VAN DER WAALS FORCES

This should prove to be the definitive work explaining van der Waals forces, how to calculate them and to take account of their impact under any circumstances and conditions. These weak intermolecular forces are of truly pervasive impact, and biologists, chemists, physicists, and engineers will profit greatly from the thorough grounding in these fundamental forces that this book offers. Parsegian has organized his book at three successive levels of sophistication to satisfy the needs and interests of readers at all levels of preparation. The Prelude and Level 1 are intended to give everyone an overview in words and pictures of the modern theory of van der Waals forces. Level 2 gives the formulae and a wide range of algorithms to let readers compute the van der Waals forces under virtually any physical or physiological conditions. Level 3 offers a rigorous basic formulation of the theory.

V. Adrian Parsegian is chief of the Laboratory of Physical and Structural Biology in the National Institute of Child Health and Human Development. He has served as Editor of the *Biophysical Journal* and President of the Biophysical Society. He is happiest when graduate students come up to him after a lecture and ask hard questions.

Van der Waals Forces

A HANDBOOK FOR BIOLOGISTS, CHEMISTS, ENGINEERS, AND PHYSICISTS

V. Adrian Parsegian
National Institutes of Health

CAMBRIDGE
UNIVERSITY PRESS

CAMBRIDGE
UNIVERSITY PRESS

32 Avenue of the Americas, New York NY 10013-2473, USA

Cambridge University Press is part of the University of Cambridge.

It furthers the University's mission by disseminating knowledge in the pursuit of education, learning and research at the highest international levels of excellence.

www.cambridge.org
Information on this title: www.cambridge.org/9780521547789

First published 2006

A catalogue record for this publication is available from the British Library

Library of Congress Cataloguing in Publication data

Parsegian, V. Adrian (Vozken, Adrian), 1939–
Van der Waals forces / V. Adrian Parsegian.
p. cm.
Includes bibliographical references and index.
ISBN 0-521-83906-8 (hardback : alk. paper) – ISBN 0-521-54778-4 (pbk. : alk. paper)
1. Van der Waals forces. I. Title.
QC175.16.M6P37 2005 20021054508
533'.7–dc22

ISBN 978-0-521-83906-8 Hardback
ISBN 978-0-521-54778-9 Paperback

CONTENTS

TABLES

Prelude

Level 1

Level 2

Tables of formulae in planar geometry

Tables of formulae in spherical geometry

Tables of formulae in cylindrical geometry

Sample spectral parameters

PREFACE

"What is this about entropy really decreasing?" I didn't know how to answer my family, worried by some preposterous news report. My best try was, "I don't know the words that you and I can use in the same way. I tell you what. Let me give you examples of where you see entropy changing, as when you put cream and sugar in coffee. You think a while about these examples. Then we can answer your question together."

That was part of the dream to which I woke the morning I was to write this welcome to readers. I connected the dream with the way my friend David Gingell came to learn about van der Waals forces 30 years ago. He began immediately by computing with previously written programs, then improved these programs to ask better questions, and finally worked back to foundations otherwise inaccessible to a zoologist.

Written using the "Gingell method," this book is an experiment in what another friend called "quantum electrodynamics for the people." First the main ideas and the general picture (Level 1); after that, practice (Level 2); then, finally, the bedrock science (Level 3), culled and rephrased from abstruse sources. This is a strategy intended to defeat the fear that stops many who need to use the theory of van der Waals forces from taking advantage of progress over the past 50 or 60 years.

Many excellent physically sophisticated texts already exist, but they remain inaccessible to too many potential users. Many popular texts simplify beyond all justification and thus deprive their readers of an exciting peek into the universe.

Although intended to be popular, the present text is not sound-bite science. There are no skimmable captions, side boxes, or section headings intended to spare the reader careful thinking. See this text as a set of conversations-at-the-blackboard to support the tables of collected or derived formulae suitable for knowing application. Peter Rand, with whom I have done more science than with any other person, says I rely heavily on the intelligence of my readers. Yes, I accept that. I hope that I can also rely on readers' motivation and pleasure in learning about a subject that reaches into all the basic sciences and into several branches of engineering.

As the book grew, I wondered if there could be more examples of applications, more details on the mechanics of computation, more exhaustive review of works in progress.

Regarding applications: I have found that many people are already eager to learn about van der Waals forces because of prior need or interest. I prefer to devote space to satisfy those needs.

Regarding computation: Spectroscopy and data processing are finally catching up with possibilities revealed by basic physical theory; any detailed How-To given here would soon be obsolete.

Regarding works in progress: "*Perfection can be achieved if a limit is accepted; without such a boundary, the end is never in sight.*" These painful phrases from Mary McCarthy's *The Stones of Florence* can burden any author who is worrying about what not to include, where to stop. The "maybe-include" list—excited states, ions in solution, atomic beams, weird geometries, etc.—grew faster than I could rationally consider. The only option was to reassure myself that, after absorbing what has been written, readers would be newly able to learn on their own. In that spirit of learning to learn, this book is designed. Through this design, I hope now to learn from my readers.

The *Prelude* gives the kind of too-brief summary and overview students might get from their pressured professors—history, principles, forms, magnitudes, examples, and measurements.

Level 1, a word-and-picture essay, tells the more motivated readers what there is for them in the modern theory. After the Prelude, it is the only part of the book best read through consecutively.

Level 2 is the doing.

Its first part, *Formulae*, examines the basic forms in a set of tables and essays that explain their versions, approximations, and elaborations. The formulae themselves are tabulated by geometry and physical properties of the interacting materials. (Take a look now. Pictures on the left; formulae on the right; occasional comments at the bottom.)

The second part, *Computation*, advises the user on algorithms as well as ways to convert experimental data into grist for the computational mill. It includes an essay on the physics of dielectric response, the aspect of van der Waals force theory that needlessly daunts potential users.

Level 3, the basic formulation, was the easiest part to write but is probably the most difficult to read. I put it last because people have a right to know what they are doing, though they need not be pushed through derivations before learning to use the theory. It is, as I imagined in the dream with my family, better to stir the coffee and have a few sips before getting into the principles of coffee making.

This brings me to think of a far more learned group of friends and fellow coffee drinkers with whom I have been lucky to study this subject (none of whom is responsible for inevitable errors or shortcomings in this text). Among them:

Barry Ninham, my original collaborator; our high moment together set our paths of learning over the next decades and founded lifelong friendship; Aharon Katzir-Katchalsky and Shneior Lifson, wise, shrewd, inspiring teachers who introduced me to this subject and who guided my early scientific life; George Weiss, my one-time "boss" who made sure that I always had complete freedom, whose corny jokes and mathematical wit have nourished me for decades; Ralph Nossal, steady friend of forty years, who has reliably provided wise advice on book writing, bike riding, and much else; Rudi Podgornik, whose "you're the one to do it" kept me doing it, and whose fertile wit made critical reading into creative science; Victor Bloomfield and Lou DeFelice, my on-line editors whose apt comments and enthusiastic encouragement came quickly and generously; Kirk Jensen, my Cambridge editor, whose deft handling of this text (and of me) earned monotonically increasing appreciation; Vicky Danahy, copy editor, who with humor, patience, and persistence demonstrated Cambridge University Press' famously fierce editing; Per Hansen and Vanik Mkrtchian, my indefatigable equation checkers who actually seemed to enjoy their days (weeks?) making sure I got it right; Luc Belloni, whose scrupulous reading of the ionic sections caught factors of 2 and inconsistencies hundreds of pages apart; David Andelman, whose love of science and teaching let him advise and read as both scientist and teacher; Sergey Bezrukov, who taught me most of what I know about noise and fluctuations; Joel Cohen, whose quest for the right word or phrase is almost as mad as my own; Roger French and Lin DeNoyer, for bringing to all of us a healthy dose of modern spectroscopy and a powerful van der Waals computation program; Dilip Asthagiri, Simon Capelin, Paul Chaikin, Fred Cohen, Milton Cole, Peter Davies, Zachary Dorsey, Michael Edidin, Evan Evans, Toni Feder, Alan Gold, Peter Gordon, Katrina Halliday, Daniel Harries, Jeff Hutter, Jacob Israelachvili, James Kiefer, Sarah Keller, Christopher Lanczycki, Laszlo Kish, Alexey Kornyshev, Nathan Kurz, Bramie Lenhoff, Graham Vaughn Lees, Sergey Leikin, Alfonso Leyva, Steve Loughin, Tom Lubensky, Elisabeth Luthanie, Jay Mann, William Marlow, Chris Miller, Eoin O'Sullivan, Nicholas Panasik, Horia Petrache, Yakov Rabinovich, Don Rau, George Rose, Wayne Saslow, Arnold Shih, Xavier Siebert, Sid Simon, Jin Wang, Lee White, Lee Young, Josh Zimmerberg, and many more (I expect I have omitted too many names and understated too many contributions) who gave me scholarly, editorial, and psychological lifts as well as criticism and stimulating ideas; Owen Rennert, Scientific Director of my day job in the National Institute of Child Health and Human Development, smart enough to direct indirectly; Aram Parsegian, whose overheard "Does Dad always write like this?" made me rethink my writing; Andrew Parsegian, Homer Parsegian, and Phyllis Kalmaz Parsegian, whose encouragement makes me such a lucky father; Valerie Parsegian, my Editor-for-Life, who deserves more credit than anyone can imagine for her witty suggestions and for unfailing encouragement; Brigitte Sitter, James Melville, and the staff of the American Embassy in Paris, who generously provided a laptop computer just after the mass murders of September 11, 2001. Thus armed, I could work in Paris while waiting almost a week to go home.

And David Gingell (1941–1995). I wish I could will myself another dream, talking with David:

Here is the book you asked me to write 30 years ago. It is not as good as it would have been after your unpredictable comments. There were not the laughs we would have had while I was writing. The book misses you. So do I. Still, it is from working with you that I wrote as I did.

From me.
For you.

Prelude

"How much does a ladybug weigh?" A straightforward question from a straightforward six-year old. I made a guess that seemed to satisfy. Thirty-five years later, I finally weighed a ladybug: 21 mg. I never doubted that she could walk on a ceiling or on the window where I caught her, but could she be holding on by the van der Waals forces about which I was writing? That 21 mg, plus a quick calculation, reassured me that these bugs might have learned some good physics a very long time ago. If so, what about other creatures? We are told that geckos might use these forces; their feet have hairy bottoms with contact areas like those of insects. They might put to good use forces whose appearance to us humans emerged from details in the pressures of gases, whose formulation resides in difficult theories, whose practicality is seen in paints and aerosols, and whose measurement requires delicate equipment. Were these same forces showing themselves to us during childhood summers?

The first clear evidence of forces between what were soon to be called molecules came from Johannes Diderik van der Waals' 1873 Ph.D. thesis formulation of the pressure p, volume V, and temperature T of dense gases. His work followed that of Robert Boyle on dilute gases of infinitesimal noninteracting particles. In high school, or even in junior high, we learn Boyle's law: pV is constant. In today's modern form, we write it as $pV = NkT$, where N is a given number of particles and k is the universal Boltzmann constant. In 1660 Boyle called his relation the "spring of air." Hold the temperature fixed; squeeze the volume; and the pressure goes up. Hold the volume fixed; heat; and the pressure goes up too.

This "spring" is still the ideal example of today's "entropic"[1] forces. Any particle or set of particles will try to realize all allowed possibilities. Particles of the ideal gas fill all the available volume V. The confining pressure depends on the volume V/N available to each particle and on kT, the vigor of thermal motion.

Boyle and Boltzmann describe most of the spring, but to describe dense-gas pressures, van der Waals found he had to replace $pV = NkT$ with $[p + (\mathrm{a}/V^2)](V - \mathrm{b}) = NkT$:

- The full volume V became $(V - \mathrm{b})$. A positive constant b accounts for that small part of the space V occupied by the gas particles themselves. Today we speak of "steric"[2] interactions in which bodies bump into each other and thereby increase the pressure needed to confine them.

- The earlier role of p_{Boyle} is now played by $p_{\mathrm{vdW}} + \mathrm{a}/V^2$. With constant positive a, $p_{\mathrm{vdW}} = p_{\mathrm{Boyle}} - \mathrm{a}/V^2$ is *less* than the ideally expected p_{Boyle} by an amount a/V^2. This difference gives evidence that the particles are attracted to each other and thus exert less outward pressure on the walls of the container. When the volume V is allowed to go to infinity, the difference disappears. The form a/V^2 tells us that this attractive correction varies with the average distance between particles. The smaller the volume V allowed to the N particles, the stronger the pressure-lowering attraction between the particles.

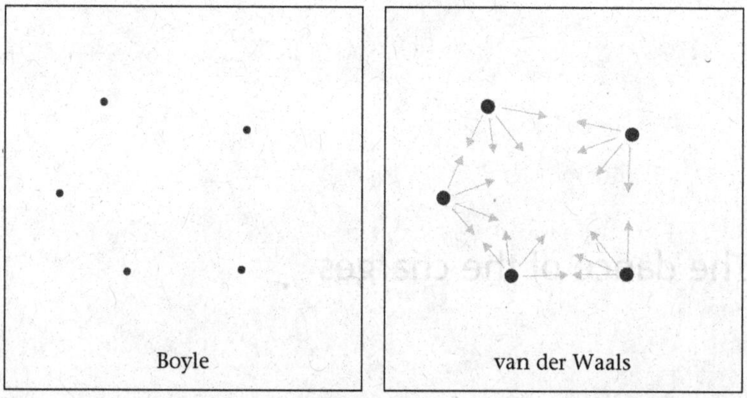

It is this attraction between molecules in a gas that most people first think of as the van der Waals interaction. These attractions in gases are so weak that they are small compared with thermal energy kT. Nevertheless, nonideal gases taught us a general truth: *Electrically neutral bodies attract.*

PR.1. The dance of the charges

In all matter there are continuous jostlings of positive and negative charges; at every point in a material body or in a vacuum, transient electric and magnetic fields arise spontaneously. These fluctuations in charge and in field occur not only because of thermal agitation but also because of inescapable quantum-mechanical uncertainties in the positions and momenta of particles and in the strengths of electromagnetic fields. The momentary positions and electric currents of moving charges act on, and react to, other charges and their fields. It is the collective coordinated interactions of moving electric charges and currents and fields, averaged over time, that create the van der Waals or "charge-fluctuation" force.

It turns out that such charge-fluctuation or "electrodynamic" forces are far more powerful within and between condensed phases—liquids and solids—than they are in gases. In fact, these forces frequently create condensed phases out of gases. The electric fields that billow out from moving charges act on many other atoms or molecules at the same time. The particles in a dilute gas are so sparsely distributed that we can safely compute the total interaction energy as the sum of interactions between the molecules considered two at a time. Rather than think in terms of a gas of pairwise particles, the modern and practical way to look at van der Waals forces is in terms of the electromagnetic properties of fully formed condensed materials. These properties can be determined from the electromagnetic absorption spectrum, i.e., from the response to externally applied electric fields.

Why? *Because the frequencies at which charges spontaneously fluctuate are the same as those at which they naturally move, or resonate, to absorb external electromagnetic waves*. This is the essence of the "fluctuation-dissipation theorem." It states that the spectrum (frequency distribution) over which charges in a material spontaneously fluctuate directly connects with the spectrum of their ability to dissipate (absorb) electromagnetic waves imposed on them. Computation of charge-fluctuation forces is essentially a conversion of observed absorption spectra. By its very nature, the measured absorption spectrum of a liquid or solid automatically includes all the interactions and couplings among constituent atoms or molecules.

Early insight

Contemporary with the 1870s formulation of the van der Waals gas equation were huge steps in electromagnetic theory. Between 1864 and 1873 James Clerk Maxwell distilled all of electricity and magnetism into four equations. In 1888 Heinrich Hertz showed how electric-charge oscillations could create and absorb (and thus detect) electromagnetic waves. In his 1894 doctoral dissertation, P. N. Lebedev saw that atoms and molecules must behave as microscopic antennae that both send and receive electrical signals. He recognized that these actions and reactions create a physical force. He saw too that in condensed materials these sendings and receivings, with their consequent forces, would have to involve many atoms or molecules at the same time. Although widely quoted, Lebedev's prescient words are worth repeating to guide us here[3]:

> Hidden in Hertz's research, in the interpretation of light oscillations as electromagnetic processes, is still another as yet undealt with question, that of the source of light emission of the processes which take place in the molecular vibrator at the time when it gives up light energy to the surrounding space; such a problem leads us, on the one hand, into the region of spectral analysis [absorption spectra], and on the other hand, quite unexpectedly as it were, to one of the most complicated problems of modern physics—the study of molecular forces.
>
> The latter circumstance follows from the following considerations: Adopting the point of view of the electromagnetic theory of light, we must state that between two radiating molecules, just as between two vibrators in which electromagnetic oscillations are excited, there exist ponderomotive [physical body] forces: They are due to the electromagnetic interaction between the alternating electric current in the molecules . . . or the alternating charges in them . . . ; we must therefore state that there exist between the molecules in such a case molecular forces whose cause is inseparably linked with the radiation processes.
>
> Of greatest interest and of greatest difficulty is the case of a physical body in which many molecules act simultaneously on one another, the vibrations of the latter not being independent owing to their close proximity.

Lebedev's program for learning was not carried out for decades. Through the 1930s, there was rapid progress in formulating interactions between particles in gases; but there was no corresponding success in formulating these forces within liquids and solids "in which many molecules act simultaneously on one another . . . owing to their close proximity." From 1894, when it was first realized that there must be a connection between absorption spectra and charge-fluctuation forces, to the 1950s when the connection was actually made, to the 1970s when it became practical to convert measured spectra to predicted forces, people still thought in terms of van der Waals forces between condensed materials as though these forces acted the same as in gases.[4]

Point–particle interactions in gases

By the end of the 1930s, the catechism of neutral-molecule interactions, at separations large compared with their size, included three kinds of dipole–dipole interactions. Their free energies—the work needed to bring them from infinite separation to finite separation r—all vary as the inverse-sixth power of distance, $-(C/r^6)$, with different positive coefficients C:

1. Keesom interactions of permanent dipoles whose mutual angles are, on average, in attractive orientations:

$$-\frac{C_{\text{Keesom}}}{r^6}$$

Dipole–dipole electrostatic interaction perturbs the randomness of orientation. If the dipole on the left is pointing "up," then there is a slightly greater chance that the dipole on the right will point "down" (or vice versa; both particles are equivalent in mutual perturbation). By increasing the chances of an attractive mutual orientation, the perturbation creates a net-attractive interaction energy.

2. Debye interactions in which a permanent dipole induces a dipole in another nonpolar molecule, with the induction necessarily in an attractive direction:

$$-\frac{C_{\text{Debye}}}{r^6}$$

The relatively sluggish permanent dipole polarizes the relatively frisky electrons on the nonpolar molecule and induces a dipole of opposite orientation. The direction of the induced dipole is such as to create attraction.

3. London dispersion interactions between transient dipoles of nonpolar but polarizable bodies:

$$-\frac{C_{\text{London}}}{r^6}$$

Here, the electrons on each molecule create transient dipoles. They couple the directions of their dipoles to lower mutual energy. "Dispersion" recognizes that natural frequencies of resonance, necessary for the dipoles to dance in step, have the same physical cause as that of the absorption spectrum—the wavelength-dependent drag on light that underlies the dispersion of white light into the spectrum of a rainbow.

There is an easy way to remember why these interaction free energies go as an inverse-sixth power. Think of the interaction between a "first" dipole pointing in a particular direction and a "second" dipole that has been oriented or induced by the $1/r^3$ electric field of the first. The degree of its orientation or induction, favorable for attraction, is proportional to the strength of the orienting or inducing electric field. The oriented or induced part of the second dipole then interacts back with the first. Because the interaction energy of two dipoles goes as $1/r^3$, we have

$$1/r^3 \text{ (for induction or orientation force)}$$
$$\times\ 1/r^3 \text{ (for interaction between the two dipoles)} = 1/r^6.$$

This is *not* an explanation of the inverse-sixth-power energy in gases; it is only a mnemonic.

In quantum mechanics, we think of each atom or molecule as having its own wave functions that describe the distribution of its electrons. The expected basis of interaction is that two atoms or molecules react to each other as dipoles, each atom's or molecule's dipolar electric field shining out as $1/r^3$ with distance r from the center. This dipole interaction averages to zero when taken over the set of electron positions predicted for the isolated atoms. However, when the isolated-atom electron distributions are themselves perturbed by each other's dipolar fields, "second-order perturbation" in the parlance, the resulting position-averaged mutual perturbation makes the extra energy go as $1/r^6$ at separations r much greater than dipole size.

Pairwise summation, lessons learned from gases applied to solids and liquids

For practical and fundamental reasons, there was a need to learn about the interactions of bodies much larger than the atoms and small molecules in gases. What interested people were systems we now call mesoscopic, with particles whose finite size Wilhelm Ostwald famously termed "the neglected dimension": 100-nm to 100-μm colloids suspended in solutions, submicrometer aerosols sprayed into air, surfaces and interfaces between condensed phases, films of nanometer to millimeter thickness. What to do?

In 1937 H. C. Hamaker,[5,6] following the work of Bradley, DeBoer, and others in the Dutch school, published an influential paper investigating the properties of van der Waals interactions between large bodies, as distinct from the small-molecule interactions that had been considered previously. Hamaker used the pairwise-summation approximation. The idea of this approximation was to imagine that incremental parts of large bodies could interact by $-C/r^6$ energies as though the remaining material were absent. The influence of the intermediate material was included as the electromagnetic equivalent of Archimedean buoyancy.

De Boer had shown that, summed over the volumes of two parallel planar blocks whose separation l was smaller than their depth and lateral extent, the $-C/r^6$ energy became an energy per area that varied as the inverse square of the separation l ($1/l^2$ for small l):

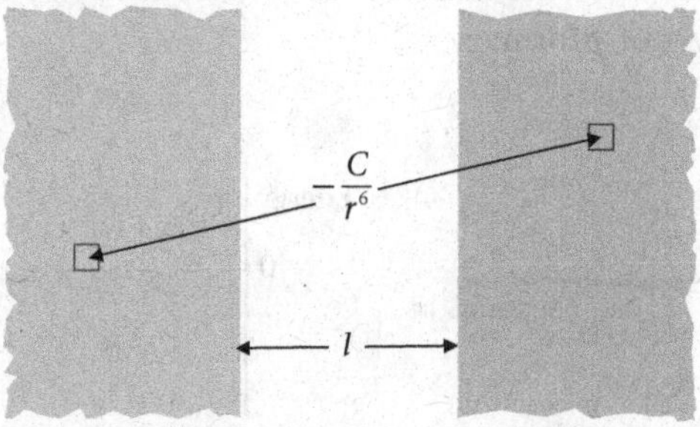

In the Hamaker summation over the volumes of two spheres, the interaction energy approaches inverse-first-power variation near contact ($1/l$ when $l \ll R$) and reverts to the expected inverse-sixth-power dependence of point particles when the spheres are widely separated compared with their size ($1/r^6$ when $r \gg R$):

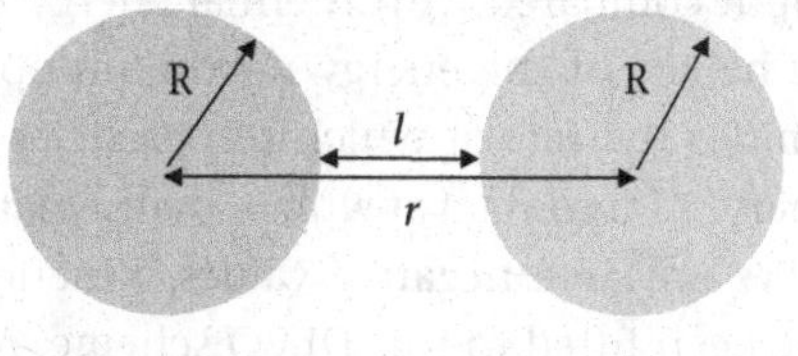

The newly recognized possibility that van der Waals forces could be of much longer range than the $1/r^6$ reach previously expected from Keesom, Debye, and London forces easily explains the well-earned influence of Hamaker's paper. The coefficient of the interaction between large objects came to be termed the "Hamaker constant" $A_{Hamaker}$, (abbreviated in this book as A_{Ham}).

Derjaguin–Landau–Verwey–Overbeek theory, colloids

There was still the problem of how to evaluate A_{Ham}. When the model was applied to colloidal interactions, this coefficient was usually fitted to data rather than estimated from any independent information. Papers written in the 1940s and early 1950s allowed "constants" that could span many orders of magnitude. This span revealed ambiguity. Nevertheless, even with such quantitative problems, van der Waals forces became recognized as the dominant long-range interaction governing the stability of colloids and often the dominant energy of interfaces. At least to first approximation, van der Waals forces followed a power law, whereas electrostatic forces, screened by the mobile ions of a salt solution, dropped off exponentially. At long enough distance a power law will always win over exponential decay. (It will also win at very short distances, but by then there are many other factors to consider.)

Neglecting all but electrostatic and van der Waals forces, and treating these two forces as though they were separable, the Derjaguin–Landau–Verwey–Overbeek (DLVO) theory created a framework (and soon a dogma) for describing the stability of colloidal suspensions.[7] In their inspiring book, published in 1948, Verwey and Overbeek make this remark: "This stability problem has been placed on a firmer physical basis by the introduction of the concept of van der Waals London dispersion forces together with the theory of the electrolytic or electro-chemical double layer."[8] Within this framework, exponentially varying electrostatic repulsion (indicated by es in the following figure) competed with power-law van der Waals attraction (indicated by vdW) to create energy versus separation curves of a form:

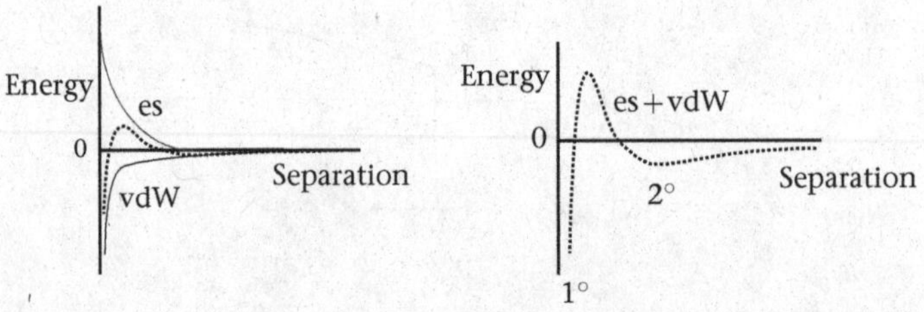

The idea has been applied to the interactions of bodies of many shapes. At long distance, the net interaction between two colloidal particles is dominated by the van der Waals attraction. This longer-range attraction creates a weak "secondary" (2°) energy minimum at a point of force balance against shorter-range electrostatic repulsion. The depth of this minimum compared with thermal energy kT allows the possibility of loose association. The height of the energy maximum opposing closer approach, again in kT units, determines the rate at which the particles might coalesce into the deep energy well of "primary" (1°) contact at which, mathematically, the inverse-power van der Waals energy takes on large negative values. Kinetic measurements, usually difficult to interpret, have been fitted to this DLVO scheme. Although more quantitative examination would require equilibrium or static measurements, and although the assumed separability of electrodynamic and electrostatic forces is incorrect, nevertheless the DLVO theory established van der Waals forces as the essential feature in the aggregation of nanometer-to-micrometer-sized particles.

The modern view

The first steps to remove the ambiguity about the magnitude of van der Waals interactions were taken almost at the same time as the development of DLVO theory, when H. B. G. Casimir formulated the interaction between two parallel metal plates. The antecedent of Casimir's view was the analysis by Max Planck in 1900 to explain the heat capacity of an empty "black box." The problem at that time was to explain the amount of heat needed to warm or cool a hollow cavity. In such a space, electromagnetic fields can exist as long as these fields satisfy the polarization properties of the vacuum in the cavity and of the bounding walls at finite temperature. In particular, the fields of this "blackbody" radiation can be expressed as oscillatory standing waves within the cavity; the energy of each wave of frequency ν changes with the emission or absorption of energy by the walls. It was Planck's famous postulate that these exchanges of energy occur as discrete units, $h\nu$, soon to be called photons, with a finite value of h, now known as Planck's constant. This proposed discreteness of energy transfer was the birth of the "quantum" theory. By summing the free energies of all allowed discrete oscillations within ideally conducting walls, and then by differentiating this total free energy with respect to temperature, Planck was able to account for the previously puzzling measured heat capacity, the energy absorbed or radiated by the space when it was heated or cooled.

In 1948 Casimir[9] used the same principle, focusing on the free energy of electromagnetic modes, to derive the force between ideally conducting walls of a box. He considered a "box" as having two of its six faces infinitely bigger than the remaining four; his box was really two large facing plates. Summing the energies of the electromagnetic modes allowed in such a box and then differentiating with respect to the distance between the two large walls, he obtained an expression for the electromagnetic pressure between the two plates. Defining the energy per area relative to a zero value when the plates are infinitely far apart then yielded the electromagnetic energy of interaction and pressures between two metal plates. Casimir's shift in perspective broke a spell. Suddenly the human race was able to step away from microscopic thinking about atoms and to survey the macroscopic whole. As Casimir told it, 50 years later[10]:

> Here is what happened. During a visit I paid to Copenhagen, it must have been in 1946 or 1947, [Niels] Bohr asked me what I had been doing and I explained our work on van der Waals forces. "That is nice," he said, "that is something new." I then explained I should like to find a simple and elegant derivation of my results. Bohr thought this over, then mumbled something like "must have something to do with the zero-point energy." That was all, but in retrospect I have to admit that I owe much to this remark.

For this reason, the Casimir formulation had another far-reaching effect. It made us recognize that "zero-point" electromagnetic-field fluctuations in a vacuum are as valid as fluctuations viewed in terms of charge motions.[11] As clearly predicted by the

Heisenberg uncertainty principle, even in a vacuum, the change ΔE in the energy of a quantum mechanically evolving system depends inversely on the duration of time Δt between observations: $\Delta E \, \Delta t \geq h/2\pi$.[12] By the inclusion of the very shortest intervals Δt, which correspond to periods of the highest frequencies or the shortest-lived fluctuations, the energy change of all possible fluctuations formally diverges to a physically impossible infinity. We are all bathed in this "vacuum infinity" of virtual electromagnetic waves, although we would never guess it from listening to a weather report.

We now recognize that "empty space" is a turmoil of electromagnetic waves of all frequencies and wavelengths. They wash through and past us in ways familiar from watching the two-dimensional version, a buoy or boat bobbing in rough water. We can turn the dancing-charges idea around. From the vacuum point of view, imagine two bodies, such as two boats in rough water or a single boat near a dock, pushed by waves from all directions except that of a wave-quelling neighbor. The net result is that the bodies are pushed together. You get close to a dock, you can stop rowing. The waves will push you in.

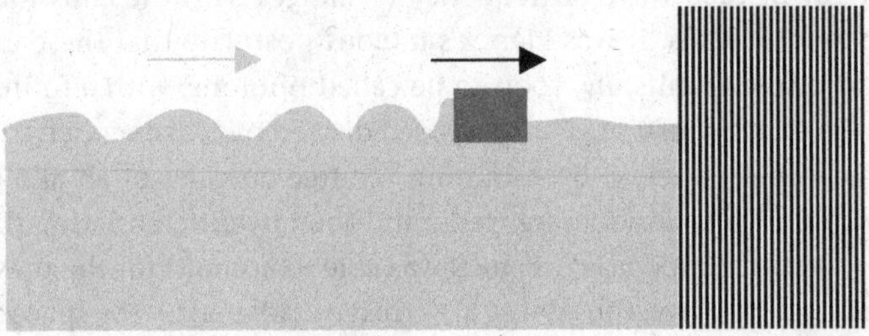

We can think of electromagnetic modes between the two bodies as the fluctuations that remain as tiny deviations from the outer turmoil. The extent of the quelling is, obviously, in proportion to the material-absorption spectra. So we can think of absorption frequencies in two ways: those at which the charges naturally dance; those at which charge polarization quells the vacuum fluctuations and stills the space between the surfaces.[13]

A careful examination[14] of forces between point particles at finite temperature and asymptotically infinite separations illustrates the value of focusing on vacuum fluctuations. These forces can be seen as driven by thermally excited, infinitely long-wavelength electromagnetic-field fluctuations that come from the surrounding vacuum. Particle polarizations are a passive response to these external fields. It is as though the particles bob on infinitely extended waves and feel each other only to the extent that their passive polarization modifies the incoming fields. Again think of the two boats on rough (read "thermally excited") water, but this time think of them as so far away as to make only the slightest perturbation on the waves.

In 1948 Casimir and Polder[15] published a second conceptual leap, showing how "retardation" changed the charge-fluctuation force between point particles. If the distance between fluctuating charges is large enough, it takes a finite amount of time for the electromagnetic field from one charge to fly across to the other. By the time the second charge responds to the field, the momentary charge configuration at the first position will have changed, and the charge fluctuations will fall out of step with each other. The strength of interaction is always weakened; its distance dependence is changed. For example, $1/r^6$ energy between small particles becomes $1/r^7$. In fact, retardation was implicitly included in the Planck and Casimir formulations; finite light velocity c means finite wavelength $\lambda = c/\nu$ for a wave of frequency ν. In fact there is no "first" sender and "second" recipient, only the coordinated fluctuation of charges at different places interacting through waves of finite wavelength. The Casimir–Polder *tour de force* ends with the observation that the simple form of the results "suggest that it might be possible to derive these expressions, perhaps apart from the numerical factors, by more elementary considerations. This would be desirable since it would also give a more physical background to our result, a result which in our opinion is rather remarkable. So far we have not been able to find such a simple argument." A few years later, the Lifshitz formulation accomplished just that by deriving the general theory, built on Casimir's earlier insight on the connection between the electromagnetic energy in a perfect-conductor Planck black box and the work of moving the walls of that box. Physicists now refer to this connection as the "Casimir effect." Some "effect"![16]

Lifshitz, using results from Rytov on the relation between absorption spectra and fluctuations, took the logical next step.[17] Conceptually he replaced the ideally conducting walls of the Planck–Casimir picture with walls of real materials.[18] Working with Dzyaloshinskii and Pitaevskii, Lifshitz replaced the intervening vacuum with real materials. The relevant electromagnetic properties of walls and intervening medium were in turn derived from the absorption spectra of the materials that composed them. By looking at all possible frequencies and all possible angles at which electromagnetic waves moved between walls, they could also see how retardation worked in the interaction between two planar parallel surfaces.

The new perspective, thinking about macroscopic interacting bodies rather than about the summed interactions of component atoms and molecules, came at a price. The theory is rigorous only at separations large enough that the materials in the bodies look like continua. This is a theory of long-range van der Waals forces in which "long-range" means separations bigger than the atomic or molecular graininess of the interacting bodies.

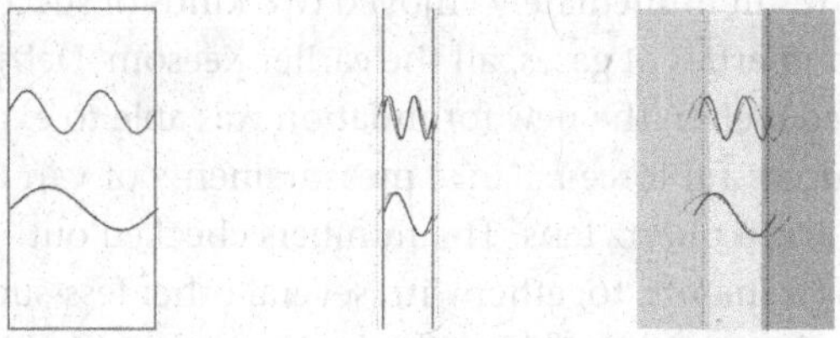

(a) Planck (b) Casimir (c) Lifshitz

Briefly, reconsider the a–b–c logical development of the macroscopic-continuum picture of van der Waals forces:

(a) Planck's analysis of a hollow black box with opaque or conducting walls. Moving electric charges in walls set up and respond to electromagnetic fields that are described as the sum of discrete waves in the cavity. The rate of change with temperature of the total wave free energy predicted the heat capacity of the box.

(b) Casimir's analysis of two parallel flat conducting surfaces, actually a rectangular box with two relatively large faces. The change of wave free energy versus separation between the large faces gives the electrodynamic force between the two plates across a vacuum.

(c) The analysis of Lifshitz, Dzyaloshinskii, and Pitaevskii of two flat surfaces of any materials facing each other across a third medium. The electrodynamic force is again a rate of change of wave free energy with separation. Unlike the previous two cases, electric and magnetic fields of "surface modes" are allowed to penetrate the outer media. Surface modes? Here and in the Casimir case, the logic is to select from all the fluctuations that spontaneously fill space only those fluctuations that depend on the location of surfaces or interfaces between materials. These are the only fluctuations whose consequence is felt as force.

The different perspectives persist in dual languages that can dichotomize researchers investigating superficially different kinds of questions. The Casimir force, with its focus on fluctuating fields, overlaps the van der Waals–Lifshitz force with its focus on the dancing charges that create or distort those fields. As the a–b–c list suggests, the Lifshitz formulation can be seen from the field as well as from the charge point of view. Because it incorporates material properties and allows the contributing fields to penetrate the walls, this formulation is not limited to ideal (infinite conductivity at all frequencies) conductors across a vacuum.

Both formulations stumble when the materials are real conductors such as salt solutions or metals. In these cases important fluctuations can occur in the limit of low frequency where we must think of long-lasting, far-reaching electric currents. Unlike brief dipolar fluctuations that can be considered to occur local to a point in a material, walls or discontinuities in conductivity at material interfaces interrupt the electrical currents set up by these longer-lasting "zero-frequency" fields. It is not enough to know finite bulk material conductivities in order to compute forces. Nevertheless, it is possible to extend the Lifshitz theory to include events such as the fluctuations of ions in salt solutions or of electrons in metals.

The 1954 Lifshitz result immediately enjoyed two kinds of success. When the materials were given the properties of gases, all the earlier Keesom–Debye–London–Casimir results readily emerged. Better, the new formulation was able to explain Derjaguin and Abrikosova's[19] first successful force-balance measurements of van der Waals forces between a quartz plate and a quartz lens. The numbers checked out.

These quartz measurements, together with several other less-successful attempts by others, had been fiercely contested.[20] Theories had been fitted to faulty measurements; there had been no adequate theory yet available for good measurements. "Measurement" drove theory. Hamaker constants (coefficients of interaction energy) were so uncertain that they were allowed to vary by factors of 100 or 1000 in order to fit the data. The Lifshitz theory put an end to all that. Disagreement meant that either there was a bad measurement or there was something acting besides a charge-fluctuation force.

It took another few years' theorizing[21] to replace the vacuum between the walls with a material. The theory of van der Waals forces was now ready for application to many kinds of experiments. Its validity was firmly enough established to justify generalization to geometries other than parallel walls. And that generalization justifies writing this book.

What are the salient properties of these forces as we now understand them? How strong are they? Where are they important? How do we compute them and measure them? What limitations must we keep in mind? At this stage, we can learn from the theory, and we can test it. So ask.

Is there a way to connect interaction energies and the shapes of interacting objects?

Already in 1934, Derjaguin[22] had proposed that many kinds of interaction, between spheres or between a sphere and a plane or between any oppositely curved surfaces near contact, could be derived from an expression for the interaction between plane-parallel facing patches. He had already foreseen how difficult it would be to derive interactions in curved coordinates compared with planar geometries. Since then, as better expressions have been discovered or developed for all kinds of interactions between planar bodies, his strategy has enjoyed ever-increasing value. However, three conditions must hold:

1. The distance of closest approach l must be small compared with the radii of curvature R_1, R_2 of the two surfaces. That is, the separation should not vary significantly over the incremental area of interaction between two facing patches.
2. The electromagnetic excitement would have to be so weak and so localized that the interaction between two facing patches would not perturb the interactions between other neighboring patches.
3. The interaction between facing patches must fall off fast enough versus patch separation that contributions only from the closest patches (lowercase θ in the next diagram) dominate.

These formal conditions are not always satisfied. In addition, real surfaces are usually rough. Real surfaces of large radius can deform under the stress of the attractive forces so that "R" itself loses meaning. Although the transform is often termed the Derjaguin approximation, its limitations are too often ignored. Nevertheless, the transform is conspicuously useful for converting van der Waals forces easily formulated in planar geometry into forces between oppositely curved, intimately close surfaces. Schematically the transform is seen as a series of steps on a curved surface. Between spheres of equal radii the steps look like this:

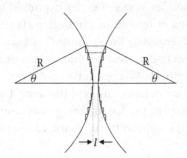

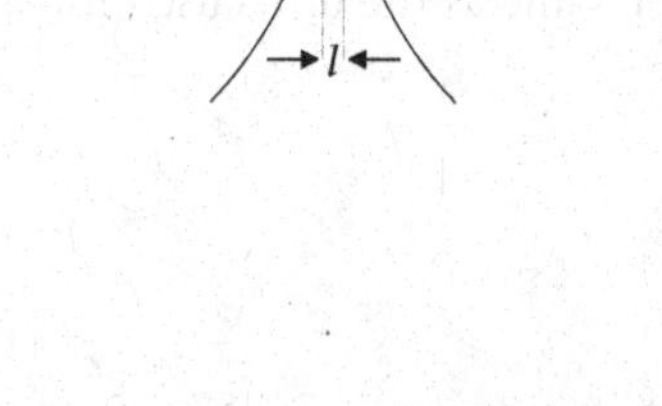

The Derjaguin idea, a mainstay in colloid science since its 1934 publication, was rediscovered by nuclear physicists in the 1970s. In the physics literature one speaks of "proximity forces," surface forces that fit the criteria already given. The "Derjaguin transformation" or "Derjaguin approximation" of colloid science, to convert parallel-surface interaction into that between oppositely curved surfaces, becomes the physicists' "proximity force theorem" used in nuclear physics and in the transformation of Casimir forces.[23]

In any language the distance between facing planar-parallel patches grows from its minimum, l, by a rate that depends on the radius of curvature. For example, from the patch summation in the stipulated small-separation limit, a planar $1/l^2$ interaction free energy per area becomes a sphere–sphere interaction free energy that goes as $1/l$ and a parallel–fat-cylinder interaction free energy that goes as $1/l^{3/2}$. Blended with the Lifshitz theory's quantitative formulation of the previously ambiguous A_{Ham} coefficient, the Derjaguin transformation confers comparable rigor on sphere–sphere and cylinder–cylinder interactions. (Warning: The spheres or cylinders must be smooth; spikes, rough spots, or dramatic deformations violate the conditions of the approximation.)

What are the powers of shapes and sizes?

Good question. Most applications and most interesting problems are for geometries different from the geometry of the simplest planar case. Different geometries change the range of the force versus separation. Being careful not to take simple limiting forms too seriously, we can glimpse the consequences of geometry beyond the large-body–small-separation regime of the Derjaguin transform. For introductory intuition, consider a few examples for cases in which there are small differences in material polarizabilities and for which we also neglect the effects of retardation. See Table Pr. 1.

Units

The coefficient A_{Ham} has units of energy. The popular form $[A_{\mathrm{Ham}}/(12\pi l^2)]$ is for the free energy or work per unit area to bring two plane-parallel infinitely extended half-spaces from infinite separation to finite separation l. For the free energy between planar faces of finite extent, multiply by facial area. Similarly, for the interaction between long parallel cylinders, the free energy is given per unit length. Between spheres and between crossed cylinders, the interaction is already in pure energy units, i.e., per pair of interacting bodies.

Put another way, the van der Waals energies expressed in these simplified forms are independent of the units of length. In physical parlance, they "scale." Consider any of the other "per-pair" energies, for example, $[-(A_{\mathrm{Ham}}/2\pi)][(\mathrm{A}_1\mathrm{A}_2)/z^4]$ between perpendicular rods. Cross-sectional areas A_1 and A_2 are in units of length squared; separation z is in units of length; $(\mathrm{A}_1\mathrm{A}_2/z^4)$ is dimensionless. As long as the ratio between the sizes of the rods and their separation are kept the same, the van der Waals energy is the same; ditto for the per-area or per-length energies as long as the length units used for the areas or lengths are the same as the units used for separations.

Table Pr.1. Idealized power-law forms of interaction free energy in various geometries

Parallel infinitely thick, infinitely extended walls ("half-spaces"), variable separation l 	$-\frac{A_{Ham}}{12\pi l^2}$ per unit area
Parallel infinitely extended slabs of fixed thickness a, variable separation l 	$-\frac{A_{Ham}}{12\pi}\left[\frac{1}{l^2}-\frac{2}{(l+a)^2}+\frac{1}{(l+2a)^2}\right]$ per unit area
Infinitely thick wall parallel to cube of finite extent or two parallel cubes of extent L 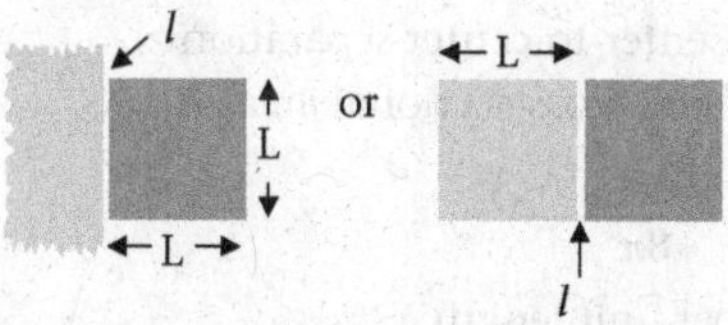	Surface separation $l \ll L$, $-\frac{A_{Ham}}{12\pi l^2}L^2$, per pair of interacting bodies
Spheres, fixed radii R_1, R_2, near contact 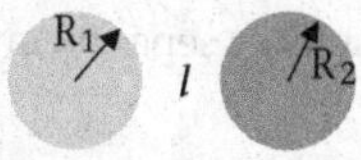	Variable surface separation $l \ll R_1, R_2$, $-\frac{A_{Ham}}{6}\frac{R_1R_2}{(R_1+R_2)l}$, per pair of spheres
Sphere near a planar thick wall 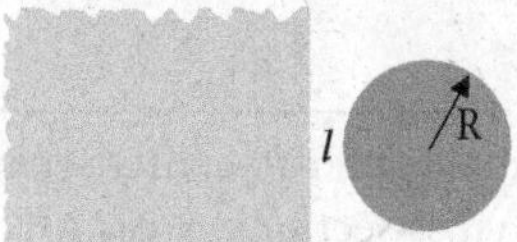	Variable surface separation $l \ll R$, $-\frac{A_{Ham}}{6}\frac{R}{l}$, per pair of bodies
Spheres, radius R, far apart 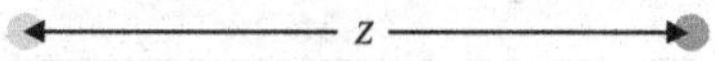	Center-to-center distance $z \approx l \gg R$, $-\frac{16}{9}\frac{R^6}{z^6}A_{Ham}$, per pair of spheres

Table Pr.1 (*cont.*)

Small sphere, radius R, far from planar surface of an infinitely thick wall 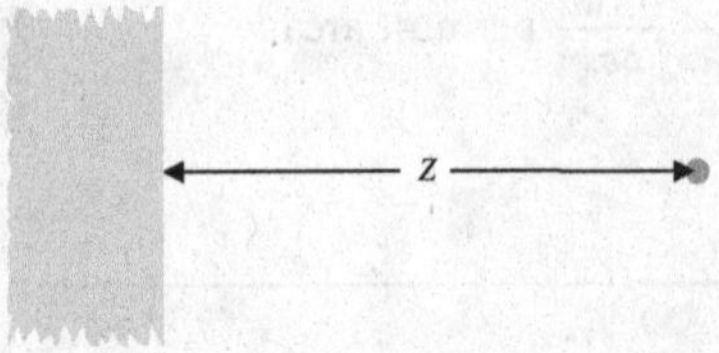	Sphere center-to-wall separation $z \gg R$, $-\frac{2A_{Ham}}{9}\left(\frac{R}{z}\right)^3$, per pair of bodies. The same relation holds for small sphere of radius of R far from a cylinder or sphere of radius much greater than z: $R_{cylinder}$ or $R_{sphere} \gg z \gg R$
Infinitely long, parallel, circular cylinders, fixed radii R, near contact 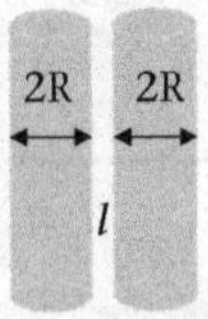	Variable surface separation $l \ll R$, $-\frac{A_{Ham}}{24 l^{3/2}} R^{1/2}$, per unit length
Infinitely long thin parallel cylinders 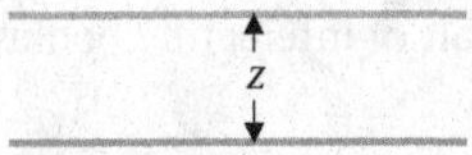	Center-to-center separation z, fixed cross-sectional areas A_1, A_2, $-\frac{3A_{Ham}}{8\pi}\frac{A_1 A_2}{z^5}$, per unit length
Infinitely long cylinders, perpendicular, near contact 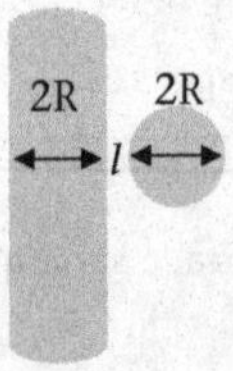	Variable surface separation $l \ll R$, $-\frac{A_{Ham}}{6}\frac{R}{l}$, per pair of cylinders
Infinitely long cylinders, perpendicular, far apart 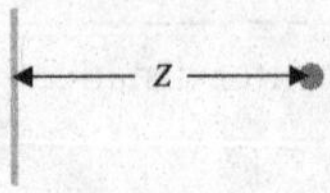	Minimal center-to-center separation z, fixed cross-sectional areas A_1, A_2, $-\frac{A_{Ham}}{2\pi}\frac{A_1 A_2}{z^4}$, per pair of cylinders

Still, units can be a nuisance. One difficulty is that much serious theoretical work is still done in centimeter–gram–second (cgs) or "Gaussian" units; such is the case with the Level 3 derivations in this text. Most students learn applications in meter–kilogram–seconds (mks) "SI" or "Système International" units. Happily, practical formulae for most van der Waals interactions look the same in both systems of units. We do not enjoy the same forgiveness with, for example, electrostatic forces for which the practical formulae in different units look different from the outset.

Not so fast! Those formulae don't look so different to me from what people have learned by using old-fashioned pairwise summation

What's different from pairwise summation? Simple: You let nature do the volume average for you and unashamedly take the electrical and magnetic behavior of the entire material. You don't try to take the properties of constituent atoms and weave them into the properties of the liquid or solid.

The forms for the limiting cases in the table are due to a happy convergence of old and new theories. These special cases

1. assume small differences in material electromagnetic properties, and
2. neglect the finite velocity of light.

In these simplified formulae, the distance dependence is what would result from pairwise summation. The huge difference is in the coefficients A_{Ham} that are now computed from whole-material properties rather than from the polarizabilities of constituent atoms or molecules. Even in formal correspondence with the old way of summing incremental contributions, the resemblance is in the distance dependence but not in the coefficient. Only in another limit, in which the media are all gases so dilute that their atoms interact two at a time as though no other particles were present, is there rigorous correspondence between old and new theories.

Think of dipoles:

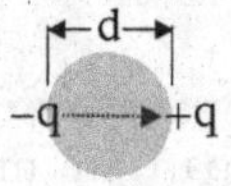

The dipole arrow goes from a negative charge $-q$ to a positive charge $+q$ of equal magnitude. Charge magnitude q times charge separation d gives the dipole moment $\mu_{dipole} = qd$. (In this text, the direction of the dipole follows physical convention: the arrow points *from the negative to the positive* charge. It points the other way in chemistry.)

Any two dipoles can interact optimally head-to-tail:

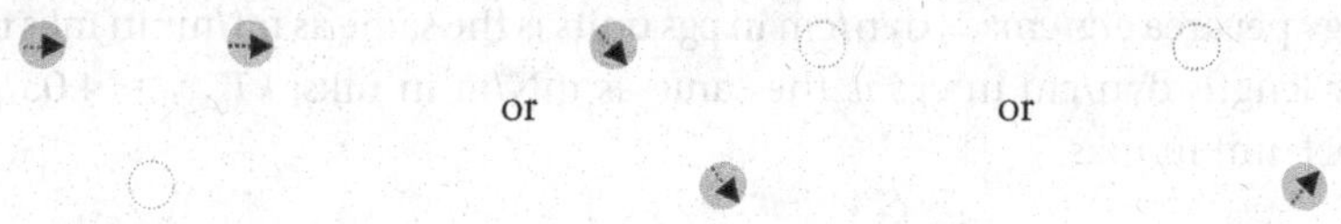

There would have to be a compromise to satisfy all three at the same time:

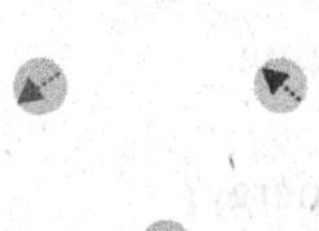

Even when dipoles are fluctuating on their own, rather than being woodenly fixed as in this picture, the net interaction cannot be as strong as the sum of interactions between isolated pairs. This holds for particles that do not bear permanent dipoles as well as for those with permanent dipoles whose interaction involves averaging over all mutual angles while jostled by temperature.

When can particles ever interact two at a time? There is a remarkably easy test. If we can apply an electric field from the outside and each atom or molecule feels that externally applied electric field as though the other atoms or molecules were not there, then we can say that the particles are so dilute that they will interact two at a time. If particles are so dense that the external field felt by each particle is distorted by the presence of other particles, then we have three's-a-crowd densities. The key variable is (particle-number density) times (individual-particle polarizability). An infinitesimal value of this product ensures that the fields of neighbors' dipole moments do not significantly contribute to the field felt by each particle. In practice, only dilute gases pass this test.

That's the nub of it. Only when the material itself behaves as the linear sum of its parts is it logical to treat the totality of its interactions as the sum of its incremental parts. How big is the correction that recognizes the simultaneity of all three of these particles fluctuating together? Formulations in the early 1940s[24] suggested an ~5%–10% correction for an isolated triplet at the density of a condensed medium. Recent formulations[25] show potentially significant three-body interactions that depend on geometry and density, as well as on the atomic context of the particles being considered three at a time. Once it is clear that three-body terms must be recognized in condensed media, we are compelled to consider yet higher-order terms. Reason? We simply do not know a priori whether the many-body series has converged after the three-body term. When practicable, the Lifshitz leap to using the properties of the whole material straightaway obviates such an indeterminate progression.

PROBLEM PR.1: On average, how far apart are molecules in a dilute gas? Show that, for a gas at 1-atm pressure at room temperature, the average interparticle distance is ~30 Å.

The use of kT_{room} as a convenient unit of energy in practical formulae does *not* mean that A_{Ham} is proportional to temperature. Except for interactions among highly polar materials, A_{Ham} depends only weakly on temperature.

The energy per area erg/cm^2 = dyn/cm in cgs units is the same as mJ/m^2 in mks units; the force per length dyn/cm in cgs is the same as mN/m in mks; $kT_{room} \approx 4.05$ pN × nm = 4.05 pN nm in mks.

How strong are long-range van der Waals interactions?

It depends on what you mean by strong.
Strong enough to measure?
Strong enough to counter thermal energy?
Strong enough to do something interesting or practical or worthwhile?

Table Pr.2. Typical estimates of Hamaker coefficients in the limit of small separation

For compactness the A_{Ham} are given here in zeptojoules (zJ): 1 zJ = 10^{-21} J = 10^{-14} ergs. A useful rule of thumb is that A_{Ham} calculated in natural units typically ranges from ~1 to ~100 times thermal energy $kT_{room} = 1.3807 \times 10^{-23}$ (J/K) × 293.15K ≈ 4.05 zJ, with T_{room} in absolute degrees Kelvin.

Material	A_{Ham} **across water (zJ)**	A_{Ham} **across vacuum (zJ)**
Organics		
Polystyrene[26]	13	79
Polycarbonate[27]	3.5	50.8
Hydrocarbon (tetradecane, Level 1)	3.8	47
Polymethyl methacrylate[27]	1.47	58.4
Protein[28, 29]	5–9, 12.6	n/a
Inorganics		
Diamond (IIa)[30]	138	296
Mica (monoclinic)[30]	13.4	98.6
Mica (Muscovite)[31]	2.9	69.6
Quartz silicon dioxide[31]	1.6	66
Aluminum oxide[31]	27.5	145
Titanium dioxide rutile[31]	60	181
Potassium chloride (cubic crystal)[30]	4.1	55.1
Water[32]	n/a	55.1
Metals		
Gold[33]	90 to 300	200 to 400
Silver[33]	100 to 400	200 to 500
Copper[33]	300	400

Measurably strong

By the criterion of measurability, even the van der Waals gas equation passes the test. It is not an easy mental journey from the van der Waals equation's a/V^2 pressure correction to a $-C/z^6$ attractive energy between atoms or molecules at separations z large compared with their size. Nevertheless, statistical mechanics shows a way to connect the interaction-perturbed randomness of the van der Waals gas with such an interaction.

The deflection of potassium, rubidium, or cesium atomic beams grazing the surface of a ~1-cm-radius gold-coated cylinder[34] clearly demonstrated the expected $1/z^3$ variation of energy with separation z. With deflections detectable for 50 nm $< z <$ 80 nm minimum approach ("impact parameter") between cesium atoms and surface, the measured coefficient K_{attr} is $7.0 \pm 0.3 \times 10^{-49}$ J m^3, and the interaction energy $-K_{attr}/z^3$ is 1.4 to 5.6 × 10^{-6} zJ, roughly one millionth room-temperature thermal energy kT_{room} ~ 4.05 zJ.

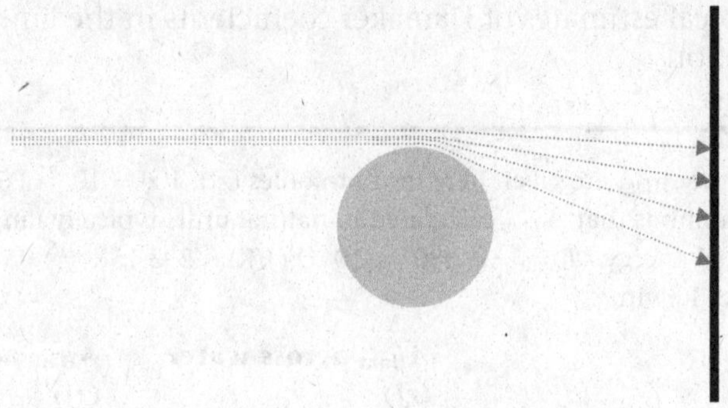

Why should the atom–cylinder interaction energy go as the energy of a point particle with a plane? With the ~1-Å-size atom, the ~50-nm impact parameter, and the ~1-cm cylinder radius, there is a clean separation of sizes. The atom sees a substrate of infinite radius along the cylinder and effectively infinite radius compared to atom–substrate distance. The interaction is then effectively that of a plane and a point particle.

PROBLEM PR.2: Calculate the effective Hamaker coefficient between the spherical atom and the gold surface.

Thermally strong

When do we think of van der Waals interactions as responsible for organizing large molecules or aggregates? One criterion is whether they create energies large compared with thermal energy kT. Then there will be forces strong enough to overcome thermal agitation. For most nonmetals interacting in liquids of comparable density, $A_{Ham} \sim 1$ to $5\,kT_{room}$; for solids or liquids interacting across vapor or vacuum, think ~10 × stronger. Conducting materials attract more strongly still.

A good rule of thumb is that *interaction energies between two bodies in any geometry will be significant compared with kT as long as their separation is less then their size.* For example, consider two square planar faces of area L^2 such that $L \gg$ separation l. The interaction free energy will go as $\{-[A_{Ham}/(12\pi l^2)]\}L^2$; as long as $A_{Ham} \geq kT$ and $l < L/6$, there will be a thermally significant interaction. Between two identical spheres of separation $l \ll$ radius R, the interaction goes as $[-(A_{Ham}/12)](R/l)$. When $A_{Ham} \geq kT$ and $l < R/12$, the interaction is greater than kT.

These two limiting forms show an equivalence between the closely approaching-sphere and the planar-block interactions. Ask, what area of the plane–plane interaction is equivalent to the sphere–sphere interaction? Equate $\{-[A_{Ham}/(12\pi l^2)]\}L^2 = [-(A_{Ham}/12)](R/l)$ to see that two spheres look like two planes of area $L^2 = R\pi l$.

In the small-sphere limit, $R \ll l \approx z$, the interaction goes as $[-(16/9)](R^6/z^6)A_{Ham}$; its strength is *never* comparable with thermal energy kT. The same holds for the interaction of two molecules at a separation large compared with their size.

PROBLEM PR.3: Show that the interaction between spheres separated by distances much greater than their radii will always be much less than thermal energy kT.

PROBLEM PR.4: Try something harder than spheres. Consider parallel cylinders of radius R, fixed length L, and surface separation l. Use the tabulated energy per unit length $[-(A_{Ham}/24l^{3/2})]R^{1/2}$ to show that, for $A_{Ham} \approx 2\,kT_{room}$, a value typical of proteins (see table in preceding section), the energy is $\gg kT$ when

1. $R = L = 1\ \mu m \gg l = 10$ nm (dimensions of colloids), and
2. $R = 1$ nm, $L = 5\ nm \gg l = 0.2$ nm (dimensions of proteins).

PROBLEM PR.5: Or easier than spheres. Consider a case of surface-shape complementarity imagined as two flat parallel surfaces. Show that the energy of interaction of two 1 nm × 1 nm patches 3 Å apart will yield an interaction energy $\sim kT$.

Mechanically strong

Van der Waals forces usually contribute significantly to the cohesion energies and interfacial energies of solids and liquids. In those cases, in which determining interactions occur at interatomic spacings, there is no doubt of their mechanical importance. However, there is still doubt about the best way to formulate and compute forces while incorporating details of molecular arrangement.

Can rigorously computed long-range forces be significant? How big can a bug be and still hold onto a ceiling by van der Waals forces?

In the spirit of pure physics, consider a cubic bug of lateral dimension L. A really juicy bug will have the density of water, $\rho \sim 1\ g/cm^3 = 1$ kg/liter. Its weight will be

$$F_{gravity} = \text{volume } (L^3) \times \text{density } (\rho) \times \text{gravitational constant } (g) = \rho L^3\, g = F_{\downarrow}.$$

A van der Waals force sufficient to hold up the bug would be the force/area $A_{Ham}/(6\pi l^3)$ between the cube and ceiling times the area of interaction L^2:

$$F_{vdW} = \frac{A_{Ham}}{6\pi l^3} L^2 = F_{\uparrow}.$$

The force to fall off grows as L^3 whereas the force to hang on grows as L^2. For big enough L, gravity will win.

How big must $L = L_{bug}$ for the two forces to be equal, for $F_{\uparrow} = F_{\downarrow}$?

Pick a weak $A_{Ham} = kT_{room}$, though for interactions in air it could be $\sim 10\ kT_{room}$. Pick a large interaction distance $l = 100$ Å $= 10$ nm, i.e., 50–100 times an interatomic distance. Still, the result is a startlingly large $L_{bug} \sim 0.02$ m $= 2$ cm or a volume of 8 cm^3.

PROBLEM PR.6: Show that van der Waals attraction at 100-Å separation is enough to hold up an ~2-cm cube even when $A_{\mathrm{Ham}} = kT_{\mathrm{room}}$:

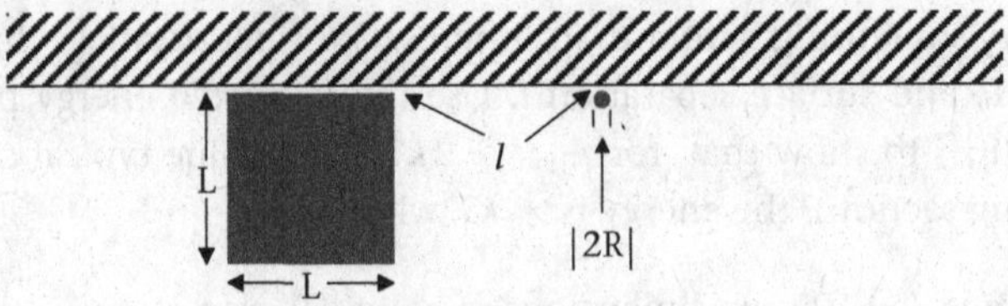

But shape makes a difference. A real physicist must consider also the well-known case of a spherically symmetric bug. With the same density, the same $A_{\mathrm{Ham}} = kT_{\mathrm{room}}$, and the same $l = 100$ Å for the minimal separation, the same $F_{\uparrow} = F_{\downarrow}$ routine gives a balancing size as a radius $R_{\mathrm{bug}} = 13 \times 10^{-4}$ cm and a not-same-at-all volume of $\sim 0.92 \times 10^{-8}\ \mathrm{cm}^3 = 9.2 \times 10^{+3} \mu\mathrm{m}^3$.

PROBLEM PR.7: Show how a change in shape makes a difference in the weight that can be maintained by a van der Waals force.

There is plenty of force in a long-range van der Waals force, at least for facultatively flat-footed bugs. There is also the important lesson to us humans that the consequences of flattening can be enormous.

PROBLEM PR.8: Show how van der Waals attraction can be a force for flattening a sphere against a flat surface.

PROBLEM PR.9: When does the van der Waals attraction between two spherical drops of water in air equal the gravity force between them? (Neglect retardation.)

PROBLEM PR.10: At what separation between two 1-μm droplets of water in air does the energy of their mutual attraction reach $-10\ kT_{\mathrm{room}}$? (Neglect retardation.)

How deeply do van der Waals forces "see"?

There is no general answer to this. To first approximation, for two bodies separated by a distance l, polarization properties to depths $\sim l$ are important. The interaction of two planar slabs of thickness a nicely illustrates this depth. Neglecting retardation and assuming that the two materials have about the same polarizability, we know that the energy will vary as shown in Table Pr.1.

$$\left[\frac{1}{l^2} - \frac{2}{(l+\mathrm{a})^2} + \frac{1}{(l+2\mathrm{a})^2}\right] = \frac{1}{l^2}\left[1 - \frac{2}{\left(1+\frac{\mathrm{a}}{l}\right)^2} + \frac{1}{\left(1+\frac{2\mathrm{a}}{l}\right)^2}\right].$$

For $l \ll \mathrm{a}$, the $1/l^2$ term dominates; the interaction is like that of two semi-infinite bodies ($\mathrm{a} \to \infty$). For separation l comparable with or larger than thickness a, the finitude of thickness asserts itself.

Measurements on interactions between singly coated bodies reveal even more about how the properties at different depths from the interacting surfaces show up as the distance between those surfaces is varied.[35] When the separation between the bodies is less than the coating thickness, the interaction approaches that of two infinitely thick bodies of the coating material. When separation is much, much greater than the coating thickness and there is not a great difference in polarizabilities of all materials, it can look as though the coating is not even there.

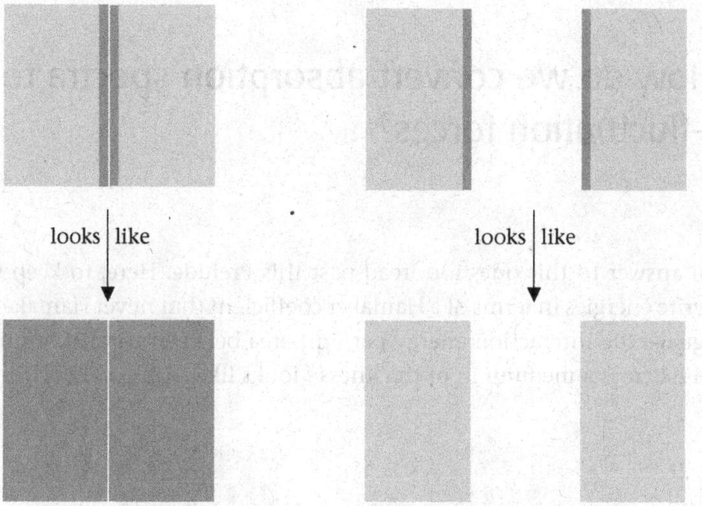

PROBLEM PR.11: Show that the forces "see" into the interacting bodies in proportion to separation.

The subject is "force" but the formulae are "energies." How to connect them?

The negative derivative of an energy with respect to distance is a force; the force per area is pressure. When the spatial variation in A_{Ham} itself (which is due to retardation screening) is neglected, the energy per unit area $-[A_{\mathrm{Ham}}/(12\pi l^2)]$ between half-spaces leads to a pressure that looks like $-[A_{\mathrm{Ham}}/6\pi l^3)]$.

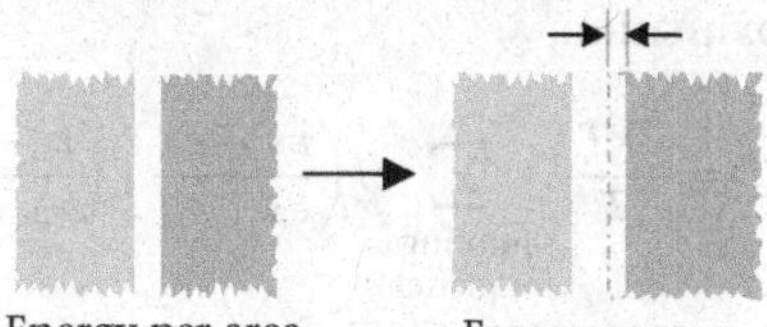

Similarly between nearby spheres at minimal separation l, energy $\sim 1/l$, the force will go as $1/l^2$:

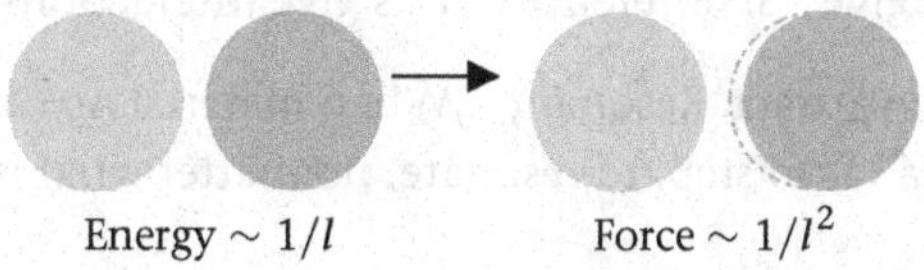

PR.2. How do we convert absorption spectra to charge-fluctuation forces?

For the real answer to this question, read past this Prelude. Here, to keep to familiar notation, write energies in terms of a Hamaker coefficient (but never Hamaker *constant*). In this language the interaction energy per unit area between parallel, infinitely thick walls A and B across a medium m of thickness l looks like $-[A_{Ham}/(12\pi l^2)]$.

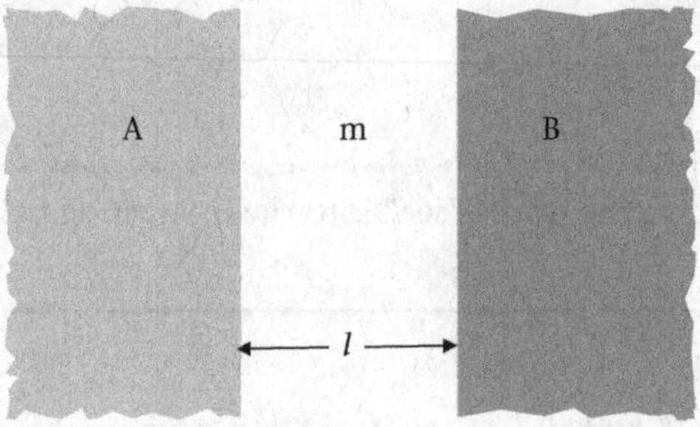

The coefficient A_{Ham} itself varies with separation l. It takes the form of a sum over all frequencies at which fluctuations can occur wherein each term depends on the frequency-dependent responses of materials A, B, and m to electromagnetic fields. These responses are written in terms of "dielectric" functions ε_A, ε_B, and ε_m that are extracted from absorption spectra. It is the *differences* in these dielectric responses that create interactions. To first approximation,

$$A_{Ham} = A_{Ham}(l) = \frac{3kT}{2} \sum_{\substack{\text{sampling} \\ \text{frequencies}}} \left(\frac{\varepsilon_A - \varepsilon_m}{\varepsilon_A + \varepsilon_m}\right)\left(\frac{\varepsilon_B - \varepsilon_m}{\varepsilon_B + \varepsilon_m}\right) \mathrm{Rel}(l).$$

For a vacuum, $\varepsilon \equiv 1$; for all materials the ε's as used here are ≥ 1. In this schematic expression, $\mathrm{Rel}(l) \leq 1$ gives the effects of "relativistic retardation," i.e., the suppression of interactions because of the finite velocity of light:

- Absorption spectra give ε's, differences in ε's give interactions.
- Like materials *always* attract. Resonance. When material A = material B their dancing charges try to fall into step, to resonate, no matter what is in between.

- Unlike materials can attract or repel depending on the comparative values of ε_A, ε_B, and ε_m at the "sampling frequencies" at which these ε's are evaluated to be used for computation.

Sampling frequencies?

Van der Waals forces result from charge and electromagnetic-field fluctuations at all possible rates. We can frequency analyze these fluctuations over the entire frequency spectrum and integrate their force consequences over the frequency continuum. Alternatively, the modern theory shows a practical way to reduce integration over all frequencies into summation by the gathering of spectral information into a set of discrete sampling frequencies or eigenfrequencies. The nature and choice of the frequencies at which dielectric functions are evaluated reveal how the modern theory combines material properties with quantum mechanics and thermodynamics.

The sampling begins at the zero frequency of static polarizability, what we think of as the "dielectric constant" of electrostatics. After zero, sampling frequencies are evenly spaced such that the photon energy (quantum mechanics!) of each frequency is a multiple of thermal energy kT (thermodynamics!). Hence, in a deft leap over everything that happens in between, the first sampling after zero corresponds to infrared (IR) frequencies; after that there are a few more samplings through the IR and the visible, but most frequencies that enter computation occur in the vastly wider ultraviolet (UV) and x-ray regions. When the photon energy is comparable with kT, which is true of the first few sampling frequencies, fluctuations will be excited by thermal agitation; when photon energy is significantly greater than kT, fluctuations flow from zero-point motions (recognized by Bohr and Casimir) demanded by the uncertainty principle.

Most intriguing to newcomers, these sampling frequencies are not expressed as ordinary sinusoidal ("real-frequency") oscillations. Instead they are crafted in the language of exponential (horribly designated as "imaginary-frequency") variations pertaining to the ways in which spontaneous charge fluctuations die away. Different kinds of fluctuations die at different rates:

- Bound electrons over periods corresponding to those of UV and higher frequencies, characteristic times $\lesssim 10^{-17}$ s;
- Vibrating molecules at frequencies corresponding to IR, characteristic times $\sim 10^{-16}$ to $\sim 10^{-12}$ s;
- Rotations at microwave frequencies, $\sim 10^{-11}$ to $\sim 10^{-6}$ s;
- Ions or conduction electrons or other mobile charges with times of displacement that extend down to the slowest frequencies and longest periods.

What does this kind of exponential frequency mean for the epsilons? Happily, the characteristic oscillations of laboratory language transform nicely into the language of exponential frequencies. After transformation, the epsilons behave smoothly, decreasing monotonically versus exponential frequency, without the spikes of absorption and dispersion seen for real oscillations.

If I take that formula seriously, then two bubbles or even two pockets of vacuum will attract across a material body. How can two nothings do anything?

If A and B are empty spaces and the medium is some material, then $\varepsilon_A = \varepsilon_B = 1$ and $\varepsilon_m \geq 1$. Except for the yet-unexplained screening factor Rel(l), the formula looks bemusingly like what you would expect if the positions were reversed, substances A and B across a vacuum m. Yes, two pockets of vacuum will enjoy van der Waals attraction.

It is not that there are actually interactions between nonexistent matter in the empty spaces, although we could argue that way from the vacuum-field point of view (see note 11 in the section titled "The modern view"). It is rather that the material between the "bubbles," or pockets A and B, likes to be with more of its own kind. If the pockets moved together, then the material in between would move out to the infinite reservoir where it could be happier with more of its kind.

What counts with van der Waals forces is *differences* in electromagnetic properties.

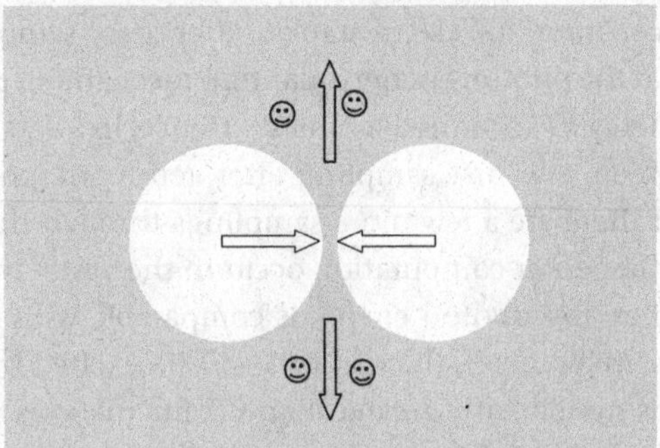

Incidentally, this bubble–bubble example illustrates the nontriviality of a theory in which the medium is *not* a vacuum. We have little trouble thinking about charges dancing in two material bodies, absorbing and emitting electromagnetic fields, sending signals to each other across empty space. It is irksomely different to think about charges dancing in the intervening medium itself, sending and receiving signals. Then energy of interaction comes from distortions in the fields at the boundaries of the medium where it comes into contact with empty space or other materials.[36]

Does charge-fluctuation resonance translate into specificity of interaction?

In a weak sense, yes, as long as we are careful to define "specificity." Imagine that A and B are two different materials. At each sampling frequency in the summation, compare the terms in the Hamaker coefficients for

$$\text{A interacting with A,}\quad A_{A-A} \sim \left(\frac{\varepsilon_A - \varepsilon_m}{\varepsilon_A + \varepsilon_m}\right)^2 \geq 0,$$

$$\text{B interacting with B,}\quad A_{B-B} \sim \left(\frac{\varepsilon_B - \varepsilon_m}{\varepsilon_B + \varepsilon_m}\right)^2 \geq 0,$$

$$\text{A interacting with B,}\quad A_{A-B} \sim \left(\frac{\varepsilon_A - \varepsilon_m}{\varepsilon_A + \varepsilon_m}\right)\left(\frac{\varepsilon_B - \varepsilon_m}{\varepsilon_B + \varepsilon_m}\right).$$

The sum of A_{A-A} and A_{B-B} is always greater than or equal to twice A_{A-B}:

$$A_{A-A} + A_{B-B} \geq 2A_{A-B}.$$

Because the van der Waals interactions go as the negative of the Hamaker coefficients, the sum of the (negative) A – A and B – B interactions at a given separation is more negative than two A – B interactions at that same separation.

In this average sense, van der Waals attraction is stronger between like than between unlike particles. This average inequality does *not* imply that either A – A or B – B attraction is alone stronger than A – B attraction. It does suggest that in a mixture of A's and B's the van der Waals forces described here will act to sort them into purified clusters of just A's and just B's.

PROBLEM PR.12: Convince yourself that $A_{A-A} + A_{B-B} \geq 2A_{A-B}$.

How does retardation come in?

For all the big words, like "relativistic," that are used here, the basic idea is simple. Van der Waals forces depend on the correlated movements or dance of charges at different places. These charges send and receive electric and magnetic signals from each other. If the charges are so close that there is essentially no time needed for the signals to travel between them, their movements and their momentary positions can nicely fall into step. If the charges are so far apart that the travel time is no longer negligible, the dancing charges will fall out of step. "Relativistic" creeps in with "retardation" because of the finite velocity of light.

When is the signal travel time long enough to worry about? Simply and intuitively, when the travel time back and forth between bodies is comparable to the length of time that the charges dwell in a particular configuration at each sampling frequency. The modern theory tells us to sum over a set of sampling frequencies that capture the important polarization–fluctuation properties of the interacting materials. Relativistic retardation screening uses the period of each sampling frequency to measure the signal travel time.

To first approximation, relativistic retardation can be expressed as the factor Rel(l) in the frequency summation $\sum_{\substack{\text{sampling}\\ \text{frequencies}}} \{[(\varepsilon_A - \varepsilon_m)/(\varepsilon_A + \varepsilon_m)][(\varepsilon_B - \varepsilon_m)/(\varepsilon_B + \varepsilon_m)]\} \times$ Rel(l) for the interaction of planar-parallel surfaces.

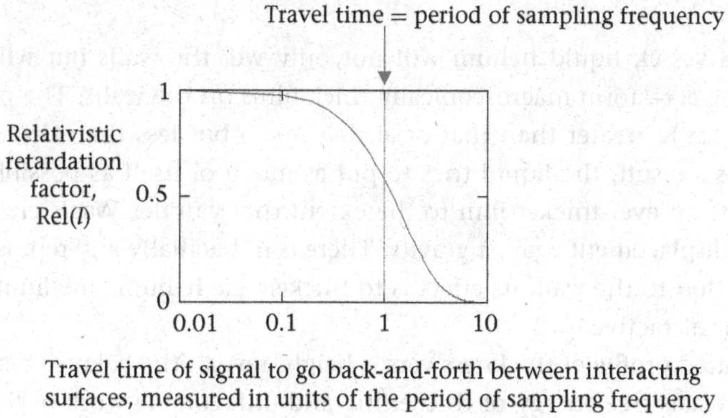

When the travel time equals the period of a sampling frequency, the contribution to the interaction is screened down to about half of its nonretarded strength. Under certain idealized conditions, where temperature-driven fluctuations are neglected and separations are greater than all wavelengths of material-absorption frequencies, retardation adds a power to the decay of interactions. The $1/r^6$ drop-off of point-dipole energies becomes $1/r^7$; the $1/l^2$ between half-spaces becomes $1/l^3$. It should be obvious just by looking at the graph of screening factor Rel versus Separation that retardation screening in fact spoils power-law variation of the van der Waals interaction.

PROBLEM PR.13: In vacuum, at what separation l does the travel time across and back equal the $\sim 10^{-14}$ s period of an IR frequency?

Can there ever be a negative Hamaker coefficient and a positive charge-fluctuation energy?

Yes. This is not surprising when you think about it. Again consider the form

$$A_{\mathrm{Ham}} = \frac{3kT}{2} \sum_{\substack{\text{sampling} \\ \text{frequencies}}} \left(\frac{\varepsilon_A - \varepsilon_m}{\varepsilon_A + \varepsilon_m}\right)\left(\frac{\varepsilon_B - \varepsilon_m}{\varepsilon_B + \varepsilon_m}\right) \times \mathrm{Rel}(l).$$

If B is more polarizable than m($\varepsilon_B > \varepsilon_m$), but m is more polarizable than A($\varepsilon_m > \varepsilon_A$), then the product $[(\varepsilon_A - \varepsilon_m)/(\varepsilon_A + \varepsilon_m)][(\varepsilon_B - \varepsilon_m)/(\varepsilon_B + \varepsilon_m)]$ is negative (rather than positive as when A and B are the same). A_{Ham} can be negative. In that case, the free energy $-[A_{\mathrm{Ham}}/(12\pi l^2)]$ would be positive. The interaction of A with B across m would look like a repulsive force in the sense that the change in energy would be favorable to thickening the film of material m. Another way to say it is that m would prefer to be near B and away from A. There's a beautiful crawling-the-walls experiment that shows just that.

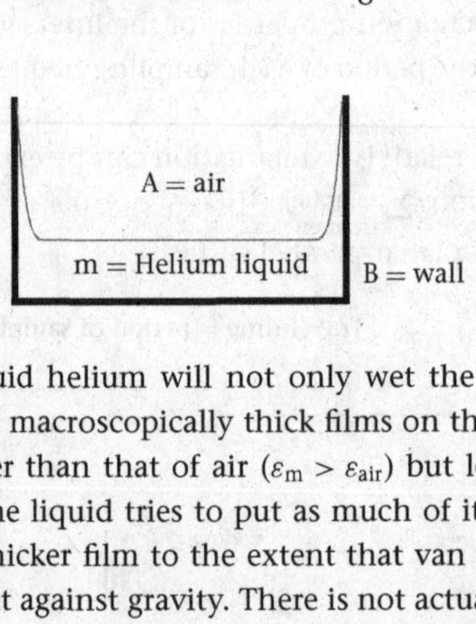

Put into a vessel, liquid helium will not only wet the walls but will also—in a gravity-defying act—form macroscopically thick films on the walls. The polarizability of liquid helium is greater than that of air ($\varepsilon_m > \varepsilon_{\mathrm{air}}$) but less than that of the walls ($\varepsilon_m < \varepsilon_{\mathrm{wall}}$). As a result, the liquid tries to put as much of itself as possible near solid walls, to create an ever-thicker film to the extent that van der Waals energy can pay for this mass displacement against gravity. There is not actually any repulsion. Rather, there is attraction to the wall. Its effect is to thicken the helium "medium" and make it move up the attractive wall.

The measured profile of thickness versus height reveals the balance between favorable van der Waals free energy of thickening and unfavorable work of lifting against gravity. Think of the helium as having equal energy everywhere on the surface, the positive work of lifting continuously compensated by the negative energy of interaction between the solid wall and the liquid helium surface. Factoring in the known weight density of the liquid gives the strength and correct form of van der Waals interaction. Helium film thicknesses, measured from 10 to 250 Å, show the coefficient of van der Waals interaction and inverse-cube variation of thickness versus height quantitatively consonant with predictions of the Lifshitz theory when the dielectric responses of air, helium, and ceramic wall are put into $-[A_{\mathrm{Ham}}/(12\pi l^2)]$.[37] Direct measurements in other asymmetric situations also show van der Waals repulsion.[38]

PR.3. How good are measurements? Do they really confirm theory?

It is curious how people working in different disciplines seek and see validation in different kinds of experiments. Only the liquid helium up-the-walls measurement, described in the preceding subsection, seems to satisfy most people that the Lifshitz theory quantitatively accounts for measured forces (see note 37 in the preceding subsection). As with the historical comments, this review of measurements is not intended to be exhaustive.[39]

Force-balance measurements and atomic-beam measurements give qualitatively or semiquantitatively satisfying results, but they present worries about the smoothness of the surfaces and about the effective "zero" of separation at contact. For example, with no fitting parameters, the atomic-beam measurements already described give the predicted $1/l^3$ power law for the energy of interaction between atom and surface, but then there is a stubborn 60–75% overestimate of the coefficient computed by Lifshitz theory fed with good spectral data versus the experiments that admit little adjustment (see note 34 in the subsection titled "Measurably strong"). This level of disagreement might be resolved by better spectra or by better modeling of the surface structure. Surface roughness is often the fingered culprit, but specific attempts to formulate and compute its consequences fail.[40] Formulations that include roughness[41] are often based on pairwise summation and must themselves be validated by systematic measurement or by comparison with exact solutions.

That's the worst. Several kinds of measurements reassure us that the Lifshitz strategy of converting spectra into forces works reliably enough to capture the key features of van der Waals forces.

The best? Physicists' awareness of the Casimir effect is growing so fast that we can expect many more careful measurements to test and to extend theory. Field fluctuations drive many kinds of events. In the physics literature, one can read heroic assertions such as (1997) "fundamental applications of the Casimir effect belong to the domains of the Kaluza-Klein supergravity, quantum chromodynamics, atomic physics, and condensed matter."[42] The purview of van der Waals = Hamaker = Casimir = Lifshitz = charge fluctuation = electrodynamic forces spans fundamental cosmology to living systems to mechanical devices and household commodities. Confidence in our prowess today allows us to say (1991) "van der Waals forces play a very important role in biology and medical sciences. They are in general particularly significant in surface phenomena, such as adhesion, colloidal stability, and foam formation. One could dare say that they are the most fundamental physical forces controlling living beings and life processes."[43] There are quasi-Casimir effects, analogous to those from fluctuating electromagnetic fields: Solid walls attract through the steric bumps of an intervening liquid-crystal[44]; spherical bubbles attract through the Bjerknes force of correlations in their spontaneously pulsing volumes.[45] The explosive growth of learning among physicists immediately translates into new possibilities for thinking and application in chemistry, biology, and engineering.

Atomic beams

Shot through a slit of varying width, a beam of sodium atoms loses intensity because of attraction to the walls of the slit. By looking at the number of atoms that make it through for a given slit separation, we can infer that the range and form of this scavenging attraction have the $1/l^4$ properties of "Casimir–Polder" or "fully retarded van der Waals" energy. Rather than the $1/l^3$ attraction energy expected when the finite velocity of light is neglected, the forces seen in this slit experiment are those that occur at much longer atom-to-surface separations.[46]

Nanoparticles

For various practical reasons, gold-coated surfaces are a favorite for measurements and for design of devices. For example, the qualitative features of the Casimir force were seen between gold-covered copper spheres at separations from 600 nm to 6 μm suspended in an electromechanical torsion pendulum.[47] Soon after, the Casimir force between gold surfaces, 0.1–1 μm apart, was used to impel a micromechanical system that might be the prototype of practical microscopic sensors or detectors.[48] The range and strength of the van der Waals interaction between surfaces might be appropriate to create switches, a desirable source of nonlinearity in a micromechanical oscillator.

Force microsopy on coated surfaces

A fussier preparation that uses "atomic force microscopy" (AFM) demonstrates that there can be a lateral force between surfaces sinusoidally coated with gold.[49] Sophisticates of earlier sections know that a probe such as AFM will "see" into a surface to a depth comparable with separation between probe tip and surface. When substrate structure varies with depth from surface, interpretation of van der Waals interactions can be problematic. Nevertheless, good success has been reported.[50,51]

Glass surfaces

Directly measured attraction between quartz surfaces in air gave the first successful quantitative test of the modern theory. In those measurements, in which the interacting bodies were a 10-cm- or 25-cm-radius sphere facing an optical flat, the theory had to be corrected by applying the Derjaguin transform. The results, for separations in the fully retarded regime, justified a triumphant, "Our experimental results agree with E. M. Lifshitz's theory. This proves P. N. Lebedeff's hypothesis concerning the electromagnetic nature of molecular forces."[52]

Quartz

The optical properties of quartz cranked into Lifshitz's formula, for plane-parallel surfaces but modified by the Derjaguin transform for a sphere and a flat, gave an attraction that fit neatly with experiments.

Mica

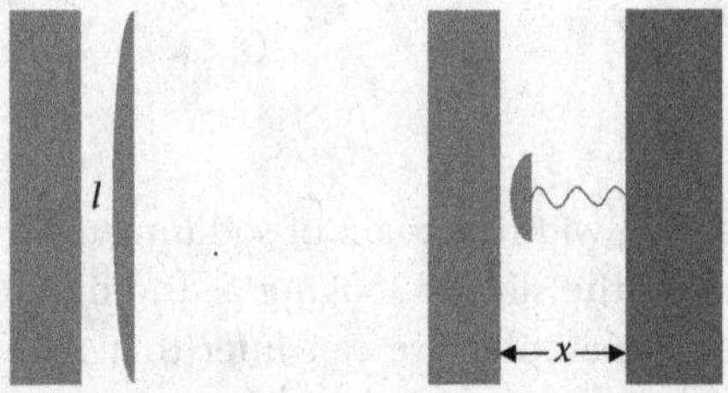

The connection between x set by the operator and l reached by the spring and the interaction across l is from a balance of forces, $F_{\text{spring}} + F_{\text{interaction}} = 0$. An applied change Δx in x evokes a measured change Δl in the separation l. Knowing the spring stiffness together with x and l gives the interaction force.

PROBLEM PR.14: Show how a Hookean spring works against an inverse-power van der Waals interaction in a force balance of a sphere and a flat surface.

Bilayers as thin films and as vesicles

There is a slight difference in the surface tension γ of the thin surfactant film and γ' of a thickened region of trapped oil film that angles away from it. From the separately measured surface tension of the flat film and the measured contact angle between the flat and the bulge, we can infer the inward pitch of the van der Waals attraction across the thin film.

PROBLEM PR.15: Convert an angle of deviation in a surface contour into an estimate of attractive energy across a film.

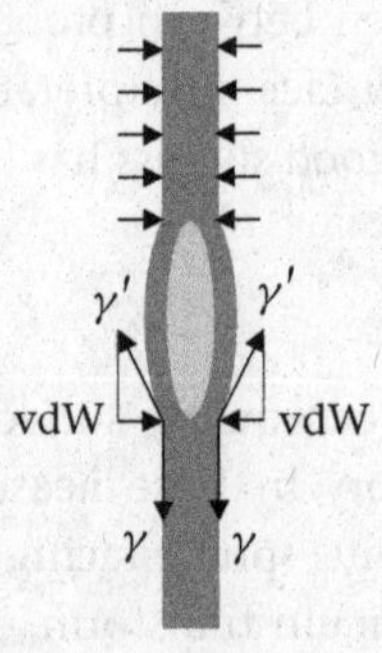

By the strength with which they flatten against each other, two juxtaposed bilayer vesicles, sucked into pipettes, also reveal van der Waals attraction.[53] The strength of flattening, corrected for thermal undulations by varying applied tension $\overline{\mathbf{T}}$ from the pipette, and combined with the interbilayer equilibrium separation measured by x-ray diffraction from multilayers,[54] produces a bilayer-across-water interaction coefficient of the same magnitude as the water-across-bilayer coefficient in the surface-tension measurement.

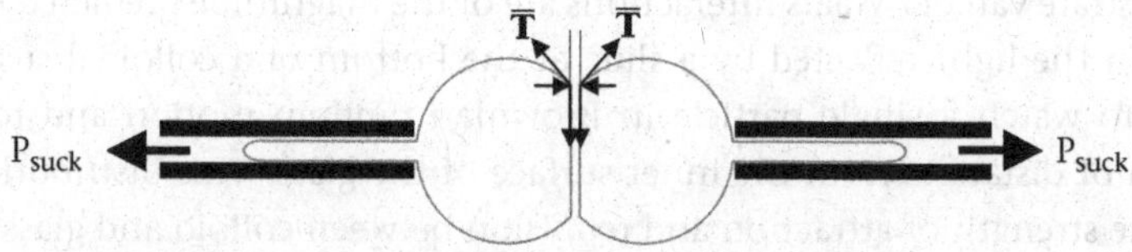

In practice, van der Waals forces appear within a mix of forces. Measured between bilayers in free suspension, they are mixed with lamellar motions as well as with repulsive hydration forces.

PROBLEM PR.16: What is the attractive energy that creates a flattening between two vesicles under tension $\bar{T}$?

Between bilayers immobilized onto substrates,[55] bilayer–bilayer interaction measurements also sense interactions involving the substrate. After compensation for these features, measurements on all three systems pleasingly agree.[56]

Cells and colloids

The corollary to the interaction of bilayers is that of lipid membranes bounding biological cells. Van der Waals forces of the magnitude seen between bilayers do not explain the strength and specificity of cell–cell or cell–substrate interaction. Still, forces inferred between gluteraldehyde-stabilized red cells and glass or metal substrates are of the magnitude predicted for van der Waals attraction balanced against electrostatic repulsion. The separation of cells from glass versus ionic strength of suspending solutions was measured by total internal reflection microscopy (TIRM).[57]

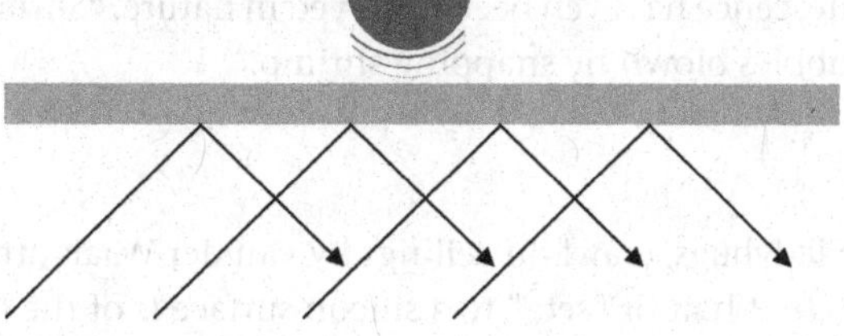

Similar to cells in size, colloidal particles such as the 1–10-μm polystyrene beads used industrially and in the lab also reveal significant van der Waals attraction. Observed with TIRM over a range of separations that cover nonretarded and retarded interactions, colloid–substrate van der Waals interactions are of the magnitude predicted by modern theory. From the light reflected by a slide at the bottom of a colloidal suspension it is possible to watch a single particle in Brownian random motion and to infer the distribution of distances from the inner surface of the glass. This distribution in turn indicates the strength of attraction and repulsion between colloid and glass or coated-glass substrate.[58]

Aerosols

Even relatively weak attraction between droplets or solid particles in aerosols suffices to create an enhanced collision rate that can change particle-size distributions and overall stability. Think in kT thermal-energy units. Alone, small suspended bodies do a Brownian bop, randomly jiggling from the kT kicks of the air. Should their mutually random paths bring two particles to separations comparable with their size, their van der Waals attraction energy also approaches kT. To previous randomness, attraction adds strength of purpose and increased chance of collision, aggregation, or fusion.[59]

From the cosmos to the kitchen: back from vacuum infinities to traditional interest in colloids and films.

There have been impressive efforts to collect spectra to compute dispersion forces for practical applications. For example, from measured absorption spectra and consequently computed forces, we learn how to design and produce thick-film resistors for computers and other electronic devices.[60]

Bright stuff. Sonoluminescence

When their sizes change rapidly, collapsing cavities burp electromagnetic waves. One suggested explanation is a "dynamical Casimir effect." The idea is that, at frequencies for which cavity and outer materials differ in dielectric response, the rapidly moving dielectric interface pushes on the virtual or zero-point electromagnetic fields to excite the blackbody spectrum of the cavity and even to create real photons.[61] Put another way, when the zero-point energy of the fields in the collapsing cavity does not have time to be dissipated as heat, the energy can be emitted as light. For the frequencies for which there is a step in dielectric response at the interface, the bursts have a spectrum similar to that of a blackbody at $\sim 10^5$ K. Well documented in the laboratory,[62] but still a puzzle,[63] sonoluminescence has even been observed in nature. "Shrimpoluminescence" comes out of the bubbles blown by snapping shrimp.[64]

Fun stuff

Geckos, and maybe ladybugs, crawl on ceilings by van der Waals attraction. The adhesive force of a gecko foot hair or "seta" to a silicon surface is of the right magnitude to lead to as much as 100-N adhesive force on one foot if all its setae act together. Geckos curl their toes to flatten and peel foot contact as though they had computed the difference between flat- and curved-surface attraction seen in the cube–sphere comparison made earlier in this chapter. It is worth thinking about these forces in terms of van der Waals interaction in order to design systematic measurements and practical materials with what we now know about dispersion forces.[65,66]

Slippery stuff, ice and water

Van der Waals interactions create a free-energy difference between material in a macroscopic phase and the same material in a film of finite thickness. This was wittily shown for pure ice in its vapor. Below freezing temperature, a thin film of water can be stable on ice. At the nominal triple point, the ~30-Å thickness of a coexisting water film exhibits significant relativistic screening. When a water film is between ice and another substance, film thickness can be a significant factor in material properties.[67] The variable energy of an oil film on water has been noted from the different spreading of different hydrocarbons on water. Small differences in polarizability show up as yes–no differences in spreading. The connection between dielectric and spreading properties of films demonstrated the utility of the modern Lifshitz theory at a critical time when it was not properly recognized.[68]

What about interfacial energies and energies of cohesion? Aren't van der Waals forces important there too, not just between bodies at a distance?

Emphatically, *yes*. Van der Waals forces probably dominate most interfacial energies and cohere most nonpolar materials. In fact, it was what he called the "continuity" of gas and liquid states that van der Waals was examining; attractive interactions in the gas suggested the existence of cohesive forces strong enough to condense them into liquids. The problem is simply that we do not have rigorous ways to compute these energies for which spacings are comparable with atomic size. Although van der Waals forces do contribute strongly to interfacial energies, rearrangements at the surface and the granularity of matter preclude use of rigorous continuum models except as qualitative guides.[69] There have been some strategies suggested to compute short-distance interactions such as self-energies, interfacial energies, and free energies of adsorption.[70] The lessons given in the present text tell you what to do within the approximations. We know that the modern theory gives an estimate of free energy between two flat surfaces of the same material brought together to the spacing of an interatomic distance, energy of the right magnitude to compare with the interfacial energies of the two interfaces that disappear on such "contact." We know that we can sum up the interaction of parts of molecules composing a liquid or solid, sum up the interaction by a $1/r^6$ law, and get an energy comparable with the cohesion energy of the material.

But we should also know that these are at best approximations and at worst only mind games. The "interfacial energy" between two surfaces that are only ~1 Å apart is an illusion. As two parallel surfaces approach such separations, atomic graininess of the surfaces becomes too important to allow us to imagine that the materials are ideally smooth continua.

Where these forces are best understood, they are also weakest. That is the regime in which they have been most studied. Fundamental work is still needed.

PROBLEM PR.17: To gauge the difference between long-range and short-range charge-fluctuation forces, compute the van der Waals attraction free energy between two flat parallel regions of hydrocarbon across 3 nm of vacuum. How does this long-range free energy compare with the $\sim$20 mJ/m^2 (= mN/m = erg/cm^2 = dyn/cm) surface tension of an oil?

PROBLEM PR.18: To get an idea about the onset of graininess, consider the interaction between one point particle and a pair of point particles at a small separation **a**; show how the interaction becomes proportional to $\mathbf{a}^2/z^2$ when the distance z between point and pair becomes much greater than **a**.

PROBLEM PR.19: Peel vs. Pull. Imagine a tape of width W with an adhesion energy G per area. Peeling off a length z removes an area of adhesion Wz and thereby incurs work GWz. Perpendicular lifting off a patch of area $A = 1$ cm^2 costs GA.

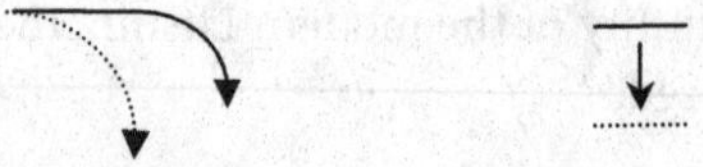

Assuming that adhesion comes from only a van der Waals attraction $G = -[A_{\mathrm{Ham}}/(12\pi l^2)]$, neglecting any balancing forces or any elastic properties of the tape, show that when tape–surface separation $l = 0.5$ nm (5 Å), $W = z = 0.01$ m (1 cm), and $G = 0.2$ mJ/m^2 (0.2 erg/cm^2), the peeling force is a tiny constant 0.002 mN = 0.2 dyn whereas the maximum perpendicular-pull-off force on this same square patch is an effortful 80 N = 8×10^6 dyn.

PR.4. What can I expect to get from this book?

The tools to think with and about charge-fluctuation forces.

Level 1, the introduction, speaks the language of the modern theory in order to help you develop intuition about the fundamental connection between material absorption spectra and charge-fluctuation forces. Its approximate formulae also show the connection between the shapes of bodies and the forces between them.

The precise tabulated formulae together with the essays on computation in Level 2 will let you compute interactions under a variety of instructive assumptions.

The conversation-in-front-of-a-blackboard style and the extensive footnotes of Level 3 can help you understand the origins of tabulated formulae and, more important, show you ways to derive new formulae. Nonphysicists need not fear to go to the back of this book. Levels 2 and 3 will give them much of what would have been learned in missed physics courses.

No excuses. Anyone who has passed a course in physical chemistry should be able to crunch this book for numbers and for equations. Engineers can treat it as a handbook with long explanations whose formulae can be plugged into original programs or packaged software (preferably both so as to understand what is being computed).

Readers with a physics background will find connections to systems that fall outside the usual purview of pure physics. The properties of these systems encourage physicists to use their prowess in new situations.

Readers from nonphysics backgrounds, for whom this text is primarily intended, will now be able to connect to the larger world of physics that has created the tools described in this book. Having digested even part of the material given here, the nonphysicist can profitably read several excellent physics texts.[71]

Unless your in-laws are physicists, you probably won't find much here for the next family reunion. However, if van der Waals forces do come up, tell the kids about geckos and ladybugs. Try to protect the children from the kind of 1930s concepts and modern elaborations that they are likely to have picked up from timid teachers or dumbed-down texts. Even if many of its instructors still work with outmoded ideas, the next generation need not be afraid to flex the muscle of modern physics.

LEVEL ONE

Introduction

Unlike deceptively simple electrostatic Coulombic interactions, van der Waals forces can appear complicated. Why? Electrostatic forces depend on only the response to constant electric fields from effectively stationary charges. Electrodynamic van der Waals forces depend on all the possible electromagnetic fields that come out of all possible modes of charges in motion. After learning the language of these fields and motions, seeing how measurements of reflection and absorption turn into calculable forces, and understanding how the electromagnetic wave equations lead us to formulate interactions for a huge variety of materials in variously shaped bodies, we are liberated. What we knew before about charge–charge interactions seems so confining once we can move into a newly accessible area.

The van der Waals interaction depends on the dielectric properties of the materials that interact and that of the medium that separates them. ("Dielectric" designates the response of material to an electric field across it: Greek δι- or διά- means "across.") The dielectric function ε can be measured experimentally by use of the reflection and transmission properties of light as functions of frequency. At low frequencies, the dielectric function ε for nonconducting materials approaches a limit that is the familiar dielectric constant. The dielectric function actually has two parts, one that measures the polarization properties and the other that measures the absorption properties of the material.

Does this sound as though all we have to do is to make some measurements and then plug into some function to calculate the van der Waals interaction? Almost, but there are a few conceptual steps. We massage reflection–absorption data on electromagnetic waves in order to speak of "imaginary frequencies." There is nothing unreal going on here, just an unfortunately chosen word intended to distinguish the oscillating form of ordinary waves from the exponential regression to equilibrium of charges that have spontaneously fluctuated to a particular transient configuration.

For the interactions themselves, we speak in terms of a free energy G: "free" to emphasize the idea that it is energy available to do physical work and "G" to indicate the Gibbs free energy for work done under conditions of controlled pressure and temperature. We think of bodies moved toward or away from each other in the usual way of measurements or experiments. Heat flows in or out to maintain temperature; volume changes to maintain pressure.[1]

L1.1. The simplest case: Material A versus material B across medium m

Begin with the interaction between two plane-parallel bodies separated by a distance l across a medium m. Each of the bodies is semi-infinite, filling the space to the left or the right of the surface (see Fig. L1.1).

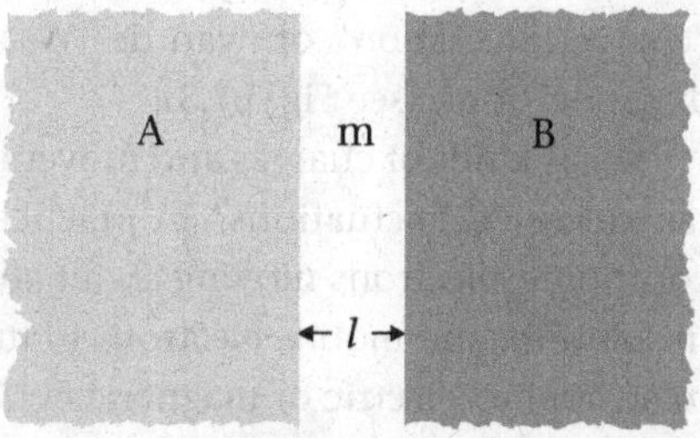

Figure L1.1

Of course, no body is infinitely or even semi-infinitely huge. The geometry here, chosen for mathematical convenience, in fact applies to cases in which the extent of the bodies is large compared with all other dimensions that are to be varied. Specifically, think of cases in which the depth away from the gap l and the lateral extent of the interface enclosing l are both much greater than the size of l itself. With changes in separation l, material m is drawn from or driven out to an external, infinitely abundant reservoir.

Schematically, think of A and B in a large medium m (see Fig. L1.2). From a distance they look like small particles; closer up they form a pair whose separation is much smaller than their sizes. At closer range, the space between them seems infinitesimal compared with their outer extent.

Although materials A, B, and m are electrically neutral, they are composed of moving charges. At any given instant there can be a net-positive or net-negative charge at a given location. Overall, there is an instantaneous configuration of charges throughout the space occupied by the bodies and a corresponding electromagnetic field throughout those bodies *and* the space around them. Moving charges also create fluctuating magnetic fields. Although they do not usually contribute as much as fluctuating electric fields, they are included in the full treatment after this Level 1 tutorial.

When two bodies are far apart, the dance of their charges and the corresponding field will depend on only their own material properties and that of the surrounding space. When they come to the finite separation l, the fields emanating from each body will act on the other body as well as on the intervening medium. If the dance of the charges and fields were to continue as if the other body were not nearby, the average effect on the energy of interaction would be zero.

In fact, the fields coming in and out of each body (and in and out of the intervening medium) distort the dance in such a way as to favor the occurrence of mutual

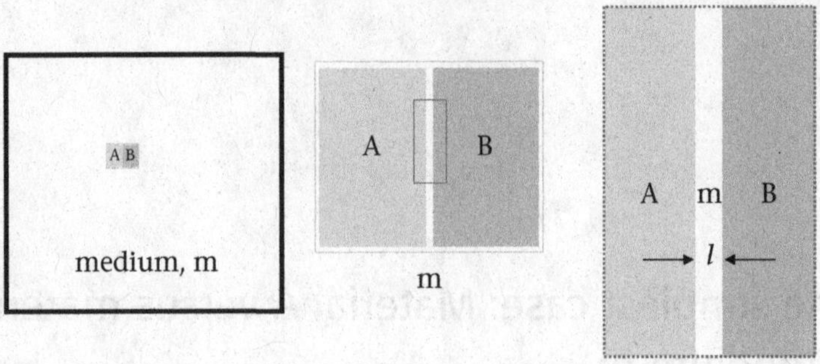

Figure L1.2

configurations and movements for a lower average electromagnetic energy. This change in probabilities, this mutual perturbation of the dance, creates the "charge fluctuation" or "van der Waals" or "electrodynamic" force (see Fig. L1.3).

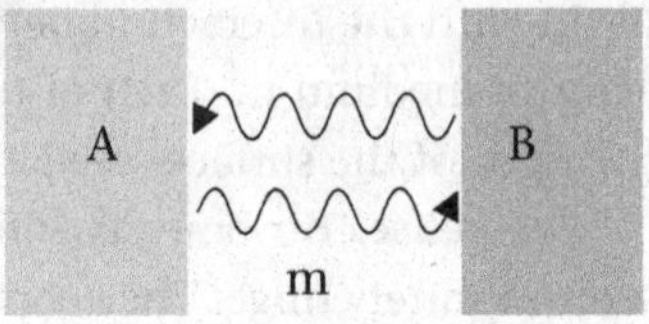

Figure L1.3

What kinds of charges and movements are important in these fluctuations? In practice, all kinds can contribute: electrons moving about atoms, vibrating and rotating dipoles, mobile ions in solution, and mobile electrons in metals. Every charge movement that can respond to an applied electric or magnetic field is a charge movement that can create transient electromagnetic fields. To visualize, we can speak of these movements through the filter of oscillations, speak of them as occurring at different "real" sinusoidal frequencies.

What drives these charge fluctuations? Slower low-frequency motions, such as ionic displacements, molecular rotations, and molecular vibrations, respond to thermal energy. The extreme of a low-frequency motion is the translation of ions or rotation of dipolar molecules, free to move under the vagaries of thermal motion, restrained by no chemical bonds but only by their relatively loose interactions with their surroundings. Time is not so important for this class of motions, collectively considered to occur in the realm of "zero frequency." The charges enjoy the leisure to sample all possible arrangements, limited only by the energetic cost of those arrangements compared with thermal-energy kT.

Faster motions correspond to higher frequencies, the UV frequencies that are typical of the absorption spectra of all materials. These motions, of electrons, are much faster than anything that occurs from thermal agitation. They are a consequence of the fundamental uncertainty in being able to specify simultaneously the position and momentum of a charged particle (or the energy and the duration of an electromagnetic field). It is these rapid motions that are an important source of charge-fluctuation forces.

Think in terms of wildly dancing charge configurations. Think of the ever-changing electric fields set up by these moving charges. Think of the spectrum of collective waves sent between the interacting bodies at different separations. The closer the separation, the stronger the coupling between fluctuations and the stronger the electrical signals between A and B across m.

The ease of time-varying charge displacement, measured as the time-dependent dielectric or magnetic permittivity (or permeability), is expressed by the dielectric function ε and magnetic function μ. Both ε and μ depend on frequency; both measure the susceptibility of a material to react to electric and magnetic fields at each frequency. For succinctness, only the dielectric function and the electrical fluctuations are described in the rest of this introductory section. The full expressions are given in the application and derivation sections of Levels 2 and 3.

Mathematical form of dependence on material properties

The electrodynamic work, or free energy, $G_{AmB}(l)$, to bring bodies A and B to a finite separation l from an infinite separation in medium m depends on there being a *difference* in the dielectric susceptibilities $(\varepsilon_A - \varepsilon_m)$ and $(\varepsilon_B - \varepsilon_m)$ of each of the bodies and the medium. If the ε's were the same for two adjacent materials, the interface between materials would be electromagnetically invisible. No electromagnetic interface, no "separation" l!

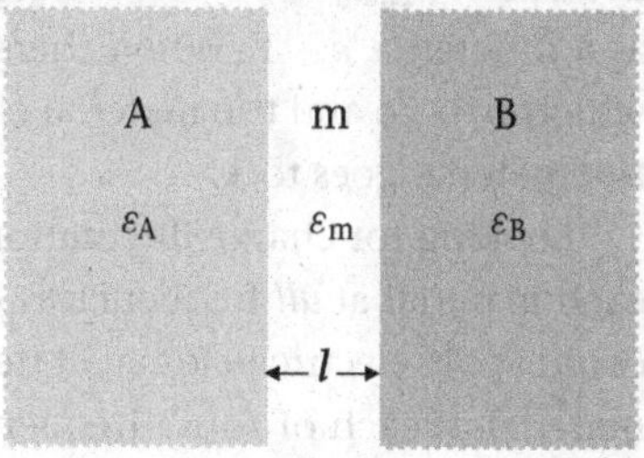

Figure L1.4

The susceptibilities ε_A, ε_B, and ε_m (see Fig. L1.4) come in as relative differences, that is, differences-over-sums, usually written as $\overline{\Delta}_{Am}$ and $\overline{\Delta}_{Bm}$:

$$\overline{\Delta}_{Am} = \frac{\varepsilon_A - \varepsilon_m}{\varepsilon_A + \varepsilon_m}, \quad \overline{\Delta}_{Bm} = \frac{\varepsilon_B - \varepsilon_m}{\varepsilon_B + \varepsilon_m}. \tag{L1.1}$$

Because the dielectric properties of each material are different functions of frequency, the $\overline{\Delta}$'s also depend on frequency. (There will be similar differences-over-sums for magnetic susceptibilities, μ_A, μ_m, and μ_B; these are usually small and are neglected in this introductory discussion.)

The dielectric response used here is a generalization of the dielectric "constant" in electrostatics. There, dielectric constants were introduced as a way to patch up Coulomb's law for charge–charge interactions in a medium other than a vacuum, where $\varepsilon_{vacuum} \equiv 1$, a way to recognize that the medium itself responds to the static or glacially changing electric fields set up by the charges. A good way to think about the response there in simple electrostatics and here in electrodynamics is to imagine the material placed between the two plates of a capacitor. Apply a known, oscillatory voltage; measure the amount of charge shifted onto the plates as the intervening material adapts to applied voltage. That amount of charge measures the capacitance, how much charge the plates can hold at a particular applied voltage; it is proportional to the polarizability or dielectric permittivity $\varepsilon(\omega)$ at frequency ω. The extra amount of charge delivered from the outside voltage source is proportional to and opposite in sign to the charge delivered by the intervening material to its outer boundaries at the plates (see Fig. L1.5).

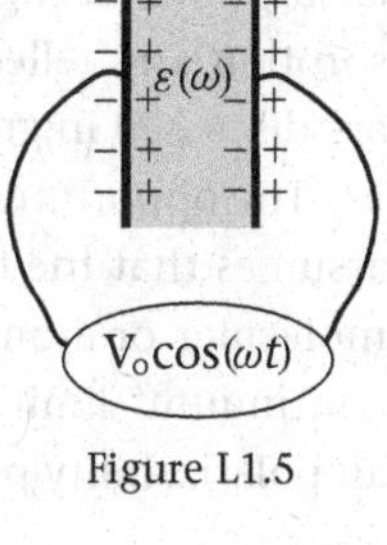

Figure L1.5

We speak here of a relative ε. If there were no material between the plates, then there would be no extra charge inside near the plates. The interesting part of $\varepsilon(\omega)$ is the extent to which

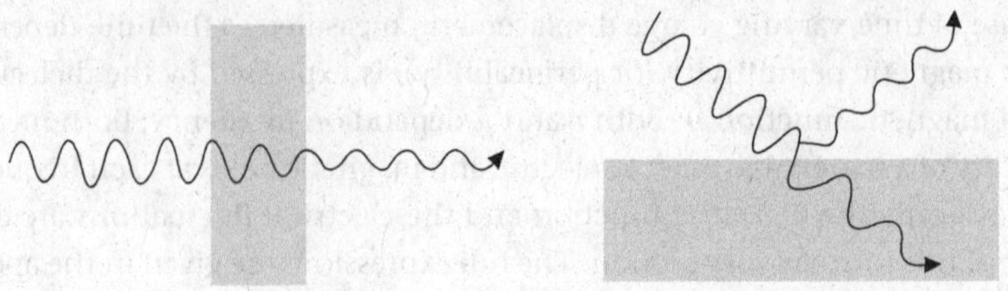

Figure L1.6

capacitance here differs from the case in which there is only vacuum between the plates. This relative $\varepsilon(\omega)$ is defined as the ratio of the capacitance to that vacuum case in which $\varepsilon_{\text{vacuum}} \equiv 1$. Alternatively we can think of the physically interesting part of ε as a difference $\varepsilon - 1$. When there is no material response, for example in the limit in which $\omega \to \infty$ and the material charges have no chance to follow the applied field, ε of the material goes to 1.

Looking for charge fluctuations (electrodynamics!), we must know the response of each material at *all* frequencies ω. It is as though we conceived an ideal capacitor that could apply electric fields at all frequencies and we could measure a susceptibility of the material at each of those frequencies. Most instructive to us, the material will absorb energy from the continually varying field and inform us of the natural frequencies at which its charges prefer to dance. When the frequency of the applied electric field is at a natural frequency, a frequency at which the material's charges would move spontaneously on their own without external disturbance, there is resonance, optimal charge displacement, and maximum absorption of energy.

Think of these natural or resonance frequencies as essentially the same as what a parent measures when determining just the right rate at which to push a kid on a swing. Pushing too fast or too slowly moves the kid plus swing but not with the same amplitude as pushing at their resonance frequency. Wrongly paced, the parent's energy is set against himself, having to slow a swing that is still moving toward him or to chase a swing that has already begun to move away. Only when pusher—the parent—and pushed—the kid plus swing—fall into rhythm do they move in smoothly periodic harmony. Only then does the outside effort turn into maximal motion; the work needed to keep pushing is only that absorbed by friction with hinges or air.

From audio frequencies (below which there are worries that are due to conductance and electrochemical reactions at the charged plates) to microwave frequencies (above which electronics are not fast enough), a capacitor is actually a practical device, not just a conceptually convenient picture for measuring dielectric response and energy absorption. At higher frequencies—IR, visible, UV—this information comes from absorption and reflection of electromagnetic waves (see Fig. L1.6). (Dielectric responses are discussed in great detail in Level 2, Subsection L2.4.A.)

The opportunity to use whole-material dielectric susceptibilities comes at a price. It assumes that the two interacting bodies A and B are so far apart that they do not see molecular or atomic features in their respective structures. This is the "macroscopic-continuum" limit: Materials are treated as macroscopic bodies on the laboratory scale; all polarizability properties are averaged out much as they average out in a capacitance measurement or in transmission–reflection measurements on macroscopic bodies. In fact, the price is not too high. Separation l (Figs. L1.1, L1.2, and L1.4) must be much greater than the graininess of the atomic packing. For most materials, separations down to $\sim$20 Å are probably allowed. This lower limit on the distance practical for computation does not prevent a very large number of important applications.[2]

Mathematical form of the charge-fluctuation free energy

For historical reasons, we write the interaction free energy $G(l)$ between the plane-parallel half-spaces of Fig. L1.1 in terms of a "Hamaker coefficient" A_{Ham} so as to put the interaction per unit area in a form

$$G(l) = -\frac{A_{\mathrm{Ham}}}{12\pi l^2}. \tag{L1.2}$$

To specify the case of material A interacting with material B across medium m of thickness l, write $G(l)$ relative to infinite separation in a subscripted form:

$$G_{\mathrm{AmB}}(l) = -\frac{A_{\mathrm{Am/Bm}}(l)}{12\pi l^2}. \tag{L1.3}$$

The qualification "relative to infinite separation" may seem academic. In spirit it is no different from the distinction between the "self-energy" of charging an isolated electrostatic charge and the Coulombic correction for its proximity to other charges. Because of the uncertainty principle and of zero-point fluctuations at all frequencies, the sum of all charge-fluctuation energies—in a medium and in vacuum—is infinite. Happily, the physically real part of this work or free energy $G_{\mathrm{AmB}}(l)$ is its derivative pressure. We integrate this pressure from infinite separation to finite l in order to extract the finite change in work done versus separation. (Similarly, it was the temperature derivative, rather than the spatial derivative, of the infinite electromagnetic energy that Planck associated with the observed energy of blackbody radiation in order to dispose of the same bothersome infinity.)

The general formula for $G_{\mathrm{AmB}}(l)$ has many pretty parts. To forestall any fear of using a complex expression, it is written here in Level 1 in a *nearly* exact version that can be easily taken apart for examination. This simplified version of the interaction formula holds for the case in which (1) relative differences in susceptibilities ε_{A}, ε_{B}, and ε_{m} are small, (2) differences in magnetic susceptibilities are neglected, and (3) the velocities of light in media A and B are set equal to its velocity in medium m. These heuristic approximations will *not* be necessary, and in fact should be avoided in actual computation.

To first approximation, the Hamaker coefficient $A_{\mathrm{Am/Bm}}(l)$ for interactions between interfaces Am and Bm is

$$A_{\mathrm{Am/Bm}}(l) \approx \frac{3kT}{2} \sum_{n=0}^{\infty}{}' \,\overline{\Delta}_{\mathrm{Am}} \overline{\Delta}_{\mathrm{Bm}} R_n(l), \tag{L1.4}$$

itself a function of distance l as well as of each material's dielectric susceptibility. The index n designates a sum over frequencies ξ_n, which are described in the next section. The prime in the summation indicates that the $n = 0$ term is to be multiplied by $1/2$. Alternatively, we can write the interaction free energy $G_{\mathrm{AmB}}(l)$ in this equal-light-velocity, small-difference-in-ε's approximation:

$$G_{\mathrm{AmB}}(l) \approx -\frac{kT}{8\pi l^2}\sum_{n=0}^{\infty}{}' \,\overline{\Delta}_{\mathrm{Am}}\overline{\Delta}_{\mathrm{Bm}} R_n(l), \tag{L1.5}$$

where the $\overline{\Delta}$'s are the relative differences in ε's already defined in Eq. (L1.1).

This variation of $A_{\mathrm{Am/Bm}}$ with separation comes from the "relativistic screening function" $R_n(l)$, which is subsequently elaborated. This factor becomes important at large distances when we must be concerned with the finite velocity of the electromagnetic wave. At short distances, $R_n(l) = 1$; the energy of interaction between two flat surfaces varies with the square of separation. At large distances any effective power-law variation of the interaction depends on the particular separation of materials and wavelengths of the operative electromagnetic waves between them.

One still hears the archaic designation "Hamaker *constant*" for $A_{\mathrm{Ham}} = A_{\mathrm{Am/Bm}}(l)$ from the time when people did not recognize that the coefficient could itself vary with separation l. In modern usage this spatially varying coefficient, evaluated at zero separation, remains a popular and useful measure of the strength of van der Waals forces.

Frequencies at which ε's, $\overline{\Delta}$'s, and R_n's are evaluated

The ε's are sampled over an infinite series of what are unfortunately called "imaginary" frequencies, a terrible name that for decades has probably stopped people from taking advantage of the modern theory. The dance of the charges is described in terms of their spontaneous fluctuation from their time-averaged positions and their gradual, exponential-in-time return to average positions. "Imaginary" frequencies describe this exponentially varying, rather than sinusoidally varying (purely oscillatory), process.

Recall that, when the imaginary number $i = \sqrt{-1}$ is used, an exponential $e^{i\theta}$ becomes

$$e^{i\theta} = \cos\theta + i\sin\theta. \tag{L1.6}$$

To use this form to describe the oscillations of a wave, we can say that θ increases with time t as $\theta = \omega t$ in terms of "radial frequency" ω that has the units of radians per second. This ω differs by a factor of 2π from usual frequency ν, whose units are cycles per second, or hertz. With

$$\omega \equiv 2\pi\nu, \tag{L1.7}$$

the oscillation of an ordinary wave $\cos(2\pi\nu t)$ reads more compactly as $\cos(\omega t)$. For an ordinary steady oscillation, we think of the mathematically real part of $e^{i\omega t}$, $\mathrm{Re}(e^{i\omega t}) = \cos(\omega t)$ that performs the usual "sinusoidal" oscillation.

Driven by uncertainty and random circumstance, responding with their own natural tendencies, electric charges move erratically. The idea of "frequency" must be generalized to include transient excursions as well as regular sinusoidal motions. The theory of charge fluctuations uses a "complex" frequency ω,

$$\omega = \omega_{\mathrm{R}} + i\xi, \tag{L1.8}$$

of real ω_{R} and imaginary $i\xi$ parts (where ω_{R} and ξ can be positive or negative but are *always* mathematically real quantities). An "oscillation" becomes

$$e^{i\omega t} = e^{i(\omega_{\mathrm{R}}+i\xi)t} = e^{-\xi t}e^{i\omega_{\mathrm{R}}t}, \tag{L1.9}$$

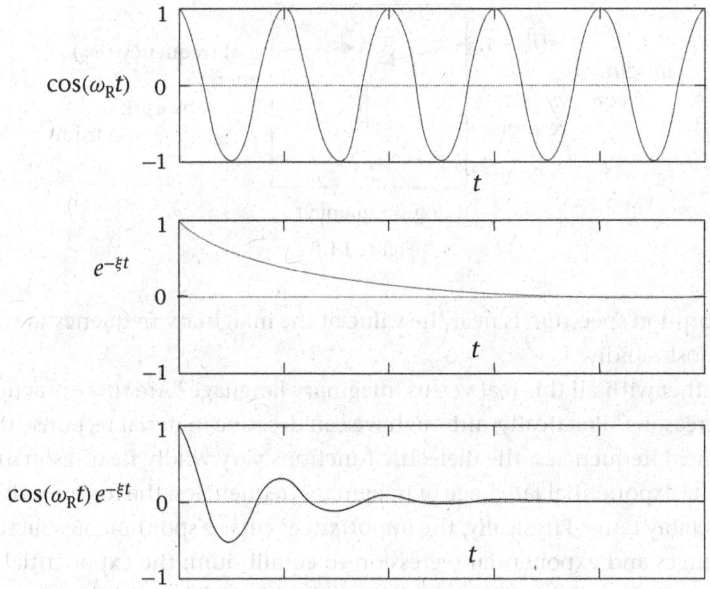

Figure L1.7

whereas $e^{i\omega t}$, formerly oscillatory with a purely real frequency, now also varies exponentially with factor $e^{-\xi t}$. There is no mystery in an "imaginary frequency." It is simply a convenient language for the average exponential variation of the electromagnetic-field and charge fluctuations that create interactions. A purely imaginary $\omega = i\xi$ with a positive ξ describes a purely decaying $e^{-\xi t}$ (see Fig. L1.7).

Remember the child on a swing. Grandma is a physicist. She gives the swing one good push, then watches. The swing moves back and forth with its own frequency; that oscillation is from ω_R. The amplitude of swinging gradually dies away; that dying away is from ξ. She watches both ω_R and ξ so as to push every once in a while at just the right moment, in phase with ω_R and before the amplitude decays to a degree that it elicits complaints.

Big brother is a lazy physics student. He decides to let his kid brother swing but expects random passersby to give the swing plus child a random push. He knows that each push will create oscillations ω_R. He hopes that these pushes will be delivered often enough compared with the decay time $1/\xi$ so that the kid plus swing will keep rocking at their natural ω_R.

Another way to think about an imaginary frequency in the context of a material property is to say that it tests the ability of the charges to follow an exponentially varying rather than an oscillating electric field. The larger ξ is and the faster the variation of the applied fields, the more difficult it is for the material charges to follow them. It is not surprising then that the dielectric response $\varepsilon(i\xi)$ to a purely imaginary positive frequency is a monotonically decreasing function of ξ down to the limiting value of 1, the response of a vacuum, when $\xi \to \infty$.

In practice, material properties $\varepsilon(\omega)$ are usually measured by use of real frequencies $\varepsilon(\omega_R)$. They are then mathematically transformed to functions $\varepsilon(i\xi)$ of positive imaginary frequencies. The value of the real frequency at which there is a maximum in the

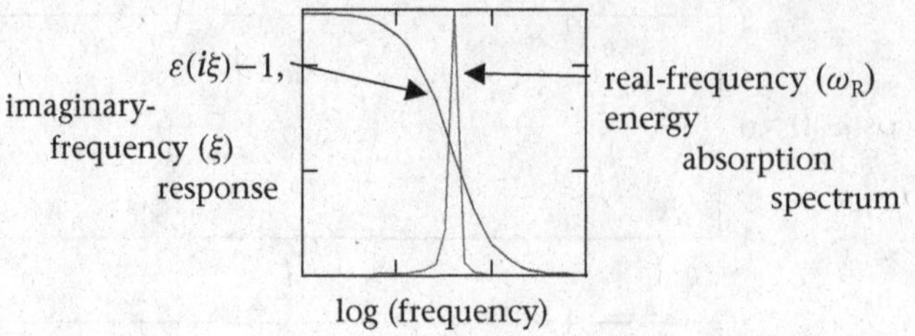

Figure L1.8

energy-absorption spectrum is near the value of the imaginary frequency at which $\varepsilon(i\xi)$ decreases most rapidly.

Why bother with all this real versus imaginary language? Are there practical as well as physical reasons? Practically, although we can describe material response in terms of oscillatory, real frequencies, the dielectric functions vary wildly near resonance; transformed to the exponential language of imaginary frequencies, the forms of the epsilons become tractably tame. Physically, the important events in spontaneous fluctuation are sudden changes and exponential regression to equilibrium; the exponential language is actually more appropriate. To compute van der Waals forces, it is not necessary but it is much easier to use the language of exponential variation.

For an idealized material with a single absorption frequency, the connection between absorption spectrum and imaginary-frequency response looks like Fig. L1.8.

In computation, susceptibilities ε are evaluated at a set of *discrete-positive* imaginary frequencies $i\xi_n$, where

$$\xi_n \equiv \frac{2\pi kT}{\hbar} n \tag{L1.10}$$

over the set of integers from $n = 0$ to infinity. These sampling frequencies come from quantum theory, where they are known as "Matsubara frequencies." In Level 3 of this text they are shown to come from properties of harmonic oscillators whose energy levels are spaced in units of $\hbar\xi_n$. (See Level 3, Subsection L3.2.A.) That is, in quantum thinking, these ξ_n sampling frequencies are those whose corresponding photon energies are proportional to thermal energy kT:

$$\hbar\xi_n \equiv 2\pi kTn. \tag{L1.11}$$

Here kT reflects the contribution of thermal agitation to charge fluctuation.

$2\pi\hbar = h$, Planck's constant ($h = 6.625 \times 10^{-34}$ J s or $= 6.625 \times 10^{-27}$ ergs s), reflects the necessity to think about $\hbar\xi_n$ in thermal units $2\pi kT$.

At room temperature, the coefficient $(2\pi kT_{\text{room}}/\hbar) = 2.41 \times 10^{14}$ rad/s so that

$$\xi_n(T_{\text{room}}) = 2.41 \times 10^{14}\, n \text{ rad/s}. \tag{L1.12}$$

In terms of photon energies $\hbar\xi_1(T_{\text{room}}) = 0.159$ eV,

$$\hbar\xi_n(T_{\text{room}}) = 0.159\, n \text{ eV}. \tag{L1.13}$$

The coefficient is big enough that $\xi_1(T_{\text{room}})$, the first sampling frequency after $\xi_0 = 0$, corresponds to the periods of IR vibrations. Its corresponding wavelength is $\lambda_1 = 2\pi c/\xi_1 = 7.82 \times 10^{-4}$ cm $= 7.82\,\mu$m $= 7.82 \times 10^4$ Å.

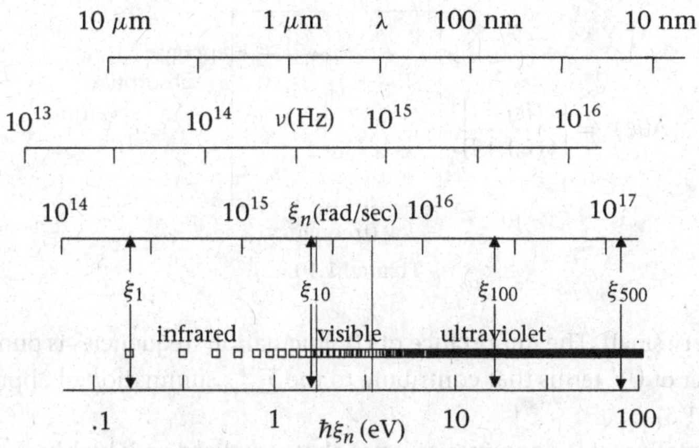

Figure L1.9

PROBLEM L1.1: How important is temperature in determining which sampling frequencies act in the charge-fluctuation force? For $n = 1$, 10, and 100, compute imaginary radial frequencies $\xi_1(T)$ at $T = 0.1$, 1.0, 10, 100, and 1000 K with corresponding frequencies $\nu_1(T)$, photon energies $\hbar\xi_1(T)$, and wavelengths λ_1.

Because of the wide range of relevant frequencies and the logarithmic way frequency is experienced, the charge-fluctuation spectrum is more efficiently plotted on a $\log_{10}$(frequency) or $\log_{10}(\hbar\xi_n)$ scale. Given the uniform spacing of the sampling frequencies ξ_n, the density of sampling frequencies goes up with frequency when plotted on a log scale. The number of sampling frequencies (indicated by the squares in Fig. L1.9, plotted for $T = T_{\text{room}}$) between $\xi = 10^{15}$ and 10^{16} rad/s is 10 times the number of sampling frequencies between $\xi = 10^{14}$ and 10^{15} rad/s.

Plotted on a log(frequency) or $\log_{10}(\hbar\xi_n)$ scale, the density of contributions to the charge-fluctuation force goes up even where the $\overline{\Delta}$'s are constant in ξ. This logarithmic property of frequency lets us see better the connection between the position of an absorption frequency and the spectrum of its contribution to charge-fluctuation forces. We can also see that details of spectra in the IR region need not be as important as those in the UV, where sampling is much denser. Watch for this density of sampling in what follows [cf. Figs. L1.22(a) and L1.22(b) in the section on the van der Waals interaction spectrum].

At room temperature, Fig. L1.9, the full sampling over the set of ξ_n involves the zero-frequency term and then a few terms in the IR followed by a very large number of terms in the visible and the UV. For this reason, most of the time, we expect UV spectral properties to dominate van der Waals interactions numerically, even though at these frequencies the magnitudes of Eq. (L1.1),

$$\overline{\Delta}_{\text{Am}} = \frac{\varepsilon_{\text{A}} - \varepsilon_{\text{m}}}{\varepsilon_{\text{A}} + \varepsilon_{\text{m}}}, \qquad \overline{\Delta}_{\text{Bm}} = \frac{\varepsilon_{\text{B}} - \varepsilon_{\text{m}}}{\varepsilon_{\text{B}} + \varepsilon_{\text{m}}},$$

$$\overline{\Delta}(i\xi)^2 = \left[\frac{\varepsilon(i\xi)-1}{\varepsilon(i\xi)+1}\right]^2$$

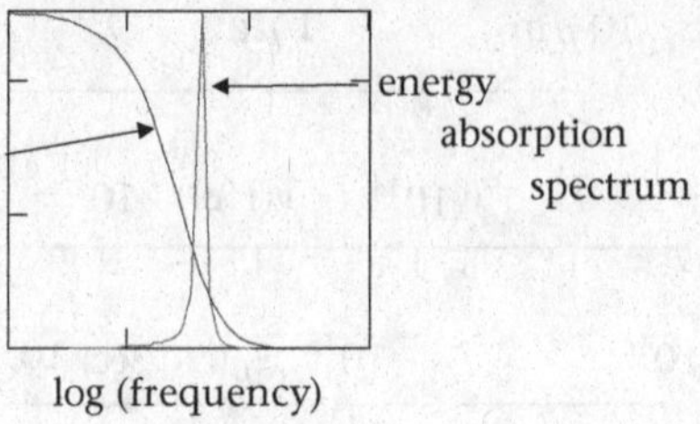

Figure L1.10

are usually very small. The importance of UV fluctuation frequencies is purely a matter of the number of UV terms that contribute to the $\sum_{n=0}^{\prime\infty}$ summation of approximations (L1.4) and (L1.5).

This expectation of UV dominance is not always realized with highly polar materials with large ε's at very low frequencies. In particular, large differences in the $\varepsilon(0)$ for water and for nonpolar materials such as hydrocarbons can create $\overline{\Delta}(n = 0)$'s whose magnitude, $\sim(80 - 2)/(80 + 2)$, is almost equal to 1. This first term in the summation over n stands out compared with its successors. At IR frequencies, the value of $\varepsilon(i\xi_1)$ for polar liquids is already down to ~ 2 or 3, and the $\overline{\Delta}$'s are already small compared with the 1.

At higher frequencies, photon energies are much larger than kT, and the corresponding charge-fluctuation excitation is not driven by temperature. At these higher frequencies too, the dielectric-response functions change so slowly that the summation over n can be accurately smoothed into an integral over frequency, an integral that shows no explicit dependence on temperature.

PROBLEM L1.2: If you take the kT factor too seriously, then it looks as though van der Waals interactions increase linearly with absolute temperature. Show that, for contributions from a sampling-frequency range $\Delta\xi$ over which $\overline{\Delta}$'s change little, there is little change in van der Waals forces with temperature, *except* for temperature-dependent changes in the component ε's themselves.

PROBLEM L1.3: If the interaction is really a free energy versus separation, then it must have energetic and entropic parts. What is the entropy of a van der Waals interaction?

Consider a case in this smoothed limit within which A and B are identical, nonpolar materials and have one absorption peak at a frequency whose photon energy is

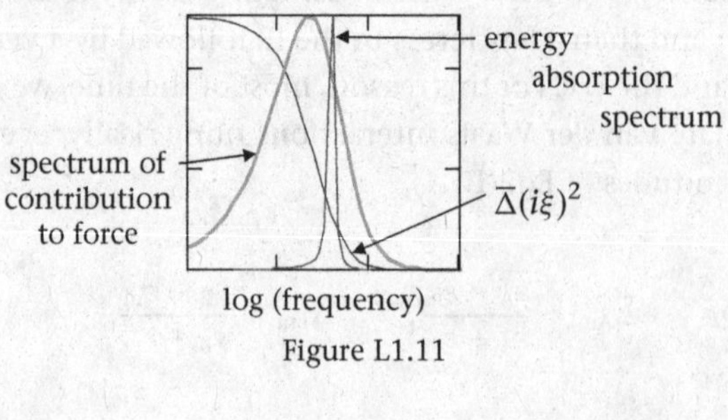

Figure L1.11

Table L1.1. Language, units, and constants

Summation $\sum_{n=0}^{\prime\infty}$ over characteristic frequencies of fluctuation.
Index $n = 0$ to ∞ (summation to infinite frequency in principle, but to a sufficiently large value in practice).
The prime indicates that the $n = 0$ term is to be multiplied by 1/2.
Characteristic frequencies $\xi_n = [(2\pi kT)/\hbar]n$.
k = Boltzmann's constant
$= 1.3807 \times 10^{-16}$ erg/K
$= 1.3807 \times 10^{-23}$ J/K
$= 8.6173 \times 10^{-5}$ eV/K.
$\hbar = 2\pi h$ (Planck's constant)
$= 1.0546 \times 10^{-27}$ erg s
$= 1.0546 \times 10^{-34}$ J s
$= 6.5821 \times 10^{-16}$ eV s.
1 electron volt (eV) $= 1.601 \times 10^{-12}$ erg $= 1.601 \times 10^{-19}$ J.
At $T = T_{\text{room}} = 20\ ^\circ$C, $[(2\pi kT_{\text{room}})/\hbar] = 2.411 \times 10^{14}$ rad/s.
$2\pi kT_{\text{room}} = 0.159$ (eV).
$kT_{\text{room}} = 4.04 \times 10^{-14}$ erg $= 4.04 \times 10^{-21}$ J.
"Imaginary" pertains to exponential variation.
Complex frequency ω has real ω_R and imaginary $i\xi$ parts: $\omega = \omega_R + i\xi$.
The time variation of a signal goes as $e^{+i\omega t} = e^{+i\omega_R t}e^{-\xi t}$.
A purely imaginary frequency goes exponentially as $e^{-\xi t}$.

much greater than kT. Let m be a vacuum with $\varepsilon_m = 1$. Then $\overline{\Delta}_{Am}(i\xi)\overline{\Delta}_{Bm}(i\xi) = \overline{\Delta}(i\xi)^2 = \{[\varepsilon(i\xi) - 1]/[\varepsilon(i\xi) + 1]\}^2$. Because the dielectric response $\varepsilon(i\xi)$ decreases monotonically in ξ, so does $\overline{\Delta}(i\xi)^2$ (see Fig. L1.10).

Because of the varying density of sampling frequencies, the spectrum of contributions to the charge-fluctuation force has a maximum. Over the frequency range at which $\overline{\Delta}(i\xi)^2$ is constant, the contribution to the force increases as a function of frequency because of the increasing density of sampling frequencies. Over the frequency range at which $\overline{\Delta}(i\xi)^2$ decreases to zero, the contribution to the force ceases (see Fig. L1.11).

About the frequency spectrum

We experience frequencies logarithmically. An "octave" is a doubling of frequency, a "decade" a factor of 10. The ranges of electromagnetic frequencies are given in terms of the log of frequency. The range of visible light covers a mere octave; other ranges of frequencies—acoustic, microwave, IR, UV, x ray, cosmic rays—cover decades. See Table L1.2.

Retardation screening from the finite velocity of the electromagnetic signal

The last factor in the summands in approximation (L1.4) for $A_{Am/Bm}$ is R_n, a dimensionless "relativistic retardation correction factor" that is due to the finite velocity of light.

Table L1.2. The frequency spectrum

Range	$\log_{10}[\nu(\text{Hz})]$	$\log_{10}[\omega(\text{rad/s})]$	λ, in vacuo	$h\nu = \hbar\omega$ (eV)
Acoustic	~1–4.3 (20 kHz)	~1–5.1	$>1.5 \times 10^6$ cm	$<10^{-10}$
Microwave	4.3–11.5 (300 GHz)	5.1–12.3	1.5×10^6–0.1 cm	$\sim10^{-10}$–$\sim10^{-3}$
IR	11.5–14.6 (4×10^{14} Hz)	12.3–15.4	0.1 cm–0.8 μm	$\sim10^{-3}$–1.6
Visible	14.6–14.9 (8×10^{14} Hz)	15.4–15.7	800–400 nm	1.6–3.3
UV	14.9–17 (10^{17} Hz)	15.7–17.8	4000–30 Å	3.3–416
X ray	17–22 (10^{22} Hz)	17.8–22.8	30–3×10^{-4} Å	416–4.2×10^7
Cosmic ray	>22	>22.8	$<3 \times 10^{-4}$ Å	$>4.2 \times 10^7$

1 eV equals the energy $h\nu$ of a photon of frequency $\nu = 2.416 \times 10^{14}$ Hz; or $\hbar\omega$ of a photon of radial frequency $\omega = 1.518 \times 10^{15}$ rad/s; or hc/λ of a photon of wavelength $\lambda = 1.242 \times 10^4$ Å $= 1242$ nm $= 1.242$ μm and equals 40 kT at room temperature.
$\omega = 10^{15}$ rad/s corresponds to a photon energy $\hbar\omega = 0.66$ eV.
$\nu = 10^{14}$ Hz (or cycles/second) corresponds to 0.414 eV.

Here is another of those complicated-sounding terms for a simple idea. Think about the dancing charges. Think of the time it takes them to do a particular step. Think of the travel time for the electric pulse created by that dancing charge to go a distance l to another body plus the time it takes for the charges on that other body to fall into step and to send back a signal. Altogether it is a round-trip that is $2l$ in length; the time it takes is this $2l$ divided by the velocity of light $c/\varepsilon_m^{1/2}$ in the intervening medium. (To keep language consistent, we write the square root of dielectric permittivity rather than the more usual index of refraction $n_{ref} = \varepsilon_m^{1/2}$. Also here in Level 1, we assume that the magnetic permeabilities μ_A, μ_m, and μ_B differ negligibly from the $\mu = 1$ of free space). This travel time,

$$\frac{2l}{c/\varepsilon_m^{1/2}}, \tag{L1.14}$$

is to be compared with the characteristic periods of charge fluctuation or lifetimes $1/\xi_n$. The characteristic frequencies ξ_n set the rhythms of dances. Interaction couples the dancers at each particular rhythm.

The pertinent ratio, r_n, the travel time relative to the fluctuation lifetime $1/\xi_n$ becomes

$$r_n = \left(\frac{2l}{c/\varepsilon_m^{1/2}}\right) \bigg/ \left(\frac{1}{\xi_n}\right) = \frac{2l\varepsilon_m^{1/2}\xi_n}{c}. \tag{L1.15}$$

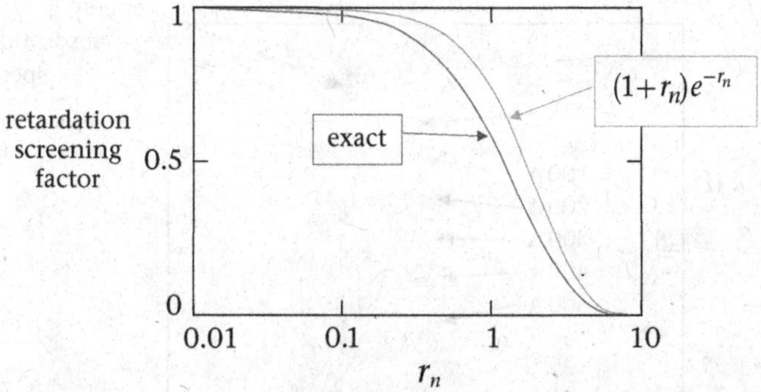

Figure L1.12

In the approximation in which the velocity of light in A and B is taken to be equal to that in medium m, the relativistic screening of a particular frequency fluctuation can be written in an intuitive form (see Fig. L1.12):

$$R_n(l;\xi_n) = [1 + r_n(l;\xi_n)]\, e^{-r_n(l;\xi_n)} \le 1. \tag{L1.16}$$

When the ratio r_n is much less than 1, that is, when the light signal travels across and back much faster than the length of time $1/\xi_n$ that a fluctuation endures, then $R_n = 1$. There is no loss of signal between the two bodies; then the finite velocity of light does not affect the van der Waals interaction from sampling frequency ξ_n. (For the "exact" screening factor see Level 2, Subsection L2.3.A.)

PROBLEM L1.4: For each sampling frequency ξ_n, or its corresponding photon energy $\hbar\xi_n$, what is the separation l_n at which $r_n = [(2l_n\varepsilon_m^{1/2}\xi_n)/c] = 1$?

When separations are small enough that effectively $r_n \to 0\,(R_n \to 1)$ for *all* contributing frequencies, the interaction between A and B across m [Eq. (L1.3)],

$$G_{AmB}(l) = -\frac{A_{Am/Bm}(l)}{12\pi l^2},$$

varies as the inverse square of separation. This is because, in the $l \to 0$ limit, the Hamaker coefficient [approximation (L1.4)],

$$A_{Am/Bm}(l) \approx \frac{3kT}{2}\sum_{n=0}^{\infty}{}'\, \overline{\Delta}_{Am}\overline{\Delta}_{Bm} R_n(l),$$

no longer depends on separation through R_n so that

$$A_{Am/Bm}(0) = \frac{3kT}{2}\sum_{n=0}^{\infty}{}'\, \overline{\Delta}_{Am}\overline{\Delta}_{Bm}. \tag{L1.17}$$

At the other limit, when $r_n \gg 1$, the time of travel of the signal across the gap l and back is longer than the lifetime of the electromagnetic fluctuation. The damping term goes to zero almost exponentially:

$$R_n = (1 + r_n)e^{-r_n} \to r_n e^{-r_n} \to 0.$$

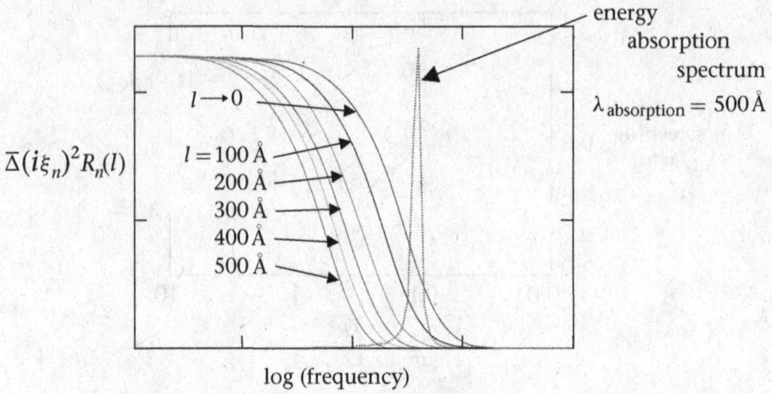

Figure L1.13

There is no longer a contribution to van der Waals forces at large distances or from high frequencies (large ξ_n, small period of fluctuation $1/\xi_n$). At such large distances the signal takes too long for the movement of charges to fall into step. The "first" charge configuration has moved too far by the time the signal comes back from the other side for it to have set up a correlation with the "second" configuration. Only the $\xi_0 = 0$ term, for which R_0 always equals one, is not subject to this loss of correlation because of the finite velocity of light.

Between these limits, the Hamaker coefficient $A_{Am/Bm}(l)$ varies with separation in a way that depends on the particular dielectric properties of the interacting materials. Beginning with contributions from the highest important frequencies, there is progressive relativistic damping of the van der Waals force. Consider again the interaction across a vacuum of two identical materials with one important absorption frequency at a wavelength $\lambda_{absorption} = 500$ Å (see Fig. L1.13).

Even at a separation of 100 Å, 1/5 the principal absorption wavelength, there is damping. By a separation $l = \lambda_{abs} = 500$ Å, practically no contribution occurs from the region of the absorption frequency. The effect of retardation screening can also be seen clearly in the changes of the density spectrum of contribution to the interaction energy at different frequencies (see Fig. L1.14).

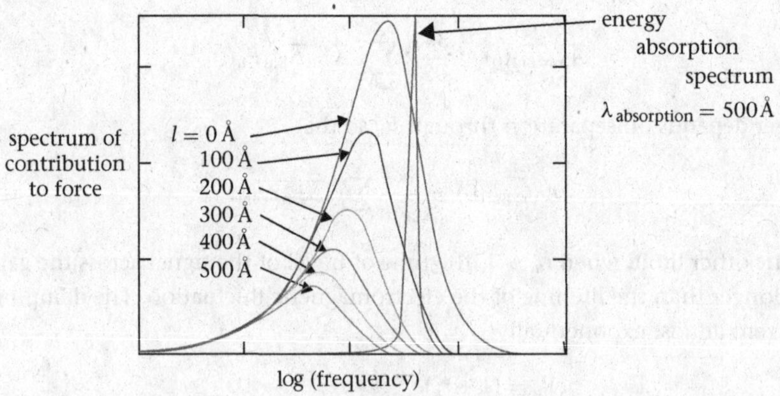

Figure L1.14

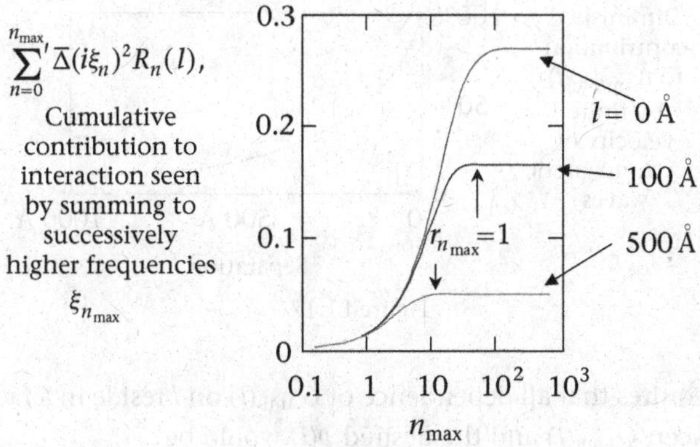

Figure L1.15

We can watch the accumulation of this spectrum by summing up to a finite maximum frequency $\xi_{n_{max}}$ as in $\sum_{n=0}^{\prime n_{max}} \overline{\Delta}(i\xi_n)^2 R_n(l)$. We can then see the full interaction develop when n_{max} grows big enough that there are no further contributions. The larger the separation l, the sooner the "big enough" occurs such that the partial sum to n_{max} is effectively to infinite frequency as in $\sum_{n=0}^{\prime\infty} \overline{\Delta}(i\xi_n)^2 R_n(l)$. This limit is reached when $r_n = r_{n_{max}} = 1$. For the example chosen with one 500-Å absorption wavelength, if $l = 100$ Å, the story is over by $n_{max} \approx 62$; if $l = 500$ Å, then by $n_{max} \approx 12$ (see Fig. L1.15).

The integrated consequence of this retardation screening shows up as a change in the contribution to the Hamaker coefficient $A_{Am/Bm}(l)$. This diminution in $A_{Am/Bm}(l)$ looks different when plotted versus log(l) (see Fig. L1.16) or plotted versus l by itself (see Fig. L1.17).

Here the Hamaker coefficient is constant only over a very small range of separations. With an absorption wavelength of 500 Å, there is a diminution in $A_{Am/Bm}(l)$ by ~50% at 100-Å separation. At a separation l equal to the absorption wavelength itself the contribution has dropped to ~25% of its $l = 0$ value.

Effective power law of van der Waals interaction versus separation

People often like to speak of the interaction energy $G_{AmB}(l) = -\{[A_{Am/Bm}(l)]/(12\pi l^2)\}$ as though it varied as a power p of separation l. If one insists on such terminology, then it is necessary to recognize that the power p must itself vary with separation. Formally,

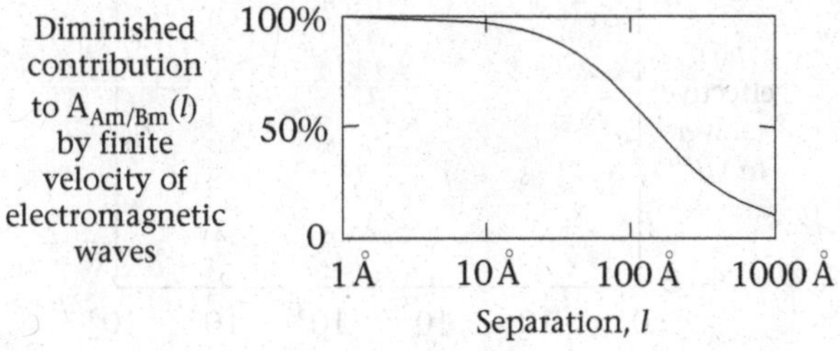

Figure L1.16

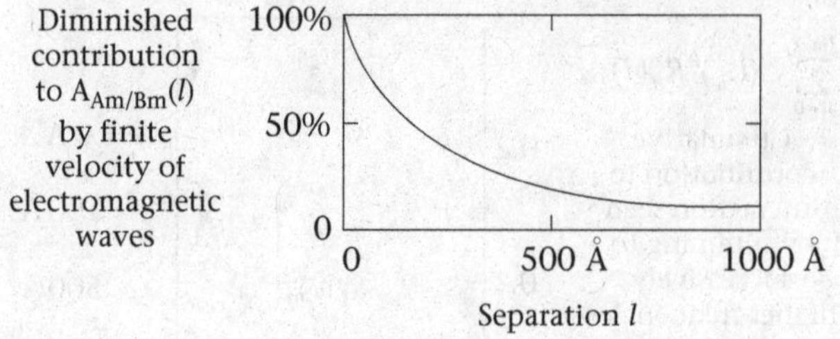

Figure L1.17

this insistence requires that all dependence of $G_{AmB}(l)$ on l reside in a factor $1/l^{p(l)}$. The connection between $G_{AmB}(l)$ and the desired $p(l)$ would be

$$p(l) = -\frac{\mathrm{d}\ln[G_{AmB}(l)]}{\mathrm{d}\ln(l)}. \tag{L1.18}$$

PROBLEM L1.5: Show how this power law emerges from free energy $G_{AmB}(l)$.

The one-absorption-wavelength example shows the relation between $p(l)$ and the computed $G_{AmB}(l)$ (see Fig. L1.18).

For l less than ~20 Å, the power $p(l)$ is almost constantly 2. When separation l approaches the 500-Å wavelength of absorption, the power grows rapidly to 3. Then the exponent remains near 3 from $l \approx 1000$ to 10,000 Å before it plunges back down to 2 in the limit of large separations. Why?

At near-zero separations, retardation screening is of no account. The summation $\sum_{n=0}^{\prime\infty} \overline{\Delta}_{Am}\overline{\Delta}_{Bm} R_n$ over $\overline{\Delta}_{Am}\overline{\Delta}_{Bm} = \overline{\Delta}(i\xi_n)^2 = \{[\varepsilon(i\xi_n) - 1]/[\varepsilon(i\xi_n + 1)]\}^2$ is complete. That is, the terms in the sum go to zero because $\overline{\Delta}(i\xi)^2$ goes to zero before retardation screening $R_n(l;\xi_n) = [1 + r_n(l;\xi_n)]e^{-r_n(l;\xi_n)} \le 1$ can act to cut down the terms any further.

At separations greater than ~20 Å, retardation screening begins progressively to snuff out the higher-frequency contributions. The terms in summation die out because both $\overline{\Delta}(i\xi_n)^2$ and $R_n(l;\xi_n)$ go to zero over the same range of ξ_n.

At yet larger separations, all frequencies over which $\overline{\Delta}(i\xi_n)^2$ decreases on its own have been screened out; for the remaining terms, $\overline{\Delta}(i\xi_n)^2$ is essentially constant. It has its zero-frequency value $\overline{\Delta}(\xi = 0)^2$, but the screening factor continues to act. The result is to make the coefficient $A(l)$ decrease as the first power in l and to make the total interaction energy vary as the inverse-third power of separation. Sometimes referred to as the purely retarded limit, this peculiar behavior is more rigorously examined in Level 2.

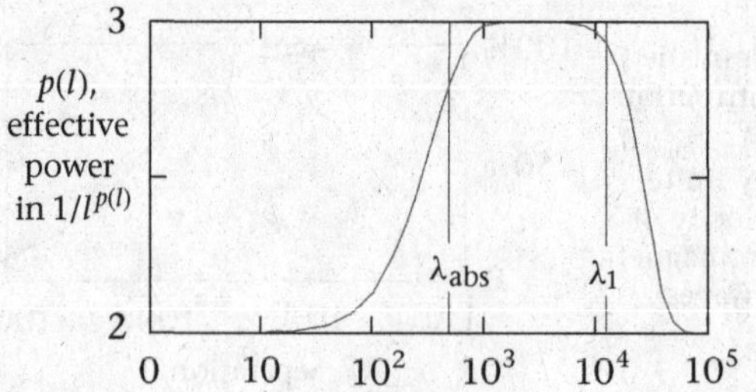

Figure L1.18

PROBLEM L1.6: How does convergence of the sum under the influence of only the retardation function $R_n(l; \xi_n)$ create the appearance of the $1/l^3$ variation of free energy?

Finally, at separations greater than the wavelength $\lambda_1 = 2\pi c/\xi_1 = 7.82 \times 10^4$ Å (at room temperature), the effective power $p(l)$ of van der Waals attraction between planar surfaces descends back to 2. Retardation has screened out even the fluctuations associated with the lowest finite sampling frequency ξ_1. The first term, $\frac{1}{2}\overline{\Delta}(\xi_{n=0})^2$, is all that remains of the summation $\sum_{n=0}^{\infty\,\prime} \overline{\Delta}_{\mathrm{Am}}\overline{\Delta}_{\mathrm{Bm}} R_n$. This $n = 0$ term endures to assert simple inverse-square variation versus separation similar in form to the power law at very small separations.

Can we say that the longest-range behavior of van der Waals forces is due to the finite velocity of light? In a negative sense, yes. Finite velocity squelches all finite-frequency contributions. But it is probably better to recognize the quantum nature of light. Its discretization of the sampling frequencies leaves us with the kT-dependent zero-frequency contribution. Except at physically unreachable $T = 0$ K, this $n = 0$ term will stand out at the largest distances. Because of its formal durability and because it can be large in polar systems, this term attracts interest. Nevertheless it too is vulnerable to its own kind of screening from the ooze of mobile-charge conductivity at very low frequencies.

The one-absorption example used for illustration here is appropriate to most van der Waals charge-fluctuation forces because of the important, usually dominant, UV-frequency range of fluctuations. Temperature is not usually a consideration; the summation over sampling frequencies ξ_n can often be smoothed into a continuous integration. However, retardation screening acts in any situation in which separations are more than a mere 20–30 Å. Only for distances less than ~20 Å and, sometimes, for distances greater than 10,000 Å, can the van der Waals interaction between parallel-planar surfaces be said to vary by the $1/l^2$ power law that it demurely reveals in its simplest representation.

Van der Waals pressure

Seeing that each frequency term in the summation of approximation (L1.5) has its own dependence on separation immediately informs us that the derivative pressure,

$$P(l) \equiv -\frac{\partial G(l)}{\partial l}, \tag{L1.19}$$

is the sum of derivatives of individual terms. With $R_n(l) = [1 + r_n(l)]e^{-r_n(l)}$ from Eq. (L1.16) and $r_n(l) = [(2\varepsilon_{\mathrm{m}}^{1/2}\xi_n)/c]l$ from Eq. (L1.15), the pressure between A and B across m is

$$P_{\mathrm{AmB}}(l) \approx -\frac{kT}{4\pi l^3}\sum_{n=0}^{\infty}{}' \overline{\Delta}_{\mathrm{Am}}\overline{\Delta}_{\mathrm{Bm}}\left(1 + r_n + \frac{r_n^2}{2}\right)e^{-r_n}, \tag{L1.20}$$

with the convention that a negative pressure denotes attraction. Derivatives of this kind will be useful for deriving forces between small particles. At very small separations such that all $r_n \to 0$ this pressure varies as l^{-3}. Otherwise, as with its integral, the interaction free energy, spatial variation is no simple power law.

PROBLEM L1.7: Take the l derivative of $G_{AmB}(l)$, approximation (L1.5), in the equal-light-velocities approximation, to obtain $P_{AmB}(l)$, approximation (L1.20).

Asymmetric systems

Specificity

Written as the sum over the products, $-\overline{\Delta}_{Am}\overline{\Delta}_{Bm}$, electrodynamic interactions show a kind of specificity. They prefer the interaction of like, resonating materials compared with the interaction of unlike species. This specificity, though, occurs as a difference between possible combinations of interactions: If A and B are made of different substances, then the sum of A–A and B–B interactions is always less than *two* interactions A–B. Term by term in the $\sum_{n=0}^{'\infty}$ sum, there is the inequality [from the inequality of geometric and arithmetic means that says ab $\leq$ (a^2 + b^2)/2]:

$$-\overline{\Delta}_{Am}\overline{\Delta}_{Am} - \overline{\Delta}_{Bm}\overline{\Delta}_{Bm} \leq -2\overline{\Delta}_{Am}\overline{\Delta}_{Bm}. \tag{L1.21}$$

This constraint translates into an inequality in free energies,

$$G_{AmA}(l) + G_{BmB}(l) \leq 2G_{AmB}(l), \tag{L1.22}$$

and in derivative pressures,

$$P_{AmA}(l) + P_{BmB}(l) \leq 2P_{AmB}(l). \tag{L1.23}$$

This inequality is not worth very much. It does *not* say that A will interact with A more strongly than with B. It says only that, given the various combinations (A–A, B–B, A–B), the strengths of interactions compete as shown. It does suggest that if there were freely mixing A's and B's in medium m, the A's and the B's will to some extent sort themselves out to make A–A and B–B rather than A–B associations if van der Waals forces alone were operating. The degree of preference depends, of course, on the magnitude of the differences in energies compared with the magnitude of kT that agitates for random mixing. (No confusion, please. Although we speak of the interaction of discrete bodies composed of A or B, we are still speaking of continuous materials A, B, and m.)

Van der Waals repulsion?

The free energies of interaction between like substances will always be attractive. Between unlike materials, when ε_m is intermediate between ε_A and ε_B, $\overline{\Delta}_{Am}$ and $\overline{\Delta}_{Bm}$ have opposite signs. The contribution of the product $-\overline{\Delta}_{Am}\overline{\Delta}_{Bm}$ to the van der Waals interaction $G_{AmB}(l) \approx -(kT/8\pi l^2)\sum_{n=0}^{'\infty}\overline{\Delta}_{Am}\overline{\Delta}_{Bm}R_n(l)$ [approximation L1.5] is actually positive.

A positive pressure between A and B in m, a positive van der Waals interaction energy, reminds us that the intrusion of an interface disrupts the interactions of materials A, m, and B with themselves. A positive pressure simply indicates a force for thickening medium m, a force to draw more material m from a region of pure m into the space between A and B. This occurs when the interaction of substance m with itself is stronger than the interaction of substance m with substance A but weaker than its interaction

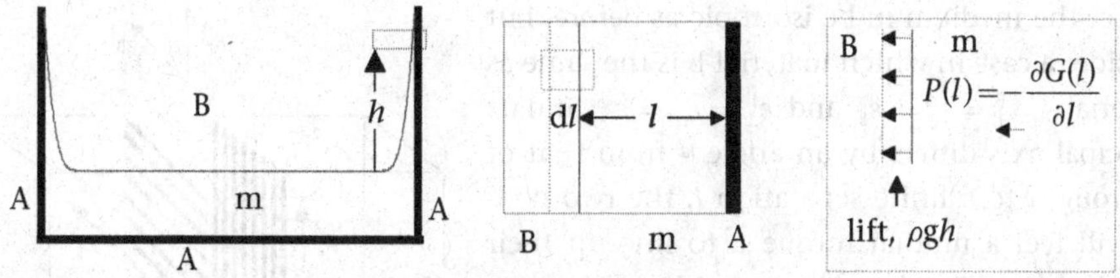

Figure L1.19

with B; so m flows to put as much of itself as possible near A rather than remain in the reservoir, where m interacts only with itself. The price is more exposure of m to B, by definition a price worth paying.

For example, consider a film of m on a substrate A under a vapor B where $\varepsilon_A > \varepsilon_m > \varepsilon_B$. The tendency will be to thicken the film by drawing from a reservoir of m. If the film is of a low-surface-tension liquid, it wets the walls of a container. Then the van der Waals force of thickening is a force to raise the liquid m of density ρ from the "reservoir" at the bottom. The cost $\rho gh\ dl$ of lifting an increment of mass $\rho\ dl$ to a height h involves work $\rho g\ dl$ against gravity (g is the gravitational constant, 9.8 N/kg). This lifting work is balanced by the electrodynamic benefit of thickening the film by an increment dl. There is balance between the change $(dG/dl)\ dl = -P(l)\ dl$ and $\rho gh\ dl$ (see Fig. L1.19):

$$\rho gh + (dG/dl) = 0 \text{ or } \rho gh = P(l). \tag{L1.24}$$

From the contour of measured thickness l versus rise h, it is possible to learn the form of $P(l)$ and $G(l)$. What is happening physically? The material in the incremental layer dl interacts more favorably with the wall a distance l away than it would interact with more of its own kind at that same distance. There is the tendency then to add more liquid to the film—until the benefit is compensated for by the gravitational work of delivery.

The total energy of the two interfaces (Am of m with the wall A and Bm of m with the vapor B) depends on the thickness l. The total energy is that of the liquid–wall and liquid–air interfaces for an infinitely thick liquid medium *plus* the free energy $G(l)$. $G(l)$ positive but decreasing with increasing l means a force to thicken the film.

PROBLEM L1.8: Can there be van der Waals repulsion between bodies separated by a vacuum? (Far-fetched? Zestfully discussed by Casimir cognoscenti.[3])

Torque

We usually think of van der Waals forces in terms of attraction or repulsion based on differences in polarizability. What if materials are anisotropic, for example, birefringent with different polarizabilities in different directions? Imagine that substance A has a principal optical axis pointing parallel to the interface between A and m, that is, there is a dielectric response coefficient $\varepsilon_{\parallel}^{A}$ parallel to the interface but a permittivity $\varepsilon_{\perp}^{A} < \varepsilon_{\parallel}^{A}$ in directions perpendicular to the principal axis (see Fig. L1.20).

Let the medium m be isotropic as before, but consider a case in which material B is the same as material A, $\varepsilon_{\parallel}^{B} = \varepsilon_{\parallel}^{A} = \varepsilon_{\parallel}$ and $\varepsilon_{\perp}^{B} = \varepsilon_{\perp}^{A} = \varepsilon_{\perp}$, and its principal axis differs by an angle θ from that of A. Brought to a finite separation l, the two bodies will feel a mutual torque τ to line up their principal axes. For the case of weak birefringence, $|\varepsilon_{\parallel} - \varepsilon_{\perp}| \ll \varepsilon_{\perp}$, with retardation neglected the free energy per unit area has the form (see Table P.9.e in Level 2).

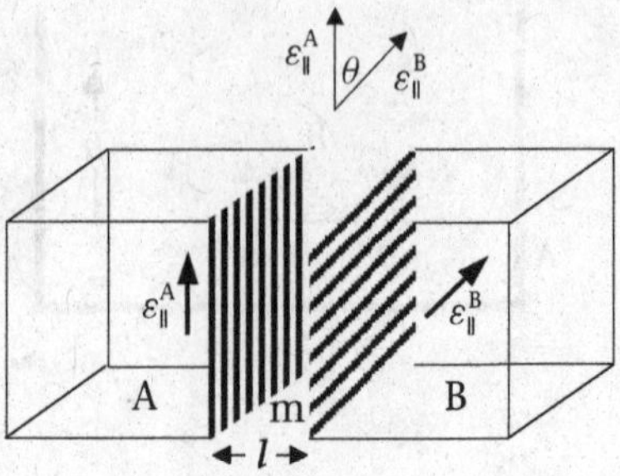

Figure L1.20

$$G(l,\theta) = -\frac{kT}{8\pi l^2}\sum_{n=0}^{\infty}{}' \left[\overline{\Delta}^2 + \gamma\overline{\Delta} + \gamma^2(1+2\cos^2\theta)\right], \tag{L1.24a}$$

where

$$\overline{\Delta} \equiv \left(\frac{\sqrt{\varepsilon_{\perp}\varepsilon_{\parallel}} - \varepsilon_{\mathrm{m}}}{\sqrt{\varepsilon_{\perp}\varepsilon_{\parallel}} + \varepsilon_{\mathrm{m}}}\right), \qquad \gamma \equiv \frac{\sqrt{\varepsilon_{\perp}\varepsilon_{\parallel}}(\varepsilon_{\perp} - \varepsilon_{\parallel})}{2\varepsilon_{\parallel}(\sqrt{\varepsilon_{\perp}\varepsilon_{\parallel}} + \varepsilon_{\mathrm{m}})} \ll 1.$$

There is a derivative torque per unit area:

$$\tau(l,\theta) = -\left.\frac{\partial G(l,\theta)}{\partial\theta}\right|_{l} = -\frac{kT}{2\pi l^2}\sum_{n=0}^{\infty}{}' \gamma^2\cos\theta\sin\theta = -\frac{kT}{4\pi l^2}\sum_{n=0}^{\infty}{}' \gamma^2\sin(2\theta). \tag{L1.24b}$$

Note that both $G(l,\theta)$ and $\tau(l,\theta)$ are doubly periodic in θ.

PROBLEM L1.9: Using the result given in Table P.9.e in Level 2, derive free energy and torque [Eqs. (L1.24a) and (L1.24b)].

PROBLEM L1.10: If $\sum_{n=0}^{\prime\infty}\gamma^2 = 10^{-2}$, how big an area L^2 of the two parallel-planar faces would suffer an energy change kT because of a 90° turn in mutual orientation?

L1.2. The van der Waals interaction spectrum

What frequencies are important to the interaction of real materials? It is best to learn from example. If we can develop an intuition to know the significant frequencies of fluctuation, our partial ignorance of absorption spectra need not daunt us in computation or, better, this intuition might give us some idea about the accuracy of computation. Differences in dielectric response create the force; sampling-frequency density weights the contribution from higher frequencies; retardation snuffs out the highest frequencies first. These general features show up in specific examples. Water, hydrocarbon (liquid tetradecane), gold, and mica are not only popular materials in van der Waals force measurement but, as a group, they also display a wide variety of dielectric response. (See, e.g., tables in Subsection L2.4.D).

Their detailed energy-absorption spectra as functions of radial frequency ω_R translate into smooth functions $\varepsilon(i\xi)$ of imaginary frequency. This blurring of details in $\varepsilon(i\xi)$ is one reason why it is often possible to compute van der Waals interactions to good accuracy without full knowledge of spectra (see Fig. L1.21).

To see how these different $\varepsilon(i\xi)$ functions combine to create an interaction, consider the case of two hydrocarbon half-spaces A = B = H across water medium m = W. First plot $\varepsilon_H(i\xi)$ and $\varepsilon_W(i\xi)$ as continuous functions [see Fig. L1.22(a)]. These are plotted at only the discrete sampling frequencies ξ_n at which they are to be evaluated; a log plot in frequency shows how compression of the arithmetically even spacing $\hbar\xi_n = 0.159\,n$ eV in index n works with the varying difference in $\varepsilon_H(i\xi_n)$ and $\varepsilon_W(i\xi_n)$ [see Fig. L1.22(b)].

The operative function $\overline{\Delta}^2_{HW}(i\xi_n) = [(\varepsilon_H - \varepsilon_W)/(\varepsilon_H + \varepsilon_W)]^2$ goes to zero when the two response functions are equal (near $\hbar\xi = 0.2$eV). This crossing of the $\varepsilon(i\xi)$ lines and the paucity of sampling frequencies in the IR region suggest that there will be little contribution to the sum over $\overline{\Delta}^2_{HW}(i\xi_n)$ that comprises the total interaction. The density of sampling frequencies ensures that the dominant contribution to the interaction energy, except for the zero-frequency contribution, comes from fluctuations having lifetimes characteristic of the periods of UV frequencies. These properties are clear in a plot of $\overline{\Delta}^2_{HW}(i\xi_n)$ (see Fig. L1.23). The comparative contributions appear clearly too in a cumulative plot of $\overline{\Delta}^2_{HW}(i\xi_n)$ up to a particular $\xi_{n_{max}}$ (see Fig. L1.24).

How important to forces are the electromagnetic fluctuations that occur in the different frequency regions?

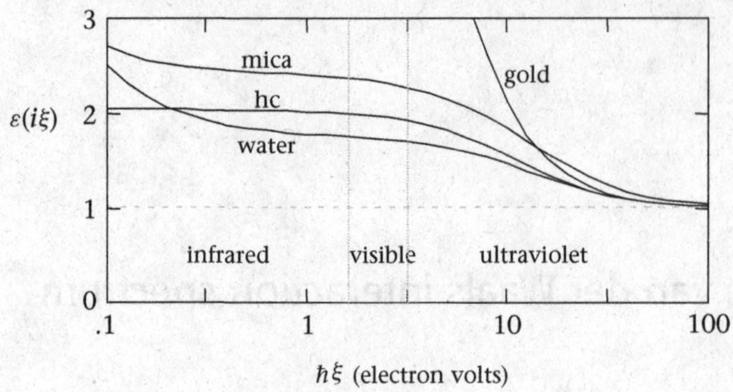

Figure L1.21

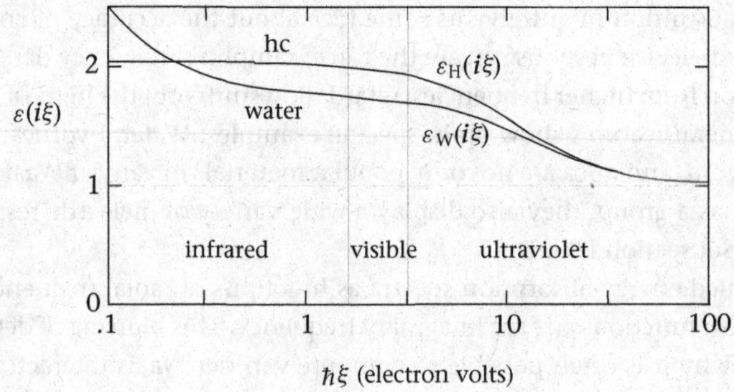

Figure L1.22a

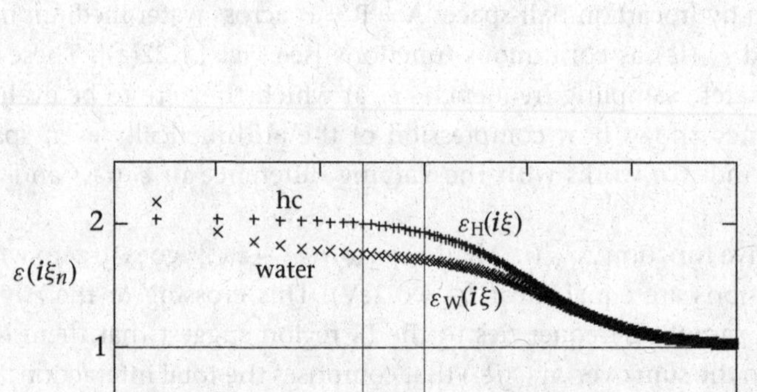

Figure L1.22b

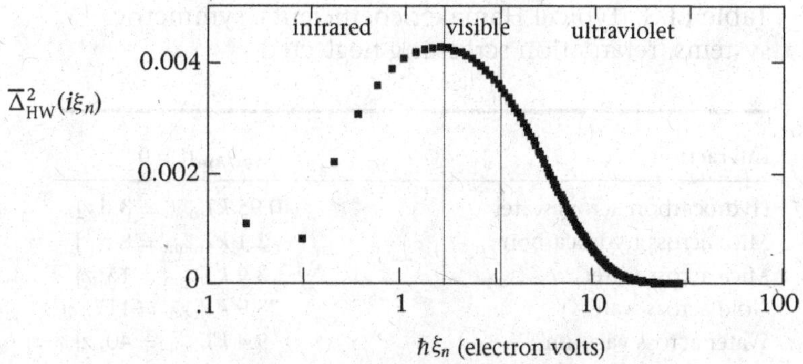

Figure L1.23

With $\varepsilon_W(0) \approx 80 \gg \varepsilon_H(0) \approx 2$, $\overline{\Delta}^2_{HW}(0) \approx (80-2/80+2)^2 = 0.905$. This contribution has only one half the weight of the finite-frequency terms (prime $'$ in summation) and weighs in as 0.452.

Over the points $n = 1$–9 that correspond to IR frequencies, the sum over frequencies $\overline{\Delta}^2_{HW}(i\xi_n)$ comes to 0.027; over the visible range, $n = 10$–20, 0.045; over the UV range, 0.107.

The finite-frequency fluctuation forces are easily dominated by action in the UV, but because of the high polarizability of water in the limit of low frequency, the 0.452 $n = 0$ term contributes most heavily of all to the total sum of $\overline{\Delta}^2_{HW}$: 0.452 ($n = 0$) + 0.0027 (IR) + 0.045 (visible) + 0.107 (UV) = 0.631 (total).

When we consider the interaction of a hydrocarbon with itself across a vacuum, $\varepsilon_m = \varepsilon_{vacuum} \equiv 1$, the magnitude of the force and the spectrum of contributions look very different. The $n = 0$ term comes out to $0.5(2.04 - 1/2.04 + 1)^2 = 0.059$. The nine IR terms sum to 1.031 followed by the visible-frequency terms that sum to 1.162, and finally a very strong 5.449 from the UV. The total summation comes to approximately 7.7, compared with 0.63 of the same hydrocarbon across water.

For water across a vacuum, the comparable figures are 0.476 from the $n = 0$ term, 0.8 from the IR, 0.782 from the visible, and 4.226 from the UV. The large 6.284 total is again dominated by UV contributions.

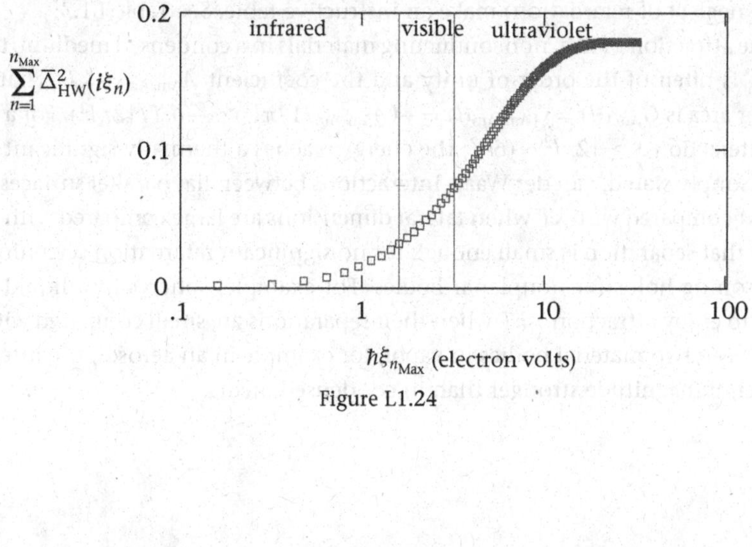

Figure L1.24

Table LI.3. Typical Hamaker coefficients, symmetric systems, retardation screening neglected

Interaction	$A_{\mathbf{Am/Am}}(l=0)^a$
Hydrocarbon across water	$0.95\ kT_{\rm room} = 3.6$ zJ.
Mica across hydrocarbon	$2.1\ kT_{\rm room} = 8.5$ zJ
Mica across water	$3.9\ kT_{\rm room} = 15.$ zJ
Gold across water	$28.9\ kT_{\rm room} = 117.$ zJ
Water across vacuum	$9.4\ kT_{\rm room} = 40.$ zJ
Hydrocarbon across vacuum	$11.6\ kT_{\rm room} = 46.9$ zJ
Mica across vacuum	$21.8\ kT_{\rm room} = 88.$ zJ
Gold across vacuum	$48.6\ kT_{\rm room} = 196$ zJ

[a] $kT_{\rm room} = 4.04 \times 10^{-14}$ ergs $= 4.04 \times 10^{-21}$ J $= 4.04$ zJ.

From these summations, neglecting any retardation or ionic screening, the unscreened Hamaker coefficients $A_{\rm Am/Bm}(l=0) \approx (3kT/2)\sum_{n=0}^{'\infty} \overline{\Delta}_{\rm Am}\overline{\Delta}_{\rm Bm}$ emerge as the following expressions:

$A_{\rm HW/HW} = 1.5 \times 0.631\ kT_{\rm room} = 0.947\ kT \approx 4 \times 10^{-14}$ ergs $= 4 \times 10^{-21}$ J $= 4$ zJ for hydrocarbon across water (or vice versa);

$A_{\rm HV/HV} = 1.5 \times 7.7\ kT_{\rm room} = 11.55\ kT \approx 5 \times 10^{-13}$ ergs $= 50$ zJ for hydrocarbon across vacuum (or vice versa);

$A_{\rm WV/WV} = 1.5 \times 6.28\ kT_{\rm room} = 9.42\ kT \approx 4 \times 10^{-13}$ ergs $= 40$ zJ for water across vacuum (or vice versa).

Magnitude of van der Waals interactions

How strong are these long-range van der Waals forces between semi-infinite bodies A and B across a medium m or across a vacuum? For the four materials whose dielectric responses are plotted in the preceding section, the corresponding Hamaker coefficients (with the neglect of retardation) make an instructive table. See Table L1.3.

For the attraction of like, nonconducting materials in a condensed medium, the sum $\sum_{n=0}^{'\infty} \overline{\Delta}_{\rm Am}^2$ is often of the order of unity and the coefficient $A_{\rm Am/Am} \sim kT$. That is, the energy per area is $G_{\rm AmA}(l) = G_{\rm Am/Am}(l) = -(A_{\rm Am/Am}/12\pi l^2) \sim -(kT/12\pi l^2)$. For a surface area of interaction $S \geq 12\pi l^2 \sim (6l)^2$, the energy reaches a thermally significant kT.

More simply stated, van der Waals interactions between flat parallel surfaces will be significant compared with kT when lateral dimensions are large compared with separation (and that separation is small enough for no significant retardation screening). The same reasoning holds for nonplanar bodies. For example, spheres in a liquid can be expected to enjoy attraction $>kT$ when their separations are small compared with their radii. Between two material bodies in vapor, for example in an aerosol, the interaction is an order of magnitude stronger than in condensed media.

L1.3. Layered planar bodies

One singly layered surface

Rather than its being a block of uniform material, let body B be coated with a layer of material B_1 of constant thickness b_1. $G_{AmB}(l)$ for the simplest AmB case becomes $G_{AmB_1B}(l; b_1)$. Because it is the difference in dielectric response that creates an electromagnetic interface, it is the distances between interfaces that appear in the van der Waals energy. Because there are now two spacings—l and $(l + b_1)$—there are two terms in the simplest-form free energy. For b_1 of fixed thickness and l of variable separation, for small differences in ε's, the free energy becomes (see Table P.2.b.2 in Level 2)

$$G_{AmB_1B}(l; b_1) = G_{Am/B_1m}(l) + G_{Am/BB_1}(l_{Am/BB_1})$$
$$= -\frac{A_{Am/B_1m}(l)}{12\pi l^2} - \frac{A_{Am/BB_1}(l + b_1)}{12\pi (l + b_1)^2} \quad \text{(L1.25)}$$

(see Fig. L1.25).

Rather than the one interaction $G_{Am/B_1m}(l) = -\{[A_{Am/B_1m(l)}]/12\pi l^2\}$ between one pair of interfaces across a distance l, there are now *two pairs* of interacting surfaces, each pair with its own coefficient and its own distance of separation. Because there are *three* materials involved, the subscripts on the free energies and on the Hamaker coefficients are written to show which surfaces are correspondingly involved. For clarity, we use an outside–inside subscripting for the materials at the different interfaces.

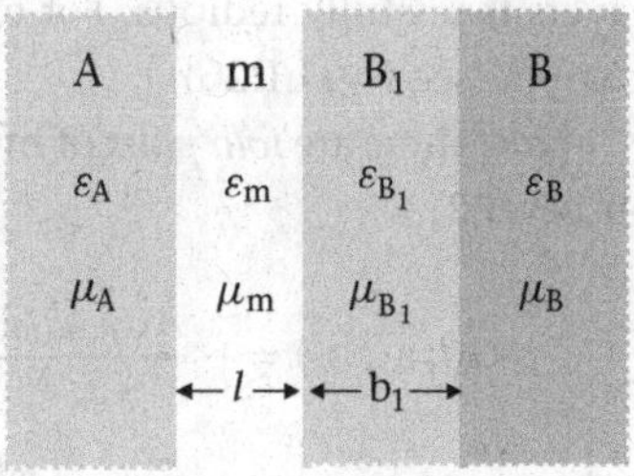

Figure L1.25

G_{Am/B_1m} and A_{Am/B_1m} designate the interaction between the Am interface and the B_1m interface separated by variable distance l (subscript Am for A "outside," to the left of m; B_1m for B_1 "outside," to the right of m), and $G_{Am/BB_1}, A_{Am/BB_1}$ go with the interaction between the Am interface and the BB_1 interface, that is, with the separation $l_{Am/BB_1} = (l + b_1)$, while thickness b_1 is kept fixed.

The evaluation of the two Hamaker coefficients is much as before with the following two exceptions:

- The ε's must match the corresponding materials at corresponding interfaces, and
- distances in the relativistic screening term are those between relevant interfaces.

$$A_{\mathrm{Am/B_1m}}(l) \approx \frac{3kT}{2} \sum_{n=0}^{\infty}{}' \overline{\Delta}_{\mathrm{Am}} \overline{\Delta}_{\mathrm{BB_1}} R_n(l), \tag{L1.26}$$

$$A_{\mathrm{Am/BB_1}}(l + \mathrm{b}_1) \approx \frac{3kT}{2} \sum_{n=0}^{\infty}{}' \overline{\Delta}_{\mathrm{Am}} \overline{\Delta}_{\mathrm{BB_1}} R_n(l + \mathrm{b}_1), \tag{L1.27}$$

with

$$\overline{\Delta}_{\mathrm{Am}} \approx \frac{\varepsilon_{\mathrm{A}} - \varepsilon_{\mathrm{m}}}{\varepsilon_{\mathrm{A}} + \varepsilon_{\mathrm{m}}}, \quad \overline{\Delta}_{\mathrm{B_1m}} \approx \frac{\varepsilon_{\mathrm{B_1}} - \varepsilon_{\mathrm{m}}}{\varepsilon_{\mathrm{B_1}} + \varepsilon_{\mathrm{m}}}, \quad \overline{\Delta}_{\mathrm{BB_1}} \approx \frac{\varepsilon_{\mathrm{B}} - \varepsilon_{\mathrm{B_1}}}{\varepsilon_{\mathrm{B}} + \varepsilon_{\mathrm{B_1}}}. \tag{L1.28}$$

Note the order of the ε's to ensure the correct sign of the interaction. The ε that comes *first* is *always* for the material that is on the *outside* of the interface looking toward the medium m.

With an important exception, R_n relativistic screening functions go as previously, $R_n = (1 + r_n)e^{-r_n}$, with the ratio r_n reflecting the distance l or $(l + \mathrm{b}_1)$. This is an approximation for which we assume that there is a negligible difference between the velocities of light in m and in B_1 at those frequencies at which relativistic retardation is occurring. The exception occurs when any region is a salt solution; then it is not always permissible to use this approximation for the $n = 0$ term. (See the subsequent section on ionic fluctuations.)

When variable separation l is much less than fixed layer thickness b_1, the interaction is dominated by the $1/l^2$ term; the interaction looks as though it occurs only between semi-infinite material A and layer material B_1 with material B forgotten. Dielectric properties of the substrate material B become important as soon as separation l and layer thickness b become comparable. To first approximation, *the van der Waals interaction "sees" into a structure to a depth comparable with the separation between structures.*

Two singly layered surfaces

At least in this simplest Hamaker-form approximation, the scheme for adding further layers is restfully tedious. For example, add a layer of material A_1 and thickness a_1 to body A (see Fig. L1.26).

Now there are *four pairs* of interfaces with four corresponding terms (see Table P.3.b.2 in Level 2):

$$\begin{aligned} G(l; \mathrm{a}_1, \mathrm{b}_1) = & -\frac{A_{\mathrm{A_1m/B_1m}}(l)}{12\pi l^2} - \frac{A_{\mathrm{A_1m/BB_1}}(l + \mathrm{b}_1)}{12\pi (l + \mathrm{b}_1)^2} \\ & - \frac{A_{\mathrm{AA_1/B_1m}}(l + \mathrm{a}_1)}{12\pi (l + \mathrm{a}_1)^2} \\ & - \frac{A_{\mathrm{AA_1/BB_1}}(l + \mathrm{a}_1 + \mathrm{b}_1)}{12\pi (l + \mathrm{a}_1 + \mathrm{b}_1)^2}. \end{aligned} \tag{L1.29}$$

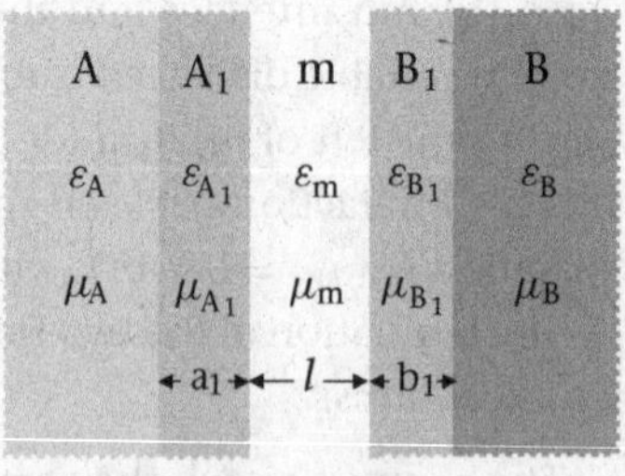

Figure L1.26

Again, the relativistic screening functions R_n in their simplest version have the same $(1 + r_n)e^{-r_n}$ form with the ratios r_n reflecting the distances l, $(l + \mathrm{a}_1)$, $(l + \mathrm{b}_1)$, and $(l + \mathrm{a}_1 + \mathrm{b}_1)$.

Interaction between two parallel slabs, in a medium versus on a substrate

If regions A and B have the same material properties as medium m, then the remaining interaction is that between two parallel slabs (see Fig. L1.27).

For example, when retardation is neglected (see Table P.3.c.3 in Level 2),

$$G(l; a_1, b_1) = -\frac{A_{A_1m/B_1m}}{12\pi}\left[\frac{1}{l^2} - \frac{1}{(l+b_1)^2} - \frac{1}{(l+a_1)^2} + \frac{1}{(l+a_1+b_1)^2}\right]. \qquad (L1.30)$$

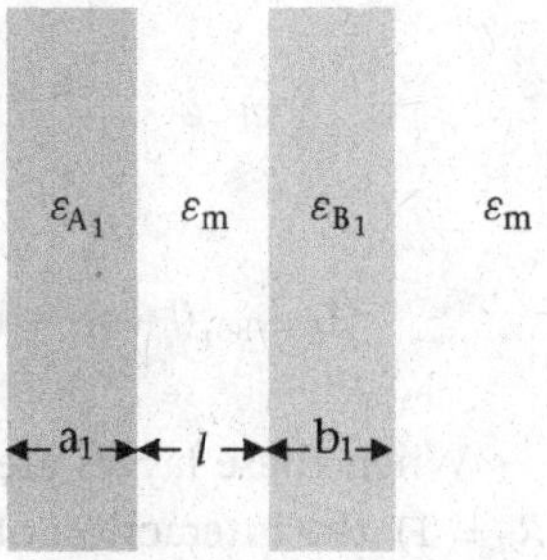

Figure L1.27

PROBLEM L1.11: Neglecting retardation, show how the interaction between two coated bodies, Eq. (L1.29) can be converted into the interaction between two parallel slabs, Eq. (L1.30).

WORKED PROBLEM: Compare the interaction of two hydrocarbon layers in solution with their interaction when immobilized on solid substrates. Show the limiting behavior with separation long or short compared with layer thickness.

SOLUTION: These slab formulae specialize further to some interesting and useful limits. For example, let A_1 and B_1 be of the same material H of constant thickness h and similarly let A, B, and m be of the same material W, where the separation is now designated as w.

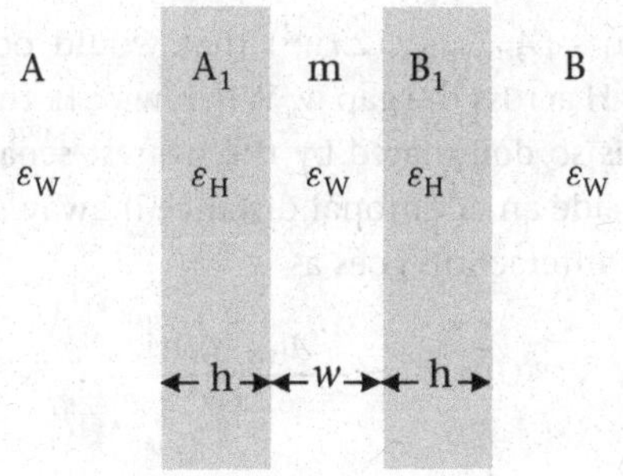

This is for an interaction between two layers (hydrocarbon, say) in a medium (water, for example). Here,

$$\overline{\Delta}_{A_1m} = \overline{\Delta}_{B_1m} = \overline{\Delta}_{HW} = \frac{\varepsilon_H - \varepsilon_W}{\varepsilon_H + \varepsilon_W}.$$

The Hamaker coefficients are similar but differ in sign:

$$A_{\mathrm{A_1m/B_1m}}(l) = A_{\mathrm{HWHW}}(w) \approx \frac{3kT}{2}\sum_{n=0}^{\infty}{}' \overline{\Delta}_{\mathrm{HW}}^2 R_n(w),$$

$$\begin{aligned} A_{\mathrm{mA_1/B_1m}}(l+\mathrm{a}_1) &= A_{\mathrm{WH/HW}}(w+\mathrm{h}) \\ &= A_{\mathrm{A_1m/BB_1}}(l+\mathrm{b}_1) = A_{\mathrm{HW/WH}}(w+\mathrm{h}) \\ &= -A_{\mathrm{HW/HW}}(w+\mathrm{h}) \approx -\frac{3kT}{2}\sum_{n=0}^{\infty}{}' \overline{\Delta}_{\mathrm{HW}}^2 R_n(w+\mathrm{h}), \end{aligned}$$

$$A_{\mathrm{AA_1/BB_1}}(l+\mathrm{a}_1+\mathrm{b}_1) = A_{\mathrm{HW/HW}}(w+2\mathrm{h}) \approx \frac{3kT}{2}\sum_{n=0}^{\infty}{}' \overline{\Delta}_{\mathrm{HW}}^2 R_n(w+2\mathrm{h}).$$

When these layers are close enough to allow the neglect of retardation terms (all $R_n = 1$), the interaction of Eq. (L1.30) achieves a particularly simple form:

$$G(w;\mathrm{h}) = -\frac{A_{\mathrm{HW/HW}}}{12\pi}\left[\frac{1}{w^2} - \frac{2}{(w+\mathrm{h})^2} + \frac{1}{(w+2\mathrm{h})^2}\right],$$

with

$$A_{\mathrm{HW/HW}} \approx \frac{3kT}{2}\sum_{n=0}^{\infty}{}' \overline{\Delta}_{\mathrm{HW}}^2.$$

$G(w;\mathrm{h})$ can also be written as

$$G(w;\mathrm{h}) = -\frac{A_{\mathrm{HW/HW}}}{12\pi w^2}\left[1 - \frac{2w^2}{(w+\mathrm{h})^2} + \frac{w^2}{(w+2\mathrm{h})^2}\right],$$

which looks like a correction factor

$$\left[1 - \frac{2w^2}{(w+\mathrm{h})^2} + \frac{w^2}{(w+2\mathrm{h})^2}\right]$$

applied to the interaction $-(A_{\mathrm{HW/HW}}/12\pi w^2)$ that would occur between two semi-infinite bodies of material H across the gap w. When $w \ll \mathrm{h}$, this leading term becomes accurate; the interaction is so dominated by the nearest separation w that the layers do not "see" to the other side an additional distance h away. When $w \gg \mathrm{h}$, the factor becomes $6(\mathrm{h}/w)^2$, and the interaction goes as

$$G(w;\mathrm{h}) = -\frac{A_{\mathrm{HW/HW}}\mathrm{h}^2}{2\pi w^4}.$$

PROBLEM L1.12: Show how the nonretarded interaction between slabs goes from inverse-square to inverse-fourth-power variation.

Here again, the extent of "seeing" into a body changes with separation. This depth of vision becomes even clearer if we imagine that the layers H sit on another material M, at the position of material A and B.

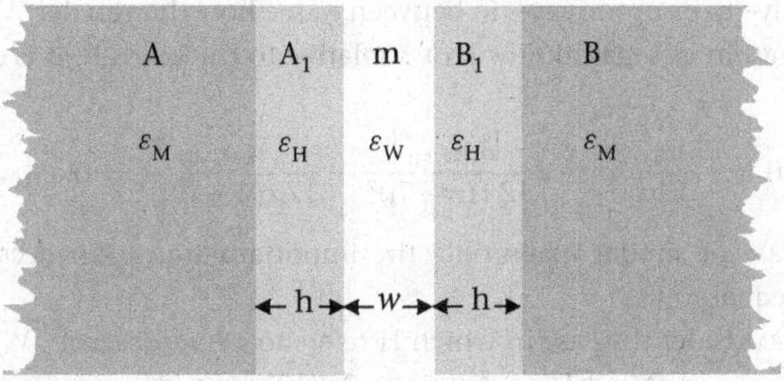

We use Eq. (L1.29) (see Table P.3.b2 in Level 2):

$$G(w;\mathrm{h}) = -\frac{A_{\mathrm{HW/HW}}}{12\pi w^2} - 2\frac{A_{\mathrm{HW/MH}}}{12\pi(w+\mathrm{h})^2} - \frac{A_{\mathrm{MH/MH}}}{12\pi(w+2\mathrm{h})^2},$$

where

$$A_{\mathrm{HW/HW}}(w) \approx \frac{3kT}{2}\sum_{n=0}^{\infty}{}' \,\overline{\Delta}_{\mathrm{HW}}^2\, R_n(w), \qquad A_{\mathrm{HW/MH}} \approx \frac{3kT}{2}\sum_{n=0}^{\infty}{}' \,\overline{\Delta}_{\mathrm{HW}}\overline{\Delta}_{\mathrm{MH}}\, R_n(w+\mathrm{h}),$$

$$A_{\mathrm{MH/MH}}(w+2\mathrm{h}) \approx \frac{3kT}{2}\sum_{n=0}^{\infty}{}' \,\overline{\Delta}_{\mathrm{MH}}^2\, R_n(w+2\mathrm{h}),$$

with

$$\overline{\Delta}_{\mathrm{HW}} = \frac{\varepsilon_{\mathrm{H}} - \varepsilon_{\mathrm{W}}}{\varepsilon_{\mathrm{H}} + \varepsilon_{\mathrm{W}}}, \qquad \overline{\Delta}_{\mathrm{MH}} = \frac{\varepsilon_{\mathrm{M}} - \varepsilon_{\mathrm{H}}}{\varepsilon_{\mathrm{M}} + \varepsilon_{\mathrm{H}}}.$$

Without retardation screening, for separation w much less than layer thickness h, interaction is dominated by the first term, $-(A_{\mathrm{HW/HW}}/12\pi w^2)$, i.e., the same as if thickness h were infinite, as though substrate M were not even there, even if it had a very high polarizability compared with W and H.

In the limit of very large separations, $w \gg \mathrm{h}$, the denominators are roughly the same, $w^2 \sim (w+\mathrm{h})^2$. Qualitatively, the interaction goes as an effective single term:

$$G(w;h) = -\frac{A_{\mathrm{eff}}}{12\pi w^2} = G_{\mathrm{W/H/W/H/W}}$$

with

$$A_{\mathrm{eff}} = A_{\mathrm{HW/HW}} + 2\,A_{\mathrm{HW/MH}} + A_{\mathrm{MH/MH}}.$$

Looking at the summations that make the Hamaker coefficients, we see that this A_{eff} has the succinct form

$$A_{\mathrm{eff}} \approx \frac{3kT}{2}\sum_{n=0}^{\infty}{}' \left(\overline{\Delta}_{\mathrm{HW}} + \overline{\Delta}_{\mathrm{MH}}\right)^2.$$

If the polarizabilities are such that $|\overline{\Delta}_{\mathrm{MH}}| \gg |\overline{\Delta}_{\mathrm{HW}}|$, it is even possible that in this limit the contribution of the coating layer becomes negligible compared with that of the substrate.

It is instructive to examine cases in between to see how the van der Waals force that is due to M comes in as separation w grows relative to thickness h. It is clear from the form

$$G(w;\mathrm{h}) = -\frac{A_{\mathrm{HW/HW}}}{12\pi w^2} - 2\frac{A_{\mathrm{HW/MH}}}{12\pi(w+\mathrm{h})^2} - \frac{A_{\mathrm{MH/MH}}}{12\pi(w+2\mathrm{h})^2} = G_{\mathrm{M/H/W/H/M}}$$

that when A's are of similar magnitude the important transition occurs when h is approximately equal to w.

Specifically, consider the case in which H refers to a hydrocarbon, W to water, and M to mica: $A_{\mathrm{HW/HW}} = 0.95\ kT_{\mathrm{room}}$, $A_{\mathrm{MH/MH}} = 2.1\ kT_{\mathrm{room}}$, and $A_{\mathrm{HW/MH}} = -0.226\ kT_{\mathrm{room}}$. Only in the limit of contact, separation $w \ll$ thickness h, can the finitude of h be ignored. The interaction ratio of hydrocarbon-coated mica compared to hydrocarbon layers alone increases monotonically with separation w and equals unity only in the limit of close approach.[4]

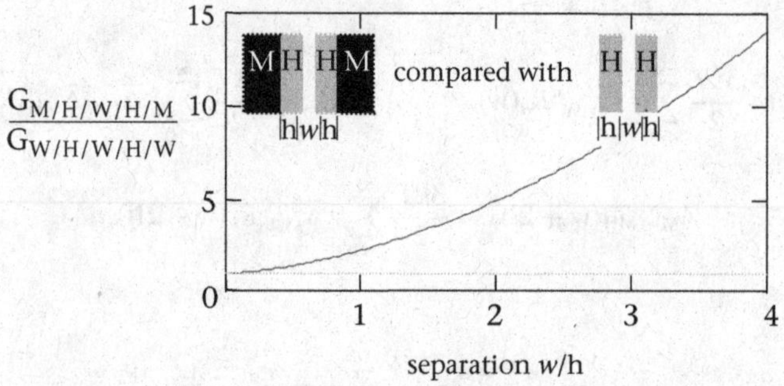

Compared with the interaction of two semi-infinite bodies of hydrocarbon, the interaction between hydrocarbon-coated mica surfaces increases steadily versus separation whereas that between two finite slabs of hydrocarbon expectably decreases. Significant deviations are already clear at $w/\mathrm{h} = 1$, when the varying separation w equals the constant layer thickness h.

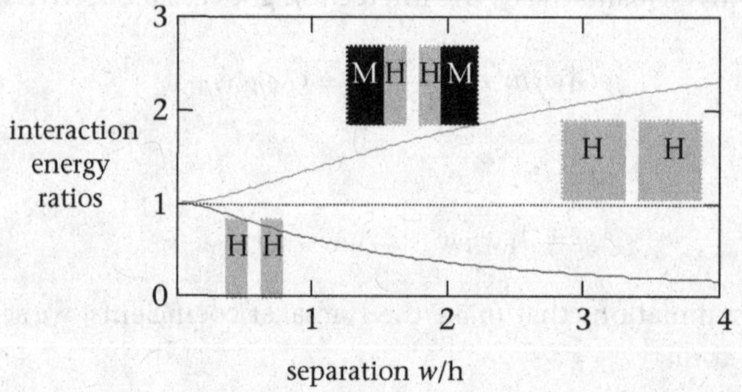

How does the interaction of hydrocarbon across water change when a solute is added to solvent water?

There is no general answer. Interactions depend on differences in ε's. Recall Figs. L1.22(a) and L1.22(b) in which $\varepsilon_{\mathrm{W}}(i\xi_n)$ and $\varepsilon_{\mathrm{H}}(i\xi_n)$ cross twice—once in the IR region, again (barely visible) in the UV. If a solute increases $\varepsilon_{\mathrm{W}}(i\xi_n)$, then the difference between ε's becomes smaller in the visible and flanking regions but larger in the far UV. The net result is that the interaction of hydrocarbon across water becomes weaker when solutes are added that increase the index of refraction. When enough solute is added, $\varepsilon_W(i\xi_n)$, or really $\varepsilon_{\text{solution}}(i\xi_n)$, can become bigger than $\varepsilon_H(i\xi_n)$ even in the visible region. The attraction of a hydrocarbon across a water solution can become stronger with further addition of solute.

Multiple layers, general scheme

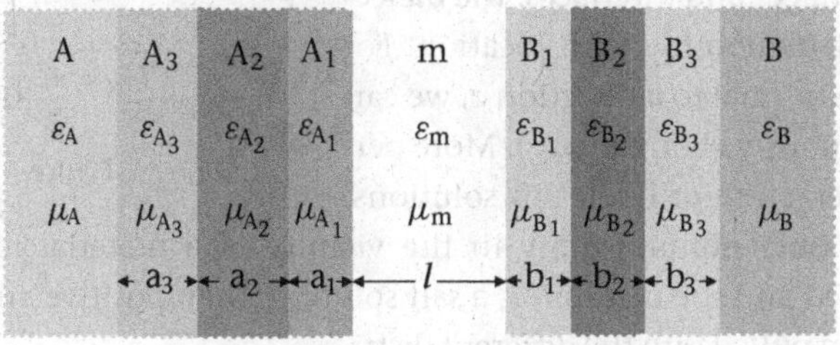

Figure L1.28

The generalization from these last two examples is straightforward. Between every pair of interfaces there is a separate term with its own Hamaker coefficient and inverse-square denominator. For example, to compute the interaction between two triply coated bodies, each with four interfaces, entails writing out $4 \times 4 = 16$ terms of the approximate form discussed so far (see Fig. L1.28). The general case of i layers on A and k layers on B entails $(i+1) \times (k+1)$ terms. Follow the same procedure for creating the Hamaker coefficient and the component $\overline{\Delta}$'s for each pair of interfaces. Always work from the outside to the inside for the sequence of ε's. To be specific, think of each pair of interfaces at a separation $l_{A'A''/B'B''}$, where each interface separates layers A′A″ and B′B″ (see Fig. L1.29).

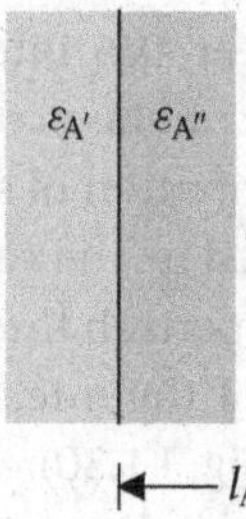

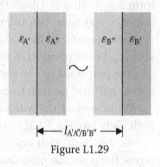

Figure L1.29

The contribution of this interaction will be

$$G_{A'A''/B'B''}(l_{A'A''/B'B''}) = -\frac{A_{A'A''/B'B''}}{12\pi\, l^2_{A'A''/B'B''}}, \tag{L1.31}$$

where, in the small-differences-in-ε's regime,

$$A_{A'A''/B'B''} \approx \frac{3kT}{2} \sum_{n=0}^{\infty}{}' \overline{\Delta}_{A'A''} \overline{\Delta}_{B'B''} R_n(l_{A'A''/B'B''}),$$

$$\overline{\Delta}_{A'A''} = \frac{\varepsilon_{A'} - \varepsilon_{A''}}{\varepsilon_{A'} + \varepsilon_{A''}}, \quad \overline{\Delta}_{B'B''} = \frac{\varepsilon_{B'} - \varepsilon_{B''}}{\varepsilon_{B'} + \varepsilon_{B''}}. \tag{L1.32}$$

Note how the sign of the $\overline{\Delta}$'s is determined by the outside–in sequence of the ε's in the numerator and how the sign of the interaction is then determined by the minus sign—in front of everything.

Smoothly varying $\varepsilon(z)$

No interface is an infinitesimally sharp mathematical step. There is a good reason. It takes an infinite amount of energy to change material properties with infinite sharpness; yet interfaces are usually modeled as steps. With electrodynamic energies, the fictitiously infinite energy it takes to make the step is recovered as a fictitiously divergent $1/l^2$ interaction energy in the limit at which separation l goes to a mathematically fictitious zero.

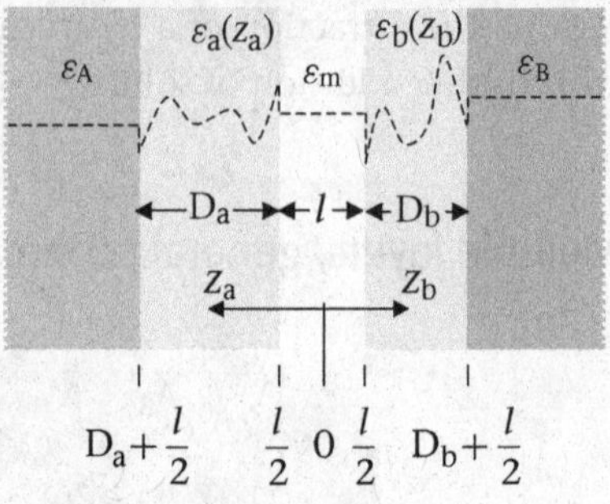

Figure L1.30

Across real surfaces and interfaces, the dielectric response varies smoothly with location. For a planar interface normal to a direction z, we can speak of a continuously changing $\varepsilon(z)$. More pertinent to the interaction of bodies in solutions, solutes will distribute nonuniformly in the vicinity of a material interface. If that interface is charged and the medium is a salt solution, then positive and negative ions will be pushed and pulled into the different distributions of an electrostatic double layer. We know that solutes visibly change the index of refraction that determines the optical-frequency contribution to the charge-fluctuation force. The nonuniform distribution of solutes thereby creates a nonuniform $\varepsilon(z)$ near the interfaces of a solution with suspended colloids or macromolecules. Conversely, the distribution of solutes can be expected to be perturbed by the very charge-fluctuation forces that they perturb through an $\varepsilon(z)$.[5]

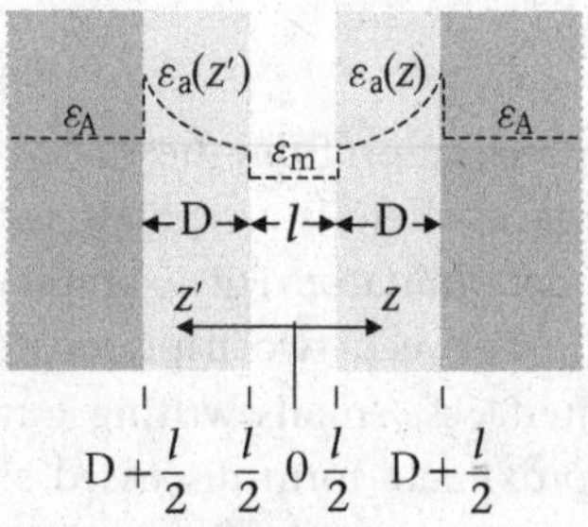

Figure L1.31

Think now of the interaction between half-spaces, where ε changes arbitrarily in each body (see Fig. L1.30).

We can deal with continuously varying $\varepsilon(z)$ as the limit of infinitesimally thin layers through the procedure for finite layers (Level 2 Formulae, section L2.3.B on continuously changing susceptibilities) or from what we know about electromagnetic fields in inhomogeneous media (such as are analyzed for wave propagation in the Earth's atmosphere) (Level 3, Subsection L3.5, on inhomogeneous media). Depending on the shape of $\varepsilon(z)$ and, more important, on the continuity of $\varepsilon(z)$ and $d\varepsilon(z)/dz$ at the interfaces with medium m, qualitatively new properties of interactions emerge in the $l \to 0$ limit of contact: Consider three cases of interactions between symmetric bodies coming into contact:

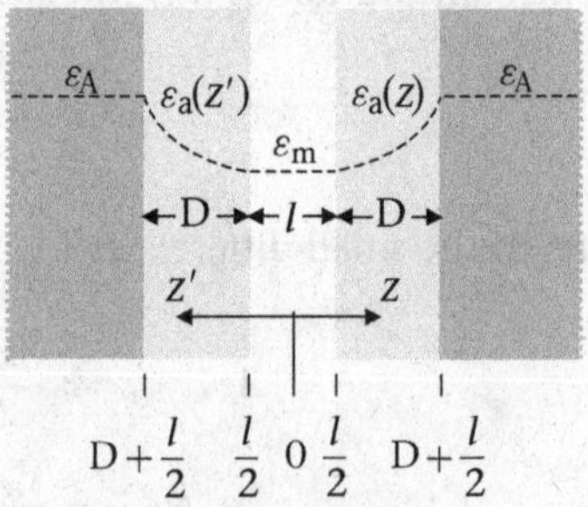

Figure L1.32

1. Discontinuity in ε at the interfaces with medium m. Imagine a symmetric juxtaposition of two exponentially varying regions of $\varepsilon_a(z)$ over thickness D with a

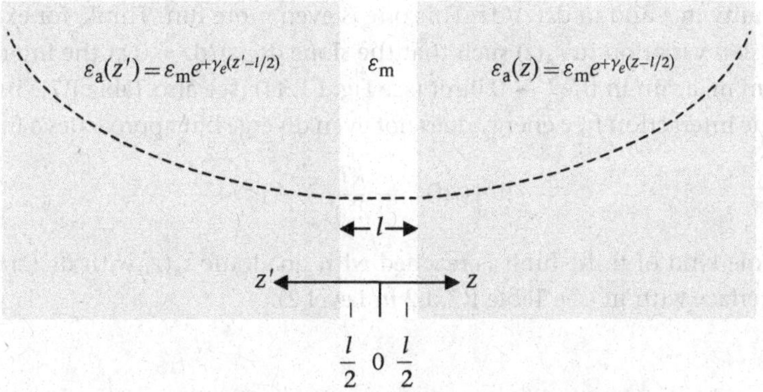

Figure L1.33

discontinuity at the intervening medium m. (see Fig. L1.31) (see also Table P.7.c.1 in Level 2).

In the limit of $l \to 0$ the dominant part of the interaction is due to the steps in ε at $z = z' = l/2$. There is the usual $1/l^2$ divergence for sharp interfaces. That is, the interaction goes over to the familiar Lifshitz form (see the footnote to Table P.7.c.1 in Level 2):

$$G(l \to 0; \mathrm{D}) = -\frac{kT}{8\pi l^2} \sum_{n=0}^{\infty}{}' \left[\frac{\varepsilon_a(l/2) - \varepsilon_m}{\varepsilon_a(l/2] + \varepsilon_m}\right]^2. \tag{L1.33}$$

2. Continuity in ε, discontinuity in $d\varepsilon(z)/dz$. Imagine the same two exponentials in $\varepsilon(z)$ but with no discontinuity at the intervening medium m (see Fig. L1.32) (see also Table P.7.c.2 in Level 2). In the $l \to 0$ limit this looks like an interaction between the variable regions only (see Fig. L1.33). The free energy varies as the log of separation (see Table P.7.c.3 in Level 2):

$$G(\gamma_e l \to 0) = \frac{kT}{32\pi} \sum_{n=0}^{\infty}{}' \gamma_e^2 \ln(\gamma_e l) \tag{L1.34}$$

[not forgetting that $\ln(\gamma_e l)$ is negative when $\gamma_e l \to 0$ as in Fig. L1.33]. The derivative pressure diverges as $1/l$.

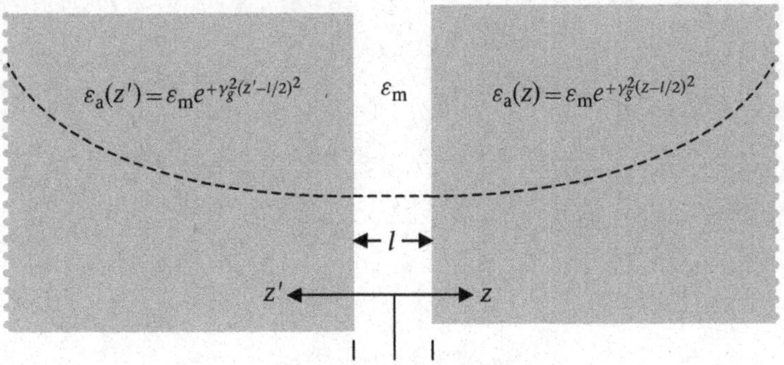

Figure L1.34

3. Continuity in ε and in $d\varepsilon(z)/dz$. This one is even more fun. Think, for example, of a Gaussian variation in $\varepsilon_a(z)$ such that the slope $d\varepsilon_a(z)/dz = 0$ at the interface with medium m, again in the $l \to 0$ limit (see Fig. L1.34) (see also Table P.7.e in Level 2). Now the interaction free energy does not even diverge but approaches a finite limit:

$$\lim_{l \to 0} G(l) = -\frac{kT}{2^8\pi} \sum_{n=0}^{\infty}{}' \gamma_g^2. \tag{L1.35}$$

The same kind of finite limit is reached with quadratic $\varepsilon_a(z)$ with $d\varepsilon_a(z)/dz = 0$ at the interface with m (see Table P.7.d.2 in Level 2).

L1.4. Spherical geometries

There is a qualitative difference in the energy versus separation between two oppositely curved bodies from the case in which their apposing surfaces are parallel planes. For one thing the underlying wave equations are much more difficult to solve than those in planar geometry. It is necessary to be satisfied with expressions obtained under restrictive conditions or in limiting cases: spheres (or cylinders) that are almost touching (separation $\ll$ diameter) or very far apart (separation $\gg$ diameter), neglected retardation screening, and small differences in polarizability.

Even with the neglect of retardation screening, the apparent power of the van der Waals interaction varies with separation itself. The reason for this power-law variation is clear when we consider the interaction of two spheres of material 1 and 2 and of radius R_1 and R_1 with a center-to-center distance $z = l + R_1 + R_2$ in a medium m (see Fig. L1.35).

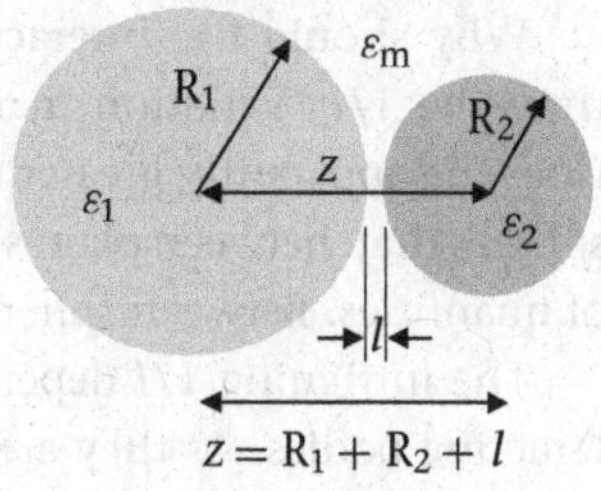

Figure L1.35

Close up, when separation $l \ll R_1$ and $l \ll R_2$, the interaction is dominated by interactions of the closest parts of the two spheres. Using the remarkable procedure derived by Derjaguin, we can express the force $F_{ss}(l; R_1, R_2)$ between two spheres in terms of the energy of interaction $G_{pp}(l)$ between two parallel planes whose separation is also l [see the Prelude and Level 2, Subsection L2.3.C on the Derjaguin transform; see also Eq. (L2.106) and Table S.1.a in Level 2]:

$$F_{ss}(l; R_1, R_2) = \frac{2\pi R_1 R_2}{(R_1 + R_2)} G_{pp}(l). \tag{L1.36}$$

In this close-approach limit, the most precise expression for the parallel-planar surfaces can be applied to the interaction of oppositely curved surfaces.

When $R_1 = R_2 = R \gg l$ [see also Eq. (L2.109) and Table S.1.c.1 in Level 2],

$$F_{ss}(l; R) = \pi R G_{pp}(l), \tag{L1.37}$$

an interaction whose strength is proportional to radius R (see Fig. L1.36).

When R_1 is infinite (i.e., the left-hand sphere in Fig. L1.35 looks flat), and $R_2 = R \gg l$, the force F_{ss} goes as [see also Eq. (L2.110) and Table S.1.c.2 in Level 2]

$$F_{ss}(l; R_1 \to \infty, R_2 = R) = F_{sp}(l; R) = 2\pi R G_{pp}(l), \tag{L1.38}$$

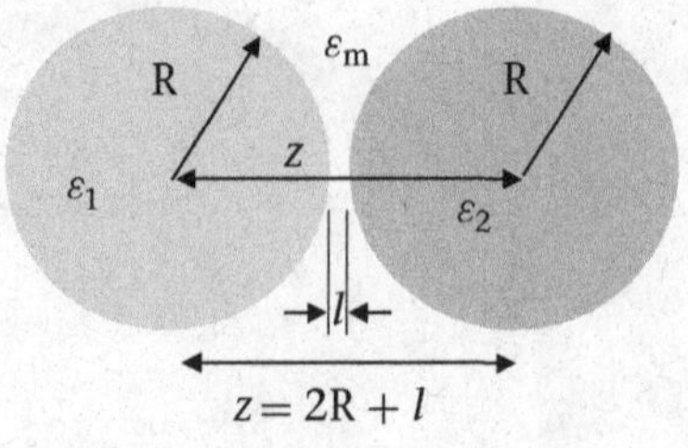

Figure L1.36

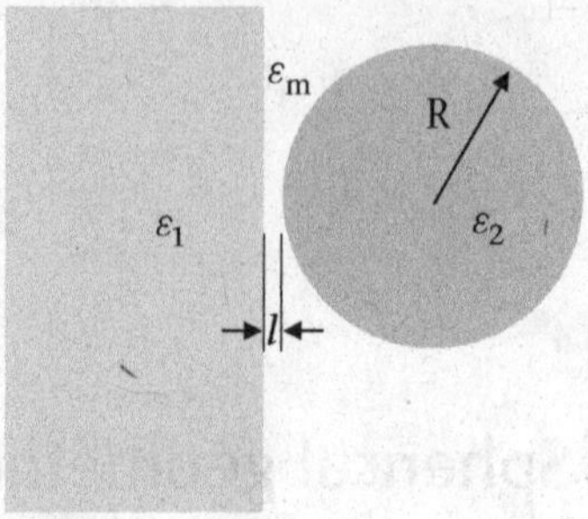

Figure L1.37

(see Fig. L1.37), proportional to R but twice as strong as the R–R equal-sphere interaction.

In the language of the Hamaker coefficient $A_{1m/2m}$, which is restricted to situations in which ε_1, ε_2, and ε_m are nearly equal, $G_{pp}(l)$ looks like the same expression already used several times for planes:

$$G_{1m/2m}(l) = -\frac{A_{1m/2m}}{12\pi l^2}, \quad A_{1m/2m} \approx \frac{3kT}{2} \sum_{n=0}^{\infty}{}' \overline{\Delta}_{1m}\overline{\Delta}_{2m} R_n(l). \qquad \text{(L1.39)}$$

With retardation neglected, $R_{n(l)} = 1$, the predicted force between two spheres is a $1/l^2$ power law; its integral, the free energy of interaction between almost-touching spheres, varies as the inverse-first power of separation.

Why should the interaction between almost-touching spheres vary more slowly than the $1/l^2$ variation in free energy between planes or slabs at close separation? Does the apparently greater reach of the $1/l$ interaction imply stronger forces between spheres than between planes? Comparison needs care. We are comparing different kinds of quantities. Between spheres the energy is per interaction; between planes, per area.

The intriguing $1/l$ dependence comes from the opposite curvature of the two interacting bodies. As they are brought together, more and more material interacts at a particular separation. The added contribution of these more distant parts of the surface, already at separation l in the planar interaction, makes the reach of the spherical interaction seem longer.

More finely stated, areas on the surfaces of the two bodies look across to each other and see bits of almost-parallel surfaces. A very small area, the smallest bit of facing surface is at separation l. Progressively large areas, in circular rings, face each other across slightly larger distances. A change in separation l means not only a change in separation between the patches but also a change in the amount of area at each distance between surfaces. The interaction between oppositely curved surfaces is of the plane–plane energies per area, weighted by the ever-larger areas, from separation l essentially to infinite separation (because the radii are so much larger than the minimum separation). Formally the integral must have a higher-power dependence on separation and consequently a longer apparent range than the original plane–plane interaction.

Does longer apparent range mean stronger actual integration? No. It can be seen as a change in the effective area of interaction at different separations.

PROBLEM L1.13: Working in the nonretarded limit and in the limit of close approach $l \ll R_1, R_2$, compare the free energy of interaction *per unit area* of planar facing surfaces, $G_{1m/2m}(l)$, with the free energy *per interaction*, that is, the integral $G_{ss}(l; R_1, R_2)$ of $F_{ss}(l; R_1, R_2)$. In particular, show that

$$G_{ss}(l; R_1, R_2) = G_{1m/2m}(l)\frac{2\pi R_1 R_2 l}{(R_1 + R_2)}.$$

It is as though the energy per interaction between spheres is the energy per area between planes of the same materials but multiplied by a continuously varying area $2\pi R_1 R_2 l/(R_1 + R_2)$ that goes to zero as the spheres are brought into contact.

Conversely, if we compare the interaction free energy $-(A_{1m/2m}/12)(R/l)$ of two equal-radius spheres with that of two circular parallel patches of radius R, area πR^2, on facing planes of the same materials in the same medium, we have $[-(A_{1m/2m}/12\pi l^2)]\pi R^2 = [-(A_{1m/2m}/12)](R/l)(R/l)$. The plane–plane interaction per unit area is stronger by a variable factor $(R/l) \gg 1$.

For small differences in ε_1, ε_2, and ε_m and with no retardation, i.e., $A_{1m/2m} \approx (3kT/2) \sum_{n=0}^{\prime\infty} \overline{\Delta}_{1m}\overline{\Delta}_{2m}$, there is a simple approximate algebraic expression, which was originally due to Hamaker, for the sphere–sphere interaction energy; at all l it has the closed form (see Table S.3.a):

$$G_{ss}(z; R_1, R_2) = -\frac{A_{1m/2m}}{3}\left[\frac{R_1 R_2}{z^2 - (R_1 + R_2)^2} + \frac{R_1 R_2}{z^2 - (R_1 - R_2)^2} + \frac{1}{2}\ln\frac{z^2 - (R_1 + R_2)^2}{z^2 - (R_1 - R_2)^2}\right]. \tag{L1.40}$$

For $R_1 = R_2 = R$, it reads (see Fig. L1.38) (also see Table S.3.b.3)

$$G_{ss}(z; R) = -\frac{A_{1m/2m}}{3}\left[\frac{R^2}{z^2 - 4R^2} + \frac{R^2}{z^2} + \frac{1}{2}\ln\left(1 - \frac{4R^2}{z^2}\right)\right]. \tag{L1.41}$$

For $R_1 \to \infty$, $R_2 = R$, in this approximate form the sphere–plane interaction goes as (see Fig. L1.39) (see Table S.5.b.1)

$$G_{sp}(l, R) = -\frac{A_{1m/2m}}{6}\left(\frac{R}{l} + \frac{R}{2R + l} + \ln\frac{l}{2R + l}\right). \tag{L1.42}$$

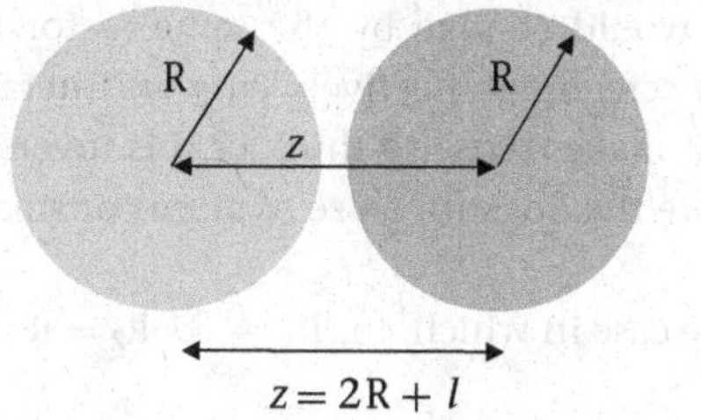

Figure L1.38

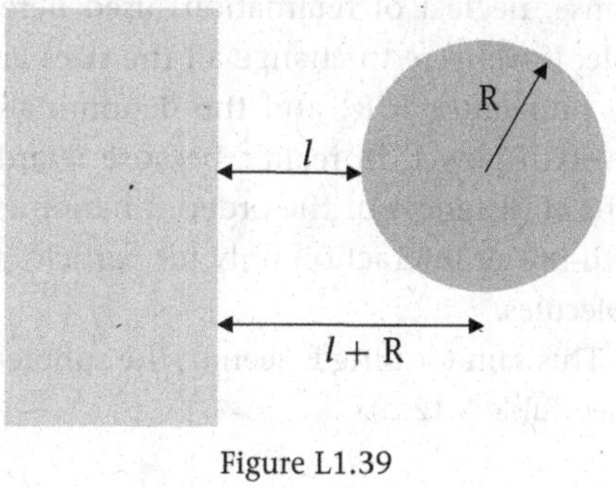

Figure L1.39

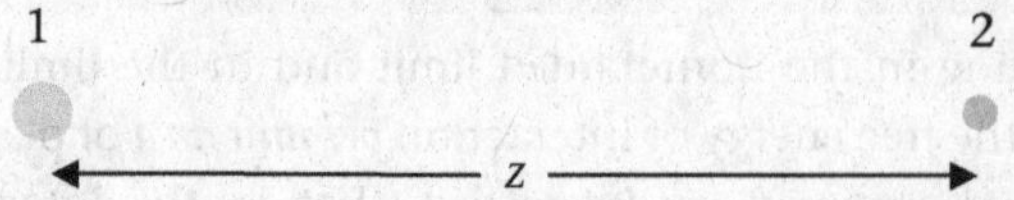

Figure L1.40

PROBLEM L1.14: Obtain Eq. (L1.42) for a sphere–plane interaction from Eq. (L1.40) for a sphere–sphere interaction.

Back to spheres, in the limit in which the spheres are almost touching, with surface-to-surface separation $l \ll R_1$ and R_2, this form of Eq. (L1.40) for $G_{ss}(z; R_1, R_2)$ goes over to an inverse-first power (see Table S.3.b.2) in minimal separation l:

$$G_{ss}(z; R_1, R_2) \to G_{ss}(l; R_1, R_2) = -\frac{A_{1m/2m}}{6}\frac{R_1 R_2}{(R_1 + R_2)\, l}. \tag{L1.43}$$

Because l is very small compared with the radii, the interaction can be thermally significant even when $A_{1m/2m}$ is of the order of kT.

When $R_1 = R_2 = R \gg l$,

$$G_{ss}(l; R) \to -\frac{A_{1m/2m}}{12}\frac{R}{l}. \tag{L1.44a}$$

For $R_1 \to \infty$, $R_2 = R \gg l$,

$$G_{sp}(l; R) = -\frac{A_{1m/2m}}{6}\frac{R}{l}. \tag{L1.44b}$$

PROBLEM L1.15: From Eq. (L1.40) obtain Eq. (L1.43) for sphere–sphere interactions in the limit of close approach, $l \ll R_1$ and R_2.

At the opposite extreme, the limit at which the center-to-center separation z is much greater than the sizes R_1 or R_2, two spheres interact (see Fig. L1.40) as $1/z^6$ (assuming no retardation!) (see Table S.3.b.1):

$$G_{ss}(z; R_1, R_2) \to -\frac{16 A_{1m/2m} R_1^3 R_2^3}{9 z^6}, \quad z \gg R_1, R_2. \tag{L1.45}$$

This approximate form of $G_{ss}(z; R_1, R_2)$ shows a general property of van der Waals interactions when formulated in the approximation (small differences in dielectric response, neglect of retardation) used here. The interaction is independent of length scale. If we were to change all the sizes and separations by any common factor, both the numerator $R_1^3 R_2^3$ and the denominator z^6 would change by the same factor to the sixth power. In reality, because retardation screening effectively cuts off interactions at distances of the order of nanometers, it makes sense to think of this inverse-sixth-power interaction only for particles that are the ångstrom size of atoms or small molecules.

This same scaling is seen in the sphere–plane case in which, for $R_1 \to \infty$, $R_2 = R \ll z$ (see Table S.12.b),

$$G_{sp}(z; R) = -\frac{2 A_{Am/sm}}{9}\frac{R^3}{z^3} \tag{L1.46}$$

Both $G_{ss}(z; R_1, R_2)$ and $G_{sp}(z; R)$ in this small-sphere limit show that, when $A_{1m/2m}$ is of the order of kT, the total interaction must be much less than thermal energy kT.

Fuzzy spheres. Radially varying dielectric response

Colloidal suspensions are often stabilized by the adsorption of polymers that are expected to exert additional configurational-steric repulsive forces. The additional, potentially significant van der Waals interactions between polymer coatings need not be neglected in stability analyses.[6] For a first glimpse of the possible attraction between layers, the Hamaker approximation formula for spheres can be easily generalized to the case in which the epsilons vary continuously, $\varepsilon_1 = \varepsilon_1(r_1)$, $\varepsilon_2 = \varepsilon_2(r_2)$ (see Fig. L1.41).

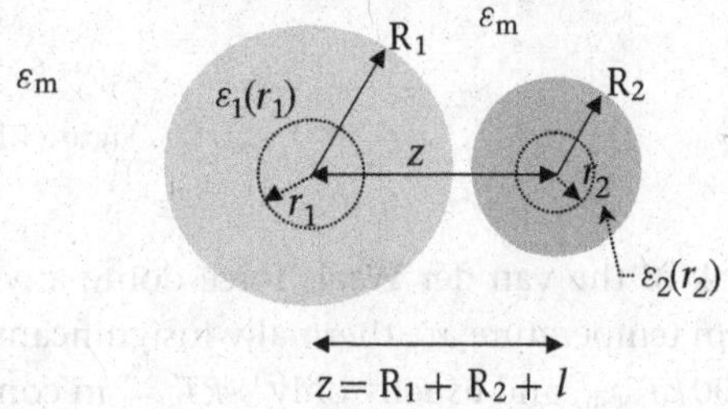

Figure L1.41

It helps our understanding if we view this interaction between inhomogeneous spheres as a sum of interactions between onionlike spheres whose shell–shell interactions add up. The discrete differences from steps in ε's are replaced with derivatives. In the case in which the ε's vary throughout the spheres and then equal ε_m of the medium at R_1 and R_2 (see Table S.4.a),

$$G_{ss}(z) = -\frac{kT}{8}\sum_{n=0}^{\infty}{}' \int_0^{R_1} dr_1 \frac{d\ln[\varepsilon_1(r_1)]}{dr_1} \int_0^{R_2} dr_2 \frac{d\ln[\varepsilon_2(r_2)]}{dr_2}$$
$$\times \left[\frac{r_1 r_2}{z^2 - (r_1 + r_2)^2} + \frac{r_1 r_2}{z^2 - (r_1 - r_2)^2} + \frac{1}{2}\ln\frac{z^2 - (r_1 + r_2)^2}{z^2 - (r_1 - r_2)^2}\right] \qquad \text{(L1.47)}$$

See how the factor within the square brackets resembles that in Eq. (L1.40) for a sphere–sphere interaction. But why the $\{d\ln[\varepsilon(r)]\}/dr$'s instead of the usual $\overline{\Delta}$ differences-over-sums? Expand

$$dr\frac{d\ln[\varepsilon(r)]}{dr} \sim dr\frac{d\varepsilon(r)}{dr}\frac{1}{\varepsilon(r)} \sim \frac{\varepsilon(r + dr) - \varepsilon(r)}{\varepsilon(r)} \sim 2\frac{\varepsilon(r + dr) - \varepsilon(r)}{\varepsilon(r + dr) + \varepsilon(r)} \leftrightarrow 2\overline{\Delta}$$

to see that even the coefficient matches that in Eq. (L1.40).

As for planar-surface interactions, the inhomogeneity of the dielectric response qualitatively changes the form of interaction. This change is especially important when the spheres near contact. In that case it is often more practical to use the transform between planar and spherical interactions rather than to suffer working in spherical coordinates.

"Point–particle" interactions

Traditionally, the inverse-sixth-power variation is regarded as the most elementary form of van der Waals interaction. The conditions of small size and large separation are those that obtain in a van der Waals gas. In fact, this condition holds only rarely when we

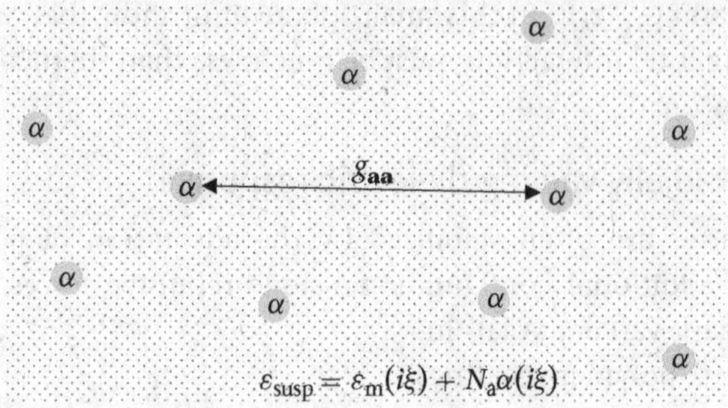

Figure L1.42a

think of the van der Waals force doing anything interesting. Interaction energies at room temperature are thermally insignificant. Hamaker coefficients are not more than $\sim 100\,kT_{room}$, and usually only $\sim kT_{room}$ in condensed media. Even for separation z only four times the radii R_1 and R_2, the sphere–sphere interaction $G_{ss}(z; R_1, R_2)$ is a puny $\sim 2 \times 100\,kT_{room} \times 1/4^6 \sim kT_{room}/20$.

What the $1/z^6$ interaction lacks in thermal interest it compensates splendidly by showing us the different ways in which it can be viewed and the different causes from which it can originate.

Particles in dilute suspension

Rather than take a limit of large separations between relatively small spheres **a** of incremental polarizability $\alpha(i\xi)$, we can think of interactions within dilute suspensions or solutions. At relatively large separations, the shape and the microscopic details of an effectively small speck become unimportant. The only feature that is of interest is that the dilute specks ever so slightly change the dielectric and ionic response of the suspension compared with that of the pure medium. When the suspension of spheres is vanishingly dilute, ε_{susp} is simply proportional to their number density $N_{\mathbf{a}}$ multiplied by $\alpha(i\xi)$, $\varepsilon_{susp} = \varepsilon_m(i\xi) + N_{\mathbf{a}}\alpha(i\xi)$ [see Fig. L1.42(a)].

In practice, thinking of small-particle interactions in terms of dilute-suspension or dilute-solution dielectrics liberates us from having to theorize about the added α. We can simply measure

$$\alpha(i\xi) \equiv \left.\frac{\partial \varepsilon_{susp}(i\xi)}{\partial N_{\mathbf{a}}}\right|_{N_{\mathbf{a}}=0},$$

the change in ε_{susp} with a change in small-particle number density $N_{\mathbf{a}}$, and use this α to compute the energy of interaction at distances large compared with particle size.

In that dilute limit, neglecting retardation, the pairwise interaction of solutes or suspended particles **a** has a form

$$g_{\mathbf{aa}}(z) = -\frac{6kT}{z^6}\sum_{n=0}^{\infty}{}' \left[\frac{\alpha(i\xi_n)}{4\pi\varepsilon_m(i\xi_n)}\right]^2 \tag{L1.48}$$

between like particles.

Between different kinds of particles, **a** and **b**, the interaction goes as (see Table S.6.b)

$$g_{\mathbf{ab}}(z) = -\frac{6kT}{z^6}\sum_{n=0}^{\infty}{}' \left\{\frac{\alpha(i\xi_n)\,\beta(i\xi_n)}{[4\pi\varepsilon_m(i\xi_n)]^2}\right\}, \tag{L1.49}$$

where $\alpha(i\xi)$ and $\beta(i\xi)$ are each the observed

$$\alpha(i\xi) \equiv \left.\frac{\partial\varepsilon_{susp}(i\xi)}{\partial N_{\mathbf{a}}}\right|_{N_{\mathbf{a}}=0, N_{\mathbf{b}}=0}, \qquad \beta(i\xi) \equiv \left.\frac{\partial\varepsilon_{susp}(i\xi)}{\partial N_{\mathbf{b}}}\right|_{N_{\mathbf{a}}=0, N_{\mathbf{b}}=0} \tag{L1.50}$$

for number densities of $N_{\mathbf{a}}$ and $N_{\mathbf{b}}$ of particles **a** and **b** [see Fig. L1.42(b)].

The interaction between unlike point particles is then the same as that between like particles except that the quantity α^2 (or β^2) is replaced with $\alpha\beta$. Each term in the sum-over-frequencies is the geometric mean of the corresponding like-particle-interaction terms.

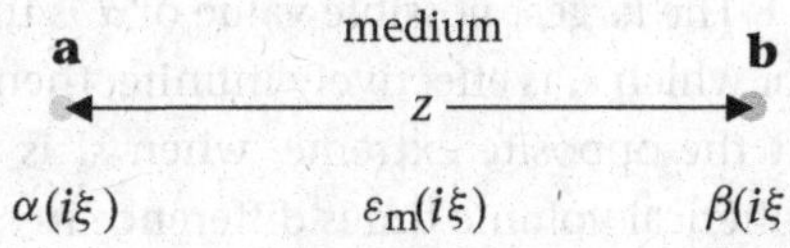

Figure L1.42b

Because $\alpha^2 + \beta^2 \geq 2\alpha\beta$, this geometric mean is the basis for an inequality, $g_{\mathbf{aa}} + g_{\mathbf{bb}} \leq 2g_{\mathbf{ab}}$. A physical interpretation of this inequality is that the total free energy would be lower if **a**-type particles were near other **a**'s and **b**'s were with **b**'s. In fact, because point–particle interaction energies are small compared with thermal energy, this is not a useful inequality when considering dilute suspensions. This kind of inequality does occur in other geometries or conditions in which van der Waals interaction energies are strong compared with thermal energy, e.g., when minimum separations are small compared with the sizes of the interacting bodies. Then these preferential energies might overcome the entropy of random mixing.

Conceptually these pairwise interactions emerge automatically from expressions for the interaction between half-spaces A and B across m. In this case, A is a suspension of particles **a** whose incremental contribution to ε_A is $\alpha(i\xi)$; B, of particles **b** whose incremental contribution to ε_B is $\beta(i\xi)$. The interaction $g_{\mathbf{ab}}(z)$ is what emerges when $\varepsilon_A = \varepsilon_m(i\xi) + N_{\mathbf{a}}\alpha(i\xi)$, $\varepsilon_B = \varepsilon_m(i\xi) + N_{\mathbf{b}}\beta(i\xi)$ so that

$$\frac{\varepsilon_A - \varepsilon_m}{\varepsilon_A + \varepsilon_m} \approx \frac{N_{\mathbf{a}}\alpha}{2\varepsilon_m} \quad \text{and} \quad \frac{\varepsilon_B - \varepsilon_m}{\varepsilon_B + \varepsilon_m} \approx \frac{N_{\mathbf{b}}\beta}{2\varepsilon_m}. \tag{L1.51}$$

In this picture, we are seeing the β-responding particles on the right interact with the α-responding particles on the left with a strength that is given in the summation by $\sum_{n=0}^{\infty}{}'\alpha\beta$. It should be obvious that this kind of pairwise summation of α/β interactions is permissible only when the suspensions are so dilute that their dielectric response is linear in particle density (see Fig. L1.43).[7]

If the particles are small spheres, then each sphere of radius a and material dielectric response ε_s boosts the total dielectric response of a dilute suspension by [see Eqs. (L2.167) and Table S.7 in Level 2]

$$\alpha = 4\pi a^3 \varepsilon_m \frac{(\varepsilon_s - \varepsilon_m)}{(\varepsilon_s + 2\varepsilon_m)} \tag{L1.52}$$

(and the equivalent for β).

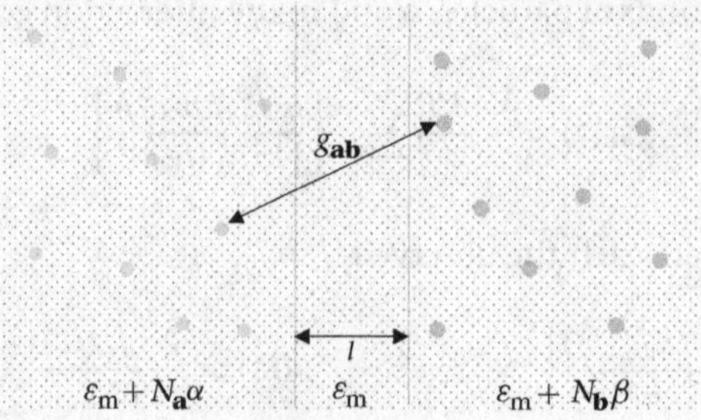

Figure L1.43

The largest possible value of α is that for a sphere of infinitely conducting material for which ε_s is effectively infinite; then $\alpha = 4\pi a^3$, three times the volume of the sphere. At the opposite extreme, when ε_s is almost equal to ε_m, $\alpha \approx [(4\pi a^3)/3](\varepsilon_s - \varepsilon_m)$, the spherical volume times difference in ε's.

Neglecting retardation, we can write the material small-sphere interaction in the form (see Table S.7.a)

$$g_{aa}(z) = -\frac{6kT}{z^6} a^6 \sum_{n=0}^{\infty}{}' \left[\frac{(\varepsilon_s - \varepsilon_m)}{(\varepsilon_s + 2\varepsilon_m)}\right]^2. \tag{L1.53}$$

Reconciliation of large- and small-particle language

Is this last result, Eq. (L1.53), compatible with the nonretarded interaction of two spheres that have a separation much greater than their radii R_1 and R_2?

In the earlier case the interaction was given as $G_{ss}(z; R_1, R_2) = \{-[(16A_{1m/2m}R_1^3R_2^3)/9z^6]\}$, Eq. (L1.45), for unlike spheres with the nonretarded Hamaker coefficient $A_{1m/2m} \approx (3kT/2)\sum_{n=0}^{\prime\infty} \overline{\Delta}_{1m}\overline{\Delta}_{2m}$, as used in Eq. (L1.40). If we match language, we put $R_1 = R_2 = a$ and $\varepsilon_1 = \varepsilon_2 = \varepsilon_s$, and Eq. (L1.45) for equal spheres becomes

$$-\frac{8kTa^6}{3z^6} \sum_{n=0}^{\infty}{}' \left(\frac{\varepsilon_s - \varepsilon_m}{\varepsilon_s + \varepsilon_m}\right)^2.$$

Why the apparent disparity with

$$-\frac{6kTa^6}{z^6} \sum_{n=0}^{\infty}{}' \left[\frac{(\varepsilon_s - \varepsilon_m)}{(\varepsilon_s + 2\varepsilon_m)}\right]^2$$

in the language of small-particle response? Their ratio is $(9/4)\{[(\varepsilon_s + \varepsilon_m)/(\varepsilon_s + 2\varepsilon_m)]^2\}$ for each term in the sum. The flaw is in using the approximate form $[(\varepsilon_s - \varepsilon_m)/(\varepsilon_s + \varepsilon_m)]^2$ in the small-dielectric-difference approximation in Eq. (L1.45). The form $[(\varepsilon_s - \varepsilon_m)/(\varepsilon_s + 2\varepsilon_m)]^2$ used in Eq. (L1.53) is accurate as long as $\alpha N \ll \varepsilon_m$ and retardation can be ignored. When there is only a small difference between ε_s and ε_m, this discrepancy factor is near 1. For example, if one material has the visible-frequency response of a hydrocarbon $\varepsilon \sim 2$ and the other that of water, $\sim 10\%$ less, ~ 1.8 at visible frequencies, the factor of disagreement amounts to $(9/4)\{[(2 + 1.8)/(2 + 2 \times 1.8)]^2\} \sim 1.04$. At the low-frequency limit, with $\varepsilon_s \sim 2$ for hydrocarbon and $\varepsilon_m \sim 80$ for water, the factor is a nontrivial $(9/4)\{[(2 + 80)/(2 + 2 \times 80)]^2\} \approx 0.58$. The difference in the two forms reminds us that these expressions rely on conceptually different assumptions in derivation and must be applied under conditions that match those assumptions.

Example: Magnitude of interaction between proteins in solution

To get a feel for using these equations as applied to macromolecules in solution, we can make a back-of-the-envelope estimate for the interaction between two proteins treated as spheres at large separation.

At visible frequencies, the change in the solution index of refraction with an added solute gives a qualitative idea of the magnitude of relative polarizability of solute and solvent. As elaborated elsewhere in the text, the complex dielectric response, $\varepsilon = \varepsilon'(\omega) + i\varepsilon''(\omega)$, is equal to the square of the complex refractive index $(n_{\text{ref}} + ik_{\text{abs}})^2 = (n_{\text{ref}}^2 - k_{\text{abs}}^2) + 2i\, n_{\text{ref}}k_{\text{abs}}$ where n_{ref} is the index of refraction and k_{abs} is the absorption coefficient. In a transparent region where $k_{\text{abs}} = 0$, the dielectric response function $\varepsilon = n_{\text{ref}}^2$. (See the Level 2.4 essay on dielectric response.)

To get the coefficient α_{solute} to be used in $\varepsilon = \varepsilon_{\text{m}} + N\alpha_{\text{solute}}$, take the derivative

$$\partial\varepsilon/\partial N = \alpha_{\text{solute}} = 2n_{\text{ref}}\ \partial n_{\text{ref}}/\partial N.$$

Recall that N is a solute *number* density. It is proportional to the weight concentration c_{wt} by a factor that is the actual molecular weight of the single protein molecule, that is, the molecular weight MW in grams per mole (or "daltons" in popular obfuscation) divided by Avogadro's number or molecules per mole, N_{Avogadro}. In grams per volume, the weight concentration $c_{\text{wt}} = (\text{MW}/N_{\text{Avogadro}})N$:

$$\alpha_{\text{solute}} = 2(\text{MW}/N_{\text{Avogadro}})\partial n_{\text{ref}}/\partial c_{\text{wt}}.$$

The quantity $\partial n_{\text{ref}}/\partial c_{\text{wt}}$ is routinely measured by differential refractometry and light scattering by nonabsorbing protein solutions. It varies slightly, from $\sim$1.7 to $\sim 2.0 \times 10^{-4}$ liters/g.[8]

At low frequencies, indices of refraction at visible frequency are no longer relevant. Dielectric-dispersion data are available for several proteins.[9] For example, for hemoglobin in the limit of low frequency, $\partial\varepsilon(0)/\partial c_{\text{wt}} = 0.3$ liter/g. The required $\alpha_{\text{solute}}(0)$ comes directly from $\partial\varepsilon(0)/\partial c_{\text{wt}}$:

$$\alpha_{\text{solute}}(0) \equiv \partial\varepsilon(0)/\partial N = [\partial\varepsilon(0)/\partial c_{\text{wt}}](\text{MW}/N_{\text{Avo}}).$$

The magnitude of interaction between small protein spheres in water can be estimated from these numbers. Take a molecule the size of hemoglobin, whose molecular weight is approximately 66,000 g/mol or

$$(\text{MW}/N_{\text{Avogadro}}) = (66{,}000/0.602 \times 10^{+24}) = 1.1 \times 10^{-19}\ \text{g}.$$

The low-frequency polarizability is

$$\begin{aligned}\alpha_{\text{solute}}(0) &= \partial\varepsilon(0)/\partial c_{\text{wt}} \times (\text{MW}/N_{\text{Avogadro}})\\ &= (0.3\ \text{liter/g}) \times (1.1 \times 10^{-19}\ \text{g}) \times (10^3\ \text{cm}^3/\text{liter})\\ &= 3.3 \times 10^{-17}\ \text{cm}^3.\end{aligned}$$

The visible-frequency polarizability is

$$\begin{aligned}\alpha_{\text{solute}} &= 2\,\partial n_{\text{ref}}/\partial c_{\text{wt}} \times (\text{MW}/N_{\text{Avogadro}})\\ &= 2 \times (1.8 \times 10^{-4}\ \text{liters/g}) \times (1.1 \times 10^{-19}\ \text{g}) \times (10^3\ \text{cm}^3/\text{liter})\\ &= 4 \times 10^{-20}\ \text{cm}^3.\end{aligned}$$

To keep things on the back of the envelope, assume that this value holds up to the beginning of UV frequencies, wavelength 2000 Å, or a cutoff frequency ξ_{uv} or $\xi_{\text{cutoff}} \sim 10^{16}$ (rad/s).

The weight density of proteins is ~4/3 that of water, i.e., ~4/3 g/cm³, so that the volume of this molecule is approximately $(1.1 \times 10^{-19}\ \text{g})/(4/3\ \text{g/cm}^3) = 8.2 \times 10^{-20}\ \text{cm}^3$. In the form of a sphere, it would have a radius of 27 Å = 2.7 nm. The point-particle formula holds then only for separations much greater than the sphere diameter, ~5 nm.

For clarity, split the summation for $g_p(z)$,

$$g_p(z) = -\frac{3kT}{8\pi^2 z^6}\sum_{n=0}^{\infty}{}' \left[\frac{\alpha_{\text{solute}}(i\xi_n)}{\varepsilon_w(i\xi_n)}\right]^2,$$

into a $n = 0$ zero-frequency term,

$$-\frac{3kT}{16\pi^2 z^6}\left[\frac{\alpha_{\text{solute}}(0)}{\varepsilon_w(0)}\right]^2,$$

plus the remaining terms,

$$-\frac{3kT}{8\pi^2 z^6}\sum_{n=1}^{\infty}\left[\frac{\alpha_{\text{solute}}(i\xi_n)}{\varepsilon_w(i\xi_n)}\right]^2.$$

Because the dielectric constant of water at low frequency is $\varepsilon_w(0) = 80$, the coefficient of z^{-6} in the zero-frequency term is (in kT_{room} units)

$$\begin{aligned}-\frac{3kT}{16\pi^2}\left[\frac{\alpha_{\text{solute}}(0)}{\varepsilon_w(0)}\right]^2 &= -\frac{3}{16\pi^2}\left(\frac{3.3 \times 10^{-17}\ \text{cm}^3}{80}\right)^2 kT_{\text{room}}\\ &= -3.2 \times 10^{-39}\ \text{cm}^6\ kT_{\text{room}}\ \text{or}\ -(3.84\ \text{nm})^6\ kT_{\text{room}}.\end{aligned}$$

For a center-to-center separation z of 7.5 nm, one radius worth of surface-to-surface separation, already violating the large-separation assumption, the interaction is

$$-(3.84\ \text{nm})^6/(7.5\ \text{nm})^6\ kT_{\text{room}} = -1.8 \times 10^{-2}\ kT_{\text{room}} \approx kT_{\text{room}}/60.$$

This computation neglects ionic-screening factors that can only make this low-frequency interaction weaker still.

For the finite-frequency charge-fluctuation energy,

$$-\frac{3kT}{8\pi^2 z^6}\sum_{n=1}^{\infty}\left[\frac{\alpha_{\text{solute}}(i\xi_n)}{\varepsilon_w(i\xi_n)}\right]^2,$$

we could again convert this into an integral as in the evaluation of the London interaction, but it is easier to keep it as a summation for which we have assumed that the summand $[\alpha_{\text{solute}}(i\xi_n)/\varepsilon_w(i\xi_n)]^2$ maintains a constant value between ξ_1 and the cutoff-frequency term corresponding to $\xi_{\text{cutoff}} = 10^{16}$ (rad/s). Because (at $T = 20$ °C) the values of ξ go as $\xi_n = [(2\pi kT)/\hbar]n = 2.411 \times 10^{14}$ (rad/s)n, there are $\approx[10^{16}(\text{rad/s})]/$

$[2.411 \times 10^{14}$ (rad/s)$] = 41$ terms in the sum. For a quick estimate, we can replace the summation $\sum_{n=1}^{\infty}\{[\alpha_{\mathrm{solute}}(i\xi_n)/\varepsilon_{\mathrm{w}}(i\xi_n)]^2\}$ by a factor of $41 \times (\alpha_{\mathrm{solute}}/\varepsilon_{\mathrm{w}})^2$.

For α_{solute} take the value $\alpha_{\mathrm{solute}} = 4 \times 10^{-20}\ \mathrm{cm}^3$ previously derived; for ε_{w} take the square of the refractive index of water, $\varepsilon_{\mathrm{w}} = 1.333^2 = 1.78$.

The coefficient of z^{-6} in the finite-frequency term is (in kT units)

$$-\frac{3}{8\pi^2}41\left(\frac{\alpha_{\mathrm{solute}}}{\varepsilon_{\mathrm{w}}}\right)^2 kT = -\frac{3}{8\pi^2}41\left(\frac{4\times 10^{-20}\ \mathrm{cm}^3}{1.78}\right)^2 kT$$
$$= -7.9 \times 10^{-40}\ \mathrm{cm}^6\ kT = -7.9 \times 10^{+2}\ \mathrm{nm}^6\ kT$$

This is smaller than the coefficient $-3.2 \times 10^{+3}\ \mathrm{nm}^6\ kT$ already estimated for the zero-frequency contribution and therefore again indicates very weak interaction energy. Inclusion of the significant degree of retardation screening expected at these separations would have yielded still smaller energies.

The inverse-sixth-power van der Waals interaction is never likely to be strong compared with thermal energy. When surface-to-surface separations are small compared with particle size, these dispersion forces can become significant factors in organizing molecules. These spectral data give $A_{\mathrm{Ham}} \sim 2.5\,kT_{\mathrm{room}} \sim 10 \times 10^{-14}$ ergs or 10^{-20} J or 10 zJ.[10] Momentarily neglecting the limitations of a continuum picture, what would this A_{Ham} mean for two parallel surfaces with an area $\mathrm{L}^2 = 1\ \mathrm{nm}^2$, $l = 0.22\ \mathrm{nm} = 2.5$ Å apart?

$$\frac{A_{\mathrm{Ham}}}{12\pi l^2} \times \mathrm{L}^2 = \frac{2.5}{12\pi}\frac{1}{0.25^2}kT_{\mathrm{room}} \sim kT_{\mathrm{room}}.$$

There is too much else going on with proteins at such separations to assume that their van der Waals energy is all that matters.

Point–particle substrate interactions

In the same spirit as that of the extraction of small-particle interactions, it is possible to specialize the general expression for the interaction of planar half-spaces in order to formulate interactions between point particles and substrates (see Fig. L1.44).

In the limiting case in which (1) retardation is neglected, (2) ε_{A} does not differ greatly from ε_{m}, and (3) the particles are in dilute suspension such that $N_{\mathbf{b}}\beta \ll \varepsilon_{\mathrm{m}}$, (see Table S.11.b.1)

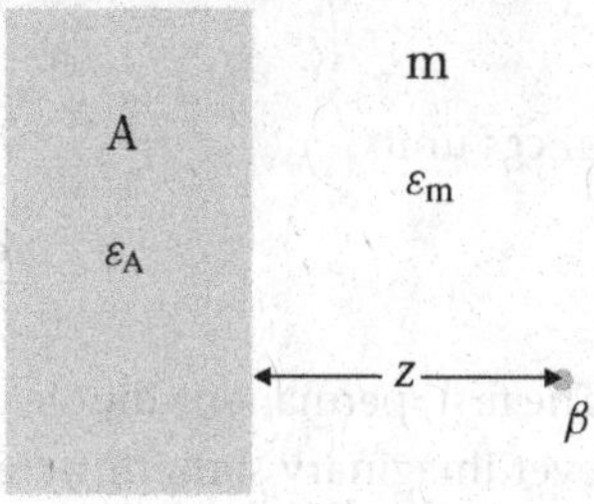

Figure L1.44

$$g_{\mathrm{p}}(z) = -\frac{kT}{2z^3}\sum_{n=0}^{\infty}{}' \left[\frac{\beta(i\xi_n)}{4\pi\varepsilon_{\mathrm{m}}(i\xi_n)}\right]\left[\frac{\varepsilon_{\mathrm{A}}(i\xi_n) - \varepsilon_{\mathrm{m}}(i\xi_n)}{\varepsilon_{\mathrm{A}}(i\xi_n) + \varepsilon_{\mathrm{m}}(i\xi_n)}\right]. \tag{L1.54}$$

For spheres of radius b and material dielectric response $\varepsilon_{\mathrm{sph}}$, β can be replaced by $4\pi \mathrm{b}^3\varepsilon_{\mathrm{m}}\frac{(\varepsilon_{\mathrm{sph}}-\varepsilon_{\mathrm{m}})}{(\varepsilon_{\mathrm{sph}}+2\varepsilon_{\mathrm{m}})}$ so that (see Table S.12.a)

$$g_{\mathrm{p}}(z) = -\frac{kT}{2z^3}\mathrm{b}^3\sum_{n=0}^{\infty}{}' \frac{\varepsilon_{\mathrm{sph}} - \varepsilon_{\mathrm{m}}}{\varepsilon_{\mathrm{sph}} + 2\varepsilon_{\mathrm{m}}}\frac{\varepsilon_{\mathrm{A}} - \varepsilon_{\mathrm{m}}}{\varepsilon_{\mathrm{A}} + \varepsilon_{\mathrm{m}}}. \tag{L1.55}$$

Particles in a dilute gas

In the context of dilute solutions, suspensions, or vapors, we can safely regard the historically earliest idealizations of inverse-sixth-power interactions. These are for the conditions that hold in a van der Waals gas. The medium is vacuum; the α's and β's of individual particles already introduced are now those of atoms or small molecules. The vapor is so dilute that its dielectric response is that of the vacuum plus very small contributions proportional to the number density of particles.

To think efficiently, though formally, again use the simplest expression for two interacting half-spaces across a medium. Imagine that regions A and B are vapors that interact across a vacuum "medium":

$$\varepsilon_A = 1 + N_A\alpha, \quad \varepsilon_B = 1 + N_B\beta, \quad N_A\alpha \ll 1, \quad N_B\beta \ll 1,$$
$$\overline{\Delta}_{Am} = \frac{\varepsilon_A - 1}{\varepsilon_A + 1} \to \frac{N_A\alpha}{2}, \quad \overline{\Delta}_{Bm} \to \frac{N_B\beta}{2}.$$

The interaction between A and B across m, $G_{Am/Bm}(l) \approx [-(kT/8\pi l^2)] \sum_{n=0}^{\prime\infty} \overline{\Delta}_{Am}\overline{\Delta}_{Bm} R_n$ becomes a product of densities $N_{\mathbf{a}}$ and $N_{\mathbf{b}}$ with responses α and β, $[-(kT/32\pi l^2)]$ $N_{\mathbf{a}}N_{\mathbf{b}} \sum_{n=0}^{\prime\infty} \alpha\beta R_n$. The procedure for extracting point-particle interactions, rigorously effected in Level 2, Subsection L2.3.E, is to find the forms $g_{\mathbf{ab}}(r)$ that add up to $G_{Am/Bm}(l)$ (see Fig. L1.45).

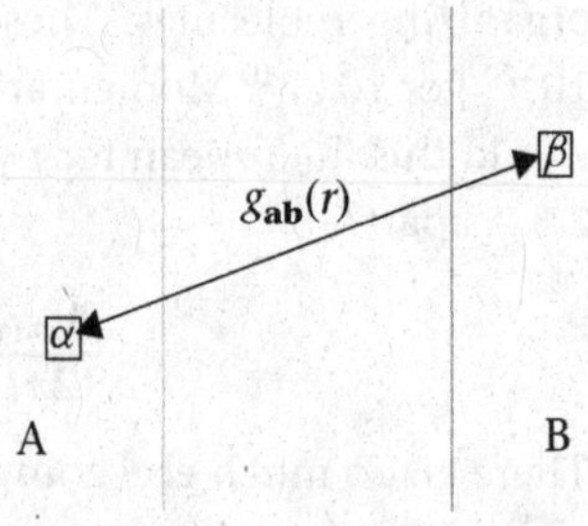

Figure L1.45

This extraction precisely reproduces the same London, Debye, and Keesom interactions, including all relativistic retardation terms that had been effortfully derived in earlier formulations. These interactions are distinguished by whether they involve the interaction of two permanent dipoles of moment μ_{dipole}, or involve an inducible polarizability α_{ind}. A water molecule, for example, has both a permanent dipole moment and inducible polarizability. The contribution of each water molecule to the total dielectric response is a sum of the form of Eqs. (L2.163) and (L2.173): in mks units,

$$\frac{\alpha(i\xi)}{4\pi} = \frac{\mu_{\text{dipole}}^2}{3kT \times 4\pi\varepsilon_0(1+\xi\tau)} + \frac{\alpha_{\text{ind}}(i\xi)}{4\pi\varepsilon_0};$$

in cgs units,

$$\frac{\alpha(i\xi)}{4\pi} = \frac{\mu_{\text{dipole}}^2}{3kT(1+\xi\tau)} + \alpha_{\text{ind}}(i\xi). \tag{L1.56}$$

The first, permanent-dipole term is important only at zero frequency in the summation over imaginary sampling frequencies ξ_n. The relaxation time τ is big enough that for $\xi_{n=1}$ the permanent-dipole term in α is effectively zero; this term counts only at zero frequency. In both mks (SI or Système International) and cgs ("Gaussian") units, the dipole moment $\mu_{\text{dipole}} = qd$ for charges $\pm q$ separated by distance d. [See table S.8 and Eq. (L2.171) in Level 2.]

PROBLEM L1.16: Show that, for $\tau = 1/1.05 \times 10^{11}$ rad/s (Table L2.1 in Level 2, Subsection L2.4.D), $\xi_{n=1}\tau \gg 1$ at room temperature.

Three traditional kinds of interactions emerge:

- "Keesom," based on the mutual alignment of permanent dipoles, at zero frequency [see Table S.8.a and Eq. (L2.177):

$$g_{\text{Keesom}}(r) = -\frac{\mu_{\text{dipole}}^4}{3(4\pi\varepsilon_0)^2 kTr^6} \text{ in mks units} = -\frac{\mu_{\text{dipole}}^4}{3kTr^6} \text{ in cgs units.} \quad \text{(L1.57)}$$

- "Debye," for the coupling between the permanent dipole and the molecular polarizability, at zero frequency [see Table S.8.b and Eq. (L2.178)]:

$$g_{\text{Debye}}(r) = -\frac{2\mu_{\text{dipole}}^2}{(4\pi\varepsilon_0)^2 r^6}\alpha_{\text{ind}}(0) \text{ in mks units} = -\frac{2\mu_{\text{dipole}}^2}{r^6}\alpha_{\text{ind}}(0) \text{ in cgs units.} \quad \text{(L1.58)}$$

- "London–Casimir–Polder," for the correlation between inducible dipoles, occurring at all frequencies (see Table S.6.a):

$$g_{\text{London}}(r) = -\frac{6kT}{(4\pi\varepsilon_0)^2 r^6}\sum_{n=0}^{\infty}{}' \alpha_{\text{ind}}^2(i\xi_n)e^{-r_n} \times \left(1 + r_n + \frac{5}{12}r_n^2 + \frac{1}{12}r_n^3 + \frac{1}{48}r_n^4\right) \text{ in mks units,}$$

$$g_{\text{London}}(r) = -\frac{6kT}{r^6}\sum_{n=0}^{\infty}{}' \alpha_{\text{ind}}^2(i\xi_n)e^{-r_n}\left(1 + r_n + \frac{5}{12}r_n^2 + \frac{1}{12}r_n^3 + \frac{1}{48}r_n^4\right) \text{ in cgs units.} \quad \text{(L1.59)}$$

PROBLEM L1.17: Show how Eqs. (L1.59) emerge from the equation of Table S.6.a.

The retardation factor $(1 + r_n + \frac{5}{12}r_n^2 + \frac{1}{12}r_n^3 + \frac{1}{48}r_n^4)e^{-r_n}$ formally creates three power-law regimes:

- The nonretarded inverse-sixth-power form [see Table S.8.c.1 and Eqs. (L2.179)]:

$$g_{\text{London}}(r) = -\frac{6kT}{(4\pi\varepsilon_0)^2 r^6}\sum_{n=0}^{\infty}{}' \alpha_{\text{ind}}(i\xi_n)^2 \text{ in mks units,}$$

$$g_{\text{London}}(r) = -\frac{6kT}{r^6}\sum_{n=0}^{\infty}{}' \alpha_{\text{ind}}(i\xi_n)^2 \text{ in cgs units,} \quad \text{(L1.60)}$$

  when $r_n = 0$.

- The much-quoted but physically impossible inverse-seventh-power interaction when the summation converges only because of the retardation factor. Impossible? Note that this form, like the inverse-third-power regime for planar interactions, holds rigorously only in the hypothetical, physically unreachable limit of $T = 0$. (see Table S.6.c):

$$g_{\text{London}}(r) = -\frac{23\hbar c}{4\pi r^7}\left[\frac{\alpha_{\text{ind}}(0)}{4\pi\varepsilon_0}\right]^2 \text{ in mks units,}$$

$$g_{\text{London}}(r) = -\frac{23\hbar c}{4\pi r^7}\alpha_{\text{ind}}(0)^2 \text{ in cgs units.} \quad \text{(L1.61)}$$

- Fully retarded inverse-sixth-power longest-range variation, $(3kT/r^6)\alpha_{\mathrm{ind}}^2(0)$ at distances so large that all finite-frequency terms are screened (see Table S.6.d):

$$g_{\mathrm{London}}(r) = -\frac{3kT}{r^6}\left[\frac{\alpha_{\mathrm{ind}}(0)}{4\pi\varepsilon_0}\right]^2 \text{ in mks units,}$$

$$g_{\mathrm{London}}(r) = -\frac{3kT}{r^6}\alpha_{\mathrm{ind}}(0)^2 \text{ in cgs units.} \tag{L1.62}$$

The variation in the apparent power law seen here is similar in spirit to what was previously illustrated for the interaction between parallel surfaces.

PROBLEM L1.18: Derive Eqs. (L1.62) from Eqs. (L1.59).

Expected magnitude of forces between point particles

Even when not reduced by retardation screening, the van der Waals interaction between point particles will be small compared with thermal energy kT. This can be seen from a few examples.

Molecules bearing a permanent-dipole moment Think of the interaction between two fairly strong dipoles, $\mu_{\mathrm{dipole}} = 2\ \mathrm{D} = 2 \times 10^{-18}$ esu cm, slightly larger than the 1.87-D moment of a water molecule.[11] Let this molecule be approximately the size of a water molecule, i.e., ~3 Å across so that the point-molecule approximation would apply at separations much greater than 3 Å. In units of kT the Keesom interaction is

$$g_{\mathrm{Keesom}}(r) = -\frac{kT}{3}\left(\frac{\mu_{\mathrm{dipole}}^2}{4\pi\varepsilon_0 kTr^3}\right)^2 \text{ in mks units,} \quad -\frac{kT}{3}\left(\frac{\mu_{\mathrm{dipole}}^2}{kTr^3}\right)^2 \text{ in cgs units.}$$

At a center-to-center separation $l = 10$ Å $= 10^{-7}$ cm, with $kT = kT_{\mathrm{room}} \approx 4 \times 10^{-14}$ ergs, this energy becomes $(kT_{\mathrm{room}}/3)\,[(4 \times 10^{-36})/(4 \times 10^{-14} \times 10^{-21})]^2 = (kT_{\mathrm{room}}/300)$, which is negligible compared with the energy of thermal excitation. At $l = 6$ Å, a surface-to-surface separation of one diameter, the interaction is $(10/6)^6 = 21$ times stronger, $\sim(kT_{\mathrm{room}}/15)$. Only if the distance between the centers of the molecules were slightly less than 4 Å, with their outer surfaces just 1 Å apart, would this energy approach kT_{room}; but by then the restriction to a separation that is large compared with size is grossly violated, and the formula for $g_{\mathrm{Keesom}}(r)$ is no longer valid.

Induced-dipole–induced-dipole interactions London forces are similarly weak. For example, assume that the interacting spheres have the maximum allowed polarizability, that of ideally conducting metallic spheres, $\alpha = 4\pi a^3 = 4\pi\alpha_{\mathrm{ind}}$ in cgs units. Let them have that polarizability up to a near-uv frequency ξ_{uv} whose value corresponds to a wavelength of 1000 Å $= 10^{-5}$ cm or a radial frequency of $(2\pi c/\lambda) \approx 2 \times 10^{16}$ rad/s. Because the low-frequency polarizability $\alpha_{\mathrm{ind}} = a^3$, $g_{\mathrm{London}}(r)$ has an appealingly simple form. The free energy of interaction between these two small spheres is

$$g_{\mathrm{London}}(r) = -\frac{6kT}{r^6}\sum_{n=0}^{\infty}{}'\,[\alpha_{\mathrm{ind}}(i\xi_n)]^2 = -\frac{3\hbar}{\pi r^6}\int_0^{\infty}\alpha_{\mathrm{ind}}(i\xi)^2\,d\xi$$

$$= -\frac{3\hbar}{\pi r^6}(a^3)^2\,\xi_{\mathrm{uv}} = -\frac{3}{\pi}\frac{a^6}{r^6}\hbar\xi_{\mathrm{uv}}.$$

The photon energy $\hbar\xi_{uv} = 1.05 \times 10^{-27}$ ergs/s $\times 2 \times 10^{16}$ rad/s $= 2.1 \times 10^{-11}$ ergs. For spheres of separation $r = 4a$, that is, one diameter's distance between their surfaces, the energy of interaction is 4.9×10^{-15} ergs $\approx kT_{room}/8$. This is much less than thermal energy even at room temperature and even at a separation that barely satisfies the requirement that separation be much greater than size. For $r = 3a$, that is, a distance of one radius between surfaces, the energy is 2.7×10^{-14} ergs $\approx kT_{room}/1.5$, roughly equal to kT_{room} but now at a separation too close to satisfy the dilute-suspension condition for valid use of the formula.

Screening of "zero-frequency" fluctuations in ionic solutions

As already emphasized, the full computation of charge-fluctuation forces necessarily includes movement by all kinds of charge. Because of their number and their ability to respond to high frequencies, electrons were traditionally thought to be the most important charges in fluctuation forces. Although this expectation is usually met in "dry" systems, it does not necessarily hold for polar liquids such as water, wherein there can be charge fluctuations from dipolar vibrations (at IR frequencies), from rotations of polar molecules (at microwave frequencies), and from ionic fluctuations (from microwave down to zero frequency). The full theory as developed in modern form implicitly includes all these contributions by using dielectric susceptibilities that can also include zero-frequency charge displacements corresponding to the currents carried by mobile charges, be they electrons in a metal or ions in a salt solution.

Salt solutions, of special interest in biological and colloidal systems, merit individual attention. Because of their capacity to form diffuse electrostatic double layers, mobile ions display a particularly seductive coupling between charge fluctuations and screening of the electric fields that come from those fluctuations. The first example of this coupling is seen as a "salt-screening" of zero frequency, $\xi_{n=0} = 0$, charge fluctuations. Imagine that medium m is a salt solution with a Debye screening length of $\lambda_{Debye} = \lambda_D$.

A low-frequency electric field emanating from body A will be screened by the salt solution with an exponential attenuation typical of double layers between parallel-planar bodies. That is, it will die as e^{-x/λ_D} versus the distance x from the interface at which the signal enters medium m from body A. By the time the signal travels the distance l to body B it will be screened to an extent e^{-l/λ_D}. The response of B back to A will also suffer a screening by a factor e^{-l/λ_D}.

The across-and-back screening of charge fluctuations has a form remarkably similar to the relativistic screening that is due to the finite velocity of light. Call the factor R_0 to emphasize that it occurs for $\xi_{n=0} = 0$, and write it in the approximate form [see Tables P.1.d.4 and P.1.d.5 and Eqs. (L3. 199) and (L3. 201)]:

$$R_0 = (1 + 2l/\lambda_D)\, e^{-2l/\lambda_D}. \tag{L1.63}$$

The Debye screening length for a 1–1 electrolyte is $\sim 3\,\text{Å}/\sqrt{I(\text{M})}$ where $I(\text{M})$ is the ionic strength in molar units (Level 2, Table P.1.d); in a 0.1 molar solution, $\lambda_D \sim 10\,\text{Å}$. The ionic screening of low-frequency fluctuations can be suffocating. For a separation $l = 10\,\text{Å}$ (already almost too small for the continuum limit in which the van der Waals theory holds) R_0 will amount to $(1 + 2)e^{-2} \approx 0.4$; for $l = 20\,\text{Å}$, $(1 + 3)e^{-3} \approx 0.2$, 80% screening.

Ionic screening of low-frequency fluctuations is even more potent when the surfaces bear permanent charge. In this case, the high average number density of the counterions creates a high concentration of salt to muffle low-frequency electric fluctuations.

Forces created by fluctuations in local concentrations of ions

The same ionic mobility that screens electric fields also allows fluctuations in ion density, transient formation of regions of net electric charge and emanation of electric fields from these regions. When medium m is saltwater, the consequence of salt is primarily to screen interactions between A and B. When semi-infinite regions A and B are ionic solutions and m is a salt-free dielectric, there are the extra interactions of mobile charge fluctuation; the extra "polarizability" conferred by ionic displacement creates an effectively very high dielectric response. Ionic fluctuation forces are important only for the zero-frequency fluctuations. Why? Because ionic displacements do not occur at or respond to the electric fields at frequencies that correspond even to the first nonzero eigenfrequency $\xi_1 = (2\pi kT/\hbar) = 2.41 \times 10^{14}$ rad/s $= 3.84 \times 10^{13}$ Hz. A quick way to see this is to recall that the diffusion constant of a typical small ion is $\sim 10^{-5}\ \text{cm}^2/\text{s} = 10^{-9}\ \text{m}^2/\text{s} = 10^{+11}$ Å^2/s. To diffuse a distance comparable to its own ~ 1 Å $= 0.1$ nm size would take 10^{-11} s, 100 times longer than the $\sim 10^{-13}$-s period of the first eigenfrequency. Conversely, in the period of the first eigenfrequency, the ion would budge only $\sim 1/100$ of its radius.[12]

Specifically, when A and B are salt solutions and medium m is a nonconducting dielectric without the high dielectric constant of water, then the $\overline{\Delta}$'s go to 1. The leading term in the $n = 0$ contribution to the interaction free energy approximation (L1.5) becomes (see Fig. L1.46)

$$G_{\text{Am/Bm}}(l) \approx -\frac{kT}{16\pi l^2}\overline{\Delta}_{\text{Am}}\overline{\Delta}_{\text{Bm}} \to -\frac{kT}{16\pi l^2}. \qquad \text{(L1.64)}$$

There is no additional double-layer screening of fluctuations correlated across the separation l.

Conversely, when A and B are pure dielectrics and m is a salt solution, the magnitudes of the $\overline{\Delta}$'s still go to 1, but there is strong salt-screening (see Fig. L1.47), so that [Eq. (L3.201) with $\kappa = 1/\lambda_D$]

$$G_{\text{AmB}}(l) \approx -\frac{kT}{16\pi l^2}(1 + 2l/\lambda_D)e^{-2l/\lambda_D}. \qquad \text{(L1.65)}$$

The interaction takes on a form much like that of relativistic screening with finite-frequency fluctuation forces.

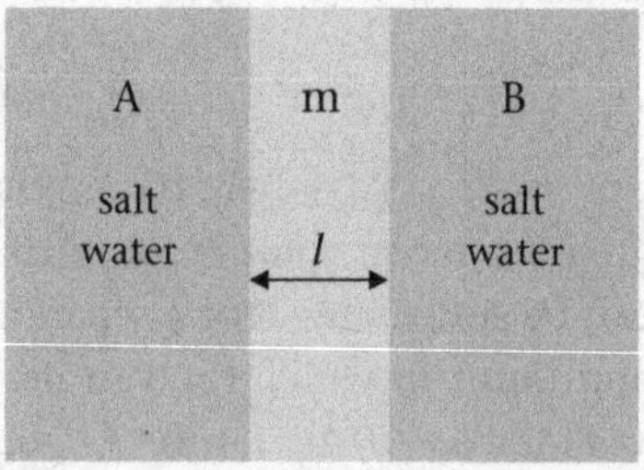

Figure L1.46

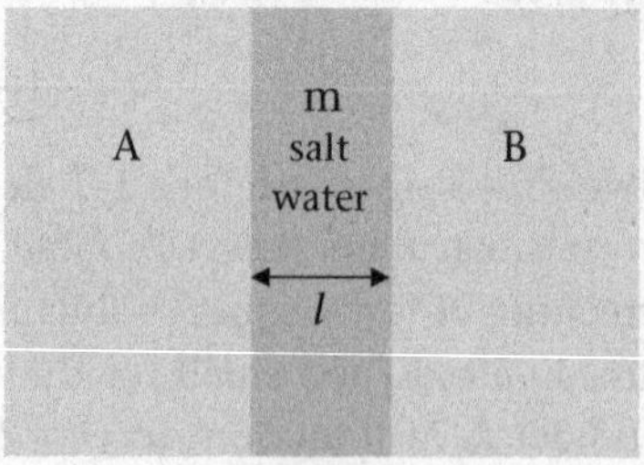

Figure L1.47

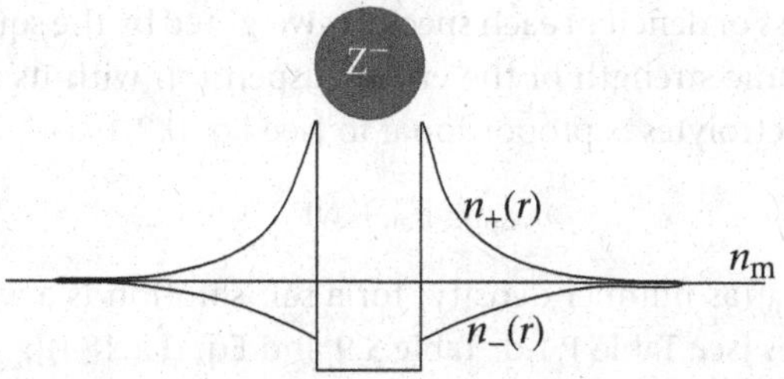

Figure L1.48

Salt confers a double feature: "Infinite" polarizability leading to large $\overline{\Delta}$'s and screening of correlated charge fluctuations.

Small-sphere ionic-fluctuation forces

Return now from gases to dilute suspensions. Imagine, in particular, a suspension of charged spherical colloidal particles or polyelectrolytes, which are large compared with the many mobile ions that compose the salt in the bathing medium, but small compared with the distance to the next colloid or polyelectrolyte. The mean distance between colloids or polyelectrolytes is also small compared with the Debye screening length λ_{D} of the bathing medium. Around each of these particles there will be an inhomogeneity in the small mobile ionic species of the bathing salt solution. The average numbers of ions of each valence ν will differ from the number-density concentrations $n_\nu(\mathrm{m})$ in the bathing medium infinitely far from the colloid or polyelectrolyte. (A singly charged positive ion has valence $\nu = +1$; a singly charged negative ion has $\nu = -1$.) If $n_\nu(r)$ is the concentration of species of valence ν at position r from the center of the sphere, then the excess of each species ν is [see Table S.10 and Eq. (L2.188)]

$$\Gamma_\nu \equiv \int_0^\infty [n_\nu(r) - n_\nu(\mathrm{m})]4\pi r^2 \mathrm{d}r. \tag{L1.66}$$

That is, the excess function Γ_ν gives the total number of ions of valence ν found in the vicinity of the spherical colloid compared with the number of such ions that would occur in the bathing solution in the absence of the colloid.

To be even more specific, solid spheres bearing negative charge Z^- suspended in a 1–1 salt solution create ionic double layers: negative mobile ions will be repelled, positive attracted. In addition, all salt will be excluded from a dielectric core of radius a. It is essential to recognize this latter exclusion of salt. The effective number of fluctuating charges will differ from Z^- and can even be negative compared with the number of mobile ions in the absence of the charged, ion-excluding colloid (see Fig. L1.48).

The fluctuations in the number of each species about each charged macroion will also differ from those in a macroion-free region. Analogous to the extra dielectric response $\alpha(i\xi)$ in a dilute gas or suspension with $\varepsilon = 1 + N\alpha$, there is an ionic response [see Table S.10 and Eq. (L2.191)]:

$$\Gamma_s \equiv \sum_{\{\nu\}} \Gamma_\nu \nu^2, \tag{L1.67}$$

where the mean excess or deficit of each species is weighted by the square of the valence. In terms of Γ_s, the ionic strength of the entire suspension with its number density of N colloids or polyelectrolytes is proportional to [see Eq. (L2.191)]

$$n_{susp} = n_m + N\Gamma_s. \tag{L1.68}$$

The ionic strength n_m (as number density) for a salt solution is a weighted average of the square of valences [see Table P.1.d, Table S.9 and Eq. (L2.184)]:

$$n_m \equiv \sum_{\{\nu\}} n_\nu(\mathrm{m})\nu^2. \tag{L1.69}$$

At separations large compared with particle size, the interaction between two charged colloids is due to a correlation between fluctuations in net charge around each of them. (At shorter distances there are multipole terms, fluctuations in potential over the space of the colloid, which lead to additional forces.) This is a monopole–monopole correlation energy [see Table S.9.c and Eq. (L2. 206)]:

$$g_{M-M}(z) = -\frac{kT}{2}\,\Gamma_s^2\,\frac{e^{-2z/\lambda_D}}{(z/\lambda_{Bj})^2}. \tag{L1.70}$$

As already emphasized, this ionic-charge-fluctuation attraction occurs only at the limit of the zero-frequency term in the summation over frequencies that compose the van der Waals force. In the exponential, the center-to-center separation z is measured in Debye lengths λ_D; in the denominator, z is measured in $\lambda_{Bjerrum}$, usually written λ_{Bj}, the Bjerrum length at which the energy of Coulombic interaction is kT (see Table S.9):

$$\lambda_{Bjerrum} \equiv e^2/4\pi\varepsilon_0\varepsilon_m kT \text{ in mks units,}$$
$$\lambda_{Bjerrum} \equiv e^2/\varepsilon_m kT \text{ in cgs units.} \tag{L1.71}$$

How to think about it?

All the ions in a salt solution undergo continuous, thermally driven fluctuations. These fluctuations can be described as deviations in number density from their average. The greater the average density, the greater the number-density fluctuation. A region containing the colloid will have an ionic number density that differs from that in a region composed only of salt solution. To an extent measured by the deviation Γ_s, the fluctuations about the charged colloid create an electrostatic potential that differs from the potential in the background salt solution. This potential decays with the $e^{-r/\lambda_D}/r$ form of a double layer around a small sphere. The potential from one center radiates to the location of another colloid at $r = z$ and perturbs the charge density there to deviate from its average charge.

The degree of perturbation is in proportion to the original radiating potential; the response is again in proportion to the extra number of charges Γ_s around the second particle. This transient perturbed charge on the second colloid radiates back to the first particle. The two fluctuations then interact as $e^{-z/\lambda_D}/z$ from the first colloid times $e^{-z/\lambda_D}/z$ back from the second colloid to give the doubly screened form $e^{-2z/\lambda_D}/z^2$.

Driving all these fluctuations is thermal energy kT, which is available to pay for transitory changes in the numbers of the various species of ions. These species are all at the same average potential, an average over a region that is small compared with the distances of variation of electrostatic fluctuation potential.

Each species of ion feels the force of an electric field (or an energy change that is due to an electric potential) in proportion to its valence, and *then* each species creates an electrical potential, again in proportion to its valence. Hence the dependence on the *square* of the valence in composing Γ_s and in the product of Γ_s's for the interaction of two particles. Separation z is measured in Debye lengths λ_D in exponential screening and in Bjerrum lengths λ_{Bj} in the Coulombic part of the potential. The idea goes back to Kirkwood and Shumaker,[13] who pointed out that fluctuations on protein titratable groups could create monopolar fluctuation forces.

There are monopolar fluctuations of the net charge on the colloid and its surrounding solution; there are dipolar fluctuations, the first moment of the ionic-charge distribution around the colloid as well as polarization of the colloid itself. Monopolar and dipolar fluctuations couple to create a hybrid interaction, g_{D-M}, again in the limit of the $n = 0$ sampling frequency at which the ions are able to fluctuate. The salt solution screens even the dipolar fluctuation the same way that the low-frequency-fluctuation term is screened in planar interactions. For dielectric spheres of radius a, ε_s whose incremental contribution to dielectric response is $\alpha = 4\pi a^3 \varepsilon_m[(\varepsilon_s - \varepsilon_m)/(\varepsilon_s + 2\varepsilon_m)]$ (see Fig. L1.49), we can idealize three forms of interaction, dipole–dipole, dipole–monopole, and monopole–monopole.

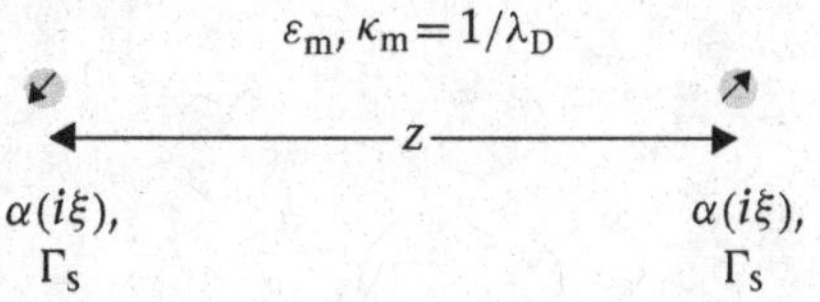

Figure L1.49

Dipole–dipole

$$g_{D-D}(z) = -3kT\frac{a^6}{z^6}\left(\frac{\varepsilon_s - \varepsilon_m}{\varepsilon_s + 2\varepsilon_m}\right)^2\left[1 + (2z/\lambda_D) + \frac{5}{12}(2z/\lambda_D)^2 + \frac{1}{12}(2z/\lambda_D)^3 + \frac{1}{96}(2z/\lambda_D)^4\right]e^{-2z/\lambda_D} \quad \text{(L1.72)}$$

(also see Table S.10.a). For parallel-plane interactions, the ionic-screening factor [Eq. (L1.63)] looked like $[1 + (2z/\lambda_D)]\,e^{-2z/\lambda_D}$, analogous to R_n for relativistic retardation. Spherical geometry creates a more complicated factor of the same general form.

Dipole–monopole

$$g_{D-M}(z) = -kT\lambda_{Bj}a^3\left(\frac{\varepsilon_s - \varepsilon_m}{\varepsilon_s + 2\varepsilon_m}\right)\Gamma_s\left[1 + (2z/\lambda_D) + \frac{1}{4}(2z/\lambda_D)^2\right]\frac{e^{-2z/\lambda_D}}{z^4} \quad \text{(L1.73)}$$

(see Table S.10.b), which includes ionic screening as well as a contribution from the ionic excess Γ_s.

Monopole–monopole

$$g_{M-M}(z) = -\frac{kT}{2}\Gamma_s^2\frac{e^{-2z/\lambda_D}}{(z/\lambda_{Bj})^2} \quad \text{(L1.74)}$$

as in the preceding subsection (see also Table S.10.c).

Again, keep in mind that (Table S.9)

$$\lambda_{Bj} \equiv e^2/4\pi\varepsilon_0\varepsilon kT \text{ in mks units,} \quad \lambda_{Bj} \equiv e^2/\varepsilon kT \text{ in cgs units; also,}$$
$$\lambda_{Bj} = \kappa_m^2/4\pi n_m = 1/4\pi n_m \lambda_D^2, \quad \text{where } \lambda_D = 1/\kappa_m.$$

The astute reader will immediately recognize a missing connection between the inhomogeneity of solute distributions that underlie ionic-fluctuation forces, the consequences of dielectric inhomogeneity [the preceding "smoothly varying $\varepsilon(z)$"], and the additional modification of high-frequency charge fluctuations that occur when there is a radially varying dielectric response in the region around suspended spheres (the preceding "fuzzy spheres").

L1.5. Cylindrical geometries

Dimensionally intermediate between spheres and planes, the interactions of cylinders reveal properties possessed by neither. In particular, there is always torque as well as force; the energy of interaction depends not only on separation z but also on mutual angle θ. For parallel cylinders of length indefinite compared with thickness and separation, the interaction is expressed as force or energy per unit length. As with spheres, there are few exact expressions for cylinder–cylinder van der Waals forces. The many approximate expressions must be used circumspectly (see Fig. L1.50).

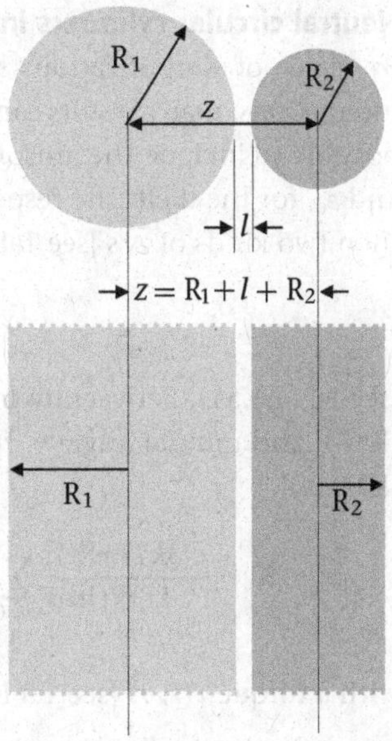

Figure L1.50

In the limit in which the dielectric responses of cylinder and medium do not greatly differ, with the neglect of retardation, at separations large compared with thickness, the interaction of parallel cylinders goes as the inverse-fifth power of the interaxial separation $z \gg R_1, R_2$ (see Table C.3.b.2):

$$G_{c\|c}(z; R_1, R_2) = -\frac{3A_{1m/2m}}{8\pi}\frac{(\pi R_1)^2(\pi R_2)^2}{z^5}. \qquad \text{(L1.74)}$$

In the opposite limit, cylinders near to touching, the interaction energy per unit length goes to an inverse 3/2-power dependence on surface-to-surface separation l ($z \to R_1 + R_2$) (see Table C.3.b.3):

$$G_{c\|c}(l; R_1, R_2) = -\sqrt{\frac{2R_1R_2}{R_1 + R_2}}\frac{A_{1m/2m}}{24l^{3/2}}. \qquad \text{(L1.75)}$$

These simplified expressions depend on the assumption of a constant Hamaker coefficient with dielectric responses ε_1, ε_2, and ε_m of similar magnitudes.

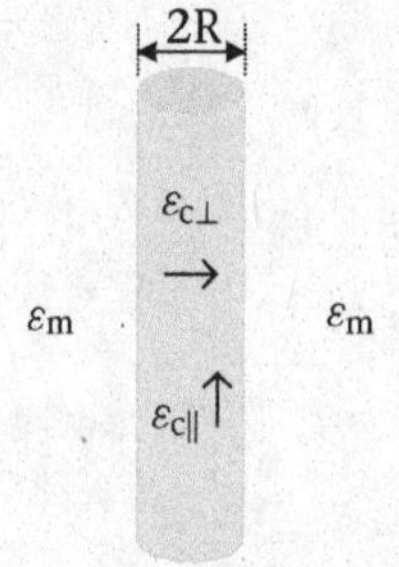

Figure L1.51

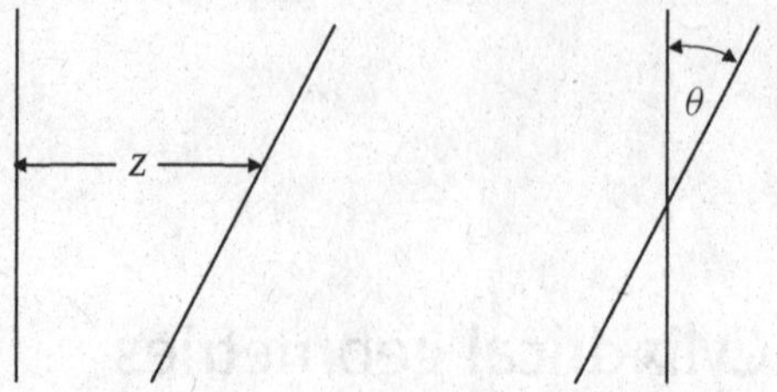

Figure L1.52

Thin cylinders, analogous to point particles

Neutral circular cylinders in salt-free solutions

By virtue of simplifications similar to those permitted in formulating point–particle interactions, many results can be derived for the interaction of thin cylinders. It is even possible to include the anisotropy of material within the rod, for example putting $\varepsilon_{c\perp}$ and $\varepsilon_{c\parallel}$ for the dielectric response perpendicular and parallel to the rod axis. There are then two kinds of $\overline{\Delta}$'s [see Table C.4 and Eqs. (L2.224)]:

$$\overline{\Delta}_\parallel = \frac{\varepsilon_{c\parallel} - \varepsilon_m}{\varepsilon_m}, \quad \overline{\Delta}_\perp = \frac{\varepsilon_{c\perp} - \varepsilon_m}{\varepsilon_{c\perp} + \varepsilon_m} \tag{L1.76}$$

(see Fig. L1.51). Between two circular cylinders of radius R, minimal interaxial separation z, and mutual angle θ, the interaction energy goes as [see Table C.4.b.1 and Eqs. (L2.234)]

$$g(z,\theta;\mathrm{R}) = -\frac{3kT(\pi \mathrm{R}^2)^2}{4\pi z^4 \sin\theta} \sum_{n=0}^{\infty}{}' \left\{ \overline{\Delta}_\perp^2 + \frac{\overline{\Delta}_\perp}{4}(\overline{\Delta}_\parallel - 2\overline{\Delta}_\perp) + \frac{2\cos^2\theta + 1}{2^7}(\overline{\Delta}_\parallel - 2\overline{\Delta}_\perp)^2 \right\}, \tag{L1.77}$$

with a torque $\tau(z,\theta)$ [see Table C.4.b.2 and Eq. (L2.235)],

$$\tau(z,\theta;\mathrm{R}) = -\left.\frac{\partial g(z,\theta;\mathrm{R})}{\partial\theta}\right|_z$$

$$= -\frac{3kT(\pi \mathrm{R}^2)^2}{4\pi z^4}\left[\frac{\cos\theta}{\sin^2\theta}\sum_{n=0}^{\infty}{}'\{\,\} + \frac{\cos\theta}{2^5}\sum_{n=0}^{\infty}{}'(\overline{\Delta}_\parallel - 2\overline{\Delta}_\perp)^2\right] \tag{L1.78}$$

toward parallel alignment (see Fig. L1.52).

Between parallel thin rods, $\theta = 0$ [see Table C.4.a and Eqs. (L2.233)],

$$g_\parallel(z;\mathrm{R}) = -\frac{9kT(\pi \mathrm{R}^2)^2}{16\pi z^5}\sum_{n=0}^{\infty}{}'\left[\overline{\Delta}_\perp^2 + \frac{\overline{\Delta}_\perp}{4}(\overline{\Delta}_\parallel - 2\overline{\Delta}_\perp) + \frac{3}{2^7}(\overline{\Delta}_\parallel - 2\overline{\Delta}_\perp)^2\right] \tag{L1.79}$$

per unit length (see Fig. L1.53).

Formal and of limited applicability as they are, these line-rod expressions nevertheless tempt us with several properties.

First, as with spheres, the strength of interaction is proportional to the product of volumes, this time the product of $\pi\mathrm{R}^2$, the volume per unit length. Superficially at least, the masses of the two rods act as polarizable units.

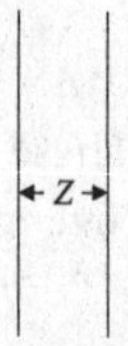

Figure L1.53

Second, to leading term, the interaction of two rods at an angle varies inversely to the sine of their mutual angle; there is a necessary divergence to infinity when two infinite rods line up parallel. But there is more to the angular dependence. Even if the rod polarizability itself is isotropic, that is, if $\varepsilon_{c\parallel} = \varepsilon_{c\perp}$, there is a slight difference between $2\overline{\Delta}_{\perp}$ and $\overline{\Delta}_{\parallel}$. This difference appears as an extra angular dependence $[(2\cos^2\theta + 1)/2^7](\overline{\Delta}_{\parallel} - 2\overline{\Delta}_{\perp})^2$ that varies by a factor of 3 in going from perpendicular to parallel rods.

Third, perhaps most intriguing but also most dangerously alluring, $\overline{\Delta}_{\parallel} = [(\varepsilon_{c\parallel} - \varepsilon_m)/\varepsilon_m]$ is not confined to values between zero and one as is the case with $\overline{\Delta}_{\perp} = [(\varepsilon_{c\perp} - \varepsilon_m)/(\varepsilon_{c\perp} + \varepsilon_m)]$ and all the other $\overline{\Delta}$'s so far encountered in van der Waals summation. Because $\overline{\Delta}_{\parallel}$ can take on very large values when $\varepsilon_{c\parallel} \gg \varepsilon_m$, the $\overline{\Delta}_{\perp}(\overline{\Delta}_{\parallel} - 2\overline{\Delta}_{\perp})$ and $(\overline{\Delta}_{\parallel} - 2\overline{\Delta}_{\perp})^2$ terms indicate the possibility of very strong van der Waals interactions to torque rods with high axial polarizability.

Imagine, for example, two metallic wires or carbon nanotubes whose intrinsic ε's can take on effectively infinite values. Formally at least, $2\overline{\Delta}_{\perp} \to 1$ whereas $\overline{\Delta}_{\parallel} \to \infty$, so that

$$g(z,\theta) = -\frac{3kT(\pi R^2)^2}{4\pi z^4 \sin\theta}\frac{2\cos^2\theta + 1}{2^7}\sum_{n=0}^{\infty}{}' \overline{\Delta}_{\parallel}^2 \text{ per interaction,} \tag{L1.80}$$

$$g_{\parallel}(z) = -\frac{9kT(\pi R^2)^2}{16\pi z^5}\frac{3}{2^7}\sum_{n=0}^{\infty}{}' \overline{\Delta}_{\parallel}^2 \text{ per unit length.} \tag{L1.81}$$

These strong interactions are a formal consequence of the infinite polarization that is possible with an infinitely long conducting rod. They can be expected to occur between a pair of isolated rods, but they have only formal significance in solutions or liquid-crystalline arrays in which there are many rods (linear molecules) interacting at the same time. In that case the "medium" is the solution itself. That is, two linear molecules "see" each other across a suspension of other linear molecules so that there is not the finite ε_m assumed in this formulation but rather a medium whose own polarizability diverges to infinity if it is populated by infinitely polarizable molecules. The $\overline{\Delta}_{\parallel}$ does not diverge in the way in which it is formally constructed here, with only two line particles in an otherwise pure medium. Foolish things have been said about "strong" rod–rod interactions in solution because this collectivity of interactions is forgotten.

Charged circular cylinders in salt solutions

Just as a charged sphere in saltwater surrounds itself with a number of mobile ions different from what would occupy the same region in its absence, so does a charged cylinder. As with spheres, there are low-frequency ionic fluctuations that create attractive forces between like cylinders. In the special case of thin cylinders whose material dielectric response is the same as that of the medium and the distance between cylinders is small compared with the Debye screening length, this ionic-fluctuation force has appealing limiting forms.

Between parallel rods ($\theta = 0$)

$$g_{\parallel}(z) = -\frac{kT\lambda_{Bj}^2}{2}\Gamma_c^2\sqrt{\pi}\frac{e^{-2z/\lambda_D}}{z(z/\lambda_D)^{1/2}} \tag{L1.82}$$

per unit length [see Table C.5.b.1 and relation (L2.257)].

Between rods at a mutual angle θ

$$g(z,\theta) = -\frac{kT\pi\lambda_{\mathrm{Bj}}^2}{\sin\theta}\Gamma_{\mathrm{c}}^2\frac{e^{-2z/\lambda_{\mathrm{D}}}}{(2z/\lambda_{\mathrm{D}})} \qquad \text{(L1.83)}$$

per interacting pair [see Table C.5.b.2 and relation (L2.256)].

In cylindrical coordinates, the excess quantities of each kind of ion are, in units per unit length [see Table C.5 and Eq. (L2.241)],

$$\Gamma_\nu \equiv \int_0^\infty [n_\nu(r) - n_\nu(\mathrm{m})]2\pi r\,\mathrm{d}r, \qquad \text{(L1.84)}$$

with the weighted sum [see Eq. (L2.241)]

$$\Gamma_{\mathrm{c}} \equiv \sum_{\{\nu\}} \Gamma_\nu \nu^2 \qquad \text{(L1.85)}$$

for the effective strength of the mobile-charge response.

LEVEL TWO

Practice

Level 2 is the doing. After reading the Prelude and Level 1, you should be able to go to any part of Levels 2 or 3 according to personal tastes or interests. Go straight to the formulae, tabulated by geometry, cross-referenced by equation numbers not only to Level 1's introduction but also to Level 2's explanatory essays and to the derivations of Level 3. The idea is to work outward from these tables—back to Level 1 or forward to Levels 2 and 3.

The opening Notation section, Section L2.1, applies not only to the tables but also to the texts of Levels 1 and 2. The Essays on formulae section, Section L2.3, which immediately follows the tables sections, reduces the results from Level 3 derivations to simpler forms. The Computation section, Section L2.4, sketches the physical foundations of the all-important dielectric-response functions and gives mathematical guidelines for calculation.

The sequence of tabulation is first to give the most exact expression available, and then to list approximations. As already described,

- When there are small differences in dielectric polarizability and also in the limits of very small and very large separation, the general *Lifshitz* formula reduces to simpler, power-law forms.
- The *Pitaevskii* density expansion specializes the Lifshitz-style formulations in order to derive interactions between point particles and thin rods in dilute suspension.
- The *Derjaguin* transform or approximation converts the interaction between plane-parallel surfaces into the interaction between oppositely curved surfaces such as spheres. This procedure and its reverse are allowed in the limit in which the closest separation is much smaller than radii of curvature.
- The *Hamaker* summation treats the form of interaction as though the dielectric responses of all media had the properties of gases. That is, the dielectric response is assumed to be proportional to the number density of component atoms or molecules.

Given the power of modern computation, it is best in principle to estimate van der Waals forces by the most exact available formulae. Still, for intuition or for ease, by choice or by necessity, we often use approximate forms. Understanding the differences among the various versions of formulae for a particular geometry, computing the consequent differences in results, instructs us (Essays on Formulae section). Similarly, computation without knowing full spectroscopic data also instructs as long as we take the trouble to compare numbers we find by using differently approximated spectra (Computation section). In fact, incomplete spectroscopic data may limit any advantage of using the most exact formulae. Beware of precision without accuracy.

L2.1. Notation and symbols

In this section the notation and symbols used throughout the book are listed alphabetically under their appropriate headings. Constants are usually given in nonitalic type and variables in italic. (This is only a general rule. By convention, Boltzmann's constant *k*, Planck's constant *h*, and other physical constants are in italic.) Boldfaced type indicates vectors and matrices. Except in section L2.4.A., the cgs (centimeter-gram-second) and the mks (meter-kilogram-second) systems of notation are used in parallel. Any symbols not listed in this section are defined where they are used or in the notation section of Level 3.

L2.1.A. Geometric quantities

a, b, c	Constant dimensions of lines, rectangles, rectangular solids.
$a_1, a_2, \ldots, b_1, b_2 \ldots$	Constant thicknesses of first, second, etc., layers on half-spaces A, B; also radii of spheres or cylinders.
A, B	Materials composing half-spaces (or L, R in Level 3 derivations for left and right convenience).
$A_1, A_2, \ldots, B_1, B_2 \ldots$	Materials composing first, second, etc., layers on half-spaces A, B.
l	Variable separation between parallel, planar surfaces; minimal separation between spheres, cylinders, oppositely curved surfaces.
m	Material in intervening medium.
$R, R_1, R_2 \ldots$	Constant radii of spheres or cylinders.
z	Variable center-to-center distance; between spheres or cylinders, $z = l + R_1 + R_2$; between sphere and cylinder and a wall, $z = l + R$.
1, 2	Material of spheres or cylinders, also as subscripts; sometimes sph or cyl, s or c.

L2.1.B. Force and energy

G	Free energy of interaction.
$G_{\mathrm{AmB}}(l)$	Planar systems between half-spaces A and B across medium m of variable thickness l.
$G_{\mathrm{Am/Bm}}(l)$	Planar systems in outside/in sequence of materials to emphasize the interaction between the Am and Bm interfaces.
$G_{\mathrm{AmB_1B}}(l; \mathrm{b_1})$	Used when there is a single layer of material $\mathrm{B_1}$ constant thickness $\mathrm{b_1}$ on half-space B or, more generally, with many layers on both A and B, use $G(l; \mathrm{a_1}, \mathrm{a_2}, \ldots, \mathrm{b_1}, \mathrm{b_2}, \ldots)$ to show variable spacing l (*italic*) with other distances constant (nonitalic).

L2.1.C. Spherical and cylindrical bodies

F	Force, negative spatial derivative of G per interaction between finite-size objects.
$g_{\mathrm{ab}}(z)$	Used between point particles a, b with a separation z.
$g_{\mathrm{p}}(z)$	Used between a point particle and a wall with a particle-to-wall separation z.
$G_{\mathrm{ss}}(l; \mathrm{R_1}, \mathrm{R_2})$ or $G_{\mathrm{cc}}(l; \mathrm{R_1}, \mathrm{R_2})$ or $G_{\mathrm{1m/2m}}(l; \mathrm{R_1}, \mathrm{R_2})$	Used between spheres or cylinders of constant radii $\mathrm{R_1}$, $\mathrm{R_2}$ of materials 1, 2.
$\hbar = \frac{h}{2\pi}, h$	Planck's constant.
k	Boltzmann's constant.
kT	Thermal energy.
kT_{room}	Thermal energy at room temperature.
P	Pressure, negative spatial derivative of G per unit area between parallel planar surfaces; negative pressure denotes attraction (a convention contrary to that in which pressure on a surface is defined in the direction of its outward normal vector).
τ	Torque, negative derivative of $G(l, \theta)$ with respect to angle θ between vectors parallel to the interfaces of half-spaces at separation l.

L2.1.D. Material properties

$c_{\mathrm{A}}, c_{\mathrm{B}}, c_{\mathrm{m}}$	Coefficients of interactions used in Hamaker summation.
i, sometimes j	Constants used to denote materials, e.g., i = A or B or $\mathrm{B_2}$ or $\mathrm{A_i}$ etc. (nonitalic)
$I(\mathrm{M})$	Ionic strength in molar units.
$J_{\mathrm{cv}} = J'_{\mathrm{cv}} + iJ''_{\mathrm{cv}}$	Interband transition strength.

$n_{\text{ref i}} = \sqrt{\varepsilon_{\text{i}}}$	Index of refraction in transparent region of material i.
$\{n_\nu^{(\text{i})}\}$ or $\{n_\nu(\text{i})\}$	Number densities of the set of ions of valence ν in region i (with ν subscript to distinguish from index in frequency summation and index of refraction).
N, N_{i}, n_{i}	Number densities of suspended particles, molecules, atoms, polarizable units.
N_e	Number density of electrons.
$\text{Rel}(r_n)$	Actual (computed) retardation screening factor for small differences in dielectric response, energy between parallel flat surfaces.
$R_n(r_n) = R_n(l;\xi_n) \equiv (1+r_n)\,e^{-r_n}$	Approximate form for retardation screening factor, interaction energy between parallel flat surfaces, also written as $R_n(l)$ to emphasize distance dependence.
$R_{\alpha\beta}(r_n) \equiv$ $e^{-r_n}\left(1+r_n+\frac{5}{12}r_n^2+\frac{1}{12}r_n^3+\frac{1}{48}r_n^4\right)$	Screening factor for point–particle interaction energy.
α	Incremental contribution of one particle to dielectric response of a dilute gas or suspension such that ε_{m} of the medium becomes $\varepsilon_{\text{m}} + \alpha N$ for a suspension.
α or α_{mks} or α_{cgs}	Coefficients for the polarization created on a single small particle a by an electric field of magnitude E. (The coefficient α is sometimes broken into a contribution that is due to the field orientation of permanent dipoles μ_{dipole} and a contribution that is due to the field's induction of a transient dipole on a polarizable particle.) Polarization $= \alpha_{\text{mks}} E_{\text{mks}}$ or $\alpha_{\text{cgs}} E_{\text{cgs}}$ in either unit system. Similarly β or β_{mks} or β_{cgs} for single small particle b.
Γ_{s}	Weighted sum of excess numbers of ions around a large charged particle or per unit length or area of an extended body, e.g., Γ_{c} per unit length of cylinder.
Γ_ν	Excess number of ions of valence ν around a large charged particle or per unit length or area of an extended body.
$\varepsilon = (n_{\text{ref}} + i\kappa_{\text{abs}})^2$	n_{ref} is the index of refraction and κ_{abs} is the absorption coefficient where $i = \sqrt{-1}$.
ε^{i}	Matrix for anisotropic relative dielectric response (of material i) with elements, e.g., ε_x^{i}, ε_y^{i}, ε_z^{i} in the x, y, or z direction; or $\varepsilon_\perp$ or $\varepsilon_\parallel$ perpendicular or parallel, respectively, to a principal axis.
$\varepsilon_{\text{i}} = \varepsilon_{\text{i}}' + i\varepsilon_{\text{i}}''$	The ε_{i}' real (elastic) and ε_{i}'' imaginary (dissipative) parts of ε_{i} for material i (often written without

	subscript when general properties are being discussed).
$\varepsilon_{\mathrm{i}}, \mu_{\mathrm{i}}$	Relative isotropic dielectric, magnetic permittivity or permeability; at zero frequency $\varepsilon(0)$ is the dielectric constant.
κ_{i}	Ionic screening constant for solution in region i, inverse to the Debye screening length.
$\lambda_{\mathrm{Bjerrum}}$ or λ_{Bj}	Bjerrum length at which the distance the energy of interaction of two univalent charges is equal to thermal energy kT.
λ_{Debye} or λ_{D}	Debye length of screening in ionic solution.
μ_{dipole}	Permanent dipole moment of small particle.
σ	Conductivity
χ	Material polarization coefficient or dielectric susceptibility such that polarization density $\mathbf{P} = \varepsilon_0 \chi^{\mathrm{mks}}\, \mathbf{E}$ in mks ("SI") units or $\chi^{\mathrm{cgs}}\, \mathbf{E}$ in cgs ("Gaussian") units.

L2.1.E. Variables to specify point positions

r	For radial distances.
x, y	For distances parallel to the faces of planes.
z	For distances perpendicular to planes (z is also used for interaxial distances between cylinders, center-to-center distances between spheres, and center-to-wall distances from sphere or cylinder-to-wall distances).

L2.1.F. Variables used for integration and summation

$\rho_{\mathrm{m}}^2 = \rho^2 + \varepsilon_{\mathrm{m}}\mu_{\mathrm{m}}\xi_n^2/c^2,\ \rho_{\mathrm{i}}^2 = \rho^2 + \varepsilon_{\mathrm{i}}\mu_{\mathrm{i}}\xi_n^2/c^2,\ \rho^2 = (u^2 + v^2),\ u,\ v.$

$x,\ x_{\mathrm{i}}^2 = x_{\mathrm{m}}^2 + \left(\frac{2l\xi_n}{c}\right)^2 (\varepsilon_{\mathrm{i}}\mu_{\mathrm{i}} - \varepsilon_{\mathrm{m}}\mu_{\mathrm{m}}),\ (x_{\mathrm{m}} = x),$

$p = x/r_n,\ r_n = r_n(l;\xi_n) = (2l\varepsilon_{\mathrm{m}}^{1/2}\mu_{\mathrm{m}}^{1/2}/c)\xi_n,$

$s_{\mathrm{i}} = \sqrt{p^2 - 1 + (\varepsilon_{\mathrm{i}}\mu_{\mathrm{i}}/\varepsilon_{\mathrm{m}}\mu_{\mathrm{m}})},\ s_{\mathrm{m}} = p.$	Components of radial-wave vectors.
ξ_n	Uniformly spaced eigenfrequencies.
$\xi_n = \frac{2\pi kT}{\hbar} n,\ n = 0, 1, 2, \ldots;\ i\xi_n$	Sometimes known as imaginary Matsubara frequencies.
$i\xi$	Continuous imaginary frequency.
$\omega = \omega_{\mathrm{R}} + i\xi$	Complex frequency with real part ω_{R}.
$\sum_{n=0}^{\prime\infty}$	Summation with $n = 0$ term multiplied by 1/2.
$\zeta(2) \equiv \sum_{q=1}^{\infty} \frac{1}{q^2},\ \zeta(3) \equiv \sum_{q=1}^{\infty} \frac{1}{q^3}$	Riemann zeta functions.

L2.1.G. Differences-over-sums for material properties

$\overline{\Delta}_{ji} = \dfrac{s_i\varepsilon_j - s_j\varepsilon_i}{s_i\varepsilon_j + s_j\varepsilon_i} = \dfrac{x_i\varepsilon_j - x_j\varepsilon_i}{x_i\varepsilon_j + x_j\varepsilon_i}$ $= \dfrac{\rho_i\varepsilon_j - \rho_j\varepsilon_i}{\rho_i\varepsilon_j + \rho_j\varepsilon_i}$	Dielectric
$\Delta_{ji} = \dfrac{s_i\mu_j - s_j\mu_i}{s_i\mu_j + s_j\mu_i} = \dfrac{x_i\mu_j - x_j\mu_i}{x_i\mu_j + x_j\mu_i}$ $= \dfrac{\rho_i\mu_j - \rho_j\mu_i}{\rho_i\mu_j + \rho_j\mu_i}$	Magnetic (In the nonretarded limit, with the finite velocity of light neglected, $\overline{\Delta}_{ji} \to \dfrac{\varepsilon_j - \varepsilon_i}{\varepsilon_j + \varepsilon_i}$, $\Delta_{ji} \to \dfrac{\mu_j - \mu_i}{\mu_j + \mu_i}$.)
$\overline{\Delta}_{Am}, \overline{\Delta}_{Bm}, \Delta_{Am}, \Delta_{Bm}$	Used for the simplest AmB planar geometry with $G_{AmB}(l)$.
$\overline{\Delta}^{eff}_{Am}, \overline{\Delta}^{eff}_{Bm}, \Delta^{eff}_{Am}, \Delta^{eff}_{Bm}$	Used for layered systems, sometimes with an argument denoting a number of layers, e.g., $\overline{\Delta}^{eff}_{Am}$ (N_A layers) for $G(l; a_1, a_2, \ldots, a_{N_A}, b_1, b_2, \ldots, b_{N_B})$.
$\varepsilon_a(z_a)$, sometimes $\varepsilon_a(z')$	Used for planar systems in which $\varepsilon(z)$ varies continuously in the direction z perpendicular to the planar interfaces, with a variation of ε in finite layer of thickness D_a next to the left-hand-side space of material A (or L in Level 3), $z_a = -z$ (measured leftward for symmetry in left-hand-side and right-hand-side notation).
$\varepsilon_b(z_b)$, sometimes $\varepsilon_b(z')$	Variation of ε in finite layer of thickness D_b next to right-hand-side space of material B (or R in Level 3), $z_b = +z$.
$\varepsilon_A, \varepsilon_B$, ($\varepsilon_{out}, \varepsilon_L, \varepsilon_R$ in Level 3)	Spatially unvarying dielectric permittivity in infinite half-spaces.
$\varepsilon_1 = \varepsilon_1(r_1), \varepsilon_2 = \varepsilon_2(r_2)$	Used for spherical or cylindrical systems in which $\varepsilon(r)$ varies with the radial position r for the different variations found in bodies 1 and 2.
$\theta(z)$, $u(z) \equiv \theta(z)e^{+2\rho(z)z}$	Used to build $\overline{\Delta}^{eff}$'s and Δ^{eff}'s for cases with continuously varying $\varepsilon(z)$.

L2.1.H. Hamaker coefficients

$A_{A'A''/B'B''}$	Used for interaction between interface A′A″ separating material A′ and material A″ and interface B′B″ between material B′ and material B″ with materials A′ and B′ on the further sides of the interfaces, constructed from $\overline{\Delta}_{A'A''}, \overline{\Delta}_{B'B''}, \Delta_{A'A''}, \Delta_{B'B''}$.
$A_{Am/Bm}$	Used for interaction between an interface Am separating material A and material m and an interface Bm separating material B and material m.

A_{Ham}	Used in generic expressions that emphasize form of equations.
$l_{A'A''/B'B''}$	The distance between these interfaces.

L2.1.I. Comparison of cgs and mks notation

Because only relative differences in dielectric response usually matter for van der Waals forces, vexing differences in centimeter-gram-second (cgs) Gaussian and meter-kilogram-second (mks) SI conventions are not major concerns. The fundamental work in van der Waals forces was, and often still is, done in cgs units. Most students learn mks units. The comparison of units summarized here is to avoid ambiguity in computation and also to allow easier access to the source literature. In what follows, the mks system is given on the left-hand side and the cgs system is given on the right-hand side.

For the force between two "point" charges, q_1 and q_2 in vacuum, Coulomb's law gives

Force $= \dfrac{q_1 q_2}{4\pi\varepsilon_0 r^2}$ N, q_1, q_2 in coulombs C, r in meters,
$\varepsilon_0 = 8.85 \times 10^{-12}\ \text{C}^2\ \text{N}^{-1}\ \text{m}^{-2}$
or
$(1/4\pi\varepsilon_0) = 8.992 \times 10^9\ \text{N m}^2/\text{C}^2$;

Force $= \dfrac{q_1 q_2}{r^2}$ dyn, q_1, q_2 in statcoulombs, r in centimeters.

For the force in a hypothetical continuum-dielectric material, introduce the dimensionless relative dielectric constant ε such that $\varepsilon_{\text{vacuum}} \equiv 1$, the same in both unit systems. Then

$$\text{Force} = [(q_1 q_2)/(4\pi\varepsilon_0\varepsilon r^2)], \qquad \text{Force} = [(q_1 q_2)/(\varepsilon r^2)].$$

The easiest way to avoid confusion here and in most situations is to keep in mind that what is ε in cgs "Gaussian" is $4\pi\varepsilon_0\varepsilon$ in mks "SI".

The electric field from a point charge q goes as

$$\mathbf{E} = \frac{q}{4\pi\varepsilon_0\varepsilon r^2}\ \text{N/C or V/m}, \qquad \mathbf{E} = \frac{q}{\varepsilon r^2}\ \text{dyn/sc or sv/cm}.$$

In vacuum the electric field that emanates from "free" or "external" electric "source" charges of density ρ_{free} obeys

$\nabla \cdot \mathbf{E} = \rho_{\text{free}}/\varepsilon_0$, with $\mathbf{E}$ in V/m and ρ_{free} in C/m^3;

$\nabla \cdot \mathbf{E} = 4\pi\rho_{\text{free}}$, with $\mathbf{E}$ in sv/cm and ρ_{free} in sc/cm^3.

For fields and source charges in a dielectric continuum, ε retains the same meaning in both systems:

$$\nabla \cdot (\varepsilon\mathbf{E}) = \rho_{\text{free}}/\varepsilon_0, \qquad \nabla \cdot (\varepsilon\mathbf{E}) = 4\pi\rho_{\text{free}}.$$

The dielectric displacement vector $\mathbf{D}$ is written as

$\mathbf{D} = \varepsilon_0\,\mathbf{E} + \mathbf{P} = \varepsilon\varepsilon_0\,\mathbf{E} = \varepsilon_0(1 + \chi^{\text{mks}})\mathbf{E}$,

$\mathbf{D} = \mathbf{E} + 4\pi\mathbf{P} = \varepsilon\,\mathbf{E} = (1 + 4\pi\chi^{\text{cgs}})\mathbf{E}$,

$\mathbf{P} = \varepsilon_0 \chi^{\text{mks}} \mathbf{E}$, $\mathbf{P} = \chi^{\text{cgs}} \mathbf{E}$,

D and **P** are in units of $\varepsilon_0 \mathbf{E}$, C/m^2, **D**, **E**, and **P** are all in sv/cm or sc/cm^2,

χ^{mks} and $\varepsilon = (1 + \chi^{\text{mks}})$ are dimensionless; χ^{cgs} and $\varepsilon = (1 + 4\pi \chi^{\text{cgs}})$ are dimensionless.

Polarization is a shift in charge (units charge × length) per unit volume (units $1/\text{length}^3$) or units of charge/length^2.

In several examples for gases and dilute suspensions, we expand the dielectric response ε around its vacuum value of 1 or around its pure-solvent value ε_m, respectively, for the suspending medium. In those cases, the dimensionless χ for the gas or for the suspension as a whole will be proportional to the number density of particles (units $1/\text{length}^3$), and the contribution to the polarizability from individual particles will have volume units (length^3).

L2.1.J. Unit conversions, mks–cgs

$e = 1.609 \times 10^{-19}$ C (mks) $= 4.803 \times 10^{-10}$ statcoulombs (cgs).

$1 \text{ sc} = (1.609 \times 10^{-19})/(4.803 \times 10^{-10})$ C $= 3.35 \times 10^{-10}$ C.

$\varepsilon_0 = 8.854\ 10^{-12}$ F/m or $C^2/(N\ m^2)$.

$4\pi\varepsilon_0 = 1.113\ 10^{-10}$ F/m or $C^2/(N\ m^2)$.

Capacitance C per area for charge Q per area.

Electric field between plates of capacitor

$$\mathbf{E} = \frac{Q}{\varepsilon_0 \varepsilon}, \qquad \mathbf{E} = \frac{4\pi Q}{\varepsilon}.$$

Separation d creates a voltage difference,

$$V = \frac{Qd}{\varepsilon_0 \varepsilon}, \qquad V = \frac{4\pi Q}{\varepsilon} d.$$

and capacitance per unit area,

$$C = \frac{Q}{V} = \frac{\varepsilon_0 \varepsilon}{d}, \qquad C = \frac{Q}{V} = \frac{\varepsilon}{4\pi d}.$$

The dielectric displacement vector varies with free charge density as

$$\text{div}\, \mathbf{D} = \rho, \qquad \text{div}\, \mathbf{D} = 4\pi\rho.$$

The relation among **D**, electric field **E**, and polarization **P** in terms of ε and ε_0 is

$$\mathbf{D} = \varepsilon_0 \varepsilon\, \mathbf{E} = \varepsilon_0 \mathbf{E} + \mathbf{P}, \qquad \mathbf{D} = \varepsilon\, \mathbf{E} = \mathbf{E} + 4\pi \mathbf{P},$$

or

$$\mathbf{D} \equiv \varepsilon_0\, \mathbf{E} + \mathbf{P} = \varepsilon \varepsilon_0 \mathbf{E} = \varepsilon_0 (1 + \chi^{\text{mks}}) \mathbf{E}, \qquad \mathbf{D} \equiv \mathbf{E} + 4\pi\, \mathbf{P} = \varepsilon\, \mathbf{E} = (1 + 4\pi \chi^{\text{cgs}}) \mathbf{E}.$$

Dielectric response of a gas For a gas of number density N of whose particles bear permanent dipole moments μ_{dipole}, for constant $\mathbf{E}$

$$\mathbf{P} = \varepsilon_0 \chi^{\text{mks}}\,\mathbf{E} \text{ yields } \chi^{\text{mks}} = N\frac{\mu^2_{\text{dipole}}}{3kT\varepsilon_0}, \qquad \mathbf{P} = \chi^{\text{cgs}}\,\mathbf{E} \text{ yields } \chi^{\text{cgs}} = N\frac{\mu^2_{\text{dipole}}}{3kT},$$

$$\varepsilon = 1 + \chi^{\text{mks}} = 1 + N\frac{\mu^2_{\text{dipole}}}{3kT\varepsilon_0}, \qquad \varepsilon = 1 + 4\pi\chi^{\text{cgs}} = 1 + N4\pi\frac{\mu^2_{\text{dipole}}}{3kT}.$$

To add polarization $\alpha\mathbf{E}$ induced on each molecule in the gas, use the form

$$\mathbf{P}_{\text{induced}} = N\alpha\mathbf{E}$$

in both unit systems.

Then

$$\mathbf{P}_{\text{induced}} = \varepsilon_0 \chi^{\text{mks}}_{\text{induced}}\mathbf{E}, \qquad \mathbf{P}_{\text{induced}} = \chi^{\text{cgs}}_{\text{induced}}\mathbf{E},$$

adds

$$N\alpha/\varepsilon_0 \text{ to } \chi^{\text{mks}} \text{ and to } \varepsilon, \qquad N\alpha \text{ to } \chi^{\text{cgs}} \text{ and } 4\pi N\alpha \text{ to } \varepsilon.$$

In this way, the relative dielectric response to static electric fields of a gas of polarizable molecules that also bear a permanent dipole moment is

$$\varepsilon_{\text{gas}} = 1 + \chi^{\text{mks}} = 1 + \frac{\mu^2_{\text{dipole}}}{3kT\varepsilon_0}N + \frac{\alpha}{\varepsilon_0}N, \qquad \varepsilon_{\text{gas}} = 1 + 4\pi\chi^{\text{cgs}} = 1 + 4\pi\frac{\mu^2_{\text{dipole}}}{3kT}N + 4\pi\alpha N.$$

Writing this with explicit imaginary frequency dependence and with Debye relaxation of the permanent dipole term of relaxation time τ gives

$$\begin{aligned}\varepsilon_{\text{gas}}(i\xi) &= 1 + \chi^{\text{mks}}(i\xi)\\ &= 1 + \frac{\mu^2_{\text{dipole}}}{3kT\varepsilon_0(1+\xi\tau)}N + \frac{\alpha^{\text{mks}}(i\xi)}{\varepsilon_0}N\\ &= 1 + N\left[\frac{\mu^2_{\text{dipole}}}{3kT\varepsilon_0(1+\xi\tau)} + \frac{\alpha^{\text{mks}}(i\xi)}{\varepsilon_0}\right]\\ &= 1 + N\frac{\alpha^{\text{mks}}_{\text{total}}(i\xi)}{\varepsilon_0},\end{aligned}$$

$$\alpha^{\text{mks}}_{\text{total}}(i\xi) \equiv \left[\frac{\mu^2_{\text{dipole}}}{3kT(1+\xi\tau)} + \alpha^{\text{mks}}(i\xi)\right];$$

$$\begin{aligned}\varepsilon_{\text{gas}}(i\xi) &= 1 + 4\pi\chi^{\text{cgs}}(i\xi)\\ &= 1 + 4\pi\frac{\mu^2_{\text{dipole}}}{3kT(1+\xi\tau)}N + 4\pi\alpha^{\text{cgs}}(i\xi)N\\ &= 1 + 4\pi N\left[\frac{\mu^2_{\text{dipole}}}{3kT(1+\xi\tau)} + \alpha^{\text{cgs}}(i\xi)\right]\\ &= 1 + 4\pi N\alpha^{\text{cgs}}_{\text{total}}(i\xi),\end{aligned}$$

$$\alpha^{\text{cgs}}_{\text{total}}(i\xi) \equiv \left[\frac{\mu^2_{\text{dipole}}}{3kT(1+\xi\tau)} + \alpha^{\text{cgs}}(i\xi)\right].$$

See also "Time out for units" page 218.

L2.2. Tables of formulae

Tables of Formulae are identified by an uppercase P, S, or C, (for planar, spherical, or cylindrical geometries), and a number (incremental, beginning with "1"). Subsets are identified by a lowercase letter, and sometimes by an additional number. Examples: Table P.2.a.1 or Table S.5.

Equations that are adapted from or are identical to those given in other parts of this book are designated by the same numbers assigned to those equations in other sections, bracketed, e.g., [L3.118]. Equations identified by a number without any letter designation (e.g.,[47]) refer to the footnote giving their original source.

L2.2.A. TABLES OF FORMULAE IN PLANAR GEOMETRY

Table P.1.a. Forms of the van der Waals interaction between two semi-infinite media

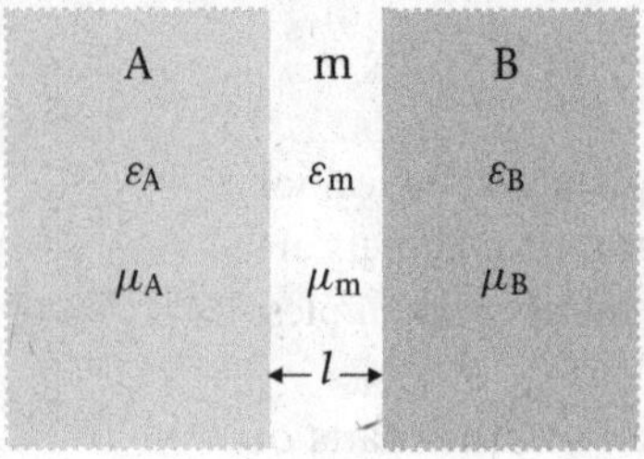

P.1.a.1. Exact, Lifshitz

$$G_{AmB}(l,T) = \frac{kT}{8\pi l^2}\sum_{n=0}^{\infty}{}' \int_{r_n}^{\infty} x \ln\left[(1-\overline{\Delta}_{Am}\overline{\Delta}_{Bm}e^{-x})(1-\Delta_{Am}\Delta_{Bm}e^{-x})\right] dx$$

$$= \frac{kT}{2\pi c^2}\sum_{n=0}^{\infty}{}' \varepsilon_m\mu_m\xi_n^2 \times \int_1^{\infty} p \ln\left[(1-\overline{\Delta}_{Am}\overline{\Delta}_{Bm}e^{-r_n p})(1-\Delta_{Am}\Delta_{Bm}e^{-r_n p})\right] dp$$

$$= -\frac{kT}{8\pi l^2}\sum_{n=0}^{\infty}{}' r_n^2 \sum_{q=1}^{\infty}\frac{1}{q} \times \int_1^{\infty} p\left[(\overline{\Delta}_{Am}\overline{\Delta}_{Bm})^q + (\Delta_{Am}\Delta_{Bm})^q\right] e^{-r_n pq} dp;$$

$$\overline{\Delta}_{ji} = \frac{s_i\varepsilon_j - s_j\varepsilon_i}{s_i\varepsilon_j + s_j\varepsilon_i},\ \Delta_{ji} = \frac{s_i\mu_j - s_j\mu_i}{s_i\mu_j + s_j\mu_i},\ s_i = \sqrt{p^2 - 1 + (\varepsilon_i\mu_i/\varepsilon_m\mu_m)},\ s_m = p,$$

$$\text{or } \overline{\Delta}_{ji} = \frac{x_i\varepsilon_j - x_j\varepsilon_i}{x_i\varepsilon_j + x_j\varepsilon_i},\ \Delta_{ji} = \frac{x_i\mu_j - x_j\mu_i}{x_i\mu_j + x_j\mu_i},$$

$$x_i^2 = x_m^2 + \left(\frac{2l\xi_n}{c}\right)^2(\varepsilon_i\mu_i - \varepsilon_m\mu_m),\ x_m = x,$$

$$p = x/r_n,\ r_n = (2l\varepsilon_m^{1/2}\mu_m^{1/2}/c)\,\xi_n.$$ [Eqs. (L3.50)–(L3.57)]

P.1.a.2. Hamaker form

$$G_{AmB}(l,T) = -\frac{A_{Am/Bm}(l,T)}{12\pi l^2}.$$ [L2.5]

P.1.a.3. Nonretarded, separations approaching contact, $l \to 0$, $r_n \to 0$

$$G_{AmB}(l\to 0,T) \to -\frac{kT}{8\pi l^2}\sum_{n=0}^{\infty}{}'\sum_{q=1}^{\infty}\frac{(\overline{\Delta}_{Am}\overline{\Delta}_{Bm})^q + (\Delta_{Am}\Delta_{Bm})^q}{q^3},$$

$$\overline{\Delta}_{ji} \to \frac{\varepsilon_j - \varepsilon_i}{\varepsilon_j + \varepsilon_i},\ \Delta_{ji} \to \frac{\mu_j - \mu_i}{\mu_j + \mu_i}.$$ [L2.8]

P.1.a.4. Nonretarded, small differences in permittivity

$$G_{AmB}(l\to 0,T) \approx -\frac{kT}{8\pi l^2}\sum_{n=0}^{\infty}{}'(\overline{\Delta}_{Am}\overline{\Delta}_{Bm} + \Delta_{Am}\Delta_{Bm}),$$

$$\overline{\Delta}_{ji} = \frac{\varepsilon_j - \varepsilon_i}{\varepsilon_j + \varepsilon_i} \ll 1,\ \Delta_{ji} = \frac{\mu_j - \mu_i}{\mu_j + \mu_i} \ll 1.$$ [L2.10]

P.1.a.5. Infinitely large separations, $l \to \infty$

$$G_{AmB}(l\to\infty,T) \to -\frac{kT}{16\pi l^2}\sum_{q=1}^{\infty}\frac{(\overline{\Delta}_{Am}\overline{\Delta}_{Bm})^q + (\Delta_{Am}\Delta_{Bm})^q}{q^3}.$$ [L2.11]

Again $\overline{\Delta}_{ji} = [(\varepsilon_j - \varepsilon_i)(\varepsilon_j + \varepsilon_i)]$, $\Delta_{ji} = [(\mu_j - \mu_i)(\mu_j + \mu_i)]$, but all ε's and μ's are evaluated only at zero frequency. This expression ignores ionic fluctuations and material conductivities.

Table P.1.b. Two half-spaces across a planar slab, separation l, zero-temperature limit

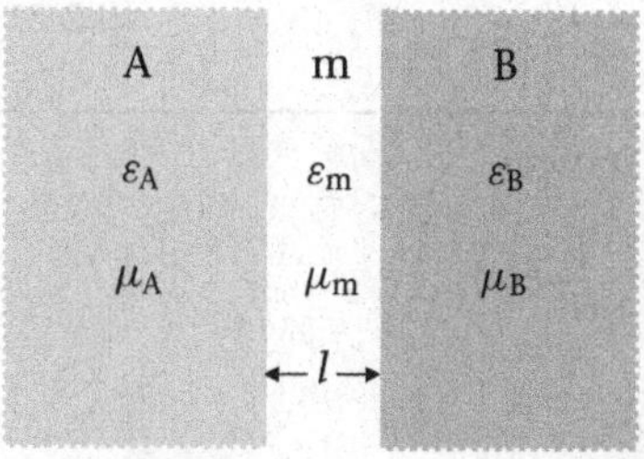

P.1.b.1. With retardation

$$G(l, T \to 0) = \frac{\hbar}{(4\pi)^2 l^2} \int_0^\infty \mathrm{d}\xi \times \int_{r_n}^\infty x \ln\left[\left(1 - \overline{\Delta}_{\mathrm{Am}}\overline{\Delta}_{\mathrm{Bm}} e^{-x}\right)\left(1 - \Delta_{\mathrm{Am}}\Delta_{\mathrm{Bm}} e^{-x}\right)\right] \mathrm{d}x$$

$$= \frac{\hbar}{(2\pi)^2 c^2} \int_0^\infty \mathrm{d}\xi\, \varepsilon_{\mathrm{m}}\mu_{\mathrm{m}}\xi^2 \times \int_1^\infty p \ln\left[\left(1 - \overline{\Delta}_{\mathrm{Am}}\overline{\Delta}_{\mathrm{Bm}} e^{-r_n p}\right)\left(1 - \Delta_{\mathrm{Am}}\Delta_{\mathrm{Bm}} e^{-r_n p}\right)\right] \mathrm{d}p. \qquad \text{[L2.12]}$$

P.1.b.2. Small-separation limit (no retardation)

$$G(l \to 0, T \to 0) = \frac{-\hbar}{(4\pi)^2 l^2} \int_0^\infty \mathrm{d}\xi \sum_{q=1}^{\infty} \frac{\left(\overline{\Delta}_{\mathrm{Am}}\overline{\Delta}_{\mathrm{Bm}}\right)^q + \left(\Delta_{\mathrm{Am}}\Delta_{\mathrm{Bm}}\right)^q}{q^3}, \qquad \text{[L2.13]}$$

$$\overline{\Delta}_{\mathrm{ji}} = \frac{\varepsilon_{\mathrm{j}} - \varepsilon_{\mathrm{i}}}{\varepsilon_{\mathrm{j}} + \varepsilon_{\mathrm{i}}}, \; \Delta_{\mathrm{ji}} = \frac{\mu_{\mathrm{j}} - \mu_{\mathrm{i}}}{\mu_{\mathrm{j}} + \mu_{\mathrm{i}}}.$$

In terms of an average photon energy $\hbar\overline{\xi}$,
$G(l \to 0, T \to 0) = [\hbar\overline{\xi}/(4\pi)^2 l^2]$,
$\overline{\xi}$ form with leading, $q = 1$, term only, $\overline{\xi} \approx \int_0^\infty (\overline{\Delta}_{\mathrm{Am}}\overline{\Delta}_{\mathrm{Bm}} + \Delta_{\mathrm{Am}}\Delta_{\mathrm{Bm}})\mathrm{d}\xi$.

P.1.b.3. Large separation limit

$l \gg$ all absorption wavelengths

$$G_{\mathrm{AmB}}(l, T \to 0) \approx -\frac{\hbar c}{8\pi^2 l^3 \varepsilon_{\mathrm{m}}^{1/2}} \overline{\Delta}_{\mathrm{Am}}\overline{\Delta}_{\mathrm{Bm}} \qquad \text{[L2.15]}$$

$$\overline{\Delta}_{\mathrm{Am}} = \frac{\sqrt{\varepsilon_{\mathrm{A}}} - \sqrt{\varepsilon_{\mathrm{m}}}}{\sqrt{\varepsilon_{\mathrm{A}}} + \sqrt{\varepsilon_{\mathrm{m}}}} = \frac{n_{\mathrm{A}} - n_{\mathrm{m}}}{n_{\mathrm{A}} + n_{\mathrm{m}}}, \; \overline{\Delta}_{\mathrm{Bm}} = \frac{\sqrt{\varepsilon_{\mathrm{B}}} - \sqrt{\varepsilon_{\mathrm{m}}}}{\sqrt{\varepsilon_{\mathrm{B}}} + \sqrt{\varepsilon_{\mathrm{m}}}} = \frac{n_{\mathrm{B}} - n_{\mathrm{m}}}{n_{\mathrm{B}} + n_{\mathrm{m}}}$$

usually evaluated through indices of refraction n_{A}, n_{m}, n_{B} in a transparent region

$$G_{\mathrm{AmB}}(l, T \to 0) \approx -\frac{\hbar c}{8\pi^2 n_{\mathrm{m}} l^3} \frac{n_{\mathrm{A}} - n_{\mathrm{m}}}{n_{\mathrm{A}} + n_{\mathrm{m}}} \frac{n_{\mathrm{B}} - n_{\mathrm{m}}}{n_{\mathrm{B}} + n_{\mathrm{m}}}.$$

Table P.1.c. Ideal conductors

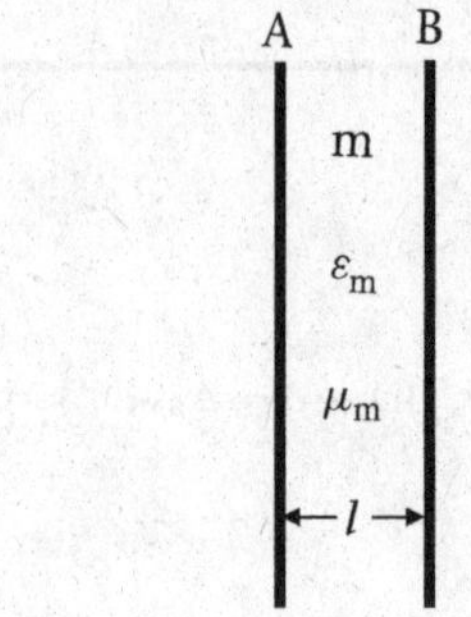

P.1.c.1. Finite temperature

$$G_{\mathrm{AmB}}(l,T) = -\frac{kT}{4\pi l^2}\sum_{n=0}^{\infty}{}' \sum_{q=1}^{\infty}\frac{(1+r_n q)e^{-r_n q}}{q^3}. \qquad \text{[L2.21]}$$

P.1.c.2. Finite temperature, long distance

$$G_{\mathrm{AmB}}\,(l\to\infty,T) \to -\frac{kT}{8\pi l^2}\sum_{q=1}^{\infty}\frac{1}{q^3} = -\frac{kT}{8\pi l^2}\zeta(3),$$

$$\zeta(3) \equiv \sum_{q=1}^{\infty}\frac{1}{q^3} \approx 1.2.$$

P.1.c.3. Zero temperature

$$G_{\mathrm{AmB}}(l,T\to 0) = -\frac{\hbar c}{8\pi^2 l^3 \varepsilon_{\mathrm{m}}^{1/2}\mu_{\mathrm{m}}^{1/2}}\zeta(4) = -\frac{\hbar c\pi^2}{720 l^3 \varepsilon_{\mathrm{m}}^{1/2}\mu_{\mathrm{m}}^{1/2}},$$

$$\zeta(4) \equiv \sum_{q=1}^{\infty}\frac{1}{q^4} = \frac{\pi^4}{90} \approx 1.1. \qquad \text{[L2.22]}$$

Energy and derivative pressure in vacuum:

$$G_{\mathrm{AmB}}(l,T\to 0) = -\frac{\hbar c\pi^2}{720 l^3},\; P(l) = -\frac{\hbar c\pi^2}{240 l^4}.$$

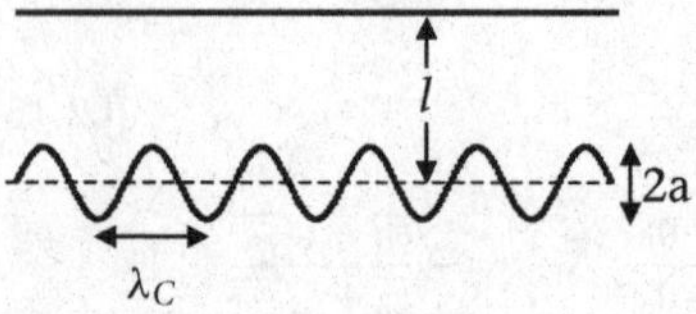

Amplitude a, period λ_C, mean separation l.

P.1.c.4. Corrugated–flat conducting surfaces, across vacuum at zero temperature

$$E_{C-s}(l;\mathrm{a}) = E_0(l) + E_{\mathrm{cf}}(l;\mathrm{a}) \quad [31],$$

$$E_0(l) = G_{\mathrm{AmB}}(l,T\to 0) = -\frac{\hbar c\pi^2}{720 l^3} \quad [32],\ [\text{P.1.c.3}]$$

$$E_{\mathrm{cf}}(l;\mathrm{a}) = -\frac{\hbar c\mathrm{a}^2}{l^5}\left[G_{\mathrm{TM}}\left(\frac{l}{\lambda_C}\right) + G_{\mathrm{TE}}\left(\frac{l}{\lambda_C}\right)\right] + \mathrm{O}(\mathrm{a}^3) \quad [32],$$

$$G_{\mathrm{TM}}(x) \equiv \frac{\pi^3 x}{480} - \frac{\pi^2 x^4}{30}\ln(1-u) + \frac{\pi}{1920x}\mathrm{Li}_2(1-u) + \frac{\pi x^3}{24}\mathrm{Li}_2(u) + \frac{x^2}{24}\mathrm{Li}_3(u)$$
$$+ \frac{x}{32\pi}\mathrm{Li}_4(u) + \frac{1}{64\pi^2}\mathrm{Li}_5(u) + \frac{1}{256\pi^3 x}\left[\mathrm{Li}_6(u) - \frac{\pi^6}{945}\right] \quad [37],$$

$$G_{\mathrm{TE}}(x) \equiv \frac{\pi^3 x}{1440} - \frac{\pi^2 x^4}{30}\ln(1-u) + \frac{\pi}{1920x}\mathrm{Li}_2(1-u) - \frac{\pi x}{48}(1+2x^2)\,\mathrm{Li}_2(u)$$
$$+ \left(\frac{x^2}{48} - \frac{1}{64}\right)\mathrm{Li}_3(u) + \frac{5x}{64\pi}\mathrm{Li}_4(u) + \frac{7}{128\pi^2}\mathrm{Li}_5(u)$$
$$+ \frac{1}{256\pi^3 x}\left[\frac{7}{2}\mathrm{Li}_6(u) - \pi^2\mathrm{Li}_4(u) + \frac{\pi^6}{135}\right] \quad [38],$$

where

$$u \equiv e^{-4\pi x},\ \mathrm{Li}_n(z) \equiv \sum_{\nu=1}^{\infty}\frac{z^\nu}{\nu^n} \quad [36].$$

Source: From T. Emig, A. Hanke, R. Golestanian, and M. Kardar, "Normal and lateral Casimir forces between deformed plates," Phys. Rev. A **67**, 022114 (2003); numbers in [ ] are equation numbers in this source paper. Separation H in that paper is replaced here with l; period of corrugation λ with λ_C; energy $\mathcal{E}$ with E. Symbols G_{TM}, G_{TE}, Li_n, a, x, z, u, and n here as in source article.

Table P.1.c (*cont.*)

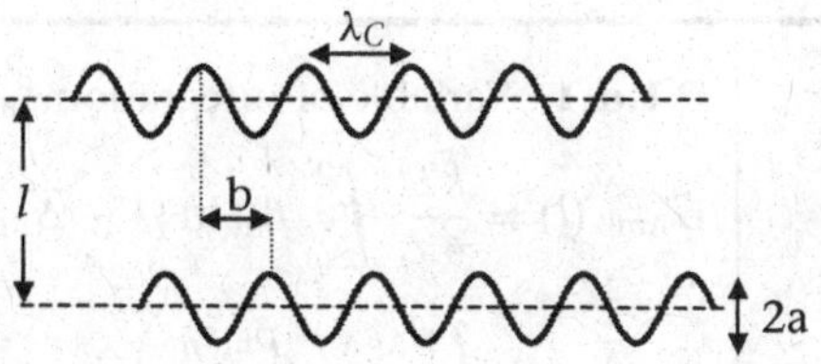

Amplitude a, period λ_C, mean separation l, lateral shift b.

P.1.c.5. Corrugated–corrugated conducting surfaces, across vacuum at zero temperature

$$E_{C-C}(l;\mathrm{a},\mathrm{b}) = E_0(l) + 2E_{\mathrm{cf}}(l;\mathrm{a}) + E_{\mathrm{cc}}(l;\mathrm{a},\mathrm{b}) \quad [44]$$

$$E_0(l) = -\frac{\hbar c\pi^2}{720 l^3} \quad [31], \quad [\text{P.1.c.3}]$$

$$E_{\mathrm{cf}}(l;\mathrm{a}) = -\frac{\hbar c \mathrm{a}^2}{l^5}\left[G_{\mathrm{TM}}\left(\frac{l}{\lambda_C}\right) + G_{\mathrm{TE}}\left(\frac{l}{\lambda_C}\right)\right] + O(\mathrm{a}^3) \quad [32];$$

$$G_{\mathrm{TM}}\left(\frac{l}{\lambda_C}\right),\ G_{\mathrm{TE}}\left(\frac{l}{\lambda_C}\right), \quad [\text{as in Table P.1.c.4}]$$

$$E_{\mathrm{cc}}(l;\mathrm{a},\mathrm{b}) = \frac{\hbar c\mathrm{a}^2}{l^5}\cos\left(\frac{2\pi \mathrm{b}}{\lambda_C}\right)\left[J_{\mathrm{TM}}\left(\frac{l}{\lambda_C}\right) + J_{\mathrm{TE}}\left(\frac{l}{\lambda_C}\right)\right] + O(\mathrm{a}^3) \quad [46],$$

$$J_{\mathrm{TM}}(x) \equiv \frac{\pi^2}{120}(16x^4-1)\,\mathrm{arctanh}(\sqrt{u}) + \sqrt{u}\left[\frac{\pi}{12}\left(x^3 - \frac{1}{80x}\right)\Phi\left(u,2,\frac{1}{2}\right) + \frac{x^2}{12}\Phi\left(u,3,\frac{1}{2}\right)\right.$$
$$\left. + \frac{x}{16\pi}\Phi\left(u,4,\frac{1}{2}\right) + \frac{1}{32\pi^2}\Phi\left(u,5,\frac{1}{2}\right) + \frac{1}{128\pi^3 x}\Phi\left(u,6,\frac{1}{2}\right)\right] \quad [50a],$$

$$J_{\mathrm{TE}}(x) \equiv \frac{\pi^2}{120}(16x^4-1)\,\mathrm{arctanh}(\sqrt{u}) + \sqrt{u}\left[-\frac{\pi}{12}\left(x^3 + \frac{x}{2} + \frac{1}{80x}\right)\Phi\left(u,2,\frac{1}{2}\right)\right.$$
$$+ \frac{1}{24}\left(x^2 - \frac{3}{4}\right)\Phi\left(u,3,\frac{1}{2}\right) + \frac{5}{32\pi}\left(x - \frac{1}{20x}\right)\Phi\left(u,4,\frac{1}{2}\right) + \frac{7}{64\pi^2}\Phi\left(u,5,\frac{1}{2}\right)$$
$$\left. + \frac{7}{256\pi^3 x}\Phi\left(u,6,\frac{1}{2}\right)\right] \quad [50b],$$

where

$$u \equiv e^{-4\pi x} \quad [36],$$

$$\Phi(z,s,\mathrm{a}) \equiv \sum_{k=0}^{\infty}\frac{z^k}{(\mathrm{a}+k)^s} \quad [49].$$

Source: From T. Emig, A. Hanke, R. Golestanian, and M. Kardar, "Normal and lateral Casimir forces between deformed plates," Phys. Rev. A **67**, 022114 (2003); numbers in [ ] are equation numbers in this source paper. Separation H in that paper is replaced here by l; period of corrugation λ with λ_C; energy $\mathcal{E}$ by E. Symbols G_{TM}, G_{TE}, J_{TM}, J_{TE}, Li_n, Φ, a, b, x, s, z, u, and n here as in source article.

Table P.1.d. Ionic solutions, zero-frequency fluctuations, two half-spaces across layer m

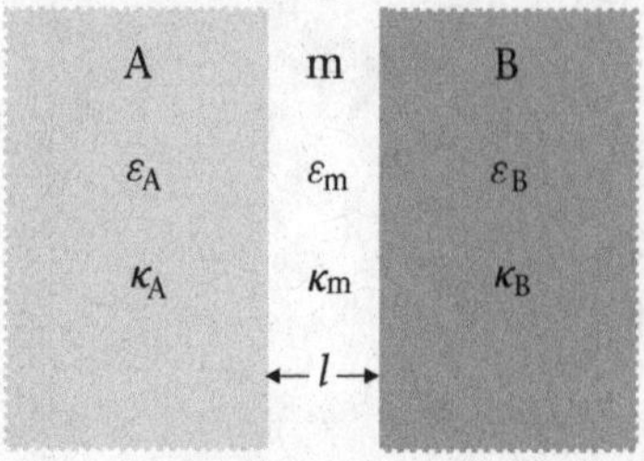

Ionic solutions in regions A, m, and B. Choose for convenience between integration over variables β_m, p, or x.

P.1.d.1. Variable of integration β_m

$$G_{AmB}(l) = \frac{kT}{4\pi} \int_{\kappa_m}^{\infty} \beta_m \ln\left(1 - \overline{\Delta}_{Am}\overline{\Delta}_{Bm} e^{-2\beta_m l}\right) d\beta_m,$$

$$\overline{\Delta}_{Am} \equiv \left(\frac{\beta_A\varepsilon_A - \beta_m\varepsilon_m}{\beta_A\varepsilon_A + \beta_m\varepsilon_m}\right), \beta_A^2 = \beta_m^2 + \left(\kappa_A^2 - \kappa_m^2\right),$$

$$\overline{\Delta}_{Bm} \equiv \left(\frac{\beta_B\varepsilon_B - \beta_m\varepsilon_m}{\beta_B\varepsilon_B + \beta_m\varepsilon_m}\right), \beta_B^2 = \beta_m^2 + \left(\kappa_B^2 - \kappa_m^2\right),$$

$$\kappa_m \le \beta_m < \infty.$$

Note form of $\beta_i\varepsilon_i$ for effective dielectric response. Double-layer screening of zero-frequency fluctuations through $e^{-2\beta_m l}$.

P.1.d.2. Variable of integration p

$$G_{AmB}(l) = \frac{kT\kappa_m^2}{4\pi} \int_1^{\infty} p \ln\left(1 - \overline{\Delta}_{Am}\overline{\Delta}_{Bm} e^{-2\kappa_m l p}\right) dp,$$

$$\overline{\Delta}_{Am} \equiv \left(\frac{s_A\varepsilon_A - p\varepsilon_m}{s_A\varepsilon_A + p\varepsilon_m}\right), s_A = \sqrt{p^2 - 1 + \kappa_A^2/\kappa_m^2},$$

$$\overline{\Delta}_{Bm} \equiv \left(\frac{s_B\varepsilon_B - p\varepsilon_m}{s_B\varepsilon_B + p\varepsilon_m}\right), s_B = \sqrt{p^2 - 1 + \kappa_B^2/\kappa_m^2},$$

$$1 \le p < \infty, \beta_m = p\kappa_m.$$

N.B.: s_L, p, s_R here multiply by ε_L, ε_m, ε_R, *not* as for the dipolar fluctuation formulae.

P.1.d.3. Variable of integration x

$$G_{AmB}(l) = \frac{kT}{16\pi l^2} \int_{2\kappa_m l}^{\infty} x\left[\ln(1 - \overline{\Delta}_{Am}\overline{\Delta}_{Bm} e^{-x})\right] dx,$$

$$\overline{\Delta}_{Am} \equiv \left(\frac{x_A\varepsilon_A - x\varepsilon_m}{x_A\varepsilon_A + x\varepsilon_m}\right), x_A = \sqrt{x^2 + \left(\kappa_A^2 - \kappa_m^2\right)(2l)^2},$$

$$\overline{\Delta}_{Bm} \equiv \left(\frac{x_B\varepsilon_B - x\varepsilon_m}{x_B\varepsilon_B + x\varepsilon_m}\right), x_B = \sqrt{x^2 + \left(\kappa_B^2 - \kappa_m^2\right)(2l)^2},$$

$$x = 2\beta_m l, \; 2\kappa_m l \le x < \infty.$$

Notes: The inverse of the Debye screening length λ_{Debye}, the screening constants κ_i, in each region i = A, m, B depend on ionic strength built from the number densities $n_\nu^{(i)}$ of mobile ions of valence ν in material i [see Eqs. (L3.176), (L2.184), and (L2.185)]: In mks units $\kappa_i^2 \equiv [e^2/(\varepsilon\varepsilon_0 kT)] \sum_{\{\nu\}} n_\nu^{(i)} \nu^2$, ion densities are per cubic meter; in cgs units $\kappa_i^2 \equiv [(4\pi e^2)/(\varepsilon kT)] \sum_{\{\nu\}} n_\nu^{(i)} \nu^2$, densities are per cubic centimeter.

The summation form $\sum_{\{\nu\}}$ takes into account the set $\{\nu\}$ of all mobile ions of all valences ν. The quantity $\sum_{\{\nu\}} n_\nu^{(i)} \nu^2$ is proportional to the ionic strength in material region i.

In molar units, the ionic strength is $I(M) \equiv \frac{1}{2} \sum_{\{\nu\}} n_\nu^{(i)}(M)\nu^2$, where number densities are concentrations expressed as moles per liter (1 mol/liter = 6.02×10^{23} particles/liter = 6.02×10^{26} particles/m^3 = 6.02×10^{20} particles/cm^3).

Table P.1.d (*cont.*)

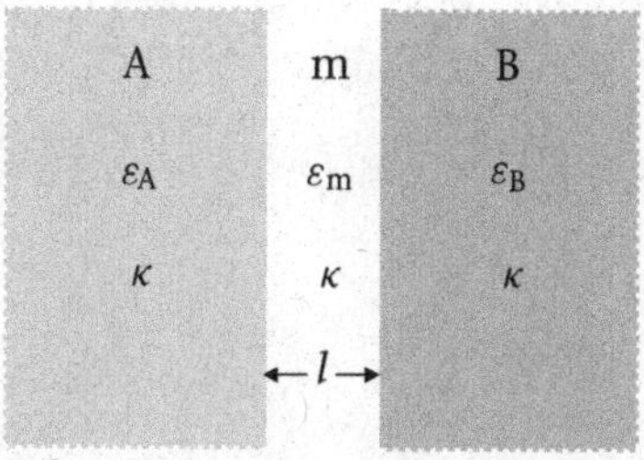

P.1.d.4. Uniform ionic strength $\kappa_A = \kappa_m = \kappa_B = \kappa$

$$G_{LmR}(l) = \frac{kT}{4\pi} \int_{\kappa}^{\infty} \beta \ln\left(1 - \overline{\Delta}_{Lm}\overline{\Delta}_{Rm} e^{-2\beta l}\right) d\beta,\ \beta_L = \beta_m = \beta_R = \beta.$$

$$\overline{\Delta}_{Lm} \equiv \left(\frac{\varepsilon_L - \varepsilon_m}{\varepsilon_L + \varepsilon_m}\right),\ \overline{\Delta}_{Rm} \equiv \left(\frac{\varepsilon_R - \varepsilon_m}{\varepsilon_R + \varepsilon_m}\right).$$

For $2\kappa l \ll 1$,

$$G_{AmB}(l) \approx -\frac{kT}{16\pi l^2}\overline{\Delta}_{Am}\overline{\Delta}_{Bm}(1 + 2\kappa l)e^{-2\kappa l}$$

$$= -\frac{kT}{16\pi l^2}\overline{\Delta}_{Am}\overline{\Delta}_{Bm}\left(1 + \frac{2l}{\lambda_D}\right)e^{-2l/\lambda_D} = -\frac{kT}{16\pi l^2}\overline{\Delta}_{Am}\overline{\Delta}_{Bm} R_0.$$

Ionic screening factor
$R_0 = (1 + 2\kappa l)e^{-2\kappa l} = [1 + (2l/\lambda_D)]e^{-2l/\lambda_D} \le 1.$

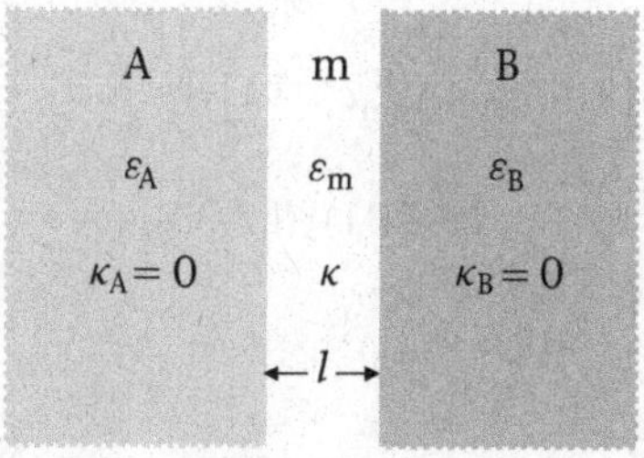

P.1.d.5. Salt solution m; pure-dielectric A, B, $\varepsilon_m \ll \varepsilon_A, \varepsilon_B$, $\kappa_A = \kappa_B = 0$

$$G_{AmB}(l) = \frac{kT}{4\pi} \int_{\kappa}^{\infty} \beta_m \ln\left(1 - \overline{\Delta}_{Am}\overline{\Delta}_{Bm} e^{-2\beta_m l}\right) d\beta_m,$$

$$\kappa_m = \kappa,\ \beta_L^2 = \beta_R^2 = \beta_m^2 - \kappa^2,$$

$$\overline{\Delta}_{Am} = \left(\frac{\varepsilon_A\beta_A - \varepsilon_m\beta_m}{\varepsilon_A\beta_A + \varepsilon_m\beta_m}\right),\ \overline{\Delta}_{Bm} = \left(\frac{\varepsilon_B\beta_B - \varepsilon_m\beta_m}{\varepsilon_B\beta_B + \varepsilon_m\beta_m}\right)\overline{\Delta}_{Am}\overline{\Delta}_{Bm} \approx 1.$$

For $2\kappa l \gg 1$,

$$G_{AmB}(l) \approx -\frac{kT}{16\pi l^2}(1 + 2\kappa l)e^{-2\kappa l} = -\frac{kT}{16\pi l^2}\left(1 + \frac{2l}{\lambda_D}\right)e^{-2l/\lambda_D}$$

$$= -\frac{kT}{16\pi l^2} R_0.$$

Ionic screening factor $R_0 = (1 + 2\kappa l)e^{-2\kappa l} = [1 + 2l/\lambda_D]e^{-2l/\lambda_D} \le 1.$

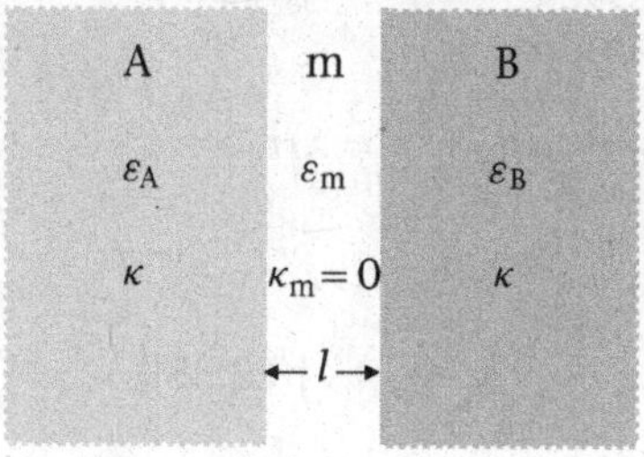

P.1.d.6. Salt solution A, B; pure-dielectric m, $\varepsilon_m \ll \varepsilon_A, \varepsilon_B$, $\kappa_A = \kappa_B = \kappa$

$$G_{AmB}(l) = \frac{kT}{4\pi} \int_{0}^{\infty} \beta_m \ln\left(1 - \overline{\Delta}_{Am}\overline{\Delta}_{Bm} e^{-2\beta_m l}\right) d\beta_m \approx -\frac{1.202kT}{16\pi l^2},$$

$$\overline{\Delta}_{Am}\overline{\Delta}_{Bm} \approx 1.$$

See also R. Netz, "Static van der Waals interaction in electrolytes," Eur. J. Phys., E 5, 189–205 (2001).

Table P.2.a. One surface singly layered

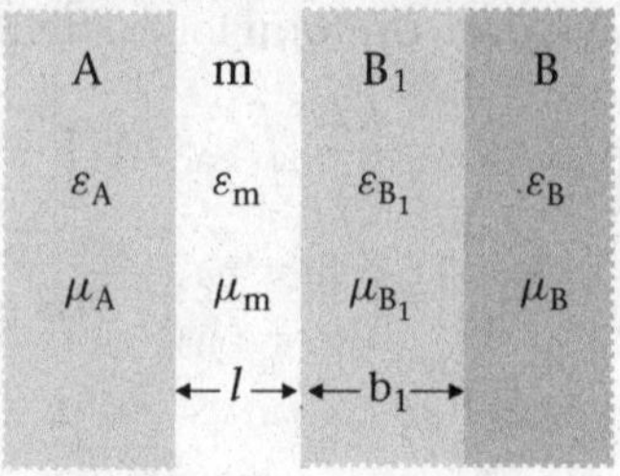

P.2.a.1. Exact, Lifshitz

$G_{\mathrm{AmB_1B}}(l;\mathrm{b}_1)$

$$= \frac{kT}{2\pi}\sum_{n=0}^{\infty}{}' \int_{\frac{\varepsilon_{\mathrm{m}}^{1/2}\mu_{\mathrm{m}}^{1/2}\xi_n}{c}}^{\infty} \rho_{\mathrm{m}} \ln\left[\left(1-\overline{\Delta}_{\mathrm{Am}}\overline{\Delta}_{\mathrm{Bm}}^{\mathrm{eff}}e^{-2\rho_{\mathrm{m}}l}\right)\left(1-\Delta_{\mathrm{Am}}\Delta_{\mathrm{Bm}}^{\mathrm{eff}}e^{-2\rho_{\mathrm{m}}l}\right)\right]\mathrm{d}\rho_{\mathrm{m}}$$

$$= \frac{kT}{8\pi l^2}\sum_{n=0}^{\infty}{}' \int_{r_n}^{\infty} x\ln\left[\left(1-\overline{\Delta}_{\mathrm{Am}}\overline{\Delta}_{\mathrm{Bm}}^{\mathrm{eff}}e^{-x}\right)\left(1-\Delta_{\mathrm{Am}}\Delta_{\mathrm{Bm}}^{\mathrm{eff}}e^{-x}\right)\right]\mathrm{d}x$$

$$= \frac{kT}{2\pi c^2}\sum_{n=0}^{\infty}{}' \varepsilon_{\mathrm{m}}\mu_{\mathrm{m}}\xi_n^2 \int_{1}^{\infty} p\ln\left[\left(1-\overline{\Delta}_{\mathrm{Am}}\overline{\Delta}_{\mathrm{Bm}}^{\mathrm{eff}}e^{-r_n p}\right)\left(1-\Delta_{\mathrm{Am}}\Delta_{\mathrm{Bm}}^{\mathrm{eff}}e^{-r_n p}\right)\right]\mathrm{d}p$$

[(L2.31)–(L2.33)]

$$\overline{\Delta}_{\mathrm{Bm}}^{\mathrm{eff}}(\mathrm{b}_1) = \frac{\left(\overline{\Delta}_{\mathrm{BB_1}}e^{-2\rho_{\mathrm{B_1}}\mathrm{b}_1}+\overline{\Delta}_{\mathrm{B_1m}}\right)}{1+\overline{\Delta}_{\mathrm{BB_1}}\overline{\Delta}_{\mathrm{B_1m}}e^{-2\rho_{\mathrm{B_1}}\mathrm{b}_1}} = \frac{\left[\overline{\Delta}_{\mathrm{BB_1}}e^{-x_{\mathrm{B_1}}(\mathrm{b}_1/l)}+\overline{\Delta}_{\mathrm{B_1m}}\right]}{1+\overline{\Delta}_{\mathrm{BB_1}}\overline{\Delta}_{\mathrm{B_1m}}e^{-x_{\mathrm{B_1}}(\mathrm{b}_1/l)}}$$

$$= \frac{\left[\overline{\Delta}_{\mathrm{BB_1}}e^{-s_{\mathrm{B_1}}r_n(\mathrm{b}_1/l)}+\overline{\Delta}_{\mathrm{B_1m}}\right]}{1+\overline{\Delta}_{\mathrm{BB_1}}\overline{\Delta}_{\mathrm{B_1m}}e^{-s_{\mathrm{B_1}}r_n(\mathrm{b}_1/l)}}, \qquad [(\mathrm{L2.36})]$$

$$\Delta_{\mathrm{Bm}}^{\mathrm{eff}}(\mathrm{b}_1) = \frac{\left(\Delta_{\mathrm{BB_1}}e^{-2\rho_{\mathrm{B_1}}\mathrm{b}_1}+\Delta_{\mathrm{B_1m}}\right)}{1+\Delta_{\mathrm{BB_1}}\Delta_{\mathrm{B_1m}}e^{-2\rho_{\mathrm{B_1}}\mathrm{b}_1}} = \frac{\left[\Delta_{\mathrm{BB_1}}e^{-x_{\mathrm{B_1}}(\mathrm{b}_1/l)}+\Delta_{\mathrm{B_1m}}\right]}{1+\Delta_{\mathrm{BB_1}}\Delta_{\mathrm{B_1m}}e^{-x_{\mathrm{B_1}}(\mathrm{b}_1/l)}}$$

$$= \frac{\left[\Delta_{\mathrm{BB_1}}e^{-s_{\mathrm{A_1}}r_n(\mathrm{b}_1/l)}+\Delta_{\mathrm{B_1m}}\right]}{1+\Delta_{\mathrm{BB_1}}\Delta_{\mathrm{B_1m}}e^{-s_{\mathrm{B_1}}r_n(\mathrm{b}_1/l)}}. \qquad [(\mathrm{L2.37})]$$

$$\rho_{\mathrm{i}}^2 = \rho^2 + \varepsilon_{\mathrm{i}}\mu_{\mathrm{i}}\xi_n^2/c^2,\; x_{\mathrm{i}} \equiv 2\rho_{\mathrm{i}}l,\; x_{\mathrm{i}}^2 = x^2 + \left(\frac{2l\xi_n}{c}\right)^2(\varepsilon_{\mathrm{i}}\mu_{\mathrm{i}}-\varepsilon_{\mathrm{m}}\mu_{\mathrm{m}}),\; p = x/r_n,$$

$$r_n \equiv \left(2l\varepsilon_{\mathrm{m}}^{1/2}\mu_{\mathrm{m}}^{1/2}/c\right)\xi_n,\; s_{\mathrm{i}} = \sqrt{p^2-1+(\varepsilon_{\mathrm{i}}\mu_{\mathrm{i}}/\varepsilon_{\mathrm{m}}\mu_{\mathrm{m}})}$$

$$\overline{\Delta}_{\mathrm{ji}} = \frac{\rho_{\mathrm{i}}\varepsilon_{\mathrm{j}}-\rho_{\mathrm{j}}\varepsilon_{\mathrm{i}}}{\rho_{\mathrm{i}}\varepsilon_{\mathrm{j}}+\rho_{\mathrm{j}}\varepsilon_{\mathrm{i}}} = \frac{x_{\mathrm{i}}\varepsilon_{\mathrm{j}}-x_{\mathrm{j}}\varepsilon_{\mathrm{i}}}{x_{\mathrm{i}}\varepsilon_{\mathrm{j}}+x_{\mathrm{j}}\varepsilon_{\mathrm{i}}} = \frac{s_{\mathrm{i}}\varepsilon_{\mathrm{j}}-s_{\mathrm{j}}\varepsilon_{\mathrm{i}}}{s_{\mathrm{i}}\varepsilon_{\mathrm{j}}+s_{\mathrm{j}}\varepsilon_{\mathrm{i}}}, \qquad [(\mathrm{L2.38})]$$

$$\Delta_{\mathrm{ji}} = \frac{\rho_{\mathrm{i}}\mu_{\mathrm{j}}-\rho_{\mathrm{j}}\mu_{\mathrm{i}}}{\rho_{\mathrm{i}}\mu_{\mathrm{j}}+\rho_{\mathrm{j}}\mu_{\mathrm{i}}} = \frac{x_{\mathrm{i}}\mu_{\mathrm{j}}-x_{\mathrm{j}}\mu_{\mathrm{i}}}{x_{\mathrm{i}}\mu_{\mathrm{j}}+x_{\mathrm{j}}\mu_{\mathrm{i}}} = \frac{s_{\mathrm{i}}\mu_{\mathrm{j}}-s_{\mathrm{j}}\mu_{\mathrm{i}}}{s_{\mathrm{i}}\mu_{\mathrm{j}}+s_{\mathrm{j}}\mu_{\mathrm{i}}}. \qquad [(\mathrm{L2.39})]$$

Table P.2.b. One surface singly layered: Limiting forms

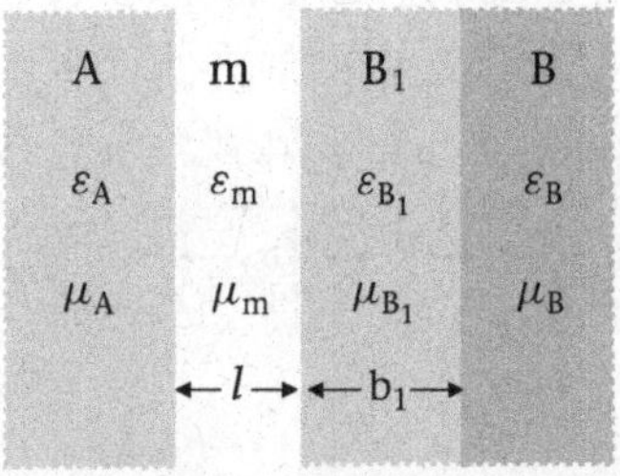

P.2.b.1. High dielectric-permittivity layer

$\varepsilon_{\mathrm{B_1}}/\varepsilon_{\mathrm{m}} \to \infty$, $\overline{\Delta}_{\mathrm{B_1m}} \to 1$, $\overline{\Delta}^{\mathrm{eff}}_{\mathrm{Bm}}(\mathrm{b_1}) \to \dfrac{(\overline{\Delta}_{\mathrm{BB_1}} e^{-2\rho_{\mathrm{B_1}}\mathrm{b_1}} + 1)}{1 + \overline{\Delta}_{\mathrm{BB_1}} e^{-2\rho_{\mathrm{B_1}}\mathrm{b_1}}} = 1$, neglecting magnetic terms;

$$G_{\mathrm{AmB_1B}}(l;\mathrm{b_1}) \to G_{\mathrm{AmB_1}}(l) = \frac{kT}{2\pi}\sum_{n=0}^{\infty}{}' \int_{\frac{\varepsilon_{\mathrm{m}}^{1/2}\mu_{\mathrm{m}}^{1/2}\xi_n}{c}}^{\infty} \rho_{\mathrm{m}} \ln\left(1 - \overline{\Delta}_{\mathrm{Am}} e^{-2\rho_{\mathrm{m}} l}\right) \mathrm{d}\rho_{\mathrm{m}}$$

$$= \frac{kT}{8\pi l^2}\sum_{n=0}^{\infty}{}' \int_{r_n}^{\infty} x \ln\left(1 - \overline{\Delta}_{\mathrm{Am}} e^{-x}\right)\mathrm{d}x$$

$$= \frac{kT}{2\pi c^2}\sum_{n=0}^{\infty}{}' \varepsilon_{\mathrm{m}}\mu_{\mathrm{m}}\xi_n^2 \int_1^{\infty} p \ln\left(1 - \overline{\Delta}_{\mathrm{Am}} e^{-r_n p}\right)\mathrm{d}p.$$

[(L2.41)]

P.2.b.2. Small differences in ε's and μ's, with retardation

$\varepsilon_{\mathrm{A}} \approx \varepsilon_{\mathrm{m}} \approx \varepsilon_{\mathrm{B}} \approx \varepsilon_{\mathrm{B}}$, $\overline{\Delta}_{\mathrm{ji}}$'s, Δ_{ij}'s $\ll 1$,

$$G_{\mathrm{AmB_1B}}(l;\mathrm{b_1}) = -\frac{kT}{2\pi}\sum_{n=0}^{\infty}{}' \int_{\frac{\varepsilon_{\mathrm{m}}^{1/2}\mu_{\mathrm{m}}^{1/2}\xi_n}{c}}^{\infty} \rho_{\mathrm{m}}\left(\overline{\Delta}_{\mathrm{Am}}\overline{\Delta}_{\mathrm{B_1m}} + \Delta_{\mathrm{Am}}\Delta_{\mathrm{B_1m}}\right) e^{-2\rho_{\mathrm{m}} l}\mathrm{d}\rho_{\mathrm{m}}$$

$$-\frac{kT}{2\pi}\sum_{n=0}^{\infty}{}' \int_{\frac{\varepsilon_{\mathrm{m}}^{1/2}\mu_{\mathrm{m}}^{1/2}\xi_n}{c}}^{\infty} \rho_{\mathrm{m}}\left(\overline{\Delta}_{\mathrm{Am}}\overline{\Delta}_{\mathrm{BB_1}} + \Delta_{\mathrm{Am}}\Delta_{\mathrm{BB_1}}\right) e^{-2\rho_{\mathrm{B_1}}\mathrm{b_1}} e^{-2\rho_{\mathrm{m}} l}\mathrm{d}\rho_{\mathrm{m}}$$

$$= -\frac{A_{\mathrm{Am/B_1m}}(l)}{12\pi l^2} - \frac{A_{\mathrm{Am/BB_1}}(l + \mathrm{b_1})}{12\pi (l + \mathrm{b_1})^2}.$$

[(L2.45)]

P.2.b.3. Small differences in ε's and μ's, without retardation

$$G_{\mathrm{AmB_1B}}(l;\mathrm{b_1}) \to -\frac{kT}{8\pi l^2}\sum_{n=0}^{\infty}{}' \left(\overline{\Delta}_{\mathrm{Am}}\overline{\Delta}_{\mathrm{B_1m}} + \Delta_{\mathrm{Am}}\Delta_{\mathrm{B_1m}}\right)$$

$$-\frac{kT}{8\pi (l + \mathrm{b_1})^2}\sum_{n=0}^{\infty}{}' \left(\overline{\Delta}_{\mathrm{Am}}\overline{\Delta}_{\mathrm{BB_1}} + \Delta_{\mathrm{Am}}\Delta_{\mathrm{BB_1}}\right).$$

[(L2.46)]

Table P.2.c. Finite planar slab with semi-infinite medium

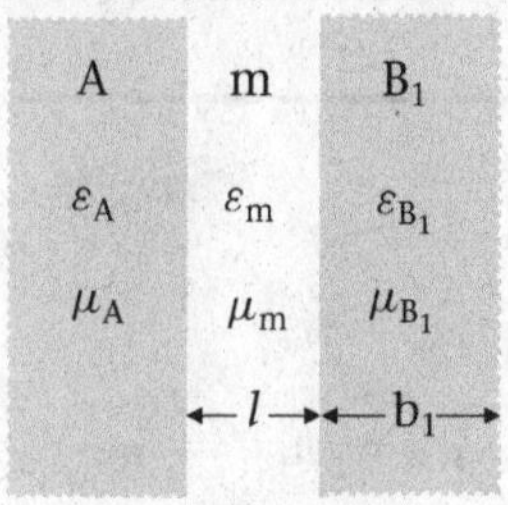

P.2.c.1. Exact, Lifshitz

$$G_{\mathrm{AmB_1}}(l;\mathrm{b}_1) = \frac{kT}{2\pi}\sum_{n=0}^{\infty}{}' \int_{\frac{\varepsilon_{\mathrm{m}}^{1/2}\mu_{\mathrm{m}}^{1/2}\xi_n}{c}}^{\infty} \rho_{\mathrm{m}} \ln\left[\left(1-\overline{\Delta}_{\mathrm{Am}}\overline{\Delta}_{\mathrm{Bm}}^{\mathrm{eff}}e^{-2\rho_{\mathrm{m}}l}\right)\left(1-\Delta_{\mathrm{Am}}\Delta_{\mathrm{Bm}}^{\mathrm{eff}}e^{-2\rho_{\mathrm{m}}l}\right)\right]\mathrm{d}\rho_{\mathrm{m}}$$

$$= \frac{kT}{8\pi l^2}\sum_{n=0}^{\infty}{}' \int_{r_n}^{\infty} x\ln\left[\left(1-\overline{\Delta}_{\mathrm{Am}}\overline{\Delta}_{\mathrm{Bm}}^{\mathrm{eff}}e^{-x}\right)\left(1-\Delta_{\mathrm{Am}}\Delta_{\mathrm{Bm}}^{\mathrm{eff}}e^{-x}\right)\right]\mathrm{d}x$$

$$= \frac{kT}{2\pi c^2}\sum_{n=0}^{\infty}{}' \varepsilon_{\mathrm{m}}\mu_{\mathrm{m}}\xi_n^2 \int_1^{\infty} p\ \ln\left[\left(1-\overline{\Delta}_{\mathrm{Am}}\overline{\Delta}_{\mathrm{Bm}}^{\mathrm{eff}}e^{-r_n p}\right)\left(1-\Delta_{\mathrm{Am}}\Delta_{\mathrm{Bm}}^{\mathrm{eff}}e^{-r_n p}\right)\right]\mathrm{d}p. \qquad \text{[(L2.51)]}$$

$$\overline{\Delta}_{\mathrm{Bm}}^{\mathrm{eff}}(\mathrm{b}_1) = \overline{\Delta}_{\mathrm{B_1m}}\frac{1-e^{-2\rho_{\mathrm{B_1}}\mathrm{b}_1}}{1-\overline{\Delta}_{\mathrm{B_1m}}^2 e^{-2\rho_{\mathrm{B_1}}\mathrm{b}_1}} = \overline{\Delta}_{\mathrm{B_1m}}\frac{1-e^{-x_{\mathrm{B_1}}(\mathrm{b}_1/l)}}{1-\overline{\Delta}_{\mathrm{B_1m}}^2 e^{-x_{\mathrm{B_1}}(\mathrm{b}_1/l)}} = \overline{\Delta}_{\mathrm{B_1m}}\frac{1-e^{-s_{\mathrm{B_1}}r_n(\mathrm{b}_1/l)}}{1-\overline{\Delta}_{\mathrm{B_1m}}^2 e^{-s_{\mathrm{B_1}}r_n(\mathrm{b}_1/l)}}.$$

$$\Delta_{\mathrm{Bm}}^{\mathrm{eff}}(\mathrm{b}_1) = \Delta_{\mathrm{B_1m}}\frac{1-e^{-2\rho_{\mathrm{B_1}}\mathrm{b}_1}}{1-\Delta_{\mathrm{B_1m}}^2 e^{-2\rho_{\mathrm{B_1}}\mathrm{b}_1}} = \Delta_{\mathrm{B_1m}}\frac{1-e^{-x_{\mathrm{B_1}}(\mathrm{b}_1/l)}}{1-\Delta_{\mathrm{B_1m}}^2 e^{-x_{\mathrm{B_1}}(\mathrm{b}_1/l)}} = \Delta_{\mathrm{B_1m}}\frac{1-e^{-s_{\mathrm{B_1}}r_n(\mathrm{b}_1/l)}}{1-\Delta_{\mathrm{B_1m}}^2 e^{-s_{\mathrm{B_1}}r_n(\mathrm{b}_1/l)}}.$$

[(L2.49), (L2.50)]

$$\rho_{\mathrm{i}}^2 = \rho^2 + \varepsilon_{\mathrm{i}}\mu_{\mathrm{i}}\xi_n^2/c^2,\ x_{\mathrm{i}} \equiv 2\rho_{\mathrm{i}}l,\ x_{\mathrm{i}}^2 = x^2 + [(2l\xi_n/c)^2](\varepsilon_{\mathrm{i}}\mu_{\mathrm{i}} - \varepsilon_{\mathrm{m}}\mu_{\mathrm{m}}),\ p = x/r_n,$$

$$r_n \equiv (2l\varepsilon_{\mathrm{m}}^{1/2}\mu_{\mathrm{m}}^{1/2}/c)\,\xi_n,\ s_{\mathrm{i}} = \sqrt{p^2 - 1 + (\varepsilon_{\mathrm{i}}\mu_{\mathrm{i}}/\varepsilon_{\mathrm{m}}\mu_{\mathrm{m}})},$$

$$\overline{\Delta}_{\mathrm{ji}} = \frac{\rho_{\mathrm{i}}\varepsilon_{\mathrm{j}} - \rho_{\mathrm{j}}\varepsilon_{\mathrm{i}}}{\rho_{\mathrm{i}}\varepsilon_{\mathrm{j}} + \rho_{\mathrm{j}}\varepsilon_{\mathrm{i}}} = \frac{x_{\mathrm{i}}\varepsilon_{\mathrm{j}} - x_{\mathrm{j}}\varepsilon_{\mathrm{i}}}{x_{\mathrm{i}}\varepsilon_{\mathrm{j}} + x_{\mathrm{j}}\varepsilon_{\mathrm{i}}} = \frac{s_{\mathrm{i}}\varepsilon_{\mathrm{j}} - s_{\mathrm{j}}\varepsilon_{\mathrm{i}}}{s_{\mathrm{i}}\varepsilon_{\mathrm{j}} + s_{\mathrm{j}}\varepsilon_{\mathrm{i}}},\ \Delta_{\mathrm{ji}} = \frac{\rho_{\mathrm{i}}\mu_{\mathrm{j}} - \rho_{\mathrm{j}}\mu_{\mathrm{i}}}{\rho_{\mathrm{i}}\mu_{\mathrm{j}} + \rho_{\mathrm{j}}\mu_{\mathrm{i}}} = \frac{x_{\mathrm{i}}\mu_{\mathrm{j}} - x_{\mathrm{j}}\mu_{\mathrm{i}}}{x_{\mathrm{i}}\mu_{\mathrm{j}} + x_{\mathrm{j}}\mu_{\mathrm{i}}} = \frac{s_{\mathrm{i}}\mu_{\mathrm{j}} - s_{\mathrm{j}}\mu_{\mathrm{i}}}{s_{\mathrm{i}}\mu_{\mathrm{j}} + s_{\mathrm{j}}\mu_{\mathrm{i}}}.$$

P.2.c.2. Small differences in ε's and μ's

$\overline{\Delta}_{\mathrm{Am}}, \Delta_{\mathrm{Am}}, \overline{\Delta}_{\mathrm{B_1m}}, \Delta_{\mathrm{B_1m}} \ll 1,$

$$G_{\mathrm{AmB_1m}}(l;\mathrm{b}_1) = -\frac{kT}{2\pi}\sum_{n=0}^{\infty}{}' \int_{\frac{\varepsilon_{\mathrm{m}}^{1/2}\mu_{\mathrm{m}}^{1/2}\xi_n}{c}}^{\infty} \rho_{\mathrm{m}}\left(\overline{\Delta}_{\mathrm{Am}}\overline{\Delta}_{\mathrm{B_1m}} + \Delta_{\mathrm{Am}}\Delta_{\mathrm{B_1m}}\right)(1-e^{-2\rho_{\mathrm{B_1}}\mathrm{b}_1})e^{-2\rho_{\mathrm{m}}l}\mathrm{d}\rho_{\mathrm{m}}$$

$$= -\frac{kT}{2\pi}\sum_{n=0}^{\infty}{}' \int_{\frac{\varepsilon_{\mathrm{m}}^{1/2}\mu_{\mathrm{m}}^{1/2}\xi_n}{c}}^{\infty} \rho_{\mathrm{m}}\left(\overline{\Delta}_{\mathrm{Am}}\overline{\Delta}_{\mathrm{B_1m}} + \Delta_{\mathrm{Am}}\Delta_{\mathrm{B_1m}}\right)e^{-2\rho_{\mathrm{m}}l}\mathrm{d}\rho_{\mathrm{m}}$$

$$+\frac{kT}{2\pi}\sum_{n=0}^{\infty}{}' \int_{\frac{\varepsilon_{\mathrm{m}}^{1/2}\mu_{\mathrm{m}}^{1/2}\xi_n}{c}}^{\infty} \rho_{\mathrm{m}}\left(\overline{\Delta}_{\mathrm{Am}}\overline{\Delta}_{\mathrm{B_1m}} + \Delta_{\mathrm{Am}}\Delta_{\mathrm{B_1m}}\right)e^{-2\rho_{\mathrm{B_1}}\mathrm{b}_1}e^{-2\rho_{\mathrm{m}}l}\mathrm{d}\rho_{\mathrm{m}}. \qquad \text{[(L2.54)]}$$

P.2.c.3. Small differences in ε's and μ's, nonretarded limit

$c \to \infty,\ \rho_{\mathrm{B_1}} \to \rho_{\mathrm{m}} \to \rho,$

$$G_{\mathrm{AmB_1m}}(l;\mathrm{b}_1) \to -\frac{kT}{8\pi}\left[\frac{1}{l^2} - \frac{1}{(l+\mathrm{b}_1)^2}\right]\sum_{n=0}^{\infty}{}'\left(\overline{\Delta}_{\mathrm{Am}}\overline{\Delta}_{\mathrm{B_1m}} + \Delta_{\mathrm{Am}}\Delta_{\mathrm{B_1m}}\right). \qquad \text{[(L2.55)]}$$

Table P.3.a. Two surfaces, each singly layered

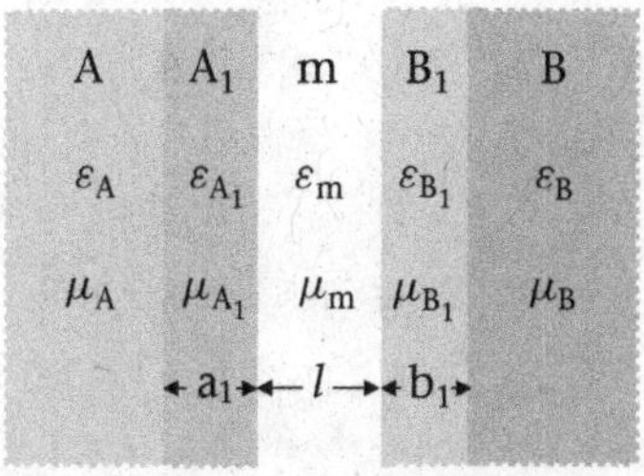

P.3.a.1. Exact, Lifshitz

$$G_{\mathrm{AA_1mB_1B}}(l;\mathrm{a_1},\mathrm{b_1}) = \frac{kT}{2\pi}\sum_{n=0}^{\infty}{}' \int_{\frac{\varepsilon_{\mathrm{m}}^{1/2}\mu_{\mathrm{m}}^{1/2}\xi_n}{c}}^{\infty} \rho_{\mathrm{m}} \ln\left[\left(1-\overline{\Delta}_{\mathrm{Am}}^{\mathrm{eff}}\overline{\Delta}_{\mathrm{Bm}}^{\mathrm{eff}}e^{-2\rho_{\mathrm{m}}l}\right)\left(1-\Delta_{\mathrm{Am}}^{\mathrm{eff}}\Delta_{\mathrm{Bm}}^{\mathrm{eff}}e^{-2\rho_{\mathrm{m}}l}\right)\right]\mathrm{d}\rho_{\mathrm{m}}$$

$$= \frac{kT}{8\pi l^2}\sum_{n=0}^{\infty}{}' \int_{r_n}^{\infty} x\ln\left[\left(1-\overline{\Delta}_{\mathrm{Am}}^{\mathrm{eff}}\overline{\Delta}_{\mathrm{Bm}}^{\mathrm{eff}}e^{-x}\right)\left(1-\Delta_{\mathrm{Am}}^{\mathrm{eff}}\Delta_{\mathrm{Bm}}^{\mathrm{eff}}e^{-x}\right)\right]\mathrm{d}x$$

$$= \frac{kT}{2\pi c^2}\sum_{n=0}^{\infty}{}' \varepsilon_{\mathrm{m}}\mu_{\mathrm{m}}\xi_n^2 \int_{1}^{\infty} p\ln\left[\left(1-\overline{\Delta}_{\mathrm{Am}}^{\mathrm{eff}}\overline{\Delta}_{\mathrm{Bm}}^{\mathrm{eff}}e^{-r_np}\right)\left(1-\Delta_{\mathrm{Am}}^{\mathrm{eff}}\Delta_{\mathrm{Bm}}^{\mathrm{eff}}e^{-r_np}\right)\right]\mathrm{d}p,$$

[(L2.56)–(L2.58)]

$$\overline{\Delta}_{\mathrm{Am}}^{\mathrm{eff}}(\mathrm{a_1}) = \frac{\left(\overline{\Delta}_{\mathrm{AA_1}}e^{-2\rho_{\mathrm{A_1}}\mathrm{a_1}}+\overline{\Delta}_{\mathrm{A_1m}}\right)}{1+\overline{\Delta}_{\mathrm{AA_1}}\overline{\Delta}_{\mathrm{A_1m}}e^{-2\rho_{\mathrm{A_1}}\mathrm{a_1}}} = \frac{\left[\overline{\Delta}_{\mathrm{AA_1}}e^{-x_{\mathrm{A_1}}(\mathrm{a_1}/l)}+\overline{\Delta}_{\mathrm{A_1m}}\right]}{1+\overline{\Delta}_{\mathrm{AA_1}}\overline{\Delta}_{\mathrm{A_1m}}e^{-x_{\mathrm{A_1}}(\mathrm{a_1}/l)}} = \frac{\left[\overline{\Delta}_{\mathrm{AA_1}}e^{-s_{\mathrm{A_1}}r_n(\mathrm{a_1}/l)}+\overline{\Delta}_{\mathrm{A_1m}}\right]}{1+\overline{\Delta}_{\mathrm{AA_1}}\overline{\Delta}_{\mathrm{A_1m}}e^{-s_{\mathrm{A_1}}r_n(\mathrm{a_1}/l)}}$$

[(L2.63) and (L2.64)]

and similarly for $\Delta_{\mathrm{Am}}^{\mathrm{eff}}(\mathrm{a_1})$, $\overline{\Delta}_{\mathrm{Bm}}^{\mathrm{eff}}(\mathrm{b_1})$, $\Delta_{\mathrm{Bm}}^{\mathrm{eff}}(\mathrm{b_1})$.

$\rho_{\mathrm{i}}^2 = \rho^2 + \varepsilon_{\mathrm{i}}\mu_{\mathrm{i}}\xi_n^2/c^2$, $x_{\mathrm{i}} \equiv 2\rho_{\mathrm{i}}l$, $x_{\mathrm{i}}^2 = x^2 + \left[(2l\xi_n/c)^2\right](\varepsilon_{\mathrm{i}}\mu_{\mathrm{i}} - \varepsilon_{\mathrm{m}}\mu_{\mathrm{m}})$, $p = x/r_n$,

$r_n \equiv (2l\varepsilon_{\mathrm{m}}^{1/2}\mu_{\mathrm{m}}^{1/2}/c)\,\xi_n$, $s_{\mathrm{i}} = \sqrt{p^2 - 1 + (\varepsilon_{\mathrm{i}}\mu_{\mathrm{i}}/\varepsilon_{\mathrm{m}}\mu_{\mathrm{m}})}$, [(L2.59)–(L2.62)]

$$\overline{\Delta}_{\mathrm{ji}} = \frac{\rho_{\mathrm{i}}\varepsilon_{\mathrm{j}}-\rho_{\mathrm{j}}\varepsilon_{\mathrm{i}}}{\rho_{\mathrm{i}}\varepsilon_{\mathrm{j}}+\rho_{\mathrm{j}}\varepsilon_{\mathrm{i}}} = \frac{x_{\mathrm{i}}\varepsilon_{\mathrm{j}}-x_{\mathrm{j}}\varepsilon_{\mathrm{i}}}{x_{\mathrm{i}}\varepsilon_{\mathrm{j}}+x_{\mathrm{j}}\varepsilon_{\mathrm{i}}} = \frac{s_{\mathrm{i}}\varepsilon_{\mathrm{j}}-s_{\mathrm{j}}\varepsilon_{\mathrm{i}}}{s_{\mathrm{i}}\varepsilon_{\mathrm{j}}+s_{\mathrm{j}}\varepsilon_{\mathrm{i}}},\quad \Delta_{\mathrm{ji}} = \frac{\rho_{\mathrm{i}}\mu_{\mathrm{j}}-\rho_{\mathrm{j}}\mu_{\mathrm{i}}}{\rho_{\mathrm{i}}\mu_{\mathrm{j}}+\rho_{\mathrm{j}}\mu_{\mathrm{i}}} = \frac{x_{\mathrm{i}}\mu_{\mathrm{j}}-x_{\mathrm{j}}\mu_{\mathrm{i}}}{x_{\mathrm{i}}\mu_{\mathrm{j}}+x_{\mathrm{j}}\mu_{\mathrm{i}}} = \frac{s_{\mathrm{i}}\mu_{\mathrm{j}}-s_{\mathrm{j}}\mu_{\mathrm{i}}}{s_{\mathrm{i}}\mu_{\mathrm{j}}+s_{\mathrm{j}}\mu_{\mathrm{i}}},$$

i, j for A, $\mathrm{A_1}$, m, $\mathrm{B_1}$, or B.

Table P.3.b. Two surfaces, each singly layered: Limiting forms

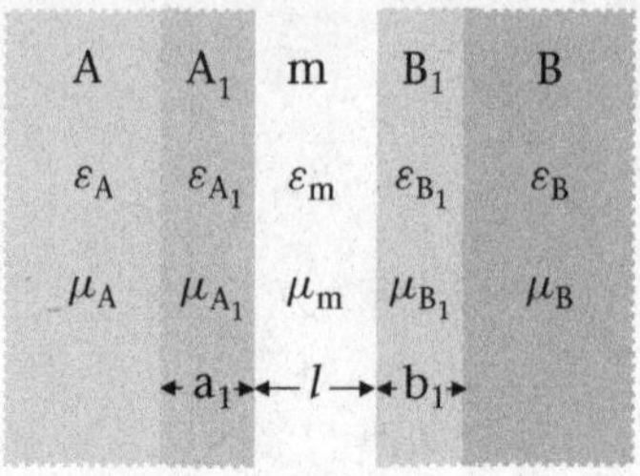

P.3.b.1. High dielectric-permittivity layer

$\varepsilon_{A_1}/\varepsilon_m \to \infty$, $\varepsilon_{B_1}/\varepsilon_m \to \infty$, neglecting magnetic terms, $\overline{\Delta}_{A_1m}, \overline{\Delta}_{B_1m} \to 1$,

$$\overline{\Delta}_{Am}^{eff}(a_1) \to \frac{\left(\overline{\Delta}_{AA_1} e^{-2\rho_{A_1} a_1} + 1\right)}{1 + \overline{\Delta}_{AA_1} e^{-2\rho_{A_1} a_1}} = 1,\ \overline{\Delta}_{Bm}^{eff}(b_1) \to \frac{\left(\overline{\Delta}_{BB_1} e^{-2\rho_{B_1} b_1} + 1\right)}{1 + \overline{\Delta}_{BB_1} e^{-2\rho_{B_1} b_1}} = 1, \qquad [(L2.67)]$$

$$G_{AA_1mB_1B}(l; a_1, b_1) \to G_{A_1mB_1}(l) = \frac{kT}{2\pi} \sum_{n=0}^{\infty}{}' \int_{\frac{\varepsilon_m^{1/2}\mu_m^{1/2}\xi_n}{c}}^{\infty} \rho_m \ln(1 - e^{-2\rho_m l}) d\rho_m \qquad [(L2.68)]$$

$$= \frac{kT}{8\pi l^2} \sum_{n=0}^{\infty}{}' \int_{r_n}^{\infty} x \ln(1 - e^{-x}) dx$$

$$= \frac{kT}{2\pi c^2} \sum_{n=0}^{\infty}{}' \varepsilon_m \mu_m \xi_n^2 \int_1^{\infty} p \ln(1 - e^{-r_n p}) dp.$$

P.3.b.2. Small differences in ε's and μ's, with retardation

$\varepsilon_A \approx \varepsilon_{A_1} \approx \varepsilon_m \approx \varepsilon_{B_1} \approx \varepsilon_B$, $\overline{\Delta}_{ji}$'s, Δ_{ij}'s $\ll 1$,

$$G_{A_1AmB_1B}(l; a_1, b_1) = -\frac{kT}{2\pi} \sum_{n=0}^{\infty}{}' \int_{\frac{\varepsilon_m^{1/2}\mu_m^{1/2}\xi_n}{c}}^{\infty} \rho_m I_{A_1AmB_1B} e^{-2\rho_m l} d\rho_m \qquad [(L2.71)]$$

$$= -\frac{A_{A_1m/B_1m}(l)}{12\pi l^2} - \frac{A_{A_1m/BB_1}(l + b_1)}{12\pi (l + b_1)^2} - \frac{A_{AA_1/B_1m}(l + a_1)}{12\pi (l + a_1)^2} - \frac{A_{AA_1/BB_1}(l + a_1 + b_1)}{12\pi (l + a_1 + b_1)^2},$$

$$I_{A_1AmB_1B} = \left(\overline{\Delta}_{A_1m}\overline{\Delta}_{B_1m} + \Delta_{A_1m}\Delta_{B_1m}\right) + \left(\overline{\Delta}_{B_1m}\overline{\Delta}_{AA_1} + \Delta_{B_1m}\Delta_{AA_1}\right) e^{-2\rho_{A_1} a_1}$$
$$+ \left(\overline{\Delta}_{A_1m}\overline{\Delta}_{BB_1} + \Delta_{A_1m}\Delta_{BB_1}\right) e^{-2\rho_{B_1} b_1} + \left(\overline{\Delta}_{BB_1}\overline{\Delta}_{AA_1} + \Delta_{BB_1}\Delta_{AA_1}\right) e^{-2\rho_{A_1} a_1} e^{-2\rho_{B_1} b_1}.$$

P.3.b.3. Small differences in ε's and μ's, without retardation

$G_{AA_1mB_1B}(l; a_1, b_1) \to$

$$-\frac{kT}{8\pi l^2} \sum_{n=0}^{\infty}{}' \left(\overline{\Delta}_{A_1m}\overline{\Delta}_{B_1m} + \Delta_{A_1m}\Delta_{B_1m}\right) - \frac{kT}{8\pi (l + a_1)^2} \sum_{n=0}^{\infty}{}' \left(\overline{\Delta}_{B_1m}\overline{\Delta}_{AA_1} + \Delta_{B_1m}\Delta_{AA_1}\right)$$
$$- \frac{kT}{8\pi (l + b_1)^2} \sum_{n=0}^{\infty}{}' \left(\overline{\Delta}_{A_1m}\overline{\Delta}_{BB_1} + \Delta_{A_1m}\Delta_{BB_1}\right) - \frac{kT}{8\pi (l + a_1 + b_1)^2} \sum_{n=0}^{\infty}{}' \left(\overline{\Delta}_{BB_1}\overline{\Delta}_{AA_1} + \Delta_{BB_1}\Delta_{AA_1}\right).$$

[(L2.72)]

Table P.3.c. Two finite slabs in medium m

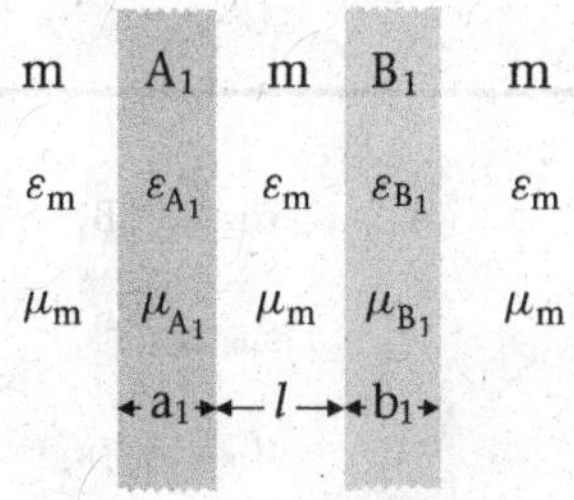

P.3.c.1. Exact, Lifshitz

$$G_{\mathrm{A_1 m B_1}}(l;\mathrm{a_1},\mathrm{b_1}) = \frac{kT}{2\pi}\sum_{n=0}^{\infty}{}' \int_{\frac{\varepsilon_{\mathrm{m}}^{1/2}\mu_{\mathrm{m}}^{1/2}\xi_n}{c}}^{\infty} \rho_{\mathrm{m}} \ln\left[\left(1-\overline{\Delta}_{\mathrm{Am}}^{\mathrm{eff}}\overline{\Delta}_{\mathrm{Bm}}^{\mathrm{eff}}e^{-2\rho_{\mathrm{m}}l}\right)\left(1-\Delta_{\mathrm{Am}}^{\mathrm{eff}}\Delta_{\mathrm{Bm}}^{\mathrm{eff}}e^{-2\rho_{\mathrm{m}}l}\right)\right]\mathrm{d}\rho_{\mathrm{m}}$$

$$= \frac{kT}{8\pi l^2}\sum_{n=0}^{\infty}{}' \int_{r_n}^{\infty} x\ln\left[\left(1-\overline{\Delta}_{\mathrm{Am}}^{\mathrm{eff}}\overline{\Delta}_{\mathrm{Bm}}^{\mathrm{eff}}e^{-x}\right)\left(1-\Delta_{\mathrm{Am}}^{\mathrm{eff}}\Delta_{\mathrm{Bm}}^{\mathrm{eff}}e^{-x}\right)\right]\mathrm{d}x$$

$$= \frac{kT}{2\pi c^2}\sum_{n=0}^{\infty}{}' \varepsilon_{\mathrm{m}}\mu_{\mathrm{m}}\xi_n^2 \int_{1}^{\infty} p\ln\left[\left(1-\overline{\Delta}_{\mathrm{Am}}^{\mathrm{eff}}\overline{\Delta}_{\mathrm{Bm}}^{\mathrm{eff}}e^{-r_n p}\right)\left(1-\Delta_{\mathrm{Am}}^{\mathrm{eff}}\Delta_{\mathrm{Bm}}^{\mathrm{eff}}e^{-r_n p}\right)\right]\mathrm{d}p.$$

[(L2.77)]

$$\overline{\Delta}_{\mathrm{Am}}^{\mathrm{eff}}(\mathrm{a_1}) = \overline{\Delta}_{\mathrm{A_1m}}\frac{1-e^{-2\rho_{\mathrm{A_1}}\mathrm{a_1}}}{1-\overline{\Delta}_{\mathrm{A_1m}}^2 e^{-2\rho_{\mathrm{A_1}}\mathrm{a_1}}} = \overline{\Delta}_{\mathrm{A_1m}}\frac{1-e^{-x_{\mathrm{A_1}}(\mathrm{a_1}/l)}}{1-\overline{\Delta}_{\mathrm{A_1m}}^2 e^{-x_{\mathrm{A_1}}(\mathrm{a_1}/l)}} = \overline{\Delta}_{\mathrm{A_1m}}\frac{1-e^{-s_{\mathrm{A_1}}r_n(\mathrm{a_1}/l)}}{1-\overline{\Delta}_{\mathrm{A_1m}}^2 e^{-s_{\mathrm{A_1}}r_n(\mathrm{a_1}/l)}},$$

$$\Delta_{\mathrm{Am}}^{\mathrm{eff}}(\mathrm{a_1}) = \Delta_{\mathrm{A_1m}}\frac{1-e^{-2\rho_{\mathrm{A_1}}\mathrm{a_1}}}{1-\Delta_{\mathrm{A_1m}}^2 e^{-2\rho_{\mathrm{A_1}}\mathrm{a_1}}} = \Delta_{\mathrm{A_1m}}\frac{1-e^{-x_{\mathrm{A_1}}(\mathrm{a_1}/l)}}{1-\Delta_{\mathrm{A_1m}}^2 e^{-x_{\mathrm{A_1}}(\mathrm{a_1}/l)}} = \Delta_{\mathrm{A_1m}}\frac{1-e^{-s_{\mathrm{A_1}}r_n(\mathrm{a_1}/l)}}{1-\Delta_{\mathrm{A_1m}}^2 e^{-s_{\mathrm{A_1}}r_n(\mathrm{a_1}/l)}},$$

$$\overline{\Delta}_{\mathrm{Bm}}^{\mathrm{eff}}(\mathrm{b_1}) = \overline{\Delta}_{\mathrm{B_1m}}\frac{1-e^{-2\rho_{\mathrm{B_1}}\mathrm{b_1}}}{1-\overline{\Delta}_{\mathrm{B_1m}}^2 e^{-2\rho_{\mathrm{B_1}}\mathrm{b_1}}} = \overline{\Delta}_{\mathrm{B_1m}}\frac{1-e^{-x_{\mathrm{B_1}}(\mathrm{b_1}/l)}}{1-\overline{\Delta}_{\mathrm{B_1m}}^2 e^{-x_{\mathrm{B_1}}(\mathrm{b_1}/l)}} = \overline{\Delta}_{\mathrm{B_1m}}\frac{1-e^{-s_{\mathrm{B_1}}r_n(\mathrm{b_1}/l)}}{1-\overline{\Delta}_{\mathrm{B_1m}}^2 e^{-s_{\mathrm{B_1}}r_n(\mathrm{b_1}/l)}},$$

$$\Delta_{\mathrm{Bm}}^{\mathrm{eff}}(\mathrm{b_1}) = \Delta_{\mathrm{B_1m}}\frac{1-e^{-2\rho_{\mathrm{B_1}}\mathrm{b_1}}}{1-\Delta_{\mathrm{B_1m}}^2 e^{-2\rho_{\mathrm{B_1}}\mathrm{b_1}}} = \Delta_{\mathrm{B_1m}}\frac{1-e^{-x_{\mathrm{B_1}}(\mathrm{b_1}/l)}}{1-\Delta_{\mathrm{B_1m}}^2 e^{-x_{\mathrm{B_1}}(\mathrm{b_1}/l)}} = \Delta_{\mathrm{B_1m}}\frac{1-e^{-s_{\mathrm{B_1}}r_n(\mathrm{b_1}/l)}}{1-\Delta_{\mathrm{B_1m}}^2 e^{-s_{\mathrm{B_1}}r_n(\mathrm{b_1}/l)}}.$$

[(L2.73)–(L2.76)]

P.3.c.2. Small differences in ε's and μ's

$$G_{\mathrm{A_1 m B_1}}(l;\mathrm{a_1},\mathrm{b_1}) = -\frac{kT}{2\pi}\sum_{n=0}^{\infty}{}' \int_{\frac{\varepsilon_{\mathrm{m}}^{1/2}\mu_{\mathrm{m}}^{1/2}\xi_n}{c}}^{\infty} \rho_{\mathrm{m}}\left(\overline{\Delta}_{\mathrm{A_1m}}\overline{\Delta}_{\mathrm{B_1m}}+\Delta_{\mathrm{A_1m}}\Delta_{\mathrm{B_1m}}\right)$$
$$\times\left(1-e^{-2\rho_{\mathrm{A_1}}\mathrm{a_1}}-e^{-2\rho_{\mathrm{B_1}}\mathrm{b_1}}+e^{-2\rho_{\mathrm{A_1}}\mathrm{a_1}}e^{-2\rho_{\mathrm{B_1}}\mathrm{b_1}}\right)e^{-2\rho_{\mathrm{m}}l}\mathrm{d}\rho_{\mathrm{m}}. \quad \text{[(L2.79)]}$$

P.3.c.3. Small differences in ε's and μ's, nonretarded limit

$$G_{\mathrm{A_1 m B_1}}(l;\mathrm{a_1},\mathrm{b_1}) \to -\frac{kT}{8\pi}\left[\frac{1}{l^2}-\frac{1}{(l+\mathrm{b_1})^2}-\frac{1}{(l+\mathrm{a_1})^2}+\frac{1}{(l+\mathrm{a_1}+\mathrm{b_1})^2}\right]\sum_{n=0}^{\infty}{}'\left(\overline{\Delta}_{A_1\mathrm{m}}\overline{\Delta}_{\mathrm{B_1m}}+\Delta_{\mathrm{A_1m}}\Delta_{\mathrm{B_1m}}\right)$$

$$= -\frac{A_{\mathrm{A_1m/B_1m}}}{12\pi}s\left[\frac{1}{l^2}-\frac{1}{(l+\mathrm{b_1})^2}-\frac{1}{(l+\mathrm{a_1})^2}+\frac{1}{(l+\mathrm{a_1}+\mathrm{b_1})^2}\right]. \quad \text{[(L2.80)]}$$

$\mathrm{a_1} = \mathrm{b_1} = \mathrm{a}$:

$$G(l;\mathrm{a},T) = -\frac{kT}{8\pi}\left[\frac{1}{l^2}-\frac{2}{(l+\mathrm{a})^2}+\frac{1}{(l+2\mathrm{a})^2}\right]\sum_{n=0}^{\infty}{}'\left(\overline{\Delta}_{\mathrm{A_1m}}\overline{\Delta}_{\mathrm{B_1m}}+\Delta_{\mathrm{A_1m}}\Delta_{\mathrm{B_1m}}\right).$$

Table P.4.a. Half-spaces, each coated with an arbitrary number of layers

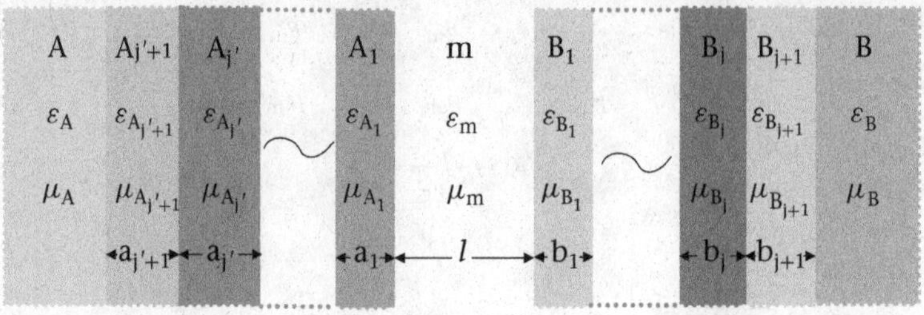

$G_{\mathrm{AmB}}(l;\mathrm{j}'+1\text{ layers on A, }\mathrm{j}+1\text{ layers on B})$

$$= \frac{kT}{2\pi}\sum_{n=0}^{\infty}{}' \int_{\frac{\varepsilon_{\mathrm{m}}^{1/2}\mu_{\mathrm{m}}^{1/2}\xi_n}{c}}^{\infty} \rho_{\mathrm{m}} \ln\left[\left(1-\overline{\Delta}_{\mathrm{Am}}^{\mathrm{eff}}(\mathrm{j}'+1\text{ layers})\overline{\Delta}_{\mathrm{Bm}}^{\mathrm{eff}}(\mathrm{j}+1\text{ layers})e^{-2\rho_{\mathrm{m}}l}\right)\right.$$

$$\left.\times\left(1-\Delta_{\mathrm{Am}}^{\mathrm{eff}}(\mathrm{j}'+1\text{ layers})\Delta_{\mathrm{Bm}}^{\mathrm{eff}}(\mathrm{j}+1\text{ layers})e^{-2\rho_{\mathrm{m}}l}\right)\right]\mathrm{d}\rho_{\mathrm{m}}$$

$$= \frac{kT}{8\pi l^2}\sum_{n=0}^{\infty}{}' \int_{r_n}^{\infty} x\ln\left[\left(1-\overline{\Delta}_{\mathrm{Am}}^{\mathrm{eff}}(\mathrm{j}'+1\text{ layers})\overline{\Delta}_{\mathrm{Bm}}^{\mathrm{eff}}(\mathrm{j}+1\text{ layers})e^{-x}\right)\right.$$

$$\left.\times\left(1-\Delta_{\mathrm{Am}}^{\mathrm{eff}}\Delta_{\mathrm{Bm}}^{\mathrm{eff}}(\mathrm{j}+1\text{ layers})e^{-x}\right)\right]\mathrm{d}x$$

$$= \frac{kT}{2\pi c^2}\sum_{n=0}^{\infty}{}' \varepsilon_{\mathrm{m}}\mu_{\mathrm{m}}\xi_n^2 \int_{1}^{\infty} p\ln\left[\left(1-\overline{\Delta}_{\mathrm{Am}}^{\mathrm{eff}}(\mathrm{j}'+1\text{ layers})\overline{\Delta}_{\mathrm{Bm}}^{\mathrm{eff}}(\mathrm{j}+1\text{ layers})e^{-r_n p}\right)\right.$$

$$\left.\times\left(1-\Delta_{\mathrm{Am}}^{\mathrm{eff}}(\mathrm{j}'+1\text{ layers})\Delta_{\mathrm{Bm}}^{\mathrm{eff}}(\mathrm{j}+1\text{ layers})e^{-r_n p}\right)\right]\mathrm{d}p.$$

$\overline{\Delta}_{\mathrm{Am}}^{\mathrm{eff}}(\mathrm{j}'+1\text{ layers})$, $\overline{\Delta}_{\mathrm{Bm}}^{\mathrm{eff}}(\mathrm{j}+1\text{ layers})$, $\Delta_{\mathrm{Am}}^{\mathrm{eff}}(\mathrm{j}'+1\text{ layers})$, $\Delta_{\mathrm{Bm}}^{\mathrm{eff}}(\mathrm{j}+1\text{ layers})$ by iteration [see expressions (L3.90)].

Table P.4.b. Addition of a layer, iteration procedure

From: $\overline{\Delta}_{Am}^{eff}(j')$ for a semi-infinite medium A coated with j' layers

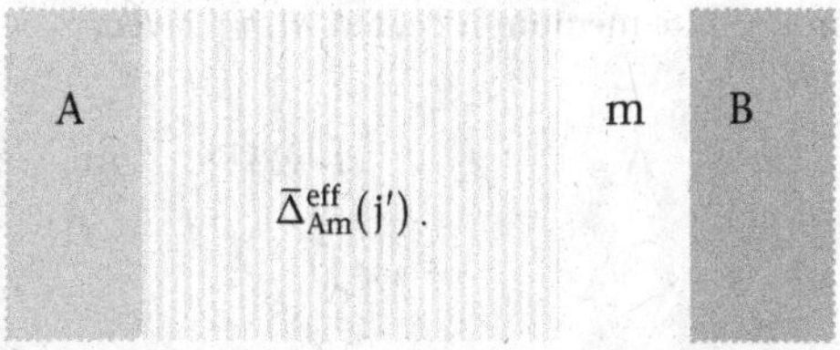

To: $\overline{\Delta}_{Am}^{eff}(j'+1)$ for A coated with $j'+1$ layers,

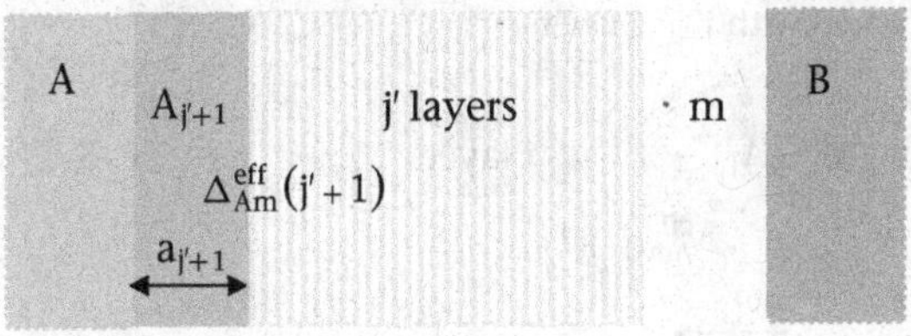

$$\overline{\Delta}_{Am}^{eff}(j'+1\ \text{layers}) = \frac{\overline{\Delta}_{AA_1}^{eff}(j'\ \text{layers})\, e^{-2\rho_{A_1} a_1} + \overline{\Delta}_{A_1m}}{\left[1+\overline{\Delta}_{AA_1}^{eff}(j'\ \text{layers})\,\overline{\Delta}_{A_1m} e^{-2\rho_{A_1} a_1}\right]}, \qquad \text{[(L2.85) and (L3.90)]}$$

with the equivalent construction for layers on body B and for magnetic terms.

For $\overline{\Delta}_{Am}^{eff}(j'+1\ \text{layers})$, replace

$$\overline{\Delta}_{AA_{j'}} = \frac{x_{A_{j'}}\varepsilon_A - x_A\varepsilon_{A_{j'}}}{x_{A_{j'}}\varepsilon_A + x_A\varepsilon_{A_{j'}}} = \frac{s_{A_{j'}}\varepsilon_A - s_A\varepsilon_{A_{j'}}}{s_{A_{j'}}\varepsilon_A + s_A\varepsilon_{A_{j'}}} \text{ in } \overline{\Delta}_{Am}^{eff}(j'\ \text{layers})$$

with

$$\frac{\overline{\Delta}_{AA_{j'+1}} e^{-x_{A_{j'+1}}[(a_{j'+1})/l]} + \overline{\Delta}_{A_{j'+1}A_{j'}}}{1+\overline{\Delta}_{AA_{j'+1}}\overline{\Delta}_{A_{j'+1}A_{j'}} e^{-x_{A_{j'+1}}[(a_{j'+1})/l]}} = \frac{\overline{\Delta}_{AA_{j'+1}} e^{-s_{A_{j'+1}} r_n[(a_{j'+1})/l]} + \overline{\Delta}_{A_{j'+1}A_{j'}}}{1+\overline{\Delta}_{AA_{j'+1}}\overline{\Delta}_{A_{j'+1}A_{j'}} e^{-s_{A_{j'+1}} r_n[(a_{j'+1})/l]}}$$

By induction, $\overline{\Delta}_{Am}^{eff}(0\ \text{layers}) = \overline{\Delta}_{Am}$ [see Eqs. (L2.81)– (L2.84)]:

$$\overline{\Delta}_{Am}^{eff}(1\ \text{layer}) = \frac{\overline{\Delta}_{AA_1} e^{-x_{A_1}(a_1/l)} + \overline{\Delta}_{A_1m}}{1+\overline{\Delta}_{AA_1}\overline{\Delta}_{A_1m} e^{-x_{A_1}(a_1/l)}} = \frac{\overline{\Delta}_{AA_1}^{eff}(0\ \text{layers})\, e^{-x_{A_1}(a_1/l)} + \overline{\Delta}_{A_1m}}{1+\overline{\Delta}_{AA_1}^{eff}(0\ \text{layers})\,\overline{\Delta}_{A_1m} e^{-x_{A_1}(a_1/l)}},$$

$$\overline{\Delta}_{Am}^{eff}(2\ \text{layers}) = \frac{\left[\dfrac{\overline{\Delta}_{AA_2} e^{-x_{A_2}(a_2/l)} + \overline{\Delta}_{A_2A_1}}{1+\overline{\Delta}_{AA_2}\overline{\Delta}_{A_2A_1} e^{-x_{A_2}(a_2/l)}}\right] e^{-x_{A_1}(a_1/l)} + \overline{\Delta}_{A_1m}}{\left\{1+\left[\dfrac{\overline{\Delta}_{AA_2} e^{-x_{A_2}(a_2/l)} + \overline{\Delta}_{A_2A_1}}{1+\overline{\Delta}_{AA_2}\overline{\Delta}_{A_2A_1} e^{-x_{A_2}(a_2/l)}}\right]\overline{\Delta}_{A_1m} e^{-x_{A_1}(a_1/l)}\right\}}$$

$$= \frac{\overline{\Delta}_{AA_1}^{eff}(1\ \text{layer})\, e^{-x_{A_1}(a_1/l)} + \overline{\Delta}_{A_1m}}{\left[1+\overline{\Delta}_{AA_1}^{eff}(1\ \text{layer})\,\overline{\Delta}_{A_1m} e^{-x_{A_1}(a_1/l)}\right]}.$$

Table P.4.c. Addition of a layer, iteration procedure for small differences in susceptibilities

From: $\overline{\Delta}_{\mathrm{Am}}^{\mathrm{eff}}(\mathrm{j}')$ for a semi-infinite medium A coated with j′ layers

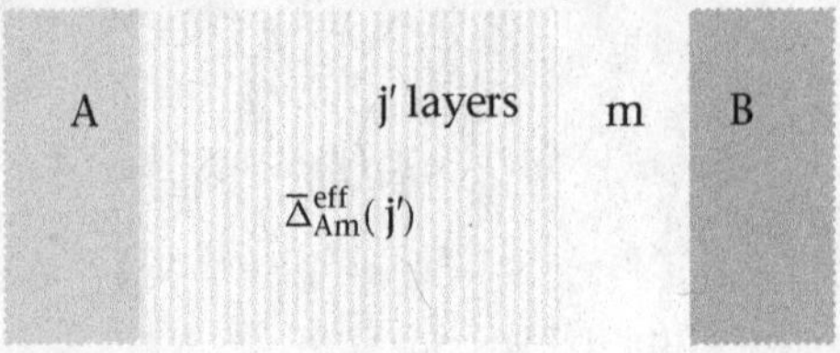

To: $\overline{\Delta}_{\mathrm{Am}}^{\mathrm{eff}}(\mathrm{j}'+1)$ for A coated with j′ + 1 layers

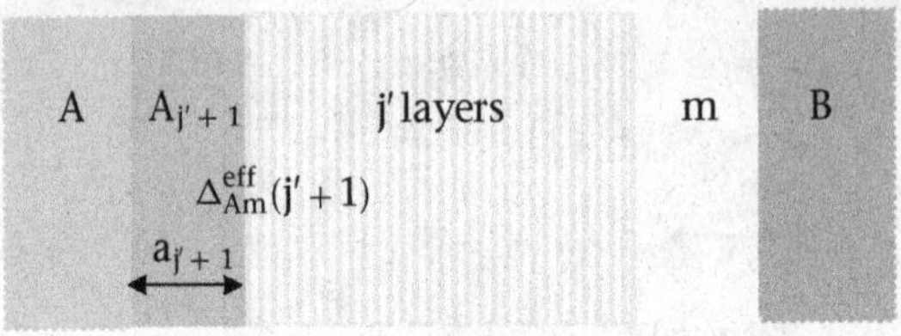

$$\overline{\Delta}_{\mathrm{Am}}^{\mathrm{eff}}(\mathrm{j}'+1\text{ layers}) = \overline{\Delta}_{\mathrm{AA_1}}^{\mathrm{eff}}(\mathrm{j}'\text{ layers})e^{-2\rho_{\mathrm{A_1}}\mathrm{a_1}} + \overline{\Delta}_{\mathrm{A_1m}},$$

with the equivalent construction for layers on body B and for magnetic terms.

For $\overline{\Delta}_{\mathrm{Am}}^{\mathrm{eff}}(\mathrm{j}'+1\text{ layers})$, replace

$$\overline{\Delta}_{\mathrm{AA_{j'}}} = \frac{x_{\mathrm{A_{j'}}}\varepsilon_{\mathrm{A}} - x_{\mathrm{A}}\varepsilon_{\mathrm{A_{j'}}}}{x_{\mathrm{A_{j'}}}\varepsilon_{\mathrm{A}} + x_{\mathrm{A}}\varepsilon_{\mathrm{A_{j'}}}} = \frac{s_{\mathrm{A_{j'}}}\varepsilon_{\mathrm{A}} - s_{\mathrm{A}}\varepsilon_{\mathrm{A_{j'}}}}{s_{\mathrm{A_{j'}}}\varepsilon_{\mathrm{A}} + s_{\mathrm{A}}\varepsilon_{\mathrm{A_{j'}}}}\cdot \text{ in } \overline{\Delta}_{\mathrm{Am}}^{\mathrm{eff}}(\mathrm{j}'\text{ layers}) \text{ with}$$

$$\overline{\Delta}_{\mathrm{AA_{j'+1}}}e^{-x_{\mathrm{A_{j'+1}}}[(a_{\mathrm{j'}}+1)/l]} + \overline{\Delta}_{\mathrm{A_{j'+1}A_{j'}}} = \overline{\Delta}_{\mathrm{AA_{j'+1}}}e^{-s_{\mathrm{A_{j'+1}}}r_n[(a_{\mathrm{j'+1}})/l]} + \overline{\Delta}_{\mathrm{A_{j'+1}A_{j'}}}.$$

By induction $\overline{\Delta}_{\mathrm{Am}}^{\mathrm{eff}}(0\text{ layers}) = \overline{\Delta}_{\mathrm{Am}}$,

$$\overline{\Delta}_{\mathrm{Am}}^{\mathrm{eff}}(1\text{ layer}) = \overline{\Delta}_{\mathrm{AA_1}}e^{-x_{\mathrm{A_1}}(\mathrm{a_1}/l)} + \overline{\Delta}_{\mathrm{A_1m}}, \qquad [(\mathrm{L2.87})]$$

$$\overline{\Delta}_{\mathrm{Am}}^{\mathrm{eff}}(2\text{ layers}) = \left[\overline{\Delta}_{\mathrm{AA_2}}e^{-x_{\mathrm{A_2}}(\mathrm{a_2}/l)} + \overline{\Delta}_{\mathrm{A_2A_1}}\right]e^{-x_{\mathrm{A_1}}(\mathrm{a_1}/l)} + \overline{\Delta}_{\mathrm{A_1m}}$$

$$= \overline{\Delta}_{\mathrm{AA_2}}e^{-x_{\mathrm{A_2}}(\mathrm{a_2}/l)}e^{-x_{\mathrm{A_1}}(\mathrm{a_1}/l)} + \overline{\Delta}_{\mathrm{A_2A_1}}e^{-x_{\mathrm{A_1}}(\mathrm{a_1}/l)} + \overline{\Delta}_{\mathrm{A_1m}} \qquad [(\mathrm{L2.88})]$$

$$\overline{\Delta}_{\mathrm{Am}}^{\mathrm{eff}}(3\text{ layers}) = \overline{\Delta}_{\mathrm{AA_3}}e^{-x_{\mathrm{A_3}}(\mathrm{a_3}/l)}e^{-x_{\mathrm{A_2}}(\mathrm{a_2}/l)}e^{-x_{\mathrm{A_1}}(\mathrm{a_1}/l)}$$

$$+ \overline{\Delta}_{\mathrm{A_3A_2}}e^{-x_{\mathrm{A_2}}(\mathrm{a_2}/l)}e^{-x_{\mathrm{A_1}}(\mathrm{a_1}/l)} + \overline{\Delta}_{\mathrm{A_2A_1}}e^{-x_{\mathrm{A_1}}(\mathrm{a_1}/l)} + \overline{\Delta}_{\mathrm{A_1m}}. \qquad [(\mathrm{L2.89})]$$

Table P.5. Multiply coated semi-infinite bodies A and B, small differences in ε's, and μ's Hamaker form

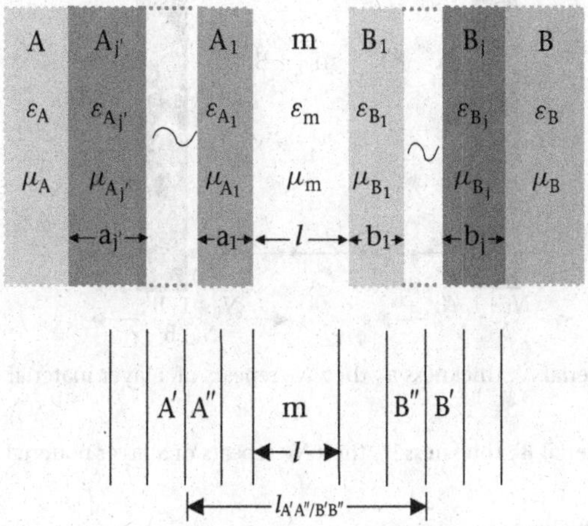

Pairs of interfaces A′A″ and B′B″ at a separation $l_{A'A''/B'B''}$; variable separation l; fixed layer thicknesses $a_1, \ldots, a_i, b_1, \ldots, b_k$.

$$G(l; a_1, a_2, \ldots, a_{j'}, b_1, b_2, \ldots, b_j) = \sum_{\substack{\text{(all pairs of}\\ \text{interfaces } A'A''/B'B'')}} G_{A'A''/B'B''}(l_{A'A''/B'B''}), \qquad [(L2.93)]$$

$$G_{A'A''/B'B''}(l_{A'A''/B'B''}) = -\frac{kT}{8\pi l^2_{A'A''/B'B''}} \sum_{n=0}^{\infty}{}' \left(\overline{\Delta}_{A'A''}\overline{\Delta}_{B'B''} + \Delta_{A'A''}\Delta_{B'B''}\right), \qquad [(L2.94)]$$

$$\overline{\Delta}_{A'A''} = \frac{\varepsilon_{A'} - \varepsilon_{A''}}{\varepsilon_{A'} + \varepsilon_{A''}}, \; \overline{\Delta}_{B'B''} = \frac{\varepsilon_{B'} - \varepsilon_{B''}}{\varepsilon_{B'} + \varepsilon_{B''}}. \qquad [(L2.92)]$$

$\overline{\Delta}_{A'A''}\overline{\Delta}_{B'B''}$, and $\Delta_{A'A''}\Delta_{B'B''}$ are written with the singly primed A′ and b′ materials on the *farther* side of the interface.

In Hamaker form,

$$G_{A'A''/B'B''}(l_{A'A''/B'B''}) = -\frac{A_{A'A''/B'B''}}{12\pi l^2_{A'A''/B'B''}}, \; A_{A'A''/B'B''} = \frac{3kT}{2}\sum_{n=0}^{\infty}{}' \overline{\Delta}_{A'A''}\overline{\Delta}_{B'B''}.$$

Table P.6.a. Multilayer-coated semi-infinite media

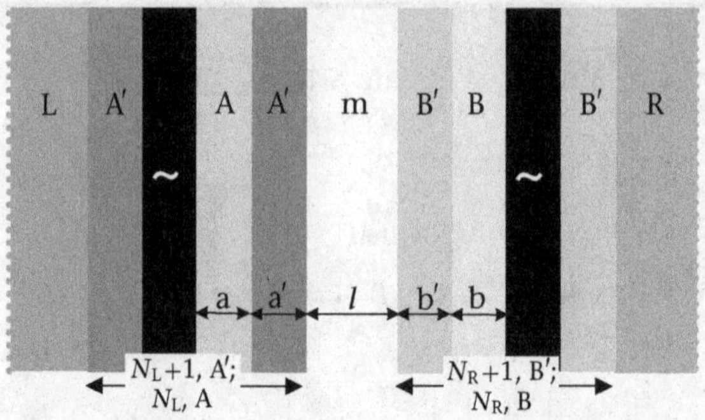

Half-space L coated with layer of material A′, thickness a′, then N_L repeats of a layer material A, thickness a, and material A′, thickness a′.

Half-space R coated with layer of material B′, thickness b′, then N_R repeats of a layer material B, thickness b, and material B′, thickness b′.

Separated by medium m of variable thickness l:

$$G_{L\sim R}(l; \mathrm{a}, \mathrm{a}', \mathrm{b}, \mathrm{b}') = \frac{kT}{2\pi} \sum_{n=0}^{\infty}{}' \int_{\frac{\varepsilon_m^{1/2}\mu_m^{1/2}\xi_n}{c}}^{\infty} \rho_m \ln\left[\left(1 - \overline{\Delta}_{Lm}^{eff} \overline{\Delta}_{Rm}^{eff} e^{-2\rho_m l}\right)\left(1 - \Delta_{Lm}^{eff} \Delta_{Rm}^{eff} e^{-2\rho_m l}\right)\right] d\rho_m. \quad \text{[(L3.50) and (L3.110)]}$$

Define Eq. (L3.106) for the Chebyshev polynomial:

$$U_{N-1}(x) = \frac{\sinh(N\zeta)}{\sinh(\zeta)} = \frac{e^{+N\zeta} - e^{-N\zeta}}{e^{+\zeta} - e^{-\zeta}}, \quad U_{N-2}(x) = \frac{\sinh[(N-1)\zeta]}{\sinh(\zeta)}, \quad x = \frac{m_{11} + m_{22}}{2} = \cosh(\zeta)$$

for $N = N_L$ or N_R, $x = x_L$ or x_R, $\zeta = \zeta_L$ or ζ_R.

Electrical terms $\overline{\Delta}_{Lm}^{eff}$, $\overline{\Delta}_{Rm}^{eff}$ Δ_{Lm}^{eff}, and Δ_{Rm}^{eff} explicitly listed in the following equations; magnetic terms are written in same form.

$$\overline{\Delta}_{Lm}^{eff} = \frac{\left[n_{22}^{(L)}\overline{\Delta}_{LA'} - n_{12}^{(L)}\right] e^{-2\rho_{A'}a'} + \left[n_{11}^{(L)} - n_{21}^{(L)}\overline{\Delta}_{LA'}\right]\overline{\Delta}_{A'm}}{\left[n_{11}^{(L)} - n_{21}^{(L)}\overline{\Delta}_{LA'}\right] + \left[n_{22}^{(L)}\overline{\Delta}_{LA'} - n_{12}^{(L)}\right]\overline{\Delta}_{A'm} e^{-2\rho_{A'}a'}}, \quad \overline{\Delta}_{Rm}^{eff} = \frac{\left[n_{22}^{(R)}\overline{\Delta}_{RB'} - n_{12}^{(R)}\right] e^{-2\rho_{B'}b'} + \left[n_{11}^{(R)} - n_{21}^{(R)}\overline{\Delta}_{RB'}\right]\overline{\Delta}_{B'm}}{\left[n_{11}^{(R)} - n_{21}^{(R)}\overline{\Delta}_{RB'}\right] + \left[n_{22}^{(R)}\overline{\Delta}_{RB'} - n_{12}^{(R)}\right]\overline{\Delta}_{B'm} e^{-2\rho_{B'}b'}}; \quad \text{[(L3.109)]}$$

$$n_{11}^{(L)} = m_{11}^{(L)} U_{N_L-1}(x_L) - U_{N_L-2}(x_L), \quad n_{12}^{(L)} = m_{12}^{(L)} U_{N_L-1}(x_L), \quad n_{11}^{(R)} = m_{11}^{(R)} U_{N_R-1}(x_R) - U_{N_R-2}(x_R), \; n_{12}^{(R)} = m_{12}^{(R)} U_{N_R-1}(x_R); \quad \text{[(L3.111)]}$$

$$n_{21}^{(L)} = m_{21}^{(L)} U_{N_L-1}(x_L), \quad n_{22}^{(L)} = m_{22}^{(L)} U_{N_L-1}(x_L) - U_{N_L-2}(x_L) \quad n_{21}^{(R)} = m_{21}^{(R)} U_{N_R-1}(x_R), \; n_{22}^{(R)} = m_{22}^{(R)} U_{N_R-1}(x_R) - U_{N_R-2}(x_R) \quad \text{[(L3.112)]}$$

$$x_L = \frac{m_{11}^{(L)} + m_{22}^{(L)}}{2}, \quad x_R = \frac{m_{11}^{(R)} + m_{22}^{(R)}}{2}, \quad \text{[(L3.113)]}$$

$$m_{11}^{(L)} = \frac{1 - \overline{\Delta}_{A'A}^2 e^{-2\rho_A a}}{\left(1 - \overline{\Delta}_{A'A}^2\right) e^{-\rho_A a} e^{-\rho_{A'} a'}}, \quad m_{11}^{(R)} = \frac{1 - \overline{\Delta}_{B'B}^2 e^{-2\rho_B b}}{\left(1 - \overline{\Delta}_{B'B}^2\right) e^{-\rho_B b} e^{-\rho_{B'} b'}},$$

$$m_{12}^{(L)} = \frac{\overline{\Delta}_{A'A}\left(1 - e^{-2\rho_A a}\right)}{\left(1 - \overline{\Delta}_{A'A}^2\right) e^{-\rho_A a} e^{-\rho_{A'} a'}}, \quad m_{12}^{(R)} = \frac{\overline{\Delta}_{B'B}\left(1 - e^{-2\rho_B b}\right)}{\left(1 - \overline{\Delta}_{B'B}^2\right) e^{-\rho_B b} e^{-\rho_{B'} b'}}, \quad \text{[(L3.114)]}$$

$$m_{21}^{(L)} = \frac{\left(e^{-2\rho_A a} - 1\right)\overline{\Delta}_{A'A} e^{-2\rho_{A'} a'}}{\left(1 - \overline{\Delta}_{A'A}^2\right) e^{-\rho_A a} e^{-\rho_{A'} a'}}, \quad m_{21}^{(R)} = \frac{\left(e^{-2\rho_B b} - 1\right)\overline{\Delta}_{B'B} e^{-2\rho_{B'} b'}}{\left(1 - \overline{\Delta}_{B'B}^2\right) e^{-\rho_B b} e^{-\rho_{B'} b'}},$$

$$m_{22}^{(L)} = \frac{\left(e^{-2\rho_A a} - \overline{\Delta}_{A'A}^2\right) e^{-2\rho_{A'} a'}}{\left(1 - \overline{\Delta}_{A'A}^2\right) e^{-\rho_A a} e^{-\rho_{A'} a'}}, \quad m_{22}^{(R)} = \frac{\left(e^{-2\rho_B b} - \overline{\Delta}_{B'B}^2\right) e^{-2\rho_{B'} b'}}{\left(1 - \overline{\Delta}_{B'B}^2\right) e^{-\rho_B b} e^{-\rho_{B'} b'}}. \quad \text{[(L3.115)]}$$

Table P.6.b. Limit of a large number of layers

Limit of large N_L:

$$\overline{\Delta}_{\mathrm{Lm}}^{\mathrm{eff}} \to \frac{\mathrm{m}_{21}^{(\mathrm{L})}\overline{\Delta}_{\mathrm{B'm}} - \left[\mathrm{m}_{22}^{(\mathrm{L})} - e^{-\zeta_{\mathrm{L}}}\right] e^{-2\rho_{\mathrm{A'}}\mathrm{a'}}}{\mathrm{m}_{21}^{(\mathrm{L})} - \left[\mathrm{m}_{22}^{(\mathrm{L})} - e^{-\zeta_{\mathrm{L}}}\right]\overline{\Delta}_{\mathrm{A'm}} e^{-2\rho_{\mathrm{A'}}\mathrm{a'}}}.$$

Limit of large N_R:

$$\overline{\Delta}_{\mathrm{Rm}}^{\mathrm{eff}} \to \frac{\mathrm{m}_{21}^{(\mathrm{R})}\overline{\Delta}_{\mathrm{B'm}} - \left[\mathrm{m}_{22}^{(\mathrm{R})} - e^{-\zeta_{\mathrm{R}}}\right] e^{-2\rho_{\mathrm{B'}}\mathrm{b'}}}{\mathrm{m}_{21}^{(\mathrm{R})} - \left[\mathrm{m}_{22}^{(\mathrm{R})} - e^{-\zeta_{\mathrm{R}}}\right]\overline{\Delta}_{\mathrm{B'm}} e^{-2\rho_{\mathrm{B'}}\mathrm{b'}}}. \qquad [(\mathrm{L3.108})]$$

Table P.6.c. Layer of finite thickness adding onto a multilayer stack

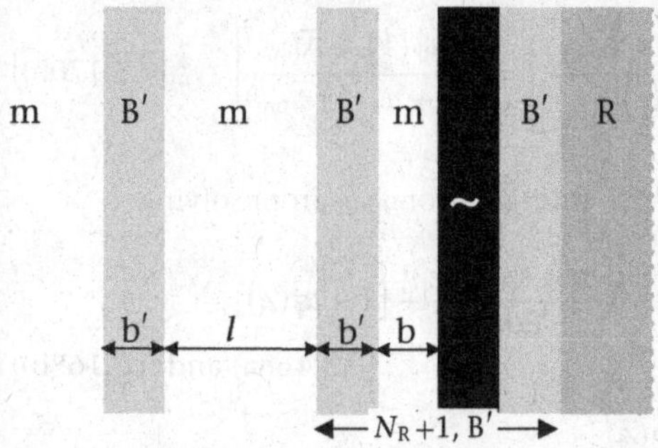

The previous result can be immediately specialized to the case of a layer of finite thickness interacting with a previously existing stack of $N+1$ layers. Let half-space L, as well as all materials B, have the same dielectric properties as medium m. Let material A′ have the same properties as material B′:

$$G_{\mathrm{B'}\sim\mathrm{R}}(l;\mathrm{b},\mathrm{b'}) = \frac{kT}{2\pi}\sum_{n=0}^{\infty}{}' \int_{\frac{\varepsilon_{\mathrm{m}}^{1/2}\mu_{\mathrm{m}}^{1/2}\xi_n}{c}}^{\infty} \rho_{\mathrm{m}} \ln\left[\left(1-\overline{\Delta}_{\mathrm{Lm}}^{\mathrm{eff}}\overline{\Delta}_{\mathrm{Rm}}^{\mathrm{eff}} e^{-2\rho_{\mathrm{m}} l}\right) \times \left(1-\Delta_{\mathrm{Lm}}^{\mathrm{eff}}\Delta_{\mathrm{Rm}}^{\mathrm{eff}} e^{-2\rho_{\mathrm{m}} l}\right)\right] \mathrm{d}\rho_{\mathrm{m}} :$$

P.6.c.1. Finite number of layers

$$\overline{\Delta}_{\mathrm{Lm}}^{\mathrm{eff}} = \overline{\Delta}_{\mathrm{B'm}} \frac{1-e^{-2\rho_{\mathrm{B'}}\mathrm{b'}}}{1-\overline{\Delta}_{\mathrm{B'm}}^2 e^{-2\rho_{\mathrm{B'}}\mathrm{b'}}}, \qquad [(\mathrm{L3.116})]$$

$$\overline{\Delta}_{\mathrm{Rm}}^{\mathrm{eff}} = \frac{\left[n_{22}^{(\mathrm{R})}\overline{\Delta}_{\mathrm{RB'}} - n_{12}^{(\mathrm{R})}\right] e^{-2\rho_{\mathrm{B'}}\mathrm{b'}} + \left[n_{11}^{(\mathrm{R})} - n_{21}^{(\mathrm{R})}\overline{\Delta}_{\mathrm{RB'}}\right]\overline{\Delta}_{\mathrm{B'm}}}{\left[n_{11}^{(\mathrm{R})} - n_{21}^{(\mathrm{R})}\overline{\Delta}_{\mathrm{RB'}}\right] + \left[n_{22}^{(\mathrm{R})}\overline{\Delta}_{\mathrm{RB'}} - n_{12}^{(\mathrm{R})}\right]\overline{\Delta}_{\mathrm{B'm}} e^{-2\rho_{\mathrm{B'}}\mathrm{b'}}}, \qquad [(\mathrm{L3.117})]$$

$$\mathrm{m}_{11}^{(\mathrm{R})} = \frac{1-\overline{\Delta}_{\mathrm{B'm}}^2 e^{-2\rho_{\mathrm{m}}\mathrm{b}}}{\left(1-\overline{\Delta}_{\mathrm{B'm}}^2\right) e^{-\rho_{\mathrm{m}}\mathrm{b}} e^{-\rho_{\mathrm{B'}}\mathrm{b'}}}, \quad \mathrm{m}_{12}^{(\mathrm{R})} = \frac{\overline{\Delta}_{\mathrm{B'm}}\left(1-e^{-2\rho_{\mathrm{m}}\mathrm{b}}\right)}{\left(1-\overline{\Delta}_{\mathrm{B'm}}^2\right) e^{-\rho_{\mathrm{m}}\mathrm{b}} e^{-\rho_{\mathrm{B'}}\mathrm{b'}}},$$

$$\mathrm{m}_{21}^{(\mathrm{R})} = \frac{\left(e^{-2\rho_{\mathrm{m}}\mathrm{b}}-1\right)\overline{\Delta}_{\mathrm{B'm}} e^{-2\rho_{\mathrm{B'}}\mathrm{b'}}}{\left(1-\overline{\Delta}_{\mathrm{B'm}}^2\right) e^{-\rho_{\mathrm{m}}\mathrm{b}} e^{-\rho_{\mathrm{B'}}\mathrm{b'}}}, \quad \mathrm{m}_{22}^{(\mathrm{R})} = \frac{\left(e^{-2\rho_{\mathrm{m}}\mathrm{b}}-\overline{\Delta}_{\mathrm{B'm}}^2\right) e^{-2\rho_{\mathrm{B'}}\mathrm{b'}}}{\left(1-\overline{\Delta}_{\mathrm{B'm}}^2\right) e^{-\rho_{\mathrm{m}}\mathrm{b}} e^{-\rho_{\mathrm{B'}}\mathrm{b'}}}. \qquad [(\mathrm{L3.118})]$$

P.6.c.2. Limit of a large number of layers

$$\overline{\Delta}_{\mathrm{Rm}}^{\mathrm{eff}} \to \frac{\mathrm{m}_{21}^{(\mathrm{R})}\overline{\Delta}_{\mathrm{B'm}} - \left[\mathrm{m}_{22}^{(\mathrm{R})} - e^{-\zeta_{\mathrm{R}}}\right] e^{-2\rho_{\mathrm{B'}}\mathrm{b'}}}{\mathrm{m}_{21}^{(\mathrm{R})} - \left[\mathrm{m}_{22}^{(\mathrm{R})} - e^{-\zeta_{\mathrm{R}}}\right]\overline{\Delta}_{\mathrm{B'm}} e^{-2\rho_{\mathrm{B'}}\mathrm{b'}}} \qquad [(\mathrm{L3.108})]$$

where $\cosh(\zeta_R) = \{[\mathrm{m}_{11}^{(\mathrm{R})} + \mathrm{m}_{22}^{(\mathrm{R})}]/2\}$, [Table P.6.a and Eq. (L3.105)], $\mathrm{m}_{11}^{(\mathrm{R})}$ and $\mathrm{m}_{22}^{(\mathrm{R})}$ still as in Table P.6.c.1 for finite number of layers.

Table P.7.a. Spatially varying dielectric responses

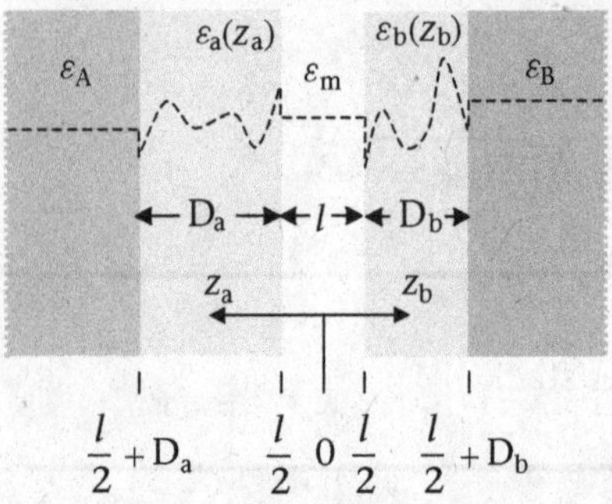

Arbitrary $\varepsilon_a(z_a)$, $\varepsilon_b(z_b)$ asymmetry about $z_a, z_b = 0$; discontinuity at $z_a(l/2)+D_a$, $z_b(l/2)+D_b$ and at $z_a, z_b = (l/2)$.

$i = A, a, m, b, B$, $\rho_i^2 = \rho^2 + \frac{\xi_n^2}{c^2}\varepsilon_i\mu_i$ with $\rho_a(z_a)$, $\rho_b(z_b)$:

$$\overline{\Delta}_{am} = \frac{\varepsilon_a\left(\frac{l}{2}\right)\rho_m - \varepsilon_m\rho_a\left(\frac{l}{2}\right)}{\varepsilon_a\left(\frac{l}{2}\right)\rho_m + \varepsilon_m\rho_a\left(\frac{l}{2}\right)},$$

[(L3.172a)]

$$\overline{\Delta}_{bm} = \frac{\varepsilon_b\left(\frac{l}{2}\right)\rho_m - \varepsilon_m\rho_b\left(\frac{l}{2}\right)}{\varepsilon_b\left(\frac{l}{2}\right)\rho_m + \varepsilon_m\rho_b\left(\frac{l}{2}\right)},$$

[(L3.172b)]

$$\overline{\Delta}_{Aa} = \frac{\varepsilon_A\rho_a\left(\frac{l}{2}+D_a\right) - \varepsilon_a\left(\frac{l}{2}+D_a\right)\rho_A}{\varepsilon_a\left(\frac{l}{2}\right)\rho_A + \varepsilon_A\rho_a\left(\frac{l}{2}\right)},$$

[(L3.167a)]

$$\overline{\Delta}_{Bb} = \frac{\varepsilon_B\rho_b\left(\frac{l}{2}+D_b\right) - \varepsilon_b\left(\frac{l}{2}+D_b\right)\rho_B}{\varepsilon_B\rho_b\left(\frac{l}{2}+D_b\right) + \varepsilon_b\left(\frac{l}{2}+D_b\right)\rho_B}.$$

[(L3.167b)]

In the absence of discontinuities $\overline{\Delta}_{am}$, $\overline{\Delta}_{bm}$, $\overline{\Delta}_{Aa}$, $\overline{\Delta}_{Bb}$ go to zero.

P.7.a.1. Spatially varying dielectric response in a finite layer, asymmetric, $\varepsilon(z)$ discontinuous at interfaces, with retardation

$$G(l; D_a, D_b) = \frac{kT}{2\pi}\sum_{n=0}^{\infty}{}' \int_0^{\infty} \rho \ln\left[\left(1-\overline{\Delta}^{eff}_{Am}\overline{\Delta}^{eff}_{Bm}e^{-2\rho l}\right)\left(1-\Delta^{eff}_{Am}\Delta^{eff}_{Bm}e^{-2\rho l}\right)\right]d\rho,$$

[(L3.165)]

$$\overline{\Delta}^{eff}_{Am} \equiv \left[\frac{\theta_a\left(\frac{l}{2}\right)e^{+\rho_a\left(\frac{l}{2}\right)l} + \overline{\Delta}_{am}}{1+\theta_a\left(\frac{l}{2}\right)e^{+\rho_a\left(\frac{l}{2}\right)l}\overline{\Delta}_{am}}\right] = \left[\frac{u_a\left(\frac{l}{2}\right)+\overline{\Delta}_{am}}{1+u_a\left(\frac{l}{2}\right)\overline{\Delta}_{am}}\right], \quad [(L3.170a)]$$

$$\overline{\Delta}^{eff}_{Bm} \equiv \left[\frac{\theta_b\left(\frac{l}{2}\right)e^{+\rho_b\left(\frac{l}{2}\right)l} + \overline{\Delta}_{bm}}{1+\theta_b\left(\frac{l}{2}\right)e^{+\rho_b\left(\frac{l}{2}\right)l}\overline{\Delta}_{bm}}\right] = \left[\frac{u_b\left(\frac{l}{2}\right)+\overline{\Delta}_{bm}}{1+u_b\left(\frac{l}{2}\right)\overline{\Delta}_{bm}}\right]. \quad [(L3.170b)]$$

$i = a, b$, $u_i(z_i) \equiv e^{2\rho_i(z_i)z_i}\theta(z_i)$; find; $u_i(\frac{l}{2})$ or $\theta_i(\frac{l}{2})$ from solving

$$\frac{du_i(z_i)}{dz_i} = +2\rho_i(z_i)u_i(z_i) - \frac{d\ln[\varepsilon_i(z_i)/\rho(z_i)]}{2dz_i}\left[1-u_i^2(z_i)\right],$$

[(L3.168a) and (L3.168b)]

$$\frac{d\theta_i(z_i)}{dz_i} = -2z_i\frac{d\rho_i(z_i)}{dz_b}\theta_i(z_i) - \frac{d\ln[\varepsilon_i(z_i)/\rho_i(z_i)]}{2dz_i}e^{-2\rho_i(z_i)z_i}\left[1-e^{+4\rho_i(z_i)z_i}\theta_i^2(z_i)\right]$$

[(L3.169a) and (L3.169b)]

whose solutions—numerically or analytically—begin at $(l/2)+D_i$:

$$\theta_a\left(\frac{l}{2}+D_a\right)e^{+\rho_a\left(\frac{l}{2}+D_a\right)(l+2D_a)} = u_a\left(\frac{l}{2}+D_a\right) = +\overline{\Delta}_{Aa}, \quad [(L3.166a)]$$

$$\theta_b\left(\frac{l}{2}+D_b\right)e^{+\rho_b\left(\frac{l}{2}+D_b\right)(l+2D_b)} = u_b\left(\frac{l}{2}+D_b\right) = +\overline{\Delta}_{Bb}, \quad [(L3.166b)]$$

at the outer interfaces.

Similarly for $\Delta^{eff}_{Am}\Delta^{eff}_{Bm}$ with magnetic $\mu_a(z_a)$, $\mu_b(z_b)$.

Table P.7.a (*cont.*)

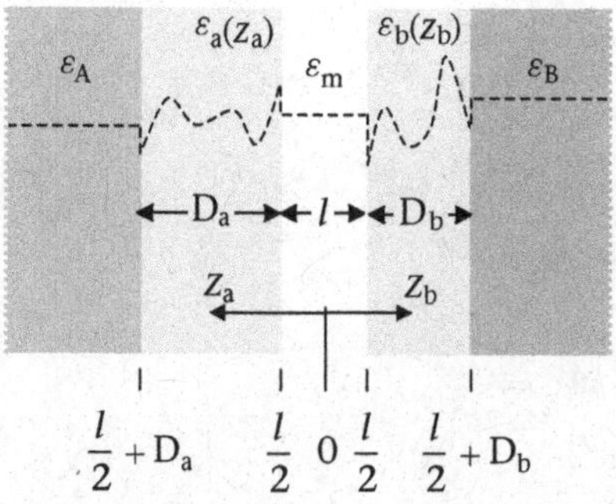

Arbitrary $\varepsilon_a(z_a)$, $\varepsilon_b(z_b)$ asymmetry about $z_a, z_b = 0$; discontinuity at $z_a(l/2) + D_a$, $z_b(l/2) + D_b$ and at $z_a, z_b = (l/2)$. $i = A, a, m, b, B$, $\rho_i^2 = \rho^2$ everywhere:

$$\overline{\Delta}_{am} = \frac{\varepsilon_a\left(\frac{l}{2}\right) - \varepsilon_m}{\varepsilon_a\left(\frac{l}{2}\right) + \varepsilon_m}, \qquad [(L3.144a)]$$

$$\overline{\Delta}_{bm} = \frac{\varepsilon_b\left(\frac{l}{2}\right) - \varepsilon_m}{\varepsilon_b\left(\frac{l}{2}\right) + \varepsilon_m}, \qquad [(L3.144b)]$$

$$\overline{\Delta}_{Aa} = \frac{\varepsilon_A - \varepsilon_a\left(\frac{l}{2} + D_a\right)}{\varepsilon_A + \varepsilon_a\left(\frac{l}{2} + D_a\right)}, \qquad [(L3.148)]$$

$$\overline{\Delta}_{Bb} = \frac{\varepsilon_B - \varepsilon_b\left(\frac{l}{2} + D_b\right)}{\varepsilon_B + \varepsilon_b\left(\frac{l}{2} + D_b\right)} \qquad [(L3.145)]$$

In the absence of discontinuities, $\overline{\Delta}_{am}$, $\overline{\Delta}_{bm}$, $\overline{\Delta}_{Aa}$, and $\overline{\Delta}_{Bb}$ go to zero.

P.7.a.2. Spatially varying dielectric response in a finite layer, asymmetric, $\varepsilon(z)$ discontinuous at inner and outer interfaces, no retardation

$$G(l; D_a, D_b) = \frac{kT}{2\pi}\sum_{n=0}^{\infty}{}' \int_0^{\infty} \rho \ln\left[\left(1 - \overline{\Delta}_{Am}^{eff}\overline{\Delta}_{Bm}^{eff}e^{-2\rho l}\right)\left(1 - \Delta_{Am}^{eff}\Delta_{Bm}^{eff}e^{-2\rho l}\right)\right]d\rho, \qquad [(L3.142)]$$

$$\overline{\Delta}_{Am}^{eff} \equiv \left[\frac{\theta_a\left(\frac{l}{2}\right)e^{+\rho l} + \overline{\Delta}_{am}}{1 + \theta_a\left(\frac{l}{2}\right)e^{+\rho l}\overline{\Delta}_{am}}\right] = \left[\frac{u_a\left(\frac{l}{2}\right) + \overline{\Delta}_{am}}{1 + u_a\left(\frac{l}{2}\right)\overline{\Delta}_{am}}\right], \qquad [(L3.143a)]$$

$$\overline{\Delta}_{Bm}^{eff} \equiv \left[\frac{\theta_b\left(\frac{l}{2}\right)e^{+\rho l} + \overline{\Delta}_{bm}}{1 + \theta_b\left(\frac{l}{2}\right)e^{+\rho l}\overline{\Delta}_{bm}}\right] = \left[\frac{u_b\left(\frac{l}{2}\right) + \overline{\Delta}_{bm}}{1 + u_b\left(\frac{l}{2}\right)\overline{\Delta}_{bm}}\right]. \qquad [(L3.143b)]$$

$i = a, b$, $u_i(z_i) \equiv e^{2\rho z_i}\theta(z_i)$; find $u_i(\frac{l}{2})$ or $\theta_i(\frac{l}{2})$ from solving

$$\frac{du_i(z_i)}{dz_i} = +2\rho u_i(z_i) - \frac{d\ln[\varepsilon_i(z_i)]}{2dz_i}\left[1 - u_i^2(z_i)\right], \qquad [(L3.147) \text{ and } (L3.150)]$$

$$\frac{d\theta_i(z_i)}{dz_i} = -\frac{e^{-2\rho z_i}}{2}\frac{d\ln[\varepsilon_i(z_i)]}{dz_i}\left[1 - e^{+4\rho z_i}\theta_i^2(z_i)\right], \qquad [(L3.146) \text{ and } (L3.149)]$$

whose solutions—numerically or analytically—begin at $(l/2) + D_i$:

$$\theta_a\left(\frac{l}{2} + D_a\right)e^{+\rho(l+2D_a)} = u_a\left(\frac{l}{2} + D_a\right) = +\overline{\Delta}_{Aa}, \qquad [(L3.145)]$$

$$\theta_b\left(\frac{l}{2} + D_b\right)e^{+\rho(l+2D_b)} = u_b\left(\frac{l}{2} + D_b\right) = +\overline{\Delta}_{Bb} \qquad [(L3.149)]$$

at the outer interfaces.

Similarly for $\Delta_{Am}^{eff}\Delta_{Bm}^{eff}$ with magnetic $\mu_a(z_a)$, $\mu_b(z_b)$.

Table P.7.b. Inhomogeneous, $\varepsilon(z)$ in finite layer, small range in ε, retardation neglected

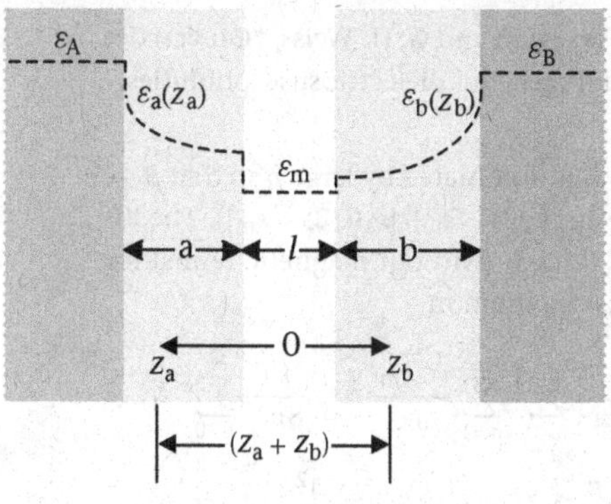

$$G(l; a, b) = -\frac{kT}{32\pi}\sum_{n=0}^{\infty}{}' \iint_{z_b, z_a} \frac{d\ln[\varepsilon_b(z_b)]}{dz_b}\frac{d\ln[\varepsilon_a(z_a)]}{dz_a}\frac{dz_a dz_b}{(z_a + z_b)^2}. \qquad [(L2.96)]$$

z_a, z_b measured outward from midpoint.

To lowest order in $(\varepsilon_A - \varepsilon_m)$, $(\varepsilon_B - \varepsilon_m)$, $[\varepsilon_a(z_a) - \varepsilon_m]$, $[\varepsilon_B(z_b) - \varepsilon_m] \ll \varepsilon_m$.

Finite steps in ε at $z_a = \frac{l}{2} + a, \frac{l}{2}$, $z_b = \frac{l}{2} + b, \frac{l}{2}$ create additional discrete terms of the same form as for cases in which ε is constant across planar regions.

Table P.7.c. Exponential $\varepsilon(z)$ infinite layer, symmetric systems

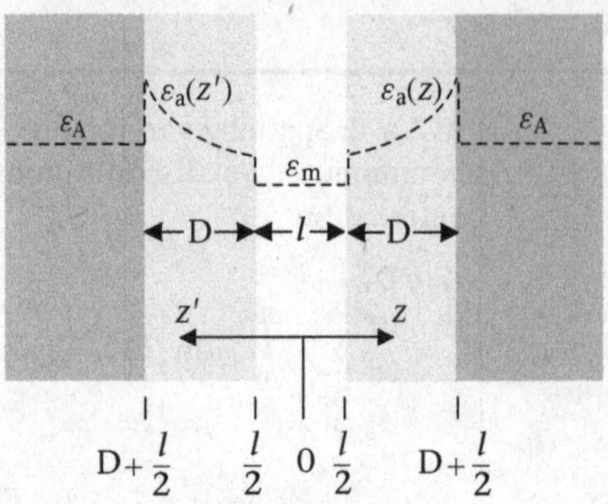

$$\varepsilon_a(z) = \Gamma_e e^{-\gamma_e z}, \frac{l}{2} \le z, z' \le D + \frac{l}{2} \quad [1],$$

$$\varepsilon_{a+} \equiv \varepsilon_a(D + l/2) = \Gamma_e e^{-\gamma_e (D+l/2)},$$

$$\varepsilon_{a-} \equiv \varepsilon_a(l/2) = \Gamma_e e^{-\gamma_e l/2} = \varepsilon_{a+} e^{+\gamma_e D} \quad [12a].$$

P.7.c.1. Two semi-infinite media A symmetrically coated with a finite layers a of thickness D with exponential variation $\varepsilon_a(z)$ perpendicular to the interface, retardation neglected

$$G(l; D) = \frac{kT}{2\pi} \sum_{n=0}^{\infty}{}' \int_0^{\infty} \rho \ln (1 - \overline{\Delta}_c^2 e^{-2\rho l}) d\rho$$

$$= \frac{kT}{8\pi l^2} \sum_{n=0}^{\infty}{}' \int_0^{\infty} x \ln (1 - \overline{\Delta}_c^2 e^{-x}) dx, \ x \equiv 2\rho l \quad [16][18],$$

$$\overline{\Delta}_c = \frac{(\alpha_+ \varepsilon_{a-} - \rho\varepsilon_m) - (\alpha_- \varepsilon_{a-} - \rho\varepsilon_m)\,\overline{\Delta}_{Aa} e^{-\beta D}}{(\alpha_+ \varepsilon_{a-} + \rho\varepsilon_m) - (\alpha_- \varepsilon_{a-} + \rho\varepsilon_m)\,\overline{\Delta}_{Aa} e^{-\beta D}} \quad [13],$$

$$\overline{\Delta}_{Aa} = \frac{\varepsilon_A \rho - \varepsilon_{a+}\alpha_+}{\varepsilon_A \rho - \varepsilon_{a+}\alpha_-}, \quad [12b]; \quad 2\alpha_\pm = -\gamma_e \pm \beta \quad [9],$$

$$\beta = \sqrt{(2\rho)^2 + \gamma_e^2} = \sqrt{(x/l)^2 + \gamma_e^2} \quad [12b],$$

$$G(l \to 0; D) \to -\frac{kT}{8\pi l^2} \sum_{n=0}^{\infty}{}' \left[\frac{\varepsilon_a(l/2) - \varepsilon_m}{\varepsilon_a(l/2) + \varepsilon_m}\right]^2.$$

Source: Equation numbers are for formulae derived in V. A. Parsegian and G. H. Weiss, "On van der Waals interactions between macroscopic bodies having inhomogeneous dielectric susceptibilities," J. Colloid Interface Sci., **40**, 35–41 (1972).

Note: In the limit $l \to 0$, where $l \ll D = $ constant, the integral is dominated by large ρ so that $\beta \to 2\rho \to \infty$, $\alpha_\pm \to \pm\rho$, $\overline{\Delta}_c \to \overline{\Delta}_{a-m} = [(\varepsilon_{a-} - \varepsilon_m)/(\varepsilon_{a-} + \varepsilon_m)] = \{[\varepsilon_a(l/2) - \varepsilon_m]/[\varepsilon_a(l/2) + \varepsilon_m]\}$. The interaction approaches the nonretarded Lifshitz form (Table P.1.a.3) (without magnetic terms) for two half-spaces of permittivity $\varepsilon_{a-} = \varepsilon_a(l/2)$ attracting across medium m,

$$G(l \to 0; D) \to \frac{kT}{8\pi l^2} \sum_{n=0}^{\infty}{}' \int_0^{\infty} x \ln \left(1 - \overline{\Delta}_{a-m}^2 e^{-x}\right) dx = -\frac{kT}{8\pi l^2} \sum_{n=0}^{\infty}{}' \sum_{q=1}^{\infty} \frac{\overline{\Delta}_{a-m}^{2q}}{q^3} \approx -\frac{kT}{8\pi l^2} \sum_{n=0}^{\infty}{}' \overline{\Delta}_{a-m}^2$$

$$= -\frac{kT}{8\pi l^2} \sum_{n=0}^{\infty}{}' \left[\frac{\varepsilon_a(l/2) - \varepsilon_m}{\varepsilon_a(l/2) + \varepsilon_m}\right]^2.$$

Table P.7.c (*cont.*)

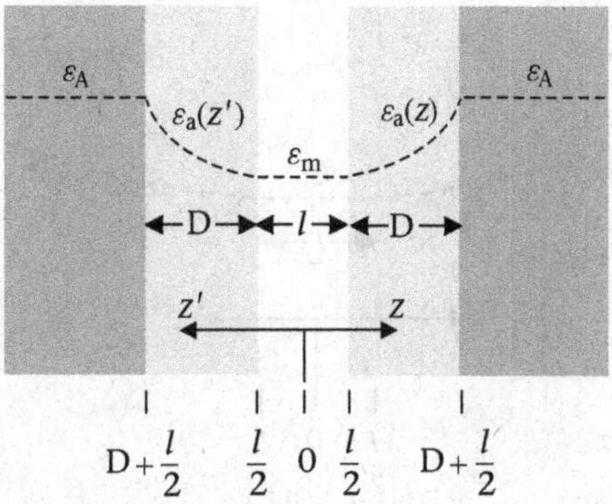

$\varepsilon_a(z) = \Gamma_e e^{-\gamma_e z} = \varepsilon_m e^{-\gamma_e(z-l/2)}$,

$\varepsilon_{a'}(z') = \Gamma_e e^{-\gamma_e z'} = \varepsilon_m e^{-\gamma_e(z'-l/2)}$, $\frac{l}{2} \le z, z' \le D + \frac{l}{2}$ [1],

$\varepsilon_{a+} \equiv \varepsilon_a(D + l/2) = \Gamma_e e^{-\gamma_e(D+l/2)} = \varepsilon_A$ [19],

$\varepsilon_{a-} \equiv \varepsilon_a(l/2) = \Gamma_e e^{-\gamma_e l/2} = \varepsilon_m = \varepsilon_A e^{+\gamma_e D}$,

$\gamma_e D = \ln\left(\frac{\varepsilon_m}{\varepsilon_A}\right)$ [20].

P.7.c.2. Exponential variation in finite layer of thickness *D*, symmetric structures, no discontinuities in ε, retardation neglected

$$G(l; D) = \frac{kT}{2\pi} \sum_{n=0}^{\infty}{}' \int_0^\infty \rho \ln\left(1 - \overline{\Delta}_c^2 e^{-2\rho l}\right) d\rho$$

$$= \frac{kT}{8\pi l^2} \sum_{n=0}^{\infty}{}' \int_0^\infty x \ln\left(1 - \overline{\Delta}_c^2 e^{-x}\right) dx, \quad x \equiv 2\rho l \quad [16]\ [18],$$

$$\overline{\Delta}_c = \frac{\gamma_e\left(1 - e^{-\beta D}\right)}{2\rho\left(1 - e^{-\beta D}\right) + \beta\left(1 + e^{-\beta D}\right)} \quad [21],$$

$$\beta = \sqrt{(2\rho)^2 + \gamma_e^2} = \sqrt{(x/l)^2 + \gamma_e^2} \quad [12b].$$

In small D/*l* limit $\overline{\Delta}_c^2 = \overline{\Delta}_{Am}^2 = [(\varepsilon_A - \varepsilon_m)/(\varepsilon_A + \varepsilon_m)]^2$ [22].

Source: Equation numbers are for formulae derived in V. A. Parsegian and G. H. Weiss, "On van der Waals interactions between macroscopic bodies having inhomogeneous dielectric susceptibilities," **40**: 35, 1972. J. Colloid Interface Sci., **40**, 35–41 (1972). From that paper, $\varepsilon_1 = \varepsilon_A$ here, $\varepsilon_3 = \varepsilon_m$, $\lambda = \gamma_e$, $\theta = \gamma_e D$, $\sqrt{\theta^2 + a^2x^2} = \beta D$, $ax = 2\rho D$ here.

In the limit where separation $l \gg$ layer thickness D, the integration in ρ is dominated by values near $\rho = 0$ where $\beta \approx \gamma_e \gg 2\rho$ so that

$$\overline{\Delta}_c \approx \frac{\gamma_e\left(1 - e^{-\beta D}\right)}{\gamma_e\left(1 + e^{-\beta D}\right)} = -\frac{\varepsilon_A - \varepsilon_m}{\varepsilon_A + \varepsilon_m}.$$

Table P.7.c (*cont.*)

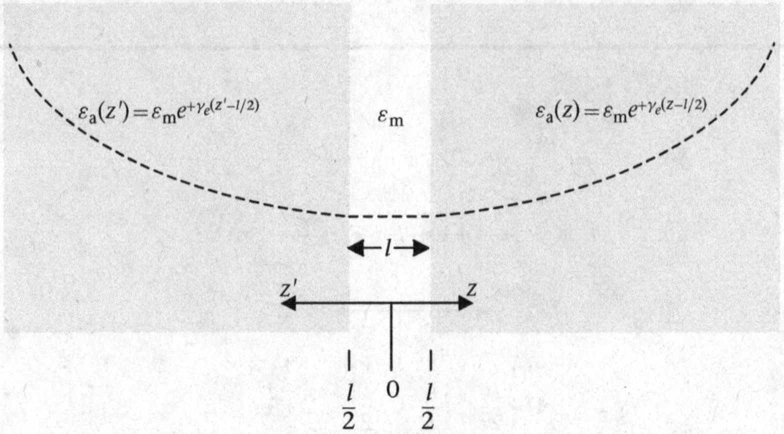

P.7.c.3. Exponential variation of dielectric response in an infinitely thick layer, no discontinuities in ε, discontinuity in $d\varepsilon(z)$ at interface, retardation neglected

$$G(l) = \frac{kT}{8\pi}\sum_{n=0}^{\infty}{}' \gamma_e^2 \int_0^{\infty} x \ln\left\{1 - \left[\frac{x-1-(x^2+1)^{1/2}}{x+1+(x^2+1)^{1/2}}\right]^2 e^{-\gamma_e l x}\right\} dx \quad [2.10], [2.11].$$

Small $\gamma_e l$ limit [$\ln(\gamma_e l)$ is negative!]:

$$G(\gamma_e l \to 0) \sim +\frac{kT}{32\pi}\sum_{n=0}^{\infty}{}' \gamma_e^2 \ln(\gamma_e l) \quad [2.14].$$

Large $\gamma_e l$ limit:

$$G(\gamma_e l \to \infty) \sim -\frac{kT}{8\pi l^2}\sum_{\mathrm{j}=1}^{\infty}\frac{1}{\mathrm{j}^3}$$

[$\gamma_e(i\xi_n) \to 0$ as $\xi_n \to \infty$, $\sum_{\xi_n=0}^{\prime\infty}$, summation inappropriate].

Source: Formulae derived in Section 2 of G. H. Weiss, J. E. Kiefer, and V. A. Parsegian, "Effects of dielectric inhomogeneity on the magnitude of van der Waals interactions," J. Colloid Interface Sci., **45**, 615–625 (1973).

Note: When $\gamma_e l \to 0$, the integral is dominated by contributions at large x, i.e., where

$$\left[\frac{x-1-(x^2+1)^{1/2}}{x+1+(x^2+1)^{1/2}}\right]^2 \to \frac{1}{4x^2}.$$

For large x the integral for each term in

$$G(l) \to -\frac{kT\gamma_e^2}{8\pi}\frac{1}{4}\int_{\sim 1}^{\infty}\frac{e^{-\gamma_e l x}}{x}dx = -\frac{kT\gamma_e^2}{32\pi}\int_{\sim\gamma_e l}^{\infty}\frac{e^{-v}}{v}dv \sim +\frac{kT\gamma_e^2}{32\pi}\ln(\gamma_e l).$$

When $l \to \infty$, the integral is dominated by values near $x = 0$, where

$$\left[\frac{x-1-(x^2+1)^{1/2}}{x+1+(x^2+1)^{1/2}}\right]^2 \to 1:$$

$$G(l \to \infty) \sim \frac{kT}{8\pi}\sum_{n=0}^{\infty}{}' \gamma_e^2 \int_0^{\infty} x \ln\left(1 - e^{-\gamma_e l x}\right)dx = -\frac{kT}{8\pi}\sum_{n=0}^{\infty}{}' \gamma_e^2 \sum_{\mathrm{j}=1}^{\infty}\frac{1}{\mathrm{j}}\int_0^{\infty} x e^{-\mathrm{j}\gamma_e l x}dx$$

$$= -\frac{kT\gamma_e^2}{8\pi}\sum_{1}^{\infty}\frac{1}{\mathrm{j}(\mathrm{j}\gamma_e l)^2} = -\frac{kT}{8\pi l^2}\sum_{\mathrm{j}=1}^{\infty}\frac{1}{\mathrm{j}^3}.$$

Table P.7.d. Power-law $\varepsilon(z)$ in a finite layer, symmetric systems

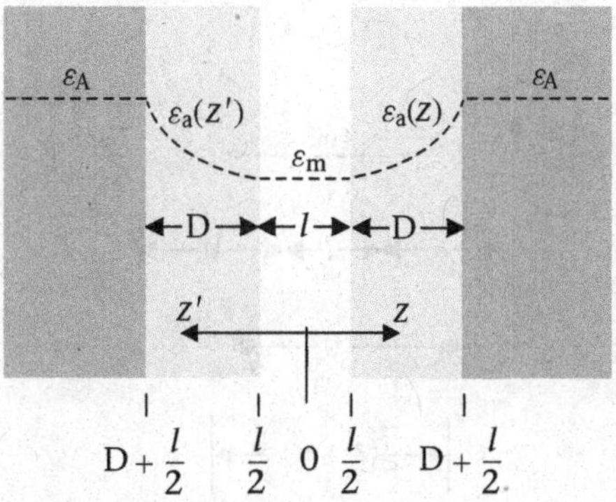

z_a, z_b measured outward from midpoint 0,

$\frac{l}{2} \le z, z' \le D + \frac{l}{2}$,

$\varepsilon_a(z) = (\alpha + \beta z)^{\gamma_p}$ [1.1], γ_p is allowed all real values,

$\alpha = \alpha(l) = \varepsilon_m^{1/\gamma_p} - \frac{l/2}{D}\left(\varepsilon_A^{1/\gamma_p} - \varepsilon_m^{1/\gamma_p}\right), \beta \equiv \frac{1}{D}\left(\varepsilon_A^{1/\gamma_p} - \varepsilon_m^{1/\gamma_p}\right)$ [5.2], for continuity at boundaries

$\psi \equiv \varepsilon_A/\varepsilon_m$, $u \equiv 1/\left|\psi^{1/\gamma_p} - 1\right|$, $v \equiv \psi^{1/\gamma_p} u$ [5.7], in arguments of functions

$\nu \equiv (1 - \gamma_p)/2$ [5.6], in (subscripted) indices of Bessel functions

P.7.d.1. Power-law variation in a finite layer of thickness D, symmetric structures, no discontinuities in ε but discontinuity in $d\varepsilon/dz$ at interfaces, retardation neglected

$$G(l; D) = \frac{kT}{2\pi} \sum_{n=0}^{\infty}{}' g_\pm(\xi_n), \; g_\pm(l; D, \xi_n) = \frac{kT}{2\pi D^2} \int_0^\infty x \ln\left[1 - \overline{\Delta}_\pm^2(x, \xi_n) e^{-2xl/D}\right] dx.$$

Use + subscript when $\beta > 0$, $\varepsilon_A^{1/n}(\xi_n) > \varepsilon_m^{1/n}(\xi_n)$.

Use − subscript when $\beta < 0$, $\varepsilon_A^{1/n}(\xi_n) < \varepsilon_m^{1/n}(\xi_n)$.

When index $\nu \equiv (1 - \gamma_p)/2$ is an integer, use [Eq. 5.7]:

$$\overline{\Delta}_+(x, \xi_n) = \frac{\eta_+(vx) - \eta_+(ux)}{\eta_+(vx) - \eta_-(ux)} \frac{1}{U(ux)}, \quad \overline{\Delta}_-(x, \xi_n) = \frac{\eta_-(vx) - \eta_-(ux)}{\eta_-(vx) - \eta_+(ux)} U(ux),$$

$\eta_\pm(x) = \frac{K_{\nu-1}(x) \mp K_\nu(x)}{I_{\nu-1}(x) \pm I_\nu(x)}$, $U(x) = \frac{I_\nu(x) - I_{\nu-1}(x)}{I_\nu(x) + I_{\nu-1}(x)}$ with the modified Bessel functions I and K.

When index $\nu \equiv (1 - \gamma_p)/2$ is *not* an integer, use:

$$\overline{\Delta}_+(x, \xi_n) = \frac{\Gamma_+(ux) - \Gamma_+(vx)}{\Gamma_-(ux) - \Gamma_+(vx)} \frac{1}{U(ux)}, \quad \overline{\Delta}_-(x, \xi_n) = \frac{\Gamma_-(ux) - \Gamma_-(vx)}{\Gamma_+(ux) - \Gamma_-(vx)} U(ux) \quad [5.11],$$

$\Gamma_\pm(x) = \frac{I_{1-\nu}(x) \pm I_\nu(x)}{I_{\nu-1}(x) \pm I_\nu(x)}$ [5.10] with only the I modified Bessel functions.

Source: Formulae derived in Section 5 of G. H. Weiss, J. E. Kiefer, and V. A. Parsegian, "Effects of dielectric inhomogeneity on the magnitude of van der Waals interactions," J. Colloid Interface Sci., **45**, 615–625 (1973); Equation numbers in [ ] from that paper. γ_p here is "n" in source paper.

Table P.7.d (*cont.*)

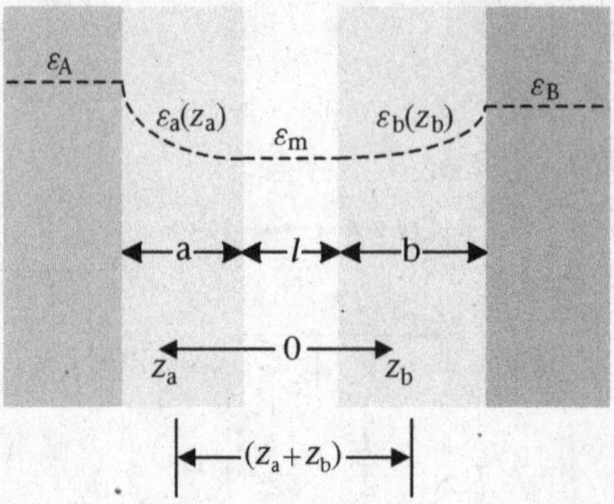

z_a, z_b measured outward from mid-point 0 $d\varepsilon(z)/dz = 0$ at interface with m

P.7.d.2. Continuously changing $\varepsilon(z)$, continuous $d\varepsilon/dz$ at inner interface; quadratic variation over finite layers, retardation neglected

$$\varepsilon_a(z_a) = \varepsilon_m + \frac{(\varepsilon_A - \varepsilon_m)}{a^2}\left(z_a - \frac{l}{2}\right)^2, \quad \varepsilon_b(z_b) = \varepsilon_m + \frac{(\varepsilon_B - \varepsilon_m)}{b^2}\left(z_b - \frac{l}{2}\right)^2,$$

[(L2.97) and (L2.98)]

$$G(l; a, b) = -\frac{kT}{16\pi}\left\{\frac{l^2}{a^2b^2}\ln\left[\frac{(l+a)(l+b)}{(l+a+b)\,l}\right] + \frac{1}{b^2}\ln\left(\frac{l+a+b}{l+a}\right)\right.$$
$$\left. + \frac{1}{a^2}\ln\left(\frac{l+a+b}{l+b}\right) - \frac{1}{ab}\right\}\sum_{n=0}^{\infty}{}' \frac{(\varepsilon_A - \varepsilon_m)(\varepsilon_B - \varepsilon_m)}{\varepsilon_m^2}, \quad \text{[(L2.99)]}$$

$$G(l \to 0; a, b) = -\frac{kT}{16\pi}\left[\frac{\ln\left(1 + \frac{b}{a}\right)}{b^2} + \frac{\ln\left(1 + \frac{a}{b}\right)}{a^2} - \frac{1}{ab}\right]\sum_{n=0}^{\infty}{}' \frac{(\varepsilon_A - \varepsilon_m)(\varepsilon_B - \varepsilon_m)}{\varepsilon_m^2},$$

[(L2.100)]

$$P(l \to 0; a, b) = -\frac{kT}{8\pi}\frac{1}{ab(a+b)}\sum_{n=0}^{\infty}{}' \frac{(\varepsilon_A - \varepsilon_m)(\varepsilon_B - \varepsilon_m)}{\varepsilon_m^2}. \quad \text{[(L2.102)]}$$

Table P.7.e. Gaussian variation of dielectric response in an infinitely thick layer, no discontinuities in ε or in $d\varepsilon/dz$, symmetric profile, retardation neglected

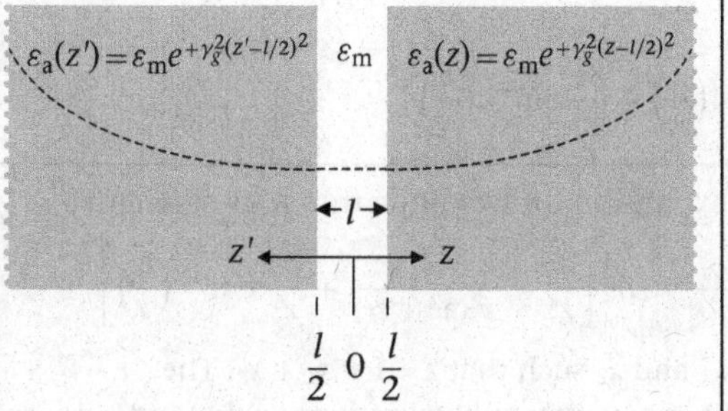

$$G(l) = \frac{2kT}{\pi}\sum_{n=0}^{\infty}{}' \gamma_g^2 \int_0^{\infty} x \ln\left\{1 - \left[\frac{x - F(x)}{x + F(x)}\right]^2 e^{-4\gamma_g l x}\right\} dx, \qquad [2.18]$$

$$F(x) = \Gamma(1 + x^2)/\Gamma\left(\frac{1}{2} + x^2\right). \qquad [2.19]$$

$\Gamma(x)$ is the gamma function (Abramowitz and Stegun, Section 6.1):

$$G(l \to 0) \sim -\frac{kT}{2^8\pi}\sum_{n=0}^{\infty}{}' \gamma_g^2$$

(finite at contact, $\gamma_g(i\xi_n) \to 0$ as $\xi_n \to \infty$).

Source: Formulae derived in section 2 of G. H. Weiss, J. E. Kiefer, and V. A. Parsegian, "Effects of dielectric inhomogeneity on the magnitude of van der Waals interactions," J. Colloid Interface Sci., **45**, 615–625 (1973).

Note: When $l \to 0$, the integral in [2.18] is dominated by the region of $x \sim \infty$. There $F(x) \sim x + (1/8x)$ [M. Abramowitz and I. A. Stegun, *Handbook of Mathematical Functions, with Formulas, Graphs and Mathematical Tables* (Dover Books, New York, 1965). Eq. (6.1.47)]:

$$\{[x - F(x)]/[x + F(x)]\}^2 \sim 1/2^8 x^4,$$

$$G(l \to 0) \sim -\frac{kT}{2^7\pi}\sum_{n=0}^{\infty}{}' \gamma_g^2 \int_{\sim 1}^{\infty} \frac{e^{-4\gamma_g l x}}{x^3} dx$$

$$= -\frac{kT}{2^7\pi}\sum_{n=0}^{\infty}{}' \gamma_g^2 (4\gamma_g l)^2 \int_{\sim 4\gamma_g l}^{\infty} \frac{e^{-q}}{q^3} dq \sim -\frac{kT}{2^8\pi}\sum_{n=0}^{\infty}{}' \gamma_g^2.$$

Table P.8.a. Edge-to-edge interaction between two thin rectangles, length a, width b, separation $l \gg$ thickness c, Hamaker limit

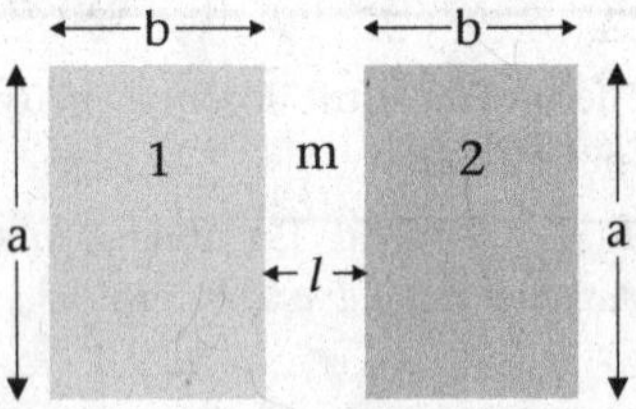

$$G(l;\mathrm{a},\mathrm{b},\mathrm{c}) = -\frac{A_{1\mathrm{m}/2\mathrm{m}}\mathrm{c}^2}{24\pi^2}\left[F(l) - 2F(l+\mathrm{b}) + F(l+2\mathrm{b})\right],$$

where

$$F(x;\mathrm{a}) \equiv -\frac{1}{x^2} + \frac{3x}{\mathrm{a}^3}\tan^{-1}\left(\frac{x}{\mathrm{a}}\right) + \frac{3\mathrm{a}}{x^3}\tan^{-1}\left(\frac{\mathrm{a}}{x}\right).$$

Think of integrating the expression for two finite thin rods of length a,

$$E(z;a) = E(z,\mathrm{a}) = -\frac{A_{1\mathrm{m}/2\mathrm{m}}}{4\pi^2}A_1A_2\left[\frac{1}{z^4} - \frac{1}{z^2(\mathrm{a}^2+z^2)} + \frac{3\mathrm{a}}{z^5}\tan^{-1}\left(\frac{\mathrm{a}}{z}\right)\right],$$

over the width variables x_1 and x_2 such that $z = l + x_1 + x_2$. The cross-sectional areas of these now-incremental rods are c dx_1 and c dx_2 so that the now-required integration is

$$-\frac{A_{1\mathrm{m}/2\mathrm{m}}\mathrm{c}^2}{4\pi^2}\int_0^{\mathrm{b}}\int_0^{\mathrm{b}}\left\{\frac{1}{(l+x_1+x_2)^4} - \frac{1}{(l+x_1+x_2)^2[\mathrm{a}^2+(l+x_1+x_2)^2]} + \frac{3\mathrm{a}}{(l+x_1+x_2)^5}\tan^{-1}\left[\frac{\mathrm{a}}{(l+x_1+x_2)}\right]\right\}dx_1dx_2.$$

To avoid tedium, define

$$H(l) = -\frac{1}{48l^2} + \frac{l}{16\mathrm{a}^3}\tan^{-1}\left(\frac{l}{\mathrm{a}}\right) + \frac{\mathrm{a}}{16l^3}\tan^{-1}\left(\frac{\mathrm{a}}{l}\right)$$

$$= \frac{1}{8}\int_0^{\infty}\int_0^{\infty}\left\{\frac{1}{(l+x_1+x_2)^4} - \frac{1}{(l+x_1+x_2)^2\left[\mathrm{a}^2+(l+x_1+x_2)^2\right]} + \frac{3\mathrm{a}}{(l+x_1+x_2)^5}\tan^{-1}\left[\frac{\mathrm{a}}{(l+x_1+x_2)}\right]\right\}dx_1dx_2.$$

[A. G. DeRocco and, W. G. Hoover, "On the interaction of colloidal particles," Proc. Natl. Acad. Sci., USA, **46**, 1057–1065 (1960)]

Here use

$$F(x) = 48\,H(x) = -\frac{1}{x^2} + \frac{3x}{\mathrm{a}^3}\tan^{-1}\left(\frac{x}{\mathrm{a}}\right) + \frac{3\mathrm{a}}{x^3}\tan^{-1}\left(\frac{\mathrm{a}}{x}\right),$$

so that the rectangle–rectangle interaction goes as $F(l) - 2F(l+\mathrm{b}) + F(l+2\mathrm{b})$.

Table P.8.b. Face-to-face interaction between two thin rectangles, length a, width b, separation l ≫ thickness c, Hamaker limit

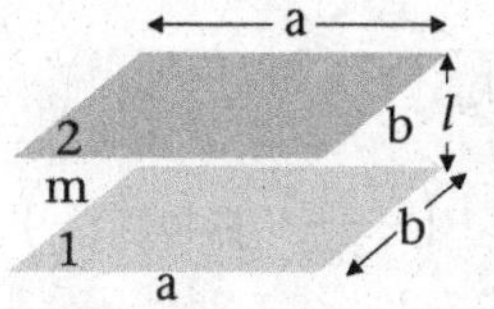

$$E(l;\mathrm{a},\mathrm{b},\mathrm{c}) = -\frac{A_{1\mathrm{m}/2\mathrm{m}}\,\mathrm{c}^2}{\pi^2}\,K_6(l;\mathrm{a},\mathrm{b}),$$

$$K_6(l;\mathrm{a},\mathrm{b}) = \left[\frac{\mathrm{b}l^2+2\mathrm{a}^2\mathrm{b}}{2l^4\,(l^2+\mathrm{a}^2)^{1/2}}\right]\tan^{-1}\left[\frac{\mathrm{b}}{(l^2+\mathrm{a}^2)^{1/2}}\right] + \left[\frac{\mathrm{a}l^2+2\mathrm{a}\mathrm{b}^2}{2l^4\,(l^2+\mathrm{b}^2)^{1/2}}\right]\tan^{-1}\left[\frac{\mathrm{a}}{(l^2+\mathrm{b}^2)^{1/2}}\right] - \frac{\mathrm{b}}{2l^3}\tan^{-1}\frac{\mathrm{b}}{l} - \frac{\mathrm{a}}{2l^3}\tan^{-1}\frac{\mathrm{a}}{l}.$$

The $1/r^6$ integration over the two sheets is

$K_6(l;\mathrm{a},\mathrm{b})$.

[A. G. DeRocco and W. G. Hoover, "On the interaction of colloidal particles," Proc. Natl. Acad. Sci., USA, **46**, 1057–1065 (1960)]. In the $l \to 0$ limit, the first two terms dominate and go to

$$\frac{\mathrm{ab}}{l^4}\left(\tan^{-1}\frac{\mathrm{b}}{\mathrm{a}} + \tan^{-1}\frac{\mathrm{a}}{\mathrm{b}}\right) = \frac{\pi\,\mathrm{ab}}{2l^4},$$

$$[\tan^{-1}(\mathrm{a}/\mathrm{b}) + \tan^{-1}(\mathrm{b}/\mathrm{a}) = \alpha+\beta = \pi/2]$$

expectably proportional to area ab and varying as the inverse-fourth power of separation appropriate for thin parallel slabs. Recall the Hamaker summation for infinitely extended thin slabs of thickness c, separation l for energy per unit area $-[(A_{\mathrm{Ham}}\mathrm{c}^2)/2\pi l^4]$. Thus $K_6(l;\mathrm{a},\mathrm{b})$ for thin rectangles given here must be multiplied by $-[(A_{\mathrm{Ham}}\,\mathrm{c}^2)/\pi^2]$ for energy of interaction:

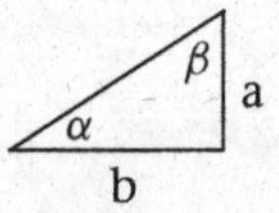

Table P.8.c. Two rectangular solids, length a, width b, height c, parallel, separated by a distance l normal to the a,b plane, Hamaker limit

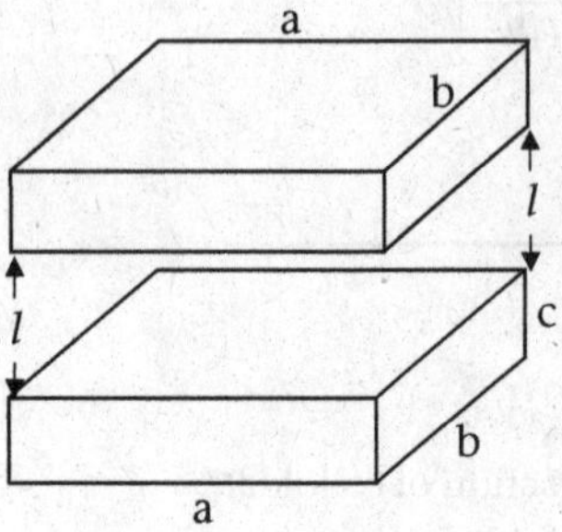

$$G_{pp}(l; \mathrm{a}, \mathrm{b}, \mathrm{c}) = -\frac{A_{1m2}}{\pi^2} K_{pp}(x)\Big|_{l+\mathrm{c},l+\mathrm{c}}^{l+2\mathrm{c},l},$$

$$K_{pp}(x)\Big|_{l+\mathrm{c},l+\mathrm{c}}^{l+2\mathrm{c},l} = K_{pp}(l+2\mathrm{c}) - 2K_{pp}(l+\mathrm{c}) + K_{pp}(l),$$

$$K_{pp}(x) = \frac{1}{4}\ln\left(\frac{x^4 + x^2\mathrm{a}^2 + x^2\mathrm{b}^2 + \mathrm{a}^2\mathrm{b}^2}{x^4 + x^2\mathrm{a}^2 + x^2\mathrm{b}^2}\right) + \left(\frac{x^2 - \mathrm{a}^2}{4\mathrm{a}x}\right)\tan^{-1}\left(\frac{\mathrm{a}}{x}\right)$$
$$+\left(\frac{x^2 - \mathrm{b}^2}{4\mathrm{b}x}\right)\tan^{-1}\left(\frac{\mathrm{b}}{x}\right) + \frac{x(\mathrm{a}^2 + \mathrm{b}^2)^{3/2}}{6\mathrm{a}^2\mathrm{b}^2}\tan^{-1}\left[\frac{x}{(\mathrm{a}^2 + \mathrm{b}^2)^{1/2}}\right]$$
$$+\left(\frac{1}{6x^2} + \frac{1}{6\mathrm{a}^2}\right)\mathrm{b}\,(x^2 + \mathrm{a}^2)^{1/2}\tan^{-1}\left[\frac{\mathrm{b}}{(x^2 + \mathrm{a}^2)^{1/2}}\right]$$
$$+\left(\frac{1}{6x^2} + \frac{1}{6\mathrm{b}^2}\right)\mathrm{a}\,(x^2 + \mathrm{b}^2)^{1/2}\tan^{-1}\left[\frac{\mathrm{a}}{(x^2 + \mathrm{b}^2)^{1/2}}\right].$$

When l goes to zero compared with a, b, and c, this interaction goes to the inverse-square dependence of two planes interacting over an area ab, $K_{pp}(x)|_{l+\mathrm{c},l+\mathrm{c}}^{l+2\mathrm{c},l} \rightarrow [(\pi \mathrm{ab})/(12l^2)]$ (too tedious to derive here). Because the interaction energy in this limit is $[(A_{1m/2m})/(12\pi l^2)] \times \mathrm{ab}$, the energy for this parallelepiped interaction for all a, b, c, goes as previously given. [A. G. DeRocco and W. G. Hoover, "On the interaction of colloidal particles," Proc. Natl. Acad. Sci., USA, **46**, 1057–1065 (1960)].

Table P.8.d. Rectangular solids, length = width = a, height c, corners are separated by the diagonal of a square of side d, Hamaker limit

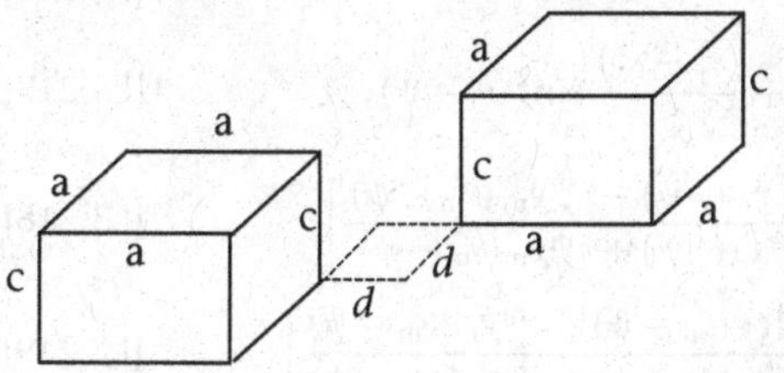

The result is so lengthy that it is not even written out in the original paper [A. G. DeRocco and W. G. Hoover, "On the interaction of colloidal particles," Proc. Natl. Acad. Sci., USA, **46**, 1057–1065 (1960)]. It consists of three expressions of the same form: one is evaluated at d as written in function $K_{\mathrm{sp}}(x;d)|_{d+\mathrm{a},d+\mathrm{a}}^{d+2\mathrm{a},d}$. The two other terms are this same function but evaluated with d replaced everywhere with $(d+\mathrm{a})$, and the resulting function $K_{\mathrm{sp}}(x;d+\mathrm{a})$ multiplied by (-2); and this same function yet again but evaluated with d replaced everywhere with $(d+2\mathrm{a})$, the resulting function $K_{\mathrm{sp}}(x;d+2\mathrm{a})$ multiplied by $(+1)$. See the following equations.

$$K_{\mathrm{sp}}(x;d)|_{d+\mathrm{a},d+\mathrm{a}}^{d+2\mathrm{a},d} = K_{\mathrm{sp}}(d+2\mathrm{a};d) - 2K_{\mathrm{sp}}(d+\mathrm{a};d) + K_{\mathrm{ap}}(d;d),$$

where

$$\begin{aligned} K_{\mathrm{sp}}(x;d) = {} & \frac{1}{8}\ln\left(\frac{d^2+x^2}{\mathrm{c}^2+d^2+x^2}\right) + \frac{1}{8}\left(\frac{x}{d}-\frac{d}{x}\right)\tan^{-1}\left(\frac{d}{x}\right) \\ & + \frac{(\mathrm{c}^2+d^2)^{3/2}x}{12\mathrm{c}^2d^2}\tan^{-1}\frac{x}{(\mathrm{c}^2+d^2)^{1/2}} + \frac{\mathrm{c}(d^2+x^2)^{1/2}}{12} \\ & \times\left(\frac{1}{d^2}+\frac{1}{x^2}\right)\tan^{-1}\frac{\mathrm{c}}{(d^2+x^2)^{1/2}} + \frac{d(\mathrm{c}^2+x^2)^{1/2}}{12} \\ & \times\left(\frac{1}{\mathrm{c}^2}+\frac{1}{x^2}\right)\tan^{-1}\frac{d}{(\mathrm{c}^2+x^2)^{1/2}}. \end{aligned}$$

In the limit at which the two bodies approach contact (again too tedious to spell out), when their separation $d^* = \sqrt{2}d$ goes to zero, the interaction approaches an inverse-first-power dependence $(\pi\mathrm{c})/(6d^*)$, similar to that seen between closely approaching spheres.

Table P.9.a. Interactions between and across anisotropic media

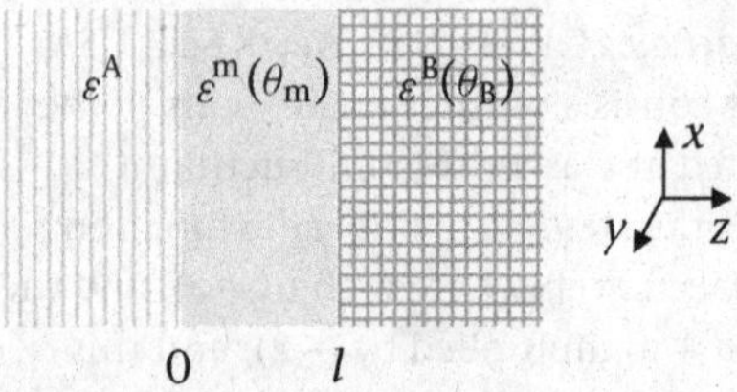

Anisotropic media, i = A, m, B
Principal axes in x, y, z directions

$$\varepsilon^{\mathrm{i}} \equiv \begin{bmatrix} \varepsilon_x^{\mathrm{i}} & 0 & 0 \\ 0 & \varepsilon_y^{\mathrm{i}} & 0 \\ 0 & 0 & \varepsilon_z^{\mathrm{i}} \end{bmatrix}.$$

Rotation of m and/or B relative to A by angles θ_{m} and/or θ_{R} about z axis creates dielectric tensors $\varepsilon^{\mathrm{m}}(\theta_{\mathrm{m}})$ and $\varepsilon^{\mathrm{B}}(\theta_{\mathrm{B}})$ ($\theta_{\mathrm{A}} \equiv 0$).

$$\varepsilon^{\mathrm{i}}(\theta_{\mathrm{i}}) = \begin{bmatrix} \varepsilon_x^{\mathrm{i}} + (\varepsilon_y^{\mathrm{i}} - \varepsilon_x^{\mathrm{i}})\sin^2(\theta_{\mathrm{i}}) & (\varepsilon_x^{\mathrm{i}} - \varepsilon_y^{\mathrm{i}})\sin(\theta_{\mathrm{i}})\cos(\theta_{\mathrm{i}}) & 0 \\ (\varepsilon_x^{\mathrm{i}} - \varepsilon_y^{\mathrm{i}})\sin(\theta_{\mathrm{i}})\cos(\theta_{\mathrm{i}}) & \varepsilon_y^{\mathrm{i}} + (\varepsilon_x^{\mathrm{i}} - \varepsilon_y^{\mathrm{i}})\sin^2(\theta_{\mathrm{i}}) & 0 \\ 0 & 0 & \varepsilon_z^{\mathrm{i}} \end{bmatrix},$$

$$G(l, \theta_{\mathrm{m}}, \theta_{\mathrm{B}}) = \frac{kT}{16\pi^2 l^2} \sum_{n=0}^{\infty}{}' \sum_{\mathrm{j}=1}^{\infty} \frac{1}{\mathrm{j}^3} \int_0^{2\pi} \frac{[\overline{\Delta}_{\mathrm{Am}}(\xi_n, \theta_{\mathrm{m}}, \psi)\overline{\Delta}_{\mathrm{Bm}}(\xi_n, \theta_{\mathrm{m}}, \theta_{\mathrm{B}}, \psi)]^{\mathrm{j}}\mathrm{d}\psi}{g_{\mathrm{m}}^2(\theta_{\mathrm{m}} - \psi)}, \quad \text{[L3.222]}$$

$$g_{\mathrm{i}}^2(\theta_{\mathrm{i}} - \psi) \equiv \frac{\varepsilon_x^{\mathrm{i}}}{\varepsilon_z^{\mathrm{i}}} + \frac{(\varepsilon_y^{\mathrm{i}} - \varepsilon_x^{\mathrm{i}})}{\varepsilon_z^{\mathrm{i}}}\sin^2(\theta_{\mathrm{i}} - \psi), \quad \text{[L3.217]}$$

$$\overline{\Delta}_{\mathrm{Am}}(\xi_n, \theta_{\mathrm{m}}, \psi) = \left[\frac{\varepsilon_z^{\mathrm{A}} g_{\mathrm{A}}(-\psi) - \varepsilon_z^{\mathrm{m}} g_{\mathrm{m}}(\theta_{\mathrm{m}} - \psi)}{\varepsilon_z^{\mathrm{A}} g_{\mathrm{A}}(-\psi) + \varepsilon_z^{\mathrm{m}} g_{\mathrm{m}}(\theta_{\mathrm{m}} - \psi)}\right], \quad \text{[L3.218]}$$

$$\overline{\Delta}_{\mathrm{Bm}}(\xi_n, \theta_{\mathrm{m}}, \theta_{\mathrm{B}}, \psi) = \left[\frac{\varepsilon_z^{\mathrm{B}} g_{\mathrm{B}}(\theta_{\mathrm{B}} - \psi) - \varepsilon_z^{\mathrm{m}} g_{\mathrm{m}}(\theta_{\mathrm{m}} - \psi)}{\varepsilon_z^{\mathrm{B}} g_{\mathrm{B}}(\theta_{\mathrm{B}} - \psi) + \varepsilon_z^{\mathrm{m}} g_{\mathrm{m}}(\theta_{\mathrm{m}} - \psi)}\right] \quad \text{[L3.219]}$$

Note: Reduction to isotropic case: $g_{\mathrm{i}}^2(\theta_{\mathrm{i}} - \psi) = 1$, $\int_0^{2\pi} \mathrm{d}\psi = 2\pi$, $G(l) = [-(kT/8\pi l^2)] \sum_{n=0}^{\prime\infty} \sum_{\mathrm{j}=1}^{\infty} [(\overline{\Delta}_{\mathrm{Am}}\overline{\Delta}_{\mathrm{Bm}})^{\mathrm{j}}/\mathrm{j}^3]$.

Table P.9.b. Interactions between anisotropic media A and B across isotropic medium m ($\varepsilon_x^{\mathrm{m}} = \varepsilon_y^{\mathrm{m}} = \varepsilon_z^{\mathrm{m}} = \varepsilon_{\mathrm{m}}$)

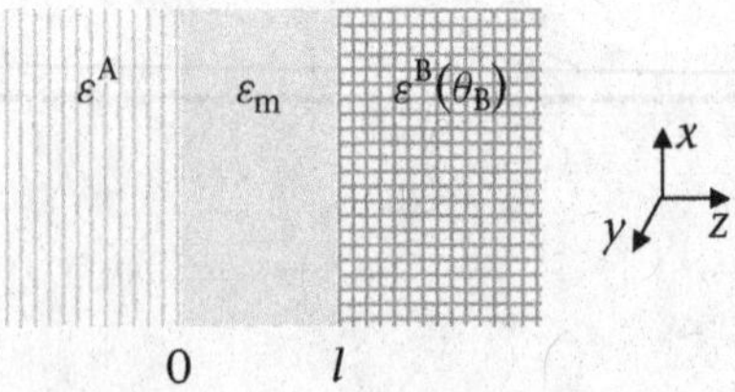

Anistropic media, i = A, B; isotropic m. Principal axes in x, y, z directions

$$\varepsilon^{\mathrm{i}} \equiv \begin{bmatrix} \varepsilon_x^{\mathrm{i}} & 0 & 0 \\ 0 & \varepsilon_y^{\mathrm{i}} & 0 \\ 0 & 0 & \varepsilon_z^{\mathrm{i}} \end{bmatrix}.$$

Rotation of B relative to A by angle θ_{B} about the z axis creates dielectric tensors $\varepsilon^{\mathrm{B}}(\theta_{\mathrm{B}})(\theta_{\mathrm{A}} \equiv 0)$:

$$\varepsilon^{\mathrm{B}}(\theta_{\mathrm{B}}) = \begin{bmatrix} \varepsilon_x^{\mathrm{B}} + (\varepsilon_y^{\mathrm{B}} - \varepsilon_x^{\mathrm{B}})\sin^2(\theta_{\mathrm{B}}) & (\varepsilon_x^{\mathrm{B}} - \varepsilon_y^{\mathrm{B}})\sin(\theta_{\mathrm{B}})\cos(\theta_{\mathrm{B}}) & 0 \\ (\varepsilon_x^{\mathrm{B}} - \varepsilon_y^{\mathrm{B}})\sin(\theta_{\mathrm{B}})\cos(\theta_{\mathrm{B}}) & \varepsilon_y^{\mathrm{B}} + (\varepsilon_x^{\mathrm{B}} - \varepsilon_y^{\mathrm{B}})\sin^2(\theta_{\mathrm{B}}) & 0 \\ 0 & 0 & \varepsilon_z^{\mathrm{B}} \end{bmatrix},$$

$$G(l, \theta_{\mathrm{B}}) = \frac{-kT}{16\pi^2 l^2} \sum_{n=0}^{\infty}{}' \sum_{\mathrm{j}=1}^{\infty} \frac{1}{\mathrm{j}^3} \int_0^{2\pi} [\overline{\Delta}_{\mathrm{Am}}(\xi_n, \psi)\overline{\Delta}_{\mathrm{Bm}}(\xi_n, \theta_{\mathrm{B}}, \psi)]^{\mathrm{j}}\mathrm{d}\psi,$$

$$\overline{\Delta}_{\mathrm{Am}}(\xi_n, \psi) = \left[\frac{\varepsilon_z^{\mathrm{A}} g_{\mathrm{A}}(-\psi) - \varepsilon_{\mathrm{m}}}{\varepsilon_z^{\mathrm{A}} g_{\mathrm{A}}(-\psi) + \varepsilon_{\mathrm{m}}}\right],$$

$$\overline{\Delta}_{\mathrm{Bm}}(\xi_n, \theta_{\mathrm{B}}, \psi) = \left[\frac{\varepsilon_z^{\mathrm{B}} g_{\mathrm{B}}(\theta_{\mathrm{B}} - \psi) - \varepsilon_{\mathrm{m}}}{\varepsilon_z^{\mathrm{B}} g_{\mathrm{B}}(\theta_{\mathrm{B}} - \psi) + \varepsilon_{\mathrm{m}}}\right],$$

$$g_{\mathrm{i}}^2(\theta_{\mathrm{i}} - \psi) \equiv \frac{\varepsilon_x^{\mathrm{i}}}{\varepsilon_z^{\mathrm{i}}} + \frac{(\varepsilon_y^{\mathrm{i}} - \varepsilon_x^{\mathrm{i}})}{\varepsilon_z^{\mathrm{i}}}\sin^2(\theta_{\mathrm{i}} - \psi), \ \mathrm{i} = \mathrm{A, B}.$$

Note: $g_{\mathrm{m}}^2 \equiv 1$.

Table P.9.c. Low-frequency ionic-fluctuation interactions between and across anisotropic media (magnetic terms neglected)

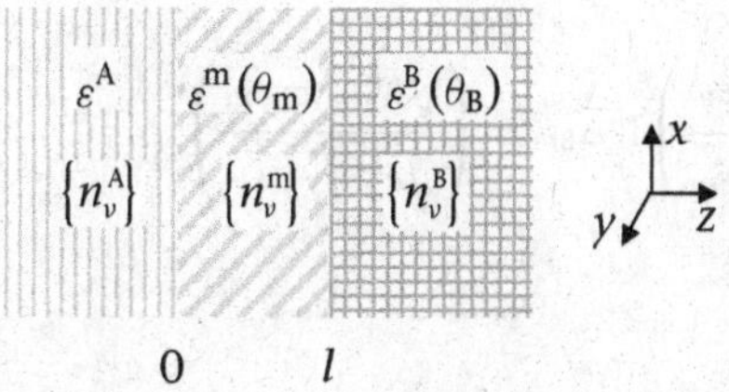

Anisotropic media, i = A, m, B
Principal axes in x, y, z directions

$$\varepsilon^i \equiv \begin{bmatrix} \varepsilon_x^i & 0 & 0 \\ 0 & \varepsilon_y^i & 0 \\ 0 & 0 & \varepsilon_z^i \end{bmatrix},$$

$\{n_\nu^i\}$, the set of ions of valence ν in regions i = A, m, B.

$$G_{n=0}(l, \theta_m, \theta_B) = \frac{kT}{8\pi^2}\int_0^{2\pi} d\psi \int_0^\infty \rho \ln[D(l, \rho, \psi, \theta_m, \theta_B)]d\rho, \quad \text{[L3.238]}$$

$$D(l, \rho, \psi, \theta_m, \theta_B) = 1 - \overline{\Delta}_{Am}(\theta_m, \psi)\overline{\Delta}_{Bm}(\theta_m, \theta_B, \psi)e^{-2\sqrt{\rho^2 g_m^2(\theta_m-\psi)+\kappa_m^2}\,l},$$

$$\overline{\Delta}_{Am}(\theta_m, \psi) = \left[\frac{\varepsilon_z^A(0)\beta_A - \varepsilon_z^m(0)\beta_m(\theta_m)}{\varepsilon_z^A(0)\beta_A + \varepsilon_z^m(0)\beta_m(\theta_m)}\right],$$

$$\overline{\Delta}_{Bm}(\theta_m, \theta_B, \psi) = \left[\frac{\varepsilon_z^B(0)\beta_B(\theta_B) - \varepsilon_z^m(0)\beta_m(\theta_m)}{\varepsilon_z^B(0)\beta_B(\theta_B) + \varepsilon_z^m(0)\beta_m(\theta_m)}\right], \quad \text{[L3.239]}$$

$$g_i^2(\theta_i - \psi) \equiv \frac{\varepsilon_x^i}{\varepsilon_z^i} + \frac{(\varepsilon_y^i - \varepsilon_x^i)}{\varepsilon_z^i}\sin^2(\theta_i - \psi); \quad \text{[L3.235]}$$

$$\beta_i^2(\theta_i) = \rho^2 g_i^2(\theta_i - \psi) + \kappa_i^2, \quad \text{[L3.234]}$$

$$\kappa_i^2 \equiv \frac{e^2}{\varepsilon_0\varepsilon_z^i kT}\sum_{\nu=-\infty}^{\nu=\infty}\nu^2 n_\nu^i \text{ mks}, \; \kappa_i^2 \equiv \frac{4\pi e^2}{\varepsilon_z^i kT}\sum_{\nu=-\infty}^{\nu=\infty}\nu^2 n_\nu^i \text{ cgs}; \quad \text{[L3.233]}$$

where n_ν^i are the mean number densities of ions of valence ν in regions i = A, m, or B; and $\varepsilon(0)$'s are the dielectric constants in the limit of zero frequency ($\xi_{n=0}$).

Table P.9.d. Birefringent media A and B across isotropic medium m, principal axes perpendicular to interface

---

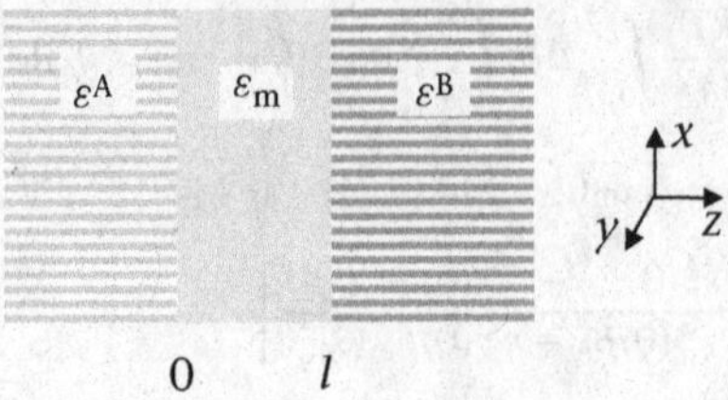

$\varepsilon_x^{\mathrm{A}} = \varepsilon_y^{\mathrm{A}} = \varepsilon_{\parallel}^{\mathrm{A}};\ \varepsilon_x^{\mathrm{B}} = \varepsilon_y^{\mathrm{B}} = \varepsilon_{\parallel}^{\mathrm{B}};\ \varepsilon_z^{\mathrm{A}} = \varepsilon_{\perp}^{\mathrm{A}};$

$\varepsilon_z^{\mathrm{B}} = \varepsilon_{\perp}^{\mathrm{B}};\ \varepsilon_x^{\mathrm{m}} = \varepsilon_y^{\mathrm{m}} = \varepsilon_z^{\mathrm{m}} = \varepsilon_{\mathrm{m}};$

$$\boldsymbol{\varepsilon}^{\mathrm{i}} = \begin{bmatrix} \varepsilon_{\parallel}^{\mathrm{i}} & 0 & 0 \\ 0 & \varepsilon_{\parallel}^{\mathrm{i}} & 0 \\ 0 & 0 & \varepsilon_{\perp}^{\mathrm{i}} \end{bmatrix},\ \mathrm{i} = \mathrm{A}, \mathrm{B}.$$

$$G(l) = -\frac{kT}{8\pi l^2} \sum_{n=0}^{\infty}{}' \sum_{\mathrm{j}=1}^{\infty} \frac{\overline{\Delta}_{\mathrm{Am}}^{\mathrm{j}} \overline{\Delta}_{\mathrm{Bm}}^{\mathrm{j}}}{\mathrm{j}^3},$$

$$\overline{\Delta}_{\mathrm{Am}} = \left(\frac{\sqrt{\varepsilon_{\perp}^{\mathrm{A}} \varepsilon_{\parallel}^{\mathrm{A}}} - \varepsilon_{\mathrm{m}}}{\sqrt{\varepsilon_{\perp}^{\mathrm{A}} \varepsilon_{\parallel}^{\mathrm{A}}} + \varepsilon_{\mathrm{m}}} \right),\quad \overline{\Delta}_{\mathrm{Bm}} = \left(\frac{\sqrt{\varepsilon_{\perp}^{\mathrm{B}} \varepsilon_{\parallel}^{\mathrm{B}}} - \varepsilon_{\mathrm{m}}}{\sqrt{\varepsilon_{\perp}^{\mathrm{B}} \varepsilon_{\parallel}^{\mathrm{B}}} + \varepsilon_{\mathrm{m}}} \right).$$

Note:

$$g_{\mathrm{i}}^2(\theta_{\mathrm{i}} - \psi) \equiv \frac{\varepsilon_x^{\mathrm{i}}}{\varepsilon_z^{\mathrm{i}}} + \frac{\left(\varepsilon_y^{\mathrm{i}} - \varepsilon_x^{\mathrm{i}}\right)}{\varepsilon_z^{\mathrm{i}}} \sin^2(\theta_{\mathrm{i}} - \psi) = \frac{\varepsilon_{\parallel}^{\mathrm{i}}}{\varepsilon_{\perp}^{\mathrm{i}}},\ \mathrm{i} = \mathrm{A}, \mathrm{B};\ g_{\mathrm{m}}^2 = 1;$$

$$\overline{\Delta}_{\mathrm{im}}(\xi_n, \theta_{\mathrm{m}}, \psi) = \left[\frac{\varepsilon_z^{\mathrm{i}} g_{\mathrm{i}}(-\psi) - \varepsilon_z^{\mathrm{m}} g_{\mathrm{m}}(\theta_{\mathrm{m}} - \psi)}{\varepsilon_z^{\mathrm{i}} g_{\mathrm{i}}(-\psi) + \varepsilon_z^{\mathrm{m}} g_{\mathrm{m}}(\theta_{\mathrm{m}} - \psi)} \right] = \left(\frac{\sqrt{\varepsilon_{\parallel}^{\mathrm{i}} \varepsilon_{\perp}^{\mathrm{i}}} - \varepsilon_{\mathrm{m}}}{\sqrt{\varepsilon_{\parallel}^{\mathrm{i}} \varepsilon_{\perp}^{\mathrm{i}}} + \varepsilon_{\mathrm{m}}} \right);$$

$$\int_0^{2\pi} \frac{\left[\overline{\Delta}_{\mathrm{Am}}(\xi_n, \theta_{\mathrm{m}}, \psi) \overline{\Delta}_{\mathrm{Bm}}(\xi_n, \theta_{\mathrm{m}}, \theta_{\mathrm{B}}, \psi)\right]^j d\psi}{g_{\mathrm{m}}^2(\theta_{\mathrm{m}} - \psi)} = \left[\overline{\Delta}_{\mathrm{Am}} \overline{\Delta}_{\mathrm{Bm}}\right]^{\mathrm{j}} \int_0^{2\pi} d\psi$$

$$G(l, \theta_{\mathrm{m}}, \theta_{\mathrm{B}}) = -\frac{kT}{16\pi^2 l^2} \sum_{n=0}^{\infty}{}' \sum_{\mathrm{j}=1}^{\infty} \frac{1}{\mathrm{j}^3} \int_0^{2\pi} \frac{\left[\overline{\Delta}_{\mathrm{Am}}(\xi_n, \theta_{\mathrm{m}}, \psi) \overline{\Delta}_{\mathrm{Bm}}(\xi_n, \theta_{\mathrm{m}}, \theta_{\mathrm{B}}, \psi)\right]^{\mathrm{j}} d\psi}{g_{\mathrm{m}}^2(\theta_{\mathrm{m}} - \psi)}$$

$$= -\frac{kT}{8\pi l^2} \sum_{n=0}^{\infty}{}' \sum_{\mathrm{j}=1}^{\infty} \frac{\left[\overline{\Delta}_{\mathrm{Am}} \overline{\Delta}_{\mathrm{Bm}}\right]^j}{\mathrm{j}^3}$$

Table P.9.e. Birefringent media A and B across isotropic medium m, principal axes parallel to interface and at a mutual angle θ

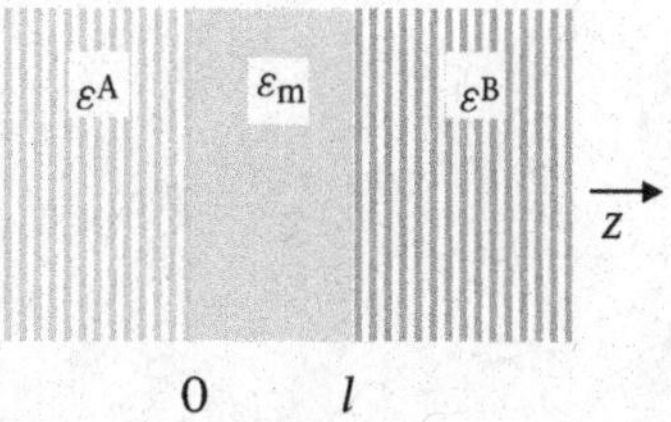

$$\varepsilon^{\mathrm{i}} = \begin{pmatrix} \varepsilon_{\parallel}^{\mathrm{i}} & 0 & 0 \\ 0 & \varepsilon_{\perp}^{\mathrm{i}} & 0 \\ 0 & 0 & \varepsilon_{\perp}^{\mathrm{i}} \end{pmatrix}, \ \mathrm{i} = \mathrm{A, B}$$

$\varepsilon_{x_A}^{A} = \varepsilon_{\parallel}^{A}, \varepsilon_{y_A}^{A} = \varepsilon_{z}^{A} = \varepsilon_{\perp}^{A};$

$\varepsilon_{x_B}^{B} = \varepsilon_{\parallel}^{B}, \varepsilon_{y_B}^{B} = \varepsilon_{z}^{B} = \varepsilon_{\perp}^{B};$

$\varepsilon_{x}^{m} = \varepsilon_{y}^{m} = \varepsilon_{z}^{m} = \varepsilon_{m};$

x_A, x_B, y_A, y_B parallel to interfaces; $\theta_A \equiv 0, \theta_B \equiv \theta$; $\theta =$ angle between principal axes of A and B.

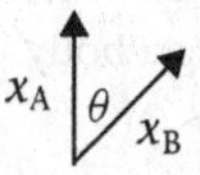

$$G(l,\theta) = -\frac{kT}{16\pi^2 l^2} \sum_{n=0}^{\infty}{}' \sum_{\mathrm{j}=1}^{\infty} \frac{1}{\mathrm{j}^3} \int_0^{2\pi} \left[\overline{\Delta}_{\mathrm{Am}}(\xi_n,\psi)\overline{\Delta}_{\mathrm{Bm}}(\xi_n,\theta,\psi)\right]^{\mathrm{j}} \mathrm{d}\psi, \qquad \text{[(L3.222)]}$$

$$\overline{\Delta}_{\mathrm{im}}(\xi_n,\theta_{\mathrm{i}},\psi) = \left\{ \frac{\sqrt{\varepsilon_{\perp}^{\mathrm{i}}\varepsilon_{\parallel}^{\mathrm{i}}}\sqrt{1+[(\varepsilon_{\perp}^{\mathrm{i}}-\varepsilon_{\parallel}^{\mathrm{i}})/\varepsilon_{\parallel}^{\mathrm{i}}]\sin^2(\theta_{\mathrm{i}}-\psi)} - \varepsilon_{\mathrm{m}}}{\sqrt{\varepsilon_{\perp}^{\mathrm{i}}\varepsilon_{\parallel}^{\mathrm{i}}}\sqrt{1+[(\varepsilon_{\perp}^{\mathrm{i}}-\varepsilon_{\parallel}^{\mathrm{i}})/\varepsilon_{\parallel}^{\mathrm{i}}]\sin^2(\theta_{\mathrm{i}}-\psi)} + \varepsilon_{\mathrm{m}}} \right\},$$

Torque: $\tau = -\partial G(l,\theta)/\partial\theta\big|_l$

Weak birefringence $|\varepsilon_{\perp}^{\mathrm{i}} - \varepsilon_{\parallel}^{\mathrm{i}}| \ll \varepsilon_{\parallel}^{\mathrm{i}}$, j = 1 term only.

$$G(l,\theta) = -\frac{kT}{8\pi l^2} \sum_{n=0}^{\infty}{}' \left[\overline{\Delta}_{\bar{\mathrm{A}}\mathrm{m}}\overline{\Delta}_{\bar{\mathrm{B}}\mathrm{m}} + \overline{\Delta}_{\bar{\mathrm{A}}\mathrm{m}}\frac{\gamma_{\mathrm{B}}}{2} + \overline{\Delta}_{\bar{\mathrm{B}}\mathrm{m}}\frac{\gamma_{\mathrm{A}}}{2} + \frac{\gamma_{\mathrm{A}}\gamma_{\mathrm{B}}}{8}(1+2\cos^2\theta) \right], \qquad \text{[L1.24]}$$

$$\overline{\Delta}_{\bar{\mathrm{i}}\mathrm{m}} \equiv \left(\frac{\sqrt{\varepsilon_{\perp}^{\mathrm{i}}\varepsilon_{\parallel}^{\mathrm{i}}} - \varepsilon_{\mathrm{m}}}{\sqrt{\varepsilon_{\perp}^{\mathrm{i}}\varepsilon_{\parallel}^{\mathrm{i}}} + \varepsilon_{\mathrm{m}}} \right), \ \gamma_{\mathrm{i}} \equiv \frac{\sqrt{\varepsilon_{\perp}^{\mathrm{i}}\varepsilon_{\parallel}^{\mathrm{i}}}\,(\varepsilon_{\perp}^{\mathrm{i}} - \varepsilon_{\parallel}^{\mathrm{i}})}{2\varepsilon_{\parallel}^{\mathrm{i}}\left(\sqrt{\varepsilon_{\perp}^{\mathrm{i}}\varepsilon_{\parallel}^{\mathrm{i}}} + \varepsilon_{\mathrm{m}}\right)} \ll 1, \mathrm{i} = \mathrm{A, B}.$$

Note:

$$g_{\mathrm{i}}^2(\theta_{\mathrm{i}}-\psi) \equiv \frac{\varepsilon_{\parallel}^{\mathrm{i}}}{\varepsilon_{\perp}^{\mathrm{i}}} + \frac{\left(\varepsilon_{\perp}^{\mathrm{i}} - \varepsilon_{\parallel}^{\mathrm{i}}\right)}{\varepsilon_{\perp}^{\mathrm{i}}}\sin^2(\theta_{\mathrm{i}}-\psi), \mathrm{i} = \mathrm{A, B}, g_{\mathrm{m}}^2 = 1, \qquad \text{[(L3.217)]}$$

$$\overline{\Delta}_{\mathrm{im}}(\xi_n,\theta_{\mathrm{m}},\theta_{\mathrm{i}},\psi) = \left[\frac{\varepsilon_z^{\mathrm{i}} g_{\mathrm{i}}(\theta_{\mathrm{i}}-\psi) - \varepsilon_z^{\mathrm{m}} g_{\mathrm{m}}(\theta_{\mathrm{m}}-\psi)}{\varepsilon_z^{\mathrm{i}} g_{\mathrm{i}}(\theta_{\mathrm{i}}-\psi) + \varepsilon_z^{\mathrm{m}} g_{\mathrm{m}}(\theta_{\mathrm{m}}-\psi)} \right], \qquad \text{[(L3.218) and (L3.219)]}$$

$|\varepsilon_{\perp}^{\mathrm{i}} - \varepsilon_{\parallel}^{\mathrm{i}}| \ll \varepsilon_{\parallel}^{\mathrm{i}}$, $\overline{\Delta}_{\mathrm{im}}(\xi_n,\theta_{\mathrm{i}},\psi) \approx \left(\frac{\sqrt{\varepsilon_{\perp}^{\mathrm{i}}\varepsilon_{\parallel}^{\mathrm{i}}} - \varepsilon_{\mathrm{m}}}{\sqrt{\varepsilon_{\perp}^{\mathrm{i}}\varepsilon_{\parallel}^{\mathrm{i}}} + \varepsilon_{\mathrm{m}}} \right) + \frac{\sqrt{\varepsilon_{\perp}^{\mathrm{i}}\varepsilon_{\parallel}^{\mathrm{i}}}(\varepsilon_{\perp}^{\mathrm{i}} - \varepsilon_{\parallel}^{\mathrm{i}})\sin^2(\theta_{\mathrm{i}}-\psi)}{2\varepsilon_{\parallel}^{\mathrm{i}}(\sqrt{\varepsilon_{\perp}^{\mathrm{i}}\varepsilon_{\parallel}^{\mathrm{i}}} + \varepsilon_{\mathrm{m}})} = \overline{\Delta}_{\bar{\mathrm{i}}\mathrm{m}} + \gamma_{\mathrm{i}}\sin^2(\theta_{\mathrm{i}}-\psi)$; use $\int_0^{2\pi}\sin^2(-\psi)\mathrm{d}\psi = \int_0^{2\pi}\sin^2(\theta-\psi)\mathrm{d}\psi = \pi$, $\int_0^{2\pi}\sin^2(-\psi)\sin^2(\theta-\psi)\mathrm{d}\psi = \sin^2(\theta)\frac{\pi}{4} + \cos^2(\theta)\frac{3}{4}\pi = \frac{\pi}{4}[1+2\cos^2(\theta)]$ (I. S. Gradshteyn & I. M. Ryzhik, *Table of Integrals, Series, and Products*, Academic Press, New York, 1965, Eqs. 2.513.7 and 2.513.21), $\int_0^{2\pi}\overline{\Delta}_{\mathrm{Am}}(\xi_n,\psi)\overline{\Delta}_{\mathrm{Bm}}(\xi_n,\theta_{\mathrm{B}},\psi)d\psi = \overline{\Delta}_{\bar{\mathrm{A}}\mathrm{m}}\overline{\Delta}_{\bar{\mathrm{B}}\mathrm{m}}2\pi + \overline{\Delta}_{\bar{\mathrm{A}}\mathrm{m}}\gamma_{\mathrm{B}}\pi + \overline{\Delta}_{\bar{\mathrm{B}}\mathrm{m}}\gamma_{\mathrm{A}}\pi + \gamma_{\mathrm{A}}\gamma_{\mathrm{B}}\frac{\pi}{4}(1+2\cos^2(\theta))$.

Table P.10.a. Sphere in a sphere, Lifshitz form, retardation neglected and magnetic terms omitted

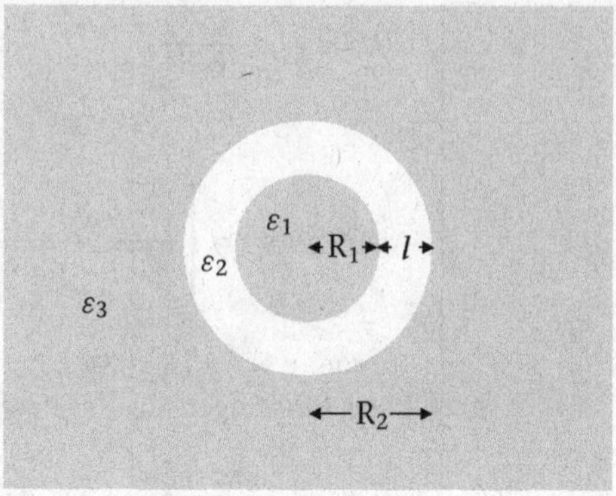

$$G_{\mathrm{sph}}(l;\mathrm{R}_1,\mathrm{R}_2) = kT\sum_{n=0}^{\infty}{}'\sum_{\mathrm{m}=1}^{\infty}(2\mathrm{m}+1)\ln\left\{1-\frac{m(m+1)(\varepsilon_1-\varepsilon_2)(\varepsilon_3-\varepsilon_2)}{[(m+1)\varepsilon_2+m\varepsilon_1]\,[m\varepsilon_2+(m+1)\varepsilon_3]}\left(\frac{\mathrm{R}_1}{\mathrm{R}_2}\right)^{2m+1}\right\}.$$

N.B. This energy is in *addition* to the interfacial energies that will change when l, R_1, and/or R_2 is varied. The energy given here is the difference in energy for making the 1–2–3 configuration shown at left *minus* the energies of making a body of substance 1, radius R_1 in medium 2,

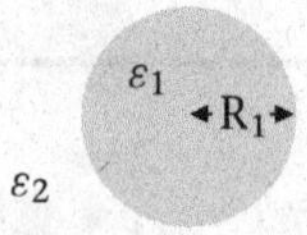

and of making a body of material 2 radius R_2 in medium 3.

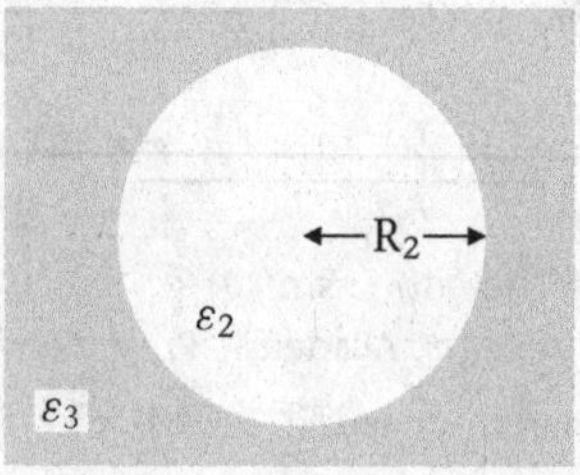

Source: $G_{\mathrm{sph}}(l;\mathrm{R}_1,\mathrm{R}_2)$ from Eq. (33) in V. A. Parsegian and G. H. Weiss, "Electrodynamic interaction between curved parallel surfaces," J. Chem. Phys. **60**, 5080–5085 (1974).

Table P.10.b. Small sphere in a concentric large sphere, special case $R_1 \ll R_2$

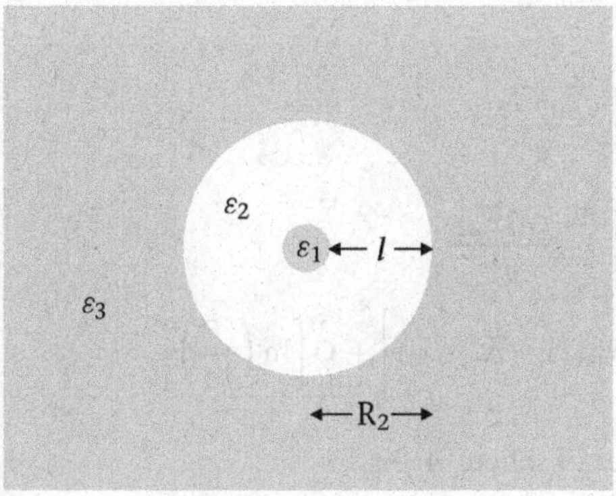

m = 1 term only:

$$G_{\mathrm{sph}}(l; R_1, R_2) = -6kT\left(\frac{R_1}{R_2}\right)^3 \sum_{n=0}^{\infty}{}' \frac{(\varepsilon_1 - \varepsilon_2)(\varepsilon_3 - \varepsilon_2)}{(\varepsilon_1 + 2\varepsilon_2)(\varepsilon_2 + 2\varepsilon_3)}$$

$$\approx -\frac{8kT}{3}\left(\frac{R_1}{R_2}\right)^3 \sum_{n=0}^{\infty}{}' \frac{(\varepsilon_1 - \varepsilon_2)(\varepsilon_3 - \varepsilon_2)}{(\varepsilon_1 + \varepsilon_2)(\varepsilon_2 + \varepsilon_3)}, \; \varepsilon_1 \approx \varepsilon_2 \approx \varepsilon_3.$$

Compare: Spherical point particle of radius R interacting with a flat surface in the limit of small differences in ε (see Table S.12.a, replace R with R_1, z with $R_2 \approx l$, ε_s with ε_1, ε_m with ε_2, ε_A with ε_3):

$$g_p(z) = -\frac{kT}{2}\left(\frac{R_1}{R_2}\right)^3 \sum_{n=0}^{\infty}{}' \frac{\varepsilon_1 - \varepsilon_2}{\varepsilon_1 + 2\varepsilon_2}\frac{\varepsilon_3 - \varepsilon_2}{\varepsilon_3 + \varepsilon_2}$$

$$\approx -\frac{kT}{3}\left(\frac{R_1}{R_2}\right)^3 \sum_{n=0}^{\infty}{}' \frac{\varepsilon_1 - \varepsilon_m}{\varepsilon_1 + \varepsilon_m}\frac{\varepsilon_3 - \varepsilon_2}{\varepsilon_3 + \varepsilon_2}.$$

Table P.10.c. Concentric parallel surfaces, special case $R_1 \approx R_2 \gg R_2 - R_1 = l$, slightly bent planes; retardation and magnetic terms neglected

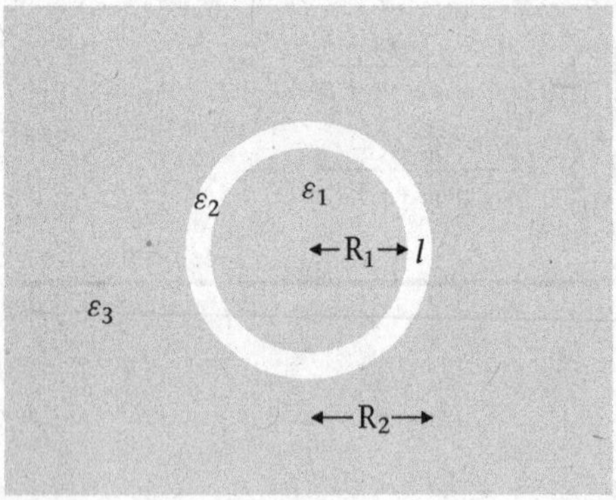

P.10.c.1. Sphere in a sphere

$$G_{sph}(l; R_1 \approx R_2) = -\frac{kTR_1^2}{2l^2}\sum_{n=0}^{\infty}{}'\sum_{q=1}^{\infty}\frac{\overline{\Delta}_{12}\overline{\Delta}_{32}}{q^3} - \frac{kTR_1}{2l}\sum_{n=0}^{\infty}{}'\left[\sum_{q=1}^{\infty}\frac{\overline{\Delta}_{12}\overline{\Delta}_{32}}{q^3} + (\overline{\Delta}_{12} - \overline{\Delta}_{32})\ln(1 - \overline{\Delta}_{12}\overline{\Delta}_{32})\right] + O\left[\ln\left(\frac{l}{R_1}\right)\right].$$

Normalized per area of sphere, $4\pi R_1^2$:

$$G_{sph}(l; R_1 \approx R_2) = -\frac{kT}{8\pi l^2}\sum_{n=0}^{\infty}{}'\sum_{q=1}^{\infty}\frac{\overline{\Delta}_{12}\overline{\Delta}_{32}}{q^3} - \frac{kT}{8\pi l R_1}\sum_{n=0}^{\infty}{}'\left[\sum_{q=1}^{\infty}\frac{\overline{\Delta}_{12}\overline{\Delta}_{32}}{q^3} + (\overline{\Delta}_{12} - \overline{\Delta}_{32})\ln(1 - \overline{\Delta}_{12}\overline{\Delta}_{32})\right] + \cdots.$$

This is the interaction of two flat parallel planes plus a correction that is a factor $\sim l/R_1$ smaller.

To lowest order in (small) $\overline{\Delta}$'s:

$$G_{sph}(l; R_1 \approx R_2) = -\frac{kT}{8\pi l^2}\left(1 + \frac{l}{R_1}\right)\sum_{n=0}^{\infty}{}'\overline{\Delta}_{12}\overline{\Delta}_{32} + \cdots + = G_{planar}(l)\left(1 + \frac{l}{R_1}\right) + \cdots.$$

P.10.c.2. Cylinder in a cylinder

$$G_{cyl}(l; R_1 \approx R_2) \sim -\frac{kTR_1}{4l^2}\sum_{n=0}^{\infty}{}'\sum_{q=1}^{\infty}\frac{\overline{\Delta}_{12}\overline{\Delta}_{32}}{q^3} + \frac{kT}{4l}(\overline{\Delta}_{32} - \overline{\Delta}_{12})\sum_{n=0}^{\infty}{}'\sum_{q=1}^{\infty}\frac{\overline{\Delta}_{12}\overline{\Delta}_{32}}{q^2}$$

per unit length.

Energy per unit area:

$$G_{cyl}(l; R_1 \approx R_2) \sim -\frac{kT}{8\pi l^2}\sum_{n=0}^{\infty}{}'\sum_{q=1}^{\infty}\frac{\overline{\Delta}_{12}\overline{\Delta}_{32}}{q^3} + \frac{kT}{8\pi l R_1}(\overline{\Delta}_{32} - \overline{\Delta}_{12})\sum_{n=0}^{\infty}{}'\sum_{q=1}^{\infty}\frac{\overline{\Delta}_{12}\overline{\Delta}_{32}}{q^2}.$$

Table P.10.c (*cont.*)

	To lowest order in (small) $\overline{\Delta}$'s: $G_{cyl}(l; R_1 \approx R_2) \rightarrow -\frac{kT}{8\pi l^2}\sum_{n=0}^{\infty}{}' \overline{\Delta}_{12}\overline{\Delta}_{32} + \frac{kT}{8\pi l R_1}\sum_{n=0}^{\infty}{}' (\overline{\Delta}_{32}-\overline{\Delta}_{12})\overline{\Delta}_{12}\overline{\Delta}_{32} = G_{planar}(l) + \frac{kT}{8\pi l R_1}\sum_{n=0}^{\infty}{}' (\overline{\Delta}_{32}-\overline{\Delta}_{12})\overline{\Delta}_{12}\overline{\Delta}_{32}.$ Material 1 same as material 3 $G_{cyl}(l; R_1 \approx R_2) = -\frac{kT}{8\pi l^2}\sum_{n=0}^{\infty}{}' \sum_{q=1}^{\infty}\frac{\overline{\Delta}_{12}\overline{\Delta}_{32}}{q^3}$ per unit area.

Note: Spheres: From Eq. 41, V. A. Parsegian and G. H. Weiss, "Electrodynamic interaction between curved parallel surfaces," J. Chem. Phys. **60**, 5080–5085 (1974):

$$\frac{G_n}{kT} = -\frac{R_1^2}{2l^2}\sum_{q=1}^{\infty}\frac{\overline{\Delta}_{12}\overline{\Delta}_{32}}{q^3} - \frac{R_1}{2l}\sum_{q=1}^{\infty}\frac{\overline{\Delta}_{12}\overline{\Delta}_{32}}{q^3} - \frac{R_1}{2l}(\overline{\Delta}_{21}+\overline{\Delta}_{32})\ln(1-\overline{\Delta}_{12}\overline{\Delta}_{32}) + O\left[\ln\left(\frac{l}{R_1}\right)\right]$$

$$= -\frac{R_1^2}{2l^2}\sum_{q=1}^{\infty}\frac{\overline{\Delta}_{12}\overline{\Delta}_{32}}{q^3} - \frac{R_1}{2l}\left[\sum_{q=1}^{\infty}\frac{\overline{\Delta}_{12}\overline{\Delta}_{32}}{q^3} + (\overline{\Delta}_{12}-\overline{\Delta}_{32})\ln(1-\overline{\Delta}_{12}\overline{\Delta}_{32})\right] + O\left[\ln\left(\frac{l}{R_1}\right)\right].$$

Divide by $4\pi R_1^2$ to create an energy per area:

$$G_n = -\frac{kT}{8\pi l^2}\sum_{q=1}^{\infty}\frac{\overline{\Delta}_{12}\overline{\Delta}_{32}}{q^3} - \frac{kT}{8\pi R_1 l}\left[\sum_{q=1}^{\infty}\frac{\overline{\Delta}_{12}\overline{\Delta}_{32}}{q^3} + (\overline{\Delta}_{12}-\overline{\Delta}_{32})\ln(1-\overline{\Delta}_{12}\overline{\Delta}_{32})\right] + O\left[\ln\left(\frac{l}{R_1}\right)\Big/R_1^2\right].$$

To lowest terms in $\overline{\Delta}$'s:

$$G_n = -\frac{kT}{8\pi l^2}\overline{\Delta}_{12}\overline{\Delta}_{32} - \frac{kT}{8\pi R_1 l}[\overline{\Delta}_{12}\overline{\Delta}_{32} + (\overline{\Delta}_{12}-\overline{\Delta}_{32})\ln(1-\overline{\Delta}_{12}\overline{\Delta}_{32})] + O\left[\ln\left(\frac{l}{R_1}\right)\Big/R_1^2\right]$$

$$\approx -\frac{kT}{8\pi l^2}\left(1+\frac{l}{R_1}\right)\overline{\Delta}_{12}\overline{\Delta}_{32}.$$

For material 1 the same as material 2:

$$G_n = -\frac{kT}{8\pi l^2}\overline{\Delta}_{12}^2 - \frac{kT}{8\pi R_1 l}\overline{\Delta}_{12}^2 = -\frac{kT}{8\pi l^2}\overline{\Delta}_{12}^2\left(1+\frac{l}{R_1}\right).$$

See also A. A. Saharian, "Scalar Casimir effect for D-dimensional spherically symmetric Robin boundaries," Phys. Rev. D, 63, 125007 (2001) and references therein.

Cylinders: From Eq. 28, V. A. Parsegian and G. H. Weiss, "Electrodynamic interaction between curved parallel surfaces," J. Chem. Phys. **60**, 5080–5085 (1974):

$$G_{cyl}(l) \sim -\frac{kTR_1}{4l^2}\sum_{n=0}^{\infty}{}'\sum_{q=1}^{\infty}\frac{\overline{\Delta}_{12}\overline{\Delta}_{32}}{q^3} + \frac{kT}{4l}(\overline{\Delta}_{32}-\overline{\Delta}_{12})\sum_{n=0}^{\infty}{}'\sum_{q=1}^{\infty}\frac{\overline{\Delta}_{12}\overline{\Delta}_{32}}{q^2}.$$

Divide by $2\pi R_1$ to create an energy per area:

$$G_{cyl}(l) \sim -\frac{kT}{8\pi l^2}\sum_{n=0}^{\infty}{}'\sum_{q=1}^{\infty}\frac{\overline{\Delta}_{12}\overline{\Delta}_{32}}{q^3} + \frac{kT}{8\pi l R_1}(\overline{\Delta}_{32}-\overline{\Delta}_{12})\sum_{n=0}^{\infty}{}'\sum_{q=1}^{\infty}\frac{\overline{\Delta}_{12}\overline{\Delta}_{32}}{q^2}.$$

See also F. D. Mazzitelli, M. J. Sanchez, N. N. Scoccala, and J. von Stecher, "Casimir interaction between two concentric cylinders: exact versus semiclassical result," Phys. Rev. A, 67, 013807 (2003) and references therein.

Table P.10.c (*cont.*)

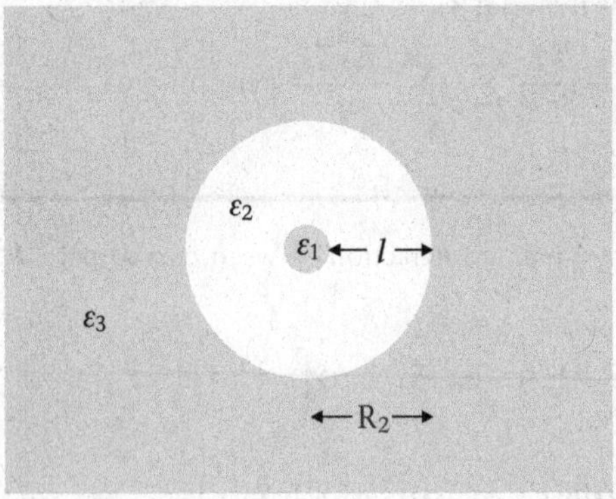

P.10.c.3. Thin cylinder in a concentric large cylinder, special case $R_1 \ll R_2$

$$G_{cyl}(l; R_1, R_2) \approx -\frac{9A_H}{4}\frac{\pi R_1^2}{R_2^2}$$

in the Hamaker approximation.

Hamaker integration over incremental interactions

$$-\frac{A_H}{\pi^2}\frac{dV_1\,dV_2}{r^6}$$

(see Section L2.3.D.).

Because the radius of the inner cylinder $R_1 \ll l \approx R_2$, the distance r between incremental volumes is $r^2 = r_2^2 + z_2^2$, $R_2 \le r_2 < \infty$, $-\infty < z_2 < +\infty$ (z_2 perpendicular to the plane of the picture), $dV_1 = \pi R_1^2$ per unit length; $dV_2 = 2\pi r_2 dr_2 dz_2$ per unit length.

The required integration is

$$-\frac{A_H}{\pi^2}\pi R_1^2 2\pi \int_{-\infty}^{\infty}\int_{l}^{\infty}\frac{r_2 dr_2 dz_2}{(r_2^2+z_2^2)^3} = -\frac{A_H}{4}3\pi R_1^2\int_{l}^{\infty}\frac{dr_2}{r_2^4}$$

$$= -\frac{9A_H}{4}\frac{\pi R_1^2}{l^3} \approx -\frac{9A_H}{4}\frac{\pi R_1^2}{R_2^2},$$

where

$$\int_{-\infty}^{\infty}\frac{dz_2}{(r_2^2+z_2^2)^3} = \frac{3\pi}{8r_2^5}.$$

(I. S. Gradshteyn and I. M. Ryzhik, *Table of Integrals, Series, and Products*, Academic Press, New York, 1965, Eq. 3.252.2).

L2.2.B. TABLES OF FORMULAE IN SPHERICAL GEOMETRY

Table S.1. Spheres at separations small compared with radius, Derjaguin transform from Lifshitz planar result, including retardation and all higher-order interactions

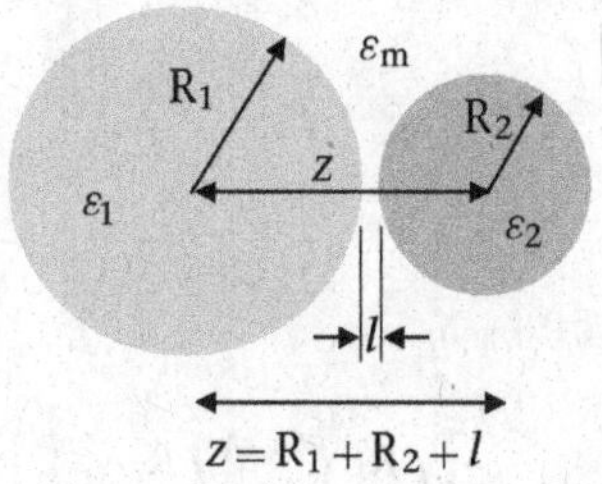

$l \ll R_1, R_2$

S.1.a. Force

$$F_{ss}(l; R_1, R_2) = \frac{2\pi R_1 R_2}{(R_1 + R_2)} G_{pp}(l)$$

$$= -\frac{kT}{4l^2}\frac{R_1 R_2}{R_1 + R_2}\sum_{n=0}^{\infty}{}' r_n^2 \sum_{q=1}^{\infty}\frac{1}{q} \times \int_1^{\infty} p[(\overline{\Delta}_{1m}\overline{\Delta}_{2m})^q + (\Delta_{1m}\Delta_{2m})^q] e^{-r_n p q} dp. \quad \text{[(L2.108)]}$$

S.1.b. Free energy of interaction

$$G_{ss}(l; R_1, R_2) = -\frac{kT}{4l}\frac{R_1 R_2}{R_1 + R_2}\sum_{n=0}^{\infty}{}' r_n \sum_{q=1}^{\infty}\frac{1}{q^2} \times \int_1^{\infty} [(\overline{\Delta}_{1m}\overline{\Delta}_{2m})^q + (\Delta_{1m}\Delta_{2m})^q] e^{-r_n p q} dp, \quad \text{[L2.113]}$$

$$\overline{\Delta}_{ji} = \frac{s_i\varepsilon_j - s_j\varepsilon_i}{s_i\varepsilon_j + s_j\varepsilon_i},\ \Delta_{ji} = \frac{s_i\mu_j - s_j\mu_i}{s_i\mu_j + s_j\mu_i},\ s_i = \sqrt{p^2 - 1 + (\varepsilon_i\mu_i/\varepsilon_m\mu_m)},$$

$$r_n = \left(2l\varepsilon_m^{1/2}\mu_m^{1/2}/c\right)\xi_n.$$

S.1.c. Nonretarded limit

$$G_{ss}(l; R_1, R_2) = -\frac{kT}{4l}\frac{R_1 R_2}{R_1 + R_2}\sum_{n=0}^{\infty}{}' \sum_{q=1}^{\infty}\frac{[(\overline{\Delta}_{1m}\overline{\Delta}_{2m})^q + (\Delta_{1m}\Delta_{2m})^q]}{q^3}. \quad \text{[L2.115]}$$

S.1.c.1. Spheres of equal radii

$$R_2 = R_1 = R,\ \frac{R_1 R_2}{R_1 + R_2} = \frac{R}{2}.$$

S.1.c.2. Sphere-with-a-plane, $R_2 \to \infty$

$$R_1 = R,\ \frac{R_1 R_2}{R_1 + R_2} = R.$$

Note: Force from free energy:

$$F_{ss}(l; R_1, R_2) = \frac{2\pi R_1 R_2}{R_1 + R_2} G_{pp}(l);\ G_{pp}(l) = -\frac{kT}{8\pi l^2}\sum_{n=0}^{\infty}{}' r_n^2 \sum_{q=1}^{\infty}\frac{1}{q}\int_1^{\infty} p[(\overline{\Delta}_{1m}\overline{\Delta}_{2m})^q + (\Delta_{1m}\Delta_{2m})^q] e^{-r_n p q} dp,$$

$$F_{ss}(l; R_1, R_2) = -\frac{kT}{4l^2}\frac{R_1 R_2}{R_1 + R_2}\sum_{n=0}^{\infty}{}' r_n^2 \sum_{q=1}^{\infty}\frac{1}{q}\int_1^{\infty} p[(\overline{\Delta}_{1m}\overline{\Delta}_{2m})^q + (\Delta_{1m}\Delta_{2m})^q] e^{-r_n p q} dp.$$

Table S.2. Sphere–sphere interactions, limiting forms

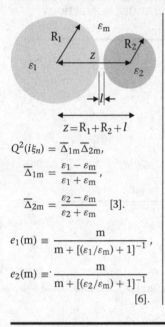

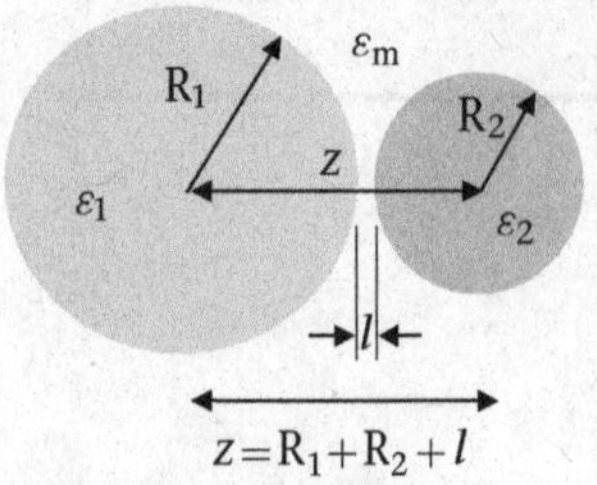

$$Q^2(i\xi_n) = \overline{\Delta}_{1m}\overline{\Delta}_{2m},$$

$$\overline{\Delta}_{1m} = \frac{\varepsilon_1 - \varepsilon_m}{\varepsilon_1 + \varepsilon_m},$$

$$\overline{\Delta}_{2m} = \frac{\varepsilon_2 - \varepsilon_m}{\varepsilon_2 + \varepsilon_m} \quad [3].$$

$$e_1(\mathrm{m}) \equiv \frac{\mathrm{m}}{\mathrm{m} + [(\varepsilon_1/\varepsilon_m) + 1]^{-1}},$$

$$e_2(\mathrm{m}) \equiv \frac{\mathrm{m}}{\mathrm{m} + [(\varepsilon_2/\varepsilon_m) + 1]^{-1}} \quad [6].$$

S.2.a. Many-body expansion to all orders, at all separations, no retardation

$$G_{ss}(z; \mathrm{R}_1, \mathrm{R}_2) = -kT \sum_{n=0}^{\infty}{}' g(z; i\xi_n) \quad [1],$$

$$g(z; i\xi_n) = \sum_{\nu=1}^{\infty} T_\nu(z; i\xi_n) \frac{Q^{2\nu}(i\xi_n)}{\nu} \quad [2],$$

$$T_\nu(z; i\xi_n) = \sum_{m_1=1}^{\infty} \ldots \sum_{n_\nu=1}^{\infty} C(m_1, n_1, m_2, n_2, \cdots m_\nu, n_\nu)$$

$$\times \prod_{i=1}^{\nu} e_1(m_i) e_2(n_i) \left(\frac{\mathrm{R}_1}{z}\right)^{2m_i+1} \left(\frac{\mathrm{R}_2}{z}\right)^{2n_i+1} \quad [4],$$

$$C(\sigma_1, \sigma_2, \sigma_3, \sigma_4, \cdots \sigma_k) = \sum_{\mu=-\infty}^{\infty} \prod_{i=1}^{k} \binom{\sigma_i + \sigma_{i+1}}{\sigma_i + \mu}, \quad \sigma_{k+1} = \sigma_1 \quad [5],$$

$$C(\sigma) = 4^\sigma, \; C(\sigma_1, \sigma_2) = \binom{2\sigma_1 + 2\sigma_2}{2\sigma_1} \quad [7].$$

Source: Original many-body formulation in Section 4.2, D. Langbein, *Van der Waals Attraction*, Springer Tracts in Modern Physics (Springer-Verlag, Berlin, 1974) (hereafter L1974) and "Non-retarded dispersion energy between macroscopic spheres," J. Phys. Chem. Solids, **32**, 1657 (1971) (hereafter L1971). Results stated here in the notation of J. E. Kiefer, V. A. Parsegian, and G. H. Weiss, "Some convenient bounds and approximations for many body van der Waals attraction between two spheres," J. Colloid Interface Sci., **63**, 140–153 (1978) (hereafter KPW 1978). Numbers in [ ] correspond to those in KPW 1978.

The coefficient for Equation [1] used here differs from that in KPW 1978 because of the substitution of summation over $\xi_n = (2\pi kT/\hbar)n$ for integration with a consequent factor $2\pi kT/\hbar$: $\frac{h}{8\pi^2}\int_{-\infty}^{\infty} d\xi = \frac{\hbar}{2\pi}\int_0^{\infty} d\xi = \frac{\hbar}{2\pi}\frac{2\pi kT}{\hbar}\int_0^{\infty} dn = kT\sum_{n=0}^{\prime\infty}$.

Equation [6] for $e_1(\mathrm{m})$, $e_2(\mathrm{m})$ here and in KPW 1978 comes from Eq. (10) in L1971 for $\eta_1(\mathrm{m})$, $\eta_2(\mathrm{m})$ by factoring out $Q^2(i\xi_n) = \overline{\Delta}_{1m}\overline{\Delta}_{2m}$: $e_1(\mathrm{m}) = \eta_1(\mathrm{m})\overline{\Delta}_{1m}$, $e_2(\mathrm{m}) = \eta_2(\mathrm{m})\overline{\Delta}_{2m}$.

The grandiose summation $\sum_{\mu=-\infty}^{\infty}$ in [5] simply means "include all values of μ that do not create the (zero value) factorial of a negative number." Expanding the product in [5],

$$\prod_{i=1}^{k}\binom{\sigma_i + \sigma_{i+1}}{\sigma_i + \mu} = \frac{(\sigma_1+\sigma_2)!}{(\sigma_1+\mu)!(\sigma_2-\mu)!}\frac{(\sigma_2+\sigma_3)!}{(\sigma_2+\mu)!(\sigma_3-\mu)!}\cdots\frac{(\sigma_k+\sigma_1)!}{(\sigma_k+\mu)!(\sigma_1-\mu)!} = \frac{\prod_{i=1}^{k}(\sigma_i+\sigma_{i+1})!}{\prod_{i=1}^{k}(\sigma_i+\mu)!(\sigma_{i+1}-\mu)!},$$

makes clear why there can be no value of μ allowed bigger than the smallest of the σ_i's.

In [7], $C(\sigma) = \sum_{\mu=-\sigma}^{+\sigma}\binom{\sigma+\sigma}{\sigma+\mu} = (2\sigma)^2 = 4^\sigma$ from the sum of the binomial coefficients; for $C(\sigma_1, \sigma_2)$, see Eq. (15) of L1971 and Eq. (4.33) of L1974.

Sphere–sphere interactions are treated in a similar spirit by J. D. Love, "On the van der Waals force between two spheres or a sphere and a wall," J. Chem. Soc. Faraday Trans. 2, **73**, 669–688 (1977).

Table S.2 (*cont.*)

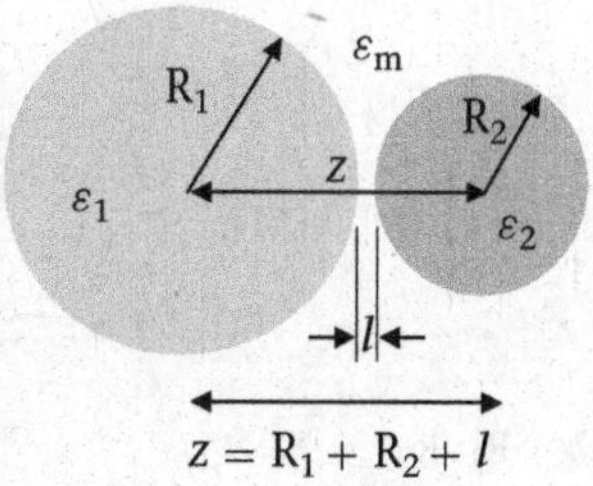

S.2.b. Sphere–sphere interaction expanded about long-distance limit, retardation neglected

$$G_{ss}(z) = -kT\sum_{n=0}^{\infty}{}' \left[\sum_{n_1=1}^{\infty} \eta_1(n_1)\left(\frac{R_1}{z}\right)^{2n_1+1} \sum_{n_2=1}^{\infty} \eta_2(n_2)\left(\frac{R_2}{z}\right)^{2n_2+1} \frac{(2n_1+2n_2)!}{(2n_1)!\,(2n_2)!}\right],$$

$$\eta_1(n_1) \equiv \frac{n_1(\varepsilon_1 - \varepsilon_m)}{n_1(\varepsilon_1 + \varepsilon_m) + \varepsilon_m};\ \eta_2(n_2) \equiv \frac{n_2(\varepsilon_2 - \varepsilon_m)}{n_2(\varepsilon_2 + \varepsilon_m) + \varepsilon_m}$$

$n_1, n_2 = 1, 2, \ldots$; ε's are $\varepsilon(i\xi_n)$.

$R_1, R_2 \ll z$ ($n_1 = n_2 = 1$ term only):

$$G_{ss}(z; R_1, R_2) \to -kT\frac{R_1^3 R_2^3}{z^6}\sum_{n=0}^{\infty}{}' \left[\frac{(\varepsilon_1 - \varepsilon_m)}{\varepsilon_1 + 2\varepsilon_m}\frac{(\varepsilon_2 - \varepsilon_m)}{\varepsilon_2 + 2\varepsilon_m}\right].$$

Note: Better than the Hamaker limit but still for small differences in susceptibilities $(\varepsilon_1 - \varepsilon_m)$ and $(\varepsilon_2 - \varepsilon_m)$, see Eqs. (4.32) and (4.43) in L1974.

Table S.2 (*cont.*)

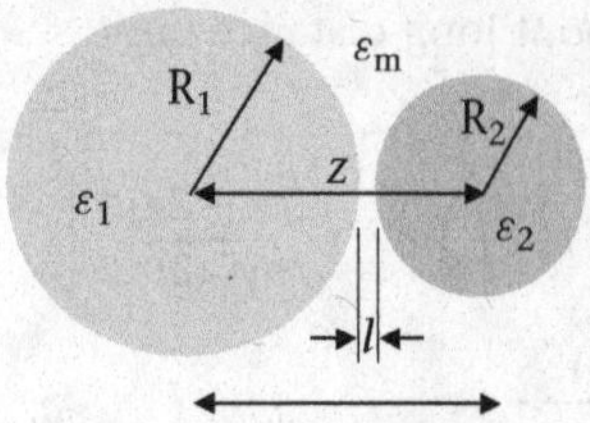

$z = R_1 + R_2 + l$

$$\cosh(2\theta_{ss}) = \frac{z^2 - R_1^2 - R_2^2}{2R_1R_2}, \quad [31]$$

$$Q^2(i\xi_n) = \overline{\Delta}_{1m}\overline{\Delta}_{2m},$$

$$\overline{\Delta}_{1m} = \frac{\varepsilon_1 - \varepsilon_m}{\varepsilon_1 + \varepsilon_m},$$

$$\overline{\Delta}_{2m} = \frac{\varepsilon_2 - \varepsilon_m}{\varepsilon_2 + \varepsilon_m}. \quad [3]$$

S.2.c. Sphere–sphere interaction, easily calculated accurate approximations to the exact, many-body form, no retardation

$$G_{ss}(z; R_1, R_2) = -kT\sum_{n=0}^{\infty}{}' g(z; i\xi_n) \quad [1],$$

$$g(z, i\xi_n) = \sum_{\nu=1}^{\infty}\left\{\frac{1}{8\nu}\left[\frac{1}{\sinh^2(\nu\theta_{ss}) + \cosh^2(\nu\theta_{ss})}\right] + P_\nu\right\}\tilde{Q}^{2\nu} \quad [51];$$

for P_ν, see subsequent equations

Easy-computation approximation: Use $\tilde{Q} = Q_E(z, i\xi_n; R_1, R_2) = (x_3/E_{22})^{1/2}Q$ [41].

Not-so-easy approximation: Use $\tilde{Q} = Q_{NSE}(z, i\xi_n; R_1, R_2) = (\tilde{T}/E_{22})^{1/2}Q$ [35].

Note:

$$E_{11} = E_{11}(z; R_1, R_2) = \frac{R_1R_2}{2}\left[\frac{1}{z^2 - (R_1 + R_2)^2} + \frac{1}{z^2 - (R_1 - R_2)^2}\right] + \frac{1}{4}\ln\left[\frac{z^2 - (R_1 + R_2)^2}{z^2 - (R_1 - R_2)^2}\right] \quad [36],$$

$$E_{22} = E_{22}(z; R_1, R_2) = \frac{R_1R_2}{2}\left[\frac{1}{z^2 - (R_1 + R_2)^2} + \frac{1}{z^2 - (R_1 - R_2)^2} - \frac{1}{z^2 - R_1^2} - \frac{1}{z^2 - R_2^2} + 1\right] \quad [33],$$

$$E_{12} = E_{12}(z; R_1, R_2) = \frac{R_1R_2}{2}\left[\frac{1}{z^2 - (R_1 + R_2)^2} + \frac{1}{z^2 - (R_1 - R_2)^2} - \frac{1}{z^2 - R_1^2}\right]$$
$$- \frac{R_2}{4z}\ln\left[\frac{(z + R_1 + R_2)(z + R_1 - R_2)(z - R_1)^2}{(z - R_1 + R_2)(z - R_1 - R_2)(z + R_1)^2}\right] \quad [37],$$

$E_{21} = E_{21}(z; R_1, R_2) = E_{12}(z; R_2, R_1)$, i.e., the same function as E_{12} $(z; R_1, R_2)$ but with the positions of R_1 and R_2 reversed.

$$x_1 = x_1(z, i\xi_n; R_1, R_2) = E_{22}\frac{\varepsilon_1 + \varepsilon_m}{\varepsilon_1 - \varepsilon_m + 2\varepsilon_m(E_{22}/E_{12})} \quad [38],$$

$$x_2 = x_2(z, i\xi_n; R_1, R_2) = E_{21}\frac{\varepsilon_1 + \varepsilon_m}{\varepsilon_1 - \varepsilon_m + 2\varepsilon_m(E_{21}/E_{11})} \quad [39],$$

$$x_3 = x_3(z, i\xi_n; R_1, R_2) = x_1\frac{\varepsilon_2 + \varepsilon_m}{\varepsilon_2 - \varepsilon_m + 2\varepsilon_m(x_1/x_2)} \quad [40],$$

$$\tilde{T} = \tilde{T}(z, i\xi_n; R) = E_{22} + \sum_{m=1}^{\infty}\sum_{m'=1}^{\infty}\binom{2m + 2m'}{2m}\frac{m}{m + [(\varepsilon_s/\varepsilon_m) + 1]^{-1}}\frac{m'}{m' + [(\varepsilon_s/\varepsilon_m) + 1]^{-1}}\left(\frac{R}{z}\right)^{2m+2m'+2} \quad [34],$$

$$P_\nu = \sum_{k=1}^{2\nu}\frac{(-1)^k}{k}[f(k, 2\nu) + g(k, 2\nu)] \quad [50],$$

where the forms of $f(k, 2\nu)$ and $g(k, 2\nu)$ depend on their arguments.

$$f(1, m) = 1/f_m \text{ and } g(1, m) = 1/g_m, \quad [46]$$

$$f_m = g_m = \frac{z}{\sqrt{R_1R_2}}\frac{\sinh[(m + 1)\theta_{ss}]}{\sinh(2\theta_{ss})}, \text{ m odd} \quad [42], [43];$$

Table S.2 (*cont.*)

$$f_m = \frac{\sinh[(m+2)\theta_{ss}]}{\sinh(2\theta_{ss})} + \frac{R_2}{R_1}\frac{\sinh(m\theta_{ss})}{\sinh(2\theta_{ss})}, \text{ m even} \quad [44];$$

$$g_m = \frac{\sinh[(m+2)\theta_{ss}]}{\sinh(2\theta_{ss})} + \frac{R_1}{R_2}\frac{\sinh[m\theta_{ss}]}{\sinh(2\theta_{ss})}, \text{ m even} \quad [45].$$

Then, for m even,

$$f(k, m) = \sum_{j=1}^{m+1-k} f(1, j)\, f(k-1, m-j) \quad [47], \quad g(k, m) = \sum_{j=1}^{m+1-k} g(1, j) g(k-1, m-j) \quad [47], [48].$$

Then, for m odd,

$$f(k, m) = g(k, m) = \sum_{j=1}^{m+1-k} f(1, j)\, f(k-1, m-j) = \sum_{j=1}^{m+1-k} g(1, j) g(k-1, m-j) \quad [49].$$

Source: From KPW 1978. Numbers in [ ] correspond to those in KPW 1978. The "E" subscript here is for the "easy approximation" in that paper, "NSE" for "not so easy." The easy approximation is good to ~1%; NSE, to ~0.2%.

Original many-body formulation in L1974 and L1971. The coefficient for Equation [1] differs because of the substitution of summation over $\xi_n = (2\pi kT/\hbar)n$ for integration with a consequent factor $2\pi kT/\hbar$.

Table S.2 (*cont.*)

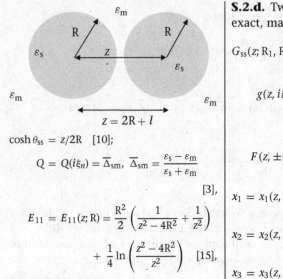

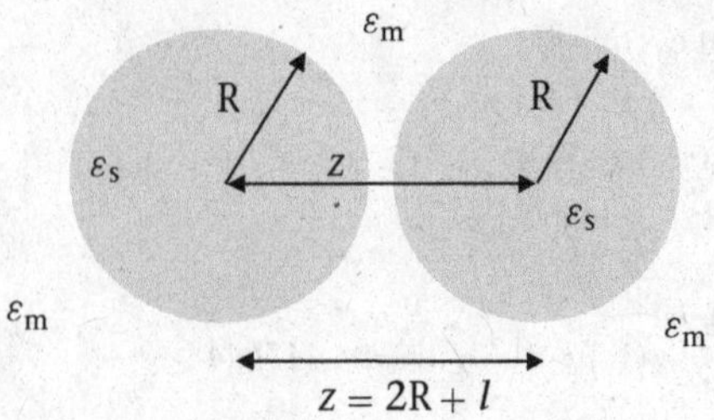

$$\cosh\theta_{ss} = z/2R \quad [10];$$

$$Q = Q(i\xi_n) = \overline{\Delta}_{sm},\ \overline{\Delta}_{sm} = \frac{\varepsilon_s - \varepsilon_m}{\varepsilon_s + \varepsilon_m} \quad [3],$$

$$E_{11} = E_{11}(z;R) = \frac{R^2}{2}\left(\frac{1}{z^2-4R^2} + \frac{1}{z^2}\right) + \frac{1}{4}\ln\left(\frac{z^2-4R^2}{z^2}\right) \quad [15],$$

$$E_{12} = E_{12}(z;R) = \frac{R^2}{2}\left(\frac{1}{z^2-4R^2} + \frac{1}{z^2} - \frac{2}{z^2-R^2}\right) - \frac{R}{4z}\ln\left[\frac{(z+2R)(z-R)^2}{(z-2R)(z+R)^2}\right] \quad [14],$$

$$E_{22} = E_{22}(z;R) = \frac{R^2}{2}\left(\frac{1}{z^2-4R^2} + \frac{3}{z^2} - \frac{4}{z^2-R^2}\right) \quad [11],$$

$$e(m) \equiv \frac{m}{m + [(\varepsilon_s/\varepsilon_m)+1]^{-1}} \quad [6].$$

S.2.d. Twin spheres, easily calculated approximations to the exact, many-body form, no retardation

$$G_{ss}(z;R_1,R_2) = -kT\sum_{n=0}^{\infty}{}' g(z;i\xi_n) \quad [1],$$

$$g(z,i\xi_n) = \frac{1}{8}\sum_{\nu=1}^{\infty}\left[\frac{1}{\sinh^2(\nu\theta_{ss})} + \frac{1}{\cosh^2(\nu\theta_{ss})}\right]\frac{\tilde{Q}^{2\nu}(z,i\xi_n)}{\nu} - \ln\{[1+F(z,\tilde{Q})][1+F(z,-\tilde{Q})]\} \quad [20],$$

$$F(z,\pm\tilde{Q}) = \sum_{m=1}^{\infty}\frac{\sinh(\theta_{ss})}{\sinh[(m+1)\theta_{ss}]}(\pm\tilde{Q})^m \quad [21],$$

$$x_1 = x_1(z,i\xi_n;R) = E_{22}\frac{\varepsilon_s+\varepsilon_m}{\varepsilon_s-\varepsilon_m+2\varepsilon_m(E_{22}/E_{12})} \quad [16],$$

$$x_2 = x_2(z,i\xi_n;R) = E_{12}\frac{\varepsilon_s+\varepsilon_m}{\varepsilon_s-\varepsilon_m+2\varepsilon_m(E_{12}/E_{11})} \quad [17],$$

$$x_3 = x_3(z,i\xi_n;R) = x_1\frac{\varepsilon_s+\varepsilon_m}{\varepsilon_s-\varepsilon_m+2\varepsilon_m(x_1/x_2)} \quad [18],$$

Easy approximation:

Use $\tilde{Q} = Q_E(z,i\xi_n;R) = (x_3/E_{22})^{1/2}Q$ [19],

NSE approximation:

Use $\tilde{Q} = Q_{NSE}(z,i\xi_n;R) = (\tilde{T}/E_{22})^{1/2}Q(i\xi_n)$ [13],

$$\tilde{T} = \tilde{T}(z,i\xi_n;R) = E_{22} + \sum_{m=1}^{\infty}\sum_{m'=1}^{\infty}\binom{2m+2m'}{2m}e(m)e(m')\left(\frac{R}{z}\right)^{2m+2m'+2} \quad [12].$$

Source: From KPW 1978. Numbers in [ ] correspond to those in source paper. The "E" subscript here is for the "easy approximation" in that paper; "NSE" for "not so easy." The coefficient for Equation [1] differs because of the substitution of summation over $\xi_n = (2\pi kT/\hbar)n$ for integration with a consequent factor $2\pi kT/\hbar$. Original many-body formulation in L1974 and L1971.

Table S.3. Sphere–sphere interaction, Hamaker hybrid form

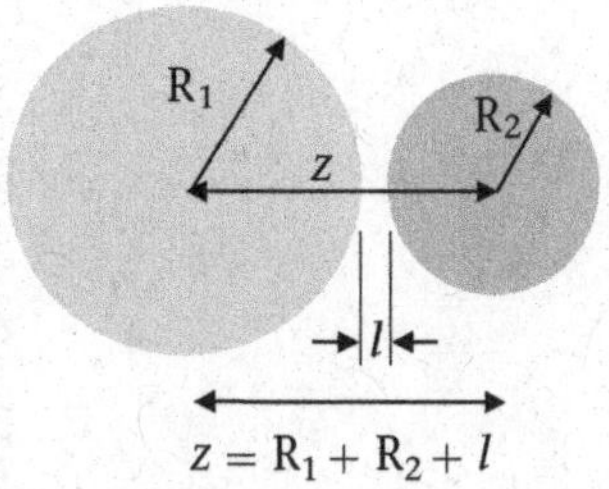

S.3.a. Hamaker summation

$$G_{ss}(z; R_1, R_2) = -\frac{A_{1m/2m}}{3}\left[\frac{R_1R_2}{z^2-(R_1+R_2)^2} + \frac{R_1R_2}{z^2-(R_1-R_2)^2} + \frac{1}{2}\ln\frac{z^2-(R_1+R_2)^2}{z^2-(R_1-R_2)^2}\right].$$

S.3.b.1. Point-particle limit

$$G_{ss}(z; R_1, R_2) \to -\frac{R_1^3R_2^3}{z^6}\frac{16}{9}A_{1m/2m} = -\frac{V_1V_2}{\pi^2 z^6}A_{1m/2m}.$$

R_1 and $R_2 \ll z \approx l$, V_1, V_2 spherical volumes.

S.3.b.2. Close-approach limit

$$G_{ss}(z; R_1, R_2) = -\frac{A_{1m/2m}}{6}\frac{R_1R_2}{(R_1+R_2)l},\ l \ll R_1 \text{ or } R_2,$$

$$A_{1m/2m} \approx \frac{3kT}{2}\sum_{n=0}^{\infty}{}' \frac{\varepsilon_1-\varepsilon_m}{\varepsilon_1+\varepsilon_m}\frac{\varepsilon_2-\varepsilon_m}{\varepsilon_2+\varepsilon_m}.$$

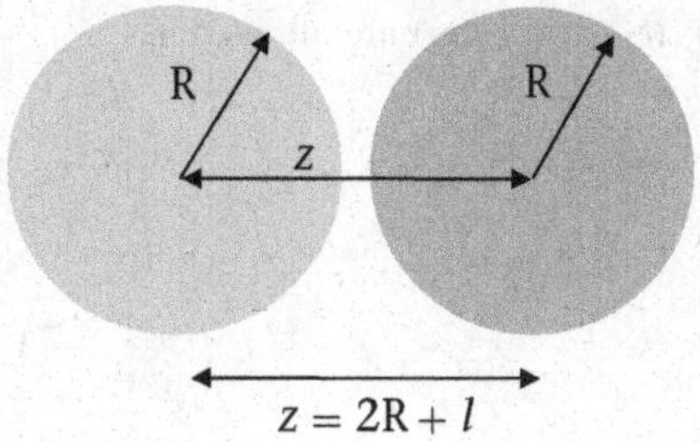

S.3.b.3. Equal-size spheres

$R_1 = R_2 = R$:

$$G_{ss}(z; R) = -\frac{A_{1m/2m}}{3}\left[\frac{R^2}{z^2-4R^2} + \frac{R^2}{z^2} + \frac{1}{2}\ln\left(1-\frac{4R^2}{z^2}\right)\right].$$

S.3.b.4. Equal-size spheres, large separation

$$R \ll z \approx l,\ G_{ss}(z; R) \to -\frac{R^6}{z^6}\frac{16}{9}A_{1m/2m} = -\frac{V_2}{\pi^2 z^6}A_{1m/2m}.$$

Note: The coefficients in the long-distance limit are easily seen for equal spheres, and almost as easily for $R_1 \neq R_2$. Let $\alpha \equiv (2R)^2/z^2$; expand

$$[\ ] = \frac{\alpha}{4}\frac{1}{1-\alpha} + \frac{\alpha}{4} + \frac{1}{2}\ln(1-\alpha) = \frac{\alpha}{4}(1+\alpha+\alpha^2+1) + \frac{1}{2}\left(-\alpha - \frac{\alpha^2}{2} - \frac{\alpha^3}{3}\right) = \left(\frac{1}{4}-\frac{1}{6}\right)\alpha^3,$$

so that

$$\frac{A}{3}\frac{\alpha^3}{12} = \frac{A}{3}\frac{(2R)^6}{12z^6} = \frac{R^6}{z^6}\frac{16}{9}A.$$

Table S.4. Fuzzy spheres, radially varying dielectric response

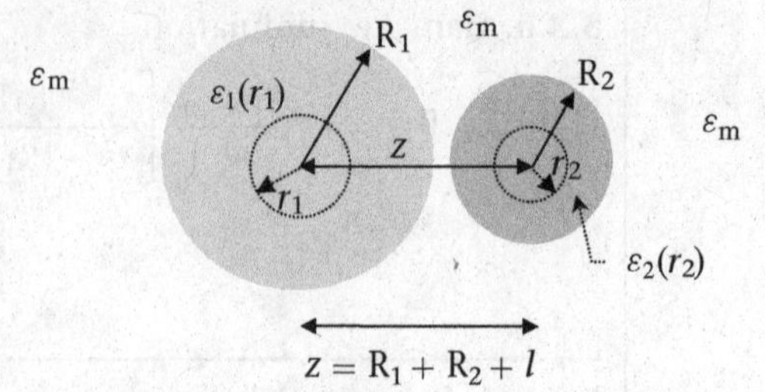

S.4.a. Small differences in ε, no retardation

$$G_{ss}(z) = -\frac{kT}{8}\sum_{n=0}^{\infty}{}' \int_0^{R_1} dr_1 \frac{d\ln[\varepsilon_1(r_1)]}{dr_1} \int_0^{R_2} dr_2 \frac{d\ln[\varepsilon_2(r_2)]}{dr_2} K(r_1, r_2),$$

$$K(r_1, r_2) = \left[\frac{r_1 r_2}{z^2 - (r_1 + r_2)^2} + \frac{r_1 r_2}{z^2 - (r_1 - r_2)^2} + \frac{1}{2}\ln\frac{z^2 - (r_1 + r_2)^2}{z^2 - (r_1 - r_2)^2}\right].$$

Note: Small-differences-in-ε's regime, summation over continuously varying dielectric response. Integration (Eq. 4.101 from L1974) of $K(r_1, r_2)$ is over the same geometric form as the sphere–sphere interaction in the Hamaker pairwise-summation limit. The conversion from zero-temperature integration in the original derivation to finite-temperature summation over frequency is effected by a factor of $2\pi kT/\hbar$ in Eq. 4.101. Discontinuities in $\varepsilon_1(r_1)$ and $\varepsilon_2(r_2)$ are allowed as delta functions in their derivatives.

Table S.4 (*cont.*)

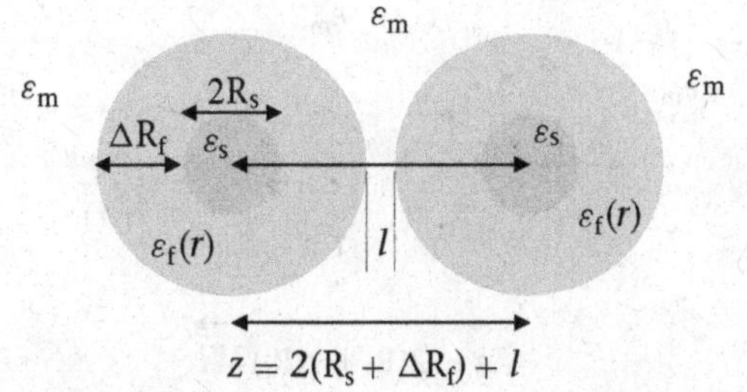

Core-sphere radius R_s, fuzzy-layer thickness ΔR_f, center-to-center distance z
Small steps allowed in ε's at R_s and $R_s + \Delta R_f$:

$r < R_{sphere} = R_s$, $\varepsilon = \varepsilon_{sphere} = \varepsilon_s$,

$R_s < r < R_s + \Delta R_f$, $\varepsilon_{fuzz} = \varepsilon_f(r)$,

$r > R_s + \Delta R_f$, $\varepsilon = \varepsilon_{medium} = \varepsilon_m$,

$$K(r_1, r_2) = \left[\frac{r_1 r_2}{z^2 - (r_1 + r_2)^2} + \frac{r_1 r_2}{z^2 - (r_1 - r_2)^2} + \frac{1}{2}\ln\frac{z^2 - (r_1 + r_2)^2}{z^2 - (r_1 - r_2)^2}\right] \quad [2].$$

S.4.b. Two like spheres, small differences in ε, no retardation

$$G_{fs/fs}(z; R_s, \Delta R_f) = -\frac{kT}{8}\sum_{n=0}^{\infty}{}' I(i\xi_n) = -\frac{kT}{8}\sum_{n=0}^{\infty}{}' \int_0^{R_1} dr_1 \frac{d\ln[\varepsilon_f(r_1)]}{dr_1}\int_0^{R_2} dr_2 \frac{d\ln[\varepsilon_f(r_2)]}{dr_2} K(r_1, r_2) \quad [1],$$

$$I(i\xi_n) = \ln^2\left[\frac{\varepsilon_f(R_s)}{\varepsilon_s}\right] K(R_s, R_s) + 2\ln\left[\frac{\varepsilon_f(R_s)}{\varepsilon_s}\right]\ln\left[\frac{\varepsilon_m}{\varepsilon_f(R_s + \Delta R_f)}\right] K(R_s + \Delta R_f, R_s)$$

$$+ \ln^2\left[\frac{\varepsilon_m}{\varepsilon_f(R_s + \Delta R_f)}\right] K(R_s + \Delta R_f, R_s + \Delta R_f) + 2\ln\left[\frac{\varepsilon_f(R_s)}{\varepsilon_s}\right]\int_{R_s}^{R_s + \Delta R_f} K(R_s, r)\frac{d\ln[\varepsilon_f(r)]}{dr} dr$$

$$+ 2\ln\left[\frac{\varepsilon_m}{\varepsilon_f(R_s + \Delta R_f)}\right]\int_{R_s}^{R_s + \Delta R_f} K(R_s + \Delta R_f, r)\frac{d\ln[\varepsilon_f(r)]}{dr} dr$$

$$+ \int_{R_s}^{R_s + \Delta R_f} dr_1 \int_{R_s}^{R_s + \Delta R_f} K(r_1, r_2)\frac{d\ln[\varepsilon_f(r_1)]}{dr_1}\frac{d\ln[\varepsilon_f(r_2)]}{dr_2} dr_2 \quad [3].$$

Source: Equation numbers in [ ] as in J. E. Kiefer, V. A. Parsegian, and G. H. Weiss, "Model for van der Waals attraction between spherical particles with nonuniform adsorbed polymer," J. Colloid Interface Sci., **51**, 543–545 (1975).

Table S.4 (*cont.*)

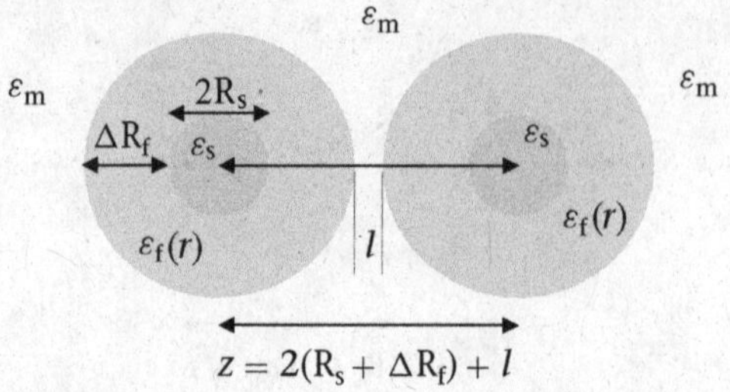

Core-sphere radius R_s, fuzzy-layer thickness ΔR_f, center-to-center distance z continuously varying ε:

$r < R_{sphere} = R_s,\ \varepsilon = \varepsilon_{sphere} = \varepsilon_s,$

$r = R_s,\ \varepsilon = \varepsilon_s = \varepsilon_f(R_s),$

$R_s < r < R_s + \Delta R_f,\ \varepsilon_{fuzz} = \varepsilon_f(r) = \varepsilon_s e^{\left(\frac{r-R}{\Delta R}\right)\ln\frac{\varepsilon_m}{\varepsilon_s}},$

$r = R_s + \Delta R_f,\ \varepsilon = \varepsilon_f(R_s + \Delta R_f) = \varepsilon_m,$

$r > R_s + \Delta R_f,\ \varepsilon = \varepsilon_{medium} = \varepsilon_m.$

S.4.c. Two like spheres with coatings of exponentially varying $\varepsilon_f(r)$: small differences in ε, no retardation

$$G_{fs/fs}(z; R_s, \Delta R_f) = -\frac{kT}{8}\sum_{n=0}^{\infty}{}' I(i\xi_n) \quad [1],$$

$$\begin{aligned} I(i\xi_n) = {} & 2\left(\alpha - \frac{\alpha^3}{3}\right) f(2\alpha) + \left(\alpha^2 + \frac{2}{3}\right) g(2\alpha) + 2\left(\beta - \frac{\beta^3}{3}\right) f(2\beta) + \left(\beta^2 + \frac{2}{3}\right) g(2\beta) \\ & - 2\left(\alpha + \beta - \frac{\alpha^3 + \beta^3}{3}\right) f(\alpha + \beta) - \left(2\alpha\beta + \frac{4}{3}\right) g(\alpha + \beta) \\ & -2\left(\beta - \alpha - \frac{\beta^3 - \alpha^3}{3}\right) f(\beta - \alpha) + \left(2\alpha\beta - \frac{4}{3}\right) g(\beta - \alpha) \quad [8], \end{aligned}$$

$$\alpha \equiv \frac{\left(1 - \frac{\Delta R_f}{R_s}\right)}{\left(2 + \frac{l}{R_s}\right)},\ \beta \equiv \frac{1}{\left(2 + \frac{l}{R_s}\right)} \quad [6],$$

$$f(x) \equiv \frac{1}{2}\ln\left[\frac{(1+x)}{(1-x)}\right],\ g(x) \equiv \frac{1}{2}\ln(1 - x^2) \quad [7].$$

Note: The condition that ε's be continuous at R_s and $R_s + \Delta R_f$ is easily removed by adding an extra sphere–sphere terms with the required discontinuity in ε. Equation numbers in [ ] as in J. E. Kiefer, V. A. Parsegian, and G. H. Weiss, "Model for van der Waals attraction between spherical particles with nonuniform adsorbed polymer," J. Colloid Interface Sci., **51**, 543–545 (1975).

Table S.5. Sphere–plane interactions

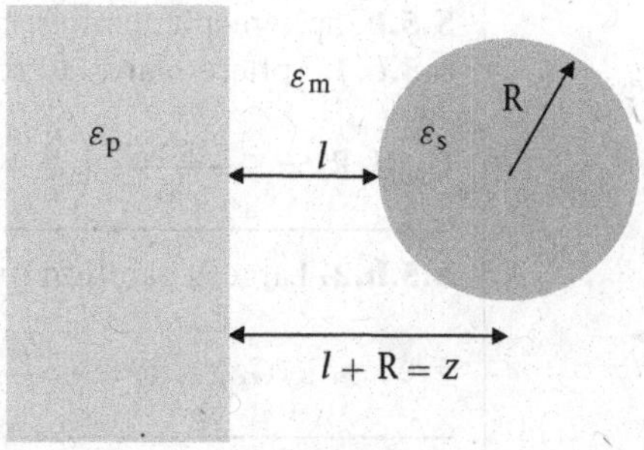

$$\cosh(2\theta_{sp}) = 1 + \frac{l}{R} = \frac{z}{R} \quad [22],$$

$$Q^2(i\xi_n) = \overline{\Delta}_{sm}\overline{\Delta}_{pm}, \ \overline{\Delta}_{sm} = \frac{\varepsilon_s - \varepsilon_m}{\varepsilon_s + \varepsilon_m}, \ \overline{\Delta}_{pm} = \frac{\varepsilon_p - \varepsilon_m}{\varepsilon_p + \varepsilon_m} \quad [3],$$

$$E_{11} = E_{11}(l; R) = \frac{R}{4}\left(\frac{1}{l} - \frac{1}{l+2R}\right) - \frac{1}{4}\ln\left(\frac{l+2R}{l}\right) \quad [26],$$

$$E_{22} = E_{22}(l; R) = \frac{R}{4}\left(\frac{1}{l} + \frac{1}{l+2R} - \frac{2}{l+R}\right) \quad [23].$$

S.5.a. Accurate approximations to the exact, many-body form, no retardation

$$G_{sp}(l; R) = G_{sp}(\theta_{sp}; R) = -kT\sum_{n=0}^{\infty}{}' g_{sp}(z; i\xi_n) \quad [1],$$

$$g(z; i\xi_n) = \frac{1}{8}\sum_{\nu=1}^{\infty}\left[\frac{1}{\sinh^2(\nu\theta_{sp})} + \frac{1}{\cosh^2(\nu\theta_{sp})}\right]\frac{\tilde{Q}^{2\nu}(z; i\xi_n)}{\nu} - \ln\{1 + F[\theta_{sp}, \tilde{Q}(z; i\xi_n)]\} \quad [29],$$

$$F(\theta_{sp}, \tilde{Q}) = \sum_{m=1}^{\infty}\frac{\sinh(2\theta_{sp})}{\sinh[(m+1)2\theta_{sp}]}\tilde{Q}^{2m} \quad [30].$$

Easy approximation:

Use $\tilde{Q} = Q_E(z, i\xi_n; R) = (x_3/E_{22})^{1/2}Q(i\xi_n)$ [28],

$$x_3 = x_3(z, i\xi_n; R) = E_{22}\frac{\varepsilon_s + \varepsilon_m}{\varepsilon_s - \varepsilon_m + 2\varepsilon_m(E_{22}/E_{11})} \quad [27].$$

NSE approximation (more accurate):

Use $\tilde{Q} = Q_{NSE}(z; i\xi_n) = (\tilde{T}/E_{22})^{1/2}Q$ [25],

$$\tilde{T} = \tilde{T}(l; R; i\xi_n) = E_{22} + \frac{1}{2}\sum_{m=1}^{\infty}\left\{\frac{m}{m + [(\varepsilon_p/\varepsilon_m) + 1]^{-1}}\right\}\left(\frac{R}{z}\right)^{2m+1} \quad [24],$$

Source: From KPW 1978. The "E" subscript here is for the "easy approximation" in that paper; "NSE" is for "not so easy." The coefficient for Eq. [1] differs because of the substitution of summation over $\xi_n = (2\pi kT/\hbar)n$ for integration with a consequent factor $2\pi kT/\hbar$. Original many-body formulation in L1974 and L1971. Numbers in [ ] correspond to those in KPW 1978. Sphere–wall interactions are also treated in J. D. Love, "On the van der Waals force between two spheres or a sphere and a wall," J. Chem. Soc. Faraday Trans. 2, **73**, 669–688 (1977).

Table S.5 (*cont.*)

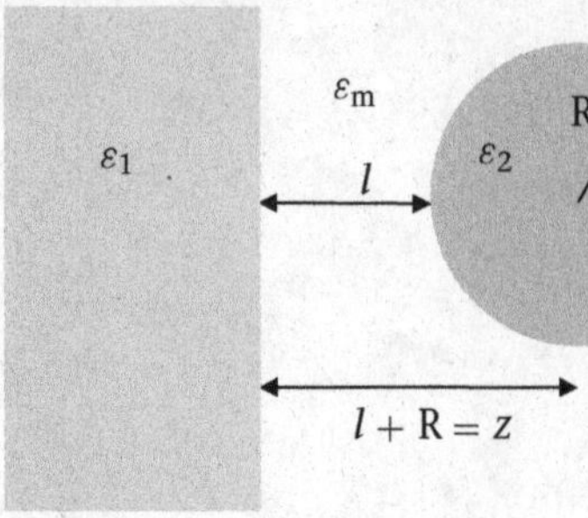

S.5.b. Sphere–plane interaction, Hamaker hybrid form

S.5.b.1. Sphere-plane, all separations

$$G_{sp}(l, R) = -\frac{A_{1m/2m}}{6}\left(\frac{R}{l} + \frac{R}{2R+l} + \ln\frac{l}{2R+l}\right).$$

S.5.b.2. Large-separation limit

$$z \approx l \gg R,\ G_{sp}(z; R) = -\frac{2A_{1m/2m}}{9}\frac{R^3}{l^3}.$$

S.5.b.3. Near contact

$$l \ll R,\ \ G_{sp}(l; R) = -\frac{A_{1m/2m}}{6}\frac{R}{l},$$

$$A_{1m/2m} \approx \frac{3kT}{2}\sum_{n=0}^{\infty}{}' \frac{\varepsilon_1 - \varepsilon_m}{\varepsilon_1 + \varepsilon_m}\frac{\varepsilon_2 - \varepsilon_m}{\varepsilon_2 + \varepsilon_m}.$$

Note: From the sphere–sphere interaction in the Hamaker approximation, $z^2 - (R_1 + R_2)^2 = [(R_1 + R_2) + l]^2 - (R_1 + R_2)^2 = 2l(R_1 + R_2) + l^2 \to 2lR_1$; $z^2 - (R_1 - R_2)^2 = [(R_1 + R_2) + l]^2 - (R_1 - R_2)^2 = 4R_1R_2 + 2l(R_1 + R_2) + l^2 \to 4R_1R_2 + 2lR_1$.

When $R_1 \ggg l$ and $R_2 = R$, R_1 cancels out of all terms in the sphere–sphere expression

$$\left[\frac{R_1R_2}{z^2 - (R_1 + R_2)^2} + \frac{R_1R_2}{z^2 - (R_1 - R_2)^2} + \frac{1}{2}\ln\frac{z^2 - (R_1 + R_2)^2}{z^2 - (R_1 - R_2)^2}\right]$$

to make this

$$\frac{1}{2}\left[\frac{R}{l} + \frac{R}{2R+l} + \ln\frac{l}{2R+l}\right].$$

When $\alpha \equiv R/l \ll 1$, [ ] expands as $\alpha + [\alpha/(1 + 2\alpha)] - \ln(1 + 2\alpha) = \alpha + \alpha(1 - 2\alpha + (2\alpha)^2 - \cdots) - [+2\alpha - (2\alpha)^2/2 + (2\alpha)^3/3 - \cdots] = (4 - 8/3)\alpha^3 = (4/3)R^3/l^3$ so that the interaction goes as

$$-\frac{A_{1m/2m}}{6}\frac{4R^3}{3l^3} = -\frac{2A_{1m2}}{9}\frac{R^3}{l^3};$$

this result can also be derived from interaction of point spheres with a plane in the small-differences-in-ε, no-retardation limit.

When $R \gg l$, the first term in [ ] dominates to give $[-(A_{1m/2m}/6)](R/l)$. This limit can also be extracted from the Derjaguin transform result for small-differences-in-ε and neglected retardation.

Table S.6. Point particles (without ionic fluctuations or ionic screening)

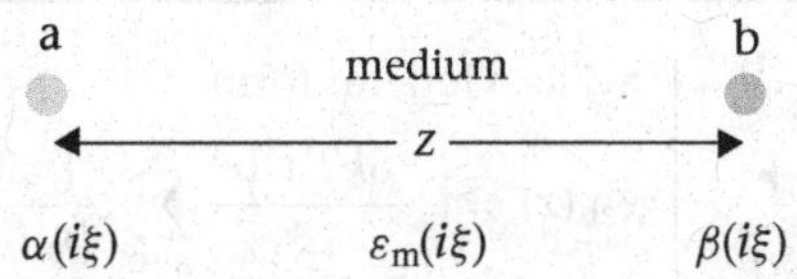

Dilute suspension or solution of colloids or macromolecules a, b at number densities N_a, N_b :

$$\alpha(i\xi) \equiv \left.\frac{\partial \varepsilon_{\text{suspension}}}{\partial N_a}\right|_{N_a, N_b=0}, \quad \beta(i\xi) \equiv \left.\frac{\partial \varepsilon_{\text{suspension}}}{\partial N_b}\right|_{N_a, N_b=0}.$$

S.6.a. General form

$$g_{ab}(z) = -\frac{6kT}{z^6}\sum_{n=0}^{\infty}{}' \left\{\frac{\alpha(i\xi_n)\beta(i\xi_n)}{[4\pi\varepsilon_m(i\xi_n)]^2}\right\}\left(1 + r_n + \frac{5}{12}r_n^2 + \frac{1}{12}r_n^3 + \frac{1}{48}r_n^4\right)e^{-r_n}.$$

[L2.150] and [L2.151]

S.6.b. Nonretarded limit

$z \ll$ all absorption wavelengths:

$$g_{ab}(z) = -\frac{6kT}{z^6}\sum_{n=0}^{\infty}{}' \frac{\alpha(i\xi_n)\beta(i\xi_n)}{[4\pi\varepsilon_m(i\xi_n)]^2}. \quad \text{[L2.152]}$$

S.6.c. Zero-temperature retarded limit

$z \gg$ all absorption wavelengths, valid only in the hypothetical $T = 0$ limit:

$$g_{ab}(z) = -\frac{23\hbar c}{(4\pi)^3 z^7}\frac{\alpha(0)\beta(0)}{\varepsilon_m(0)^{5/2}}. \quad \text{[L2.154]}$$

S.6.d. Fully retarded finite-temperature low-frequency limit

$z \gg \lambda_1$ corresponding to wavelength of first finite sampling frequency ξ_1:

$$g_{ab}(z) = -\frac{3kT}{z^6}\frac{\alpha(0)\beta(0)}{[4\pi\varepsilon_m(0)]^2}. \quad \text{[L2.155]}$$

Note: Generic α, β can be connected with particle polarizabilities α_{mks}, β_{mks} or α_{cgs}, β_{cgs} in mks or cgs units [Eq. (L2.162)–(L2.164), (L2.169)]:

generic	mks	cgs
$\varepsilon_{susp} = \varepsilon_m + N_a\alpha + N_b\beta$,	$\varepsilon_{susp} = \varepsilon_m + N_a(\alpha_{mks}/\varepsilon_0) + N_b(\beta_{mks}/\varepsilon_0)$,	$\varepsilon_{susp} = \varepsilon_m + N_a(4\pi\alpha_{cgs}) + N_b(4\pi\beta_{cgs})$,
$\dfrac{\alpha\beta}{(4\pi\varepsilon_m)^2}$	$\dfrac{\alpha_{mks}\beta_{mks}}{(4\pi\varepsilon_0\varepsilon_m)^2}$	$\dfrac{\alpha_{cgs}\beta_{cgs}}{\varepsilon_m^2}$

Table S.7. Small spheres (without ionic fluctuations or ionic screening)

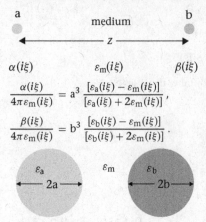

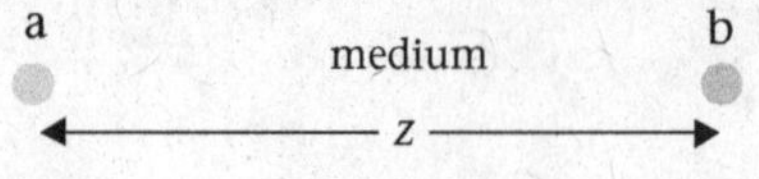

$\alpha(i\xi)$ $\quad\varepsilon_{\mathrm{m}}(i\xi)$ $\quad\beta(i\xi)$

$$\frac{\alpha(i\xi)}{4\pi\varepsilon_{\mathrm{m}}(i\xi)} = \mathrm{a}^3\frac{[\varepsilon_{\mathrm{a}}(i\xi)-\varepsilon_{\mathrm{m}}(i\xi)]}{[\varepsilon_{\mathrm{a}}(i\xi)+2\varepsilon_{\mathrm{m}}(i\xi)]},$$

$$\frac{\beta(i\xi)}{4\pi\varepsilon_{\mathrm{m}}(i\xi)} = \mathrm{b}^3\frac{[\varepsilon_{\mathrm{b}}(i\xi)-\varepsilon_{\mathrm{m}}(i\xi)]}{[\varepsilon_{\mathrm{b}}(i\xi)+2\varepsilon_{\mathrm{m}}(i\xi)]}.$$

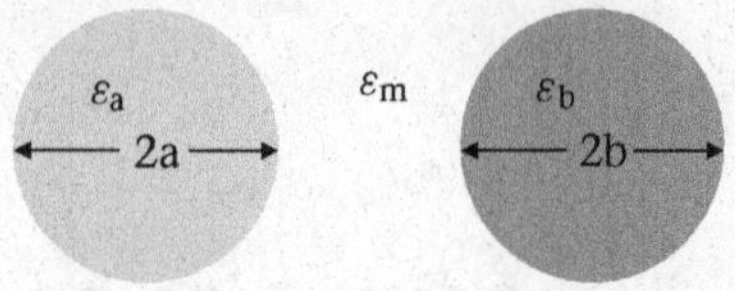

See Eqs. (L2.166)–(L2.169).

S.7.a. General form

$$g_{\mathrm{ab}}(z) = -\frac{6kT\mathrm{a}^3\mathrm{b}^3}{z^6}{\sum_{n=0}^{\infty}}'\frac{[\varepsilon_{\mathrm{a}}(i\xi_n)-\varepsilon_{\mathrm{m}}(i\xi_n)]}{[\varepsilon_{\mathrm{a}}(i\xi_n)+2\varepsilon_{\mathrm{m}}(i\xi_n)]}\frac{[\varepsilon_{\mathrm{b}}(i\xi_n)-\varepsilon_{\mathrm{m}}(i\xi_n)]}{[\varepsilon_{\mathrm{b}}(i\xi_n)+2\varepsilon_{\mathrm{m}}(i\xi_n)]}$$
$$\times\left(1+r_n+\frac{5}{12}r_n^2+\frac{1}{12}r_n^3+\frac{1}{48}r_n^4\right)e^{-r_n} \qquad \text{[L2.168]}$$

S.7.b. Nonretarded limit

$z \ll$ all absorption wavelengths:

$$g_{\mathrm{ab}}(z) = -\frac{6kT\mathrm{a}^3\mathrm{b}^3}{z^6}{\sum_{n=0}^{\infty}}'\frac{[\varepsilon_{\mathrm{a}}(i\xi_n)-\varepsilon_{\mathrm{m}}(i\xi_n)]}{[\varepsilon_{\mathrm{a}}(i\xi_n)+2\varepsilon_{\mathrm{m}}(i\xi_n)]}\frac{[\varepsilon_{\mathrm{b}}(i\xi_n)-\varepsilon_{\mathrm{m}}(i\xi_n)]}{[\varepsilon_{\mathrm{b}}(i\xi_n)+2\varepsilon_{\mathrm{m}}(i\xi_n)]}.$$

S.7.c. Zero-temperature retarded limit, $T = 0$

$z \gg$ all absorption wavelengths, valid only in the hypothetical $T = 0$ limit:

$$g_{\mathrm{ab}}(z) = -\frac{23\hbar c}{4\pi\varepsilon_{\mathrm{m}}^{1/2}(0)}\frac{\mathrm{a}^3\mathrm{b}^3}{z^7}\frac{[\varepsilon_{\mathrm{a}}(0)-\varepsilon_{\mathrm{m}}(0)]}{[\varepsilon_{\mathrm{a}}(0)+2\varepsilon_{\mathrm{m}}(0)]}\frac{[\varepsilon_{\mathrm{b}}(0)-\varepsilon_{\mathrm{m}}(0)]}{[\varepsilon_{\mathrm{b}}(0)+2\varepsilon_{\mathrm{m}}(0)]}.$$

S.7.d. Fully retarded finite-temperature low-frequency limit

$z \gg \lambda_1$ corresponding to wavelength of first finite sampling frequency ξ_1:

$$g_{\mathrm{ab}}(z) = -\frac{3kT\mathrm{a}^3\mathrm{b}^3}{z^6}{\sum_{n=0}^{\infty}}'\frac{[\varepsilon_{\mathrm{a}}(0)-\varepsilon_{\mathrm{m}}(0)]}{[\varepsilon_{\mathrm{a}}(0)+2\varepsilon_{\mathrm{m}}(0)]}\frac{[\varepsilon_{\mathrm{b}}(0)-\varepsilon_{\mathrm{m}}(0)]}{[\varepsilon_{\mathrm{b}}(0)+2\varepsilon_{\mathrm{m}}(0)]}.$$

Table S.8. Point–particle interaction in vapor, like particles without retardation screening

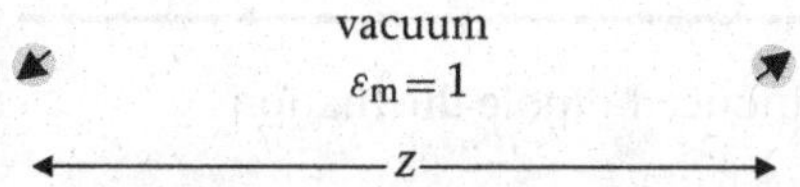

$\varepsilon_{vapor}(i\xi) = 1 + \alpha_{total}(i\xi)N$ (particle number density N), $\alpha_{total} = \alpha_{permanent} + \alpha_{induced}$, $\alpha_{permanent}(i\xi) = \frac{\mu^2_{dipole}}{3kT(1+\xi\tau)}$, dipole moment μ_{dipole}.

Units: Use

$\alpha = \alpha_{mks}/\varepsilon_0 = 4\pi\alpha_{cgs}$; $\alpha_{cgs} = \alpha_{mks}/(4\pi\varepsilon_0)$.

S.8.a. "Keesom" energy, mutual alignment of permanent dipoles

$$g_{Keesom}(z) = -\frac{\mu^4_{dipole}}{3(4\pi\varepsilon_0)^2 kTz^6}\ \text{(mks)} = -\frac{\mu^4_{dipole}}{3kTz^6}\ \text{(cgs)}. \quad \text{[L2.177]}$$

S.8.b. "Debye" interaction, permanent dipole and inducible dipole

$\alpha_{induced}(0)$ zero-frequency polarizability:

$$g_{Debye}(z) = -\frac{2\mu^2_{dipole}}{(4\pi\varepsilon_0)^2 z^6}\alpha_{ind}(0)\ \text{(mks)} = -\frac{2\mu^2_{dipole}}{z^6}\alpha_{ind}(0)\ \text{(cgs)}. \quad \text{[L2.178]}$$

S.8.c. "London" energy between mutually induced dipoles

1. *Finite temperature:*

$$g_{London}(z) = -\frac{6kT}{(4\pi\varepsilon_0)^2 z^6}{\sum_{n=0}^{\infty}}' \alpha_{ind}(i\xi_n)^2\ \text{(mks)},$$

$$g_{London}(z) = -\frac{6kT}{(z^6)}{\sum_{n=0}^{\infty}}' \alpha_{ind}(i\xi_n)^2\ \text{(cgs)}. \quad \text{[L2.179]}$$

2. *Low temperature:*

$$g_{London}(z, T \to 0) = -\frac{3\hbar}{\pi(4\pi\varepsilon_0)^2 z^6}\int_0^\infty \alpha^2_{ind}(i\xi)d\xi\ \text{(mks)},$$

$$= -\frac{3\hbar}{\pi z^6}\int_0^\infty \alpha^2_{ind}(i\xi)d\xi\ \text{(cgs)}. \quad \text{[L2.180]}$$

Note: Dipole moment μ_{Dipole} = charge × distance: in mks units, coulombs × meters; in cgs units, statcoulombs × centimeters.

For historical reasons, dipole moment or strength is often stated in Debye units (P. Debye, *Polar Molecules*, Dover, New York, 1929),

$$1 \text{ Debye unit} = 10^{-18}\ \text{sc} \times \text{cm}.$$

For example, a dipole pair of charges $+q$ and $-q$ each of elementary-charge magnitude $e = 4.803 \times 10^{-10}$ sc, separated by $d = 1$ Å $= 10^{-8}$ cm has a moment $\mu_{dipole} = 4.803$ Debye units.

Table S.9. Small, charged particles in saltwater, zero-frequency fluctuations only, ionic screening

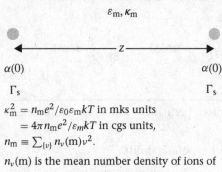

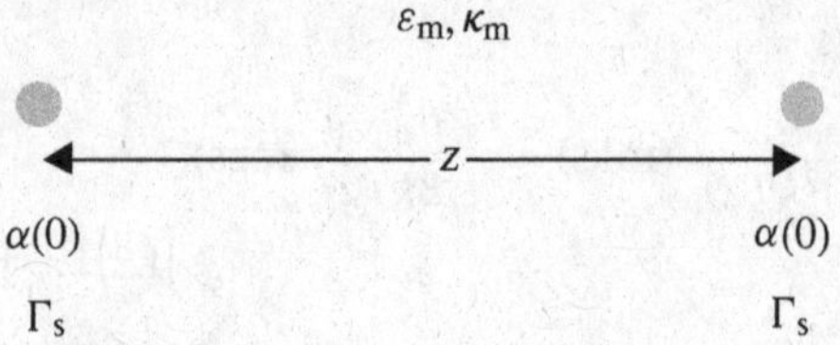

$\kappa_m^2 = n_m e^2/\varepsilon_0\varepsilon_m kT$ in mks units
$= 4\pi n_m e^2/\varepsilon_m kT$ in cgs units,
$n_m \equiv \sum_{\{\nu\}} n_\nu(m)\nu^2$.
$n_\nu(m)$ is the mean number density of ions of valence ν in the bathing solution (not including charge on small charged particles);

$\lambda_{Debye} = 1/\kappa_m$;

$\Gamma_s \equiv \sum_{\{\nu\}} \Gamma_\nu \nu^2$, Γ_ν is the mean excess in the number of mobile ions of valence ν around the small charged particle.

$\lambda_B \equiv e^2/4\pi\varepsilon_0\varepsilon_m kT$ (mks),

$\lambda_B \equiv [e^2/(\varepsilon_m kT)]$ (cgs).

$[\kappa_m^2/(4\pi n_m)] = \lambda_B$ in *either* unit system:

$$\left(\frac{\alpha}{4\pi\varepsilon_m}\right) = \frac{\alpha_{mks}}{4\pi\varepsilon_0\varepsilon_m} \text{ (mks)},$$

$$\frac{\alpha}{4\pi\varepsilon_m} = \frac{\alpha_{cgs}}{\varepsilon_m} \text{ (cgs)}.$$

S.9.a. Induced-dipole–induced-dipole fluctuation correlation

$$g_{D\text{-}D}(z) = -3kT\left[\frac{\alpha(0)}{4\pi\varepsilon_m(0)}\right]^2 \times\left[1+(2\kappa_m z)+\frac{5}{12}(2\kappa_m z)^2+\frac{1}{12}(2\kappa_m z)^3 + \frac{1}{96}(2\kappa_m z)^4\right]\frac{e^{-2\kappa_m z}}{z^6}. \quad \text{[L2.200]}$$

S.9.b. Induced-dipole–monopole fluctuation correlation

$$g_{D\text{-}M}(z) = -\frac{kT\kappa_m^2}{4\pi}\left[\frac{\alpha(0)}{4\pi\varepsilon_m(0)}\right]\left(\frac{\Gamma_s}{n_m}\right) \times\left[1+(2\kappa_m z)+\frac{1}{4}(2\kappa_m z)^2\right]\frac{e^{-2\kappa_m z}}{z^4}$$

or

$$g_{D\text{-}M}(l) = -kT\lambda_{Bjerrum}\left[\frac{\alpha(0)}{4\pi\varepsilon_m(0)}\right] \times\Gamma_s\left[1+(2\kappa_m z)+\frac{1}{4}(2\kappa_m z)^2\right]\frac{e^{-2\kappa_m z}}{z^4}. \quad \text{[L2.204]}$$

S.9.c. Monopole–monopole fluctuation correlation

$$g_{M\text{-}M}(l) = -\frac{kT\kappa_m^4}{2}\left(\frac{\Gamma_s}{4\pi n_m}\right)^2\frac{e^{-2\kappa_m z}}{z^2} = -\frac{kT}{2}\Gamma_s^2\frac{e^{-2z/\lambda_{Debye}}}{(z/\lambda_{Bj})^2} \quad \text{[L2.206]}$$

Formulae are valid only in the "dilute-suspension" limit, wherein particle number density N is low enough that
$N|\Gamma_s| \ll n_m$,
$N|\alpha| \ll \varepsilon_m$, and
$\varepsilon_{suspension} = \varepsilon_m + (\alpha_{mks}/\varepsilon_0)N$ or $\varepsilon_{suspension} = \varepsilon_m + 4\pi\alpha_{cgs}N$.

Table S.10. Small charged spheres in saltwater, zero-frequency fluctuations only, ionic screening

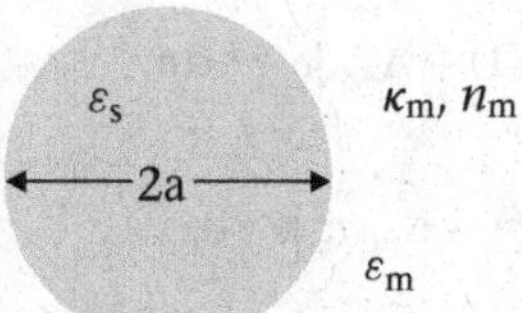

$$\frac{\alpha}{4\pi\varepsilon_m} = a^3\left(\frac{\varepsilon_s - \varepsilon_m}{\varepsilon_s + 2\varepsilon_m}\right) \quad [(L2.166)–(L2.169)]$$

Sphere of radius a, dielectric ε_s, in medium m, dielectric ε_m, mean ion density n_m, ionic screening constant κ_m.

Charge on the sphere redistributes mobile ions in surroundings:

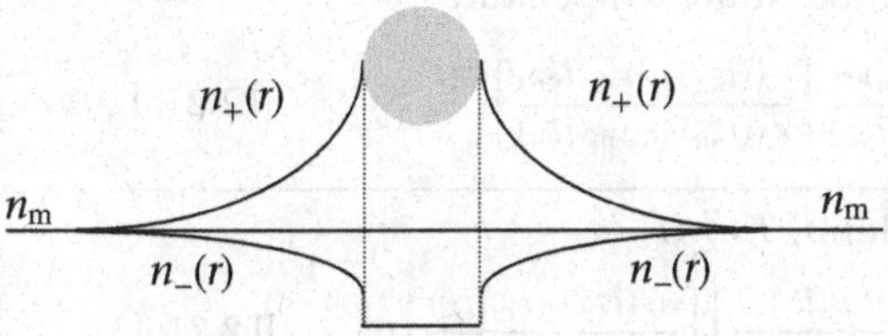

For a 1–1 salt bathing solution,

$$\Gamma_s = \Gamma_{+1} + \Gamma_{-1}$$
$$\equiv \int_0^\infty \{(n_+(r) - n_m) + [n_-(r) - n_m]\} 4\pi r^2\, dr.$$

S.10.a. Induced-dipole–induced-dipole fluctuation correlation

$$g_{D\text{-}D}(z) = -3kTa^6\left(\frac{\varepsilon_s - \varepsilon_m}{\varepsilon_s + 2\varepsilon_m}\right)^2\left[1 + (2\kappa_m z) + \frac{5}{12}(2\kappa_m z)^2 + \frac{1}{12}(2\kappa_m z)^3 + \frac{1}{96}(2\kappa_m z)^4\right]\frac{e^{-2\kappa_m z}}{z^6}.$$

S.10.b. Induced-dipole–monopole fluctuation correlation

$$g_{D\text{-}M}(z) = -kT\lambda_{Bj}a^3\left(\frac{\varepsilon_s - \varepsilon_m}{\varepsilon_s + 2\varepsilon_m}\right) \times \Gamma_s\left[1 + (2\kappa_m z) + \frac{1}{4}(2\kappa_m z)^2\right]\frac{e^{-2\kappa_m z}}{z^4}.$$

S.10.c. Monopole–monopole fluctuation correlation

$$g_{M\text{-}M}(z) = -\frac{kT\lambda_{Bj}^2}{2}\Gamma_s^2\frac{e^{-2z/\lambda_D}}{z^2} = -\frac{kT}{2}\Gamma_s^2\frac{e^{-2z/\lambda_D}}{(z/\lambda_{Bj})^2},$$

$$\lambda_{Bjerrum} = \lambda_{Bj} = \kappa_m^2/4\pi n_m,\ \lambda_{Debye} = \lambda_D = 1/\kappa_m,$$

$$\lambda_{Bj} \equiv e^2/4\pi\varepsilon_0\varepsilon_m kT \text{ (mks)},\ \lambda_{Bj} \equiv e^2/\varepsilon_m kT \text{ (cgs)},$$

$$\Gamma_s \equiv \sum_{\{\nu\}}\Gamma_\nu \nu^2, \quad \Gamma_\nu \equiv \int_0^\infty [n_\nu(r) - n_\nu(m)]4\pi r^2\, dr.$$

Table S.11. Point–particle substrate interactions

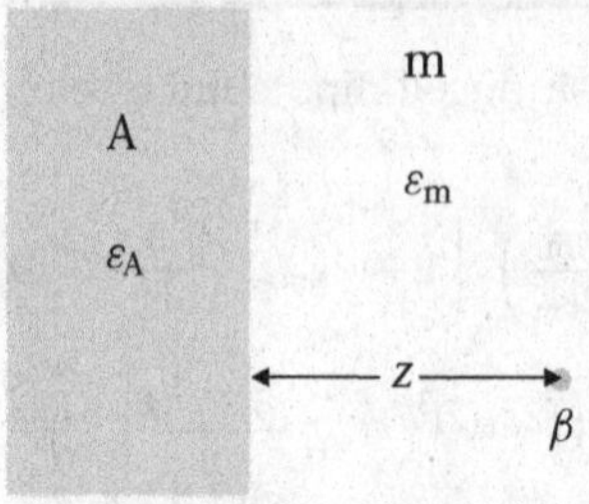

separation $z \gg$ particle size

$$\overline{\Delta}_{\mathrm{Am}} = \frac{p\varepsilon_{\mathrm{A}} - s_{\mathrm{A}}\varepsilon_{\mathrm{m}}}{p\varepsilon_{\mathrm{A}} + s_{\mathrm{A}}\varepsilon_{\mathrm{m}}},$$

$$\Delta_{\mathrm{Am}} = \frac{p - s_{\mathrm{A}}}{p + s_{\mathrm{A}}}.$$

β as in Tables S.6 and S.7 and Eqs. (L2.166)–(L2.169).

S.11.a.1. General case

$$g_p(z) = -\frac{kT}{8z^3}\sum_{n=0}^{\infty}{}' \left[\frac{\beta(i\xi_n)}{4\pi\varepsilon_{\mathrm{m}}(i\xi_n)}\right] r_n^3 \int_1^{\infty} [\overline{\Delta}_{\mathrm{Am}}(2p^2-1) - \Delta_{\mathrm{Am}}]e^{-r_n p}\,\mathrm{d}p$$

$$= -\frac{kT}{8z^3}\sum_{n=0}^{\infty}{}' \left[\frac{\beta(i\xi_n)}{4\pi\varepsilon_{\mathrm{m}}(i\xi_n)}\right] \int_{r_n}^{\infty} [\overline{\Delta}_{\mathrm{Am}}(2x^2 - r_n^2) - \Delta_{\mathrm{Am}} r_n^2]e^{-x}\mathrm{d}x.$$

[L2.211]

S.11.a.2. Small- $\overline{\Delta}_{\mathrm{Am}}$ limit

$$g_p(z) = -\frac{kT}{2z^3}\sum_{n=0}^{\infty}{}' \left[\frac{\beta(i\xi_n)}{4\pi\varepsilon_{\mathrm{m}}(i\xi_n)}\right]\left[\frac{\varepsilon_{\mathrm{A}}(i\xi_n) - \varepsilon_{\mathrm{m}}(i\xi_n)}{\varepsilon_{\mathrm{A}}(i\xi_n) + \varepsilon_{\mathrm{m}}(i\xi_n)}\right]\left(1 + r_n + \frac{r_n^2}{4}\right)e^{-r_n}.$$

[L2.212]

S.11.b.1. Nonretarded limit, finite temperature

$$g_p(z) = -\frac{kT}{2z^3}\sum_{n=0}^{\infty}{}' \frac{\beta(i\xi_n)}{4\pi\varepsilon_{\mathrm{m}}(i\xi_n)}\left[\frac{\varepsilon_{\mathrm{A}}(i\xi_n) - \varepsilon_{\mathrm{m}}(i\xi_n)}{\varepsilon_{\mathrm{A}}(i\xi_n) + \varepsilon_{\mathrm{m}}(i\xi_n)}\right]. \quad \text{[L2.215]}$$

S.11.b.2. Nonretarded limit, $T \to 0$

$$g_{p,T\to 0}(z) = -\frac{\hbar}{4\pi z^3}\int_0^{\infty}\left[\frac{\beta(i\xi)}{4\pi\varepsilon_{\mathrm{m}}(i\xi)}\right]\left[\frac{\varepsilon_{\mathrm{A}}(i\xi) - \varepsilon_{\mathrm{m}}(i\xi)}{\varepsilon_{\mathrm{A}}(i\xi) + \varepsilon_{\mathrm{m}}(i\xi)}\right]\mathrm{d}\xi. \quad \text{[L2.216]}$$

S.11.c. Fully retarded limit

$T = 0$ and $r_n = (2l\xi_n\varepsilon_{\mathrm{m}}^{1/2})/c \to \infty$ for $\xi_n \ll$ absorption frequencies:

$$g_p(z) = -\frac{3\hbar c}{8\pi z^4}\frac{(\beta/4\pi)}{\varepsilon_{\mathrm{m}}^{3/2}}\Theta(\varepsilon_{\mathrm{A}}/\varepsilon_{\mathrm{m}}), \quad \text{[L2.217]}$$

$$\Theta(\varepsilon_{\mathrm{A}}/\varepsilon_{\mathrm{m}}) \equiv \frac{1}{2}\int_1^{\infty}\{[\overline{\Delta}_{\mathrm{Am}}(2p^2-1) - \Delta_{\mathrm{Am}}]/p^4\}\mathrm{d}p.$$

$$\varepsilon_{\mathrm{A}} \gg \varepsilon_{\mathrm{m}}, \quad \Theta(\varepsilon_{\mathrm{A}}/\varepsilon_{\mathrm{m}}) = 1, \quad \varepsilon_{\mathrm{A}} \approx \varepsilon_{\mathrm{m}},$$

$$\Theta(\varepsilon_{\mathrm{A}}/\varepsilon_{\mathrm{m}}) \approx \frac{23}{30}\left(\frac{\varepsilon_{\mathrm{A}} - \varepsilon_{\mathrm{m}}}{\varepsilon_{\mathrm{A}} + \varepsilon_{\mathrm{m}}}\right). \quad \text{[L2.220]}$$

Table S.12. Small-sphere substrate interactions

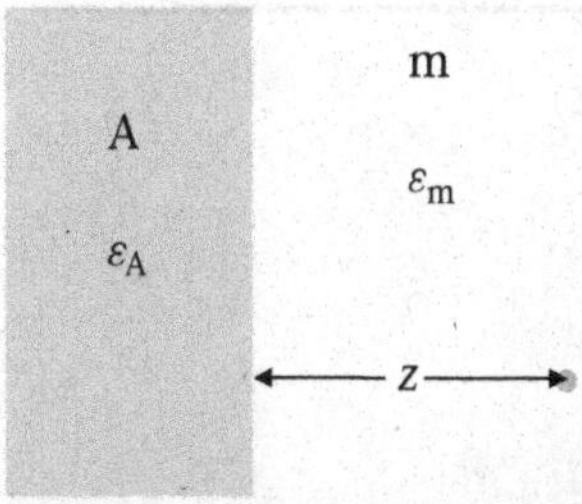

Separation z measured from A/m interface to center of sphere

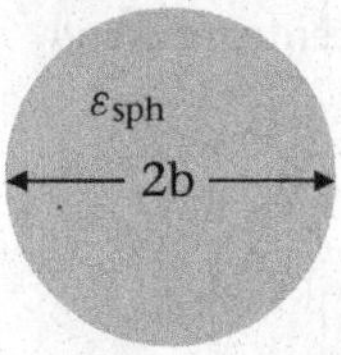

Sphere material dielectric response $\varepsilon_{\mathrm{sph}}$, sphere radius b

S.12.a. Spherical point particle of radius b in the limit of small differences in ε

$$g_p(z) = -\frac{kT}{2z^3}\mathrm{b}^3 \sum_{n=0}^{\infty}{}' \frac{\varepsilon_{\mathrm{sph}} - \varepsilon_{\mathrm{m}}}{\varepsilon_{\mathrm{sph}} + 2\varepsilon_{\mathrm{m}}} \frac{\varepsilon_{\mathrm{A}} - \varepsilon_{\mathrm{m}}}{\varepsilon_{\mathrm{A}} + \varepsilon_{\mathrm{m}}}$$

$$\approx -\frac{kT}{3z^3}\mathrm{b}^3 \sum_{n=0}^{\infty}{}' \frac{\varepsilon_{\mathrm{sph}} - \varepsilon_{\mathrm{m}}}{\varepsilon_{\mathrm{sph}} + \varepsilon_{\mathrm{m}}} \frac{\varepsilon_{\mathrm{A}} - \varepsilon_{\mathrm{m}}}{\varepsilon_{\mathrm{A}} + \varepsilon_{\mathrm{m}}}.$$

S.12.b. Hamaker form for large separations

$$G_{\mathrm{sp}}(z;\mathrm{b}) = -\frac{2A_{\mathrm{Am/sm}}}{9}\frac{\mathrm{b}^3}{z^3}, \quad A_{\mathrm{Am/sm}} = \frac{3kT}{2}\sum_{n=0}^{\infty}{}' \frac{\varepsilon_{\mathrm{sph}} - \varepsilon_{\mathrm{m}}}{\varepsilon_{\mathrm{sph}} + \varepsilon_{\mathrm{m}}} \frac{\varepsilon_{\mathrm{A}} - \varepsilon_{\mathrm{m}}}{\varepsilon_{\mathrm{A}} + \varepsilon_{\mathrm{m}}}.$$

S.12.c. Small sphere of radius b concentric within a large sphere of radius $\mathrm{R}_2 \approx z$

[See Table (P.10.b); replace R_1 with b, $\mathrm{R}_2 \approx l$ with z, ε_1 with $\varepsilon_{\mathrm{sph}}$, ε_2 with ε_{m}, ε_3 with ε_{A}]:

$$G_{\mathrm{sph}}(z;\mathrm{b}) \to -\frac{6kT}{z^3}\mathrm{b}^3 \sum_{n=0}^{\infty}{}' \frac{(\varepsilon_{\mathrm{sph}} - \varepsilon_{\mathrm{m}})(\varepsilon_{\mathrm{A}} - \varepsilon_{\mathrm{m}})}{(\varepsilon_{\mathrm{sph}} + 2\varepsilon_{\mathrm{m}})(\varepsilon_{\mathrm{m}} + 2\varepsilon_{\mathrm{A}})}$$

$$\approx -\frac{8kT}{3z^3}\mathrm{b}^3 \sum_{n=0}^{\infty}{}' \frac{\varepsilon_{\mathrm{sph}} - \varepsilon_{\mathrm{m}}}{\varepsilon_{\mathrm{sph}} + \varepsilon_{\mathrm{m}}} \frac{\varepsilon_{\mathrm{A}} - \varepsilon_{\mathrm{m}}}{\varepsilon_{\mathrm{A}} + \varepsilon_{\mathrm{m}}}.$$

Note: For a sphere of radius b, material dielectric response ε_{s}, in a medium m, $(\beta/4\pi\varepsilon_{\mathrm{m}})$ for point particles becomes $\mathrm{b}^3[(\varepsilon_{\mathrm{sph}} - \varepsilon_{\mathrm{m}})/(\varepsilon_{\mathrm{sph}} + 2\varepsilon_{\mathrm{m}})]$. When $\varepsilon_{\mathrm{s}} \approx \varepsilon_{\mathrm{m}}$, $(\varepsilon_{\mathrm{sph}} + 2\varepsilon_{\mathrm{m}}) \approx (3/2)(\varepsilon_{\mathrm{sph}} + \varepsilon_{\mathrm{m}})$, so that

$$\sum_{n=0}^{\infty}{}' \frac{(\varepsilon_{\mathrm{sph}} - \varepsilon_{\mathrm{m}})}{(\varepsilon_{\mathrm{sph}} + 2\varepsilon_{\mathrm{m}})}\left(\frac{\varepsilon_{\mathrm{A}} - \varepsilon_{\mathrm{m}}}{\varepsilon_{\mathrm{A}} + \varepsilon_{\mathrm{m}}}\right) \approx \frac{2}{3}\sum_{n=0}^{\infty}{}' \left(\frac{\varepsilon_{\mathrm{sph}} - \varepsilon_{\mathrm{m}}}{\varepsilon_{\mathrm{sph}} + \varepsilon_{\mathrm{m}}}\right)\left(\frac{\varepsilon_{\mathrm{A}} - \varepsilon_{\mathrm{m}}}{\varepsilon_{\mathrm{A}} + \varepsilon_{\mathrm{m}}}\right).$$

Table S.13. Two point particles in a vapor, near or touching a substrate (nonretarded limit)

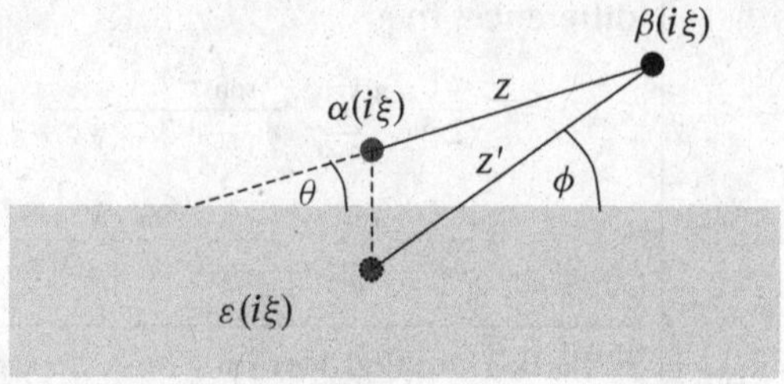

z = distance from center of α to center of β

z' = distance from center of image of α to center of β

The interface between substrate ε and the vapor above is midway between center of α and image of α.

S.13.a. Near

Points of polarizability $\alpha(i\xi)$, $\beta(i\xi)$:

$$g(z, z') = -\frac{3kT}{8\pi^2 z^6}\sum_{n=0}^{\infty}{}' \alpha(i\xi_n)\beta(i\xi_n) - \frac{3kT}{8\pi^2 z'^6}\sum_{n=0}^{\infty}{}' \alpha(i\xi_n)\beta(i\xi_n)\left[\frac{\varepsilon(i\xi_n)-1}{\varepsilon(i\xi_n)+1}\right]^2$$

$$+\frac{[2+3\cos(2\theta)+3\cos(2\phi)]kT}{16\pi^2 z^3 z'^3}\sum_{n=0}^{\infty}{}' \alpha(i\xi_n)\beta(i\xi_n)\left[\frac{\varepsilon(i\xi_n)-1}{\varepsilon(i\xi_n)+1}\right].$$

S.13.b. Touching

Points of polarizability $\alpha(i\xi)$, $\beta(i\xi)$ effectively on the interface: $z \to z'$; $\theta, \phi \to 0$

$$g(z) = -\frac{kT}{4\pi^2 z^6}\sum_{n=0}^{\infty}{}' \alpha(i\xi_n)\beta(i\xi_n)\frac{\varepsilon(i\xi_n)^2+5}{(\varepsilon(i\xi_n)+1)^2}.$$

Source: From Eq. 1.4 of A. D. McLachlan, "Van der Waals forces between an atom and a surface," Mol. Phys., **7**, 381–388 (1964). Note that the $\alpha(i\xi_n)$, $\beta(i\xi_n)$ in that paper differ by factors of 4π from the same symbols as used here and that the substitution $\xi = \xi_n = [2\pi kT/\hbar]n$ has been introduced to include the effects of finite temperature; specifically replace $\hbar d\xi$ in the published formula by

$$\frac{2\pi kT}{(4\pi)^2} = \frac{kT}{8\pi}$$

and replace integration with summation.

L2.2.C. TABLES OF FORMULAE IN CYLINDRICAL GEOMETRY

Table C.1. Parallel cylinders at separations small compared with radius, Derjaguin transform from full Lifshitz result, including retardation

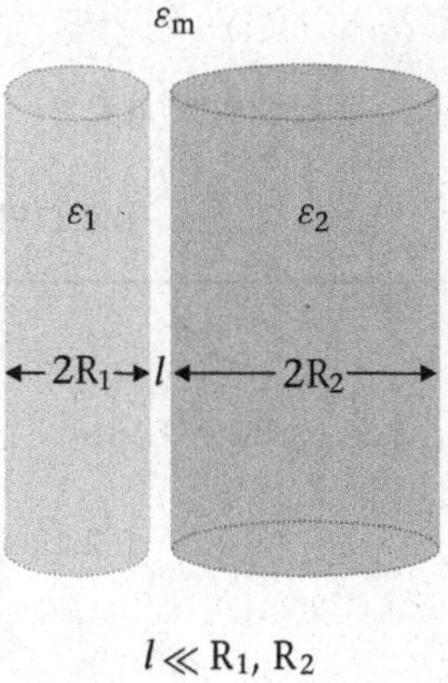

C.1.a. Force per unit length

$$F_{c\|c}(l; R_1, R_2) = -\sqrt{\frac{2\pi R_1 R_2}{R_1 + R_2}} \frac{kT}{8\pi l^{5/2}} \sum_{n=0}^{\infty}{}' r_n^{5/2} \sum_{q=1}^{\infty} \frac{1}{q^{1/2}}$$
$$\times \int_1^{\infty} p^{3/2}[(\overline{\Delta}_{Am}\overline{\Delta}_{Bm})^q + (\Delta_{Am}\Delta_{Bm})^q] e^{-r_n pq}\, dp. \quad \text{[L2.117]}$$

C.1.b. Free energy of interaction per unit length

$$G_{c\|c}(l; R_1, R_2) = -\sqrt{\frac{2\pi R_1 R_2}{R_1 + R_2}} \frac{kT}{8\pi l^{3/2}} \sum_{n=0}^{\infty}{}' r_n^{3/2} \sum_{q=1}^{\infty} \frac{1}{q}$$
$$\times \int_1^{\infty} p\left[(\overline{\Delta}_{Am}\overline{\Delta}_{Bm})^q + (\Delta_{Am}\Delta_{Bm})^q\right] \frac{e^{-r_n pq}}{\sqrt{pq}}\, dp, \quad \text{[L2.116]}$$

$$\overline{\Delta}_{ji} = \frac{s_i\varepsilon_j - s_j\varepsilon_i}{s_i\varepsilon_j + s_j\varepsilon_i},\ \Delta_{ji} = \frac{s_i\mu_j - s_j\mu_i}{s_i\mu_j + s_j\mu_i},$$

$$s_i = \sqrt{p^2 - 1 + (\varepsilon_i\mu_i/\varepsilon_m\mu_m)},\ r_n = (2l\varepsilon_m^{1/2}\mu_m^{1/2}/c)\,\xi_n.$$

C.1.c.1. Nonretarded (infinite light velocity) limit

$$G_{c\|c}(l; R_1, R_2) = -\sqrt{\frac{2R_1 R_2}{R_1 + R_2}} \frac{kT}{16 l^{3/2}} \sum_{n=0}^{\infty}{}' \sum_{q=1}^{\infty} \frac{[(\overline{\Delta}_{1m}\overline{\Delta}_{2m})^q + (\Delta_{1m}\Delta_{2m})^q]}{q^3}. \quad \text{[L2.118]}$$

C.1.c.2. Cylinders of equal radii

$$R_2 = R_1 = R,\ \sqrt{\frac{2\pi R_1 R_2}{R_1 + R_2}} = \sqrt{\pi R}.$$

C.1.c.3. Cylinder with a plane

$$R_2 \to \infty,\ R_1 = R,\ \sqrt{\frac{2\pi R_1 R_2}{R_1 + R_2}} = \sqrt{2\pi R}.$$

Note: Nonretarded limit, $r_n \to 0$, integral dominated by large p where $s_i = s_2 = p$

$$r_n^{3/2} \sum_{q=1}^{\infty} \frac{1}{q} \int_1^{\infty} p[(\overline{\Delta}_{1m}\overline{\Delta}_{2m})^q + (\Delta_{1m}\Delta_{2m})^q] \frac{e^{-r_n pq}}{\sqrt{pq}} dp \to \sum_{q=1}^{\infty} \frac{[(\overline{\Delta}_{1m}\overline{\Delta}_{2m})^q + (\Delta_{1m}\Delta_{2m})^q]}{q^3} \int_{r_n q}^{\infty} \sqrt{r_n pq}\, e^{-r_n pq} d(r_n pq)$$

$$\to \frac{\pi^{1/2}}{2} \sum_{q=1}^{\infty} \frac{[(\overline{\Delta}_{1m}\overline{\Delta}_{2m})^q + (\Delta_{1m}\Delta_{2m})^q]}{q^3},\ \int_{r_n q \to 0}^{\infty} \sqrt{r_n pq}\, e^{-r_n pq} d(r_n pq) \to \int_0^{\infty} \sqrt{x} e^{-x} dx = \Gamma\left(\frac{3}{2}\right) = \frac{\pi^{1/2}}{2};$$

$$G_{c\|c}(l; R_1, R_2) = -\sqrt{\frac{2\pi R_1 R_2}{R_1 + R_2}} \frac{kT}{8\pi l^{3/2}} \frac{\pi^{1/2}}{2} \sum_{n=0}^{\infty}{}' \sum_{q=1}^{\infty} \frac{[(\overline{\Delta}_{1m}\overline{\Delta}_{2m})^q + (\Delta_{1m}\Delta_{2m})^q]}{q^3}$$

$$= -\sqrt{\frac{2R_1 R_2}{R_1 + R_2}} \frac{kT}{16 l^{3/2}} \sum_{q=1}^{\infty} \frac{[(\overline{\Delta}_{1m}\overline{\Delta}_{2m})^q + (\Delta_{1m}\Delta_{2m})^q]}{q^3}.$$

Table C.2. Perpendicular cylinders, $R_1 = R_2 = R$, Derjaguin transform from full Lifshitz planar result, including retardation

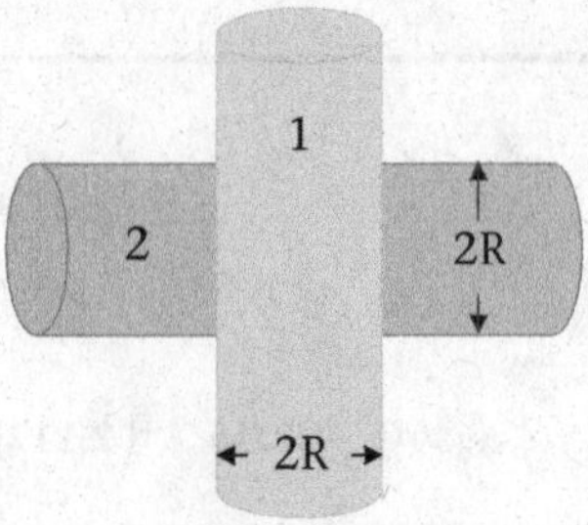

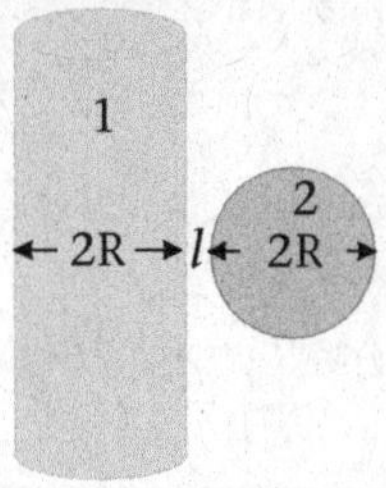

C.2.a. Force

$$F_{c\perp c}(l;R) = 2\pi R G_{pp}(l)$$

$$= -\frac{kTR}{4l^2}\sum_{n=0}^{\infty}{}' r_n^2 \sum_{q=1}^{\infty}\frac{1}{q}\int_1^{\infty} p[(\overline{\Delta}_{1m}\overline{\Delta}_{2m})^q + (\Delta_{1m}\Delta_{2m})^q]e^{-r_n pq}\,dp.$$

[L2.119]

C.2.b. Free energy per interaction

$$G_{c\perp c}(l;R) = -\frac{kTR}{4l}\sum_{n=0}^{\infty}{}' r_n \sum_{q=1}^{\infty}\frac{1}{q^2}\int_1^{\infty} [(\overline{\Delta}_{1m}\overline{\Delta}_{2m})^q + (\Delta_{1m}\Delta_{2m})^q]e^{-r_n pq}\,dp,$$

[L2.120]

$$\overline{\Delta}_{ji} = \frac{s_i\varepsilon_j - s_j\varepsilon_i}{s_i\varepsilon_j + s_j\varepsilon_i},\quad \Delta_{ji} = \frac{s_i\mu_j - s_j\mu_i}{s_i\mu_j + s_j\mu_i},$$

$$s_i = \sqrt{p^2 - 1 + (\varepsilon_i\mu_i/\varepsilon_m\mu_m)},\quad r_n = (2l\varepsilon_m^{1/2}\mu_m^{1/2}/c)\,\xi_n.$$

C.2.c. Nonretarded (infinite light velocity) limit

$$G_{c\perp c}(l;R) = -\frac{kTR}{4l}\sum_{n=0}^{\infty}{}' \sum_{q=1}^{\infty}\frac{[(\overline{\Delta}_{1m}\overline{\Delta}_{2m})^q + (\Delta_{1m}\Delta_{2m})^q]}{q^3}.$$

C.2.d. Light velocities taken everywhere equal to that in the medium, small $\overline{\Delta}_{ji}$, Δ_{ji}, $q = 1$

$$G_{c\perp c}(l;R) = -\frac{kT}{4}\frac{R}{l}\sum_{n=0}^{\infty}{}' (\overline{\Delta}_{1m}\overline{\Delta}_{2m} + \Delta_{1m}\Delta_{2m})e^{-r_n}.$$

C.2.e. Hamaker–Lifshitz hybrid form

$$G_{c\perp c}(l;R) = -\frac{A_{Am/Bm}}{6}\frac{R}{l},\quad A_{1m/2m}(l) = \frac{3kT}{2}\sum_{n=0}^{\infty}{}' (\overline{\Delta}_{1m}\overline{\Delta}_{2m} + \Delta_{1m}\Delta_{2m}).$$

Note: Nonretarded limit, $r_n \to 0$, integral dominated by large p, where $s_1 = s_2 = p$:

$$r_n\sum_{q=1}^{\infty}\frac{1}{q^2}\int_1^{\infty}[(\overline{\Delta}_{1m}\overline{\Delta}_{2m})^q + (\Delta_{1m}\Delta_{2m})^q]e^{-r_n pq}\,dp \to \sum_{q=1}^{\infty}\frac{[(\overline{\Delta}_{1m}\overline{\Delta}_{2m})^q + (\Delta_{1m}\Delta_{2m})^q]}{q^3}\int_{r_n q\to 0}^{\infty} e^{-r_n pq}\,d(r_n pq)$$

$$= \sum_{q=1}^{\infty}\frac{[(\overline{\Delta}_{1m}\overline{\Delta}_{2m})^q + (\Delta_{1m}\Delta_{2m})^q]}{q^3}.$$

Equal velocities, small-delta's approximation: use $q = 1$ term only, $s_i = s_j = p$:

$$\sum_{n=0}^{\infty}{}' r_n \sum_{q=1}^{\infty}\frac{1}{q^2}\int_1^{\infty}[(\overline{\Delta}_{1m}\overline{\Delta}_{2m})^q + (\Delta_{1m}\Delta_{2m})^q]e^{-r_n pq}\,dp \to \sum_{n=0}^{\infty}{}' (\overline{\Delta}_{1m}\overline{\Delta}_{2m} + \Delta_{1m}\Delta_{2m})r_n\int_1^{\infty} e^{-r_n p}\,dp$$

$$= \sum_{n=0}^{\infty}{}' (\overline{\Delta}_{1m}\overline{\Delta}_{2m} + \Delta_{1m}\Delta_{2m})e^{-r_n}.$$

Table C.3. Two parallel cylinders

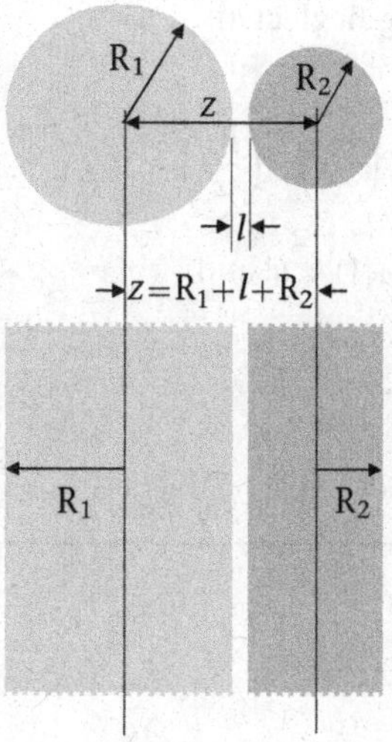

C.3.a. Two parallel cylinders, retardation screening neglected, solved by multiple reflection

$$G_{c\|c}(z; \mathrm{R}_1, \mathrm{R}_2) = -\frac{kT}{\pi z}\sum_{n=0}^{\infty}{}' \sum_{q=1}^{2} \frac{(\overline{\Delta}_{1m}\overline{\Delta}_{2m})^q}{q} \sum_{i,j=1}^{\infty} A(q, i, j)\left(\frac{\mathrm{R}_1}{z}\right)^{2i}\left(\frac{\mathrm{R}_2}{z}\right)^{2j}$$

per unit length. For coefficients $A(q, i, j)$ see following table. Radii R_1, R_2, interaxial separation z.

	$q = 1$						$q = 2$				
i	$j = 1$	2	3	4	5	i	$j = 1$	2	3	4	5
1	5.5517	17.35	35.42	59.77	90.40	1	0	0	0	0	0
2	17.35	106.3	358.6	904.1	1910.0	2	0	6.731	31.25	90.48	206.9
3	35.42	358.6	1808.0	6366.0	17904.0	3	0	31.25	212.2	847.7	2556.0
4	59.77	904.1	6366.0	29841.0	107799	4	0	90.48	847.7	2556.0	6439.0
5	90.40	1910.0	17904.0	107799.0	486443.0	5	0	206.9	2556.0	4468.0	17216.0
6	127.3	3581.0	43119.0	257629.0		6	0	408.4	6439.0	17216.0	
7	170.5	6160.0	92654.0			7	0	728.7	14274.0		
8	220.0	9928.0				8	0	1207.0			

Source: Taken from D. Langbein, Phys. Kondens. Mat., **15**, 61–86 (1972) [Eqs. (41), p. 71, and Table 2, p. 79] for the energy of interaction between two parallel cylinders of length L much greater than their radii and separation. The expression given on p. 79 apparently lacked a factor π in the denominator and should have read

$$\Delta E_{12}(z; \mathrm{R}_1, \mathrm{R}_2) = -\frac{\hbar L}{4\pi^2 z}\sum_{q=1}^{2}\frac{\Omega_q}{q}\sum_{i,j=1}^{\infty} A(q, i, j)\left(\frac{\mathrm{R}_1}{z}\right)^{2i}\left(\frac{\mathrm{R}_2}{z}\right)^{2j}.$$

[To verify this typographical correction, compare with the limit for thin rods, Eqs. (58) & (59).] Notation has been changed slightly to conform to that used in this text. The coefficients $A(q, i, j)$ are solved numerically. $\Omega_q \equiv \int_{-\infty}^{\infty} (\overline{\Delta}_{1m}\overline{\Delta}_{2m})^q d\xi$ (Eq. (59), p. 73, op cit.].

Here, the product $\hbar d\xi$ to be used in the hypothetical limit of low temperature is converted back to the finite-temperature form by $\hbar\xi_n = 2\pi kTn$. (Recall that, in the limit of low temperature, ξ_n takes on continuous values although n is the set of discrete integers.) At finite temperature the discrete ξ_n creates a summation rather than integration:

$$\hbar\Omega_q = \hbar\int_{\xi=-\infty}^{\infty}(\overline{\Delta}_{1m}\overline{\Delta}_{2m})^q d\xi \to 2\pi kT\int_{n=-\infty}^{\infty}(\overline{\Delta}_{1m}\overline{\Delta}_{2m})^q dn \to 2\pi kT\sum_{n=-\infty}^{\infty}(\overline{\Delta}_{1m}\overline{\Delta}_{2m})^q = 4\pi kT\sum_{n=0}^{\infty}{}'(\overline{\Delta}_{1m}\overline{\Delta}_{2m})^q.$$

In this summation then $\hbar\Omega_q$ is replaced with $4\pi kT\sum_{n=0}^{\prime\infty}(\overline{\Delta}_{1m}\overline{\Delta}_{2m})^q$ and the energy E_{12} is seen as a work or free energy of interaction between parallel cylinders $G_{c\|c}$:

$$G_{c\|c}(z; \mathrm{R}_1, \mathrm{R}_2) = -\frac{kTL}{\pi z}\sum_{n=0}^{\infty}{}'\sum_{q=1}^{2}\frac{(\overline{\Delta}_{1m}\overline{\Delta}_{2m})^q}{q}\sum_{i,j=1}^{\infty} A(q, i, j)\left(\frac{\mathrm{R}_1}{z}\right)^{2i}\left(\frac{\mathrm{R}_2}{z}\right)^{2j}.$$

Table C.3 (cont.)

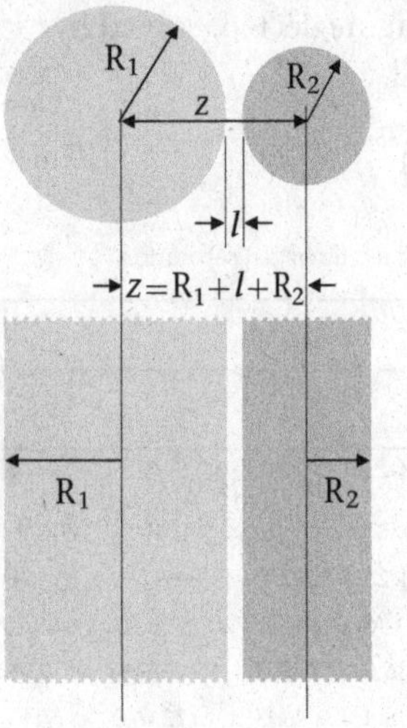

radii R_1, R_2, separation l

C.3.b. Two parallel cylinders, pairwise-summation approximation, Hamaker–Lifshitz hybrid, retardation screening neglected

C.3.b.1. All separations

$$G_{c\|c}\,(z;R_1,R_2) = -\frac{2A_{1m/2m}}{3z}\sum_{i,j=1}^{\infty}\frac{\Gamma^2\left(i+j+\frac{1}{2}\right)}{i!\,j!(i-1)!(j-1)!}\left(\frac{R_1}{z}\right)^{2i}\left(\frac{R_2}{z}\right)^{2j},$$

free energy per unit length. For an integer n, $\Gamma(n+1) = n!$ and

$$\Gamma\left(n+\frac{1}{2}\right) = \frac{1\times3\times5\times7\times\cdots\times(2n-1)}{2^n}\pi^{1/2},$$

$$A_{1m/2m} \approx \frac{3kT}{2}\sum_{n=0}^{\infty}{}'\,\overline{\Delta}_{1m}\overline{\Delta}_{2m}.$$

C.3.b.2. Large separations

$z \gg R_1$ and R_2, $i = j = 1$ term only, $\Gamma^2(5/2) = (9\pi/16)$:

$$G_{c\|c}(z;R_1,R_2) = -\frac{3A_{1m/2m}}{8\pi}\frac{(\pi R_1)^2(\pi R_2)^2}{z^5}.$$

C.3.b.3. Small separations

$l \ll R_1$ and R_2:

$$G_{c\|c}(l;R_1,R_2) = -\sqrt{\frac{2R_1R_2}{R_1+R_2}}\frac{A_{1m/2m}}{24l^{3/2}}.$$

Note: See Eqs. (4.62) and (4.63) in D. Langbein, *Van der Waals Attraction*, Springer Tracts in Modern Physics (Springer-Verlag, Berlin, 1974) or Eqs. (4) and (10) in D. Langbein, "Van der Waals attraction between cylinders, rods or fibers," Phys. Kondens. Mat., **15**, 61–86 (1972). If we translate to the geometric and summation variables used here, the coefficient $\hbar\langle\omega_1\rangle$ in the form

$$\frac{\hbar\langle\omega_1\rangle}{4\pi z}\sum_{i=1}^{\infty}\sum_{j=1}^{\infty}\frac{\Gamma^2(i+j+\frac{1}{2})}{i!\,j!(i-1)!(j-1)!}\left(\frac{R_1}{z}\right)^{2i}\left(\frac{R_2}{z}\right)^{2j}$$

(Eq. 4.62, op. cit.) is converted from integration to summation by the contour-integral procedure used in the Level 3 derivation of the original Lifshitz result. Specifically, from Eq. 4.63 (op. cit.),

$$\hbar\langle\omega_1\rangle = \hbar\int_{\xi=-\infty}^{\infty}d\xi\,\coth\left(\frac{\hbar\xi}{kT}\right)\left(\frac{\varepsilon_1-\varepsilon_m}{\varepsilon_1+\varepsilon_m}\right)\left(\frac{\varepsilon_2-\varepsilon_m}{\varepsilon_2+\varepsilon_m}\right) \to 2\pi kT\sum_{n=-\infty}^{\infty}\overline{\Delta}_{1m}\overline{\Delta}_{2m} = 4\pi kT\sum_{n=0}^{\infty}{}'\,\overline{\Delta}_{1m}\overline{\Delta}_{2m},$$

with

$$A_{1m/2m} = \frac{3kT}{2}\sum_{n=0}^{\infty}{}'\,\overline{\Delta}_{1m}\overline{\Delta}_{2m},\quad \frac{\hbar\langle\omega_1\rangle}{4\pi z} = \frac{kT}{z}\sum_{n=0}^{\infty}{}'\,\overline{\Delta}_{1m}\overline{\Delta}_{2m} \to \frac{2A_{1m/2m}}{3z}$$

or

$$G_{c\|c}(z;R_1,R_2) = -\frac{2A_{1m/2m}}{3z}\sum_{i,j=1}^{\infty}\frac{\Gamma^2(i+j+\frac{1}{2})}{i!\,j!(i-1)!(j-1)!}\left(\frac{R_1}{z}\right)^{2i}\left(\frac{R_2}{z}\right)^{2j}.$$

Langbein gives a more elaborate summation to include higher-order terms in $\overline{\Delta}_{1m}$ and $\overline{\Delta}_{2m}$.

Table C.4. "Thin" dielectric cylinders; parallel and at all angles, interaxial separation $z \ll$ radius R; Lifshitz form; retardation, magnetic, and ionic fluctuation terms not included

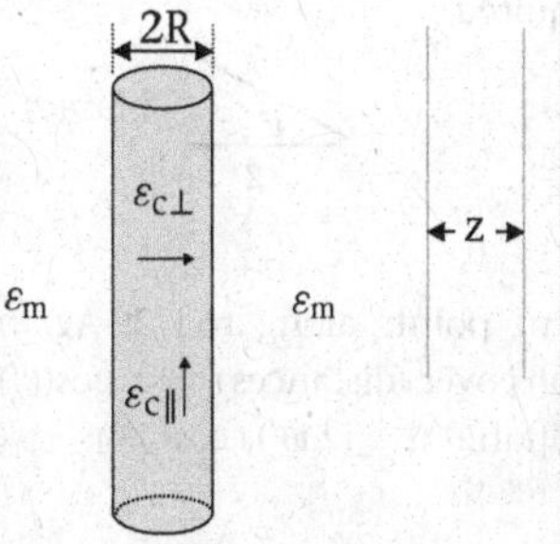

$$\overline{\Delta}_{\parallel} \equiv \frac{\varepsilon_{c\parallel} - \varepsilon_m}{\varepsilon_m},$$

$$\overline{\Delta}_{\perp} \equiv \frac{\varepsilon_{c\perp} - \varepsilon_m}{\varepsilon_{c\perp} + \varepsilon_m}$$

C.4.a. Parallel, interaxial separation z

$$g_{\parallel}(z;\mathrm{R}) = -\frac{9kT(\pi \mathrm{R}^2)^2}{16\pi z^5} \sum_{n=0}^{\infty}{}' \left[\overline{\Delta}_{\perp}^2 + \frac{\overline{\Delta}_{\perp}}{4}(\overline{\Delta}_{\parallel} - 2\overline{\Delta}_{\perp}) + \frac{3}{2^7}(\overline{\Delta}_{\parallel} - 2\overline{\Delta}_{\perp})^2 \right] \quad \text{[L2.233]}$$

free energy per unit length.

C.4.b.1. At an angle θ, minimal interaxial separation z

$$g(z,\theta;\mathrm{R}) = -\frac{3kT(\pi \mathrm{R}^2)^2}{4\pi z^4 \sin\theta} \sum_{n=0}^{\infty}{}' \left\{ \overline{\Delta}_{\perp}^2 + \frac{\overline{\Delta}_{\perp}}{4}(\overline{\Delta}_{\parallel} - 2\overline{\Delta}_{\perp}) + \frac{2\cos^2\theta + 1}{2^7}(\overline{\Delta}_{\parallel} - 2\overline{\Delta}_{\perp})^2 \right\}$$

[L2.234]

free energy per interaction.

C.4.b.2. Torque $\tau(z,\theta)$.

$$\tau(z,\theta;\mathrm{R}) = -\left.\frac{\partial g(z,\theta;\mathrm{R})}{\partial \theta}\right|_z$$

$$= -\frac{3kT(\pi \mathrm{R}^2)^2}{4\pi z^4} \left[\frac{\cos\theta}{\sin^2\theta} \sum_{n=0}^{\infty}{}' \{\} + \frac{\cos\theta}{2^5} \sum_{n=0}^{\infty}{}' (\overline{\Delta}_{\parallel} - 2\overline{\Delta}_{\perp})^2 \right]. \quad \text{[L2.235]}$$

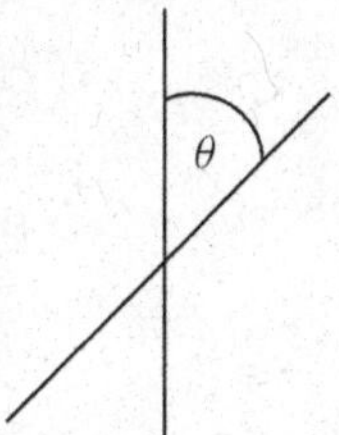

C.4.c. Hamaker hybrid form (small-delta limit with $\varepsilon_{c\perp} = \varepsilon_{c\parallel}$)

$$g_{\parallel}(z) = -\frac{3A_{1m/2m}(\pi \mathrm{R}^2)^2}{8\pi z^5}, \quad g(z,\theta) \cong -\frac{A_{1m/2m}(\pi \mathrm{R}^2)^2}{2z^4\pi \sin\theta},$$

$$A_{1m/2m} \approx \frac{3kT}{2} \sum_{n=0}^{\infty}{}' \overline{\Delta}_{1m}\overline{\Delta}_{2m} = \frac{3kT}{2} \sum_{n=0}^{\infty}{}' \overline{\Delta}^2, \quad \overline{\Delta} = \overline{\Delta}_{\perp} = \overline{\Delta}_{\parallel}$$ (second and third terms in [ ] neglected).

Note: Hamaker summation, actually integration, over two volumes goes as

$$-\frac{A_{\mathrm{Ham}}}{\pi^2} \iint_{V_1, V_2} \frac{\mathrm{d}V_1 \mathrm{d}V_2}{r^6} \quad \text{[Eq. (L2.125)].}$$

For two *parallel* thin cylinders of cross section A_1 and A_2, $\mathrm{d}V_1 = A_1 \mathrm{d}y_1$, $\mathrm{d}V_2 = A_2 \mathrm{d}y_2$ and $r^2 = z^2 + (y_2 - y_1)^2$. For an energy of interaction per unit length, the required integral is

$$-\frac{A_{\mathrm{Ham}} A_1 A_2}{\pi^2} \int_{-\infty}^{\infty} \frac{\mathrm{d}y_2}{r^6}$$

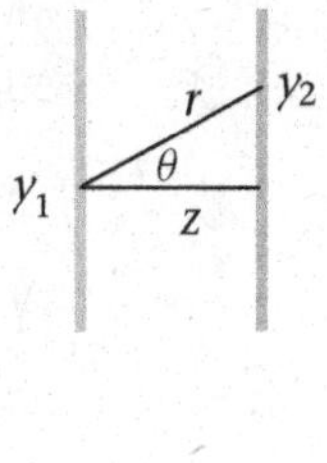

where we can use $\mathrm{d}y_2 = z\mathrm{d}[\tan(\theta)] = [z\mathrm{d}\theta/\cos^2(\theta)]$; $r = z/\cos(\theta)$, so that

$$\int_{-\infty}^{\infty} \frac{\mathrm{d}y_2}{r^6} = \frac{1}{z^5} \int_{-\frac{\pi}{2}}^{\frac{\pi}{2}} \cos^4(\theta)\mathrm{d}\theta = \frac{3\pi}{8z^5}$$

(Gradshteyn and Ryzhik, p. 369, Eq. 3.621.3).

Table C.4 (cont.)

For two *perpendicular* thin cylinders of cross section A_1 and A_2, $dV_1 = A_1 dx_1$, $dV_2 = A_2 dy_2$ and $r^2 = x_1^2 + z^2 + y_2^2$, with $-\infty < x_1, y_2 < +\infty$. For an interaction per pair of rods, the required integral is

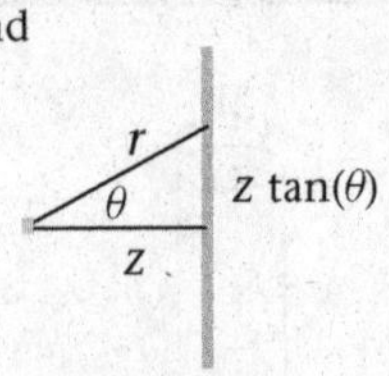

$$-\frac{A_{\mathrm{Ham}} A_1 A_2}{\pi^2} \int_{-\infty}^{\infty} dx_1 \int_{-\infty}^{\infty} \frac{dy_2}{r^6}.$$

Think first of the interaction between the $x_1 = 0$ position on rod 1 with different points along rod 2: Again $\int_{-\infty}^{\infty} (dy_2/r^6) = (3\pi/8z^5)$. Next integrate over all x_1 positions along rod 1. The integration covers distances $r = z/\cos(\theta')$ from each point on rod 1 to the closest point on rod 2 (now drawn end on). $dx_1 = z d[\tan(\theta')] = (z d\theta')/\cos^2(\theta')$:

$$\int_{-\infty}^{\infty} \frac{dx_1}{r^5} = \frac{1}{z^4} \int_{-\frac{\pi}{2}}^{\frac{\pi}{2}} \cos^3(\theta)\, d\theta = \frac{4}{3z^4}$$

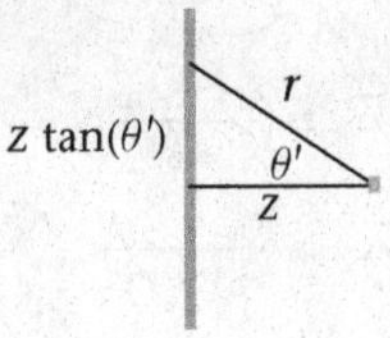

(Gradshteyn and Ryzhik, p. 369, Eq. 3.621.4).

The full integration gives

$$-\frac{A_{\mathrm{Ham}} A_1 A_2}{\pi^2} \int_{-\infty}^{\infty} dx_1 \int_{-\infty}^{\infty} \frac{dy_2}{r^6} = -\frac{A_{\mathrm{Ham}} A_1 A_2}{\pi^2} \frac{4}{3z^4} \frac{3\pi}{8} = -\frac{A_{\mathrm{Ham}} A_1 A_2}{2\pi z^4}.$$

Table C.5.a. Thin dielectric cylinders in saltwater, parallel and at an angle, low-frequency ($n = 0$) dipolar and ionic fluctuations

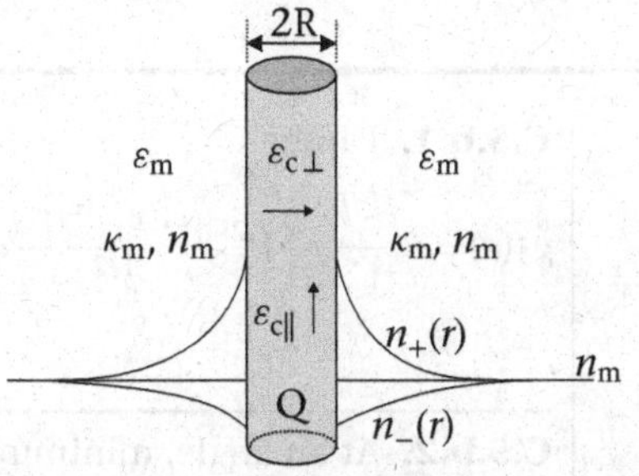

Cylinder bearing negative fixed charge of magnitude Qe per unit length. Net surrounding mobile charges neutralize that on cylinder.

Γ_c, total excess number of mobile charges apparent in charge fluctuations, $\Gamma_c \equiv \sum_{\{\nu\}} \Gamma_\nu \nu^2$, per unit length sum of all mobile charges ν minus the number of mobile charges in the salt solution if the cylinder were not present (figure drawn for 1–1 electrolyte). $\Gamma_\nu \equiv \int_0^\infty [n_\nu(r) - n_\nu(\mathrm{m})] 2\pi r\,\mathrm{d}r$, per unit length.

C.5.a.1. Parallel, center-to-center separation z

$$g_\parallel(z) = -\frac{2kT\kappa_m^5(\pi R^2)^2}{\pi^2}\{\,\};\quad \{\,\} = \left[\bar{\Delta}_\perp^2 + \frac{\bar{\Delta}_\perp(\bar{\Delta}_\parallel - 2\bar{\Delta}_\perp)}{4} + \frac{3(\bar{\Delta}_\parallel - 2\bar{\Delta}_\perp)^2}{2^7}\right]\int_1^\infty K_0(2\kappa_m p z)p^4\,\mathrm{d}p \qquad \text{[L2.254]}$$

$$+\left[\begin{array}{l}\left(\dfrac{\Gamma_c}{\pi a^2 n_m}\right)\dfrac{\bar{\Delta}_\perp}{2} + \left(\dfrac{\Gamma_c}{\pi a^2 n_m}\right)\dfrac{(\bar{\Delta}_\parallel - 2\bar{\Delta}_\perp)}{16} \\ -\bar{\Delta}_\perp^2 - \dfrac{3\bar{\Delta}_\perp(\bar{\Delta}_\parallel - 2\bar{\Delta}_\perp)}{8} - \dfrac{3(\bar{\Delta}_\parallel - 2\bar{\Delta}_\perp)^2}{2^6}\end{array}\right]\int_1^\infty K_0(2\kappa_m p z)p^2\,\mathrm{d}p$$

$$+\left[\begin{array}{l}\left(\dfrac{\Gamma_c}{\pi a^2 n_m}\right)^2\dfrac{1}{16} - \left(\dfrac{\Gamma_c}{\pi a^2 n_m}\right)\dfrac{\bar{\Delta}_\perp}{4} - \left(\dfrac{\Gamma_c}{\pi a^2 n_m}\right)\dfrac{(\bar{\Delta}_\parallel - 2\bar{\Delta}_\perp)}{16} \\ +\dfrac{\bar{\Delta}_\perp^2}{4} + \dfrac{\bar{\Delta}_\perp(\bar{\Delta}_\parallel - 2\bar{\Delta}_\perp)}{8} + \dfrac{3(\bar{\Delta}_\parallel - 2\bar{\Delta}_\perp)^2}{2^7}\end{array}\right]\int_1^\infty K_0(2\kappa_m p z)\,\mathrm{d}p.$$

For $x \to 0$, $K_0(x) \to -\ln x$, M. Abramowitz and I. A. Stegun, *Handbook of Mathematical Functions, with Formulas, Graphs and Mathematical Tables*, p. 375, Dover Books, New York (1965), Eq. 9.6.8. For $x \to \infty$, $K_0(x) \to \sqrt{\frac{\pi}{2x}}e^{-x}$, op. cit., p. 378, Eq. 9.7.2.

C.5.a.2. At an angle θ with minimum center-to-center separation z

$$g(z,\theta) = -\frac{kT\kappa_m^4(\pi R^2)^2}{\pi \sin\theta}\{\,\};$$

$$\{\,\} = \left[\bar{\Delta}_\perp^2 + \frac{\bar{\Delta}_\perp(\bar{\Delta}_\parallel - 2\bar{\Delta}_\perp)}{4} + \frac{(\bar{\Delta}_\parallel - 2\bar{\Delta}_\perp)^2}{2^7}(1 + 2\cos^2\theta)\right]\frac{6e^{-2\kappa_m z}}{(2\kappa_m z)^4}\left[1 + 2\kappa_m z + \frac{(2\kappa_m z)^2}{2} + \frac{(2\kappa_m z)^3}{6}\right]$$

$$+\left[\begin{array}{l}\left(\dfrac{\Gamma_c}{\pi R^2 n_m}\right)\dfrac{\bar{\Delta}_\perp}{2} + \left(\dfrac{\Gamma_c}{\pi R^2 n_m}\right)\dfrac{(\bar{\Delta}_\parallel - 2\bar{\Delta}_\perp)}{16} \\ -\dfrac{\bar{\Delta}_\perp^2}{2} - \dfrac{3\bar{\Delta}_\perp(\bar{\Delta}_\parallel - 2\bar{\Delta}_\perp)}{8} - \dfrac{(\bar{\Delta}_\parallel - 2\bar{\Delta}_\perp)^2}{2^6}(1 + 2\cos^2\theta)\end{array}\right]\frac{1}{(2\kappa_m z)^2}e^{-2\kappa_m z}(1 + 2\kappa_m z)$$

$$+\left[\begin{array}{l}\left(\dfrac{\Gamma_c}{\pi R^2 n_m}\right)^2\dfrac{1}{16} - \left(\dfrac{\Gamma_c}{\pi R^2 n_m}\right)\dfrac{\bar{\Delta}_\perp}{4} - \left(\dfrac{\Gamma_c}{\pi R^2 n_m}\right)\dfrac{(\bar{\Delta}_\parallel - 2\bar{\Delta}_\perp)}{16} \\ +\dfrac{\bar{\Delta}_\perp^2}{4} + \dfrac{\bar{\Delta}_\perp(\bar{\Delta}_\parallel - 2\bar{\Delta}_\perp)}{8} + \dfrac{(\bar{\Delta}_\parallel - 2\bar{\Delta}_\perp)^2}{2^7}(1 + 2\cos^2\theta)\end{array}\right]E_1(2\kappa_m z). \qquad \text{[L2.252]}$$

The exponential integral $E_1(2\kappa_m z) = -\gamma - \ln(2\kappa_m z) - \sum_{n=1}^\infty \frac{(-2\kappa_m z)^n}{nn!}$, $\gamma = 0.5772156649$. For large arguments, $E_1(2\kappa_m z) \to [(e^{-2\kappa_m z})/(2\kappa_m z)]$.

Table C.5.b. Thin cylinders in saltwater, parallel and at an angle, ionic fluctuations only, at separations $\gg$ Debye length

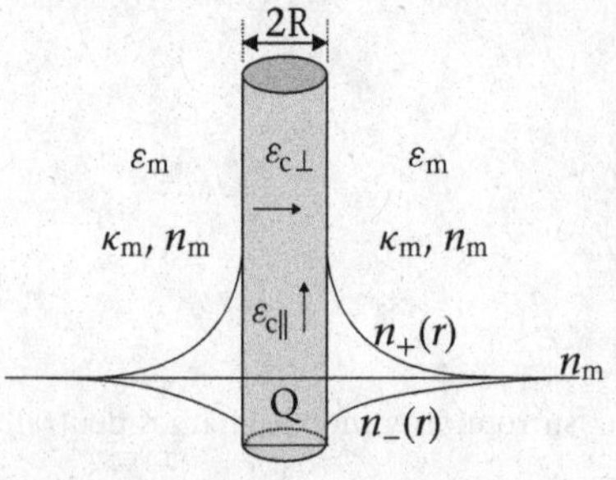

Cylinder bearing negative fixed charge of magnitude Qe per unit length. Net surrounding mobile charges neutralize those on cylinder.

Total excess number of mobile charges apparent in charge fluctuations is the valence-squared weighted sum of all mobile charges minus the number of mobile charges in the salt solution if the cylinder were not present.

C.5.b.1. Parallel

$$g_{\|}(z) = -\frac{kT\lambda_{\mathrm{Bj}}^2}{2}\Gamma_c^2\sqrt{\pi}\frac{e^{-2\kappa_m z}}{\kappa_m^{1/2} z^{3/2}} = -\frac{kT\lambda_{\mathrm{Bj}}^2}{2}\Gamma_c^2\sqrt{\pi}\frac{e^{-2z/\lambda_D}}{z(z/\lambda_D)^{1/2}}.$$

[L2.257]

C.5.b.2. At an angle, minimum separation z

$$g(z,\theta) = -\frac{kT\lambda_{\mathrm{Bj}}^2}{\sin\theta}\Gamma_c^2\frac{e^{-2\kappa_m z}}{2\kappa_m z} = -\frac{kT\pi\lambda_{\mathrm{Bj}}^2}{\sin\theta}\Gamma_c^2\frac{e^{-2z/\lambda_D}}{(2z/\lambda_D)},$$

[L2.256]

$$\Delta_{\|} \to 0,\ \Delta_{\perp} \to 0,\ z \gg \lambda_{\mathrm{Debye}} = \lambda_D = 1/\kappa_m.$$

Table C.6. Parallel, coterminous thin rods, length a, interaxial separation z, Hamaker form

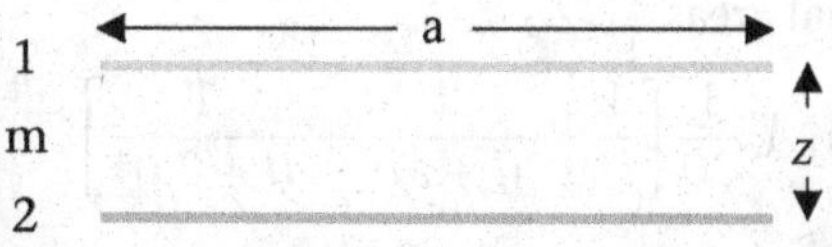

C.6.a. Cross-sectional areas A_1, A_2

$$G(z;a) = -\frac{A_{1m/2m}}{4\pi^2} A_1 A \left[\frac{1}{z^4} - \frac{1}{z^2(a^2+z^2)} + \frac{3a}{z^5}\tan^{-1}\left(\frac{a}{z}\right)\right]$$

per interaction,

$$A_{1m/2m} \approx \frac{3kT}{2}\sum_{n=0}^{\infty}{}' \overline{\Delta}_{1m}\overline{\Delta}_{2m}.$$

C.6.b. Circular rods of radii R_1, R_2

$$G(z;a) = -\frac{A_{1m/2m}}{4}\pi R_1^2 R_2^2 \left[\frac{1}{z^4} - \frac{1}{z^2(a^2+z^2)} + \frac{3a}{z^5}\tan^{-1}\left(\frac{a}{z}\right)\right].$$

Limit $a/z \to \infty$, $\tan^{-1}(a/z) \to (\pi/2)$.

Energy of interaction per unit length

$$\frac{G(z;a)}{a} \to -\frac{3A_{1m/2m}}{8\pi}\frac{A_1 A_2}{z^5}.$$

Note: Coefficients can be confusing. These formulae are per total interaction, not divided to be per rod. Recall the Hamaker form for incremental interactions,

$$-\frac{A_{Ham}}{\pi^2}\frac{dV_1\, dV_2}{r^6}.$$

Between the two thin rods here, the incremental volumes are $dV_1 = A_1\, dx_1$, $dV_2 = A_2\, dx_2$; their separation r varies as $r^2 = [z^2 + (x_2 - x_1)^2]$. Between infinitely long rods, the energy of interaction per unit length requires one integration only, over $x = (x_2 - x_1)$ from $x = -\infty$ to $+\infty$,

$$-\frac{A_{Ham}}{\pi^2}A_1 A_2 \int_{-\infty}^{\infty}\frac{dx}{(z^2+x^2)^3} = -\frac{A_{Ham}}{\pi^2}A_1 A_2 \frac{3\pi}{8z^5} = -\frac{3A_{Ham}}{8\pi}\frac{A_1 A_2}{z^5}.$$

This is identical to the result $[G(z;a)]/a$ for $a \to \infty$ just above and similar in form to the Lifshitz result for two like parallel cylinders of radius R, Table C.4.a. The $1/r^6$ integrations are taken from A. G. DeRocco and W. G. Hoover, "On the interaction of colloidal particles," Proc. Natl. Acad. Sci. USA, **46**, 1057–1065 (1960).

Table C.7. Coaxial thin rods, minimum separation l, length a, Hamaker form

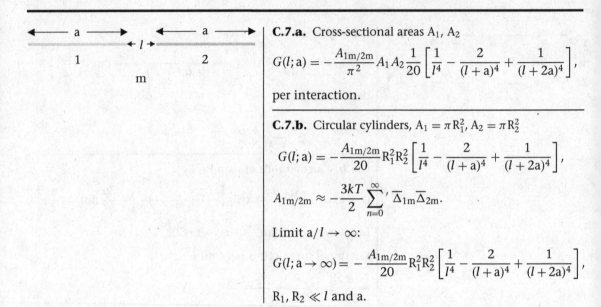

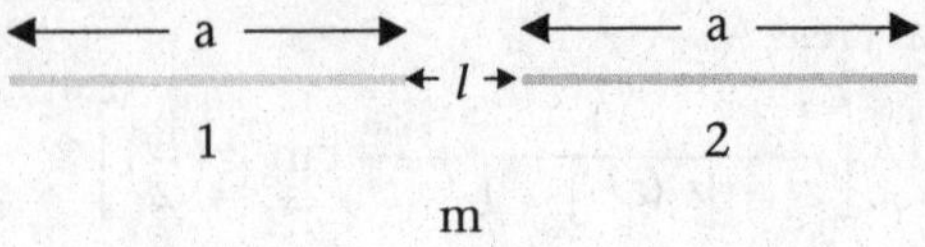

C.7.a. Cross-sectional areas A_1, A_2

$$G(l;\mathrm{a}) = -\frac{A_{1\mathrm{m}/2\mathrm{m}}}{\pi^2} A_1 A_2 \frac{1}{20}\left[\frac{1}{l^4} - \frac{2}{(l+\mathrm{a})^4} + \frac{1}{(l+2\mathrm{a})^4}\right],$$

per interaction.

C.7.b. Circular cylinders, $A_1 = \pi R_1^2$, $A_2 = \pi R_2^2$

$$G(l;\mathrm{a}) = -\frac{A_{1\mathrm{m}/2\mathrm{m}}}{20} R_1^2 R_2^2 \left[\frac{1}{l^4} - \frac{2}{(l+\mathrm{a})^4} + \frac{1}{(l+2\mathrm{a})^4}\right],$$

$$A_{1\mathrm{m}/2\mathrm{m}} \approx -\frac{3kT}{2} \sum_{n=0}^{\infty}{}' \overline{\Delta}_{1\mathrm{m}} \overline{\Delta}_{2\mathrm{m}}.$$

Limit $\mathrm{a}/l \to \infty$:

$$G(l;\mathrm{a}\to\infty) = -\frac{A_{1\mathrm{m}/2\mathrm{m}}}{20} R_1^2 R_2^2 \left[\frac{1}{l^4} - \frac{2}{(l+\mathrm{a})^4} + \frac{1}{(l+2\mathrm{a})^4}\right],$$

$R_1, R_2 \ll l$ and a.

Note:

$$-\frac{A_{\mathrm{Ham}}}{\pi^2} A_1 A_2 \int_0^{\mathrm{a}} \int_0^{\mathrm{a}} \frac{dz_1 dz_2}{(z_1 + z_2 + l)^6} = -\frac{A_{\mathrm{Ham}}}{\pi^2} \frac{A_1 A_2}{20} \left[\frac{1}{l^4} - \frac{2}{(l+\mathrm{a})^4} + \frac{1}{(l+2\mathrm{a})^4}\right].$$

Table C.8. Circular disks and rods

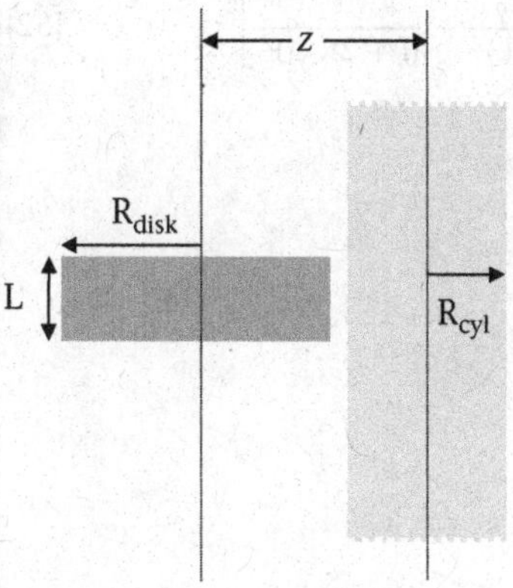

Disk thickness L, radius R_{disk}, cylinder radius R_{cyl}, interaxial spacing z

C.8.a. Circular disk or rod of finite length, with axis parallel to infinitely long cylinder, pairwise-summation form

$$G_{disk/cylinder}(z; R_{disk}, R_{cyl}, L) = -\frac{A_{Ham} L}{4}\{\};$$

$$\begin{aligned}
\{\} = {} & \frac{\pi}{2}\left[\frac{-1}{z+R_{disk}-R_{cyl}} - \frac{R_{cyl}}{2(z+R_{disk}-R_{cyl})^2} + \frac{1}{z-R_{disk}-R_{cyl}} + \frac{R_{cyl}}{2(z-R_{disk}-R_{cyl})^2}\right] \\
& - \frac{\pi}{2}\left[\frac{-1}{z+R_{disk}+R_{cyl}} + \frac{R_{cyl}}{2(z+R_{disk}+R_{cyl})^2} + \frac{1}{z-R_{disk}+R_{cyl}} + \frac{R_{cyl}}{2(z-R_{disk}+R_{cyl})^2}\right] \\
& + \frac{R_{disk}^2 - z^2}{2z}\left[\frac{-1}{2(z+R_{disk}-R_{cyl})^2} + \frac{1}{2(z-R_{disk}-R_{cyl})^2}\right] \\
& + \frac{z^2 - R_{disk}^2}{2z}\left[\frac{-1}{2(z+R_{disk}+R_{cyl})^2} + \frac{1}{2(z-R_{disk}+R_{cyl})^2}\right] \\
& - \frac{1}{2z}\left[\ln\left(\frac{z+R_{disk}-R_{cyl}}{z-R_{disk}-R_{cyl}}\right) - \frac{2R_{cyl}}{z+R_{disk}-R_{cyl}} + \frac{2R_{cyl}}{z-R_{disk}-R_{cyl}} - \frac{R_{cyl}^2}{2(z+R_{disk}-R_{cyl})^2} + \frac{R_{cyl}^2}{2(z-R_{disk}-R_{cyl})^2}\right] \\
& + \frac{1}{2z}\left[\ln\left(\frac{z+R_{disk}+R_{cyl}}{z-R_{disk}+R_{cyl}}\right) - \frac{2R_{cyl}}{z+R_{disk}+R_{cyl}} - \frac{2R_{cyl}}{z-R_{disk}+R_{cyl}} - \frac{R_{cyl}^2}{2(z+R_{disk}+R_{cyl})^2} + \frac{R_{cyl}^2}{2(z-R_{disk}+R_{cyl})^2}\right] \quad [41].
\end{aligned}$$

Source: Results translated from S. W. Montgomery, M. A. Franchek, and V. W. Goldschmidt, "Analytical dispersion force calculations for nontraditional geometries," J. Colloid Interface Sci., **227**, 567–587 (2000).

Notation: $U_{P/P} = (-\beta/l^6)$, Eq. [3] in that paper, analogous to $-(c_A c_B/r^6)$ [from expression (L2.121)] for two particles attracting across a vacuum $dU = [-Q/\beta/l^6)]dV$, Eq. [4], for volume integration; Q represents the number density per unit volume, the same as N in this text. $A_{Ham} = \pi^2 Q_A Q_B \beta$ there, $A_{Ham} = \pi^2 N_A c_A N_B c_B$ across vacuum of $A_{Ham} = \pi^2(N_A c_A - N_m c_m)(N_B c_B - N_m c_m)$ [Eq. (L2.124)] across medium m here. $R_2 = R_{cyl}$; $R_3 = R_{disk}$; $H_2 = z$ = interaxial distance; $D = l = z - R_{cyl} - R_{disk}$.

Table C.8 (cont.)

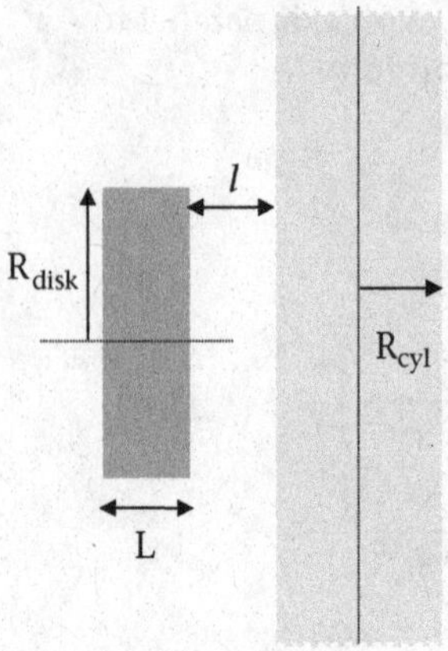

Disk thickness L, radius R_{disk}, cylinder radius R_{cyl}, nearest separation l.

C.8.b. Circular disk with axis perpendicular to axis of infinite cylinder, pairwise-summation form

$$G_{perp-disk/cylinder}(l; R_{disk}, R_{cyl}, L)$$
$$= -\frac{A_{Ham}\pi R_{disk}^3}{8}\left[\frac{1}{l^2} + \frac{1}{(l+L+2R_{cyl})^2} - \frac{1}{(l+L)^2} - \frac{1}{(l+2R_{cyl})^2}\right] \quad [52].$$

Source: S. W. Montgomery, M. A. Franchek, and V. W. Goldschmidt, "Analytical dispersion force calculations for nontraditional geometries," J. Colloid Interface Sci., **227**, 567–584 (2000).

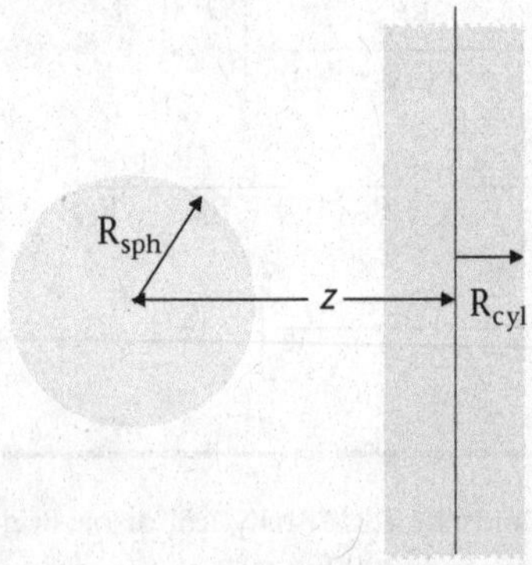

Sphere radius R_{sph}, cylinder radius R_{cyl}, sphere center to cylinder axis distance z.

C.8.c. Sphere with infinite cylinder, pairwise-summation form

$$G_{sphere/cylinder}(z; R_{sph}, R_{cyl}) = -\frac{A_{Ham}\pi}{8}\{\};$$

$$\begin{aligned}\{\} = {} & \frac{R_{sph}^3 - (z+R_{cyl})^2}{2(z+R_{cyl}+R_{sph})^2} + \frac{(z-R_{cyl})^2 - R_{sph}^3}{2(z-R_{cyl}+R_{sph})^2} \\ & - \frac{(z-R_{cyl})^2 - R_{sph}^3}{2(z-R_{cyl}-R_{sph})^2} + \frac{(z+R_{cyl})^2 - R_{sph}^3}{2(z+R_{cyl}-R_{sph})^2} \\ & + \frac{2z+2R_{cyl}}{z+R_{cyl}+R_{sph}} - \frac{2z+2R_{cyl}}{z+R_{cyl}-R_{sph}} \\ & + \frac{2R_{cyl}-2z}{z-R_{cyl}+R_{sph}} + \frac{2z-2R_{cyl}}{z-R_{cyl}-R_{sph}} \\ & + \ln\left(\frac{z+R_{cyl}+R_{sph}}{z+R_{cyl}-R_{sph}}\right) + \ln\left(\frac{z-R_{cyl}-R_{sph}}{z-R_{cyl}+R_{sph}}\right)\end{aligned}$$

[60].

Source: Equation from S. W. Montgomery, M. A. Franchek, and V. W. Goldschmidt, "Analytical dispersion force calculations for nontraditional geometries," J. Colloid Interface Sci., **227**, 567–587 (2000).

L2.3. Essays on formulae

for·mu·lary \ ˈfȯr-myə-ˌler-ē \ *n, pl* -lar·ies (1541) **1:** A book or other collection of stated and fixed forms, such as prayers **2:** A statement expressed in formulas **3:** A fixed form or pattern; a formula **4:** A book containing a list of pharmaceutical substances along with their formulas, uses, and methods of preparation—**formulary** *adj*

The American Heritage Dictionary of the English Language, Fourth Edition Copyright © 2000 by Houghton Mifflin Company.

Between the ecclesiastic (definition 1) and the operative (definition 4), between origin (Level 3) and result (Level 2 tables), is exegesis. Different minds will find different ways to work from Level 3 Foundations to Level 2 tabulations. Each of the following sections describes not simply one particular set of steps but, more important, the kinds of steps to take:

- Examining the full expression for the interaction between half-spaces to see which features are revealed in its specialized limiting forms (Section L2.3.A);
- Generalizing the original half-space geometry to layered and inhomogeneous planar bodies (Section L2.3.B);
- Converting to cases of curved structures (Section L2.3.C);
- Reducing to Hamaker theory (Section L2.3.D) for gases and dilute suspensions, but now
- Incorporating fluctuations and screening in ionic solutions (Section L2.3.E), and then
- Extending to include interactions between small particles and substrates (Section L2.3.F) as well as between one-dimensional linear bodies (Section L2.3.G).

In what follows, the numbers in parentheses next to or below the equations are the actual equation numbers. Numbers given in square brackets correspond to the tables of Level 2 (prefaced by P, S, or C) or to equations and expressions given elsewhere in this book (prefaced with the level number).

This is the place to get the pencils moving, to become familiar with the manipulations that bring out instructive or critical features of interactions, to be able to think about forces beyond the automated conversion of formulae into numbers.

L2.3.A. Interactions between two semi-infinite media

Exact, Lifshitz

The general formula for the electrodynamic free energy per unit area between two semi-infinite media A and B separated by a planar slab of material m, thickness l is

$$G_{\mathrm{AmB}}(l,T) = \frac{kT}{8\pi l^2}\sum_{n=0}^{\infty}{}' \int_{r_n}^{\infty} x \ln\left[(1-\overline{\Delta}_{\mathrm{Am}}\overline{\Delta}_{\mathrm{Bm}}e^{-x})(1-\Delta_{\mathrm{Am}}\Delta_{\mathrm{Bm}}e^{-x})\right]\mathrm{d}x$$

$$= \frac{kT}{2\pi c^2}\sum_{n=0}^{\infty}{}' \varepsilon_{\mathrm{m}}\mu_{\mathrm{m}}\xi_n^2 \int_{1}^{\infty} p \ln\left[(1-\overline{\Delta}_{\mathrm{Am}}\overline{\Delta}_{\mathrm{Bm}}e^{-r_n p})(1-\Delta_{\mathrm{Am}}\Delta_{\mathrm{Bm}}e^{-r_n p})\right]\mathrm{d}p$$

$$= -\frac{kT}{8\pi l^2}\sum_{n=0}^{\infty}{}' r_n^2 \sum_{q=1}^{\infty}\frac{1}{q}\int_{1}^{\infty} p[(\overline{\Delta}_{\mathrm{Am}}\overline{\Delta}_{\mathrm{Bm}})^q + (\Delta_{\mathrm{Am}}\Delta_{\mathrm{Bm}})^q]e^{-r_n p q}\mathrm{d}p, \qquad \text{(L2.1)}$$

[P.1.a.1]

where the prime in the summation $\sum_{n=0}^{\prime\infty}$ stipulates that the $n = 0$ term is to be multiplied by 1/2 and

$$\overline{\Delta}_{\mathrm{ji}} = \frac{s_{\mathrm{i}}\varepsilon_{\mathrm{j}} - s_{\mathrm{j}}\varepsilon_{\mathrm{i}}}{s_{\mathrm{i}}\varepsilon_{\mathrm{j}} + s_{\mathrm{j}}\varepsilon_{\mathrm{i}}}, \quad \Delta_{\mathrm{ji}} = \frac{s_{\mathrm{i}}\mu_{\mathrm{j}} - s_{\mathrm{j}}\mu_{\mathrm{i}}}{s_{\mathrm{i}}\mu_{\mathrm{j}} + s_{\mathrm{j}}\mu_{\mathrm{i}}}, \quad s_{\mathrm{i}} = \sqrt{p^2 - 1 + (\varepsilon_{\mathrm{i}}\mu_{\mathrm{i}}/\varepsilon_{\mathrm{m}}\mu_{\mathrm{m}})},$$
$$s_{\mathrm{m}} = p \qquad \text{(L2.2)}$$

or

$$\overline{\Delta}_{\mathrm{ji}} = \frac{x_{\mathrm{i}}\varepsilon_{\mathrm{j}} - x_{\mathrm{j}}\varepsilon_{\mathrm{i}}}{x_{\mathrm{i}}\varepsilon_{\mathrm{j}} + x_{\mathrm{j}}\varepsilon_{\mathrm{i}}}, \quad \Delta_{\mathrm{ji}} = \frac{x_{\mathrm{i}}\mu_{\mathrm{j}} - x_{\mathrm{j}}\mu_{\mathrm{i}}}{x_{\mathrm{i}}\mu_{\mathrm{j}} + x_{\mathrm{j}}\mu_{\mathrm{i}}}, \quad x_{\mathrm{i}}^2 = x_{\mathrm{m}}^2 + \left(\frac{2l\xi_n}{c}\right)^2(\varepsilon_{\mathrm{i}}\mu_{\mathrm{i}} - \varepsilon_{\mathrm{m}}\mu_{\mathrm{m}}),$$
$$x_{\mathrm{m}} = x, \qquad \text{(L2.3)}$$

where

$$p = x/r_n, \quad r_n = (2l\varepsilon_{\mathrm{m}}^{1/2}\mu_{\mathrm{m}}^{1/2}/c)\,\xi_n. \qquad \text{(L2.4)}$$

The eigenfrequencies ξ_n at which $\varepsilon_{\mathrm{i}}(i\xi_n)$ and $\mu_{\mathrm{i}}(i\xi_n)$ are evaluated for materials i = A, m, and B are uniformly spaced at

$$\xi_n \frac{2\pi kT}{\hbar} n, \; n = 0, 1, 2 \ldots. \qquad \text{(L2.5)}$$

Physically, r_n is the ratio of the time for an electromagnetic signal to travel and return across the gap of medium m and thickness l divided by the characteristic time $1/\xi_n$ of the particular electromagnetic fluctuation.

This formula can also be written in Hamaker form with the Hamaker coefficient $A_{\mathrm{AmB}}(l,T)$ (see Table P.1.a.2):

$$G_{\mathrm{AmB}}(l,T) = -\frac{A_{\mathrm{AmB}}(l,T)}{12\pi l^2}. \qquad \text{(L2.6)}$$

The quantities ε_i, ε_j and μ_i, μ_j are dielectric- and magnetic-response functions of the individual materials. Indices i and j refer to materials A, B, or m; x and p are variables of integration. The variable x has a physical meaning: $x = 2\rho_m l$, where

$$\rho_m^2 = \rho^2 + \varepsilon_m \mu_m \xi_n^2 / c^2 \tag{L2.7}$$

and $\rho^2 = (u^2 + v^2)$ is the sum of squares of the radial components of the surface-wave vector. It is the perturbation of these surface waves (or surface modes) by varied separation l that creates the van der Waals force. In principle, the dielectric and magnetic responses ε and μ depend on the wave vector ρ, but this dependence is neglected in the macroscopic continuum theory. (See Section L2.4, pp. 258–260, for a discussion of this neglect.)

Nonretarded, separations approaching contact, $l \to 0$, $r_n \to 0$

$$G_{AmB}(l \to 0, T) \to \frac{kT}{8\pi l^2} \sum_{n=0}^{\infty}{}' \int_0^{\infty} x \left[\ln\left(1 - \overline{\Delta}_{Am}\overline{\Delta}_{Bm} e^{-x}\right)\left(1 - \Delta_{Am}\Delta_{Bm} e^{-x}\right)\right] dx$$

$$= -\frac{kT}{8\pi l^2} \sum_{n=0}^{\infty}{}' \sum_{q=1}^{\infty} \frac{(\overline{\Delta}_{Am}\overline{\Delta}_{Bm})^q + (\Delta_{Am}\Delta_{Bm})^q}{q^3}. \tag{L2.8}$$

[P.1.a.3]

In this case, the deltas can be put into a simpler form:

$$\overline{\Delta}_{ji} = \frac{\varepsilon_j - \varepsilon_i}{\varepsilon_j + \varepsilon_i} \quad \Delta_{ji} = \frac{\mu_j - \mu_i}{\mu_j + \mu_i}. \tag{L2.9}$$

There are at least two ways to see the reason for this simplification of deltas: First, because $p = x/r_n$, p is effectively infinite in the integration over x. With infinitely large p, the s_i become equal to p, and the s's and p's cancel in the numerator and denominator of the deltas. Alternatively, consider the second form of integration, in p:

$$\int_1^{\infty} p\left[\ln\left(1 - \overline{\Delta}_{Am}\overline{\Delta}_{Bm} e^{-r_n p}\right)\left(1 - \Delta_{Am}\Delta_{Bm} e^{-r_n p}\right)\right] dp.$$

For this integral to converge when $r_n \to 0$, the important contributions to the integrand must be in the limit where $p \to \infty$.

Nonretarded, small differences in permittivity With $\overline{\Delta}_{ji} = [(\varepsilon_j - \varepsilon_i)/(\varepsilon_j + \varepsilon_i)] \ll 1$, $\Delta_{ji} = [(\mu_j - \mu_i)/(\mu_j + \mu_i)] \ll 1$, only the leading term in q is significant:

$$G_{AmB}(l \to 0, T) \approx -\frac{kT}{8\pi l^2} \sum_{n=0}^{\infty}{}' \left(\overline{\Delta}_{Am}\overline{\Delta}_{Bm} + \Delta_{Am}\Delta_{Bm}\right). \tag{L2.10}$$

[P.1.a.4]

Infinitely large separations, $l \to \infty$ Here all $r_n \to \infty$, except for $r_{n=0} = 0$. Except for the $n = 0$ term, all integrands are driven to zero by virtue of the factor $e^{-x} \to 0$. Only the $n = 0$ term remains in the summation $\sum_{n=0}^{\prime\infty}$, so that the interaction free energy takes the form

$$G_{AmB}(l \to \infty, T) \to -\frac{kT}{16\pi l^2} \sum_{q=1}^{\infty} \frac{(\overline{\Delta}_{Am}\overline{\Delta}_{Bm})^q + (\Delta_{Am}\Delta_{Bm})^q}{q^3}. \tag{L2.11}$$

[P.1.a.5]

Here again $\overline{\Delta}_{ji} = [(\varepsilon_j - \varepsilon_i)/(\varepsilon_j + \varepsilon_i)]$, $\Delta_{ji} = [(\mu_j - \mu_i)/(\mu_j + \mu_i)]$, but all ε's and μ's are evaluated at zero frequency. (This expression ignores ionic fluctuations and material conductivities.)

"Low" temperatures, with retardation

When kT is much less than photon energy $\hbar\xi$, charge fluctuations are no longer driven by thermal fluctuations. Only zero-point uncertainty-principle fluctuations remain. In the low-temperature regime, the spacing between successive eigenfrequencies, $\xi_n = [(2\pi kT)/\hbar]n$, becomes small enough for these frequencies to be considered a continuum rather than discrete, separated values. Because the frequencies ξ_n are closely spaced, the successive terms in the summation $\sum_{n=0}^{\prime\infty}$ over discrete n can be considered a smoothly varying function over a smoothly varying n. That is, we can view $\sum_{n=0}^{\prime\infty}$ as an integral $\int_0^\infty \mathrm{d}n$. Replacing differential $\mathrm{d}n$ with $(\hbar/2\pi kT)\mathrm{d}\xi$ converts the sum into an integral in frequency ξ:

$$\frac{kT}{8\pi l^2}\sum_{n=0}^{\infty}{}' \rightarrow \frac{kT}{8\pi l^2}\int_0^\infty \mathrm{d}n = \frac{kT}{8\pi l^2}\frac{\hbar}{2\pi kT}\int_0^\infty \mathrm{d}\xi = \frac{\hbar}{(4\pi)^2 l^2}\int_0^\infty \mathrm{d}\xi.$$

The interaction free energy becomes

$$G(l, T \rightarrow 0) = \frac{\hbar}{(4\pi)^2 l^2}\int_0^\infty \mathrm{d}\xi \int_{r_n}^\infty x \ln\left[(1 - \overline{\Delta}_{\mathrm{Am}}\overline{\Delta}_{\mathrm{Bm}}e^{-x})(1 - \Delta_{\mathrm{Am}}\Delta_{\mathrm{Bm}}e^{-x})\right]\mathrm{d}x$$

$$= \frac{\hbar}{(2\pi)^2 c^2}\int_0^\infty \mathrm{d}\xi\, \varepsilon_{\mathrm{m}}\mu_{\mathrm{m}}\xi^2 \int_1^\infty p \ln\left[(1 - \overline{\Delta}_{\mathrm{Am}}\overline{\Delta}_{\mathrm{Bm}}e^{-r_n p})(1 - \Delta_{\mathrm{Am}}\Delta_{\mathrm{Bm}}e^{-r_n p})\right]\mathrm{d}p.$$

(L2.12)
[P.1.b.1]

Low-temperature, small-separation limits (no retardation)

With the further restriction that all retardation factors be ignored,

$$G(l \rightarrow 0, T \rightarrow 0) = \frac{\hbar}{(4\pi)^2 l^2}\int_0^\infty \mathrm{d}\xi \sum_{q=1}^{\infty}\frac{(\overline{\Delta}_{\mathrm{Am}}\overline{\Delta}_{\mathrm{Bm}})^q + (\Delta_{\mathrm{Am}}\Delta_{\mathrm{Bm}})^q}{q^3}, \qquad \text{(L2.13)}$$

[P.1.b.2]

where again $\overline{\Delta}_{ji} = [(\varepsilon_j - \varepsilon_i)/(\varepsilon_j + \varepsilon_i)]$, $\Delta_{ji} = [(\mu_j - \mu_i)/(\mu_j + \mu_i)]$.

This low-temperature, small-separation form can be succinctly written as though there were an average photon energy $\hbar\overline{\xi}$:

$$G(l \rightarrow 0, T \rightarrow 0) = \frac{-\hbar\overline{\xi}}{(4\pi)^2 l^2}. \qquad \text{(L2.14)}$$

$\overline{\xi}$ is usually approximated by its leading, $q = 1$, term

$$\overline{\xi} \approx \int_0^\infty (\overline{\Delta}_{\mathrm{Am}}\overline{\Delta}_{\mathrm{Bm}} + \Delta_{\mathrm{Am}}\Delta_{\mathrm{Bm}})\mathrm{d}\xi.$$

Low-temperature, large-separation limits

In formulating this limit it is usually assumed that the ε_i and μ_i responses are effectively constant, kept at their low-frequency values during the integration over all frequencies ξ. In practice, the true low-frequency values are ignored. Rather, the important values of ε_i and μ_i are those that pertain to a finite-frequency region where they are effectively constant. This is usually in the visible-frequency region where $\varepsilon_i = n^2_{\text{ref i}}$, the square of the index of refraction. There the magnetic responses μ_i are effectively equal to 1, and the dielectric responses ε_i are given by the square of the refractive index. The distance is taken to be large enough that the integration in frequency converges purely from retardation screening. The differences in ε_i are taken to be small enough that the $\overline{\Delta}$'s and Δ's are $\ll 1$:

$$G_{AmB}(l, T \to 0) = \frac{\hbar}{(4\pi)^2 l^2}\int_0^\infty d\xi \int_{r_n}^\infty x \ln\left[(1-\overline{\Delta}_{Am}\overline{\Delta}_{Bm}e^{-x})\,(1-\Delta_{Am}\Delta_{Bm}e^{-x})\right]dx$$

$$\approx -\frac{\hbar c}{8\pi^2 l^3 \varepsilon_m^{1/2}}(\overline{\Delta}_{Am}\overline{\Delta}_{Am}), \quad \text{(L2.15)}$$

[P.1.b.3]

where now

$$\overline{\Delta}_{Am} = \frac{\sqrt{\varepsilon_A}-\sqrt{\varepsilon_m}}{\sqrt{\varepsilon_A}+\sqrt{\varepsilon_m}} = \frac{n_A - n_m}{n_A + n_m}, \quad \overline{\Delta}_{Bm} = \frac{\sqrt{\varepsilon_B}-\sqrt{\varepsilon_m}}{\sqrt{\varepsilon_B}+\sqrt{\varepsilon_m}} = \frac{n_B - n_m}{n_B + n_m}, \quad \text{(L2.16)}$$

so that

$$G_{AmB}(l, T \to 0) \approx -\frac{\hbar c}{8\pi^2 n_m l^3}\frac{n_A - n_m}{n_A + n_m}\frac{n_B - n_m}{n_B + n_m}. \quad \text{(L2.17)}$$

The derivation of this highly specialized but surprisingly popular formula deserves comment. It involves taking several different limits with severe assumptions about the important values of variables.

When the integral is kept over the frequencies and all r_n are allowed to go to infinity, there is no longer a residual $1/l^2$ contribution from an $n = 0$ term as there was in the "infinite" separation case. The integrals over x and p converge so rapidly that they are dominated by $x \sim r_n$ and $p \sim 1$, respectively. Put another way, the integrands are driven down so rapidly by the exponentials e^{-x} and $e^{-r_n p}$ that all other terms are effectively constant. With $p = 1$, $s_i = \sqrt{p^2 - 1 + (\varepsilon_i\mu_i/\varepsilon_m\mu_m)} = \sqrt{(\varepsilon_i\mu_i/\varepsilon_m\mu_m)}$, so that $\overline{\Delta}_{ji} = [(s_i\varepsilon_j - s_j\varepsilon_i)/(s_i\varepsilon_j + s_j\varepsilon_i)]$ and $\overline{\Delta}_{ji} = [(s_i\mu_j - s_j\mu_i)/(s_i\mu_j + s_j\mu_i)]$ are no longer functions of p or x:

$$\Delta_{Am} = \frac{\sqrt{\varepsilon_m}-\sqrt{\varepsilon_A}}{\sqrt{\varepsilon_m}+\sqrt{\varepsilon_A}} = -\overline{\Delta}_{Am}, \quad \Delta_{Am} = \frac{\sqrt{\varepsilon_m}-\sqrt{\varepsilon_B}}{\sqrt{\varepsilon_m}+\sqrt{\varepsilon_B}} = -\overline{\Delta}_{Bm}.$$

Because of retardation screening, the integrals converge so rapidly that there is no chance for the material responses to vary, the response functions are taken to be effectively constant in frequency.

For integration over x, expand

$$\int_{r_n}^\infty x\left[\ln(1-\overline{\Delta}_{Am}\overline{\Delta}_{Bm}e^{-x})(1-\Delta_{Am}\Delta_{Bm}e^{-x})\right]dx$$

$$= -\int_{r_n}^\infty x\left[\sum_{q=1}^\infty \frac{(\overline{\Delta}_{Am}\overline{\Delta}_{Bm}e^{-x})^q + (\Delta_{Am}\Delta_{Bm}e^{-x})^q}{q}\right]dx$$

$$= -\sum_{q=1}^\infty\left[\frac{(\overline{\Delta}_{Am}\overline{\Delta}_{Bm})^q}{q} + \frac{(\Delta_{Am}\Delta_{Bm})^q}{q}\right]\int_{r_n}^\infty xe^{-qx}dx$$

$$= -\sum_{q=1}^\infty\left[(\overline{\Delta}_{Am}\overline{\Delta}_{Bm})^q + (\Delta_{Am}\Delta_{Bm})^q\right]\frac{(1+qr_n)e^{-qr_n}}{q^2}. \quad \text{(L2.18)}$$

Integration over frequencies ξ is effectively integration over $r_n = (2l\varepsilon_m^{1/2}\mu_m^{1/2}/c)\xi_n$, where the discrete ξ_n have now become continuous and r_n too can be treated as continuous, going from 0 to infinity; $\int_0^\infty d\xi \to [c/(2l\varepsilon_m^{1/2}\mu_m^{1/2})]\int_0^\infty dr_n$ creates integrals of the form $\int_0^\infty (1+qr_n)e^{-qr_n}dr_n = (2/q^2)$.

With these dubious devices, and with all μ_i taken to be equal, the interaction free energy per unit area is written in an approximate form:

$$\begin{aligned}
&G(l \to \infty, T \to 0)\\
&= \frac{\hbar}{(4\pi)^2 l^2}\int_0^\infty d\xi \int_{r_n}^\infty x\left[\ln\left(1-\overline{\Delta}_{Am}\overline{\Delta}_{Bm}e^{-x}\right)\left(1-\Delta_{Am}\Delta_{Bm}e^{-x}\right)\right]dx\\
&= -\frac{\hbar}{(4\pi)^2 l^2}\frac{c}{2l\varepsilon_m^{1/2}\mu_m^{1/2}}\int_0^\infty dr_n \sum_{q=1}^\infty\left[(\overline{\Delta}_{Am}\overline{\Delta}_{Bm})^q + (\Delta_{Am}\Delta_{Bm})^q\right]\frac{(1+qr_n)e^{-qr_n}}{q^2}\\
&= -\frac{\hbar c}{16\pi^2 l^3\varepsilon_m^{1/2}\mu_m^{1/2}}\sum_{q=1}^\infty\left[\frac{(\overline{\Delta}_{Am}\overline{\Delta}_{Bm})^q + (\overline{\Delta}_{Am}\overline{\Delta}_{Bm})^q}{q^4}\right]\\
&\approx -\frac{\hbar c}{16\pi^2 l^3\varepsilon_m^{1/2}\mu_m^{1/2}}\left[(\overline{\Delta}_{Am}\overline{\Delta}_{Bm}) + (\overline{\Delta}_{Am}\overline{\Delta}_{Bm})\right] \approx -\frac{\hbar c}{8\pi^2 l^3\varepsilon_m^{1/2}\mu_m^{1/2}}(\overline{\Delta}_{Am}\overline{\Delta}_{Bm}).
\end{aligned} \tag{L2.19}$$

Ideal conductors

When it is assumed that at all frequencies bodies A and B are ideally conducting zero resistance materials, their interaction across a nonconductor goes to a particularly simple form. Let $\varepsilon_A = \varepsilon_B = \varepsilon_C \to \infty$ while ε_m remains finite and make $\mu_A = \mu_m = \mu_B$. Then $s_A = s_B = s_c = \sqrt{p^2 - 1 + (\varepsilon_C/\varepsilon_m)} \to \sqrt{(\varepsilon_C/\varepsilon_m} \to \infty$:

$$\overline{\Delta}_{Am} = \overline{\Delta}_{Bm} = \overline{\Delta}_{Cm} \to \frac{\varepsilon_C p - \varepsilon_m s_C}{\varepsilon_C p + \varepsilon_m s_C} = \frac{\sqrt{\varepsilon_C}p - \sqrt{\varepsilon_m}}{\sqrt{\varepsilon_C}p + \sqrt{\varepsilon_m}} \to 1,$$

$$\Delta_{Am} = \Delta_{Bm} = \Delta_{Cm} \to \frac{p - s_C}{p + s_C} = \frac{p - \sqrt{\varepsilon_C/\varepsilon_m}}{p + \sqrt{\varepsilon_C/\varepsilon_m}} \to -1.$$

$$\begin{aligned}
G_{AmB}(l, T) &= -\frac{kT}{8\pi l^2}\sum_{n=0}^{\infty}{}' r_n^2 \sum_{q=1}^\infty \frac{1}{q}\int_1^\infty p\left[(\overline{\Delta}_{Am}\overline{\Delta}_{Bm})^q + (\Delta_{Am}\Delta_{Bm})^q\right]e^{-r_n pq}dp\\
&\to -\frac{kT}{4\pi l^2}\sum_{n=0}^{\infty}{}' r_n^2 \sum_{q=1}^\infty \frac{1}{q}\int_1^\infty p e^{-r_n pq}dp\\
&= -\frac{kT}{4\pi l^2}\sum_{n=0}^{\infty}{}' \sum_{q=1}^\infty \frac{(1+r_n q)e^{-r_n q}}{q^3}.
\end{aligned} \tag{L2.20}$$

[P.1.c.1]

In the limit of infinite separation at which $r_1 \gg 1$, only the $n = 0$ term survives retardation screening:

$$G_{AmB}(l \to \infty, T) \to -\frac{kT}{8\pi l^2}\sum_{q=1}^\infty \frac{1}{q^3} = -\frac{kT}{8\pi l^2}\zeta(3) \tag{L2.21}$$

[P.1.c.2]

where $\zeta(3) \equiv \sum_{q=1}^\infty (1/q^3) \sim 1.2$ is the Riemann zeta function.

When finite temperature is neglected, $\sum_{n=0}^{\prime\infty}(1+r_n q)\,e^{-r_n q} \to \int_{n=0}^{\infty}(1+r_n q)e^{-r_n q}\,dn$, where $dn = [\hbar c/(4\pi kTl\varepsilon_m^{1/2}\mu_m^{1/2}q)]d(r_n q)$ so that

$$G_{AmB}(l, T\to 0) \to -\frac{kT}{4\pi l^2}\sum_{n=0}^{\infty}{}' \sum_{q=1}^{\infty}\frac{(1+r_n q)e^{-r_n q}}{q^3} = -\frac{2\hbar c}{16\pi^2 l^3 \varepsilon_m^{1/2}\mu_m^{1/2}}\sum_{q=1}^{\infty}\frac{1}{q^4}$$

$$= -\frac{\hbar c}{8\pi^2 l^3 \varepsilon_m^{1/2}\mu_m^{1/2}}\zeta(4) = -\frac{\hbar c\pi^2}{720 l^3 \varepsilon_m^{1/2}\mu_m^{1/2}} \tag{L2.22}$$

[P.1.c.3]

Here $\zeta(4) \equiv \sum_{q=1}^{\infty}(1/q^4) = (\pi^4/90) \approx 1.1$ confers compact form.

In the case in which the intermediate m is a vacuum, $\varepsilon_m = \mu_m = 1$, $G_{AmB}(l, T\to 0)$ coincides with the Casimir interaction energy whose derivative pressure between metallic plates is

$$P(l) = -\frac{\hbar c\pi^2}{240 l^4}. \tag{L2.23}$$

Warning: These aesthetically pleasing limiting forms should not be used to compute the interaction of real metals. Except at low frequencies, the dielectric response of a real metal is not that of an ideal conductor. Only when $T \to 0$ does the summation over eigenfrequencies become the integration that is needed to derive these seductive popular forms.

Pause for retardation

Except for the fact that the variables s_i are functions of p, the integration in x or p for $G_{AmB}(l, T)$ can be done analytically. Several properties of the ε_i and μ_i often allow an approximation that all the $s_i = \sqrt{p^2 - 1 + (\varepsilon_i\mu_i/\varepsilon_m\mu_m)}$ and p are practically equal. In this way the s's and p's cancel in the numerator and denominator of $\overline{\Delta}_{ji} = [(s_i\varepsilon_j - s_j\varepsilon_i)/(s_i\varepsilon_j + s_j\varepsilon_i)]$ and $\Delta_{ji} = [(s_i\mu_j - s_j\mu_i)/(s_i\mu_j + s_j\mu_i)]$.

The relevant properties are:

1. Except possibly at zero frequency ($\xi_n = 0$), magnetic susceptibilities μ_i are close to unity for most materials. For visible and higher frequencies, the dielectric permittivities ε are also close to each other. In these cases the ratio $(\varepsilon_i\mu_i/\varepsilon_m\mu_m)$ is not very different from 1.

2. Because of the form of the integrals $\int_{r_n}^{\infty} x\ln[(1-\overline{\Delta}_{Am}\overline{\Delta}_{Bm}e^{-x})(1-\Delta_{Am}\Delta_{Bm}e^{-x})]dx$ and $\int_1^{\infty} p\ln[(1-\overline{\Delta}_{Am}\overline{\Delta}_{Bm}e^{-r_n p})(1-\Delta_{Am}\Delta_{Bm}e^{-r_n p})]dp$ (Table P.1.a.1), the integrand takes on its greatest values in the vicinity of $x \sim 1$ or $p \sim 1/r_n = (c/2l\xi_n\varepsilon_m^{1/2}\mu_m^{1/2})$.
   a. For $\xi_{n=0} = 0$, this means that the important contributions occur when $p \to \infty$. In this case the quantities s_i are all rigorously equal to p.
   b. For finite frequencies, when $r_n \ll 1$, the dominant contribution still comes from regions of integration where $p \gg 1$, and still all $s_i \approx p$ in the region of important contributions [as long as $(\varepsilon_i\mu_i/\varepsilon_m\mu_m)$ is not very different from 1].
   c. In the limit of very high (x-ray) frequencies, all ε's and μ's go to 1 so that the ratio $(\varepsilon_i\mu_i/\varepsilon_m\mu_m)$ also goes to 1. Again $s_i \approx p$. In this $s_i = p$ approximation, then, $\overline{\Delta}_{ji} \approx [(\varepsilon_j - \varepsilon_i)/(\varepsilon_j + \varepsilon_i)]$ and $\Delta_{ji} \approx [(\mu_j - \mu_i)/(\mu_j + \mu_i)]$.

PROBLEM L2.1: Show that this simple form of $\overline{\Delta}_{ji}$ and Δ_{ji} emerges immediately from assuming that the velocity of light is finite but everywhere equal.

Further, except near zero frequency ($\xi_{n=0} = 0$), magnetic susceptibilities are usually nearly equal to 1, so that this approximate Δ_{ji} is often set equal to zero; differences in ε's are usually small enough that $\overline{\Delta}_{ji} \ll 1$. Then the logarithms in the integrand are expanded only to leading terms: $\ln(1 - \overline{\Delta}_{Am}\overline{\Delta}_{Bm}e^{-x}) \approx -\overline{\Delta}_{Am}\overline{\Delta}_{Bm}e^{-x}$. Separating out the $\xi_n = 0$ term from the rest of the summation $\sum_{n=0}^{\prime\infty}$ gives

$$G_{AmB}(l, T) \approx \frac{kT}{8\pi l^2}\left\{\frac{1}{2}\int_0^\infty x\ln\left[\left(1-\overline{\Delta}_{Am}\overline{\Delta}_{Bm}e^{-x}\right)\left(1-\Delta_{Am}\Delta_{Bm}e^{-x}\right)\right]dx - \sum_{n=1}^{\infty}\overline{\Delta}_{Am}\overline{\Delta}_{Bm}\int_{r_n}^{\infty} xe^{-x}dx\right\}$$

$$= -\frac{kT}{8\pi l^2}\left\{\frac{1}{2}\sum_{q=1}^{\infty}\frac{(\overline{\Delta}_{Am}\overline{\Delta}_{Bm})^q + (\Delta_{Am}\Delta_{Bm})^q}{q^3} + \sum_{n=1}^{\infty}\overline{\Delta}_{Am}\overline{\Delta}_{Bm}(1+r_n)e^{-r_n}\right\} \tag{L2.24}$$

The first term in { } is a sum of rapidly converging series $(\ \ )^q/q^3$, where the ε's and μ's are evaluated at zero frequency. Because the $\overline{\Delta}_{ji}$ and Δ_{ji} functions can never be greater than 1, the summation in q converges rapidly and is usually dominated by the leading term. The second term in { } is the sum over finite frequencies,

$$\xi_n = \frac{2\pi kT}{\hbar} n \text{ for } n = 1, 2 \ldots.$$

Neglecting all magnetic contributions, i.e., taking all μ's = 1, and including only the leading term in the zero-frequency q summation, gives

$$G_{AmB}(l, T) \approx -\frac{kT}{8\pi l^2}\left[\frac{\overline{\Delta}_{Am}(0)\overline{\Delta}_{Bm}(0)}{2} + \sum_{n=1}^{\infty}\overline{\Delta}_{Am}(\xi_n)\overline{\Delta}_{Bm}(\xi_n)(1+r_n)e^{-r_n}\right]$$

$$= -\frac{kT}{8\pi l^2}\sum_{n=0}^{\infty}{}' \overline{\Delta}_{Am}(\xi_n)\overline{\Delta}_{Bm}(\xi_n)(1+r_n)e^{-r_n}$$

$$= -\frac{kT}{8\pi l^2}\sum_{n=0}^{\infty}{}' \overline{\Delta}_{Am}(\xi_n)\overline{\Delta}_{Bm}(\xi_n)R_n(\xi_n), \tag{L2.25}$$

with

$$R_n(r_n) \equiv (1+r_n)e^{-r_n}. \tag{L2.26}$$

In the Hamaker form,

$$G_{AmB}(l) = -\frac{A_{Ham}}{12\pi l^2}, \quad A_{Ham} \approx +\frac{3kT}{2}\sum_{n=0}^{\infty}{}' \overline{\Delta}_{Am}(\xi_n)\overline{\Delta}_{Bm}(\xi_n)R_n(\xi_n). \tag{L2.27}$$

[P.1.a.2]

But how reliable is it to assume that the velocity of light is everywhere equal to that in medium m? The assumption is only qualitatively correct. It should *not* be used in careful computation. For example, set $\mu_i = \mu_m$ and consider the case in which $\varepsilon_A = \varepsilon_B = \varepsilon$, $s_A = s_B = s = \sqrt{p^2 - 1 + \varepsilon/\varepsilon_m}$ so that $\overline{\Delta}_{Am} = \overline{\Delta}_{Bm} = \overline{\Delta} = [(p\varepsilon - s\varepsilon_m)/(p\varepsilon + s\varepsilon_m)]$, $\Delta_{Am} = \Delta_{Bm} = \Delta = [(p - s)/(p + s)]$. How does the integral $\int_{r_n}^{\infty} x(\overline{\Delta}^2 + \Delta^2)e^{-x}dx$ differ

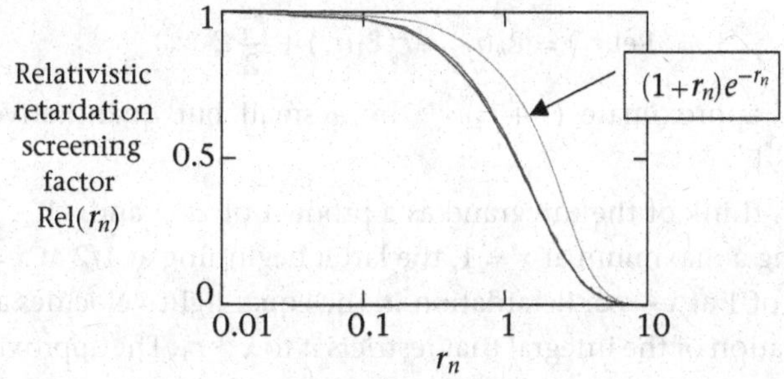

Figure L2.1

from the form $[(\varepsilon - \varepsilon_m)/(\varepsilon + \varepsilon_m)]^2(1 + r_n)e^{-r_n}$ that emerges from the light-velocities-everywhere-equal assumption?

To facilitate comparison, define a screening factor Rel(r_n) acting on $[(\varepsilon - \varepsilon_m)/(\varepsilon + \varepsilon_m)]^2$ such that

$$\text{Rel}(r_n)\left(\frac{\varepsilon - \varepsilon_m}{\varepsilon + \varepsilon_m}\right)^2 \equiv \int_{r_n}^{\infty} x(\overline{\Delta}^2 + \Delta^2)e^{-x}\,\mathrm{d}x. \tag{L2.28}$$

Computed screening is essentially the same for $\varepsilon = 1.001\varepsilon_m$ and $\varepsilon = 2\varepsilon_m$ (two nearly indistinguishable lower curves in Fig. L2.1). The approximate form $(1 + r_n)e^{-r_n}$ is only qualitatively like the numerically computed Rel(r_n).

Why does such a small difference in epsilons still lead to stronger screening than described by the equal-velocity form? We dissect the integral $\int_{r_n}^{\infty} x(\overline{\Delta}^2 + \Delta^2)e^{-x}\mathrm{d}x$ by again setting $\varepsilon_A = \varepsilon_B = \varepsilon$, $\mu_i = \mu_m$ and writing $(\varepsilon/\varepsilon_m) = 1 + \eta$, $s_A = s_B = s = \sqrt{p^2 + \eta}$.

For small η,

$$\overline{\Delta} = \frac{p\varepsilon - s\varepsilon_m}{p\varepsilon + s\varepsilon_m} = \frac{(1+\eta) - \sqrt{1 + \eta/p^2}}{(1+\eta) + \sqrt{1+\eta/p^2}} \approx \frac{\eta}{2}(1 - 1/2p^2) \text{ or } \frac{2}{\eta}\overline{\Delta} = 1 - \frac{1}{2p^2};$$

$$\Delta = \frac{p - s}{p + s} = \frac{1 - \sqrt{1 + \eta/p^2}}{1 + \sqrt{1 + \eta/p^2}} \approx -\frac{\eta}{4p^2} \text{ or } \frac{2}{\eta}\Delta = -\frac{1}{2p^2} \text{ and } \frac{\varepsilon - \varepsilon_m}{\varepsilon + \varepsilon_m} = \frac{\eta}{2 + \eta} \approx \frac{\eta}{2}.$$

In this approximation,

$$(\overline{\Delta}^2 + \Delta^2)\left(\frac{2}{\eta}\right)^2 = \left(1 - \frac{1}{2p^2}\right)^2 + \left(\frac{1}{2p^2}\right)^2 = 1 - \frac{r_n^2}{x^2} + \frac{r_n^4}{2x^4},$$

so that

$$\int_{r_n}^{\infty} x\left(\overline{\Delta}^2 + \Delta^2\right)e^{-x}\mathrm{d}x = \left(\frac{\varepsilon - \varepsilon_m}{\varepsilon + \varepsilon_m}\right)^2 \int_{r_n}^{\infty} x\left(1 - \frac{r_n^2}{x^2} + \frac{r_n^4}{2x^4}\right)e^{-x}\mathrm{d}x.$$

Integration of the first term over xe^{-x} gives the screening factor

$$R_n(r_n) = (1 + r_n)e^{-r_n}. \tag{L2.29}$$

The second and third integrations are exponential integrals:

$$-r_n^2 \int_{r_n}^{\infty} \frac{e^{-x}}{x}\mathrm{d}x = -r_n^2 E_1(r_n),$$

$$\frac{r_n^4}{2}\int_{r_n}^{\infty} \frac{e^{-x}}{x^3}\mathrm{d}x = \frac{r_n^4}{2}\int_{r_n}^{\infty} \frac{e^{-r_n p}}{(r_n p)^3}\mathrm{d}(r_n p) = \frac{r_n^2}{2}\int_1^{\infty} \frac{e^{-r_n p}}{p^3}\mathrm{d}p = \frac{r_n^2}{2}E_3(r_n).$$

The actual retardation factor

$$\mathrm{Rel}(r_n) = \mathrm{R}_n(r_n) - r_n^2\mathrm{E}_1(r_n) + \frac{r_n^2}{2}E_3(r_n) \tag{L2.30}$$

differs from the approximate $(1+r_n)e^{-r_n}$ by a small but quantitatively significant $-r_n^2[E_1(r_n) - \frac{E_3(r_n)}{2}]$.

Alternatively, think of the integrand as a product of xe^{-x} and $(1 - \frac{r_n^2}{x^2} + \frac{r_n^4}{2x^4})$, with the former having a maximum at $x = 1$, the latter beginning at 1/2 at $x = r_n$ and going to its maximum of 1 at $x \to \infty$. Retardation in the equal-light-velocities approximation is simply a truncation of the integral that restricts it to $x \geq r_n$. The approximate function $R(r_n) = (1 + r_n)e^{-r_n}$ captures enough of the main features to allow its use in examining the main features of retardation.

L2.3.B. Layered systems

SEMI-INFINITE MEDIUM AND A SINGLY COATED SEMI-INFINITE MEDIUM

Exact

The general formula (Table P.2.a.1) for the interaction of half-space A and half-space B coated with a layer of material B_1 of thickness b_1 has the same outward form as the original Lifshitz formula for the interaction of two half-spaces. To recognize its inner possibilities, consider the single-layer interaction in terms of different variables of integration:

1. With the variable of integration ρ_m

$$G_{\mathrm{AmB_1B}}(l;\mathrm{b}_1) = \frac{kT}{2\pi}\sum_{n=0}^{\infty}{}' \int_{\frac{\varepsilon_m^{1/2}\mu_m^{1/2}\xi_n}{c}}^{\infty} \rho_m \times \ln\left[\left(1 - \overline{\Delta}_{\mathrm{Am}}\overline{\Delta}_{\mathrm{Bm}}^{\mathrm{eff}}e^{-2\rho_m l}\right)\left(1 - \Delta_{\mathrm{Am}}\Delta_{\mathrm{Bm}}^{\mathrm{eff}}e^{-2\rho_m l}\right)\right] d\rho_m. \tag{L2.31}$$

2. With $x = 2\rho_m\, l$ and $r_n \equiv (2l\varepsilon_m^{1/2}\mu_m^{1/2}/c)\xi_n$,

$$G_{\mathrm{AmB_1B}}(l;\mathrm{b}_1) = \frac{kT}{8\pi l^2}\sum_{n=0}^{\infty}{}' \int_{r_n}^{\infty} x\ln\left[\left(1 - \overline{\Delta}_{\mathrm{Am}}\overline{\Delta}_{\mathrm{Bm}}^{\mathrm{eff}}e^{-x}\right)\left(1 - \Delta_{\mathrm{Am}}\Delta_{\mathrm{Bm}}^{\mathrm{eff}}e^{-x}\right)\right] dx. \tag{L2.32}$$

[See (L3.83) and (L3.84)]

3. With $p = x/r_n$

$$G_{\mathrm{AmB_1B}}(l;\mathrm{b}_1) = \frac{kT}{2\pi c^2}\sum_{n=0}^{\infty}{}' \varepsilon_m\mu_m\xi_n^2 \times \int_1^{\infty} p\ln\left[\left(1 - \overline{\Delta}_{\mathrm{Am}}\overline{\Delta}_{\mathrm{Bm}}^{\mathrm{eff}}e^{-r_n p}\right)\left(1 - \Delta_{\mathrm{Am}}\Delta_{\mathrm{Bm}}^{\mathrm{eff}}e^{-r_n p}\right)\right] dp. \tag{L2.33}$$

[L3.85]

The variables

$$\rho_i^2 = \rho^2 + \varepsilon_i\mu_i\xi_n^2/c^2 \tag{L2.34}$$

used in derivation are conveniently transformed to

$$x_i \equiv 2\rho_i l = s_i r_n,\ x_i^2 = x_m^2 + \left(\frac{2l\xi_n}{c}\right)^2(\varepsilon_i\mu_i - \varepsilon_m\mu_m),$$
$$s_i = \sqrt{p^2 - 1 + (\varepsilon_i\mu_i/\varepsilon_m\mu_m)}. \tag{L2.35}$$

The functions $\overline{\Delta}_{Bm}$, Δ_{Bm} in the simplest AmB case are replaced by

$$\overline{\Delta}_{Bm}^{eff}(b_1) = \frac{\left(\overline{\Delta}_{BB_1} e^{-2\rho_{B_1} b_1} + \overline{\Delta}_{B_1 m}\right)}{1 + \overline{\Delta}_{BB_1}\overline{\Delta}_{B_1 m} e^{-2\rho_{B_1} b_1}} = \frac{\left(\overline{\Delta}_{BB_1} e^{-x_{B_1}(b_1/l)} + \overline{\Delta}_{B_1 m}\right)}{1 + \overline{\Delta}_{BB_1}\overline{\Delta}_{B_1 m} e^{-x_{B_1}(b_1/l)}}$$

$$= \frac{\left(\overline{\Delta}_{BB_1} e^{-s_{B_1} r_n (b_1/l)} + \overline{\Delta}_{B_1 m}\right)}{1 + \overline{\Delta}_{BB_1}\overline{\Delta}_{B_1 m} e^{-s_{B_1} r_n (b_1/l)}}, \tag{L2.36}$$

$$\Delta_{Bm}^{eff}(b_1) = \frac{\left(\Delta_{BB_1} e^{-2\rho_{B_1} b_1} + \Delta_{B_1 m}\right)}{1 + \Delta_{BB_1}\Delta_{B_1 m} e^{-2\rho_{B_1} b_1}} = \frac{\left(\Delta_{BB_1} e^{-x_{B_1}(b_1/l)} + \Delta_{B_1 m}\right)}{1 + \Delta_{BB_1}\Delta_{B_1 m} e^{-x_{B_1}(b_1/l)}}$$

$$= \frac{\left(\Delta_{BB_1} e^{-s_{A_1} r_n (b_1/l)} + \Delta_{B_1 m}\right)}{1 + \Delta_{BB_1}\Delta_{B_1 m} e^{-s_{B_1} r_n (b_1/l)}}. \tag{L2.37}$$

The elemental $\overline{\Delta}_{ji}$'s and Δ_{ij}'s are still

$$\overline{\Delta}_{ji} = \frac{\rho_i \varepsilon_j - \rho_j \varepsilon_i}{\rho_i \varepsilon_j + \rho_j \varepsilon_i} = \frac{x_i \varepsilon_j - x_j \varepsilon_i}{x_i \varepsilon_j + x_j \varepsilon_i} = \frac{s_i \varepsilon_j - s_j \varepsilon_i}{s_i \varepsilon_j + s_j \varepsilon_i}, \tag{L2.38}$$

$$\Delta_{ji} = \frac{\rho_i \mu_j - \rho_j \mu_i}{\rho_i \mu_j + \rho_j \mu_i} = \frac{x_i \mu_j - x_j \mu_i}{x_i \mu_j + x_j \mu_i} = \frac{s_i \mu_j - s_j \mu_i}{s_i \mu_j + s_j \mu_i}. \tag{L2.39}$$

The quantities ε_i, ε_j and μ_i, μ_j are dielectric- and magnetic-response functions of the individual materials. Indices i and j now refer to four materials: A, B, B_1, or m.

Limiting forms

$B_1 = B$ It is obvious that, when material B_1 is the same as material B, then $\overline{\Delta}_{BB_1}$ and Δ_{BB_1} equal zero; $\overline{\Delta}_{Bm}^{eff}(b_1)$ and $\Delta_{Bm}^{eff}(b_1)$ revert to $\overline{\Delta}_{Bm}$ and $\overline{\Delta}_{Bm}$ to make the interaction that between B and A across m at separation l.

$B_1 = m$ It is almost as obvious that, when material B_1 is the same as material m, then

$$\overline{\Delta}_{Bm}^{eff}(b_1) \to \overline{\Delta}_{BB_1} e^{-2\rho_{B_1} b_1} = \overline{\Delta}_{Bm} e^{-2\rho_m b_1}$$

so that

$$\overline{\Delta}_{Am}\overline{\Delta}_{Bm}^{eff} e^{-2\rho_m l} \to \overline{\Delta}_{Am}\overline{\Delta}_{Bm} e^{-2\rho_m b_1} e^{-2\rho_m l} = \overline{\Delta}_{Am}\overline{\Delta}_{Bm} e^{-2\rho_m (b_1 + l)} \tag{L2.40}$$

(and the same for $\Delta_{Am}\Delta_{Bm}^{eff} e^{-2\rho_m l}$). Then the interaction is still A with B across m, but this time across a distance $(b_1 + l)$.

$b_1 \gg l$ When thickness $b_1 \gg l$, $\overline{\Delta}_{Bm}^{eff}(b_1) \to \overline{\Delta}_{B_1 m}$, $\Delta_{Bm}^{eff}(b_1) \to \Delta_{B_1 m}$, material B disappears from the interaction. We have B_1 interacting with A across m at a separation l.

High dielectric-permittivity layer When $\varepsilon_{B_1}/\varepsilon_m \to \infty$, neglecting magnetic terms, $\overline{\Delta}_{B_1m} \to 1$,

$$\overline{\Delta}_{Bm}^{eff}(b_1) \to \frac{\left(\overline{\Delta}_{BB_1} e^{-2\rho_{B_1} b_1} + 1\right)}{1 + \overline{\Delta}_{BB_1} e^{-2\rho_{B_1} b_1}} = 1,$$

$$G_{AmB_1B}(l; b_1) \to G_{AmB_1}(l) = \frac{kT}{2\pi} \sum_{n=0}^{\infty}{}' \int_{\frac{\varepsilon_m^{1/2}\mu_m^{1/2}\xi_n}{c}}^{\infty} \rho_m \ln\left(1 - \overline{\Delta}_{Am} e^{-2\rho_m l}\right) d\rho_m. \quad \text{(L2.41)}$$

[P.2.b.1]

Semi-infinite material B is screened out by ideally metallic B_1.

Small differences in ε's and μ's When $\overline{\Delta}_{ji}$'s and Δ_{ij}'s are $\ll 1$, we can restrict ourselves to leading terms:

$$\overline{\Delta}_{Bm}^{eff}(b_1) \to \left(\overline{\Delta}_{BB_1} e^{-2\rho_{B_1} b_1} + \overline{\Delta}_{B_1m}\right)$$
$$= \left(\overline{\Delta}_{BB_1} e^{-x_{B_1}(b_1/l)} + \overline{\Delta}_{B_1m}\right) = \left(\overline{\Delta}_{BB_1} e^{-s_{B_1} r_n (b_1/l)} + \overline{\Delta}_{B_1m}\right) \ll 1, \quad \text{(L2.42)}$$

$$\Delta_{Bm}^{eff}(b_1) \to \left(\Delta_{BB_1} e^{-2\rho_{B_1} b_1} + \Delta_{B_1m}\right)$$
$$= \left(\Delta_{BB_1} e^{-x_{B_1}(b_1/l)} + \Delta_{B_1m}\right) = \left(\Delta_{BB_1} e^{-s_{B_1} r_n (b_1/l)} + \Delta_{B_1m}\right) \ll 1, \quad \text{(L2.43)}$$

$$\ln\left[\left(1 - \overline{\Delta}_{Am}\overline{\Delta}_{Bm}^{eff} e^{-2\rho_m l}\right)\left(1 - \Delta_{Am}\Delta_{Bm}^{eff} e^{-2\rho_m l}\right)\right]$$
$$\to -\overline{\Delta}_{Am}\overline{\Delta}_{Bm}^{eff} e^{-2\rho_m l} - \Delta_{Am}\Delta_{Bm}^{eff} e^{-2\rho_m l}$$
$$\to -\overline{\Delta}_{Am}\left(\overline{\Delta}_{BB_1} e^{-2\rho_{B_1} b_1} + \overline{\Delta}_{B_1m}\right) e^{-2\rho_m l} - \Delta_{Am}\left(\Delta_{BB_1} e^{-2\rho_{B_1} b_1} + \Delta_{B_1m}\right) e^{-2\rho_m l}$$
$$= -(\overline{\Delta}_{Am}\overline{\Delta}_{B_1m} + \Delta_{Am}\Delta_{B_1m}) e^{-2\rho_m l} - (\overline{\Delta}_{Am}\overline{\Delta}_{BB_1} + \Delta_{Am}\Delta_{BB_1}) e^{-2\rho_{B_1} b_1} e^{-2\rho_m l}.$$

(L2.44)

The interaction free energy reduces to two terms:

$$G_{AmB_1B}(l; b_1) = -\frac{kT}{2\pi} \sum_{n=0}^{\infty}{}' \int_{\frac{\varepsilon_m^{1/2}\mu_m^{1/2}\xi_n}{c}}^{\infty} \rho_m (\overline{\Delta}_{Am}\overline{\Delta}_{B_1m} + \Delta_{Am}\Delta_{B_1m}) e^{-2\rho_m l} d\rho_m$$
$$- \frac{kT}{2\pi} \sum_{n=0}^{\infty}{}' \int_{\frac{\varepsilon_m^{1/2}\mu_m^{1/2}\xi_n}{c}}^{\infty} \rho_m (\overline{\Delta}_{Am}\overline{\Delta}_{BB_1} + \Delta_{Am}\Delta_{BB_1}) e^{-2\rho_{B_1} b_1} e^{-2\rho_m l} d\rho_m.$$

(L2.45)

[P.2.b.2]

The first term is the Lifshitz interaction of A with B_1 across m at separation l; the second is the interaction of B with A across m and B_1. Because of the difference in the velocity of light in materials B_1 and m, there is a difference in the ρ_{B_1} and ρ_m that measures thicknesses b_1 and l. The second term has almost, but not exactly, the simplest Lifshitz form.

Small differences in ε's and μ's, nonretarded limit For small differences in susceptibilities, with $c \to \infty$, $\rho_{B_1} \to \rho_m \to \rho$,

$$G_{AmB_1B}(l; b_1) \to -\frac{kT}{2\pi} \sum_{n=0}^{\infty}{}' (\overline{\Delta}_{Am}\overline{\Delta}_{B_1m} + \Delta_{Am}\Delta_{B_1m}) \int_0^{\infty} \rho\, e^{-2\rho l} d\rho$$
$$- \frac{kT}{2\pi} \sum_{n=0}^{\infty}{}' (\overline{\Delta}_{Am}\overline{\Delta}_{BB_1} + \Delta_{Am}\Delta_{BB_1}) \int_0^{\infty} \rho\, e^{-2\rho(l+b_1)} d\rho$$

$$= -\frac{kT}{8\pi l^2}\sum_{n=0}^{\infty}{}'\left(\overline{\Delta}_{\mathrm{Am}}\overline{\Delta}_{\mathrm{B_1m}} + \Delta_{\mathrm{Am}}\Delta_{\mathrm{B_1m}}\right)$$

$$-\frac{kT}{8\pi (l+\mathrm{b}_1)^2}\sum_{n=0}^{\infty}{}'\left(\overline{\Delta}_{\mathrm{Am}}\overline{\Delta}_{\mathrm{BB_1}} + \Delta_{\mathrm{Am}}\Delta_{\mathrm{BB_1}}\right). \quad \text{(L2.46)} \; \text{[P.2.b.3]}$$

Nonretarded limit When the velocity of light is imagined infinite, all $r_n \to 0$, $\rho_{\mathrm{i}} \to \rho_{\mathrm{m}} \to \rho$, $x_{\mathrm{i}} \to x_{\mathrm{m}} = x\, s_{\mathrm{i}} \to p$,

$$\overline{\Delta}_{\mathrm{Am}}\overline{\Delta}_{\mathrm{Bm}}^{\mathrm{eff}}e^{-2\rho_{\mathrm{m}}l} \to \overline{\Delta}_{\mathrm{Am}}\frac{\left(\overline{\Delta}_{\mathrm{BB_1}}e^{-2\rho\mathrm{b}_1} + \overline{\Delta}_{\mathrm{B_1m}}\right)}{1+\overline{\Delta}_{\mathrm{BB_1}}\overline{\Delta}_{\mathrm{B_1m}}e^{-2\rho\mathrm{b}_1}}e^{-2\rho l} = \overline{\Delta}_{\mathrm{Am}}\frac{\left(\overline{\Delta}_{\mathrm{BB_1}}e^{-x(\mathrm{b}_1/l)} + \overline{\Delta}_{\mathrm{B_1m}}\right)}{1+\overline{\Delta}_{\mathrm{BB_1}}\overline{\Delta}_{\mathrm{B_1m}}e^{-x(\mathrm{b}_1/l)}}e^{-x}$$

$$= \overline{\Delta}_{\mathrm{Am}}\frac{\left(\overline{\Delta}_{\mathrm{BB_1}}e^{-pr_n(\mathrm{b}_1/l)} + \overline{\Delta}_{\mathrm{B_1m}}\right)}{1+\overline{\Delta}_{\mathrm{BB_1}}\overline{\Delta}_{\mathrm{B_1m}}e^{-pr_n(\mathrm{b}_1/l)}}e^{-pr_n}, \quad \text{(L2.47)}$$

and similarly for the magnetic terms. These can be expanded as an infinite series:

$$\overline{\Delta}_{\mathrm{Am}}\overline{\Delta}_{\mathrm{Bm}}^{\mathrm{eff}}e^{-2\rho_{\mathrm{m}}l} \to \overline{\Delta}_{\mathrm{Am}}\left(\overline{\Delta}_{\mathrm{B_1m}} + \overline{\Delta}_{\mathrm{BB_1}}e^{-2\rho\mathrm{b}_1}\right)e^{-2\rho l}\sum_{j=0}^{\infty}\left(-\overline{\Delta}_{\mathrm{BB_1}}\overline{\Delta}_{\mathrm{B_1m}}\right)^j e^{-2\rho j\mathrm{b}_1}. \quad \text{(L2.48)}$$

From this expansion we see that the exponential $e^{-2\rho l}$ factor that led to a $1/l^2$ factor in the simplest AmB interaction is now replaced by an infinite set of factors that correspond to distances $l, l+\mathrm{b}_1, l+2\mathrm{b}_1, \ldots, l+j\mathrm{b}_1, \ldots,$ successively multiplied by powers of $-\overline{\Delta}_{\mathrm{BB_1}}\overline{\Delta}_{\mathrm{B_1m}}$. The leading terms, valid for small $\overline{\Delta}_{\mathrm{ji}}$'s, are interactions of $\mathrm{B_1}$ with A across m at a distance l, $\overline{\Delta}_{\mathrm{Am}}\overline{\Delta}_{\mathrm{B_1m}}e^{-2\rho l}$, and B with A across $\mathrm{B_1}$ and m at a distance $l+\mathrm{b}_1$, $\overline{\Delta}_{\mathrm{Am}}\overline{\Delta}_{\mathrm{BB_1}}e^{-2\rho(l+\mathrm{b}_1)}$.

Because $\overline{\Delta}_{\mathrm{Am}}\overline{\Delta}_{\mathrm{Bm}}^{\mathrm{eff}}$ and $\Delta_{\mathrm{Am}}\Delta_{\mathrm{Bm}}^{\mathrm{eff}}$ themselves act within logarithms $\ln[(1-\overline{\Delta}_{\mathrm{Am}}\overline{\Delta}_{\mathrm{Bm}}^{\mathrm{eff}}e^{-2\rho_{\mathrm{m}}l})(1-\Delta_{\mathrm{Am}}\Delta_{\mathrm{Bm}}^{\mathrm{eff}}e^{-2\rho_{\mathrm{m}}l})]$ that can be expanded with their own summation, a too-literal pursuit of individual terms is an easy route to insanity. It suffices to point out that even this first elaboration, i.e., adding one layer, leads to complex behaviors.

SEMI-INFINITE MEDIUM AND A SLAB OF FINITE THICKNESS

Exact

Make material B the same as medium material m. $\overline{\Delta}_{\mathrm{BB_1}} = -\overline{\Delta}_{\mathrm{B_1m}}$:

$$\overline{\Delta}_{\mathrm{Bm}}^{\mathrm{eff}}(\mathrm{b}_1) = \overline{\Delta}_{\mathrm{B_1m}}\frac{1-e^{-2\rho_{\mathrm{B_1}}\mathrm{b}_1}}{1-\overline{\Delta}_{\mathrm{B_1m}}^2e^{-2\rho_{\mathrm{B_1}}\mathrm{b}_1}} = \overline{\Delta}_{\mathrm{B_1m}}\frac{1-e^{-x_{\mathrm{B_1}}(\mathrm{b}_1/l)}}{1-\overline{\Delta}_{\mathrm{B_1m}}^2e^{-x_{\mathrm{B_1}}(\mathrm{b}_1/l)}}$$

$$= \overline{\Delta}_{\mathrm{B_1m}}\frac{1-e^{-s_{\mathrm{B_1}}r_n(\mathrm{b}_1/l)}}{1-\overline{\Delta}_{\mathrm{B_1m}}^2e^{-s_{\mathrm{B_1}}r_n(\mathrm{b}_1/l)}}, \quad \text{(L2.49)}$$

$$\Delta_{\mathrm{Bm}}^{\mathrm{eff}}(\mathrm{b}_1) = \Delta_{\mathrm{B_1m}}\frac{1-e^{-2\rho_{\mathrm{B_1}}\mathrm{b}_1}}{1-\Delta_{\mathrm{B_1m}}^2e^{-2\rho_{\mathrm{B_1}}\mathrm{b}_1}} = \Delta_{\mathrm{B_1m}}\frac{1-e^{-x_{\mathrm{B_1}}(\mathrm{b}_1/l)}}{1-\Delta_{\mathrm{B_1m}}^2e^{-x_{\mathrm{B_1}}(\mathrm{b}_1/l)}}$$

$$= \Delta_{\mathrm{B_1m}}\frac{1-e^{-s_{\mathrm{B_1}}r_n(\mathrm{b}_1/l)}}{1-\Delta_{\mathrm{B_1m}}^2e^{-s_{\mathrm{B_1}}r_n(\mathrm{b}_1/l)}} \quad \text{(L2.50)}$$

$$G_{\mathrm{AmB_1m}}(l;\mathrm{b}_1)$$
$$= \frac{kT}{2\pi}\sum_{n=0}^{\infty}{}' \int_{\frac{\varepsilon_{\mathrm{m}}^{1/2}\mu_{\mathrm{m}}^{1/2}\xi_n}{c}}^{\infty} \rho_{\mathrm{m}}\left[\ln\left(1-\overline{\Delta}_{\mathrm{Am}}\overline{\Delta}_{\mathrm{Bm}}^{\mathrm{eff}}e^{-2\rho_{\mathrm{m}}l}\right)\left(1-\Delta_{\mathrm{Am}}\Delta_{\mathrm{Bm}}^{\mathrm{eff}}e^{-2\rho_{\mathrm{m}}l}\right)\right]\mathrm{d}\rho_{\mathrm{m}}$$
$$= \frac{kT}{8\pi l^2}\sum_{n=0}^{\infty}{}' \int_{r_n}^{\infty} x\left[\ln\left(1-\overline{\Delta}_{\mathrm{Am}}\overline{\Delta}_{\mathrm{Bm}}^{\mathrm{eff}}e^{-x}\right)\left(1-\Delta_{\mathrm{Am}}\Delta_{\mathrm{Bm}}^{\mathrm{eff}}e^{-x}\right)\right]\mathrm{d}x$$
$$= \frac{kT}{2\pi c^2}\sum_{n=0}^{\infty}{}' \varepsilon_{\mathrm{m}}\mu_{\mathrm{m}}\xi_n^2 \int_{1}^{\infty} p\left[\ln\left(1-\overline{\Delta}_{\mathrm{Am}}\overline{\Delta}_{\mathrm{Bm}}^{\mathrm{eff}}e^{-r_n p}\right)\left(1-\Delta_{\mathrm{Am}}\Delta_{\mathrm{Bm}}^{\mathrm{eff}}e^{-r_n p}\right)\right]\mathrm{d}p.$$

(L2.51)
[P.2.c.1]

Small differences in ε's and μ's

When $\overline{\Delta}_{\mathrm{Am}}, \Delta_{\mathrm{Am}}, \overline{\Delta}_{\mathrm{B_1m}}, \Delta_{\mathrm{B_1m}} \ll 1$,

$$\overline{\Delta}_{\mathrm{Bm}}^{\mathrm{eff}}(\mathrm{b}_1) \to \overline{\Delta}_{\mathrm{B_1m}}\left(1-e^{-2\rho_{\mathrm{B_1}}\mathrm{b}_1}\right) = \overline{\Delta}_{\mathrm{B_1m}}\left(1-e^{-x_{\mathrm{B_1}}(\mathrm{b}_1/l)}\right) = \overline{\Delta}_{\mathrm{B_1m}}\left(1-e^{-s_{\mathrm{B_1}}r_n(\mathrm{b}_1/l)}\right), \quad \text{(L2.52)}$$

$$\Delta_{\mathrm{Bm}}^{\mathrm{eff}}(\mathrm{b}_1) \to \Delta_{\mathrm{B_1m}}\left(1-e^{-2\rho_{\mathrm{B_1}}\mathrm{b}_1}\right) = \Delta_{\mathrm{B_1m}}\left(1-e^{-x_{\mathrm{B_1}}(\mathrm{b}_1/l)}\right) = \Delta_{\mathrm{B_1m}}\left(1-e^{-s_{\mathrm{B_1}}r_n(\mathrm{b}_1/l)}\right), \quad \text{(L2.53)}$$

the free energy reduces to two terms:

$$G_{\mathrm{AmB_1m}}(l;\mathrm{b}_1)$$
$$= -\frac{kT}{2\pi}\sum_{n=0}^{\infty}{}' \int_{\frac{\varepsilon_{\mathrm{m}}^{1/2}\mu_{\mathrm{m}}^{1/2}\xi_n}{c}}^{\infty} \rho_{\mathrm{m}}(\overline{\Delta}_{\mathrm{Am}}\overline{\Delta}_{\mathrm{B_1m}}+\Delta_{\mathrm{Am}}\Delta_{\mathrm{B_1m}})(1-e^{-2\rho_{\mathrm{B_1}}\mathrm{b}_1})e^{-2\rho_{\mathrm{m}}l}\mathrm{d}\rho_{\mathrm{m}}$$
$$= -\frac{kT}{2\pi}\sum_{n=0}^{\infty}{}' \int_{\frac{\varepsilon_{\mathrm{m}}^{1/2}\mu_{\mathrm{m}}^{1/2}\xi_n}{c}}^{\infty} \rho_{\mathrm{m}}(\overline{\Delta}_{\mathrm{Am}}\overline{\Delta}_{\mathrm{B_1m}}+\Delta_{\mathrm{Am}}\Delta_{\mathrm{B_1m}})e^{-2\rho_{\mathrm{m}}l}\mathrm{d}\rho_{\mathrm{m}}$$
$$+ \frac{kT}{2\pi}\sum_{n=0}^{\infty}{}' \int_{\frac{\varepsilon_{\mathrm{m}}^{1/2}\mu_{\mathrm{m}}^{1/2}\xi_n}{c}}^{\infty} \rho_{\mathrm{m}}(\overline{\Delta}_{\mathrm{Am}}\overline{\Delta}_{\mathrm{B_1m}}+\Delta_{\mathrm{Am}}\Delta_{\mathrm{B_1m}})e^{-2\rho_{\mathrm{B_1}}\mathrm{b}_1}e^{-2\rho_{\mathrm{m}}l}\mathrm{d}\rho_{\mathrm{m}}. \quad \text{(L2.54)}$$

[P.2.c.2]

Small differences in ε's and μ's, nonretarded limit

In the further limit at which $c \to \infty$ and $\rho_{\mathrm{B_1}} \to \rho_{\mathrm{m}} \to \rho$,

$$G_{\mathrm{AmB_1m}}(l;\mathrm{b}_1) \to -\frac{kT}{8\pi}\left[\frac{1}{l^2}-\frac{1}{(l+\mathrm{b}_1)^2}\right]\sum_{n=0}^{\infty}{}'(\overline{\Delta}_{\mathrm{Am}}\overline{\Delta}_{\mathrm{B_1m}}+\Delta_{\mathrm{Am}}\Delta_{\mathrm{B_1m}}). \quad \text{(L2.55)}$$

[P.2.c.3]

TWO SINGLY COATED SEMI-INFINITE MEDIA

Exact

The interaction of half-space A coated with a layer of material A_1 of thickness a_1 and half-space B coated with a layer of material B_1 of thickness b_1 again follows the Lifshitz form [see Table P.3.a.1 and Eq. (L3.87)]:

1. As the integral used in derivation,

$$G_{\mathrm{AA_1mB_1B}}(l;\mathrm{a_1},\mathrm{b_1})$$
$$=\frac{kT}{2\pi}\sum_{n=0}^{\infty}{}'\int_{\frac{\varepsilon_\mathrm{m}^{1/2}\mu_\mathrm{m}^{1/2}\xi_n}{c}}^{\infty}\rho_\mathrm{m}\ln\left[\left(1-\overline{\Delta}_\mathrm{Am}^\mathrm{eff}\overline{\Delta}_\mathrm{Bm}^\mathrm{eff}e^{-2\rho_\mathrm{m}l}\right)\left(1-\Delta_\mathrm{Am}^\mathrm{eff}\Delta_\mathrm{Bm}^\mathrm{eff}e^{-2\rho_\mathrm{m}l}\right)\right]\mathrm{d}\rho_\mathrm{m}. \tag{L2.56}$$

2. With $x=2\rho_\mathrm{m}l$ and $r_n\equiv(2l\varepsilon_\mathrm{m}^{1/2}\mu_\mathrm{m}^{1/2}/c)\xi_n$,

$$G_{\mathrm{AA_1mB_1B}}(l;\mathrm{a_1},\mathrm{b_1})$$
$$=\frac{kT}{8\pi l^2}\sum_{n=0}^{\infty}{}'\int_{r_n}^{\infty}x\ln\left[\left(1-\overline{\Delta}_\mathrm{Am}^\mathrm{eff}\overline{\Delta}_\mathrm{Bm}^\mathrm{eff}e^{-x}\right)\left(1-\Delta_\mathrm{Am}^\mathrm{eff}\Delta_\mathrm{Bm}^\mathrm{eff}e^{-x}\right)\right]\mathrm{d}x. \tag{L2.57}$$

3. With $p=x/r_n$,

$$G_{\mathrm{AA_1mB_1B}}(l;\mathrm{a_1},\mathrm{b_1})=\frac{kT}{2\pi c^2}\sum_{n=0}^{\infty}{}'\varepsilon_\mathrm{m}\mu_\mathrm{m}\xi_n^2\int_1^{\infty}p$$
$$\times\ln\left[\left(1-\overline{\Delta}_\mathrm{Am}^\mathrm{eff}\overline{\Delta}_\mathrm{Bm}^\mathrm{eff}e^{-r_np}\right)\left(1-\Delta_\mathrm{Am}^\mathrm{eff}\Delta_\mathrm{Bm}^\mathrm{eff}e^{-r_np}\right)\right]\mathrm{d}p, \tag{L2.58}$$

$$\rho_\mathrm{i}^2=\rho^2+\varepsilon_\mathrm{i}\mu_\mathrm{i}\xi_n^2/c^2,\ x_\mathrm{i}\equiv2\rho_\mathrm{i}l,\ x_\mathrm{i}^2=x_\mathrm{m}^2+\left(\frac{2l\xi_n}{c}\right)^2(\varepsilon_\mathrm{i}\mu_\mathrm{i}-\varepsilon_\mathrm{m}\mu_\mathrm{m}), \tag{L2.59}$$

$$s_\mathrm{i}=\sqrt{\mathrm{p}^2-1+(\varepsilon_\mathrm{i}\mu_\mathrm{i}/\varepsilon_\mathrm{m}\mu_\mathrm{m})}, \tag{L2.60}$$

$$\overline{\Delta}_\mathrm{ji}=\frac{\rho_\mathrm{i}\varepsilon_\mathrm{j}-\rho_\mathrm{j}\varepsilon_\mathrm{i}}{\rho_\mathrm{i}\varepsilon_\mathrm{j}+\rho_\mathrm{j}\varepsilon_\mathrm{i}}=\frac{x_\mathrm{i}\varepsilon_\mathrm{j}-x_\mathrm{j}\varepsilon_\mathrm{i}}{x_\mathrm{i}\varepsilon_\mathrm{j}+x_\mathrm{j}\varepsilon_\mathrm{i}}=\frac{s_\mathrm{i}\varepsilon_\mathrm{j}-s_\mathrm{j}\varepsilon_\mathrm{i}}{s_\mathrm{i}\varepsilon_\mathrm{j}+s_\mathrm{j}\varepsilon_\mathrm{i}}, \tag{L2.61}$$

$$\Delta_\mathrm{ji}=\frac{\rho_\mathrm{i}\mu_\mathrm{j}-\rho_\mathrm{j}\mu_\mathrm{i}}{\rho_\mathrm{i}\mu_\mathrm{j}+\rho_\mathrm{j}\mu_\mathrm{i}}=\frac{x_\mathrm{i}\mu_\mathrm{j}-x_\mathrm{j}\mu_\mathrm{i}}{x_\mathrm{i}\mu_\mathrm{j}+x_\mathrm{j}\mu_\mathrm{i}}=\frac{s_\mathrm{i}\mu_\mathrm{j}-s_\mathrm{j}\mu_\mathrm{i}}{s_\mathrm{i}\mu_\mathrm{j}+s_\mathrm{j}\mu_\mathrm{i}}. \tag{L2.62}$$

The functions $\overline{\Delta}_\mathrm{Am}$, Δ_Am in the A/m/B and A/m/B$_1$/B cases have been replaced with

$$\overline{\Delta}_\mathrm{Am}^\mathrm{eff}(\mathrm{a_1})=\frac{\left(\overline{\Delta}_\mathrm{AA_1}e^{-2\rho_\mathrm{A_1}\mathrm{a_1}}+\overline{\Delta}_\mathrm{A_1m}\right)}{1+\overline{\Delta}_\mathrm{AA_1}\overline{\Delta}_\mathrm{A_1m}e^{-2\rho_\mathrm{A_1}\mathrm{a_1}}}=\frac{\left[\overline{\Delta}_\mathrm{AA_1}e^{-x_\mathrm{A_1}(\mathrm{a_1}/l)}+\overline{\Delta}_\mathrm{A_1m}\right]}{1+\overline{\Delta}_\mathrm{AA_1}\overline{\Delta}_\mathrm{A_1m}e^{-x_\mathrm{A_1}(\mathrm{a_1}/l)}}$$
$$=\frac{\left(\overline{\Delta}_\mathrm{AA_1}e^{-s_\mathrm{A_1}r_n(\mathrm{a_1}/l)}+\overline{\Delta}_\mathrm{A_1m}\right)}{1+\overline{\Delta}_\mathrm{AA_1}\overline{\Delta}_\mathrm{A_1m}e^{-s_\mathrm{A_1}r_n(\mathrm{a_1}/l)}}, \tag{L2.63}$$

[L3.86]

$$\Delta_\mathrm{Am}^\mathrm{eff}(\mathrm{a_1})=\frac{\left[\Delta_\mathrm{AA_1}e^{-2\rho_\mathrm{A_1}\mathrm{a_1}}+\Delta_\mathrm{A_1m}\right]}{1+\Delta_\mathrm{AA_1}\Delta_\mathrm{A_1m}e^{-2\rho_\mathrm{A_1}\mathrm{a_1}}}=\frac{\left(\Delta_\mathrm{AA_1}e^{-x_\mathrm{A_1}(\mathrm{a_1}/l)}+\Delta_\mathrm{A_1m}\right)}{1+\Delta_\mathrm{AA_1}\Delta_\mathrm{A_1m}e^{-x_\mathrm{A_1}(\mathrm{a_1}/l)}}$$
$$=\frac{\left[\Delta_\mathrm{AA_1}e^{-s_\mathrm{B_1}r_n\mathrm{a_1}/l}+\Delta_\mathrm{A_1m}\right]}{1+\Delta_\mathrm{AA_1}\Delta_\mathrm{A_1m}e^{-s_\mathrm{A_1}r_n(\mathrm{a_1}/l)}}. \tag{L2.64}$$

Indices i and j now refer to five materials: A, A$_1$, B, B$_1$, or m.

Limiting forms

$A_1 = A$, $B_1 = B$ It is obvious that when A_1 is the same as A and B_1 is the same as B, then $\overline{\Delta}_{Am}^{eff}(a_1)$, $\Delta_{Am}^{eff}(a_1)$ and $\overline{\Delta}_{Bm}^{eff}(b_1)$, $\Delta_{Bm}^{eff}(b_1)$ revert to $\overline{\Delta}_{Am}$, $\overline{\Delta}_{Am}$ and $\overline{\Delta}_{Bm}$, $\overline{\Delta}_{Bm}$ to recreate the interaction between B and A across m at separation l.

$A_1 = B_1 = m$ Here

$$\overline{\Delta}_{Am}^{eff}(a_1) \to \overline{\Delta}_{AA_1} e^{-2\rho_{A_1} a_1} = \overline{\Delta}_{Am} e^{-2\rho_m a_1},\ \overline{\Delta}_{Bm}^{eff}(b_1) \to \overline{\Delta}_{BB_1} e^{-2\rho_{B_1} b_1} = \overline{\Delta}_{Bm} e^{-2\rho_m b_1}, \tag{L2.65}$$

so that

$$\overline{\Delta}_{Am}^{eff}\overline{\Delta}_{Bm}^{eff} e^{-2\rho_m l} \to \overline{\Delta}_{Am}\overline{\Delta}_{Bm} e^{-2\rho_m a_1} e^{-2\rho_m b_1} e^{-2\rho_m l} = \overline{\Delta}_{Am}\overline{\Delta}_{Bm} e^{-2\rho_m (a_1+b_1+l)}$$

(and the same for $\Delta_{Am}^{eff}\Delta_{Bm}^{eff} e^{-2\rho_m l}$). Then the interaction is still A with B across m, but this time across a distance $(a_1 + b_1 + l)$.

$a_1, b_1 \gg l$ When thicknesses a_1 and $b_1 \gg l$,

$$\overline{\Delta}_{Am}^{eff}(a_1) \to \overline{\Delta}_{A_1m},\ \Delta_{Am}^{eff}(a_1) \to \Delta_{A_1m},\ \overline{\Delta}_{Bm}^{eff}(b_1) \to \overline{\Delta}_{B_1m},\ \Delta_{Bm}^{eff}(b_1) \to \Delta_{B_1m}, \tag{L2.66}$$

materials A and B disappear from the interaction. B_1 interacts with A_1 across m at a separation l.

High dielectric-permittivity layers When $\varepsilon_{A_1}/\varepsilon_m \to \infty$, $\varepsilon_{B_1}/\varepsilon_m \to \infty$, neglecting magnetic terms, $\overline{\Delta}_{A_1m}$, $\overline{\Delta}_{B_1m} \to 1$,

$$\overline{\Delta}_{Am}^{eff}(a_1) \to \frac{\left(\overline{\Delta}_{AA_1} e^{-2\rho_{A_1} a_1} + 1\right)}{1 + \overline{\Delta}_{AA_1} e^{-2\rho_{A_1} a_1}} = 1,\quad \overline{\Delta}_{Bm}^{eff}(b_1) \to \frac{\left(\overline{\Delta}_{BB_1} e^{-2\rho_{B_1} b_1} + 1\right)}{1 + \overline{\Delta}_{BB_1} e^{-2\rho_{B_1} b_1}} = 1, \tag{L2.67}$$

$$\begin{aligned} G_{AA_1mB_1B}(l; a_1, b_1) &\to G_{A_1mB_1}(l) \\ &= \frac{kT}{2\pi} \sum_{n=0}^{\infty}{}' \int_{\frac{\varepsilon_m^{1/2}\mu_m^{1/2}\xi_n}{c}}^{\infty} \rho_m \ln(1 - e^{-2\rho_m l})\, d\rho_m \\ &= \frac{kT}{8\pi l^2} \sum_{n=0}^{\infty}{}' \int_{r_n}^{\infty} x \ln(1 - e^{-x}) dx \\ &= \frac{kT}{2\pi c^2} \sum_{n=0}^{\infty}{}' \varepsilon_m \mu_m \xi_n^2 \int_1^{\infty} p \ln(1 - e^{-r_n p}) dp. \end{aligned} \tag{L2.68}$$

[P.3.b.1]

Semi-infinite materials A and B are screened out by ideally metallic A_1 and B_1.

Small differences in ε's and μ's When $\overline{\Delta}_{ji}$'s and Δ_{ij}'s are $\ll 1$, we can restrict ourselves to leading terms:

$$\overline{\Delta}_{\mathrm{Am}}^{\mathrm{eff}}(\mathrm{a}_1) \to \left(\overline{\Delta}_{\mathrm{AA}_1} e^{-2\rho_{\mathrm{A}_1}\mathrm{a}_1} + \overline{\Delta}_{\mathrm{A}_1\mathrm{m}}\right)$$
$$= \left(\overline{\Delta}_{\mathrm{AA}_1} e^{-x_{\mathrm{A}_1}(\mathrm{a}_1/l)} + \overline{\Delta}_{\mathrm{A}_1\mathrm{m}}\right) = \left(\overline{\Delta}_{\mathrm{AA}_1} e^{-s_{\mathrm{A}_1} r_n(\mathrm{a}_1/l)} + \overline{\Delta}_{\mathrm{A}_1\mathrm{m}}\right) \ll 1, \quad \text{(L2.69)}$$

and similarly for $\Delta_{\mathrm{Am}}^{\mathrm{eff}}(\mathrm{a}_1)$, $\overline{\Delta}_{\mathrm{Bm}}^{\mathrm{eff}}(\mathrm{b}_1)$, $\Delta_{\mathrm{Bm}}^{\mathrm{eff}}(\mathrm{b}_1)$ so that

$$\ln\left[\left(1 - \overline{\Delta}_{\mathrm{Am}}^{\mathrm{eff}}\overline{\Delta}_{\mathrm{Bm}}^{\mathrm{eff}} e^{-2\rho_{\mathrm{m}}l}\right)\left(1 - \Delta_{\mathrm{Am}}^{\mathrm{eff}}\Delta_{\mathrm{Bm}}^{\mathrm{eff}} e^{-2\rho_{\mathrm{m}}l}\right)\right]$$
$$\to -\overline{\Delta}_{\mathrm{Am}}^{\mathrm{eff}}\overline{\Delta}_{\mathrm{Bm}}^{\mathrm{eff}} e^{-2\rho_{\mathrm{m}}l} - \Delta_{\mathrm{Am}}^{\mathrm{eff}}\Delta_{\mathrm{Bm}}^{\mathrm{eff}} e^{-2\rho_{\mathrm{m}}l}$$
$$\to -(\overline{\Delta}_{\mathrm{A}_1\mathrm{m}}\overline{\Delta}_{\mathrm{B}_1\mathrm{m}} + \Delta_{\mathrm{A}_1\mathrm{m}}\Delta_{\mathrm{B}_1\mathrm{m}})e^{-2\rho_{\mathrm{m}}l}$$
$$-(\overline{\Delta}_{\mathrm{A}_1\mathrm{m}}\overline{\Delta}_{\mathrm{BB}_1} + \Delta_{\mathrm{A}_1\mathrm{m}}\Delta_{\mathrm{BB}_1})e^{-2\rho_{\mathrm{B}_1}\mathrm{b}_1}e^{-2\rho_{\mathrm{m}}l}$$
$$-(\overline{\Delta}_{\mathrm{B}_1\mathrm{m}}\overline{\Delta}_{\mathrm{AA}_1} + \Delta_{\mathrm{B}_1\mathrm{m}}\Delta_{\mathrm{AA}_1})e^{-2\rho_{\mathrm{A}_1}\mathrm{a}_1}e^{-2\rho_{\mathrm{m}}l}$$
$$-(\overline{\Delta}_{\mathrm{BB}_1}\overline{\Delta}_{\mathrm{AA}_1} + \Delta_{\mathrm{BB}_1}\Delta_{\mathrm{AA}_1})e^{-2\rho_{\mathrm{A}_1}\mathrm{a}_1}e^{-2\rho_{\mathrm{B}_1}\mathrm{b}_1}e^{-2\rho_{\mathrm{m}}l}. \quad \text{(L2.70)}$$

The interaction free energy reduces to four terms:

$$-\frac{kT}{2\pi}\sum_{n=0}^{\infty}{}' \int_{\frac{\varepsilon_{\mathrm{m}}^{1/2}\mu_{\mathrm{m}}^{1/2}\xi_n}{c}}^{\infty} \rho_{\mathrm{m}}(\overline{\Delta}_{\mathrm{A}_1\mathrm{m}}\overline{\Delta}_{\mathrm{B}_1\mathrm{m}} + \Delta_{\mathrm{A}_1\mathrm{m}}\Delta_{\mathrm{B}_1\mathrm{m}})e^{-2\rho_{\mathrm{m}}l}d\rho_{\mathrm{m}},$$
$$-\frac{kT}{2\pi}\sum_{n=0}^{\infty}{}' \int_{\frac{\varepsilon_{\mathrm{m}}^{1/2}\mu_{\mathrm{m}}^{1/2}\xi_n}{c}}^{\infty} \rho_{\mathrm{m}}(\overline{\Delta}_{\mathrm{A}_1\mathrm{m}}\overline{\Delta}_{\mathrm{BB}_1} + \Delta_{\mathrm{A}_1\mathrm{m}}\Delta_{\mathrm{BB}_1})e^{-2\rho_{\mathrm{B}_1}\mathrm{b}_1}e^{-2\rho_{\mathrm{m}}l}d\rho_{\mathrm{m}},$$
$$-\frac{kT}{2\pi}\sum_{n=0}^{\infty}{}' \int_{\frac{\varepsilon_{\mathrm{m}}^{1/2}\mu_{\mathrm{m}}^{1/2}\xi_n}{c}}^{\infty} \rho_{\mathrm{m}}(\overline{\Delta}_{\mathrm{B}_1\mathrm{m}}\overline{\Delta}_{\mathrm{AA}_1} + \Delta_{\mathrm{B}_1\mathrm{m}}\Delta_{\mathrm{AA}_1})e^{-2\rho_{\mathrm{A}_1}\mathrm{a}_1}e^{-2\rho_{\mathrm{m}}l}d\rho_{\mathrm{m}},$$
$$-\frac{kT}{2\pi}\sum_{n=0}^{\infty}{}' \int_{\frac{\varepsilon_{\mathrm{m}}^{1/2}\mu_{\mathrm{m}}^{1/2}\xi_n}{c}}^{\infty} \rho_{\mathrm{m}}(\overline{\Delta}_{\mathrm{BB}_1}\overline{\Delta}_{\mathrm{AA}_1} + \Delta_{\mathrm{BB}_1}\Delta_{\mathrm{AA}_1})e^{-2\rho_{\mathrm{A}_1}\mathrm{a}_1}e^{-2\rho_{\mathrm{B}_1}\mathrm{b}_1}e^{-2\rho_{\mathrm{m}}l}d\rho_{\mathrm{m}}. \quad \text{(L2.71)}$$

[P.3.b.2]

The first term is the Lifshitz interaction of A with B_1 across m at separation l; the second is the interaction of B with A_1 across m and B_1. The next two terms, corresponding to interactions between the other two pairs of interfaces across m, have almost the Lifshitz form. They differ slightly from it because of the different velocities of light through each medium.

Small differences in ε's and μ's, nonretarded limit For small differences in susceptibilities, with $c \to \infty$, $\rho_{\mathrm{i}} \to \rho_{\mathrm{m}} \to \rho$,

$$G_{\mathrm{AA}_1\mathrm{mB}_1\mathrm{B}}(l; \mathrm{a}_1, \mathrm{b}_1) \to -\frac{kT}{8\pi l^2}\sum_{n=0}^{\infty}{}'(\overline{\Delta}_{\mathrm{A}_1\mathrm{m}}\overline{\Delta}_{\mathrm{B}_1\mathrm{m}} + \Delta_{\mathrm{A}_1\mathrm{m}}\Delta_{\mathrm{B}_1\mathrm{m}})$$
$$-\frac{kT}{8\pi(l+\mathrm{a}_1)^2}\sum_{n=0}^{\infty}{}'(\overline{\Delta}_{\mathrm{B}_1\mathrm{m}}\overline{\Delta}_{\mathrm{AA}_1} + \Delta_{\mathrm{B}_1\mathrm{m}}\Delta_{\mathrm{AA}_1})$$
$$-\frac{kT}{8\pi(l+\mathrm{b}_1)^2}\sum_{n=0}^{\infty}{}'(\overline{\Delta}_{\mathrm{A}_1\mathrm{m}}\overline{\Delta}_{\mathrm{BB}_1} + \Delta_{\mathrm{A}_1\mathrm{m}}\Delta_{\mathrm{BB}_1})$$
$$-\frac{kT}{8\pi(l+\mathrm{a}_1+\mathrm{b}_1)^2}\sum_{n=0}^{\infty}{}'(\overline{\Delta}_{\mathrm{BB}_1}\overline{\Delta}_{\mathrm{AA}_1} + \Delta_{\mathrm{BB}_1}\Delta_{\mathrm{AA}_1}). \quad \text{(L2.72)}$$

[P.3.b.3]

TWO SLABS OF FINITE THICKNESS

Exact

Let materials A = B = m, $\overline{\Delta}_{AA_1} = -\overline{\Delta}_{A_1m}$, $\overline{\Delta}_{BB_1} = -\overline{\Delta}_{B_1m}$:

$$\overline{\Delta}_{Am}^{eff}(a_1) = \overline{\Delta}_{A_1m}\frac{1-e^{-2\rho_{A_1}a_1}}{1-\overline{\Delta}_{A_1m}^2 e^{-2\rho_{A_1}a_1}} = \overline{\Delta}_{A_1m}\frac{1-e^{-x_{A_1}(a_1/l)}}{1-\overline{\Delta}_{A_1m}^2 e^{-x_{A_1}(a_1/l)}}$$

$$= \overline{\Delta}_{A_1m}\frac{1-e^{-s_{A_1}r_n(a_1/l)}}{1-\overline{\Delta}_{A_1m}^2 e^{-s_{A_1}r_n(a_1/l)}}, \tag{L2.73}$$

$$\Delta_{Am}^{eff}(a_1) = \Delta_{A_1m}\frac{1-e^{-2\rho_{A_1}a_1}}{1-\Delta_{A_1m}^2 e^{-2\rho_{A_1}a_1}} = \Delta_{A_1m}\frac{1-e^{-x_{A_1}(a_1/l)}}{1-\Delta_{A_1m}^2 e^{-x_{A_1}(a_1/l)}}$$

$$= \Delta_{A_1m}\frac{1-e^{-s_{A_1}r_n(a_1/l)}}{1-\Delta_{A_1m}^2 e^{-s_{A_1}r_n(a_1/l)}}, \tag{L2.74}$$

$$\overline{\Delta}_{Bm}^{eff}(b_1) = \overline{\Delta}_{B_1m}\frac{1-e^{-2\rho_{B_1}b_1}}{1-\overline{\Delta}_{B_1m}^2 e^{-2\rho_{B_1}b_1}} = \overline{\Delta}_{B_1m}\frac{1-e^{-x_{B_1}(b_1/l)}}{1-\overline{\Delta}_{B_1m}^2 e^{-x_{B_1}(b_1/l)}}$$

$$= \overline{\Delta}_{B_1m}\frac{1-e^{-s_{B_1}r_n(b_1/l)}}{1-\overline{\Delta}_{B_1m}^2 e^{-s_{B_1}r_n(b_1/l)}}, \tag{L2.75}$$

$$\Delta_{Bm}^{eff}(b_1) = \Delta_{B_1m}\frac{1-e^{-2\rho_{B_1}b_1}}{1-\Delta_{B_1m}^2 e^{-2\rho_{B_1}b_1}} = \Delta_{B_1m}\frac{1-e^{-x_{B_1}(b_1/l)}}{1-\Delta_{B_1m}^2 e^{-x_{B_1}(b_1/l)}}$$

$$= \Delta_{B_1m}\frac{1-e^{-s_{B_1}r_n(b_1/l)}}{1-\Delta_{B_1m}^2 e^{-s_{B_1}r_n(b_1/l)}}, \tag{L2.76}$$

$$G_{A_1mB_1}(l; a_1, b_1)$$

$$= \frac{kT}{2\pi}\sum_{n=0}^{\infty}{}' \int_{\frac{\varepsilon_m^{1/2}\mu_m^{1/2}\xi_n}{c}}^{\infty} \rho_m \ln\left[\left(1-\overline{\Delta}_{Am}^{eff}\overline{\Delta}_{Bm}^{eff}e^{-2\rho_m l}\right)\left(1-\Delta_{Am}^{eff}\Delta_{Bm}^{eff}e^{-2\rho_m l}\right)\right] d\rho_m$$

$$= \frac{kT}{8\pi l^2}\sum_{n=0}^{\infty}{}' \int_{r_n}^{\infty} x \ln\left[\left(1-\overline{\Delta}_{Am}^{eff}\overline{\Delta}_{Bm}^{eff}e^{-x}\right)\left(1-\Delta_{Am}^{eff}\Delta_{Bm}^{eff}e^{-x}\right)\right] dx$$

$$= \frac{kT}{2\pi c^2}\sum_{n=0}^{\infty}{}' \varepsilon_m\mu_m\xi_n^2 \int_1^{\infty} p \ln\left[\left(1-\overline{\Delta}_{Am}^{eff}\overline{\Delta}_{Bm}^{eff}e^{-r_n p}\right)\left(1-\Delta_{Am}^{eff}\Delta_{Bm}^{eff}e^{-r_n p}\right)\right] dp. \tag{L2.77}$$

[P.3.c.1]

Small differences in ε's and μ's When $\overline{\Delta}_{Am}, \Delta_{Am}, \overline{\Delta}_{B_1m}, \Delta_{B_1m} \ll 1$,

$$\overline{\Delta}_{Am}^{eff}(a_1) \to \overline{\Delta}_{A_1m}(1-e^{-2\rho_{A_1}a_1}) = \overline{\Delta}_{A_1m}\left(1-e^{-x_{A_1}(a_1/l)}\right) = \overline{\Delta}_{A_1m}\left(1-e^{-s_{A_1}r_n(a_1/l)}\right), \tag{L2.78}$$

and similarly for $\Delta_{Am}^{eff}(a_1)$, $\overline{\Delta}_{Bm}^{eff}(b_1)$, $\Delta_{Bm}^{eff}(b_1)$. The free energy reduces to one integral with four terms:

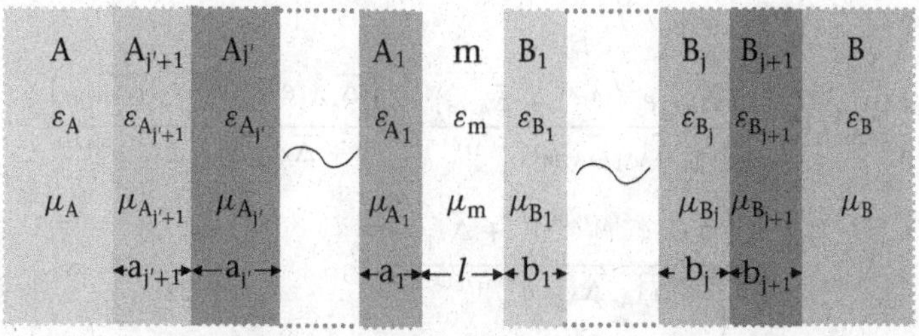

Figure L2.2

$$G_{A_1mB_1}(l; a_1, b_1) = -\frac{kT}{2\pi}\sum_{n=0}^{\infty}{}' \int_{\frac{\varepsilon_m^{1/2}\mu_m^{1/2}\xi_n}{c}}^{\infty} \rho_m(\overline{\Delta}_{A_1m}\overline{\Delta}_{B_1m} + \Delta_{A_1m}\Delta_{B_1m})$$
$$\times (1 - e^{-2\rho_{A_1}a_1} - e^{-2\rho_{B_1}b_1} + e^{-2\rho_{A_1}a_1}e^{-2\rho_{B_1}b_1})e^{-2\rho_m l}d\rho_m. \quad \text{(L2.79)}$$
[P.3.c.2]

Small differences in ε's and μ's, nonretarded limit In the further limit at which $c \to \infty$, $\rho_{A_1}, \rho_{B_1} \to \rho_m \to \rho$:

$$G_{A_1mB_1}(l; a_1, b_1) \to -\frac{kT}{8\pi}\left[\frac{1}{l^2} - \frac{1}{(l+b_1)^2} - \frac{1}{(l+a_1)^2} + \frac{1}{(l+a_1+b_1)^2}\right]$$
$$\times \sum_{n=0}^{\infty}{}' (\overline{\Delta}_{A_1m}\overline{\Delta}_{B_1m} + \Delta_{A_1m}\Delta_{B_1m}). \quad \text{(L2.80)}$$
[P.3.c.3]

TWO MULTIPLY COATED SEMI-INFINITE MEDIA

Exact

The addition of successive layers involves the generation of successive forms for $\overline{\Delta}_{Am}^{eff}$, $\overline{\Delta}_{Bm}^{eff}$, Δ_{Am}^{eff}, Δ_{Bm}^{eff} to go into $\int_{\frac{\varepsilon_m^{1/2}\mu_m^{1/2}\xi_n}{c}}^{\infty} \rho_m \ln[(1-\overline{\Delta}_{Am}^{eff}\overline{\Delta}_{Bm}^{eff}e^{-2\rho_m l})(1-\Delta_{Am}^{eff}\Delta_{Bm}^{eff}e^{-2\rho_m l})]d\rho_m$ or its variants. Either by matrix multiplication or by induction, the $\overline{\Delta}_{Am}^{eff}$ (j′) for j′ layers can be converted into $\overline{\Delta}_{Am}^{eff}$ (j′ + 1) (see Fig. L2.2).

Use the convention that material $A_{j'}$ is next to half-space A and that a j′ + 1st layer of thickness $a_{j'+1}$ is inserted between $A_{j'}$ and A. Then the difference in susceptibilities embodied in $\overline{\Delta}_{AA_{j'}}$ is replaced with

$$\frac{\overline{\Delta}_{AA_{j'+1}}e^{-2\rho_{A_{j'+1}}a_{j'+1}} + \overline{\Delta}_{A_{j'+1}A_{j'}}}{1+\overline{\Delta}_{AA_{j'+1}}\overline{\Delta}_{A_{j'+1}A_{j'}}e^{-2\rho_{A_{j'+1}}a_{j'+1}}} = \frac{\overline{\Delta}_{AA_{j'+1}}e^{-x_{A_{j'+1}}(a_{j'+1}/l)} + \overline{\Delta}_{A_{j'+1}A_{j'}}}{1+\overline{\Delta}_{AA_{j'+1}}\overline{\Delta}_{A_{j'+1}A_{j'}}e^{-x_{A_{j'+1}}(a_{j'+1}/l)}}$$
$$= \frac{\overline{\Delta}_{AA_{j'+1}}e^{-s_{A_{j'+1}}r_n(a_{j'+1}/l)} + \overline{\Delta}_{A_{j'+1}A_{j'}}}{1+\overline{\Delta}_{AA_{j'+1}}\overline{\Delta}_{A_{j'+1}A_{j'}}e^{-s_{A_{j'+1}}r_n(a_{j'+1}/l)}}. \quad \text{(L2.81)}$$
[L3.90]

In this notation, the unlayered half-space A is represented by

$$\overline{\Delta}_{Am}^{eff}(0) = \overline{\Delta}_{Am}, \quad \text{(L2.82)}$$

the singly layered A by

$$\overline{\Delta}_{\mathrm{Am}}^{\mathrm{eff}}(1)=\frac{\left(\overline{\Delta}_{\mathrm{AA_1}}e^{-2\rho_{\mathrm{A_1}}a_1}+\overline{\Delta}_{\mathrm{A_1m}}\right)}{1+\overline{\Delta}_{\mathrm{AA_1}}\overline{\Delta}_{\mathrm{A_1m}}e^{-2\rho_{\mathrm{A_1}}a_1}}=\frac{\left(\overline{\Delta}_{\mathrm{AA_1}}e^{-x_{\mathrm{A_1}}(a_1/l)}+\overline{\Delta}_{\mathrm{A_1m}}\right)}{1+\overline{\Delta}_{\mathrm{AA_1}}\overline{\Delta}_{\mathrm{A_1m}}e^{-x_{\mathrm{A_1}}(a_1/l)}}$$

$$=\frac{\left(\overline{\Delta}_{\mathrm{AA_1}}e^{-s_{\mathrm{A_1}}r_n(a_1/l)}+\overline{\Delta}_{\mathrm{A_1m}}\right)}{1+\overline{\Delta}_{\mathrm{AA_1}}\overline{\Delta}_{\mathrm{A_1m}}e^{-s_{\mathrm{A_1}}r_n(a_1/l)}},\tag{L2.83}$$

the doubly layered A by

$$\overline{\Delta}_{\mathrm{Am}}^{\mathrm{eff}}(2)=\frac{\left(\dfrac{\overline{\Delta}_{\mathrm{AA_2}}e^{-2\rho_{\mathrm{A_2}}a_2}+\overline{\Delta}_{\mathrm{A_2A_1}}}{1+\overline{\Delta}_{\mathrm{AA_2}}\overline{\Delta}_{\mathrm{A_2A_1}}e^{-2\rho_{\mathrm{A_2}}a_2}}\right)e^{-2\rho_{\mathrm{A_1}}a_1}+\overline{\Delta}_{\mathrm{A_1m}}}{\left[1+\left(\dfrac{\overline{\Delta}_{\mathrm{AA_2}}e^{-2\rho_{\mathrm{A_2}}a_2}+\overline{\Delta}_{\mathrm{A_2A_1}}}{1+\overline{\Delta}_{\mathrm{AA_2}}\overline{\Delta}_{\mathrm{A_2A_1}}e^{-2\rho_{\mathrm{A_2}}a_2}}\right)\overline{\Delta}_{\mathrm{A_1m}}e^{-2\rho_{\mathrm{A_1}}a_1}\right]}$$

$$=\frac{\overline{\Delta}_{\mathrm{AA_1}}^{\mathrm{eff}}(1)e^{-2\rho_{\mathrm{A_1}}a_1}+\overline{\Delta}_{\mathrm{A_1m}}}{\left[1+\overline{\Delta}_{\mathrm{AA_1}}^{\mathrm{eff}}(1)\overline{\Delta}_{\mathrm{A_1m}}e^{-2\rho_{\mathrm{A_1}}a_1}\right]},\tag{L2.84}$$

and the general form by

$$\overline{\Delta}_{\mathrm{Am}}^{\mathrm{eff}}(j'+1)=\frac{\overline{\Delta}_{\mathrm{AA_1}}^{\mathrm{eff}}(j')e^{-2\rho_{\mathrm{A_1}}a_1}+\overline{\Delta}_{\mathrm{A_1m}}}{\left[1+\overline{\Delta}_{\mathrm{AA_1}}^{\mathrm{eff}}(j')\overline{\Delta}_{\mathrm{A_1m}}e^{-2\rho_{\mathrm{A_1}}a_1}\right]}.\tag{L2.85}$$

[P.4.b]

Keep in mind the polyglot possibilities, epitomized in $e^{-2\rho_{\mathrm{A_{j'}}}a_{j'}}$ or $e^{-x_{\mathrm{A_{j'}}}(a_{j'}/l)}$ or $e^{-s_{\mathrm{A_{j'}}}r_n(a_{j'}/l)}$, and use the same generalization rules for $\overline{\Delta}_{\mathrm{Bm}}^{\mathrm{eff}}$, $\Delta_{\mathrm{Am}}^{\mathrm{eff}}$, $\Delta_{\mathrm{Bm}}^{\mathrm{eff}}$ as given here for $\overline{\Delta}_{\mathrm{Am}}^{\mathrm{eff}}$. Then generate and manipulate expressions for any set of layers on facing surfaces.

Small differences in ε's and μ's When there are only small differences in susceptibilities in that all $\overline{\Delta}_{\mathrm{ij}}$'s and Δ_{ij}'s at material interfaces are $\ll 1$, only leading terms need be retained. $\overline{\Delta}_{\mathrm{AA_{j'}}}$, between A and its adjacent material layer j′, can now be replaced with

$$\overline{\Delta}_{\mathrm{AA_{j'+1}}}e^{-2\rho_{\mathrm{A_{j+1}}}a_{j+1}}+\overline{\Delta}_{\mathrm{A_{j'+1}A_{j'}}}=\overline{\Delta}_{\mathrm{AA_{j'+1}}}e^{-x_{\mathrm{A_{j'+1}}}(a_{j'+1}/l)}+\overline{\Delta}_{\mathrm{A_{j'+1}A_{j'}}}$$

$$=\overline{\Delta}_{\mathrm{AA_{j'+1}}}e^{-s_{\mathrm{A_{j'+1}}}r_n(a_{j'+1}/l)}+\overline{\Delta}_{\mathrm{A_{j'+1}A_{j'}}}\tag{L2.86}$$

to create sums

$$\overline{\Delta}_{\mathrm{Am}}^{\mathrm{eff}}(1)\to\left(\overline{\Delta}_{\mathrm{AA_1}}e^{-2\rho_{\mathrm{A_1}}a_1}+\overline{\Delta}_{\mathrm{A_1m}}\right),\tag{L2.87}$$

$$\overline{\Delta}_{\mathrm{Am}}^{\mathrm{eff}}(2)\to\overline{\Delta}_{\mathrm{AA_2}}e^{-2\rho_{\mathrm{A_2}}a_2}e^{-2\rho_{\mathrm{A_1}}a_1}+\overline{\Delta}_{\mathrm{A_2A_1}}e^{-2\rho_{\mathrm{A_1}}a_1}+\overline{\Delta}_{\mathrm{A_1m}},\tag{L2.88}$$

$$\overline{\Delta}_{\mathrm{Am}}^{\mathrm{eff}}(3)\to\overline{\Delta}_{\mathrm{AA_3}}e^{-2\rho_{\mathrm{A_3}}a_3}e^{-2\rho_{\mathrm{A_2}}a_2}e^{-2\rho_{\mathrm{A_1}}a_1}+\overline{\Delta}_{\mathrm{A_3A_2}}e^{-2\rho_{\mathrm{A_2}}a_2}e^{-2\rho_{\mathrm{A_1}}a_1}+\overline{\Delta}_{\mathrm{A_2A_1}}e^{-2\rho_{\mathrm{A_1}}a_1}+\overline{\Delta}_{\mathrm{A_1m}}.\tag{L2.89}$$

Multiplication of $\overline{\Delta}_{\mathrm{Am}}^{\mathrm{eff}}$ and $\overline{\Delta}_{\mathrm{Bm}}^{\mathrm{eff}}$ creates pairs of terms that correspond to combinations of all interfaces between layers on A and layers on B. The integration

$$\int_{\frac{\varepsilon_m^{1/2}\mu_m^{1/2}\xi_n}{c}}^{\infty}\rho_m\ln\left[\left(1-\overline{\Delta}_{\mathrm{Am}}^{\mathrm{eff}}\overline{\Delta}_{\mathrm{Bm}}^{\mathrm{eff}}e^{-2\rho_m l}\right)\left(1-\Delta_{\mathrm{Am}}^{\mathrm{eff}}\Delta_{\mathrm{Bm}}^{\mathrm{eff}}e^{-2\rho_m l}\right)\right]\mathrm{d}\rho_m$$

creates a sum of terms of the form

$$\int_{\frac{\varepsilon_m^{1/2}\mu_m^{1/2}\xi_n}{c}}^{\infty} \rho_m \overline{\Delta}_{A_{k'}A_{k'-1}} \overline{\Delta}_{B_kB_{k-1}} e^{-2\left(\sum_{g=1}^{k'} \rho_{A_g} a_g + \sum_{h=1}^{k} \rho_{B_h} b_h + \rho_m l\right)} d\rho_m, \tag{L2.90}$$

[P.4.c]

where $1 \le k \le j$, $1 \le k' \le j'$. What this superficially cumbersome expression states is that the individual interfaces interact across their separation distance $\sum_{g=1}^{k'} a_g + \sum_{h=1}^{k} b_h + l$ whose segments, the layer thicknesses a_g and b_h are weighted by ρ's that reflect the local velocities of light.

Small differences in ε's and μ's, nonretarded limit In the limit of infinite velocity of light, all $\rho_i^2 = \rho^2 + \varepsilon_i\mu_i\xi_n^2/c^2 \to \rho^2$, the integral for the interaction between pairs of interfaces at a separation $l_{k'/k} = \sum_{g=1}^{k'} a_g + \sum_{h=1}^{k} b_h + l$ now becomes the far more intuitive

$$\overline{\Delta}_{A_{k'}A_{k'-1}} \overline{\Delta}_{B_kB_{k-1}} \int_0^{\infty} \rho_m e^{-2\rho_m l_{k'/k}} d\rho_m. \tag{L2.91}$$

Returning to the form of the interaction free energy,

$$\int_{\frac{\varepsilon_m^{1/2}\mu_m^{1/2}\xi_n}{c}}^{\infty} \rho_m \ln\left[\left(1 - \overline{\Delta}_{Am}^{\rm eff}\overline{\Delta}_{Bm}^{\rm eff} e^{-2\rho_m l}\right)\left(1 - \Delta_{Am}^{\rm eff}\Delta_{Bm}^{\rm eff} e^{-2\rho_m l}\right)\right] d\rho_m,$$

the individual pair interactions emerge in the familiar form

$$-\frac{kT}{2\pi}\sum_{n=0}^{\infty}{}' \overline{\Delta}_{A_{k'}A_{k'-1}} \overline{\Delta}_{B_kB_{k-1}} \int_0^{\infty} \rho_m e^{-2\rho_m l_{k'/k}} d\rho_m = -\frac{kT}{8\pi l_{k'/k}^2}\sum_{n=0}^{\infty}{}' \overline{\Delta}_{A_{k'}A_{k'-1}} \overline{\Delta}_{B_kB_{k-1}}, \tag{L2.92}$$

with $\overline{\Delta}_{ji} = [(\varepsilon_j - \varepsilon_i)/(\varepsilon_j + \varepsilon_i)^2]$, and similarly for magnetic terms.

The total interaction is a sum of interactions between all pairs of interfaces across the variable part of the separation l. The result is so intuitively clear but the notation so disconcertingly messy that it is better to write in commonsense terms. Replace "$l_{k/k'}$." Instead, write $l_{A'A''/B'B''}$ for the distance between an interface on the A side and an interface on the B side. The interaction between j′-layered A and j-layered B is a sum:

$$G(l; a_1, a_2, \ldots a_{j'}, b_1, b_2, \ldots b_j) = \sum_{\substack{\text{(all pairs of}\\ \text{interfaces } A'A''/B'B'')}} G_{A'A''/B'B''}(l_{A'A/B'B''}), \tag{L2.93}$$

where each term has the form of a nonretarded small $\overline{\Delta}_{ij}$, Δ_{ij} interaction between half-spaces:

$$G_{A'A''/B'B''}(l_{A'A''/B'B''}) = -\frac{kT}{8\pi l_{A'A''/B'B''}^2}\sum_{n=0}^{\infty}{}' \left(\overline{\Delta}_{A'A''}\overline{\Delta}_{B'B''} + \Delta_{A'A''}\Delta_{B'B''}\right). \tag{L2.94}$$

[P.5]

Reminder: You ensure the correct sign of the interaction by remembering that the $\overline{\Delta}_{A'A''}\overline{\Delta}_{B'B''}$ and $\Delta_{A'A''}\Delta_{B'B''}$ are written with the singly primed A′ and B′ materials on the *farther* side of the interface.

CONTINUOUSLY CHANGING SUSCEPTIBILITIES, INHOMOGENEOUS MEDIA

When ε varies in the direction perpendicular to the planar interfaces, many of the resulting properties can be seen in terms of planar-layer formulae. Although it is best to compute using the tabulated general formulae derived in Level 3, it is instructive to see the behaviors that emerge in special limits of slowly varying $\varepsilon(z)$.

Nonretarded limit

Imagine $\varepsilon(z)$ as changing in infinitesimal steps $d\varepsilon_b = [d\varepsilon_b(z_b)/dz_b]dz_b$, and $d\varepsilon_a = [d\varepsilon_a(z_a)/dz_a]dz_a$, where the two position variables z_a and z_b are measured from the middle of the medium of width l. There is a separation $(z_a + z_b)$ between positions z_a and z_b in the transition regions. There can be steps in ε at the interfaces in addition to the changes in ε in the transition regions. The contribution of these steps to the total interaction energy is the same as for the steplike changes of the kind considered so far (see Fig. L2.3).

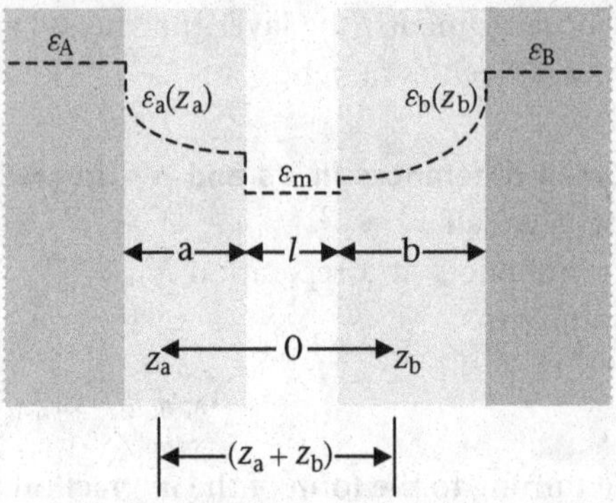

Figure L2.3

When the differences in all epsilons are all much smaller than ε_m and retardation screening is neglected, the contribution of the continuous regions can be imagined as coming from the sum of infinitesimal steps, $d\varepsilon_a(z_a)$ and $d\varepsilon_b(z_b)$ (see Fig. L2.4). The contribution from this continuously varying region is an integral over incremental energies of the form

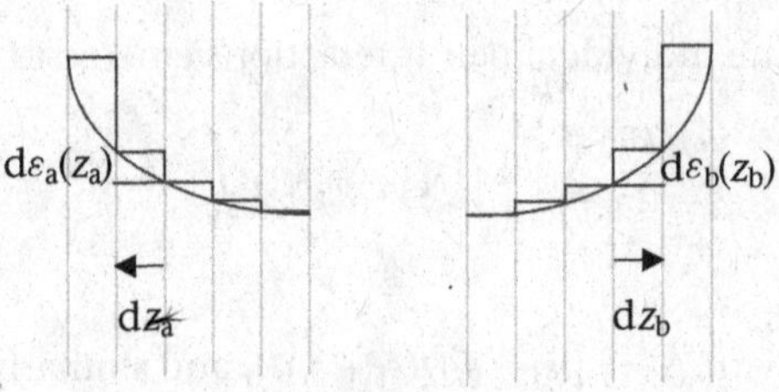

Figure L2.4

$$-\frac{kT}{8\pi}\frac{\sum_{n=0}^{\prime\infty}\overline{\Delta}(z_a)\overline{\Delta}(z_b)}{(z_a+z_b)^2}.$$

The continuously varying $\overline{\Delta}$'s go as

$$\overline{\Delta}_a(z_a) = \left\{\frac{[\varepsilon_a(z_a) + d\varepsilon_a(z_a)] - \varepsilon_a(z_a)}{[\varepsilon_a(z_a) + d\varepsilon_a(z_a)] + \varepsilon_a(z_a)}\right\} = \frac{d\ln[\varepsilon_a(z_a)]}{2dz_a}dz_a,$$

$$\overline{\Delta}_b(z_b) = \frac{d\ln[\varepsilon_b(z_b)]}{2dz_b}dz_b, \qquad \text{(L2.95)}$$

so that the energy of interaction, from integration over two layers of continuous variation in dielectric permittivity, has the form

$$G(l; a, b) = -\frac{kT}{32\pi}\sum_{n=0}^{\infty}{}' \iint_{z_b, z_a} \frac{d\ln[\varepsilon_b(z_b)]}{dz_b}\frac{d\ln[\varepsilon_a(z_a)]}{dz_a}\frac{dz_a dz_b}{(z_a+z_b)^2}. \qquad \text{(L2.96)}$$

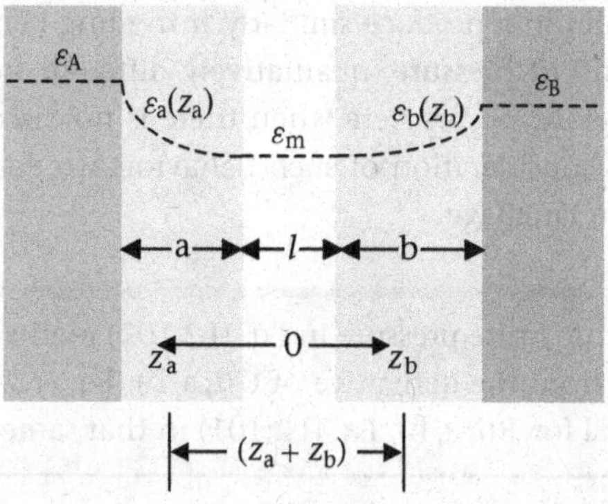

Figure L2.5

For bemusement, consider a smooth quadratic transition in ε from ε_m to the ε_A and ε_B of semi-infinite regions A and B:

$$\varepsilon_a(z_a) = \varepsilon_m + \frac{(\varepsilon_A - \varepsilon_m)}{a^2}\left(z_a - \frac{l}{2}\right)^2, \quad \frac{l}{2} \le z_a \le a + \frac{l}{2}, \tag{L2.97}$$

$$\varepsilon_b(z_b) = \varepsilon_m + \frac{(\varepsilon_B - \varepsilon_m)}{b^2}\left(z_b - \frac{l}{2}\right)^2, \quad \frac{l}{2} \le z_a \le b + \frac{l}{2}. \tag{L2.98}$$

In this particular case, there are no steps in the dielectric profile, and there is no change in slope at the interfaces with the medium (see Fig. L2.5).

In this case, with neglect of retardation and considering only the leading term in the dielectric differences, the integration over layers becomes

$$G(l; a, b) = -\frac{kT}{8\pi a^2 b^2} \sum_{n=0}^{\infty}{}' \frac{(\varepsilon_A - \varepsilon_m)(\varepsilon_B - \varepsilon_m)}{\varepsilon_m^2} \int_{l/2}^{b+l/2} \int_{l/2}^{a+l/2} \frac{z_a z_b}{(z_a + z_b)^2} dz_a dz_b$$

$$= -\frac{kT}{16\pi}\left\{\frac{l^2}{a^2 b^2} \ln\left[\frac{(l+a)(l+b)}{(l+a+b)\,l}\right] + \frac{1}{b^2}\ln\left(\frac{l+a+b}{l+a}\right) + \frac{1}{a^2}\ln\left(\frac{l+a+b}{l+b}\right) - \frac{1}{ab}\right\} \sum_{n=0}^{\infty}{}' \frac{(\varepsilon_A - \varepsilon_m)(\varepsilon_B - \varepsilon_m)}{\varepsilon_m^2}. \tag{L2.99}$$

[P.7.d.2]

At separations l much greater than layer thicknesses a and b, $G(l;a, b)$ reduces to the usual result for step-function changes in ε. More intriguing, in the $l \to 0$ limit of contact, the energy goes to a finite value:

$$G(l \to 0; a, b) = -\frac{kT}{16\pi}\left[\frac{\ln\left(1 + \frac{b}{a}\right)}{b^2} + \frac{\ln\left(1 + \frac{a}{b}\right)}{a^2} - \frac{1}{ab}\right] \sum_{n=0}^{\infty}{}' \frac{(\varepsilon_A - \varepsilon_m)(\varepsilon_B - \varepsilon_m)}{\varepsilon_m^2}. \tag{L2.100}$$

The pressure, the negative derivative with respect to separation l, goes as

$$P(l; a, b) = \left.\frac{\partial G(l; a, b)}{\partial l}\right|_{a,b}$$

$$= -\frac{kT}{4\pi a^2 b^2} \sum_{n=0}^{\infty}{}' \frac{(\varepsilon_A - \varepsilon_m)(\varepsilon_B - \varepsilon_m)}{\varepsilon_m^2} \int_{l/2}^{b+l/2} \int_{l/2}^{a+l/2} \frac{z_b z_a}{(z_a + z_b)^3} dz_a dz_b. \tag{L2.101}$$

In the $l \to 0$ limit this pressure goes to a finite value:

$$P(l \to 0; a, b) = -\frac{kT}{8\pi}\frac{1}{ab(a+b)} \sum_{n=0}^{\infty}{}' \frac{(\varepsilon_A - \varepsilon_m)(\varepsilon_B - \varepsilon_m)}{\varepsilon_m^2}. \tag{L2.102}$$

Other continuous profiles in ε produce similarly intriguing behaviors. The nondivergence of free energy and of pressure, qualitatively different from the power-law divergences in Lifshitz theory, occurs here when there is no discontinuity in ε itself or its z derivative. Deeper consideration of such behaviors would require going beyond macroscopic-continuum language.

PROBLEM L2.2: The limiting finite pressure in Eq. (L2.102) merits further consideration. Show that it comes (1) from the derivative of $G(l; \mathrm{a}, \mathrm{b})$, Eq. (L2.99), in the $l \to 0$ limit and (2) from the integral for $P(l; \mathrm{a}, \mathrm{b})$, Eq. (L2.101) in that same zero-l limit.

L2.3.C. The Derjaguin transform for interactions between oppositely curved surfaces

In 1934, B. V. Derjaguin showed how the interaction between two spheres or between a sphere and a plane near contact could be derived from the interaction between facing plane-parallel surfaces.[1] There were two conditions:

- Distance of closest approach l had to be small compared with the radii of curvature R_1, R_2. That is, the separation should not vary significantly over an area of interaction.
- Interaction energy had to be localized enough that the interactions in one patch did not perturb other places on the surface (see Fig. L2.6).

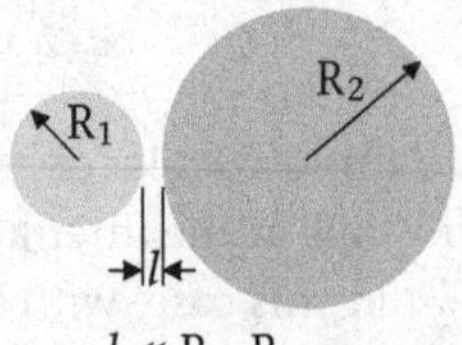

Figure L2.6

Schematically the transform is seen as a series of steps on a curved surface (see Fig. L2.7). The distance between facing patches grows from its minimum l by a rate that depends on the radius of curvature (see Fig. L2.8).

Specifically, write the distance between patches as $h = l + R_1(1 - \cos\theta_1) + R_2(1 - \cos\theta_2)$. Because the radii are much greater than l, and because the interaction between planar surfaces decays at a rate greater than or equal to $1/l^2$, there will be important contributions to the interaction only for small θ_1, θ_2. The operative range of the cosine functions is such that they can be approximated by $\cos\theta = 1 - \theta^2/2$.

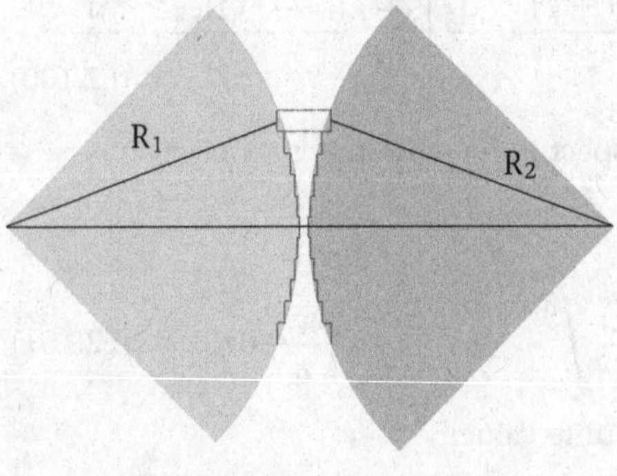

Figure L2.7

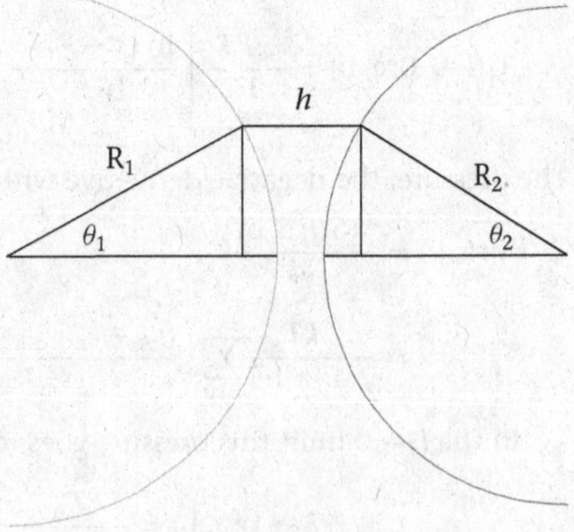

Figure L2.8

Sphere–sphere interactions

The area of a patch at a distance h goes as $(2\pi \text{ R} \sin\theta)\text{R}\, d\theta$ on each of the spheres. Adding up the interactions between facing patches involves an integration over the two angles θ_1 and θ_2 constrained by the requirement that facing patches be at the same distance, $(\text{R}_1 \sin\theta_1) = (\text{R}_2 \sin\theta_2)$, from the main axis connecting the centers of the two spheres. Again, because θ is small, $\sin\theta$ can be approximated by θ. The constraint becomes $\text{R}_1\,\theta_1 = \text{R}_2\,\theta_2$ or $\theta_2 = (\text{R}_1/\text{R}_2)\,\theta_1$. In this small-angle limit, the areas of the two patches are necessarily the same, $(2\pi\text{ R}_1 \sin\theta_1)\text{R}_1 d\theta_1 = (2\pi \text{ R}_2 \sin\theta_2)\text{R}_2\, d\theta_2$. The separation between patches can be written as

$$\begin{aligned} h &= l + \text{R}_1(1-\cos\theta_1) + \text{R}_2(1-\cos\theta_2) \approx l + \text{R}_1\,(\theta_1^2/2) + \text{R}_2\,(\theta_2^2/2) \\ &= l + \text{R}_1\,(\theta_1^2/2) + (\text{R}_1^2/\text{R}_2)\,(\theta_1^2/2) \\ &= l + (\text{R}_1/2)[1 + (\text{R}_1/\text{R}_2)]\theta_1^2. \end{aligned} \quad \text{(L2.103)}$$

The integration over oppositely curved spherical surfaces is the per-unit-area interaction energy $G_{\text{pp}}(h)$ between planes weighted by the area

$$\begin{aligned} (2\pi\text{ R}_1 \sin\theta_1)\,(\text{R}_1\, d\theta_1) &= 2\pi\text{ R}_1^2\,\theta_1\, d\theta_1 \\ &= \pi\text{ R}_1^2\, d\theta_1^2 \text{ or } \pi\text{ R}_2^2\, d\theta_2^2. \end{aligned} \quad \text{(L2.104)}$$

Because of the presumed rapid convergence of $G_{\text{pp}}(h)$ in h, the angular variable of integration, written here as $t = \theta_1^2$, can be taken over 0 to ∞. For succinctness write

$$h = l + \alpha t, \quad \alpha = (\text{R}_1/2)[1 + (\text{R}_1/\text{R}_2)]. \quad \text{(L2.105)}$$

The energy of interaction between spheres becomes $\pi\text{R}_1^2 \int_0^\infty G_{\text{pp}}(l+\alpha t)\text{d}t$. The *force* between spheres, $F(l;\text{R}_1,\text{R}_2)$, is the negative derivative of energy with respect to separation l. This is still an integral in t. For this reason there is a factor $1/\alpha$ introduced by integration:

$$F_{\text{ss}}(l;\text{R}_1,\text{R}_2) = -\pi\text{R}_1^2\int_0^\infty G'_{\text{pp}}(l+\alpha t)\text{d}t = \frac{\pi\text{R}_1^2}{\alpha}G_{\text{pp}}(l) = \frac{2\pi\text{R}_1\text{R}_2}{(\text{R}_1+\text{R}_2)}G_{\text{pp}}(l). \quad \text{(L2.106)}$$

With the Lifshitz expression

$$G_{\text{pp}}(l) = -\frac{kT}{8\pi l^2}\sum_{n=0}^{\infty}{}' r_n^2 \sum_{q=1}^{\infty}\frac{1}{q}\int_1^\infty p\left[(\overline{\Delta}_{\text{Am}}\overline{\Delta}_{\text{Bm}})^q + (\Delta_{\text{Am}}\Delta_{\text{Bm}})^q\right]e^{-r_n pq}\text{d}p, \quad \text{(L2.107)}$$

[P.1.a.1]

the sphere–sphere force becomes

$$F_{\text{ss}}(l;\text{R}_1,\text{R}_2) = -\frac{kT}{4l^2}\frac{\text{R}_1\text{R}_2}{\text{R}_1+\text{R}_2}\sum_{n=0}^{\infty}{}' r_n^2\sum_{q=1}^{\infty}\frac{1}{q}\int_1^\infty p\left[(\overline{\Delta}_{\text{Am}}\overline{\Delta}_{\text{Bm}})^q + (\Delta_{\text{Am}}\Delta_{\text{Bm}})^q\right]e^{-r_n pq}\text{d}p.$$

(L2.108)

[S.1.a]

When $\text{R}_2 = \text{R}_1 = \text{R}$, this force goes to a simple form:

$$F_{\text{ss}}(l;\text{R}) = \pi\text{R}G_{\text{pp}}(l). \quad \text{(L2.109)}$$

When $\text{R}_2 \to \infty$, $\text{R}_1 = \text{R}$, the case of a sphere interacting with a plane, the force doubles to

$$F_{\text{sp}}(l;\text{R}) = 2\pi\text{R}G_{\text{pp}}(l) \quad \text{(L2.110)}$$

Remarkably, the connection between a force between spheres and an energy between planes holds for the fullest expression of $G_{pp}(l)$. In the regime of large radii compared with minimal separation, this relation can also hold for other curved surfaces or protrusions having gradual local spherical curvature.

In the particular case of spheres, the force $F_{ss}(l;\mathrm{R}_1,\mathrm{R}_2)$ can be integrated to give the free energy $G_{ss}(l;\mathrm{R}_1,\mathrm{R}_2)$ in this same limit of $l \ll \mathrm{R}_1,\mathrm{R}_2$:

$$G_{ss}(l;\mathrm{R}_1,\mathrm{R}_2) = -\int_\infty^l F_{ss}(l;\mathrm{R}_1,\mathrm{R}_2)\mathrm{d}l = -\frac{2\pi \mathrm{R}_1\mathrm{R}_2}{\mathrm{R}_1+\mathrm{R}_2}\int_\infty^l G_{pp}(l)\mathrm{d}l. \tag{L2.111}$$

By use of the form

$$G_{pp}(l) = G_{\mathrm{AmB}}(l)$$

$$= -\frac{kT}{8\pi l^2}\sum_{n=0}^{\infty}{}' r_n^2 \sum_{q=1}^{\infty}\frac{1}{q}\int_1^\infty p\left[(\overline{\Delta}_{\mathrm{Am}}\overline{\Delta}_{\mathrm{Bm}})^q + (\Delta_{\mathrm{Am}}\Delta_{\mathrm{Bm}})^q\right]e^{-r_n pq}\mathrm{d}p, \tag{L2.112}$$

[P.1.a.1]

$r_n = (2l\varepsilon_{\mathrm{m}}^{1/2}\mu_{\mathrm{m}}^{1/2}/c)\xi_n$, the only dependence on separation l is in the exponential factor $e^{-r_n pq}$. The necessary integration over separation l amounts only to

$$\int_\infty^l e^{-\gamma q l}\mathrm{d}l = -\frac{e^{-\gamma q l}}{\gamma q} = -\frac{e^{-r_n pq}}{r_n pq}l \quad [\text{with } \gamma \equiv r_n p/l = (2\varepsilon_{\mathrm{m}}^{1/2}\mu_{\mathrm{m}}^{1/2}/c)\xi_n p].$$

Then, with subscripts 1 and 2 rather than A and B, take

$$\int_\infty^l G_{pp}(l)\mathrm{d}l = \frac{kT}{8\pi l}\sum_{n=0}^{\infty}{}' r_n \sum_{q=1}^{\infty}\frac{1}{q^2}\int_1^\infty \left[(\overline{\Delta}_{1\mathrm{m}}\overline{\Delta}_{2\mathrm{m}})^q + (\Delta_{1\mathrm{m}}\Delta_{2\mathrm{m}})^q\right]e^{-r_n pq}\mathrm{d}p,$$

and the sphere–sphere interaction free energy becomes

$$G_{ss}(l;\mathrm{R}_1,\mathrm{R}_2) = -\frac{kT}{4l}\frac{\mathrm{R}_1\mathrm{R}_2}{\mathrm{R}_1+\mathrm{R}_2}\sum_{n=0}^{\infty}{}' r_n \sum_{q=1}^{\infty}\frac{1}{q^2}\int_1^\infty \left[(\overline{\Delta}_{1\mathrm{m}}\overline{\Delta}_{2\mathrm{m}})^q + (\Delta_{1\mathrm{m}}\Delta_{2\mathrm{m}})^q\right]e^{-r_n pq}\mathrm{d}p. \tag{L2.113}$$

[S.1.b]

In the limit at which the velocity of light is effectively infinite, where r_n is effectively zero, the integral does not converge until p goes to infinitely large values. The [ ] term comes out of the p integration because $s_{\mathrm{i}} = \sqrt{p^2 - 1 + (\varepsilon_{\mathrm{i}}\mu_{\mathrm{i}}/\varepsilon_{\mathrm{m}}\mu_{\mathrm{m}})} \to p$ to make $\overline{\Delta}_{\mathrm{ji}} = [(\varepsilon_{\mathrm{j}} - \varepsilon_{\mathrm{i}})/(\varepsilon_{\mathrm{j}} + \varepsilon_{\mathrm{i}})]$, $\Delta_{\mathrm{ji}} = [(\mu_{\mathrm{j}} - \mu_{\mathrm{i}})/(\mu_{\mathrm{j}} + \mu_{\mathrm{i}})]$. From the corresponding form of the Lifshitz result,

$$G_{pp}(l \to 0, T) \to \frac{kT}{8\pi l^2}\sum_{n=0}^{\infty}{}' \sum_{q=1}^{\infty}\frac{(\overline{\Delta}_{1\mathrm{m}}\overline{\Delta}_{2\mathrm{m}})^q + (\Delta_{1\mathrm{m}}\Delta_{2\mathrm{m}})^q}{q^3}$$

$$G_{ss}(l;\mathrm{R}_1,\mathrm{R}_2) = -\frac{2\pi\mathrm{R}_1\mathrm{R}_2}{\mathrm{R}_1+\mathrm{R}_2}\int_\infty^l G_{pp}(l)\mathrm{d}l \tag{L2.114}$$

[P.1.a.3]

yields

$$G_{ss}(l;\mathrm{R}_1,\mathrm{R}_2) = -\frac{kT}{4l}\frac{\mathrm{R}_1\mathrm{R}_2}{\mathrm{R}_1+\mathrm{R}_2}\sum_{n=0}^{\infty}{}' \sum_{q=1}^{\infty}\frac{(\overline{\Delta}_{1\mathrm{m}}\overline{\Delta}_{2\mathrm{m}})^q + (\Delta_{1\mathrm{m}}\Delta_{2\mathrm{m}})^q}{q^3}. \tag{L2.115}$$

[S.1.c]

Parallel cylinders

The linear "areas" of facing patches at separation h go as $R_1\, d\theta_1 = R_2\, d\theta_2$ per unit length on each of the cylinders. As with spheres, we have $(R_1 \sin\theta_1) = (R_2 \sin\theta_2)$ and, writing $\theta = \theta_1$, $h = l + \alpha\theta^2$ with $\alpha = (R_1/2)[1 + (R_1/R_2)]$.

The integration over oppositely curved cylindrical surfaces is the energy of interaction $G_{pp}(h)$ per unit area between planes, weighted here by $R_1\, d\theta$, where θ^2 can be written as though going over an infinite range from $-\infty$ to $+\infty$. The energy of interaction between parallel cylinders becomes

$$G_{c\|c}(l; R_1, R_2) = R_1 \int_{-\infty}^{\infty} G_{pp}(l + \alpha\theta^2)\, d\theta.$$

Again as with spheres, use the form

$$G_{pp}(h) = -\frac{kT}{8\pi h^2} \sum_{n=0}^{\infty}{}' r_n^2 \sum_{n=0}^{\infty} \frac{1}{q} \int_1^{\infty} p\left[(\overline{\Delta}_{1m}\overline{\Delta}_{2m})^q + (\Delta_{1m}\Delta_{2m})^q\right] e^{-r_n p}\, dp.$$

[P.1.a.1]

The only dependence on separation h is in the exponential factor $e^{-r_n pq}$, through $r_n = (2l\varepsilon_m^{1/2}\mu_m^{1/2}/c)\xi_n$. Integration over θ,

$$R_1 \int_{-\infty}^{\infty} e^{-\gamma q h(\theta)}\, d\theta = R_1 \int_{-\infty}^{\infty} e^{-\gamma q(l+\alpha\theta^2)}\, d\theta = R_1 e^{-\gamma q l} \sqrt{\frac{\pi}{\alpha\gamma q}} = \sqrt{\frac{2\pi R_1 R_2}{R_1 + R_2}}\, l^{1/2} \frac{e^{-r_n pq}}{\sqrt{r_n pq}},$$

gives the energy of interaction per unit length between closely approaching parallel cylinders:

$$G_{c\|c}(l; R_1, R_2) = -\sqrt{\frac{2\pi R_1 R_2}{R_1 + R_2}} \frac{kT}{8\pi l^{3/2}} \sum_{n=0}^{\infty}{}' r_n^{3/2} \sum_{q=1}^{\infty} \frac{1}{q} \int_1^{\infty} p\left[(\overline{\Delta}_{1m}\Delta_{2m})^q + (\overline{\Delta}_{1m}\Delta_{2m})^q\right] \frac{e^{-r_n pq}}{\sqrt{pq}}\, dp.$$

(L2.116)
[C.1.b]

Because all actual l dependence resides only in the exponential $e^{-r_n pq} = e^{-\gamma q l}$, the force per unit length $F_{c\|c}(l; R_1, R_2)$ between cylinders is only a negative derivative that introduces a factor $\gamma q = r_n pq/l$ so that

$$F_{c\|c}(l; R_1, R_2) = -\sqrt{\frac{2\pi R_1 R_2}{R_1 + R_2}} \frac{kT}{8\pi l^{5/2}} \sum_{n=0}^{\infty}{}' r_n^{5/2} \sum_{q=1}^{\infty} \frac{1}{q^{1/2}} \times \int_1^{\infty} p^{3/2}\left[(\overline{\Delta}_{1m}\Delta_{2m})^q + (\overline{\Delta}_{1m}\Delta_{2m})^q\right] e^{-r_n pq}\, dp. \qquad (L2.117)$$

[C.1.a]

When $R_2 = R_1 = R$, $\{[(2\pi R_1 R_2)/(R_1 + R_2)]^{1/2}\} = \sqrt{\pi R}$. When $R_2 \to \infty$, $R_1 = R$, cylinder-with-a-plane, $\{[(2\pi R_1 R_2)/(R_1 + R_2)]^{1/2}\} = \sqrt{2\pi R}$.

In the nonretarded limit $G_{c\|c}(l; R_1, R_2) = R_1 \int_{-\infty}^{\infty} G_{pp}(l + \alpha\theta^2)\, d\theta$, $\int_{-\infty}^{\infty}[dx/(1 + x^2)^2] = (\pi/2)$, $(\alpha/R_1^2) = [(R_1 + R_2)/(2R_1R_2)]$,

$$G_{pp}(l \to 0, T) = -\frac{kT}{8\pi l^2} \sum_{n=0}^{\infty}{}' \sum_{q=1}^{\infty} \frac{(\overline{\Delta}_{1m}\overline{\Delta}_{2m})^q + (\Delta_{1m}\Delta_{2m})^q}{q^3} \qquad [P.1.a.3]$$

give

$$G_{c\|c}(l;R_1,R_2) = -\sqrt{\frac{2R_1R_2}{R_1+R_2}}\frac{kT}{16l^{3/2}}\sum_{q=1}^{\infty}\frac{\left[(\overline{\Delta}_{1m}\overline{\Delta}_{2m})^q + (\Delta_{1m}\Delta_{2m})^q\right]}{q^2} \qquad \text{(L2.118) [C.1.c.1]}$$

Perpendicular cylinders of equal radius R

The contour of the two surfaces is such that the distance h between them is the same function of angle as that between a sphere of radius R and a flat surface. Thus the force between two perpendicular cylinders is

$$F_{c\perp c}(l;R) = 2\pi R G_{pp}(l)$$

$$= -\frac{kTR}{4l^2}\sum_{n=0}^{\infty}{}' r_n^2 \sum_{q=1}^{\infty}\frac{1}{q}\int_1^{\infty} p\left[(\overline{\Delta}_{1m}\overline{\Delta}_{2m})^q + (\Delta_{1m}\Delta_{2m})^q\right]e^{-r_n p q}\,\mathrm{d}p. \qquad \text{(L2.119) [C.2.a]}$$

The free energy of their interaction is

$$G_{c\perp c}(l;R) = -\frac{kTR}{4l}\sum_{n=0}^{\infty}{}' r_n \sum_{q=1}^{\infty}\frac{1}{q^2}\int_1^{\infty}\left[(\overline{\Delta}_{1m}\overline{\Delta}_{2m})^q + (\Delta_{1m}\Delta_{2m})^q\right]e^{-r_n p q}\,\mathrm{d}p. \qquad \text{(L2.120) [C.2.b]}$$

L2.3.D. Hamaker approximation: Hybridization to modern theory

Even today, many people speak casually of Hamaker "constants" and about van der Waals forces in the pairwise-summation language of H. C. Hamaker's influential 1937 paper.[2] Fortunately the modern theory shows us the conditions under which some of that appealing language can be preserved as we accurately estimate force magnitudes. When differences in material susceptibilities are small and when separations are small enough to ignore retardation screening, then pairwise-summation language can be grafted onto modern thinking.

In fact, the graft is exceedingly helpful for geometries in which field equations of the modern theory are too difficult to solve but pairwise summation (actually integration) can be effected. The distance dependence of the interaction is taken from summation whereas the Hamaker *coefficient* is estimated with modern theory. To see how to connect old and new, consider the formal procedure for summation, then see its equivalence to a much-reduced version of the general theory.

Imagine the same two semi-infinite planar bodies for which the Lifshitz formulation was first carried out (see Fig. L2.9).

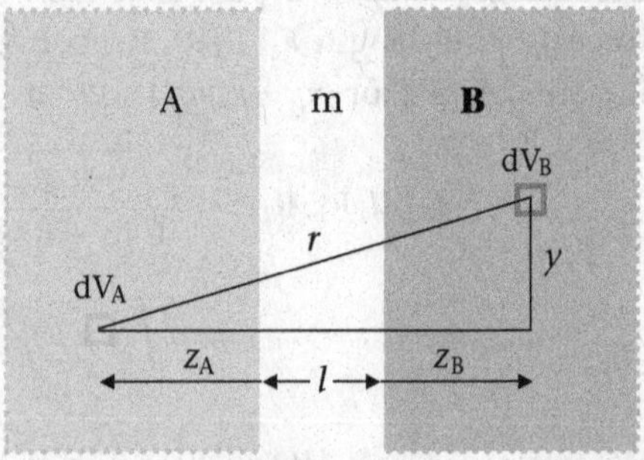

Figure L2.9

The Hamaker idea is to sum the interactions between atoms in the two incremental volumes $\mathrm{d}V_A$ and $\mathrm{d}V_B$. If these atoms are packed at number densities N_A and N_B, then there will be $N_A\mathrm{d}V_A \times N_B\mathrm{d}V_B$ individual interactions at the separation r. These individual *atomic* interactions go as

$$-\frac{c_A c_B}{r^6}. \qquad \text{(L2.121)}$$

The minus sign in $-(c_A c_B/r^6)$ makes explicit that the energy of interaction between like particles is negative. The coefficients in the numerator are those appropriate to the interaction of point particles in vacuum. Because $-(c_A c_B/r^6)$ is an energy of interaction, the coefficient $c_A c_B$ has units of *energy* × *length*^{6}.

The intent of summing individual r^{-6} interactions is to obtain the net interaction energy per unit area between the two half-spaces A and B across the gap l. For this reason we add up interactions between a patch of unit area on the face of A and integrate into A over the range $0 \le z_A < \infty$ but integrate over all of B, $0 \le z_B < \infty, 0 \le y < \infty$. Separation r varies as $r^2 = y^2 + (z_A + l + z_B)^2$. These z_A, z_B, y coordinates are chosen here to have positive values ranging from 0 to ∞ as shown in Fig. L2.9. For each value of y there is a circle of locations of circumference $2\pi y$ that corresponds to the same value of r; each value of y is weighted by a factor $2\pi y$. The integral then has the form

$$\int_0^\infty \int_0^\infty \int_0^\infty \frac{dz_A\, dz_B\, 2\pi y\, dy}{r^6}. \qquad \text{(L2.122)}$$

The interesting part of the interaction is of course its change with separation l. When l increases, a region of space occupied by material A or B is instead occupied by the medium m whose atoms experience their own van der Waals interactions with the atoms of A, B and m. Buoyancy! For interactions versus separation l, the material in each volume dV_A and dV_B in A and B is interesting only to the extent that it differs from the medium. If the medium is a vacuum, then the integral is taken over the quantity $(c_A N_A c_B N_B)/r^6$. If the medium is a material of atomic density N_m whose atom–atom interactions go as $c_m c_m/r^6$ with itself and as $c_A c_m/r^6$ and $c_B c_m/r^6$ with atoms of materials A and B, then it is necessary to subtract $N_m c_m$ from $N_A c_A$ and from $N_B c_B$. The effective coefficient of $1/r^6$ in the integral becomes the product of differences:

$$(N_A c_A - N_m c_m)(N_B c_B - N_m c_m) = N_A c_A N_B c_B - N_m c_m (N_A c_A + N_B c_B) + (N_m c_m)^2. \qquad \text{(L2.123)}$$

If the interaction density $N_m c_m$ of the medium were equal to *either* $N_A c_A$ or $N_B c_B$, there would be no interaction between the bodies A and B. If $N_m c_m$ were greater than either $N_A c_A$ or $N_B c_B$, the interaction would change sign. And if $N_m c_m$ were greater than both $N_A c_A$ and $N_B c_B$, the sign of the interaction would change back again.

If regions A and B were vacuum and region m not vacuum, then there would still be a finite interaction between A and B, an attraction just as in the Lifshitz formulation. This attraction indicates the preference of material m to be expelled from between two vacua, to be in an infinite medium of its own kind rather than to remain in a slab of finite thickness l.

It became customary to define a "Hamaker constant" A_{Ham} (sometimes also written in this text as A_H) which in the language of pairwise summation is

$$A_{Ham} = \pi^2 (N_A c_A - N_m c_m)(N_B c_B - N_m c_m). \qquad \text{(L2.124)}$$

In terms of this "constant," the effective incremental interaction between bits of each body is

$$-\frac{A_{Ham}}{\pi^2} \frac{dV_1\, dV_2}{r^6}. \qquad \text{(L2.125)}$$

Integration over half-spaces A and B,

$$\int_0^\infty \int_0^\infty \int_0^\infty \frac{dz_A\, dz_B 2\pi y dy}{r^6} = \pi \int_0^\infty dz_A \int_0^\infty dz_B \int_0^\infty \frac{dy^2}{\left[y^2 + (z_A + z_B + l)^2\right]^3}$$

$$= \frac{\pi}{2} \int_0^\infty dz_A \int_0^\infty \frac{dz_B}{(z_A + z_B + l)^4} = \frac{\pi}{12l^2},$$

gives an energy per area

$$E(l) = -\frac{A_{Ham}}{12\pi l^2}, \quad \text{(L2.126)}$$

with A_{Ham} in units of energy and the minus sign written separately to make explicit that there is always attraction between like materials. (Infinitely extended parallel bodies will necessarily have infinite energies. Energy per area or per length is the logical measure of strength.)

The pressure, or force per unit area or energy per displaced volume, is the derivative

$$P(l) = -\frac{\partial E(l)}{\partial l} = -\frac{A_{Ham}}{6\pi l^3}, \quad \text{(L2.127)}$$

negative, attractive between like bodies.

Because of the form $A_{Ham} = \pi^2(N_A c_A - N_m c_m)(N_B c_B - N_m c_m)$ [Eq. (L2.124)], it is tempting to think of the interaction between unlike materials A and B across m as though it were the geometric mean of A–A and B–B interactions:

$$A_{Am/Am} = \pi^2(N_A c_A - N_m c_m)^2 \quad \text{for A to A across m;}$$
$$A_{Bm/Bm} = \pi^2(N_B c_B - N_m c_m)^2 \quad \text{for B to B across m;}$$
$$A_{Am/Bm} = \pi^2(N_A c_A - N_m c_m)(N_B c_B - N_m c_m) \quad \text{for A to B across m;} \quad \text{(L2.128)}$$

so that

$$A_{Am/Bm} = (A_{Am/Am} A_{Bm/Bm})^{1/2}. \quad \text{(L2.129)}$$

Because a geometric mean is always less than or equal to an arithmetic mean,

$$(A_{Am/Am} A_{Bm/Bm})^{1/2} \leq (A_{Am/Am} + A_{Bm/Bm})/2, \quad \text{(L2.130)}$$

two A–B attractions will always be weaker than or equal to an A–A plus a B–B attraction:

$$2E_{Am/Bm} = -\frac{A_{Am/Bm}}{12\pi l^2} \geq -\frac{A_{Am/Am}}{12\pi l^2} - \frac{A_{Bm/Bm}}{12\pi l^2} = (E_{Am/Am} + E_{Bm/Bm}). \quad \text{(L2.131)}$$

In fact, $E_{Am/Bm}$ can be repulsive when $N_A c_A > N_m c_m > N_B c_B$ or $N_A c_A < N_m c_m < N_B c_B$. $E_{Am/Am}$ and $E_{Bm/Bm}$ are *always* negative.

Connection between the Hamaker pairwise-summation picture and the modern theory

When modern theory is restricted to the limits at which all relativistic retardation is neglected and differences in the dielectric susceptibilities are small, the interaction between half-spaces (omitting magnetic terms) goes as

$$G_{Am/Bm}(l, T) \approx -\frac{kT}{8\pi l^2} \sum_{n=0}^{\infty}{}' \overline{\Delta}_{Am} \overline{\Delta}_{Bm}. \quad \text{(L2.132)}$$

[P.1.a.4]

In the Hamaker form,

$$G_{\mathrm{Am/Bm}}(l,T) = -\frac{A_{\mathrm{Am/Bm}}}{12\pi l^2} \tag{L2.133}$$
[P.1.a.4]

$$A_{\mathrm{Am/Bm}} \approx +\frac{3kT}{2}\sum_{n=0}^{\infty}{}' \overline{\Delta}_{\mathrm{Am}}\overline{\Delta}_{\mathrm{Bm}}. \tag{L2.134}$$

How does Hamaker pairwise summation emerge from reduction of the Lifshitz theory?

At low atomic densities N, the dielectric response of a medium can be written in the popular Clausius–Mossotti or Lorentz–Lorenz form as a function of N and a coefficient α that includes atomic or molecular polarizability:

$$\varepsilon \approx \frac{1+2N\alpha/3}{1-N\alpha/3}. \tag{L2.135}$$

This form is valid for gases up to high pressure. For dilute gases $N\alpha/3$ is so small that the ε can be approximated by a linear form $\varepsilon \to 1 + N\alpha$. In this case, in which $N\alpha \ll 1$, it is possible to write $\overline{\Delta}_{\mathrm{Am}} = [(\varepsilon_{\mathrm{A}} - \varepsilon_{\mathrm{m}})/(\varepsilon_{\mathrm{A}} + \varepsilon_{\mathrm{m}})]$, $\overline{\Delta}_{\mathrm{Bm}} = [(\varepsilon_{\mathrm{B}} - \varepsilon_{\mathrm{m}})/(\varepsilon_{\mathrm{B}} + \varepsilon_{\mathrm{m}})$ as

$$\overline{\Delta}_{\mathrm{Am}} = \frac{N_{\mathrm{A}}\alpha_{\mathrm{A}} - N_{\mathrm{m}}\alpha_{\mathrm{m}}}{2}, \quad \overline{\Delta}_{\mathrm{Bm}} = \frac{N_{\mathrm{B}}\alpha_{\mathrm{B}} - N_{\mathrm{m}}\alpha_{\mathrm{m}}}{2}. \tag{L2.136}$$

In this limit and *only* in this limit in which each of the materials A, m and B can be considered a dilute gas, does the Hamaker pairwise-summation limit agree rigorously with the modern theory. Compare [Eqs. (L2.123) and (L2.124)].

$$\begin{aligned} A_{\mathrm{Am/Bm}} &\equiv \pi^2(N_{\mathrm{A}}c_{\mathrm{A}} - N_{\mathrm{m}}c_{\mathrm{m}})(N_{\mathrm{B}}c_{\mathrm{B}} - N_{\mathrm{m}}c_{\mathrm{m}}) \\ &= \pi^2\left[N_{\mathrm{A}}c_{\mathrm{A}}N_{\mathrm{B}}c_{\mathrm{B}} - N_{\mathrm{m}}c_{\mathrm{m}}(N_{\mathrm{A}}c_{\mathrm{A}} + N_{\mathrm{B}}c_{\mathrm{B}}) + (N_{\mathrm{m}}c_{\mathrm{m}})^2\right] \end{aligned}$$

with

$$\begin{aligned} A_{\mathrm{Am/Bm}} &\approx +\frac{3kT}{2}\sum_{n=0}^{\infty}{}' \frac{N_{\mathrm{A}}\alpha_{\mathrm{A}} - N_{\mathrm{m}}\alpha_{\mathrm{m}}}{2}\frac{N_{\mathrm{B}}\alpha_{\mathrm{B}} - N_{\mathrm{m}}\alpha_{\mathrm{m}}}{2}, \\ &= +\frac{3kT}{8}\sum_{n=0}^{\infty}{}' (N_{\mathrm{A}}N_{\mathrm{B}}\alpha_{\mathrm{A}}\alpha_{\mathrm{B}} - N_{\mathrm{A}}N_{\mathrm{m}}\alpha_{\mathrm{A}}\alpha_{\mathrm{m}} - N_{\mathrm{B}}N_{\mathrm{m}}\alpha_{\mathrm{B}}\alpha_{\mathrm{m}} + N_{\mathrm{m}}^2\alpha_{\mathrm{m}}^2). \end{aligned} \tag{L2.137}$$

The two coincide when the pairwise interaction coefficients $c_{\mathrm{A}}c_{\mathrm{B}}$, $c_{\mathrm{A}}c_{\mathrm{m}}$, $c_{\mathrm{B}}c_{\mathrm{m}}$, and c_{m}^2 are evaluated as $c_{\mathrm{i}}c_{\mathrm{j}} = (3kT/8\pi^2)\sum_{n=0}^{\prime\infty} \alpha_{\mathrm{i}}\alpha_{\mathrm{j}}$. The inequality $2E_{\mathrm{Am/Bm}} \geq (E_{\mathrm{Am/Am}} + E_{\mathrm{Bm/Bm}})$ of the pure Hamaker form is preserved, but the geometric mean that creates the inequality holds only for the individual terms in the summation over frequencies $\sum_{n=0}^{\prime\infty}$. The total free energy of interaction $G_{\mathrm{Am/Bm}}(l,T)$ is *not* the geometric mean of $G_{\mathrm{Am/Am}}(l,T)$ and $G_{\mathrm{Bm/Bm}}(l,T)$.

PROBLEM L2.3: Instead of the limiting form $\varepsilon \to 1 + N\alpha$, use the Clausius–Mossotti expression $\varepsilon \approx [(1+2N\alpha/3)/(1-N\alpha/3)]$ [approximation (L2.135)] in expression (L2.138) for the interaction of two condensed gases across a vacuum $\varepsilon_{\mathrm{m}} = 1$. Then, $\overline{\Delta}_{\mathrm{Am}} = \overline{\Delta}_{\mathrm{Bm}} = [(\varepsilon - 1)/(\varepsilon + 1)$. Show that the result is a power series in density N in which the corrections to the $N^2\alpha^2$ leading term come in as successive factors $N\alpha/3$, and then $49N^2\alpha^2/288$.

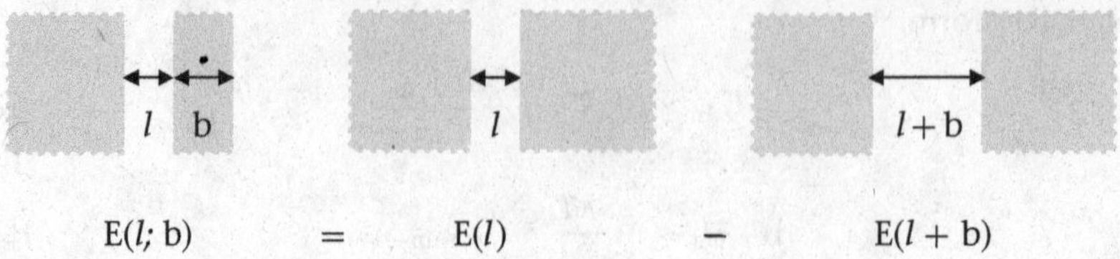

Figure L2.10

Hybridization of the Hamaker and Lifshitz formulations

Rather than think in terms of joining the two formulations in the dilute-gas limit, it is possible to reduce the general form to one that superficially looks like that limit. Recall that when retardation and magnetic susceptibilities are ignored the Lifshitz result (omitting magnetic terms) becomes

$$G_{\mathrm{Am/Bm}}(l, T) \to -\frac{kT}{8\pi l^2} \sum_{n=0}^{\infty}{}' \sum_{q=1}^{\infty} \frac{\left(\overline{\Delta}_{\mathrm{Am}}\overline{\Delta}_{\mathrm{Bm}}\right)^q}{q^3}. \tag{L2.138}$$

[P.1.a.3]

When the differences $\varepsilon_A - \varepsilon_m$ and $\varepsilon_B - \varepsilon_m$ are much less than ε_m, then only the $q = 1$ term in the summation is significant and

$$\overline{\Delta}_{\mathrm{Am}} = \frac{\varepsilon_A - \varepsilon_m}{\varepsilon_A + \varepsilon_m} \approx \frac{\varepsilon_A - \varepsilon_m}{2\varepsilon_m}, \quad \overline{\Delta}_{\mathrm{Bm}} = \frac{\varepsilon_B - \varepsilon_m}{\varepsilon_B + \varepsilon_m} \approx \frac{\varepsilon_B - \varepsilon_m}{2\varepsilon_m}. \tag{L2.139}$$

The causative part of $G_{\mathrm{Am/Bm}}(l, T)$ is merely the sum of products $(\varepsilon_A - \varepsilon_m)(\varepsilon_B - \varepsilon_m)$ formally (but only formally!) like the product of differences $(N_A c_A - N_m c_m)(N_B c_B - N_m c_m)$ in pairwise summation. It is as though the electromagnetic waves that constitute the electrodynamic force were waves in a medium m that suffered only small perturbations because of the small difference between ε_A, ε_B and ε_m. It is *not* because the atoms in the different media see each other individually, as imagined in pairwise summation; the ε_i's are not proportional to the respective number density N_i.

In this small-difference limit then, at which it is accurate to compute the Hamaker coefficient as

$$A_{\mathrm{Am/Bm}} \approx +\frac{3kT}{2} \sum_{n=0}^{\infty}{}' \overline{\Delta}_{\mathrm{Am}}\overline{\Delta}_{\mathrm{Bm}}, \tag{L2.140}$$

this coefficient can be grafted into the form of geometrical variation that comes from Hamaker summation. This assumption can be rigorously evaluated in those cases in which a full Lifshitz solution exists. One such case is the interaction of planar slabs.

Hamaker summation for the case of a half-space A interacting with a finite slab of material B

For the interaction of planar slabs, the Hamaker approach entails integration over finite ranges of z_A or z_B. For the interaction between a half-space A and a parallel slab of B of finite thickness b, this procedure is equivalent to subtracting from $E(l) = -(A_{\mathrm{Ham}}/12\pi l^2)$ an amount $-[A_{\mathrm{Ham}}/12\pi(l + b)^2]$ (see Fig. L2.10). This subtraction yields a form equivalent to the equation of Table P.2.b.3 (see Fig. L2.11):

$$E(l; b) = -\frac{A_{\mathrm{Ham}}}{12\pi}\left[\frac{1}{l^2} - \frac{1}{(l + b)^2}\right] \tag{L2.141}$$

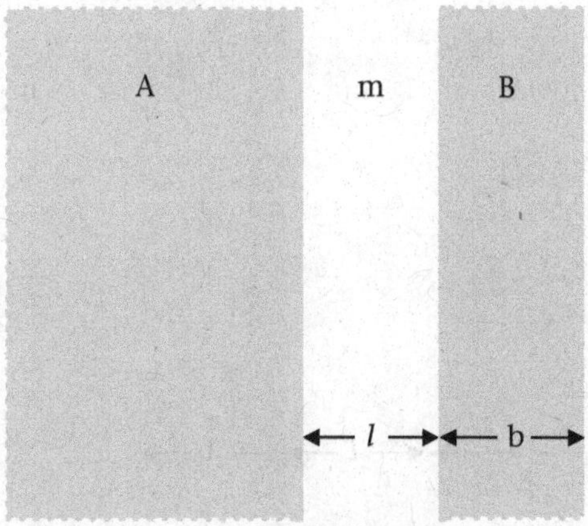

Figure L2.11

When the slab is thin compared with separation, $b \ll l$, $E(l; b)$ goes to an inverse-cube form that is linear in thickness b:

$$E(l; b) \approx -\frac{A_{Ham}\, b}{6\pi\, l^3}. \tag{L2.142}$$

PROBLEM L2.4: Show how thin-body formulae can often be derived either as expansions or as derivatives.

Hamaker summation for the case of a finite slab of material A interacting with a finite slab of material B

For the interaction of *two slabs of finite thickness*, a and b, the easiest procedure is to subtract again, this time taking $E(l; b) - E(l + a; b)$ (see Fig. L2.12):

$$E(l; a, b) = E(l; a, b) = -\frac{A_{Ham}}{12\pi}\left[\frac{1}{l^2} - \frac{1}{(l+b)^2} - \frac{1}{(l+a)^2} + \frac{1}{(l+a+b)^2}\right] \tag{L2.143}$$

(compare with the equation of Table P.3.c.3).

For slabs of equal thickness, $a = b$, this energy per area is

$$E(l; b) = -\frac{A_{Ham}}{12\pi}\left[\frac{1}{l^2} - \frac{2}{(l+b)^2} + \frac{1}{(l+2b)^2}\right]. \tag{L2.144}$$

When thickness b is much less than separation l, this turns into a fourth-power interaction whose magnitude goes as b^2, i.e., the product of the interacting masses:

$$E(l; b) \approx -\frac{A_{Ham}\, b^2}{2\pi\, l^4}. \tag{L2.145}$$

PROBLEM L2.5: Derive approximation (L2.145) by expansion of Eq. (L2.144) and by differentiation of $-[A_{Ham}/12\pi l^2]$ for the interaction of half-spaces.

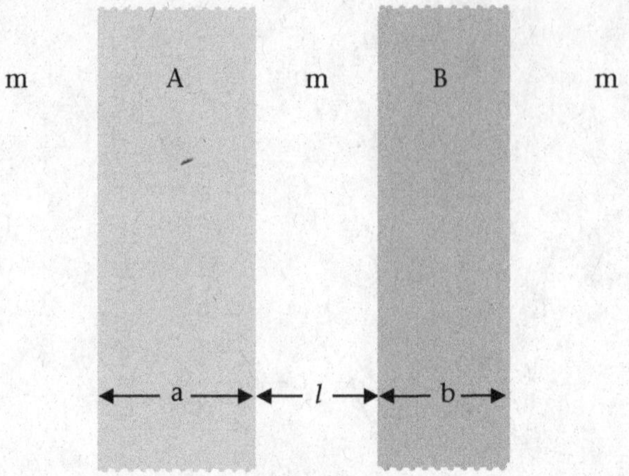

Figure L2.12

In most cases it is in fact much easier to proceed further in deriving results by means of the general theory rather than by pairwise summation. This is the place to see, though, that there is an easy merger of old and new languages. The Hamaker "constant" of fictitious pairwise summation metamorphoses, under strictly limited conditions, to the Hamaker "coefficient" of the Lifshitz theory. This is allowed as long as there are no great differences in dielectric susceptibilities and as long as relativistic retardation can be ignored. This Hamaker coefficient can be used as a prefactor for the spatially varying function that goes with the geometry of each interaction. It is in this spirit of combining old and new that we can write down a catalog of results for van der Waals interactions between particles of different shape.

L2.3.E. Point particles in dilute gases and suspensions

DIPOLAR INTERACTIONS, REDUCTION OF THE LIFSHITZ RESULT

It is of practical as well as ideological importance that the modern theory of van der Waals forces reduces to the older forms derived for the interaction of individual small molecules in dilute gases. The modern approach can in fact be used to derive new expressions for the interaction between pairs of solutes in dilute solutions. The essential property of ε in the dilute-gas or dilute-solution limit is that the dielectric response is strictly proportional to the number density of gas or solute molecules. That is, an electric field applied to a dilute gas or solution acts on each dilute species without distortion of the field by other gas or solute molecules.

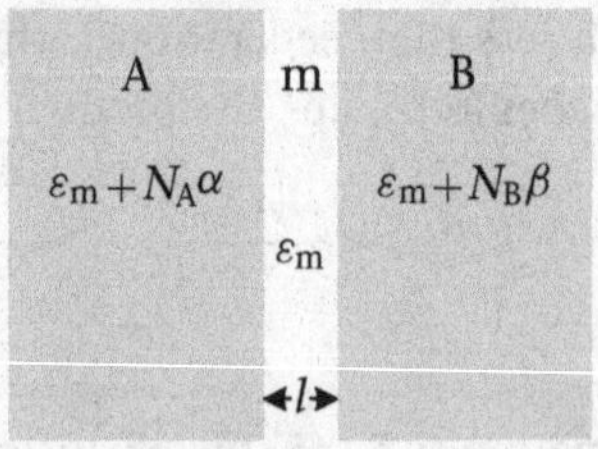

Imagine media A and B as dilute suspensions or gases ($\varepsilon_m = 1$) whose particle number densities N_A, N_B and polarizabilities are so small that susceptibilities can be written as $\varepsilon_A = \varepsilon_m + N_A\alpha$ and $\varepsilon_B = \varepsilon_m + N_B\beta$, quantities that deviate very slightly from ε_m of the pure medium. The incremental susceptibilities α and β are further specified later. Only proportionality to number densities N_A and N_B is important here.

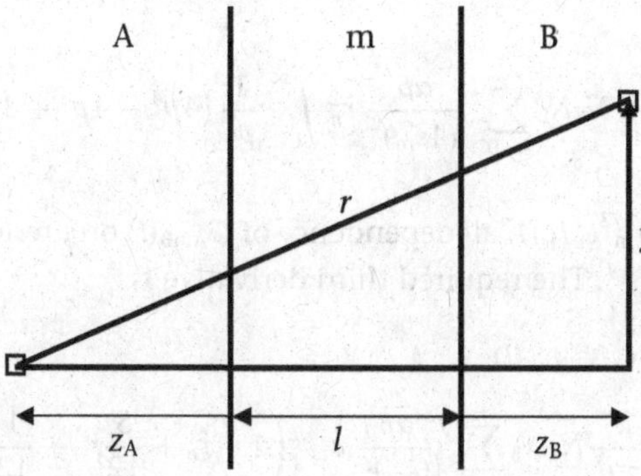

In this limit we can reverse the Hamaker summation procedure by reference to the previous section and regard the interaction energy per area $G_{AmB}(l)$ between the two regions as an integral over $g_{\alpha\beta}(r)$, the individual interaction between suspended particles. The separation between incremental volumes in regions A and B is $r = \sqrt{(z_A + l + z_B)^2 + {y_B}^2}$. The distance z_A is measured to the left from the A/m boundary and z_B is measured to the right from the B/m boundary. By pairwise summation,

$$G_{AmB}(l) = N_A N_B \int_0^\infty \int_0^\infty \int_0^\infty g_{\alpha\beta}\left(\sqrt{(z_A + l + z_B)^2 + y_B^2}\right) 2\pi\, y_B\, dy_B dz_B dz_A. \tag{L2.146}$$

Because $G_{AmB}(l)$ is an energy of interaction per unit area, the variables of integration sweep over all positions (y_B, z_B) on one side but sweep only in the z_A direction on the other side. The factor $2\pi y_B$ is there to include all positions that are equally distant from z_A at a given z_B and y_B.

To extract the function $g_{\alpha\beta}(r)$ buried under the three integrals, take three corresponding derivatives. There are a couple of neat maneuvers to do this:

- To extract a function $f(l)$ from under an integral of the form $\int_0^\infty f(l + x)dx$, take the derivative $(d/dl)\int_0^\infty f(l + x)dx = (d/dl)[F(\infty) - F(l)] = -f(l)$.
- To extract from under an integral of the form $\int_0^\infty f(l^2 + y^2)2y dy$, again take a derivative, $(d/dl)\int_0^\infty f(l^2 + y^2)2y dy = (d/dl)\int_0^\infty f(l^2 + q)dq = (d/dl)[F(\infty) - F(l^2)] = -2l f(l^2)$, and note the factor $2l$.

The right-hand side of Eq. (L2.146) becomes $2\pi l g_{\alpha\beta}(l)$, whereas the left-hand side becomes the third derivative of $G_{AmB}(l)$ with respect to l:

$$-G'''_{AmB}(l) = 2\pi l N_A N_B\, g_{\alpha\beta}(l).$$

Before taking derivatives, simplify the full Lifshitz expression,

$$G_{AmB}(l, T) = \frac{kT}{2\pi c^2} \sum_{n=0}^{\infty}{}' \varepsilon_m \mu_m \xi_n^2 \int_1^\infty p \ln\left[\left(1 - \overline{\Delta}_{Am}\overline{\Delta}_{Bm} e^{-r_n p}\right)\left(1 - \Delta_{Am}\Delta_{Bm} e^{-r_n p}\right)\right] dp,$$

in order to put it in tractable form and to reveal its newly pairwise summable nature.

Define the quantities $\overline{\Delta}_{\mathrm{Am}}$, $\overline{\Delta}_{\mathrm{Bm}}$ and Δ_{Am}, Δ_{Bm} using

$$s_{\mathrm{A}} \approx p + \frac{N_{\mathrm{A}}\alpha}{2p\varepsilon_{\mathrm{m}}}, \qquad s_{\mathrm{B}} \approx p + \frac{N_{\mathrm{B}}\beta}{2p\varepsilon_{\mathrm{m}}}, \tag{L2.147}$$

so that

$$G_{\mathrm{AmB}}(l, T) = -\frac{kT}{8\pi l^2} N_{\mathrm{A}} N_{\mathrm{B}} \sum_{n=0}^{\infty}{}' \frac{\alpha\beta}{(4\varepsilon_{\mathrm{m}})^2} r_n^2 \int_1^{\infty} \frac{1}{p^3}(4p^4 - 4p^2 + 2)e^{-r_n p}\mathrm{d}p. \tag{L2.148}$$

Because $r_n = (2\varepsilon_{\mathrm{m}}^{1/2}\mu_{\mathrm{m}}^{1/2}\xi_n/c)\,l$, all dependence of $G_{\mathrm{AmB}}(l)$ on l resides in the exponential $e^{-r_n p} = e^{-(2\varepsilon_{\mathrm{m}}^{1/2}\mu_{\mathrm{m}}^{1/2}\xi_n/c)\,pl}$. The required third derivative is

$$\begin{aligned} -G'''_{\mathrm{AmB}}(l) &= 2\pi l N_{\mathrm{A}} N_{\mathrm{B}}\, g_{\alpha\beta}(l) \\ &= -12\frac{kT}{\pi l^5} N_{\mathrm{A}} N_{\mathrm{B}} \sum_{n=0}^{\infty}{}' \frac{\alpha\beta}{(\varepsilon_{\mathrm{m}})^2} e^{-r_n}\left(1 + r_n + \frac{5}{12}r_n^2 + \frac{1}{12}r_n^3 + \frac{1}{48}r_n^4\right). \end{aligned} \tag{L2.149}$$

The interaction between pointlike particles emerges as

$$\begin{aligned} g_{\alpha\beta}(l) &= -\frac{6kT}{l^6}\sum_{n=0}^{\infty}{}' \frac{\alpha(i\xi_n)\beta(i\xi_n)}{[4\pi\varepsilon_{\mathrm{m}}(i\xi_n)]^2} e^{-r_n}\left(1 + r_n + \frac{5}{12}r_n^2 + \frac{1}{12}r_n^3 + \frac{1}{48}r_n^4\right) \\ &= -\frac{6kT}{l^6}\sum_{n=0}^{\infty}{}' \frac{\alpha(i\xi_n)\beta(i\xi_n)}{[4\pi\varepsilon_{\mathrm{m}}(i\xi_n)]^2} R_{\alpha\beta}(r_n), \end{aligned} \tag{L2.150}$$

[S.6.a]

with the screening function

$$R_{\alpha\beta}(r_n) \equiv e^{-r_n}\left(1 + r_n + \frac{5}{12}r_n^2 + \frac{1}{12}r_n^3 + \frac{1}{48}r_n^4\right). \tag{L2.151}$$

In the limit of low temperature, the summation in n is replaced with an integration in frequency and a factor $\hbar/(2\pi kT)$.

It is remarkable that the distance dependence of this result is that derived from the full quantum theory in 1948 by Casimir and Polder.[3]

By writing $\mu_{\mathrm{A}} = \mu_{\mathrm{m}} + N_{\mathrm{A}}\alpha_{\mathrm{M}}$ and $\mu_{\mathrm{B}} = \mu_{\mathrm{m}} + N_{\mathrm{B}}\beta_{\mathrm{M}}$ (subscript M for magnetic) as well as $\varepsilon_{\mathrm{A}} = \varepsilon_{\mathrm{m}} + N_{\mathrm{A}}\alpha_{\mathrm{E}}$ and $\varepsilon_{\mathrm{B}} = \varepsilon_{\mathrm{m}} + N_{\mathrm{B}}\beta_{\mathrm{E}}$, and by proceeding with the same expansion, we can extend the reduction of the Lifshitz result to the interaction of point particles to include magnetic susceptibilities.

PROBLEM L2.6: Because of the number of ways they can be used elsewhere, it is worth exercising the manipulations used to extract Eqs. (L2.150) and (L2.151) from the general form for $G_{\mathrm{AmB}}(l, T)$:

1. Ignore differences in magnetic susceptibilities, feed $\varepsilon_{\mathrm{A}} = \varepsilon_{\mathrm{m}} + N_{\mathrm{A}}\alpha_{\mathrm{E}}$ and $\varepsilon_{\mathrm{B}} = \varepsilon_{\mathrm{m}} + N_{\mathrm{B}}\beta_{\mathrm{E}}$ to $s_{\mathrm{A}} = \sqrt{p^2 - 1 + (\varepsilon_{\mathrm{A}}/\varepsilon_{\mathrm{m}})}$, $s_{\mathrm{B}} = \sqrt{p^2 - 1 + (\varepsilon_{\mathrm{B}}/\varepsilon_{\mathrm{m}})}$; expand to lowest powers in number densities so as to verify approximations (L2.147).

2. Similarly, introduce approximations (L2.147) into $\overline{\Delta}_{\mathrm{Am}}$, $\overline{\Delta}_{\mathrm{Bm}}$ and Δ_{Am}, Δ_{Bm} and expand in densities to verify Eq. (L2.148).

3. From here it is an easy trip, differentiating with respect to l and then integrating with respect to p so as to achieve Eqs. (L2.150) and (L2.151).

Nonretarded limit

In the limit $r_n = 0$, at which there are no retardation effects because of the finite velocity of light, the screening factor $R_{\alpha\beta}(r_n)$ goes to 1 and the small-particle interaction becomes

$$g_{\alpha\beta}(l) = -\frac{6kT}{l^6}\sum_{n=0}^{\infty}{}' \frac{\alpha(i\xi_n)\beta(i\xi_n)}{[4\pi\varepsilon_m(i\xi_n)]^2}, \tag{L2.152}$$

[S.6.b][L1.49]

with the familiar inverse-sixth power variation that one expects for van der Waals forces between point dipoles.

Fully retarded limit

In the opposite limit, at which distances are so large that $r_n = (2\varepsilon_m^{1/2}\mu_m^{1/2}\xi_n/c)\,l \gg 1$ and the wavelengths of the important fluctuating electric fields are small compared with particle separation, that is, where the frequency dependence in $\alpha(i\xi_n)\beta(i\xi_n)/\varepsilon_m(i\xi_n)^2$ is negligibly slow compared with the rate of change of the screening factor $R_{\alpha\beta}(r_n)$, $\alpha(i\xi_n)\beta(i\xi_n)/\varepsilon_m(i\xi_n)^2$ is treated as a constant, to be evaluated at zero frequency, so that the summation over frequencies has the form

$$\frac{6kT}{l^6}\frac{\alpha(0)\beta(0)}{[4\pi\varepsilon_m(0)]^2}\sum_{n=0}^{\infty}{}' R_{\alpha\beta}(r_n). \tag{L2.153}$$

[S.6.d]

At $T = 0$, but *only* at $T = 0$, the sum smoothes out into an integral over the index n to give

$$g_{\alpha\beta}(l) = -\frac{23\hbar c}{(4\pi)^3 l^7}\frac{\alpha(0)\beta(0)}{\varepsilon_m(0)^{5/2}}, \tag{L2.154}$$

[S.6.c]

which has an inverse-seventh-power variation.

PROBLEM L2.7: Show that, in the highly idealized limit of zero temperature, the terms in the sum $\sum_{n=0}^{\prime\infty} e^{-r_n}(\cdots)$ [Eq. (L2.151) and expression (L2.153)] change so slowly with respect to index n that the sum can be approximated by an integral $\int_0^{\infty}(\cdots)e^{-r_n}\,dn$. In this limit, derive Eq. (L2.154) with its apparently-out-of-nowhere factor of 23.

Separations greater than wavelength of first finite sampling frequency ξ_1

Set equal to zero all but the first term in $g_{\alpha\beta}(l) = -\frac{6kT}{l^6}\sum_{n=0}^{\prime\infty}\frac{\alpha(i\xi_n)\beta(i\xi_n)}{[4\pi\varepsilon_m(i\xi_n)]^2}\,e^{-r_n}(1 + r_n + \frac{5}{12}r_n^2 + \frac{1}{12}r_n^3 + \frac{1}{48}r_n^4)$ and multiply by 1/2 to account for the prime in summation:

$$g_{\alpha\beta}(l) = -\frac{3kT}{l^6}\frac{\alpha(0)\beta(0)}{[4\pi\varepsilon_m(0)]^2} \tag{L2.155}$$

[S.6.d]

If ever needed, it is trivial to add magnetic-fluctuation terms to all these point-particle results.

Specific cases

Time out for units Now we can consider more carefully the incremental change in the response of a medium to which small particles are added in dilute suspension. Units can be a nuisance, so put the facts first. *Focus on the physically important change in polarization $\partial\mathbf{P}/\partial N$ with the addition of one particle.* Then it is worthwhile to work slowly through the steps in translation.

In either unit system, define an incremental change in polarization by particle addition in the dilute, $N \to 0$ limit,

$$\left.\frac{\partial\mathbf{P}}{\partial N}\right|_{N=0} = \alpha_{\text{mks}}\mathbf{E}, \quad \left.\frac{\partial\mathbf{P}}{\partial N}\right|_{N=0} = \alpha_{\text{cgs}}\mathbf{E}. \tag{L2.156}$$

The dielectric response of the medium ε_{m} relative to the value 1 for a vacuum is written

$$\varepsilon_{\text{m}} = 1 + \chi_{\text{m}}^{\text{mks}} \text{ in mks}, \quad \varepsilon_{\text{m}} = 1 + 4\pi\chi_{\text{m}}^{\text{cgs}} \text{ in cgs}. \tag{L2.157}$$

The total dielectric displacement vector is then

$$\mathbf{D} = \varepsilon_0\varepsilon_{\text{m}}\mathbf{E} = \varepsilon_0\mathbf{E} + \mathbf{P} = \varepsilon_0\left(1 + \chi_{\text{m}}^{\text{mks}}\right)\mathbf{E},$$

$$\mathbf{D} = \varepsilon_{\text{m}}\mathbf{E} = \mathbf{E} + 4\pi\mathbf{P} = \left(1 + 4\pi\chi_{\text{m}}^{\text{cgs}}\right)\mathbf{E}, \tag{L2.158}$$

such that the material polarization density is written as

$$\mathbf{P} = \varepsilon_0\chi_{\text{m}}^{\text{mks}}\mathbf{E} \text{ in mks}, \quad \mathbf{P} = \chi_{\text{m}}^{\text{cgs}}\mathbf{E} \text{ in cgs}. \tag{L2.159}$$

The additional induced polarization that is due to an added dilute suspension of number density N becomes

$$N\alpha_{\text{mks}}\mathbf{E} = \varepsilon_0\chi_{\text{induced}}^{\text{mks}}\mathbf{E}, \quad N\alpha_{\text{cgs}}\mathbf{E} = \chi_{\text{induced}}^{\text{cgs}}\mathbf{E}, \tag{L2.160}$$

so that

$$\chi_{\text{induced}}^{\text{mks}} = (\alpha_{\text{mks}}/\varepsilon_0)N, \quad \chi_{\text{induced}}^{\text{cgs}} = \alpha_{\text{cgs}}N, \tag{L2.161}$$

and the relative dielectric response that we need in formulation plays out to

$$\varepsilon_{\text{suspension}} = \varepsilon_{\text{m}} + \chi_{\text{induced}}^{\text{mks}} = \varepsilon_{\text{m}} + (\alpha_{\text{mks}}/\varepsilon_0)\,N,$$

$$\varepsilon_{\text{suspension}} = \varepsilon_{\text{m}} + 4\pi\,\chi_{\text{induced}}^{\text{cgs}} = \varepsilon_{\text{m}} + 4\pi\alpha_{\text{cgs}}N. \tag{L2.162}$$

That is, our generic proportionality α connects with the individual particle polarizabilities α_{mks} and α_{cgs} as

$$\alpha = \alpha_{\text{mks}}/\varepsilon_0, \quad \alpha = 4\pi\alpha_{\text{cgs}}, \tag{L2.163}$$

and for the coefficient of particle–particle interaction as [S.6]

$$\frac{\alpha\beta}{(4\pi\varepsilon_{\text{m}})^2} = \frac{\alpha_{\text{mks}}\beta_{\text{mks}}}{(4\pi\varepsilon_0\varepsilon_{\text{m}})^2}, \quad \frac{\alpha\beta}{(4\pi\varepsilon_{\text{m}})^2} = \frac{\alpha_{\text{cgs}}\beta_{\text{cgs}}}{\varepsilon_{\text{m}}^2}. \tag{L2.164}$$

Incidentally, the familiar difference-over-sum ratio used in van der Waals formulations twiddles out to

$$\frac{\varepsilon_{\text{suspension}} - \varepsilon_{\text{m}}}{\varepsilon_{\text{suspension}} + \varepsilon_{\text{m}}} \sim \frac{\alpha N}{2\varepsilon_{\text{m}}} = \frac{\alpha_{\text{mks}}}{2\varepsilon_0\varepsilon_{\text{m}}}N, \quad \frac{\varepsilon_{\text{suspension}} - \varepsilon_{\text{m}}}{\varepsilon_{\text{suspension}} + \varepsilon_{\text{m}}} \sim \frac{\alpha N}{2\varepsilon_{\text{m}}} = \frac{2\pi\alpha_{\text{cgs}}}{\varepsilon_{\text{m}}}N. \tag{L2.165}$$

Conversion between systems of units often follows this rule-of-thumb: $4\pi\varepsilon_0\varepsilon_m$ *in mks versus* ε_m *in cgs.*

Small spheres A dilute suspension of spheres of material with a radius a and *volume fraction* $v_{sph} = N_A V_A = N(4\pi/3)a^3$ has a composite dielectric response[4]:

$$\varepsilon_{suspension} = \varepsilon_m + 3v_{sph}\varepsilon_m(\varepsilon_{sph} - \varepsilon_m)/(\varepsilon_{sph} + 2\varepsilon_m)N$$
$$= \varepsilon_m + 4\pi a_{sph}^3\varepsilon_m(\varepsilon_{sph} - \varepsilon_m)/(\varepsilon_{sph} + 2\varepsilon_m)N, \quad \text{(L2.166)}$$

so that

$$\frac{\alpha}{4\pi\varepsilon_m} = a^3\frac{(\varepsilon_a - \varepsilon_m)}{(\varepsilon_a + 2\varepsilon_m)}, \text{ and similarly } \frac{\beta}{4\pi\varepsilon_m} = b^3\frac{(\varepsilon_b - \varepsilon_m)}{(\varepsilon_b + 2\varepsilon_m)} \quad \text{(L2.167)}$$

for spheres of material b with radius b.

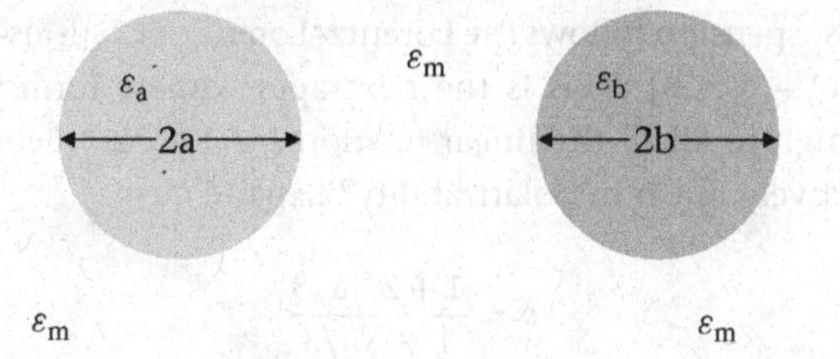

The $[\alpha(i\xi_n)\beta(i\xi_n)]/\{[4\pi\varepsilon_m(i\xi_n)]^2\}$ in expressions for point particles becomes $a^3b^3\frac{(\varepsilon_a-\varepsilon_m)}{(\varepsilon_a+2\varepsilon_m)}\frac{(\varepsilon_b-\varepsilon_m)}{(\varepsilon_b+2\varepsilon_m)}$. Irksome mks–cgs worries disappear because of canceling ε_0's and 4π's. From $g_{\alpha\beta}(l)$ derived for point–particle interactions, the interaction between spheres at relatively large center-to-center separation $z \gg$ a, b becomes

$$g_{ab}(z) = -\frac{6kT a^3b^3}{z^6}\sum_{n=0}^{\infty}{}' \frac{(\varepsilon_a - \varepsilon_m)}{(\varepsilon_a + 2\varepsilon_m)}\frac{(\varepsilon_b - \varepsilon_m)}{(\varepsilon_b + 2\varepsilon_m)}e^{-r_n}\left(1 + r_n + \frac{5}{12}r_n^2 + \frac{1}{12}r_n^3 + \frac{1}{48}r_n^4\right).$$

(L2.168)
[S.7.a]

For the record, because

$$\alpha = 4\pi a_{sph}^3\varepsilon_m(\varepsilon_{sph} - \varepsilon_m)/(\varepsilon_{sph} + 2\varepsilon_m),$$
$$\alpha_{mks} = \varepsilon_0\alpha$$
$$= 4\pi\varepsilon_0\varepsilon_m a_{sph}^3(\varepsilon_{sph} - \varepsilon_m)/(\varepsilon_{sph} + 2\varepsilon_m),$$
$$\alpha_{cgs} = \alpha/4\pi$$
$$= \varepsilon_m a_{sph}^3(\varepsilon_{sph} - \varepsilon_m)/(\varepsilon_{sph} + 2\varepsilon_m), \quad \text{(L2.169)}$$

the inducible extra polarizability of the individual spheres in the two unit systems. The dielectric response of spheres is so pretty and so fundamental that it is worthwhile to elaborate a couple of points.

Polarizability goes as the cube of radius The electrostatic potential set up by a dipole of moment μ_{dipole} has the form $(\mu_{dipole}/r^2)\cos\vartheta$, where ϑ is the angle between the dipole direction and the line to the position where the dipole potential is being sensed.[5] For example, a metallic sphere of radius a placed in a constant external electric field $\mathbf{E}_0$

is polarized and modifies the potential $-\mathbf{E}_0 r\cos\vartheta$ of the field $\mathbf{E}_0$ with an additional potential $(b/r^2)\cos\vartheta$ outside the sphere. This additional potential can be considered the dipole potential that is set up by the polarized metallic sphere. Taking the total potential to be zero at the center of the sphere, recognizing that the potential is constant throughout the conducting sphere, the potential $-E_0 \mathrm{a}\cos\vartheta + (b/\mathrm{a}^2)\cos\vartheta$ just at the radius $\mathrm{r} = \mathrm{a}$ of the sphere must be zero as well. (Don't confuse *italic* coefficient b used here with nonitalic spherical radii a, b!) In this way see that the coefficient b is equal to $E_0\,\mathrm{a}^3$. This b is a dipole moment equivalent in form to μ_{dipole} used in formulae. Because the dipole moment μ_{dipole} is the polarizability times the applied electric field ($\mathbf{E}_0$), the polarizability of the metallic sphere goes as its radius cubed, a^3.

The dilute limit emerges when $\mathrm{a}^3/\mathrm{z}^3 \ll 1$, i.e., when the spheres occupy a minor fraction of the volume To see the emergence of the dilute limit, assume that the dielectric response of dense suspension follows the Lorentz–Lorenz or Clausius–Mossotti relation[6] $\varepsilon = [(1+2N\alpha/3)/(1-N\alpha/3)]$ (This is the next approximate form when the number density N is too high to allow the linear relation $\varepsilon = 1 + N\alpha$.) Below what density N will this ε be effectively linear in polarizability? Expand

$$\varepsilon = \frac{1+2N\alpha/3}{1-N\alpha/3}$$

in powers of $N\alpha/3$:

$$\begin{aligned}\varepsilon &= (1+2N\alpha/3)(1+N\alpha/3+(N\alpha/3)^2+(N\alpha/3)^3+\cdots+)\\ &= 1+N\alpha+(N\alpha)^2/3+(N\alpha)^3/9+\cdots.\end{aligned}$$

The nonlinear third term can be neglected only if it is negligible compared with the linear second term $N\alpha$, i.e., if $N\alpha \ll 3$.

PROBLEM L2.8: Assuming the worst-case situation, a metallic sphere for which $\alpha = 4\pi\mathrm{a}^3$, and using the center-to-center distance z between spheres as a measure of number density, N is one sphere per cubic volume z^3, show that the inequality condition $N\alpha \ll 3$ becomes $4\pi\mathrm{a}^3 \ll 3z^3$.

For $\mathrm{z} = 4\mathrm{a}$, with a diameter's worth of separation between spheres, show that the inequality between $N\alpha$ and $(N\alpha)^2/3$ is a factor of $\sim 1/16$.

Atoms or molecules in a dilute gas: Keesom, Debye, London forces In a gas the "medium" is a vacuum, $\varepsilon_m = 1$, and the dielectric response varies in proportion to molecular density N. To understand the interaction between particles in a gas, it is worth considering this response more carefully than the simple coefficient of proportionality written so far. We want to include molecules that bear a permanent dipole moment as well as an ability to be polarized.

To think intuitively about "point" dipoles and their dipole moments, start with the Coulomb energy between point charges Q and q at separation z, $(Qq/4\pi\varepsilon_0\varepsilon z)$ (here in mks units).

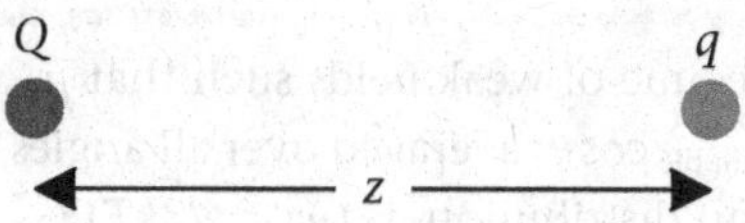

Next, write the sum of interactions between Q and q at separation z as well as Q and $-q$ at separation $z+d$, where $d <<< z$ and d is in the same direction as z:

$$\frac{Qq}{4\pi\varepsilon_0\varepsilon}\left(\frac{1}{z}-\frac{1}{z+d}\right)\approx\frac{Qq}{4\pi\varepsilon_0\varepsilon z}\left[1-\left(1-\frac{d}{z}\right)\right]=\frac{Q}{4\pi\varepsilon_0\varepsilon}\frac{qd}{z^2}, \tag{L2.170}$$

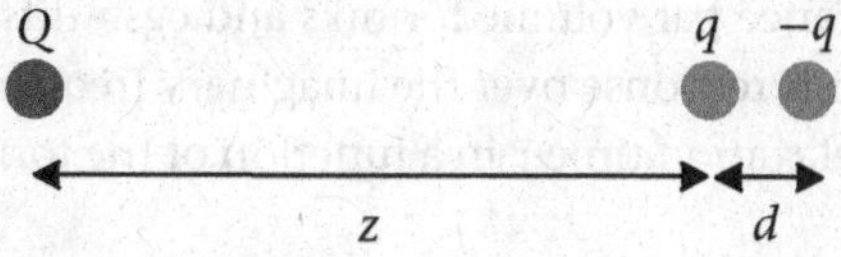

where q/z is replaced with qd/z^2.

In cgs units these interactions are written as $(Qq/\varepsilon z)$ and $(Q/\varepsilon)/(qd/z^2)$, respectively. In either case, the operative part of the $(q,-q)$ pair or dipole is its moment of magnitude:

$$\mu_{\text{dipole}}\equiv qd. \tag{L2.171}$$

Its units are charge × length (in mks units, coulombs × meters or C m; in cgs units statcoulombs × centimeters or sc cm). In physics and in this text, it is a vector that points from $+q$ to $-q$ (in chemistry, the convention is to point from $+q$ to $-q$).

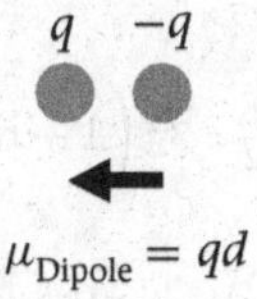

How would such a dipole respond to an applied electric field? The energy of a dipole μ_{dipole} pointing at an angle θ to the direction of an electric field $\mathbf{E}$ is $-\mathbf{E}\cdot\mu_{\text{dipole}}=-\mu_{\text{dipole}}E\cos\theta$. (Remember that an electric field pushes the positive charge and pulls the negative.)

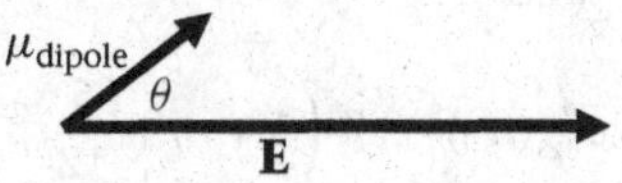

The extent of orientation polarization of the point dipole is $\mu_{\text{dipole}}\cos\theta$ in the direction of the electric field. In the limit of linear response to the electric field, the net average polarization of a freely rotating permanent dipole is

$$\frac{\mu_{\text{dipole}}^2}{3kT}E$$

with the same charge × distance units of μ_{dipole}.

PROBLEM L2.9: Under the regime of weak fields such that $\mu_{\text{dipole}} E \ll kT$, show that the orientation polarization $\mu_{\text{dipole}} \cos\theta$, averaged over all angles and weighted by energies $\mu_{\text{dipole}} E \cos\theta$ in a Boltzmann distribution, is $(\mu^2_{\text{dipole}}/3kT)E$.

The polarization density of a gas of permanent dipoles of number density N in a constant electric field of magnitude E is then

$$\mathbf{P}_{\text{dipole}} = N \frac{\mu^2_{\text{dipole}}}{3kT} \mathbf{E} = N \alpha_{\text{perm}} \mathbf{E} \tag{L2.172}$$

in units of charge × distance per volume for mks and cgs units.

The dependence of this response over the imaginary frequencies ξ of van der Waals interaction brings in a relaxation time τ in a function of the form that is due to Debye[7]:

$$\alpha_{\text{perm}}(i\xi) = \frac{\mu^2_{\text{dipole}}}{3kT(1+\xi\tau)}. \tag{L2.173}$$

This response is that of a permanent dipole that is partly oriented by a weak electrostatic field. "Weak" means that the energy put into orientation is much less than thermal energy; the field gently perturbs otherwise random orientation. This response is slow; τ is so large that the contribution to forces from dipole orientation "counts" only in the $n = 0$ limit of low frequency.

To the permanent-dipole response α_{perm} add the coefficient α_{ind} for a dipole induced by the electric field on a polarizable particle for a total polarizability $\alpha_{\text{perm}} + \alpha_{\text{ind}}$. The interaction of molecules in a gas can then be written immediately from the general form:

$$g_{\alpha\beta}(l) = -\frac{6kT}{l^6} \sum_{n=0}^{\infty}{}' \frac{\alpha(i\xi_n)\beta(i\xi_n)}{[4\pi\varepsilon_m(i\xi_n)]^2} e^{-r_n}\left(1 + r_n + \frac{5}{12}r_n^2 + \frac{1}{12}r_n^3 + \frac{1}{48}r_n^4\right). \tag{L2.174}$$

[S.6.a]

With $\varepsilon_m = 1$ and for the interaction of two like particles $\alpha = \beta$, recalling that

$$\alpha = \alpha_{\text{mks}}/\varepsilon_0, \qquad \alpha = 4\pi\alpha_{\text{cgs}},$$

we have

$$g_{\alpha\alpha}(l) = -\frac{6kT}{l^6} \sum_{n=0}^{\infty}{}' \frac{\alpha^2_{\text{mks}}(i\xi_n)}{(4\pi\varepsilon_0)^2} e^{-r_n}\left(1 + r_n + \frac{5}{12}r_n^2 + \frac{1}{12}r_n^3 + \frac{1}{48}r_n^4\right),$$

$$g_{\alpha\alpha}(l) = -\frac{6kT}{l^6} \sum_{n=0}^{\infty}{}' \alpha^2_{\text{cgs}}(i\xi_n)^2 e^{-r_n}\left(1 + r_n + \frac{5}{12}r_n^2 + \frac{1}{12}r_n^3 + \frac{1}{48}r_n^4\right). \tag{L2.175}$$

[S.6.a]

The pairwise interaction in a gas consists of three terms corresponding to $\alpha^2_{\text{total}} = \alpha^2_{\text{perm}} + 2\alpha_{\text{perm}}\alpha_{\text{ind}} + \alpha^2_{\text{ind}}$:

$$\frac{1}{2}\left(\frac{\mu^2_{\text{dipole}}}{3kT}\right)^2 + \left(\frac{\mu^2_{\text{dipole}}}{3kT}\right)\alpha_{\text{ind}}(0) + \sum_{n=0}^{\infty}{}' \alpha_{\text{ind}}(i\xi)^2, \tag{L2.176}$$

where, as usual, the factor 1/2 on the $n = 0$ term reflects the prime in summation. These terms are multiplied by

$$-\frac{6kT}{(4\pi\varepsilon_0)^2 l^6} \text{ in mks and by } -\frac{6kT}{l^6} \text{ in cgs units.}$$

Each term has its historical antecedent:

- The *Keesom* energy that is due to the mutual alignment of permanent dipoles:

$$g_{\text{Keesom}}(l) = -\frac{\mu_{\text{dipole}}^4}{3(4\pi\varepsilon_0)^2 kTl^6} \text{ in mks} = -\frac{\mu_{\text{dipole}}^4}{3kTl^6} \text{ in cgs units.} \quad \text{(L2.177) [S.8.a]}$$

- The *Debye* interaction between a permanent dipole μ_{dipole} and an inducible dipole, built on the polarizability $\alpha_{\text{ind}}(0)$ in the limit of zero frequency:

$$g_{\text{Debye}}(l) = -\frac{2\mu_{\text{dipole}}^2}{(4\pi\varepsilon_0)^2 l^6}\alpha_{\text{ind}}(0) \text{ in mks} = -\frac{2\mu_{\text{dipole}}^2}{l^6}\alpha_{\text{ind}}(0) \text{ in cgs units.} \quad \text{(L2.178) [S.8.b]}$$

- The *London* energy between two inducible dipoles at separations small compared with the wavelengths of their fluctuating electric fields (Table S.8.c, the first equation):

$$g_{\text{London}}(l) = -\frac{6kT}{(4\pi\varepsilon_0)^2 l^6}\sum_{n=0}^{\infty}{}' \alpha_{\text{ind}}^2(i\xi_n)^2 \text{ in mks units,}$$

$$g_{\text{London}}(l) = -\frac{6kT}{l^6}\sum_{n=0}^{\infty}{}' \alpha_{\text{ind}}^2(i\xi_n)^2 \text{ in cgs units.} \quad \text{(L2.179)}$$

Because London forces depend on frequencies whose photon energies are much greater than thermal energy kT, the sum $\sum_{n=0}^{\prime\infty} \alpha_{\text{ind}}^2(i\xi)$ is usually converted into an integral[8] (Table S.8.c, the second equation):

$$g_{\text{London}}(l, \mathrm{T} \to 0) = -\frac{3\hbar}{\pi(4\pi\varepsilon_0)^2 l^6}\int_0^{\infty} \alpha_{\text{ind}}^2(i\xi)\, d\xi \text{ in mks and}$$

$$-\frac{3\hbar}{\pi l^6}\int_0^{\infty} \alpha_{\text{ind}}^2(i\xi)\, d\xi \text{ in cgs.} \quad \text{(L2.180)}$$

MONOPOLAR INTERACTIONS, IONIC-FLUCTUATION FORCES, BETWEEN SMALL CHARGED PARTICLES

Following the strategy for extracting small-particle van der Waals interactions from the interaction between semi-infinite media, we can specialize the general expression for ionic-fluctuation forces to derive these forces between particles in salt solutions. Because of the low frequencies at which ions respond, only the $n = 0$ or zero-frequency terms contribute. In addition to ionic screening of dipolar fluctuations, there are ionic fluctuations that are due to the excess number of ions associated with each particle.

Imagine the interaction between two dilute suspensions of charged or neutral particles in a salt solution across a particle-free salt solution with which they are in

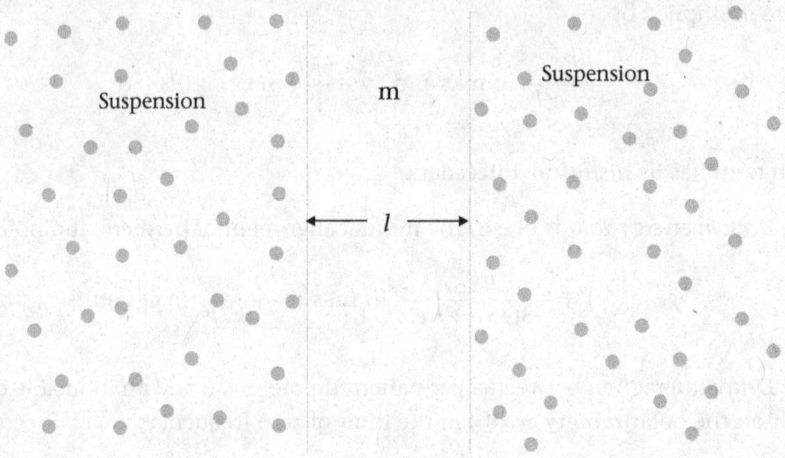

Figure L2.13

electrochemical equilibrium (see Fig. L2.13). Exclusive of conductance, the dielectric response of the suspensions is

$$\varepsilon_{\mathrm{susp}} = \varepsilon_{\mathrm{m}} + N\alpha, \tag{L2.181}$$

and their mean ionic strength (as an ion number density) is

$$n_{\mathrm{susp}} = n_{\mathrm{m}} + N\Gamma_{\mathrm{s}}. \tag{L2.182}$$

The incremental quantities α and Γ_s are the addition by each particle to the dielectric response and to the mean number of ions beyond that of the medium m, respectively. In this construction, the density N of spheres in the suspension is high enough compared with the wavelengths of important fluctuation that the suspensions can be treated as continuous media. At the same time, these suspended particles are dilute enough that the dielectric response of the left and the right regions can be expanded accurately to lowest terms in sphere number density. The conditions

$$N|\Gamma_{\mathrm{s}}| \ll n_{\mathrm{m}}, \quad N|\alpha| \ll \varepsilon_{\mathrm{m}} \tag{L2.183}$$

define "dilute" (see Fig. L2.13).

Not only are there fluctuations in the electric fields that create the dipolar fluctuations of most van der Waals forces but there are also fluctuations in electric potential with concomitant fluctuations in the number density of ions and the net charge on and around these small spheres. Monopolar charge-fluctuation forces occur when ion fluctuations in the spheres differ from ion fluctuations in the medium. Perhaps it is better to say that these forces occur when ion fluctuations around the suspended particles differ from what they would have been in the solution in the absence of particles. To formulate these interactions, we allow the ionic population of the spheres to equilibrate with the surrounding salt solution and to exchange ions with that surrounding solution. Then we compare the ionic fluctuations that occur from the presence of the small spheres with those in their absence. To do this we must have a way to count the number of extra ions associated with each sphere compared with the number of ions in their absence.

Let region m be a pure salt solution with dielectric-response functions ε_{m} and with a valence-weighted sum of concentrations (as number densities)

$$n_{\mathrm{m}} \equiv \sum_{\{\nu\}} n_\nu(\mathrm{m})\nu^2 \tag{L2.184}$$

and Debye constant

$$\kappa_{\mathrm{m}}^2 = n_{\mathrm{m}} e^2/\varepsilon_0\varepsilon_{\mathrm{m}} kT \text{ in mks} = 4\pi n_{\mathrm{m}} e^2/\varepsilon_{\mathrm{m}}\, kT \text{ in cgs}$$
$$= 4\pi n_{\mathrm{m}} \lambda_{\mathrm{Bj}} \text{ (in either unit system).} \tag{L2.185}$$

Spheres

For illustration, consider for a moment the particular case of a suspension of charged spheres of radius a and volume fraction $N(4\pi/3)\mathrm{a}^3$ (see Fig. L2.14).

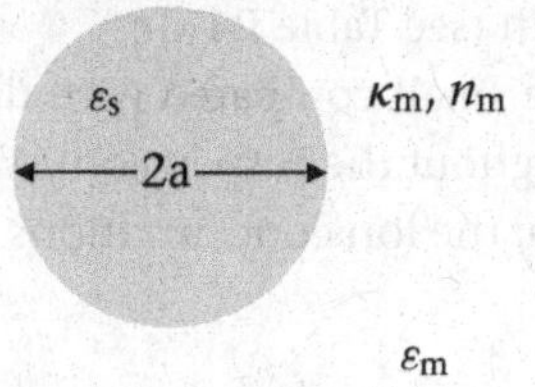

Figure L2.14

The composite dielectric response, $\varepsilon_{\mathrm{susp}} = \varepsilon_{\mathrm{m}} + N\alpha$, becomes

$$\varepsilon_{\mathrm{m}} + N\alpha = \varepsilon_{\mathrm{m}} + 4\pi \mathrm{a}^3 N\varepsilon_{\mathrm{m}} \frac{(\varepsilon_{\mathrm{s}} - \varepsilon_{\mathrm{m}})}{(\varepsilon_{\mathrm{s}} + 2\varepsilon_{\mathrm{m}})}, \tag{L2.186}$$

where ε_{s} is the dielectric response of the material within the sphere. The dilute condition is

$$4\pi \mathrm{a}^3 N \frac{(\varepsilon_{\mathrm{s}} - \varepsilon_{\mathrm{m}})}{(\varepsilon_{\mathrm{s}} + 2\varepsilon_{\mathrm{m}})} \ll 1, \tag{L2.187}$$

as in the case of dipolar fluctuations.

A sphere, charged or uncharged, immersed with its counterions into a salt solution accumulates or repels mobile ions to create an excess or deficit of each species. We use the symbol Γ_ν to designate the mean excess number of ions of valence ν per sphere. The number excess is defined as

$$\Gamma_\nu \equiv \int_0^\infty [n_\nu(r) - n_\nu(\mathrm{m})] 4\pi r^2 \mathrm{d}r, \tag{L2.188}$$

where r is measured from the center of the sphere and $n_\nu(r)$ is the mean number density of ions of valence ν at position r. Note that these "excess" quantities can be negative. The material in the sphere can exclude ions so that in the overall integration there can be fewer ions than would be present were the spheres not displacing them. A neutral sphere that does nothing but exclude ions and solvent creates a deficit

$$\Gamma_\nu = -n_\nu(\mathrm{m})(4\pi \mathrm{a}^3/3). \tag{L2.189}$$

General case

It is not necessary to consider spheres only. For any particular system the dielectric-response functions and excess ion densities can be measured or formulated as functions of the size, shape, and charge of the suspended particles as well as from ionic properties of the bathing solution. It is a separate procedure to compute the excess numbers Γ_ν from mean-field theory. They are used here as given quantities.

Over all, the mean ion concentrations of species of valence ν in the suspensions differ from that in region m by Γ_ν per sphere times sphere number density N:

$$n_\nu(\mathrm{A}) = n_\nu(\mathrm{B}) = n_\nu(\mathrm{m}) + N\Gamma_\nu. \tag{L2.190}$$

Correspondingly, the mean ionic strength differs from that in region m as

$$n_{\mathrm{susp}} = \sum_{\{\nu\}} [n_\nu(\mathrm{m}) + N\Gamma_\nu]\nu^2 = n_{\mathrm{m}} + N\Gamma_{\mathrm{s}},\ \Gamma_{\mathrm{s}} \equiv \sum_{\{\nu\}} \Gamma_\nu \nu^2. \tag{L2.191}$$

Γ_{s}, the mean excess in the number of mobile ions around each small charged particle, is the physical quantity driving an ionic-fluctuation force. It is the sum of the numbers of all excess ions of valence ν weighted by the square of valence. In this ν^2 weighting Γ_{s} is analogous to ionic strength (see Table P.1.d).

The suspended particles are small compared with distances $\sim l$ over which the potential $\phi(x, y, z)$ will vary throughout the suspension. When there is a weak fluctuation ϕ in potential from mean value, the ion concentrations deviate as in linearized Debye–Hückel theory:

$$n_\nu(\mathrm{s}, \phi) - n_\nu(\mathrm{s}) = n_\nu(\mathrm{s})[e^{-e\nu\phi/kT} - 1] = -\frac{e\nu n_\nu(\mathrm{s})}{kT}\phi. \tag{L2.192}$$

For the wave equation, $\nabla^2\phi = \kappa^2\phi$, we take a volume average over the ionic strengths of the suspending medium and of the material composing the sphere. To see why this is so, recall that the $\kappa^2\phi$ term is that part of the charge density that depends on the value of potential ϕ. This charge density is proportional to the mean ionic strength $\Sigma_\nu n_\nu \nu^2$ of the entire suspension. The waves here extend over many suspended particles, over a volume average (proportional to the total ionic strength),

$$n_{\mathrm{susp}} = n_{\mathrm{m}} + N\Gamma_{\mathrm{s}}, \tag{L2.193}$$

that gives the average ionic response of the composite suspension. With the average dielectric response, $\varepsilon_{\mathrm{susp}} = \varepsilon_{\mathrm{m}} + N\alpha$,

$$\nabla^2\phi = \kappa_{\mathrm{susp}}^2\phi \tag{L2.194}$$

where

$$\kappa_{\mathrm{susp}}^2 = \kappa_{\mathrm{m}}^2 \frac{\left(1 + N\frac{\Gamma_{\mathrm{s}}}{n_{\mathrm{m}}}\right)}{\left(1 + N\frac{\alpha}{\varepsilon_{\mathrm{m}}}\right)} \approx \kappa_{\mathrm{m}}^2 \left[1 + N\left(\frac{\Gamma_{\mathrm{s}}}{n_{\mathrm{m}}} - \frac{\alpha}{\varepsilon_{\mathrm{m}}}\right)\right] \tag{L2.195}$$

for the suspension's composite Debye constant.

In terms of the variable of integration p, $1 \le p < \infty$, *without* the magnetic-field fluctuations that necessarily go with ionic currents from the displacement of ionic charge, the $n = 0$ term in the van der Waals interaction between the two suspensions across m is [see Eqs. (L3.192)–(L3.194)]

$$G_{\mathrm{SmS}}(l) = \frac{kT\kappa_{\mathrm{m}}^2}{4\pi}\int_1^\infty p\ln(1 - \overline{\Delta}_{\mathrm{Sm}}^2 e^{-2\kappa_{\mathrm{m}} lp})\mathrm{d}p \approx -\frac{kT\kappa_{\mathrm{m}}^2}{4\pi}\int_1^\infty p\,\overline{\Delta}_{\mathrm{Sm}}^2 e^{-2\kappa_{\mathrm{m}} lp}\mathrm{d}p, \tag{L2.196}$$

where (NB: difference in location of p and s in $\overline{\Delta}_{\mathrm{Sm}}$ here vs. in nonionic cases)

$$\overline{\Delta}_{\mathrm{Sm}} = \left(\frac{s\varepsilon_{\mathrm{susp}} - p\varepsilon_{\mathrm{m}}}{s\varepsilon_{\mathrm{susp}} + p\varepsilon_{\mathrm{m}}}\right), \qquad \Delta_{\mathrm{Sm}} = \left(\frac{s - p}{s + p}\right),$$

$$s = \left[p^2 - 1 + \kappa_{\mathrm{susp}}^2/\kappa_{\mathrm{m}}^2\right]^{\frac{1}{2}} \approx \left[p^2 + N\left(\frac{\Gamma_{\mathrm{s}}}{n_{\mathrm{m}}} - \frac{\alpha}{\varepsilon_{\mathrm{m}}}\right)\right]^{\frac{1}{2}}. \tag{L2.197}$$

Recall that extraction of the pairwise interaction $g_p(l)$ between dilute particles requires a third derivative of G_{SmS} with respect to separation l [see Eq. (L2.149)]:

$$-G'''_{SmS}(l) = 2\pi l N^2 g_p(l) \approx -\frac{2kT\kappa_m^5}{\pi}\int_1^\infty p^4 \overline{\Delta}_{Sm}^2 e^{-2\kappa_m l p}\, dp. \qquad \text{(L2.198)}$$

Expanding the integrand to lowest order in N and then integrating in p yields a point-particle interaction that consists of three terms,

$$g_p(l) = g_{D-D}(l) + g_{D-M}(l) + g_{M-M}(l), \qquad \text{(L2.199)}$$

with each term reflecting dipole–dipole, dipole–monopole, and monopole–monopole correlation interactions. Because these three terms correspond to three different kinds of small-sphere interactions, they are written separately.

Dipole–dipole fluctuation correlation

$$g_{D-D}(l) = -3kT\left(\frac{\alpha}{4\pi\varepsilon_m}\right)^2\left[1 + (2\kappa_m l) + \frac{5}{12}(2\kappa_m l)^2 + \frac{1}{12}(2\kappa_m l)^3 + \frac{1}{96}(2\kappa_m l)^4\right]\frac{e^{-2\kappa_m l}}{l^6}. \qquad \text{(L2.200)}$$

[S.9.a]

This interaction is like the first $n = 0$ term for small-particle van der Waals forces but with ionic rather than retardation screening. In the limit of low salt concentration, $\kappa_m \to 0$, it has the familiar $1/l^6$ form:

$$g_{D-D}(l) \to -\frac{3kT}{l^6}\left(\frac{\alpha}{4\pi\varepsilon_m}\right)^2. \qquad \text{(L2.201)}$$

For $\kappa_m l \gg 1$, it is dominated by an exponential:

$$g_{D-D}(l) = -\frac{kT\kappa_m^4}{2}\left(\frac{\alpha}{4\pi\varepsilon_m}\right)^2\frac{e^{-2\kappa_m l}}{l^2}. \qquad \text{(L2.202)}$$

Recall that $\kappa_m^2/n_m = 4\pi\lambda_{Bj}$, where λ_{Bj} is the Bjerrum length of the medium; the coefficient can be rewritten to give

$$g_{D-D}(l) = -\frac{kT}{2}\left(\frac{n_m\alpha}{\varepsilon_m}\right)^2\frac{e^{-2\kappa_m l}}{(l/\lambda_{Bj})^2}. \qquad \text{(L2.203)}$$

Here, as in the formulae that follow next, there is a double screening in this high salt limit: the interaction across the distance l with a screening factor $e^{-\kappa_m l}/l$ and then another interaction across l again with $e^{-\kappa_m l}/l$ screening to effect the correlation in mutual perturbation.

Dipole–monopole correlation

$$g_{D-M}(l) = -\frac{kT\kappa_m^2}{4\pi}\left(\frac{\alpha}{4\pi\varepsilon_m}\right)\left(\frac{\Gamma_s}{n_m}\right)\left[1 + (2\kappa_m l) + \frac{1}{4}(2\kappa_m l)^2\right]\frac{e^{-2\kappa_m l}}{l^4} \qquad \text{(L2.204)}$$

[S.9.b]

or

$$g_{D-M}(l) = -kT\lambda_{Bj}\left(\frac{\alpha}{4\pi\varepsilon_m}\right)\Gamma_s\left[1 + (2\kappa_m l) + \frac{1}{4}(2\kappa_m l)^2\right]\frac{e^{-2\kappa_m l}}{l^4}.$$

In the limit of large $\kappa_m l$,

$$g_{D-M}(l) = -kT\lambda_{Bj}\left(\frac{\alpha}{4\pi\varepsilon_m}\right)\Gamma_s\kappa_m^2\frac{e^{-2\kappa_m l}}{l^2} = -kT\left(\frac{n_m\alpha}{\varepsilon_m}\right)\Gamma_s\frac{e^{-2\kappa_m l}}{(l/\lambda_{Bj})^2} \qquad \text{(L2.205)}$$

Monopole–monopole correlation

$$g_{\text{M-M}}(l) = -\frac{kT\kappa_{\text{m}}^4}{2}\left(\frac{\Gamma_{\text{s}}}{4\pi n_{\text{m}}}\right)^2 \frac{e^{-2\kappa_{\text{m}}l}}{l^2}$$

$$= -\frac{kT}{2}\Gamma_{\text{s}}^2 \frac{e^{-2\kappa_{\text{m}}l}}{(l/\lambda_{\text{Bj}})^2} \text{ or } -\frac{kT}{2}\Gamma_{\text{s}}^2 \frac{e^{-2l/\lambda_{\text{Debye}}}}{(l/\lambda_{\text{Bj}})^2} \qquad \text{(L2.206)}$$

[S.9.c]

Driven by a Boltzmann thermal-energy source kT and measured in kT units, coupled by a number of effective mobile ionic charges Γ_{s} from Gibbs, the monopole–monopole correlation force is screened by λ_{Debye} across the length $2l$, back and forth between point particles. At the same time its power-law dependence on length is measured in the natural thermal unit λ_{Bj}. Boltzmann, Gibbs, Debye, Bjerrum—all at the same time. Can it get any prettier?

These small-particle formulae have only symbolic validity in the zero-salt $\kappa_{\text{m}}, n_{\text{m}} \to 0$ limit. Why? Because even though κ_{m} and n_{m} need not appear explicitly, the results have been derived under the condition $N|\Gamma_{\text{s}}| \ll n_{\text{m}}$. They can be applied only under conditions for which this stringent inequality is satisfied. In the limit of small n_{m}, the ratio $|\Gamma_{\text{s}}|/n_{\text{m}}$ diverges for any finite Γ_{s}; this divergence violates the very condition under which these expressions are derived.

PROBLEM L2.10: Collecting Eqs. (L2.181), (L2.182), (L2.195), and (L2.196), expanding everything to lowest terms in particle number density, and using Eqs. (L2.198) and (L2.199), derive Eqs. (L2.200), (L2.204), and (L2.206) for $g_{\text{D-D}}(l)$, $g_{\text{D-M}}(l)$, and $g_{\text{M-M}}(l)$, respectively.

L2.3.F. Point particles and a planar substrate

General form

Begin with the general form for interaction energy per unit area between half-spaces A and B:

$$G_{\text{AmB}}(l,T) = \frac{kT}{2\pi c^2}\sum_{n=0}^{\infty}{}' \varepsilon_{\text{m}}\mu_{\text{m}}\xi_n^2 \int_1^{\infty} p \ln\left[(1-\overline{\Delta}_{\text{Am}}\overline{\Delta}_{\text{Bm}}e^{-r_np})(1-\Delta_{\text{Am}}\Delta_{\text{Bm}}e^{-r_np})\right] dp.$$

[P.1.a.1]

Treat one of the spaces, A, as occupied by a condensed material of susceptibility ε_{A} while taking the other space, B, to be occupied by a dilute gas (or solution) whose dielectric susceptibility is $\varepsilon_{\text{B}} = \varepsilon_{\text{m}} + N_{\text{B}}\beta$ (see Fig. L2.15). The interaction G_{AmB} between the solid or liquid A and the dilute cloud B is now a sum of individual interactions $g_{\text{p}}(r)$ between molecules on the right and the substrate on the left.

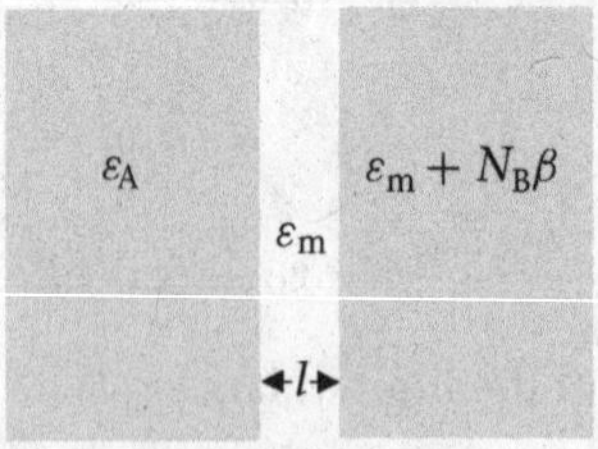

Figure L2.15

If we imagine increasing the separation l by an incremental amount dl, it is equivalent to removing an interaction $N\,dl\,g_{\text{p}}(l)$ per unit area, that is, effecting a change $-N\,dl\,g_{\text{p}}(l)$ in the energy of interaction per unit area.

Because N is a particle density per volume, $N\mathrm{d}l$ is a number of particles per area at the distance l in the "slice" dl. If the interaction is attractive, $g_p(l)$ is negative and the change in energy from increasing l by $+\mathrm{d}l$ is positive. Hence the minus sign. At the same time, in macroscopic-continuum language, the change in energy for shifting the separation from l to $l+\mathrm{d}l$ is $[\mathrm{d}G_{\mathrm{AmB}}(l)/\mathrm{d}l]\mathrm{d}l$ (see Fig. L2.16).

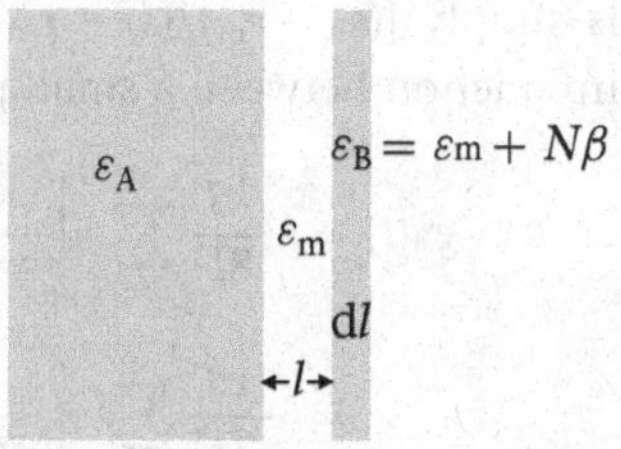

Figure L2.16

Formally, $-N\,\mathrm{d}l\; g_p(l) = [\mathrm{d}G_{\mathrm{AmB}}(l)/\mathrm{d}l]\mathrm{d}l$, or

$$N\,g_p(l) = -[\mathrm{d}G_{\mathrm{AmB}}(l)/\mathrm{d}l]. \qquad \text{(L2.207)}$$

That is, to obtain the function $g_p(l)$, take only one derivative of $G_{\mathrm{AmB}}(l)$. As when extracting point–particle–particle interactions, first simplify the full expression, but this time keep the unrestricted general forms for $\overline{\Delta}_{\mathrm{Am}}$ and Δ_{Am}:

$$\overline{\Delta}_{\mathrm{Am}} = \frac{p\varepsilon_{\mathrm{A}} - s_{\mathrm{A}}\varepsilon_{\mathrm{m}}}{p\varepsilon_{\mathrm{A}} + s_{\mathrm{A}}\varepsilon_{\mathrm{m}}};\ \Delta_{\mathrm{Am}} = \frac{p - s_{\mathrm{A}}}{p + s_{\mathrm{A}}};\ s_{\mathrm{A}} = \sqrt{p^2 - 1 + (\varepsilon_{\mathrm{A}}/\varepsilon_{\mathrm{m}})}. \qquad \text{(L2.208)}$$

(Here neglect magnetic-susceptibility differences; they can be restored with no trouble.)

Because the absolute values of the dilute suspension, $\overline{\Delta}_{\mathrm{Bm}}$ and Δ_{Bm} are much less than unity, it is possible to expand $G_{\mathrm{AmB}}(l, T)$ accurately to its lowest term in N. When

$$s_{\mathrm{B}} \approx p + \frac{N\beta}{2p\varepsilon_{\mathrm{m}}},\ \overline{\Delta}_{\mathrm{Bm}} \approx \left(\frac{N\beta}{4\varepsilon_{\mathrm{m}}}\right)\frac{1}{p^2}(2p^2 - 1),\ \Delta_{\mathrm{Bm}} \approx -\left(\frac{N\beta}{4\varepsilon_{\mathrm{m}}}\right)\left(\frac{1}{p^2}\right) \qquad \text{(L2.209)}$$

are used, this free energy of interaction per area is

$$G_{\mathrm{AmB}}(l, T) \approx -N\frac{kT}{8l^2}\sum_{n=0}^{\infty}{}' \left[\frac{\beta(i\xi_n)}{4\pi\varepsilon_{\mathrm{m}}(i\xi_n)}\right] r_n^2 \int_1^{\infty} p\left[\overline{\Delta}_{\mathrm{Am}}\left(\frac{2p^2 - 1}{p^2}\right) - \Delta_{\mathrm{Am}}\frac{1}{p^2}\right] e^{-r_n p}\,\mathrm{d}p. \qquad \text{(L2.210)}$$

From Eq. (L2.207) the derivative $[\mathrm{d}G_{\mathrm{AmB}}(l)/\mathrm{d}l]$ yields a particle–substrate interaction:

$$g_p(l) = -\frac{kT}{8l^3}\sum_{n=0}^{\infty}{}' \left[\frac{\beta(i\xi_n)}{4\pi\varepsilon_{\mathrm{m}}(i\xi_n)}\right] r_n^3 \int_1^{\infty} \left[\overline{\Delta}_{\mathrm{Am}}(2p^2 - 1) - \Delta_{\mathrm{Am}}\right] e^{-r_n p}\,\mathrm{d}p \qquad \text{(L2.211) [S.11.a.1]}$$

or

$$-\frac{kT}{8l^3}\sum_{n=0}^{\infty}{}' \left[\frac{\beta(i\xi_n)}{4\pi\varepsilon_{\mathrm{m}}(i\xi_n)}\right] \int_{r_n}^{\infty} \left[\overline{\Delta}_{\mathrm{Am}}\,(2x^2 - r_n^2) - \Delta_{\mathrm{Am}} r_n^2\right] e^{-x}\,\mathrm{d}x.$$

This is the most general form at distances l long compared with particle size. Any further simplification requires specializing assumptions. For careful computation it is usually safest to integrate this complete expression numerically. To see the qualitative features of the interaction, it helps to examine special limiting forms.

PROBLEM L2.11: Take the l derivative required for Eq. (L2.211).

Small-$\overline{\Delta}_{\mathrm{Am}}$ limit

In the case in which retardation is included, but with the approximation that $\overline{\Delta}_{\mathrm{Am}}$ is simply $[(\varepsilon_{\mathrm{A}}-\varepsilon_{\mathrm{m}})/(\varepsilon_{\mathrm{A}}+\varepsilon_{\mathrm{m}})]$ so that it can be taken out of the p integration, the interaction between a small particle and a wall is[9]

$$g_p(l) = -\frac{kT}{8l^3}\sum_{n=0}^{\infty}{}' \left[\frac{\beta(i\xi_n)}{4\pi\varepsilon_{\mathrm{m}}(i\xi_n)}\right]\overline{\Delta}_{\mathrm{Am}} r_n^3 \int_1^{\infty}(2p^2-1)e^{-r_n p}\,\mathrm{d}p$$

$$= -\frac{kT}{2l^3}\sum_{n=0}^{\infty}{}' \left[\frac{\beta(i\xi_n)}{4\pi\varepsilon_{\mathrm{m}}(i\xi_n)}\right]\left[\frac{\varepsilon_{\mathrm{A}}(i\xi_n)-\varepsilon_{\mathrm{m}}(i\xi_n)}{\varepsilon_{\mathrm{A}}(i\xi_n)+\varepsilon_{\mathrm{m}}(i\xi_n)}\right]\left(1+r_n+\frac{r_n^2}{4}\right)e^{-r_n}, \qquad \text{(L2.212) [S.11.a.2]}$$

with a contribution of similar form for magnetic interactions.

Nonretarded limit

In the limit where there is *no retardation screening*, $r_n = 0$, only the first term in the integrand endures. In that case, the general interaction reduces to

$$g_p(l) = -\frac{kT}{4l^3}\sum_{n=0}^{\infty}{}' \left[\frac{\beta(i\xi_n)}{4\pi\varepsilon_{\mathrm{m}}(i\xi_n)}\right]\int_0^{\infty}\overline{\Delta}_{\mathrm{Am}}x^2e^{-x}\,\mathrm{d}x. \qquad \text{(L2.213) [S.11.b]}$$

This is a cleanly inverse-cube interaction whose coefficient can be integrated in closed form even though the function $\overline{\Delta}_{\mathrm{Am}}$ itself depends on x (through p), a worry in cases where ε_{A} takes on very large values. In fact, this is not a serious limitation because, for $\varepsilon_A \gg \varepsilon_{\mathrm{m}}$, $\overline{\Delta}_{\mathrm{Am}}$ still goes to unity (as does $-\Delta_{\mathrm{Am}}$). Specifically, for $\varepsilon_{\mathrm{A}}/\varepsilon_{\mathrm{m}} \to \infty$, $s_{\mathrm{i}} = \sqrt{p^2-1+(\varepsilon_{\mathrm{i}}\mu_{\mathrm{i}}/\varepsilon_{\mathrm{m}}\mu_{\mathrm{m}})}$,

$$\overline{\Delta}_{\mathrm{Am}} = \frac{p\varepsilon_{\mathrm{A}}-s_{\mathrm{A}}\varepsilon_{\mathrm{m}}}{p\varepsilon_{\mathrm{A}}+s_{\mathrm{A}}\varepsilon_{\mathrm{m}}} \to \frac{\sqrt{\varepsilon_{\mathrm{A}}}-\sqrt{\varepsilon_{\mathrm{m}}/p}}{\sqrt{\varepsilon_{\mathrm{A}}}+\sqrt{\varepsilon_{\mathrm{m}}/p}} \to 1,$$

$$\Delta_{\mathrm{Am}} = \frac{p-s_{\mathrm{A}}}{p+s_{\mathrm{A}}} \to -\frac{\sqrt{\varepsilon_{\mathrm{A}}}-p\sqrt{\varepsilon_{\mathrm{m}}}}{\sqrt{\varepsilon_{\mathrm{A}}}+p\sqrt{\varepsilon_{\mathrm{m}}}} \to -1. \qquad \text{(L2.214)}$$

Warning: In these high-ε limits, computation with simplified forms is particularly risky. These Δ limits are written here only to show that the inverse-cube, nonretarded form can be realized even for very large ε_{A}.

When $r_n \to 0$ here, the variable of integration $x = r_n p$ takes on its important values at very large p so that $\overline{\Delta}_{\mathrm{Am}}$ is simply $[(\varepsilon_{\mathrm{A}}-\varepsilon_{\mathrm{m}})/(\varepsilon_{\mathrm{A}}+\varepsilon_{\mathrm{m}})]$ ($s_{\mathrm{A}} \to p$ for very large p as long as ε_{A} does not approach the effectively infinite values of a conductor).

In this simpler no-retardation limit

$$g_p(l) = -\frac{kT}{4l^3}\sum_{n=0}^{\infty}{}' \left[\frac{\beta(i\xi_n)}{4\pi\varepsilon_{\mathrm{m}}(i\xi_n)}\right]\int_0^{\infty}\overline{\Delta}_{\mathrm{Am}}x^2e^{-x}\,\mathrm{d}x$$

$$\to -\frac{kT}{2l^3}\sum_{n=0}^{\infty}{}' \frac{\beta(i\xi_n)}{4\pi\varepsilon_{\mathrm{m}}(i\xi_n)}\left[\frac{\varepsilon_{\mathrm{A}}(i\xi_n)-\varepsilon_{\mathrm{m}}(i\xi_n)}{\varepsilon_{\mathrm{A}}(i\xi_n)+\varepsilon_{\mathrm{m}}(i\xi_n)}\right]. \qquad \text{(L2.215) [S.11.b.1]}$$

When temperature is effectively zero, the discrete values of $\xi_{\mathrm{n}} = (2\pi kT/\hbar)n$ merge into a continuum; the interaction energy in the absence of ionic or retardation screening is an integral over the frequency ξ,

$$g_{p,T\to 0}(l) = -\frac{\hbar}{4\pi l^3}\int_0^\infty \left[\frac{\beta(i\xi)}{4\pi\varepsilon_m(i\xi)}\right]\left[\frac{\varepsilon_A(i\xi)-\varepsilon_m(i\xi)}{\varepsilon_A(i\xi)+\varepsilon_m(i\xi)}\right]d\xi. \tag{L2.216}$$
[S.11.b.2]

All these expressions go as $1/l^3$, the same algebraic power law the energy would follow if the van der Waals energy were the sum of inverse-sixth-power interactions between the suspended particle and the incremental pieces of the substrate. One should not use this resemblance to think that one can apply pairwise-sum reasoning for computation. That reasoning would require that the substrate "A" and the medium "m" have the density dependence of a dilute gas. The coefficients would be wrong.

Fully retarded, zero-temperature limit

In the *fully retarded limit* at which temperature is effectively zero but the velocity of light is sufficiently slow that the ratios $r_n = (2l\xi_n\varepsilon_m^{1/2})/c$ go to infinity for ξ_n less than the important absorption frequencies, β and the ε's are treated as constants (in principle evaluated at zero frequency). Then there is a fourth-power dependence on separation:

$$g_p(l) = -\frac{3\hbar c}{8\pi l^4}\frac{(\beta/4\pi)}{\varepsilon_m^{3/2}}\Theta(\varepsilon_A/\varepsilon_m), \tag{L2.217}$$
[S.11.c]

where

$$\Theta(\varepsilon_A/\varepsilon_m) \equiv \frac{1}{2}\int_1^\infty \{[\overline{\Delta}_{Am}(2p^2-1)-\Delta_{Am}]/p^4\}dp. \tag{L2.218}$$

If $\varepsilon_A \gg \varepsilon_m$, $\overline{\Delta}_{Am}$ goes to $+1$ and Δ_{Am} goes to -1:

$$\Theta(\varepsilon_A/\varepsilon_m) \to \frac{1}{2}\int_1^\infty \{(2p^2)/p^4\}dp = \int_1^\infty \{1/p^2\}dp = 1. \tag{L2.219}$$

If $\varepsilon_A \approx \varepsilon_m$,

$$\Theta(\varepsilon_A/\varepsilon_m) \approx \frac{23}{30}\left(\frac{\varepsilon_A-\varepsilon_m}{\varepsilon_A+\varepsilon_m}\right). \tag{L2.220}$$

PROBLEM L2.12: Beginning with Eq. (L2.211), convert summation to integration for the zero-temperature limit of Eqs. (L2.217) and (L2.218).

PROBLEM L2.13: Expanding Eq. (L2.218) for small differences in $\varepsilon_A \approx \varepsilon_m$, show how the 23/30 comes into approximation (L2.220).

It is easy and instructive to verify that, under the conditions of validity of these equations, the interaction energy is not great compared with kT. Use the same parameters for particle polarizability as those used in estimating particle–particle interactions.

For careful work, sum and integrate the most general form of the interaction including the p dependence of the difference-over-sum functions $\overline{\Delta}_{Am}$ and Δ_{Am}. The limiting forms are given here mainly to build intuition and to show quick preliminary ways to estimate energies.

L2.3.G. Line particles in dilute suspension

DIPOLAR FLUCTUATION FORCES BETWEEN THIN CYLINDERS

Begin by considering the interaction between two like anisotropic media across an isotropic medium m (see Fig. L2.17) (see also Section L3.7). There is an angle θ between the principal axes of the two anisotropic regions (both regions point in the x direction when $\theta = 0$).

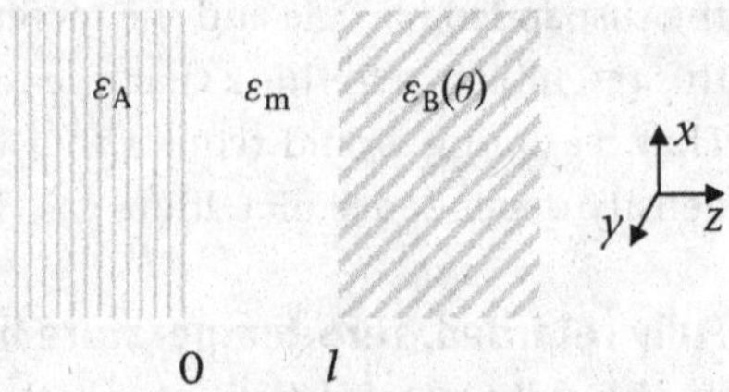

Figure L2.17

In the present case, the anisotropy of the two media A and B is due to the presence of cylindrical rods (which may themselves have anisotropic susceptibilities). The intervening space is filled with an isotropic medium with (scalar) susceptibility ε_m. This same m permeates A and B to fill the space between rods. The susceptibility of the rod *material* is $\varepsilon_{c\perp}$ in the z direction (perpendicular to the A/m/B interfaces) and $\varepsilon_{c\parallel}$ in the direction of the rod axis (see Fig. L2.18).

The rods, parallel in each region, have an average cross-sectional number density N. The volume fraction for rods of radius a is

$$v = \pi a^2 N. \tag{L2.221}$$

At low density, $v \ll 1$, the uniaxial susceptibility *tensor*

$$\begin{pmatrix} \varepsilon_\parallel & 0 & 0 \\ 0 & \varepsilon_\perp & 0 \\ 0 & 0 & \varepsilon_\perp \end{pmatrix}$$

for each anisotropic medium has diagonal elements

$$\varepsilon_\parallel = \varepsilon_m(1-v) + v\varepsilon_{c\parallel} = \varepsilon_m(1 + v\overline{\Delta}_\parallel) \tag{L2.222}$$

parallel to the rod axes and

$$\varepsilon_\perp = \varepsilon_m\left(1 + \frac{2v\overline{\Delta}_\perp}{1 - v\overline{\Delta}_\perp}\right) = \varepsilon_m\left(\frac{1 + v\overline{\Delta}_\perp}{1 - v\overline{\Delta}_\perp}\right) \approx \varepsilon_m(1 + 2v\overline{\Delta}_\perp) \tag{L2.223}$$

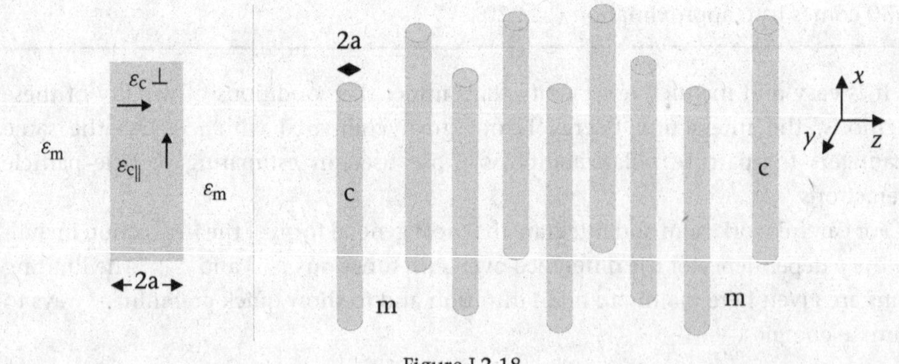

Figure L2.18

perpendicular to the rod axis, where[10]

$$\overline{\Delta}_{\parallel} \equiv \frac{\varepsilon_{c\parallel} - \varepsilon_m}{\varepsilon_m}, \quad \overline{\Delta}_{\perp} \equiv \frac{\varepsilon_{c\perp} - \varepsilon_m}{\varepsilon_{c\perp} + \varepsilon_m}. \tag{L2.224}$$

The interaction between uniaxial A and B at a mutual angle θ across isotropic region m is [Eq. (L3.222)]

$$G_{AmB}(l,\theta) = -\frac{kT}{16\pi^2 l^2} \sum_{n=0}^{\infty}{}' \sum_{j=1}^{\infty} \frac{1}{j^3} \int_0^{2\pi} \left[\overline{\Delta}_{Am}(\xi_n,\psi)\overline{\Delta}_{Bm}(\xi_n,\theta,\psi)\right]^j d\psi, \tag{L2.225}$$

where

$$\overline{\Delta}_{Am}(\xi_n,\psi) = \left[\frac{\varepsilon_{\perp} g_A(-\psi) - \varepsilon_m}{\varepsilon_{\perp} g_A(-\psi) + \varepsilon_m}\right], \quad \overline{\Delta}_{Bm}(\xi_n,\theta,\psi) = \left(\frac{\varepsilon_{\perp} g_B(\theta-\psi) - \varepsilon_m}{\varepsilon_{\perp} g_B(\theta-\psi) + \varepsilon_m}\right), \tag{L2.226}$$

when [Eq. (L3.217)]

$$g_A^2(-\psi) = 1 + \gamma\cos^2(-\psi), \quad g_B^2(\theta-\psi) = 1 + \gamma\cos^2(\theta-\psi), g_m = 1, \tag{L2.227}$$

$$\gamma \equiv \frac{(\varepsilon_{\parallel} - \varepsilon_{\perp})}{\varepsilon_{\perp}} \approx v(\overline{\Delta}_{\parallel} - 2\overline{\Delta}_{\perp}) \ll 1,$$

$$g_A(-\psi) \approx 1 + (\gamma/2)\cos^2(-\psi), \qquad g_B(\theta-\psi) = 1 + (\gamma/2)\cos^2(\theta-\psi).$$

To leading terms in rod density,

$$\overline{\Delta}_{Am}(\xi_n,\psi) = \left\{\frac{\varepsilon_m(1+2v\overline{\Delta}_{\perp})[1+(\gamma/2)\cos^2(-\psi)] - \varepsilon_m}{\varepsilon_m(1+2v\overline{\Delta}_{\perp})[1+(\gamma/2)\cos^2(-\psi)] + \varepsilon_m}\right\}$$

$$= v\left[\overline{\Delta}_{\perp} + \frac{(\overline{\Delta}_{\parallel} - 2\overline{\Delta}_{\perp})}{4}\cos^2(-\psi)\right], \tag{L2.228}$$

$$\overline{\Delta}_{Bm}(\xi_n,\theta,\psi) = v\left[\overline{\Delta}_{\perp} + \frac{(\overline{\Delta}_{\parallel} - 2\overline{\Delta}_{\perp})}{4}\cos^2(\theta-\psi)\right].$$

In this dilute-rod limit,

$$G_{AmB}(l,\theta) \approx -\frac{kT}{16\pi^2 l^2} \sum_{n=0}^{\infty}{}' \int_0^{2\pi} \overline{\Delta}_{Am}(\xi_n,\psi)\overline{\Delta}_{Bm}(\xi_n,\theta,\psi)\, d\psi$$

$$= -\frac{kT}{8\pi l^2}(\pi a^2)^2 N^2 \sum_{n=0}^{\infty}{}' \left[\overline{\Delta}_{\perp}^2 + \frac{\overline{\Delta}_{\perp}}{4}(\overline{\Delta}_{\parallel} - 2\overline{\Delta}_{\perp}) + \frac{2\cos^2(\theta)+1}{2^7}(\overline{\Delta}_{\parallel} - 2\overline{\Delta}_{\perp})^2\right]. \tag{L2.229}$$

PROBLEM L2.14: Beginning with the leading, $j = 1$, term in Eq. (L2.225) for $G_{AmB}(l,\theta)$ and introducing lowest-order terms for $\overline{\Delta}_{Am}(\xi_n,\psi)$ and $\overline{\Delta}_{Bm}(\xi_n,\theta,\psi)$ from Eqs. (L2.228), derive Eq. (L2.229).

Pairwise interaction of thin rods

Extraction of the pairwise interaction potential follows one of two procedures, depending on whether the rods in A and B are parallel ($\theta = 0$) or are skewed. As with the reduction of the Lifshitz result for planar isotropic bodies to give the interaction between point particles, here we connect the per-unit-area interaction $G_{AmB}(l, \theta)$ between planar regions of embedded cylinders with the pair interaction potential $g(l, \theta)$ between cylinders at an angle θ or the pair interaction energy per unit length $g(l, \theta = 0)$ between parallel cylinders. A sufficient condition for this connection is that $G_{AmB}(l, \theta)$ be accurately expressed by the first, quadratic term in a series expansion in density N or volume fraction v.

For parallel cylinders, the connection is an integral:

$$\lim_{N\to 0} \frac{d^2 G_{AmB}(l, \theta = 0)}{dl^2} = N^2 \int_{-\infty}^{+\infty} g(\sqrt{l^2 + y^2}, \theta = 0) dy + O(N^3). \qquad \text{(L2.230)}$$

Here $g(l, \theta = 0)$ is an energy per unit length; each rod in A, say, interacts with all rods in B. Think of the second derivative as the interaction between two infinitesimally thin slabs at separation l. If we say that these slabs are parallel to the x, y plane and that the parallel rods all point in the direction x, then we must integrate over direction y parallel in order to collect all interactions between a rod in slab A with rods in the apposing slab B.

For cylinders at a mutual angle θ, the connection is

$$\lim_{N\to 0} \frac{d^2 G_{AmB}(l, \theta)}{dl^2} = N^2 \sin\theta \, g(l, \theta) + O(N^3). \qquad \text{(L2.231)}$$

Think of the interaction per unit area between two infinitesimally thin layers of A and B. $N^2 \sin\theta$ is the number of rod–rod interactions per unit area in the two thin layers. Because $g(l, \theta)$ is the energy per interacting pair, $N^2 \sin\theta \, g(l, \theta)$ is an energy per area.

In both cases l is the minimal separation between the cylinders. Here $G_{AmB}(l, \theta)$ has the pleasingly tractable form

$$G_{AmB}(l, \theta) \approx -N^2 \frac{C(\theta)}{l^2},$$

where

$$C(\theta) = \frac{kT}{8\pi}(\pi a^2)^2 {\sum_{n=0}^{\infty}}' \left[\overline{\Delta}_\perp^2 + \frac{\overline{\Delta}_\perp}{4}(\overline{\Delta}_\parallel - 2\overline{\Delta}_\perp) + \frac{2\cos^2(\theta) + 1}{2^7}(\overline{\Delta}_\parallel - 2\overline{\Delta}_\perp)^2 \right]. \qquad \text{(L2.232)}$$

The interaction $g(l, \theta = 0) = g_\parallel(l)$ per unit length, between two parallel thin rods, is succinctly written as

$$g(l, \theta = 0) = g_\parallel(l) = -\frac{c_\parallel}{l^5},$$

$$c_\parallel = \frac{9kT}{16\pi}(\pi a^2)^2 {\sum_{n=0}^{\infty}}' \left[\overline{\Delta}_\perp^2 + \frac{\overline{\Delta}_\perp}{4}(\overline{\Delta}_\parallel - 2\overline{\Delta}_\perp) + \frac{3}{2^7}(\overline{\Delta}_\parallel - 2\overline{\Delta}_\perp)^2 \right]. \qquad \text{(L2.233)}$$

[L1.79][C.4.a]

The interaction $g(l, \theta)$ of thin rods at an angle becomes

$$g(l, \theta) = -\frac{6C(\theta)}{l^4 \sin\theta} = -\frac{c(\theta)}{l^4},$$

$$c(\theta) = \frac{6C(\theta)}{\sin\theta}$$
$$= \frac{3kT(\pi a^2)^2}{4\pi \sin\theta} \sum_{n=0}^{\infty}{}' \left\{ \overline{\Delta}_\perp^2 + \frac{\overline{\Delta}_\perp}{4}(\overline{\Delta}_\| - 2\overline{\Delta}_\perp) + \frac{2\cos^2(\theta)+1}{2^7}(\overline{\Delta}_\| - 2\overline{\Delta}_\perp)^2 \right\}. \tag{L2.234}$$

[L1.77][C.4.b.1]

There is a torque $\tau(l, \theta)$ to this interaction,

$$\tau(l,\theta) = -\left.\frac{\partial g(l,\theta)}{\partial\theta}\right|_l$$
$$= -\frac{3kT\,(\pi a^2)^2}{4\pi l^4}\left[\frac{\cos\theta}{\sin^2\theta}\sum_{n=0}^{\infty}{}'\{\,\} + \frac{\cos\theta}{2^5}\sum_{n=0}^{\infty}{}'(\overline{\Delta}_\| - 2\overline{\Delta}_\perp)^2\right], \tag{L2.235}$$

[L1.78][C.4.b.2]

that causes the rods to twist to align parallel.

PROBLEM L2.15: From Eq. (L2.229), putting Eq. (L2.230) into the form of an Abel transform (see, e.g., Section 8.11 in the *Transforms and Applications Handbook*, Alexander D. Poularikas, ed., CRC Press, Boca Raton, FL, 1996), $h(l) = \int_{-\infty}^{\infty} g(l^2 + y^2)\mathrm{d}y$, use the inverse Abel transform

$$g(l) = -\frac{1}{\pi}\int_l^\infty \frac{h'(y)}{\sqrt{y^2+l^2}}\mathrm{d}y$$

to derive Eq. (L2.233) for the attraction of parallel thin rods.

Breakdown of pairwise additivity at higher density or strong polarizability

We can ask, how do cylinders respond to electric fields when they are not infinitely dilute? For simplicity, think of an array of parallel rods. An electric field perpendicular to the array elicits a collective dielectric response,

$$\varepsilon_{\mathrm{array}}^{\perp} = \varepsilon_{\mathrm{m}}\left(1 + 2\frac{N\pi a^2\overline{\Delta}_\perp}{1 - N\pi a^2\overline{\Delta}_\perp}\right), \tag{L2.236}$$

plus higher-order terms in cross-sectional density.

N is the number of rods per unit area, and $N\pi a^2$ is the volume fraction of rods in the array. Only when $N\pi a^2\overline{\Delta}_\perp \ll 1$ is the response of the array simply proportional to their density N. Otherwise, each rod feels an electric field that is distorted by its neighbors. In the same way, the electric fields created and received by electrodynamically interacting rods are in fact felt by many rods at the same time.

Electric fields parallel to the array experience a response that is exactly a volume average of the rod and medium response:

$$\varepsilon_{\mathrm{array}}^{\|} = \varepsilon_{\mathrm{m}}(1 + N\pi a^2\overline{\Delta}_\|) \tag{L2.237}$$

(think of a capacitor with different materials reaching across at different locations between its plates). Although $\varepsilon_{\mathrm{array}}^{\|}$ is proportional to N, $\overline{\Delta}_\|$ is not like $\overline{\Delta}_\perp$ in that $\overline{\Delta}_\|$ is allowed to take on values much greater than one. Even at low density the rods can dominate the net response $\varepsilon_{\mathrm{array}}^{\|}$. The composite medium between two rods separated

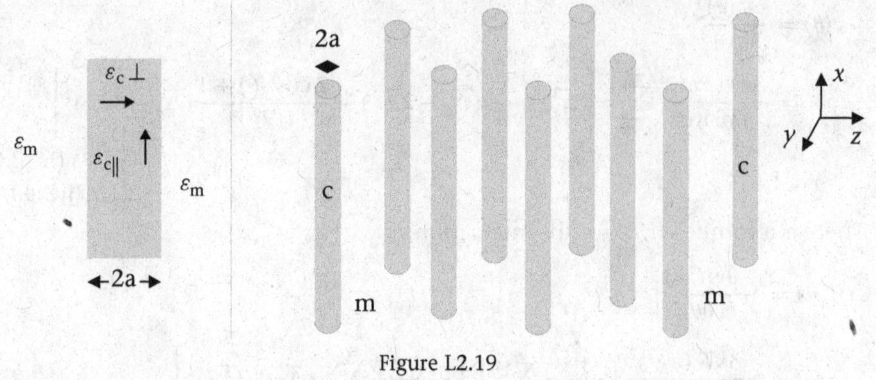

Figure L2.19

by a space containing other rods will not look like the pure ε_m of an infinitely dilute suspension. Just when $\overline{\Delta}_{\|}$ takes on the large values that make us think there will be strong interactions of the form of Eq. (L2.234)

$$g(l,\theta) = -\frac{3kT(\pi a^2)^2}{4\pi l^4 \sin\theta}\frac{2\cos^2\theta + 1}{2^7}\sum_{n=0}^{\infty}{}' \overline{\Delta}_{\|}^2$$

or of Eq. (L2.233)

$$g_{\|}(l) = -\frac{9kT(\pi a^2)^2}{16\pi l^5}\frac{3}{2^7}\sum_{n=0}^{\infty}{}' \overline{\Delta}_{\|}^2,$$

there is a breakdown of the infinite-dilution assumptions on which these expressions are built.

MONOPOLAR IONIC-FLUCTUATION FORCES BETWEEN THIN CYLINDERS

Following the Pitaevskii strategy for extracting small-particle van der Waals interactions for the interaction between suspensions, we specialize the general expression for ionic-fluctuation forces to derive forces between cylinders (Level 3). As with the extraction of dipolar forces between rods, consider two regions A and B, dilute suspensions of parallel rods immersed in salt solution interacting across a region of salt solution m (see Fig. L2.19).

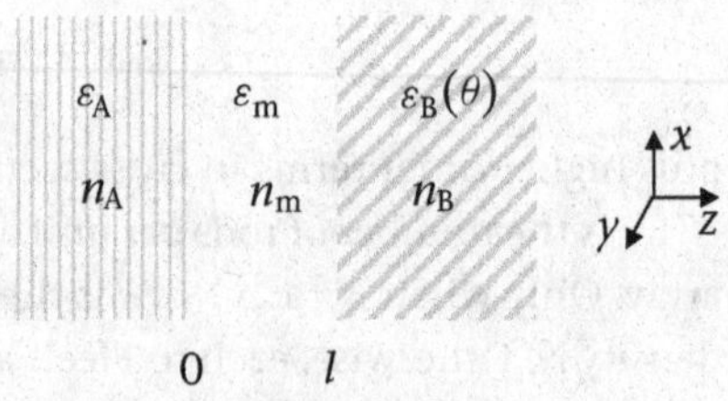

Figure L2.20

Monopolar fluctuation forces occur because of the excess number of ions associated with each cylinder. Only the $n = 0$, zero frequency, terms contribute (see Fig. L2.20).

Region m has an ionic strength as valence-weighted number density

$$n_m = \sum_{\{\nu\}} n_\nu(\mathrm{m})\nu^2 \tag{L2.238}$$

with a corresponding Debye constant

$$\begin{aligned}\kappa_m^2 &= n_m e^2/\varepsilon_0\varepsilon_m kT \text{ (in mks)} = 4\pi n_m e^2/\varepsilon_m kT \text{ (in cgs)}\\ &= 4\pi n_m \lambda_{Bj} \text{ (in either unit system).}\end{aligned} \tag{L2.239}$$

The equivalent for dilute suspensions A and B is an additional Γ_c per unit length per cylinder:

$$n_A = n_B = n_{susp} = n_m + N\Gamma_c, \tag{L2.240}$$

$$\Gamma_c \equiv \sum_{\{\nu\}} \Gamma_\nu \nu^2, \; \Gamma_\nu \equiv \int_0^\infty [n_\nu(r) - n_\nu(m)] 2\pi r \, dr, \tag{L2.241}$$

where $N\Gamma_c \ll n_m$. Γ_ν is the excess number *per unit length* of ions of valence ν that is due to the presence of each rod.

The dielectric response of the suspensions, exclusive of ionic conductance, is ε_m in region m and the uniaxial tensor

$$\begin{pmatrix} \varepsilon_\parallel & 0 & 0 \\ 0 & \varepsilon_\perp & 0 \\ 0 & 0 & \varepsilon_\perp \end{pmatrix},$$

with elements parallel or perpendicular to the rods in regions A and B. As in the case in which ionic fluctuations are ignored, and at low volume fraction of rods, $v = \pi a^2 N$,

$$\varepsilon_\parallel = \varepsilon_m(1 + v\overline{\Delta}_\parallel), \; \varepsilon_\perp \approx \varepsilon_m(1 + 2v\overline{\Delta}_\perp), \; \overline{\Delta}_\parallel \equiv \frac{\varepsilon_{c\parallel} - \varepsilon_m}{\varepsilon_m}, \; \overline{\Delta}_\perp \equiv \frac{\varepsilon_{c\perp} - \varepsilon_m}{\varepsilon_{c\perp} + \varepsilon_m}. \tag{L2.242}$$

The wave equations follow

$$\nabla \cdot (\varepsilon_i \nabla \phi) = k_i^2 \phi, \tag{L2.243}$$

with

$$k_i^2 = n_i e^2/\varepsilon_0 kT \text{ in mks} \quad \text{or} \quad k_i^2 = 4\pi n_i e^2/kT \text{ in cgs units.} \tag{L2.244}$$

In each region i = A, m, B, except for the missing ε in the denominator, these are the same as the Debye κ^2. Because ε can now be a tensor, it is not possible to bring it out of the $\nabla\cdot$ operation and divide it into $k_i^2\phi$. The key feature is that, through n_i, k_i^2 depends on volume average of the ionic strengths.

Because solutions of the wave equation require continuity of $\varepsilon\nabla\phi$ perpendicular to the interfaces at $z = 0$ and l, it is the perpendicular response $\varepsilon_\perp$ of ε_B and ε_B that couples with $k_A^2 = k_B^2$ to create the ionic screening lengths in both semi-infinite regions. For small enough volume fractions $v = N\pi a^2$ and added ionic strengths $N\Gamma_c/n_m$, define

$$\kappa_A^2 = \kappa_B^2 \equiv \frac{k_A^2}{\varepsilon_\perp} = \frac{k_B^2}{\varepsilon_\perp} \approx \frac{k_m^2(1 + N\Gamma_c/n_m)}{\varepsilon_m(1 + 2v\overline{\Delta}_\perp)}$$
$$\approx \kappa_m^2[1 + N(\Gamma_c/n_m - 2\pi a^2\overline{\Delta}_\perp)], \tag{L2.245}$$

so that

$$\kappa_A = \kappa_B \approx \kappa_m[1 + (N/2)(\Gamma_c/n_m - 2\pi a^2\overline{\Delta}_\perp)]. \tag{L2.246}$$

Interaction energy between A and B From the zero-frequency part of the van der Waals interaction free energy per unit area between the two anisotropic semi-infinite regions A and B across a slab m of thickness l [see Eqs. (L3.237) and (L3.238)],

$$G_{AmB}(l,\theta)=\frac{kT}{8\pi^2}\int_0^{2\pi}\mathrm{d}\psi\int_0^{\infty}\ln[D(\rho,\psi,l,\theta)]\rho\mathrm{d}\rho. \tag{L2.247}$$

The ρ are radial wave vectors within the x, y plane; ψ is for angular integration over all directions in ρ to take account of anisotropy.

In the present case, the secular determinant of Eq. (L3.237) for the relevant modes can be written in the form

$$D(\rho,\psi,l,\theta)=1-\overline{\Delta}_{Am}(\psi)\overline{\Delta}_{Bm}(\theta-\psi)e^{-2\sqrt{\rho^2+\kappa_m^2}l}, \tag{L2.248}$$

where $\overline{\Delta}_{Am}(\psi)\overline{\Delta}_{Bm}(\theta-\psi)\ll 1$.

Extraction of rod–rod interactions

For small enough values of $N\pi a^2\overline{\Delta}_{\|}$, $N\pi a^2\overline{\Delta}_{\perp}$, and $N\Gamma_c/n_m$, $\overline{\Delta}_{Am}$ and $\overline{\Delta}_{Bm}$ can be expanded to terms linear in number density N. The interaction energy then goes as N^2. In this dilute limit, rods in the two media interact pairwise across the gap l,[11]

$$G_{AmB}(l,\theta)\approx N^2\frac{kT}{4\pi}\kappa_m^2\int_1^{\infty}f(p,\theta)e^{-2p\kappa_m l}\,p\mathrm{d}p, \tag{L2.249}$$

where $f(p,\theta)$ is independent of separation. As with the extraction for dipolar fluctuation forces, the interaction $g(l,\theta)$ for thin rods *at an angle* is taken from the second derivative of G_{AmB} with respect to separation,

$$\lim_{N\to 0}\frac{\mathrm{d}^2G_{AmB}(l,\theta)}{\mathrm{d}l^2}=N^2\sin\theta\, g(l,\theta)+\mathrm{O}(N^3), \tag{L2.250}$$

where now

$$\frac{\mathrm{d}^2G_{AmB}(l,\theta)}{\mathrm{d}l^2}=-N^2\frac{kT}{\pi}\kappa_m^4\int_1^{\infty}f(p,\theta)e^{-2p\kappa_m l}p^3\mathrm{d}p \tag{L2.251}$$

so that

$$g(l,\theta)=-\frac{kT\kappa_m^4(\pi a^2)^2}{\pi\sin\theta}\{\,\}, \tag{L2.252}$$

[C.5.a.2]

where the integration in p produces this lengthy result[12]:

$$\{\,\}=\left[\overline{\Delta}_{\perp}^2+\frac{\overline{\Delta}_{\perp}(\overline{\Delta}_{\|}-2\overline{\Delta}_{\perp})}{4}+\frac{(\overline{\Delta}_{\|}-2\overline{\Delta}_{\perp})^2}{2^7}(1+2\cos^2\theta)\right]$$

$$\times\frac{6e^{-2\kappa_m l}}{(2\kappa_m l)^4}\left[1+2\kappa_m l+\frac{(2\kappa_m l)^2}{2}+\frac{(2\kappa_m l)^3}{6}\right]$$

$$+\left[\begin{array}{l}\left(\dfrac{\Gamma_c}{\pi a^2 n_m}\right)\dfrac{\overline{\Delta}_{\perp}}{2}+\left(\dfrac{\Gamma_c}{\pi a^2 n_m}\right)\dfrac{(\overline{\Delta}_{\|}-2\overline{\Delta}_{\perp})}{16}\\ -\overline{\Delta}_{\perp}^2-\dfrac{3\overline{\Delta}_{\perp}(\overline{\Delta}_{\|}-2\overline{\Delta}_{\perp})}{8}-\dfrac{(\overline{\Delta}_{\|}-2\overline{\Delta}_{\perp})^2}{2^6}(1+2\cos^2\theta)\end{array}\right]$$

$$\times \frac{1}{(2\kappa_m l)^2} e^{-2\kappa_m l}\left[1 + 2\kappa_m l\right]$$

$$+\left[\begin{array}{l} \left(\frac{\Gamma_c}{\pi a^2 n_m}\right)^2 \frac{1}{16} - \left(\frac{\Gamma_c}{\pi a^2 n_m}\right)\frac{\overline{\Delta}_\perp}{4} - \left(\frac{\Gamma_c}{\pi a^2 n_m}\right)\frac{(\overline{\Delta}_\parallel - 2\overline{\Delta}_\perp)}{16} \\ +\frac{\overline{\Delta}_\perp^2}{4} + \frac{\overline{\Delta}_\perp(\overline{\Delta}_\parallel - 2\overline{\Delta}_\perp)}{8} + \frac{(\overline{\Delta}_\parallel - 2\overline{\Delta}_\perp)^2}{2^7}(1 + 2\cos^2\theta) \end{array}\right] E_1(2\kappa_m l).$$

Here we use the exponential integral $E_1(2\kappa_m l) = -\gamma - \ln(2\kappa_m l) - \sum_{n=1}^{\infty}[(-2\kappa_m l)^n/nn!)]$, with $\gamma = 0.5772156649$; for large arguments $E_1(2\kappa_m l) \rightarrow [(e^{-2\kappa_m l})/(2\kappa_m l)]$.

For *parallel* rods, we require a more complicated procedure that leads to the form[13]

$$g(l, \theta = 0) = g_\parallel(l) = -\frac{2kT}{\pi^2}\kappa_m^5 \int_1^\infty f(p, \theta = 0) K_0(2\kappa_m pl) p^4 \mathrm{d}p. \qquad \text{(L2.253)}$$

Here $K_0(x)$ is the modified Bessel cylindrical function of order 0:

$$g_\parallel(l) = -\frac{2kT\kappa_m^5(\pi a^2)^2}{\pi^2}\{\,\}, \qquad \text{(L2.254)}$$

[C.5.a.1]

where

$$\{\,\} = \left[\overline{\Delta}_\perp^2 + \frac{\overline{\Delta}_\perp(\overline{\Delta}_\parallel - 2\overline{\Delta}_\perp)}{4} + \frac{3(\overline{\Delta}_\parallel - 2\overline{\Delta}_\perp)^2}{2^7}\right]\int_1^\infty K_0(2\kappa_m pl) p^4 \mathrm{d}p$$

$$+\left[\begin{array}{l} \left(\frac{\Gamma_c}{\pi a^2 n_m}\right)\frac{\overline{\Delta}_\perp}{2} + \left(\frac{\Gamma_c}{\pi a^2 n_m}\right)\frac{(\overline{\Delta}_\parallel - 2\overline{\Delta}_\perp)}{16} \\ -\overline{\Delta}_\perp^2 - \frac{3\overline{\Delta}_\perp(\overline{\Delta}_\parallel - 2\overline{\Delta}_\perp)}{8} - \frac{3(\overline{\Delta}_\parallel - 2\overline{\Delta}_\perp)^2}{2^6} \end{array}\right]\int_1^\infty K_0(2\kappa_m pl) p^2 \mathrm{d}p$$

$$+\left[\begin{array}{l} \left(\frac{\Gamma_c}{\pi a^2 n_m}\right)^2 \frac{1}{16} - \left(\frac{\Gamma_c}{\pi a^2 n_m}\right)\frac{\overline{\Delta}_\perp}{4} - \left(\frac{\Gamma_c}{\pi a^2 n_m}\right)\frac{(\overline{\Delta}_\parallel - 2\overline{\Delta}_\perp)}{16} \\ +\frac{\overline{\Delta}_\perp^2}{4} + \frac{\overline{\Delta}_\perp(\overline{\Delta}_\parallel - 2\overline{\Delta}_\perp)}{8} + \frac{3(\overline{\Delta}_\parallel - 2\overline{\Delta}_\perp)^2}{2^7} \end{array}\right]\int_1^\infty K_0(2\kappa_m pl)\mathrm{d}p.$$

The integrations in p are, unfortunately, not neatly solvable for all values of $2\kappa_m l$. For $2\kappa_m l \gg 1$, there are exponential forms[14]:

$$\int_1^\infty K_0(2\kappa_m pl) p^{2q} \mathrm{d}p \sim \sqrt{\frac{\pi}{2}}\frac{e^{-2\kappa_m l}}{(2\kappa_m l)^{3/2}}. \qquad \text{(L2.255)}$$

Purely ionic fluctuations

In the case in which there are only ionic fluctuations, $\Delta_\parallel \rightarrow 0$ and $\Delta_\perp \rightarrow 0$, and when $\kappa_m l \gg 1$,

$$g(l, \theta) \rightarrow -\frac{kT\kappa_m^4(\pi a^2)^2}{\pi \sin\theta}\left[\left(\frac{\Gamma_c}{\pi a^2 n_m}\right)^2 \frac{1}{16}\right] E_1(2\kappa_m l)$$

$$= -\frac{kT(4\pi\lambda_{Bj} n_m)^2}{16\pi \sin\theta}\left[\left(\frac{\Gamma_c}{n_m}\right)^2\right]\frac{e^{-2\kappa_m l}}{2\kappa_m l} = -\frac{kT\pi\lambda_{Bj}^2}{\sin\theta}\Gamma_c^2\frac{e^{-2\kappa_m l}}{2\kappa_m l}, \qquad \text{(L2.256)}$$

[C.5.b.2]

$$g_{\|}(l) \to -\frac{2kT\kappa_{\mathrm{m}}^5(\pi a^2)^2}{\pi^2}\left\{\left[\left(\frac{\Gamma_{\mathrm{c}}}{\pi a^2 n_{\mathrm{m}}}\right)^2\frac{1}{16}\right]\int_1^\infty K_0(2\kappa_{\mathrm{m}}pl)\mathrm{d}p\right\}$$

$$\to -\frac{2kT\kappa_{\mathrm{m}}^5}{16\pi^2}\left(\frac{\Gamma_{\mathrm{c}}}{n_{\mathrm{m}}}\right)^2\sqrt{\frac{\pi}{2}}\frac{e^{-2\kappa_{\mathrm{m}}l}}{(2\kappa_{\mathrm{m}}l)^{3/2}} = -\frac{kT\lambda_{\mathrm{Bj}}^2}{2}\Gamma_{\mathrm{c}}^2\sqrt{\pi}\frac{e^{-2\kappa_{\mathrm{m}}l}}{\kappa_{\mathrm{m}}^{1/2}l^{3/2}}. \quad \text{(L2.257)} \; \text{[C.5.b.1]}$$

Again, as with ionic-fluctuation forces in planar and spherical geometries, the exponential shows the effect of double screening, the only difference being the distance dependence in the denominator.

L2.4. Computation

Attitude

It is remarkable how many people think it difficult to compute van der Waals forces directly by using the Lifshitz theory. Invariably, after a few minutes' instruction, there is the reaction "I didn't know how easy it was." Essentially it is a matter of introducing tabulated experimental information for ε's and numerically summing or integrating for the interaction energy.

Thanks to rapid progress in spectroscopy we can soon expect to compute van der Waals forces by direct conversion of material responses to applied electromagnetic fields. The future of precise computation relies on measuring these responses on the same materials as those used to measure forces. This is because small changes—in composition of materials, dopants that confer conductance, solutes that modify spectra, even in atomic arrangement that create anisotropies—all have quantitative consequences. Whether the unintended result of handling materials or artful modifications to create forces, spectral details merit respect.

Combining these data with the theory of dielectrics helps us to think about the connection between specific features of absorption spectra and forces in order to create strategies for designing materials or to find reasons to explain measured forces. Outside the purview of this text, the actual practice of assembling data into usable $\varepsilon(\omega)$'s and $\varepsilon(i\xi)$'s has been well described in the physics and the engineering literature. To connect readers with that literature, this chapter first introduces the essential physical features and language used to create ε's of computation. It then gives examples of different ways to compute forces with different approximations to ε.

L.2.4.A. Properties of dielectric response

As emphasized elsewhere in this text, the physical act constituting an electrodynamic force is the correlated time-varying fluctuation of all component electric charges and electromagnetic fields in each material composing a system. Charge fluctuations at each point are either spontaneous or are in response to electric fields set up by fluctuations elsewhere. The dielectric permittivity is an experimental quantity that codifies not only the response of a material to an applied electric field but also the magnitude of spontaneous fluctuations.

Much of the unnecessary failure to use the easier, modern theory of van der Waals forces comes from its language, the uncommon form in which the dielectric permittivity is employed. For many people "complex dielectric permittivity" and "imaginary frequency" are terms in a strange language. Dielectric permittivity describes what a material does when exposed to an electric field. An imaginary-frequency field is one that varies exponentially versus time rather than as oscillatory sinusoidal waves.

In practice we usually think about the response of a material to oscillatory fields—absorption, reflection, transmission, refraction, etc. We learn to connect the frequencies at which electromagnetic waves are absorbed with the natural motions of the material. If necessary, we can use oscillatory-field responses to know what the material would do in nonoscillatory fields.

We can run the cause–effect connection the other way. The natural motions of the charges within a material will necessarily create electric fields whose time-varying spectral properties are those known from how the materials absorb the energy of applied fields (the "fluctuation–dissipation theorem"). It is the correlations between these spontaneously occurring electric fields and their source charges that create van der Waals forces. At a deeper level, we can even think of all these charge or field fluctuations as results or distortions of the electromagnetic fields that would occur spontaneously in vacuum devoid of matter.

Strategy

How do we see the conversion of spectra into charge fluctuation? It comes down to creating appropriate language to store the information:

1. Recognize causality. An effect (charge displacement) must come after a cause (applied electric field).
2. Frequency-analyze cause and effect so as to extract a response function to electromagnetic waves over the full range of frequencies.
3. Use this full-frequency material response to describe how materials respond to (and create) any time-varying electromagnetic field.

In practice, even the requirement for a "full range of frequencies" is sometimes eased because we can identify the particular features of response important to a particular force computation. The principal limitation to the language of dielectric response is its restriction to electric fields weak enough to provoke only a linear response. This weak-field condition poses no limitation to computing forces between materials in their thermal-equilibrium states.

Some elementary definitions

"Dielectric" designates the response of material to an electric field applied across it ($\delta\iota$ or $\delta\iota\alpha$, Greek *di* or *dia* means "across"). Recall that a "dielectric constant," ε, is a coefficient of proportionality relating a constant electric field **E** in a material to the electric polarization **P** of the material in response to that field (note the use of cgs Gaussian units here).

$$\mathbf{E} + 4\pi\mathbf{P} = \varepsilon\mathbf{E} \qquad [\text{or } 4\pi\mathbf{P} = \mathbf{E}(\varepsilon - 1)], \tag{L2.258}$$

where $\mathbf{E} + 4\pi\mathbf{P}$ is the dielectric induction

$$\mathbf{D} = \mathbf{E} + 4\pi\mathbf{P}. \tag{L2.259}$$

If the medium is a vacuum, then no polarization is possible, $\mathbf{E} = \varepsilon\mathbf{E}$ and $\varepsilon \equiv 1$. But in a material substance there will be a shifting of positive and negative charges proportional to the electric field. The polarizability coefficient χ is defined by

$$\mathbf{P} = \chi\mathbf{E}, \text{ so that } \varepsilon = 1 + 4\pi\chi. \tag{L2.260}$$

The material properties of any substance are measured by a deviation of ε from unity. **E**, **P**, and **D** as used here are averages over a small volume inside the material, a volume large enough compared with molecular sizes and spacings so as to be able to treat the material content as a macroscopic continuum.

When measurement involves application of a known outside field $\mathbf{E}_{\text{outside}}$ to the material sample, this $\mathbf{E}_{\text{outside}}$ is related to the $\mathbf{E}_{\text{inside}}$ inside by the two boundary conditions at a material interface bearing no free charge:

1. Components of $\mathbf{E}_{\text{outside}}$ and $\mathbf{E}_{\text{inside}}$ parallel to the surface are equal,
2. Components perpendicular to the surface are related by $\varepsilon_{\text{outside}}\mathbf{E}_{\text{outside}} = \varepsilon_{\text{inside}}\mathbf{E}_{\text{inside}}$.

Causality in polarization response

Now ask about the response to an electric field $\mathbf{E}(t)$ that varies with time. The material will have some memory of the applied field so that its polarization $\mathbf{P}(t)$ at time t will reflect the electric field applied at time t as well as at previous times. Imagine a memory function $f(\tau)$ that gives a $4\pi\mathbf{P}(t)$ at time t that is due to electric fields $\mathbf{E}(t-\tau)$ applied at times τ earlier than t. As τ goes to infinity, $f(\tau)$ must go to zero because polarization cannot reflect the effect of fields applied an infinitely long time before. (Neglecting hysteresis.) The function $f(\tau)$ cannot go to infinite value because this would imply an infinitely rapid or infinitely strong response to a finite field. Because $\mathbf{E}(t)$ and $\mathbf{P}(t)$ are physically real quantities, $f(\tau)$ is a mathematically real quantity.

This $\mathbf{P}(t)$, $\mathbf{E}(t-\tau)$ relation can be written as an integral that accumulates the experience of all previous times:

$$4\pi\mathbf{P}(t) = \int_0^\infty f(\tau)\mathbf{E}(t-\tau)\mathrm{d}\tau. \tag{L2.261}$$

It seems obvious, but it is worth stating explicitly, that because of causality we are allowed to think only about *positive* values of τ. An effect (**P**) can come only *after* a cause (**E**).

In the language of frequencies

A time-varying electric field can be expressed as the sum of Fourier components $\mathbf{E}_\omega$ multiplied by time-varying factors $e^{-i\omega t}$:

$$\mathbf{E}(t) = \int_{-\infty}^{\infty} \mathbf{E}_\omega e^{-i\omega t}\mathrm{d}\omega/2\pi. \tag{L2.262}$$

The ω is the radial frequency in radians per second. Experimentalists usually think of frequency in hertz or, more descriptively, in "cycles per second," ν. The conversion is $\omega = 2\pi\nu$. Radial frequency has literary advantages that allow us to write $\cos(\omega t)$ rather than $\cos(2\pi\nu t)$. (The 2π comes from the circumference of the unit circle by which the unit of radians is defined.)

Displacement $\mathbf{D}(t)$ and polarization $\mathbf{P}(t)$ can be similarly frequency resolved. The connection between each $\mathbf{D}_\omega$ and $\mathbf{E}_\omega$ is written as a coefficient $\varepsilon(\omega)$:

$$\mathbf{D}_\omega = \mathbf{E}_\omega + 4\pi\mathbf{P}_\omega = \varepsilon(\omega)\mathbf{E}_\omega, \tag{L2.263}$$

where we now show that

$$\varepsilon(\omega) \equiv 1 + \int_0^\infty f(\tau)e^{i\omega\tau}\mathrm{d}\tau. \tag{L2.264}$$

It is worth spelling this out. Think of

$$\mathbf{E}_\omega \equiv \int_{-\infty}^{\infty} \mathbf{E}(t)e^{i\omega t}\mathrm{d}t;\quad \mathbf{P}_\omega \equiv \int_{-\infty}^{\infty} \mathbf{P}(t)e^{i\omega t}\mathrm{d}t;\quad \mathbf{D}_\omega \equiv \int_{-\infty}^{\infty} \mathbf{D}(t)e^{i\omega t}\mathrm{d}t, \tag{L2.265}$$

but with the connection [Eq. (L2.261)]

$$4\pi\mathbf{P}(t) = \int_0^\infty f(\tau)\mathbf{E}(t-\tau)\,\mathrm{d}\tau. \tag{L2.266}$$

Then

$$\begin{aligned} 4\pi\mathbf{P}_\omega &\equiv \int_{-\infty}^{\infty}\int_0^\infty f(\tau)\mathbf{E}(t-\tau)e^{i\omega t}\,\mathrm{d}\tau\,\mathrm{d}t \\ &= \int_{-\infty}^{\infty}\int_0^\infty f(\tau)\,e^{i\omega\tau}\mathrm{d}\tau\;\mathbf{E}(t-\tau)e^{i\omega(t-\tau)}\,\mathrm{d}(t-\tau) \qquad \text{(L2.267)} \\ &= \left[\int_0^\infty f(\tau)e^{i\omega\tau}\mathrm{d}\tau\right]\times\left[\int_{-\infty}^{\infty}\mathbf{E}(t')e^{i\omega t'}\,\mathrm{d}t'\right] = \int_0^\infty f(\tau)\,e^{i\omega\tau}\mathrm{d}\tau\times\mathbf{E}_\omega, \qquad \text{(L2.268)} \end{aligned}$$

from which Eq. (L2.264) follows.

We see in this relation that $\varepsilon(\omega)$ picks out the frequency characteristics of the memory function $f(\tau)$. We take advantage of this mathematical form to regard $\varepsilon(\omega)$ purely mathematically as a complex function

$$\varepsilon(\omega) = \varepsilon'(\omega) + i\varepsilon''(\omega) \tag{L2.269}$$

with real and imaginary parts, $\varepsilon'(\omega)$ and $\varepsilon''(\omega)$. This separation of $\varepsilon(\omega)$ follows automatically from the fact that the factor $e^{i\omega t}$, used in Eq. (L2.264) to create $\varepsilon(\omega)$, is a sum of real and imaginary parts.

The frequency ω itself can be regarded as a complex variable. We are not restricted to thinking about only sinusoidal oscillations of fields but we can use frequency language to talk about fields that shrink or grow exponentially. The language must be able to talk about real materials in which electric fields or charge fluctuations occur, oscillate with natural frequencies of the substance, and die away over time.

For this reason, we think of ω as a combination of "real" ω_R and "imaginary" ξ parts:

$$\omega = \omega_R + i\xi \tag{L2.270}$$

As described in Level 1, the function $e^{i\omega t}$ factors into two exponentials $e^{i\omega t} = e^{i\omega_R t} e^{-\xi t}$. The "complex-frequency" language now describes oscillations, $e^{i\omega_R t}$, and exponential change, $e^{-\xi t}$. In this way, when we speak of $\varepsilon(\omega)$ we think of a function of two real variables, ω_R and ξ. The response $\varepsilon(\omega)$ is conveniently plotted on a complex-frequency plane with axes ω_R and ξ. The dielectric response $\varepsilon(\omega)$ of a material is a function of these two variables (see Fig. L2.21).

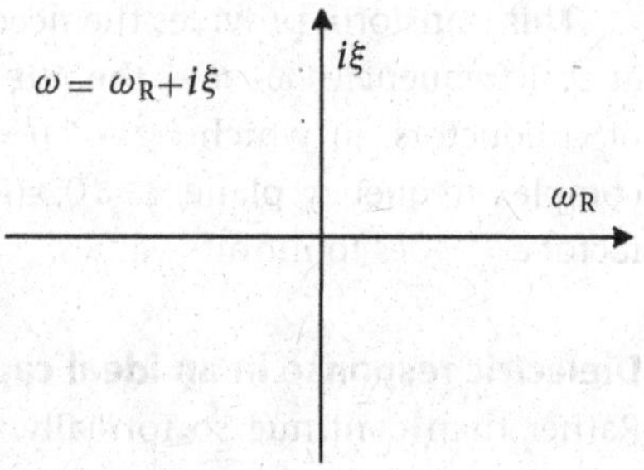

Figure L2.21

Inescapable properties of the response function

Expand $\varepsilon(\omega) = \varepsilon'(\omega) + i\varepsilon''(\omega)$ in terms of ω_R and ξ:

$$\varepsilon(\omega) = \varepsilon'(\omega) + \varepsilon''(\omega) = 1 + \int_0^\infty f(\tau) e^{i\omega\tau} d\tau = 1 + \int_0^\infty f(\tau) e^{i\omega_R \tau} e^{-\xi\tau} d\tau. \quad \text{(L2.271)}$$

From this form alone we know that $\varepsilon(\omega)$ has the following properties:

1. For $\xi > 0$, that is, on the *upper* half of the complex-frequency plane, $\varepsilon(\omega)$ must remain finite because $f(\tau)$ is finite. (Causality again! Only positive τ allowed.) The positive-ξ factor, $e^{-\xi\tau} \le 1$, ensures that the integral never blows up.

2. On the real-frequency axis, where $\xi = 0$ and $e^{i\omega_R \tau} = \cos(\omega_R \tau) + i \sin(\omega_R \tau)$, the real part of $\varepsilon(\omega_R)$ is an even function of frequency,

$$\varepsilon'(\omega_R) = 1 + \int_0^\infty f(\tau) \cos(\omega_R \tau) d\tau, \quad \varepsilon'(\omega_R) = \varepsilon'(-\omega_R); \quad \text{(L2.272)}$$

and the imaginary part of $\varepsilon(\omega_R)$ is an odd function,

$$\varepsilon''(\omega_R) = \int_0^\infty f(\tau) \sin(\omega_R \tau) d\tau, \quad \varepsilon''(\omega_R) = -\varepsilon''(-\omega_R). \quad \text{(L2.273)}$$

3. On the imaginary-frequency axis, where $\omega_R = 0$,

$$\varepsilon(i\xi) = 1 + \int_0^\infty f(\tau) e^{-\xi\tau} d\tau \quad \text{(L2.274)}$$

is a purely real function, $\varepsilon''(i\xi) = 0$. For positive ξ, $\varepsilon(i\xi)$ decreases monotonically versus ξ.

Because the essential properties are in $f(\tau)$, there is necessarily a connection—the Kramers–Kronig relations[1]—among $\varepsilon'(\omega_R)$, $\varepsilon''(\omega_R)$, and, important because of its use in computation, $\varepsilon(i\xi)$. For positive ξ,

$$\varepsilon(i\xi) = 1 + \frac{2}{\pi} \int_0^\infty \frac{\omega_R\, \varepsilon''(\omega_R)}{\omega_R^2 + \xi^2} d\omega_R \quad \text{(L2.275)}$$

This transform provides the needed conversion between measurements conducted at real frequencies ω_R and the $\varepsilon(i\xi)$ used in force computation. Except for the case of conductors, in which $\varepsilon(\xi \to 0) = \infty$, $\varepsilon(0)$ is finite. In the entire lower half of the complex-frequency plane, $\xi < 0$, $\varepsilon(i\xi)$ can take on infinitely high values because the factor $e^{-\xi\tau}$ goes to infinite values.

Dielectric response in an ideal capacitor

Rather than continue so formally, consider dielectric susceptibilities in terms of illustrative models. Conceptually the simplest picture of a dielectric response is that in an electric circuit. Think about a capacitor as a sandwich of interesting material between two parallel conducting plates (see Fig. L2.22).

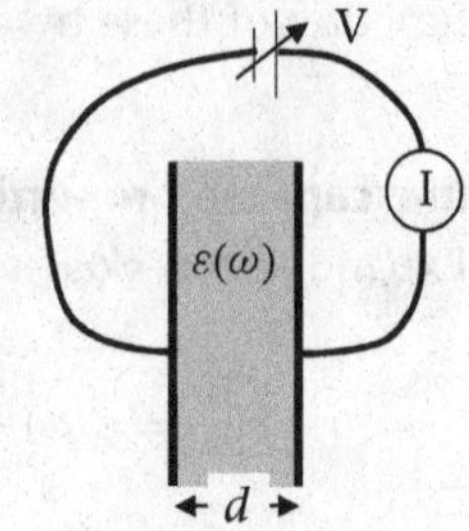

Figure L2.22

Here this capacitor has parallel plates of unit area separated by distance d. The space between them is filled with a uniform substance. For this example, thickness d is taken to be thin compared with the wavelength of any applied electric field. (Thickness d is also very small compared with the lateral dimensions so that there are no edge effects.)

In the linear-response regime treated here, an oscillatory voltage difference V across the substance can be written in the form

$$V(t) = V_\omega \,\mathrm{Re}(e^{-i\omega t}) = V_\omega \cos(\omega t). \tag{L2.276}$$

This applied voltage is written in such a long-winded way to keep the language of a complex-number oscillation $e^{-i\omega t}$ while reminding us that a voltage is a mathematically real quantity. Here, as this capacitor model is described, V_ω and ω are also defined as real quantities (with the cumbersome subscript R temporarily omitted from ω_R for the sinusoidal frequency applied to or detected between the capacitor plates).

The capacitance of this system is defined as the magnitude of charge Q_ω that can be deposited on each of the plates at a given voltage V_ω:

$$C(\omega) = Q_\omega / V_\omega. \tag{L2.277}$$

The interesting part of this capacitance is that which is due to the presence of the material between the plates. Write C_0 for the relatively uninteresting case in which the plates are separated only by vacuum. The dielectric susceptibility $\varepsilon(\omega)$ of the intervening substance is defined as the ratio of the measured capacitance $C(\omega)$ compared with C_0:

$$\varepsilon(\omega) \equiv C(\omega)/C_0. \tag{L2.278}$$

The substance is able to respond to the applied voltage by shifting its own constituent charges. A positive charge will move toward the negative plate; a negative charge toward the positive plate. If the applied voltage is oscillating, the charge must be able to respond quickly enough to follow the voltage variation.

In the limit of very high applied frequencies, the constituent material charges can no longer keep up with the applied field. The material response will be the same as that of a vacuum:

$$\varepsilon(\omega \to \infty) \to 1. \tag{L2.279}$$

The capacitance per unit area of a vacuum space is (in cgs units)

$$C_0 = 1/4\pi d. \tag{L2.280}$$

The capacitance with intervening material is

$$C(\omega) = \varepsilon(\omega)/4\pi d. \tag{L2.281}$$

To develop intuition, consider the electric current $I(t)$ flowing in response to the applied voltage $V(t)$. From elementary circuit theory, the capacitive impedance, sometimes "reactance," against this voltage is

$$Z_\omega = \frac{-1}{i\omega\, C(\omega)} = \frac{-4\pi d}{i\omega\, \varepsilon(\omega)}. \tag{L2.282}$$

The resulting current $I(t)$ in response to oscillatory $V(t)$ is also oscillatory but suffers a change in phase. We can still write $I(t)$ at frequency ω as

$$I(t) = \mathrm{Re}(I_\omega \mathrm{e}^{-i\omega t}), \tag{L2.283}$$

but we must remember that I_ω can be complex. Formally this is because the impedance is complex; physically this is because I_ω is shifted in phase from the applied voltage component V_ω that was defined as a real quantity in order to let voltage set the reference point for the phase:

$$I_\omega = \frac{V_\omega}{Z_\omega} = -V_\omega \frac{i\omega\, \varepsilon(\omega)}{4\pi d} = +\frac{\omega\, V_\omega}{4\pi d}[-i\varepsilon'(\omega) + \varepsilon''(\omega)]. \tag{L2.284}$$

Then, explicitly putting $e^{-i\omega t} = \cos(\omega t) - i\sin(\omega t)$, we find that the electric current is

$$I(t) = \mathrm{Re}(I_\omega e^{-i\omega t}) = +\frac{\omega\, V_\omega}{4\pi d}\{-i\varepsilon'(\omega)[-i\sin(\omega t)] + \varepsilon''(\omega)\cos(\omega t)\}$$

$$= +\frac{\omega\, V_\omega}{4\pi d}[\varepsilon''(\omega)\cos(\omega t) - \varepsilon'(\omega)\sin(\omega t)]. \tag{L2.285}$$

One part of this current, $\varepsilon''(\omega)\cos(\omega t)$, changes simultaneously with $V_\omega \cos(\omega t)$ whereas the second part, $-\varepsilon'(\omega)\sin(\omega t)$, is 90° out of phase with voltage.

From this distinction, we see an instructive physical difference between the real and the imaginary parts of $\varepsilon(\omega)$. Examine the average power required for maintaining the oscillatory current, that is, the time average of current × voltage. Two features stand out:

1. The 90°-out-of-phase term averages to zero because

$$\lim_{T\to\infty} \frac{1}{2T}\int_{-T}^{T} \sin(\omega t)\cos(\omega t)\mathrm{d}(\omega t) = 0. \tag{L2.286}$$

2. The in-phase term gives a nonzero average because

$$\lim_{T\to\infty} \frac{1}{2T}\int_{-T}^{T} \cos^2(\omega t)\mathrm{d}(\omega t) = 1/2. \tag{L2.287}$$

With this factor 1/2 for the time average, the average dissipation of energy, from current × driving voltage, is

$$\frac{1}{2}\mathrm{Re}(I_\omega V_\omega) = \frac{1}{2}\mathrm{Re}\left(\frac{V_\omega}{Z_\omega} V_\omega\right) = \frac{1}{2}\mathrm{Re}\left[-V_\omega^2 \frac{i\omega\, \varepsilon(\omega)}{4\pi d}\right] = \frac{1}{2}\frac{V_\omega^2}{4\pi d}\,\omega\varepsilon''(\omega). \tag{L2.288}$$

This result, well known from electric circuitry, conveys essential information. Because on the real-frequency axis $\varepsilon''(\omega_R)$ is an odd function [Eq. (L2.273)], the product $\omega_R\varepsilon''(\omega_R)$ is always positive. An applied real sinousoidal frequency can only dissipate energy. Proportional to the "oscillator strength" of the material, $\omega\varepsilon''(\omega)$ measures the dissipation, the absorption of electrical energy from an oscillating field. This ability to absorb electromagnetic energy is the material property that is converted into the imaginary-frequency susceptibility $\varepsilon(i\xi)$ used in force computation.

At those frequencies at which $\varepsilon''(\omega)$ is zero, it takes no work to maintain an applied voltage. The electric field does work on the material during half the cycle; restoring forces in the material return that work during the other half. The real part of the susceptibility, $\varepsilon'(\omega)$, measures the springiness of the material, its ability to store and give back electrical energy.

At "resonance frequencies" at which the real part of $\varepsilon(\omega)$, $\varepsilon'(\omega)$, is at or near zero, $\varepsilon''(\omega)$ is comparatively large. There the charges in the system move in step with the applied voltage; little or no work is returned to the external driving voltage source. The work done by the field degenerates into heat within the sample.

Resonance frequencies or absorption frequencies occur when the natural frequencies of charge motion are close to the frequencies of the applied fields. It is no surprise that absorption frequencies are what show up in forces that depend on spontaneous charge fluctuation (see Fig. L2.23).

Now consider the converse of capacitive response. Remove the driving voltage source $V(t)$; replace the old current meter with an ideally sensitive device that measures the electric currents created by spontaneous charge fluctuations. Because the material is at a finite temperature, charges move in thermal agitation. The frequencies of such fluctuations will be those corresponding to the natural motions of the constituent charges.[2] Because of the uncertainty in simultaneous position and motion, even at very low temperature, uncertainty-principle zero-point motions will set up electric and magnetic fields at those natural frequencies.

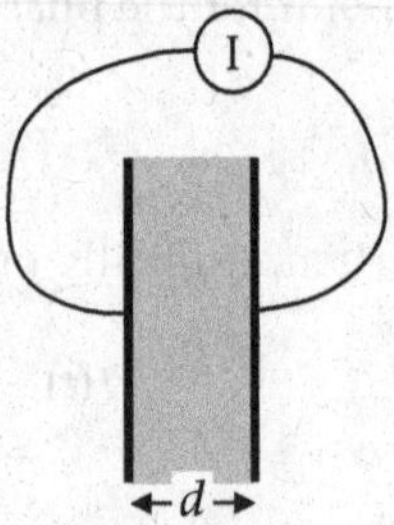

Figure L2.23

From the general theorem for current fluctuations in any circuit, the mean-square current at a radial frequency ω goes as[3]

$$\langle I^2\rangle_\omega = \frac{\hbar\omega}{2\pi}\coth\left(\frac{\hbar\omega}{2kT}\right)\mathrm{Re}\left(\frac{1}{Z_\omega}\right). \tag{L2.289}$$

Here, $\hbar\omega$ is the photon energy at frequency ω and kT is thermal energy.

For frequencies such that $\hbar\omega \ll kT$, the classical limit,

$$\frac{\hbar\omega}{2\pi}\coth\left(\frac{\hbar\omega}{2kT}\right) \to \frac{kT}{\pi}, \tag{L2.290}$$

current fluctuations become proportional to temperature.

For the quantum limit, $\hbar\omega \gg kT$,

$$\frac{\hbar\omega}{2\pi}\coth\left(\frac{\hbar\omega}{2kT}\right) \to \frac{\hbar\omega}{2\pi}, \tag{L2.291}$$

the intensity of current fluctuations goes as the photon energy $\hbar\omega$.

With

$$Z_\omega = \frac{-1}{i\omega\, C(\omega)} = \frac{-4\pi d}{i\omega\, \varepsilon(\omega)}, \qquad \mathrm{Re}\left[\frac{1}{Z_\omega}\right] = \frac{\omega\varepsilon''(\omega)}{4\pi d},$$

the square of the current fluctuation at frequency ω can be written as

$$\langle I^2\rangle_\omega = \frac{\hbar\omega}{2\pi}\coth\left(\frac{\hbar\omega}{2kT}\right)\frac{\omega\varepsilon''(\omega)}{4\pi d}, \tag{L2.292}$$

with fluctuation explicitly dependent on $\omega\varepsilon''(\omega)$, as was dissipation.

For example, at high temperatures the current fluctuations in our ideal device go as

$$\frac{kT}{\pi}\frac{\omega\varepsilon''(\omega)}{4\pi d} = kT\frac{\omega\varepsilon''(\omega)}{(2\pi)^2 d}. \tag{L2.293}$$

It is only a small step from here to the equivalent statement, known as the Nyquist theorem, for current noise of thermal fluctuations in a circuit with resistance R.[4]

Fluctuations may be expressed in terms of charge or electric field as well as of current.

Charge fluctuations: I_ω is the time derivative of the charge Q_ω on the faces of the capacitor. That is, $I(t) = (\partial Q/\partial t)$, where $Q(t)$ is $\mathrm{Re}(Q_\omega e^{-i\omega t})$. For a given frequency component, $|I_\omega| = \omega|Q_\omega|$, and Eq. (L2.292) gives

$$\langle Q^2\rangle_\omega = \frac{\hbar}{8\pi^2 d}\coth\left(\frac{\hbar\omega}{2kT}\right)\varepsilon''(\omega). \tag{L2.294}$$

Although described here in terms of a capacitor model, these connections between the fluctuation and the dissipative part $\varepsilon''(\omega)$ of the susceptibility are general. They hold well beyond the illustrative example of an idealized capacitor used here to describe them. It is not always possible to express fluctuations in so many equivalent terms—current, voltage, charge, field—as is done here because of the explicit connection enforced by the circuitry equations. Nevertheless, fluctuations of one kind can often be transformed from one variable to another.

Which of the fluctuations discussed here is really the fundamental one? Conceptually, for theory, it is probably easiest to think of charge fluctuations. Operationally, in experiments, it is probably a fluctuation in detected voltage or current that is most accessible. We learn every way we can.

Susceptibility as seen at optical frequencies: At optical frequencies we usually think of an index of refraction $n_{\mathrm{ref}}(\omega_{\mathrm{R}})$ and an absorption coefficient $\kappa_{\mathrm{abs}}(\omega_{\mathrm{R}})$ for the propagation and attenuation of light of radial frequency ω_{R}. These are related to susceptibility ε as

$$\varepsilon = (n_{\mathrm{ref}} + i\kappa_{\mathrm{abs}})^2, \tag{L2.295}$$

so that

$$\varepsilon'(\omega) = n_{\mathrm{ref}}^2 - \kappa_{\mathrm{abs}}^2, \tag{L2.296}$$

$$\varepsilon''(\omega) = 2n_{\mathrm{ref}}\kappa_{\mathrm{abs}}. \tag{L2.297}$$

Measured n_{ref} and κ_{abs} yield $\varepsilon'(\omega)$ and $\varepsilon''(\omega)$. For example, they can be obtained by the reflection of light where the reflection coefficient of light incident perpendicular to a planar surface is

$$\text{Reflectivity} = \frac{(n_{\mathrm{ref}} - 1)^2 + \kappa_{\mathrm{abs}}^2}{(n_{\mathrm{ref}} + 1)^2 + \kappa_{\mathrm{abs}}^2} \tag{L2.298}$$

when the substance is in a vacuum.

In a transparent medium, $\kappa_{\text{abs}} = 0$, a plane wave has the form $e^{i\kappa_{\text{abs}} n_{\text{ref}} x}$ where $\kappa_{\text{abs}} = (2\pi/\lambda_0)$, where λ_0 is the wavelength in vacuum.

In an absorbing medium, $\kappa_{\text{abs}} \neq 0$, this wave will be attenuated as $e^{i\kappa_{\text{abs}} n_{\text{ref}} x}$. The imaginary part κ of the complex index of refraction, $n_{\text{ref}} + i\kappa_{\text{abs}}$, is a measure of the attenuation of an electromagnetic wave as seen, for example, in a spectrophotometer. We speak of $\kappa_{\text{abs}}(\omega)$ as the absorption spectrum of light. In the limit of high absorption ($\kappa_{\text{abs}} \to \infty$), for example a perfect metal, a body will show maximum reflectivity (Reflectivity $\to$ 1) (see Fig. L2.24).

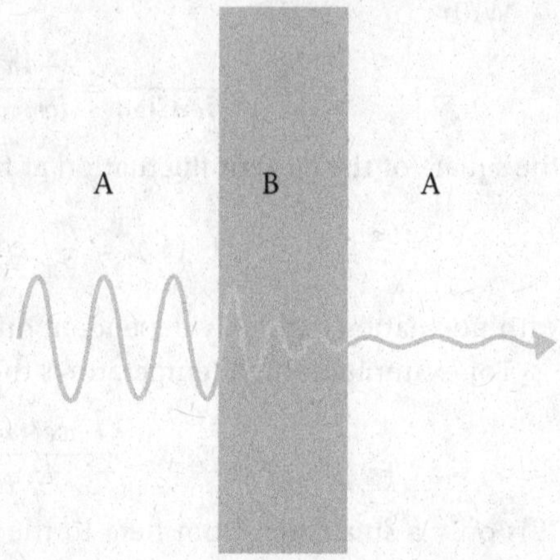

Figure L2.24

Energy-loss spectra

Rather than restricting ourselves to the electromagnetic field coming from light, think of a charged particle moving through a dielectric medium. The material around the moving charge experiences a time-varying electric field $\mathbf{E}(t)$ at each location whose frequency distribution depends on particle velocity. This field does work in proportion to $\varepsilon''(\omega)$, or more precisely $\{[\omega\varepsilon''(\omega)]/[\varepsilon'(\omega)^2 + \varepsilon''(\omega)^2]\}$, where $\omega = \omega_R$,[5] a dissipative work that slows the charged particle. Probing the energy loss of particles, or "stopping power," at differen velocities reveals the material's response spectrum (see Fig. L2.25).

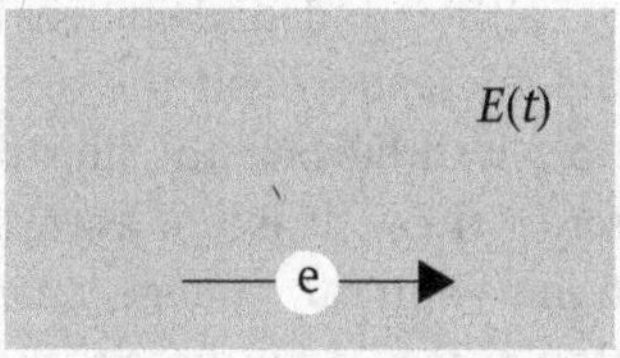

Figure L2.25

Because of the wide range of possible velocities, valence electron energy loss spectra have proven most useful for gathering data used in van der Waals force computation.

Numerical storage of data

In modern work the combined information of the several spectroscopies are often stored numerically. For the response of electrons, the inaccessible parts of the spectrum as well as differences in the kinds of information given by the different forms of measurements can often be reconciled by "sum rules" that constrain the total action shown by all electrons.[6]

In practical applications an "interband transition strength" for electrons of mass m_e,

$$J_{\text{cv}}(\omega) = J'_{\text{cv}}(\omega) + iJ''_{\text{cv}}(\omega) = \frac{m_e^2}{e^2}\frac{\omega^2}{8\pi^2}[\varepsilon''(\omega) + i\varepsilon'(\omega)] = -\frac{m_e^2}{e^2}\frac{\omega^2}{8\pi^2}\varepsilon^*(\omega), \quad \text{(L2.299)}$$

where $\varepsilon^*(\omega) = \varepsilon'(\omega) - i\varepsilon'(\omega)$, measured at real frequencies $\omega = \omega_R$, compactly catalogues the electronic response to the various probes.[7,8] In this J_{cv} language, data are stored, transformed, and applied numerically.

Approximate oscillator forms for ε

A major strength of the modern theory of electrodynamic forces is its empirical–agnostic attitude toward dielectric and magnetic susceptibilities. A rigorous physical theory explaining the behavior of ε for real materials is usually incomplete and dauntingly technical. But for computation we need only to be able to summarize electromagnetic data in a convenient form. To this end it is practical to use the experience of spectroscopists, whose language is usually adequate for fitting or, at least, for storing data.[9] In practice, it is often preferable to store raw data or processed data rather than force the data to a particular form.

Putting data in model mathematical forms, though, often helps us think about the sources of the charge fluctuations that create forces. Simplified forms often help us to think about the relevant behavior of materials; they also can ensure that our interpretation of incomplete data satisfies the fundamental properties of dielectric response.

Happily, the qualitative mathematical form for $\varepsilon(\omega)$, or more specifically for $\varepsilon(i\xi)$, is virtually imposed by its physical definition. The essential features are that

1. ε is a linear coefficient between the electric field and the polarization [Eqs. (L2.263)–(L2.268)].
2. ε cannot violate the direction of causality [Eq. (L2.261)].

A general linear relation between time-varying polarization and applied electric field could be written as a sum of derivatives:

$$\sum_k \mathrm{a}_k \frac{\mathrm{d}^k \mathbf{P}(t)}{\mathrm{d}t^k} = \mathbf{E}(t). \tag{L2.300}$$

At each particular frequency, $\mathbf{P}(t) \sim \mathbf{P}_\omega e^{-i\omega t}$, $\mathbf{E}(t) \sim \mathbf{E}_\omega e^{-i\omega t}$, the derivatives create factors $(-i\omega)^k$:

$$\sum_k (-i\omega)^k \mathrm{a}_k \mathbf{P}_\omega = \mathbf{E}_\omega. \tag{L2.301}$$

By causality, $\varepsilon(i\xi)$ [Eq. (L2.275)] must be positive and decrease monotonically in positive ξ. For $\omega = i\xi$ with $\xi > 0$, the form

$$\sum_k (-i\omega)^k \mathrm{a}_k = \sum_k \xi^k \mathrm{a}_k \tag{L2.302}$$

requires that all the a_k's must be real and positive.

The denominator $\sum_{k=0}^{\infty} (i\omega)^k \mathrm{a}_k$ can go to zero on the real-frequency, $\omega = \omega_R$, axis and on the lower half of the complex-frequency plane, $\xi \leq 0$. There $\varepsilon(\omega)$ can take on formally infinite values corresponding to resonance conditions under which an applied oscillating field invokes a large displacement of charges.

In practice, the polynomials of Eqs. (L2.301) and (L2.302) go too far. Why? Because of its finite mass, there is a limit to how far we can wiggle an electron, the lightest of the field-responsive charges, in a changing electric field.

Electron oscillator model For intuition, think about charge displacement in terms of a driven oscillator: a negative charge $-e$ of mass m_e, restrained from moving too far in displacement $x = x(t)$ because of a Hooke's law restoring force, with spring constant k_h, and because of viscous drag. In this language, polarization $p(t) = -ex(t)$. The spring force on the particle is $-k_h x$. (The negative sign reminds us that the spring force pulls toward negative x when displacement x is positive.) We want a relation between oscillatory $p(t) = -ex(t)$ and the force $-eE(t)$ of the oscillating applied electric field $\mathbf{E}(t)$. The electric and mechanical forces balance.

In this displacement language the p–E connection [Eq. (L2.300)] becomes

$$-e\sum_{k=0}^{2} \mathrm{a}_k \frac{\mathrm{d}^k x(t)}{\mathrm{d}t^k} = E(t), \qquad \text{(L2.303)}$$

or, thinking about the electric-field force $-eE(t)$ on the electron,

$$e^2\sum_{k=0}^{2} \mathrm{a}_k \frac{\mathrm{d}^k x(t)}{\mathrm{d}t^k} = -eE(t). \qquad \text{(L2.304)}$$

The terms can be associated with physical images. The driving force $-eE(t)$ is equal in magnitude and opposite in sign to the sum of mechanical, viscous, and inertial forces. That is, $-eE(t)$ equals the sum of the Hookean restoring force, the viscous drag, and the acceleration:

1. The Hooke's law force corresponds to $\mathrm{a}_0 x(t)$, with $\mathrm{a}_0 \propto -k_h$; a shift to the right, $x > 0$, encounters a spring force to the left.

2. Stokes-like viscous drag is proportional to velocity $\mathrm{d}x(t)/\mathrm{d}t$ in such a way as to slow the particle, $\mathrm{a}_1 \propto -\mathrm{b}_d$, velocity to the right ($[\mathrm{d}x(t)/\mathrm{d}t] > 0$) encounters drag toward the left.

3. Newtonian acceleration, force = mass × acceleration, with acceleration $\mathrm{d}^2x(t)/\mathrm{d}t^2$, and coefficient $\mathrm{a}_2 \propto m_e$, the particle mass.

Combined, the balance of the forces accelerating the electron is

$$m_e \frac{\mathrm{d}^2 x(t)}{\mathrm{d}t^2} = -eE(t) - \mathrm{b}_d \frac{\mathrm{d}x(t)}{\mathrm{d}t} - k_h x(t). \qquad \text{(L2.305)}$$

It should be obvious, but bears repeating, that this intuitive language does not constitute a physical theory. It is simply a convenient way to think about charge displacement in a time-varying electric field.

At a particular frequency of the applied field, $E(t) = E_\omega e^{-i\omega t}$, displacement $x(t)$ will oscillate,

$$x(t) = \mathrm{Re}\left(x_\omega e^{-i\omega t}\right) \qquad \text{(L2.306)}$$

to create an x–E connection in frequency,

$$m_e(-i\omega)^2 x_\omega = -eE_\omega - \mathrm{b}_d(-i\omega)x_\omega - k_h x_\omega, \qquad \text{(L2.307)}$$

$$(k_h - i\omega \mathrm{b}_d - \omega^2 m_e)x_\omega = -eE_\omega, \qquad \text{(L2.308)}$$

or, with $p_\omega = -ex_\omega$,

$$p_\omega = \left(\frac{e^2}{k_h - i\omega \mathrm{b}_d - \omega^2 m_e}\right) E_\omega. \qquad \text{(L2.309)}$$

Although this is phrased in the language of a single oscillating particle, it is equivalent to the equation for normal modes, where k_{h}, $\mathrm{b_d}$, m_e, and e are effective quantities describing entire collective oscillations. There are several special cases of $e^2/(k_{\mathrm{h}} - i\omega \mathrm{b_d} - \omega^2 m_e)$ for certain further idealized forms of response.

Resonant electrons: At IR and higher frequencies, oscillator language takes on its most appealing form, with a resonant frequency of magnitude $\sqrt{k_{\mathrm{h}}/m_e}$. Speak of the total response $\varepsilon(\omega)$ as the sum or integral of individual resonant responses. The total polarization is a sum over terms that look like p_ω in Eq. (L2.309) weighted by a number density n_j of electrons of mass m_e that resonate at each particular frequency ω_j, replacing the single resonance at $\sqrt{k_{\mathrm{h}}/m_e}$ and the single drag term with ad hoc coefficients γ_j. Combined with Eq. (L2.309), $E_\omega + 4\pi P_\omega = D_\omega = \varepsilon(\omega) E_\omega$ from Eq. (L2.263) gives

$$\varepsilon(\omega) = 1 + \frac{4\pi e^2}{m_e} \sum_j \frac{n_j}{\omega_j^2 - i\omega\gamma_j - \omega^2}. \tag{L2.310}$$

In this language, the oscillator or dissipative strength of the material varies with frequency as

$$\omega\varepsilon''(\omega) = \frac{4\pi e^2}{m_e} \sum_j \frac{n_j \gamma_j \omega^2}{\left(\omega_j^2 - \omega^2\right)^2 + (\omega\gamma_j)^2}.$$

From this form we can see that the dissipative response at one frequency ω is the sum or integral over all oscillators j as they report at that particular frequency.

Very high frequencies: When frequency $\omega \gg$ all resonant frequencies ω_j, this relation for electron dispersion converges to Eq. (L2.310) with its total electron density governed by a sum rule $N_e = \sum_j n_j$. Only the lightest particles, electrons of mass m_e, can follow rapidly varying fields. The $\omega^2 m_e$ term in the denominator dominates the dielectric response. If N_e is the total number density of electrons in the entire material, then the polarization response per volume is $N_e p_\omega$, and the dielectric susceptibility is

$$\varepsilon(\omega) = 1 - \frac{4\pi N_e e^2}{m_e \omega^2}. \tag{L2.311}$$

This $\varepsilon(\omega)$ is a purely real quantity with no dissipation of electric energy. It is the response at "hard" or highest-frequency x-ray frequencies at which the proportionality to electron density is the reason why x-ray diffraction gives electron distributions. In fact, the truth of Eq. (L2.311) in the x-ray region justifies dropping the higher powers in the polynomials of Eqs. (L2.303) and (L2.304). This limiting behavior occurs at $\omega > 10^{17}$ rad/s (far ultraviolet to x ray and above) and provides a tidy way to finish off $\varepsilon(\omega)$ at frequencies at which it is hard to measure. Compute the electron density N_e for lighter elements easily known from the weight density.

With $\omega = i\xi$, Equation (L2.311) also explains the good behavior of

$$\varepsilon(i\xi) = 1 + \frac{4\pi N_e e^2}{m_e \xi^2} \tag{L2.312}$$

in the limit of infinite frequency. From this form of $\varepsilon(i\xi)$, $\sum_{n=0}^{\prime\infty}$ is guaranteed to have a soft landing in the summation–integration over ξ_n when going to the "infinite" frequencies involved in force computation.

Metals: Ideally, conduction electrons can move with no restoring force, that is, $k_h = 0$. The electric field acts to accelerate the charges and, most important, to work against some general drag so that m and b_d are still finite. The form of response is then

$$\frac{-e^2}{\omega(\omega m_e + i b_d)}. \tag{L2.313a}$$

At low frequencies, $\omega \ll b_d/m_e$, this response has the pure-dissipation form of a conductor $(-e^2/i b_d \omega) = (ie^2/b_d \omega)$ and goes to infinity as frequency goes to zero as σ/ω for conductance σ. At high frequencies, $\omega \gg b_d/m_e$, the charges are pulled back and forth too fast to conduct electricity. The dielectric response goes to zero as $-e^2/m_e\omega^2$ with no dissipation of energy.

Nonideally, physicists are still wrestling with real conductors as opposed to infinite-ε ideal metals. The real problem is the $n = 0$ term in the summation for the van der Waals free energy. Its character differs from those at finite frequency and must therefore be trusted only in specifically validated cases. For real conductors, properties of the form $-e^2/[\omega(\omega m_e + i\gamma)]$ are expressed explicitly in terms of conductivity σ as a term

$$\frac{4\pi i \sigma}{\omega(1 - i\omega b)}. \tag{L2.313b}$$

This is valid only in the case of an effectively infinite medium in which no walls limit the flow of charges. Conductors must be considered case-by-case under the limitations imposed by boundary surfaces. See, for example, the treatment of ionic solutions (Level 1, Ionic fluctuation forces; Tables P.1.d, P.9.c, S.9, S.10, and C.5; Level 2, Sections L2.3.E L2.3.G; and Level 3, Sections L3.6 and L3.7).

Permanent dipoles: For particles whose acceleration is a negligible part of the balance of forces governing oscillation, there is only a restoring force (from rotational diffusion) and a drag term. The polarizability is of the Debye form

$$\frac{e^2}{k_h - i\omega b_d}. \tag{L2.314a}$$

In notation with dipole moment μ_{dipole} and relaxation time τ,[10] this becomes

$$\frac{\mu^2_{dipole}}{3kT(1 - i\omega\tau)}, \tag{L2.314b}$$

as written in Eqs. (L1.56) and (L2.173) and in Table S.8.

Vapors dilute and not-so-dilute In a dilute vapor, the external electric field polarizing any one particle—atom or molecule—is unchanged by electric fields emanating from dipoles induced on the other particles. (These dipolar fields drop off as the inverse cube of the distance from the particle.) The total polarization per unit volume of the dilute gas is the sum of individual particle dipoles. If $\alpha(\omega)$ denotes the single-particle polarizability and N is the number of particles per unit volume, then for a vapor

$$\varepsilon_{vap}(\omega) = 1 + 4\pi N\alpha(\omega). \tag{L2.315}$$

For these same particles at the density of a liquid or a solid, it is necessary to recognize their interaction with dipoles that are simultaneously induced on neighboring particles. In some cases, it may be possible and convenient to express the dielectric susceptibility $\varepsilon(\omega)$ of the solid or liquid in terms of previously measured individual atomic or molecular polarizabilities $\alpha(\omega)$. The Lorentz–Lorenz model imagines that each particle sits at the center of a spherical hollow carved out of the continuous medium without disturbing the uniform polarization of the surrounding medium. The electric field at the center of this cavity differs by a factor of $[\varepsilon(\omega) + 2]/3$ from the average field in the whole substance. The polarization of each atom or molecule will be $\alpha(\omega)\{[\varepsilon(\omega) + 2]/3\}E_\omega$ rather than $\alpha(\omega)E_\omega$. The sum of polarizations of all particles will be

$$P_\omega = \alpha(\omega)N\frac{\varepsilon(\omega)+2}{3}E_\omega. \tag{L2.316}$$

Because P_ω is already related to E_ω by the definition $\varepsilon(\omega)E_\omega = 1 + 4\pi P_\omega$, the relation between $\varepsilon(\omega)$ and $\alpha(\omega)$ is

$$\varepsilon(\omega) = \frac{1+2\left[\frac{4\pi N\alpha(\omega)}{3}\right]}{1-\left[\frac{4\pi N\alpha(\omega)}{3}\right]} \quad \text{or} \quad \alpha(\omega) = \frac{3}{4\pi N}\frac{\varepsilon(\omega)-1}{\varepsilon(\omega)+2}. \tag{L2.317}$$

PROBLEM L2.16: Show that if $\alpha(\omega)$ for an isolated particle has the form of a resonant oscillator, $\alpha(\omega) = [f_\alpha/(\omega_\alpha^2 - \omega^2 - i\omega\gamma_\alpha)]$, so does $\varepsilon(\omega)$ when we use the Lorentz–Lorenz transform $\varepsilon(\omega) = \{[1 + 2N\alpha(\omega)/3]/[1 - N\alpha(\omega)/3]\}$ for particles at number density N. The strength of response from the total number of particles is preserved through replacing f_α with Nf_α; the resonance frequency ω_α^2 is shifted to $\omega_\alpha^2 - Nf_\alpha/3$; the width parameter γ_α remains the same.

PROBLEM L2.17: How dilute is dilute? Use $\varepsilon(\omega) = \{[1 + 2N\alpha(\omega)/3]/[1 - N\alpha(\omega)/3]\}$ to show how deviation from dilute-gas pairwise additivity of energies creeps in with increasing number density N. Ignoring retardation, imagine two like nondilute gases with $\varepsilon_A = \varepsilon_B = \varepsilon = [(1 + 2N\alpha/3)/(1 - N\alpha/3)]$ interacting across a vacuum $\varepsilon_m = 1$. Expand this $\varepsilon(\omega)$ beyond the linear term in N, feed the result to the difference-over-sum $\overline{\Delta}^2 = [(\varepsilon - 1)/(\varepsilon + 1)]^2$ (Table P.1.a.4) used to compute forces.

Apply the result to metal spheres of radius a, $\alpha/4\pi = a^3$ [Table S.7 and Eqs. (L2.166)–(L2.169)] occupying an average volume $(1/N) = (4\pi/3)\rho^3$ per particle. Show that, for an average distance $z \sim 2\rho$ between particle centers, the condition of diluteness becomes $z \gg 2a$.

Practical working form of $\varepsilon(\omega)$

Models are nice. They give us pretty pictures, language, and occasional intuition. But data come first. For present purposes, we can flesh out the definition of $\varepsilon(\omega)$ with a sum (or integral if we choose) of terms with the dipole and the resonant-damped-oscillator forms:

$$\begin{aligned}\varepsilon(\omega) &= 1 + \sum_j \frac{d_j}{1 - i\omega\tau_j} + \sum_j \frac{f_j}{\omega_j^2 + g_j(-i\omega) + (-i\omega)^2} \\ &= 1 + \sum \frac{c_d}{1 - i\omega\tau_d} + \sum \frac{c_j\omega_j^2}{\omega_j^2 - i\omega\gamma_j - \omega^2}\end{aligned} \tag{L2.318}$$

or

$$\varepsilon(\omega) = 1 + \sum \frac{c_d}{1 - i\omega\tau_d} + \sum \frac{c_j}{1 - i\left(\omega\gamma_j/\omega_j^2\right) - \left(\omega^2/\omega_j^2\right)} \tag{L2.319}$$

where the coefficients c_d and c_j reflect the strength of response and the ω_j resonance frequencies.

In this language, the various f's, c's, ω_j's, and τ's are fitting parameters written to coincide with expressions used by spectroscopists to summarize their data. Built on the principle of a linear response only to events past, they reflect only the consequence of that principle. They do not become a theory of the response or even an adequate representation of the actual material for which data are being summarized. There has been some unfortunate confusion on this point because people have taken too seriously the language of models used to codify. The goal is to represent data in the most accurate, tractable, convenient form.

With increasing ability to process data numerically, there will be progressively less need to represent it in any mathematical form. At the same time, there is the danger of relying too much on numbers and too little on understanding the sources of the fluctuation that create forces; it will probably always be good practice to write analytic approximations for spectroscopic data to make clear to ourselves how spectra couple with forces. For this reason, it is still worthwhile to keep the intuition given us by considering spectra in traditional forms.

The terms $\sum[c_d/(1 - i\omega\tau_d)]$ occur at microwave frequencies (up to 10^{11} Hz, or $\sim 10^{12}$ rad/s, or relaxation times τ longer than $\sim 10^{-12}$ s). We often replace the sum by adding an extra parameter α to a single term,

$$\frac{c_d}{(1 - i\omega\tau_d)^{1-\alpha_d}}, \tag{L2.320}$$

to cover a range of responses centered about one average relaxation time. The parameters are extracted through "Cole–Cole" plots of $\varepsilon''(\omega)$ versus $\varepsilon'(\omega)$.[11]

Examination of the form

$$\varepsilon(\omega) = 1 + \sum \frac{c_d}{1 - i\omega\tau_d} + \sum \frac{c_j\omega_j^2}{\omega_j^2 - i\omega\gamma_j - \omega^2} \tag{L2.321}$$

shows that the denominators go to zero only when ξ, the imaginary part of $\omega = \omega_R + i\xi$, is less than or equal to zero. The function diverges to infinity in only the lower half of the frequency plane. For the c_d term, there are infinite values on only the imaginary-frequency axis, $\omega = -i/\tau_d$ or $\xi = -1/\tau_d$.

For the resonant-oscillator terms, the denominator

$$\omega_j^2 - i\omega\gamma_j - \omega^2 = \omega_j^2 - i(\omega_R + i\xi)\gamma_j - (\omega_R + i\xi)^2 \tag{L2.322}$$

has an imaginary part, $-i\omega_R\gamma_j - 2i\omega_R\xi$, that can be zero only for $\xi = -\gamma_j/2$.

Split $\varepsilon(\omega) = \varepsilon'(\omega) + i\varepsilon''(\omega)$ into its two parts:

$$\varepsilon'(\omega) = 1 + \sum \frac{c_d}{1 + (\omega\tau_d)^2} + \sum \frac{c_j\omega_j^2\left(\omega_j^2 - \omega^2\right)}{\left(\omega_j^2 - \omega^2\right)^2 + (\omega\gamma_j)^2}, \tag{L2.323}$$

$$\varepsilon''(\omega) = \sum \frac{c_d\omega\tau_d}{(1 + \omega\tau_d)^2} + \sum \frac{c_j\omega_j^2\,\gamma_j\omega}{\left(\omega_j^2 - \omega^2\right)^2 + (\omega\gamma_j)^2}. \tag{L2.324}$$

The resonant-frequency terms take on large values at or near $\omega = \omega_j$, where the frequencies ω_j are real and detected as maxima in the dissipative part of $\varepsilon''(\omega)$ on the real-frequency axis (see Fig. L2.26).

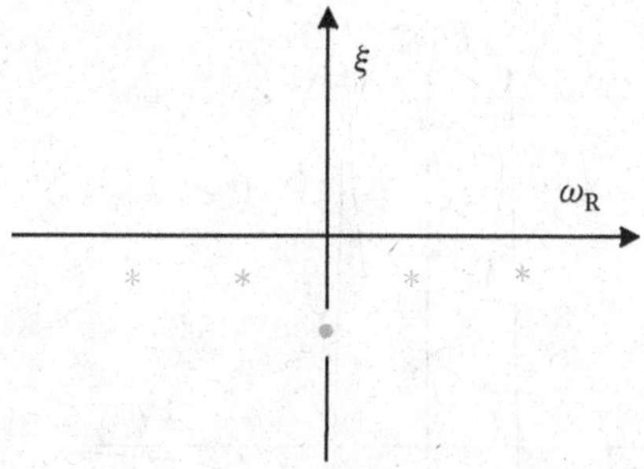

Figure L2.26

The infinite values of $\varepsilon(\omega) = \varepsilon(\omega_R + i\xi)$ occur on only one half of the frequency plane ($*$ symbols in Fig. L2.26). For infinitely sharp resonances ($\gamma_j \to 0$), the poles approach the ω_R, $\xi = 0$, axis. For dipolar "resonance," the pole occurs on the negative-ξ axis (symbol $\bullet$). (Locations of $*$ and $\bullet$ in Fig. L2.26 are drawn for illustration and are not in proportion to where resonances occur for real materials.)

For $\omega = \omega_j$ at a particular ω_j,

$$\frac{c_j\omega_j^2\left(\omega_j^2 - \omega^2\right)}{\left(\omega_j^2 - \omega^2\right)^2 + (\omega\gamma_j)^2} \to 0 \text{ in } \varepsilon'(\omega), \tag{L2.325}$$

$$\frac{c_j\omega_j^2\gamma_j\omega}{\left(\omega_j^2 - \omega^2\right)^2 + (\omega\gamma_j)^2} \to \frac{c_j\,\omega_j}{\gamma_j} \text{ in } \varepsilon''(\omega). \tag{L2.326}$$

For a well-defined, sharp resonance, ω_j is much greater than γ_j, and $(c_j\,\omega_j)/\gamma_j$ goes to very high values.

Plots of $\varepsilon'(\omega_R)$ and $\varepsilon''(\omega_R)$ show the relation between parameters (see Figs. L2.27 and L2.28).

On the positive-ξ axis used in force computation, $\varepsilon(i\xi)$ is a reassuringly sedate function:

$$\varepsilon(i\xi) = 1 + \sum_j \frac{d_j}{1 + \xi\tau_j} + \sum_j \frac{f_j}{\omega_j^2 + g_j\,\xi + \xi^2}. \tag{L2.327}$$

As prescribed by general principles [Eqs. (L2.274) and (L2.275)], $\varepsilon(i\xi)$ decreases monotonically on the positive-ξ axis, where it exhibits no sudden infinite values. Because γ_j is less than ω_j at a well-defined resonance, the γ_j terms in $\varepsilon(i\xi)$ are not always important. In practice, γ_j is used as a parameter to average over a range of resonant frequencies much as the exponent α_d is used in the microwave dipolar relaxation

$$\frac{c_d}{(1 - i\omega\tau_d)^{1-\alpha_d}}. \tag{L2.328}$$

NB: On the negative-ξ axis, it is another story. Because of the infinite values that $\varepsilon(i\xi)$ takes on for $\xi < 0$, the heuristic derivation (Level 3, Subsection L3.3) of the van der Waals force loses rigor. The derivation requires assuming symmetry about the real-frequency axis [Eqs. (L3.46) and (L3.47)]. In practice, because the eigenfrequencies ξ_n used in summing the forces are far from the $\xi = -1/\tau_j$, where $\varepsilon(i\xi)$ misbehaves by taking on infinite values, this is not a real problem. In the van der Waals force summation the first term is at zero frequency. The next, at $(2\pi kT/\hbar) \approx 2.411 \times 10^{14}$ rad/s at room temperature, is at much higher frequency than the location of the singular point at $\xi = -1/\tau_j$. For this reason, dipolar relaxation contributes to only the zero-frequency term (see Fig. L2.29).

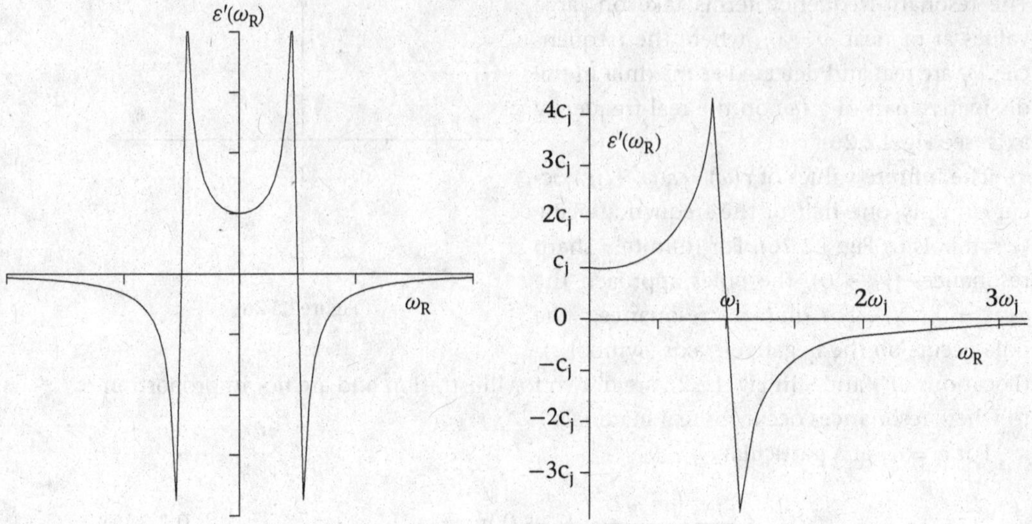

Figure L2.27

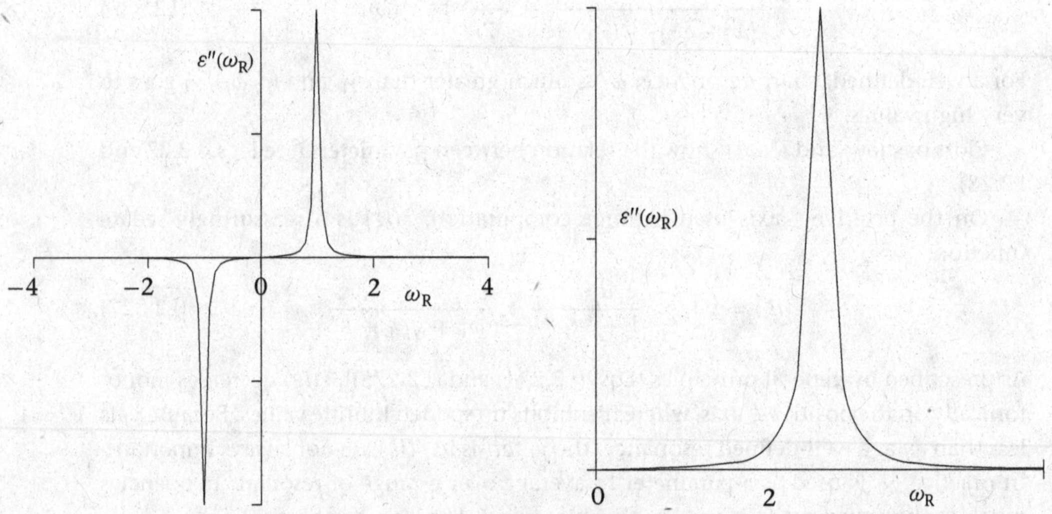

Figure L2.28

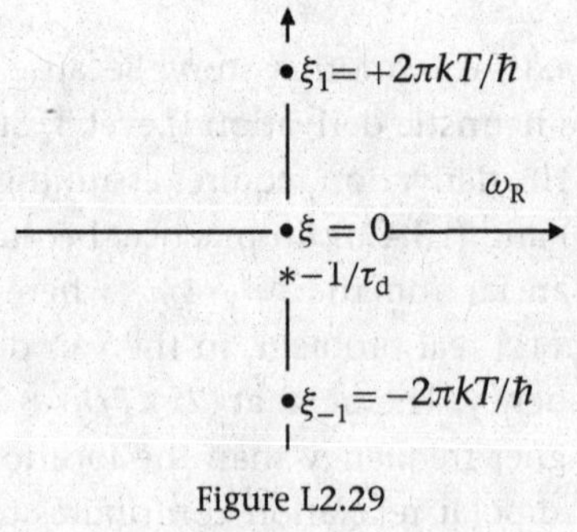

Figure L2.29

In discussing the response of electrons, we commonly speak of an oscillator strength $f(\omega_j)$:

$$\int_0^\infty f(\omega_j)\mathrm{d}\omega_j = N_e. \tag{L2.329}$$

Nonlocal dielectric response

The Lifshitz theory uses only the so-called "local" dielectric and magnetic responses. That is to say, the electric field at a place polarizes that place and that place only. What if the field is from a wave sinusoidally oscillating in space? Then the material polarization must oscillate in space to follow the field. What if that oscillation in space is of such a short wavelength that the structure of the material cannot accommodate the spatial variation of the wave? We are confronted with what is referred to as a "nonlocal" response: a polarization at a particular place is constrained by polarizations and electric fields at other places.

Instead of depending only on frequency, the response also depends on the spatial wave vector $\mathbf{k}$ of magnitude $2\pi/\lambda$, where λ is the wavelength of the spatial variation of the applied field traveling through the material: $\varepsilon(\omega)$ becomes $\varepsilon(\omega; \mathbf{k})$. The wave vector $\mathbf{k}$ in epsilon immediately brings us to think about the structure of the material. The limit in which we ordinarily speak corresponds to $\lambda = \infty$ or $\mathbf{k} = 0$, that is, what occurs when λ effectively equals infinity.

X-ray diffraction is an instructive example of such a nonlocal response. The material is polarizable in proportion to the local density of electrons. It is not polarizable at all points along the sinusoidal wave. The structure factor of x-ray diffraction describes the nonlocal response to a wave that is only weakly absorbed but that is strongly bent by the way its spatial variation couples with that of the sample to which it is exposed. Reradiation from the acceleration of the electrons creates waves that reveal the electron distribution. In no way can the scattering of the original wave be described or formulated in the continuum limit of featureless dielectric response. Because x-ray frequencies are often so high that the material absorbs little energy, it is possible to interpret x-ray scattering to infer molecular structure.

Time has no boundary. We can speak of the effect at a time t as the result of causes accumulated over all time before t. Space has boundaries. Except in the continuum $\mathbf{k} = 0$ limit, the location of an interface affects the way we use $\varepsilon(\omega; \mathbf{k})$. For this reason, formulation of van der Waals forces including fluctuations with a finite-$\mathbf{k}$ $\varepsilon(\omega; \mathbf{k})$ response has so far proven difficult.[12]

Noncontinuous media

Because it is the atomic or molecular feature that is ignored in Lifshitz theory, the separations between interacting bodies cannot be so small that these features are "seen" between them. In the macroscopic-continuum Lifshitz regime, computation is restricted to distances large compared with interatomic spacing or compared with molecular structure. Qualitatively speaking, between planar surfaces of separation l the error that is due to the continuum assumption comes in as terms of the order of $\sim(a/l)^2$, where a is a characteristic length or atomic spacing in the interacting bodies.

There is an easy but crude way to see why this is the form of the first finite-**k** correction. Imagine we expanded the dielectric response in powers of **k**:

$$\varepsilon(\omega, \mathbf{k}) = \varepsilon(\omega, \mathbf{k} = 0) + \alpha\mathbf{k} + \beta\mathbf{k}^2 + \cdots$$

Because it doesn't matter to the velocity of light whether it travels from left to right or from right to left through an isotropic material, there cannot be a linear $\alpha\mathbf{k}$ term. The first important term for small **k** comes in as $\beta\mathbf{k}^2$.

More, because **k** has units of 1/length and ε is dimensionless, the coefficient of $\mathbf{k}^2$ must have units of length squared. What length to associate with this coefficient? The only length available is some characteristic spacing in the material. Think, for example, of an interatomic spacing a:

$$\varepsilon(\omega, \mathbf{k}) \sim \varepsilon(\omega, \mathbf{k} = 0) \pm \mathrm{a}^2\mathbf{k}^2.$$

What if this $\mathrm{a}^2\mathbf{k}^2$ term were in ε in the integration of the Lifshitz expression

$$G(l, T) = \frac{kT}{8\pi l^2} \sum_{n=0}^{\infty}{}' \int_{r_n}^{\infty} x \ln[(1 - \overline{\Delta}_{\mathrm{Am}} \overline{\Delta}_{\mathrm{Bm}} e^{-x})(1 - \Delta_{\mathrm{Am}} \Delta_{\mathrm{Bm}} e^{-x})]\, dx?$$

Simplify to essentials. Set $r_n = 0$ to ignore retardation, retain dielectric terms only, and imagine small $\overline{\Delta}_{\mathrm{Am}} = \overline{\Delta}_{\mathrm{Bm}} = \overline{\Delta}$ for two like materials interacting across a vacuum:

$$\overline{\Delta}^2 = \left[\frac{\varepsilon(\omega, \mathbf{k}) - 1}{\varepsilon(\omega, \mathbf{k}) + 1}\right]^2 \sim \left[\frac{\varepsilon(\omega, \mathbf{k} = 0) - 1 + \beta\mathbf{k}^2}{2}\right]^2 \sim \left(\frac{\varepsilon - 1}{2}\right)^2 \pm \left(\frac{\varepsilon - 1}{2}\right)\mathrm{a}^2\mathbf{k}^2,$$

with $\varepsilon \equiv \varepsilon\,(\omega, \mathbf{k} = 0)$. In this crude first-correction-in-**k** description, the magnitude of the wave vector **k** is written as the radial component ρ vector of the surface modes, related to integration variable x as $x = 2\rho l = 2kl$, i.e., we replace $\mathrm{a}^2\mathbf{k}^2$ with $(\mathrm{a}/2l)^2x^2$. The remaining integral $\int_0^\infty \overline{\Delta}^2 e^{-x} x\, dx$ now has an x dependence in

$$\overline{\Delta}^2 \approx \left(\frac{\varepsilon - 1}{2}\right)^2 \pm \left(\frac{\varepsilon - 1}{2}\right)\left(\frac{\mathrm{a}}{2l}\right)^2 x^2$$

that gives an additional term in x^2.

The consequence is two integrals:

$$\left(\frac{\varepsilon - 1}{2}\right)^2 \int_0^\infty e^{-x} x\, dx = \left(\frac{\varepsilon - 1}{2}\right)^2$$

for the $\mathbf{k} = 0$ Lifshitz limit and

$$\left(\frac{\varepsilon - 1}{2}\right)\left(\frac{\mathrm{a}}{2l}\right)^2 \int_0^\infty x^3 e^{-x}\, dx = \frac{3}{2}\left(\frac{\varepsilon - 1}{2}\right)\left(\frac{\mathrm{a}}{l}\right)^2$$

for the finite-**k** correction. The integrand in $\int_0^\infty x^3 e^{-x}\, dx$ takes on its greatest value near $x = 2\rho l = 3$. With $\mathrm{a} \ll l$, this integral achieves most of its value long before x corresponds to $\mathrm{a}^2\mathbf{k}^2 = (\mathrm{a}/2l)^2 x^2$ for which higher-order terms in **k** expansion would be noticed.

PROBLEM L2.18: Show that, in the regime of pairwise summability, the continuum limit is violated by terms of the order of $(\mathrm{a}/z)^2$ where, just here, a is atomic spacing.

L.2.4.B. Integration algorithms

There are several sweet choices among computation algorithms, with the usual trade-off between human wit and computer labor. It is perfectly respectable nowadays to chop integration into the tiniest little bits and to sum these integrals over ridiculously large numbers of sampling frequencies and wave vectors. It is even fashionable, as it once was for mothers to give birth while under sedation, to feed an integral to an all-purpose program and expect good numbers to emerge.

Still, thought has its virtues. Computation becomes more efficient. Better yet, some recognition of the nature of the required integrals teaches us to think about the quantity being computed. Even package programs sometimes offer choices of integration algorithm; it helps to know what to tell the program.

Regard the object of interest, a sum of integrals in the form for the interaction free energy:

$$G(l, T) = \frac{kT}{8\pi l^2} \sum_{n=0}^{\infty}{}' \int_{r_n}^{\infty} x \ln[D(x, \xi_n)]\, dx \qquad \text{(L2.330)}$$

The integrands $x \ln[D(x, \xi_n)]$ can encompass all the geometries for which formulae have been derived as well as all temperatures at which spectral data are collected and converted for computation.

The range of each wave-vector x integration is infinite as is the set of frequency indices n in the primed sum. $D(x, \xi_n)$ has the well-behaved form

$$D(x, \xi_n) = \left(1 - \overline{\Delta}_{\mathrm{Am}}^{\mathrm{eff}} \overline{\Delta}_{\mathrm{Bm}}^{\mathrm{eff}} e^{-x}\right)\left(1 - \Delta_{\mathrm{Am}}^{\mathrm{eff}} \Delta_{\mathrm{Bm}}^{\mathrm{eff}} e^{-x}\right). \qquad \text{(L2.331)}$$

The delta's depend on x and ξ_n but their magnitudes vary smoothly between 0 and 1. They always go to zero for large ξ_n. For all ξ_n, $x \ln[D(x, \xi_n)]$ tails off exponentially at large x and goes emphatically to zero at $x = 0$. The maximum value of $x \ln[D(x, \xi_n)]$ occurs at $x \sim 1$:

- The troublesome part of wave-vector x integration is only the extent to which the x dependence of $x \ln[D(x, \xi_n)]$ deviates from the trivially integrable xe^{-x}. The deviation becomes negligible not only for large x but also for large ξ_n.
- The troublesome part of frequency n summation is to find the number of terms after which the deltas have settled down to a form that allows the sum to be finished off as an integral.

The Laguerre form of Gaussian integration (cf. Section 25.4.45 and Table 25.9, Abramowitz and Stegun, 1965) evaluates integrals of the form $\int_0^\infty I(y)e^{-y}dy$ through a summation $\sum_{j=1}^{J} w_j I(y_j)$. The weightings w_j and choice of evaluation points y_j are tabulated for different choices of the number of terms J. Summation returns an exact integral for the polynomial of degree J that is best fit to the function $I(y)$. (For the limiting case of deltas that do not depend on y, the "polynomial" is just a constant times y.) Rather than being Simpson's rule donkeys, chopping y into small equal increments and evaluating the integrand $I(y)e^{-y}$ at hundreds of points from $y = 0$ to $y \gg 1$, we can evaluate the integral using only a dozen or fewer terms. The only problem is that in our case the actual range of integration goes from r_n to ∞ rather than the 0 to ∞ for which the procedure is designed. It is possible to set $I(y) = 0$ for $0 \le y \le r_n$ but this maneuver would create a step in $I(y)$ that would drive a best-fit polynomial crazy. It makes far better sense to create a new variable of integration $(y - r_n)$ to preserve the designed distribution of weightings and evaluation points. At the same time, the exponential factor e^{-y} does not come out as a clean factor in the integral needed for computation. So it is easier to think of the integrand as a function $K(y) = I(y)e^{-y}$ such that the required integral can be evaluated as

$$\int_0^\infty K(y)\mathrm{d}y = \int_0^\infty I(y)e^{-y}\mathrm{d}y = \sum_{j=1}^{J} w_j I(y_j)$$
$$= \sum_{j=1}^{J} (w_j e^{+y_j})(I(y_j)e^{-y_j}) = \sum_{j=1}^{J} w_j e^{+y_j} K(y_j). \qquad \text{(L2.332)}$$

Specifically, consider $\int_{r_n}^\infty x \ln[D(x, \xi_n)]\,\mathrm{d}x$ as a weighted sum whose terms $x_j \ln[D(x_j, \xi_n)]$ are evaluated as a set of values x_j from tabulated y_j by $x_j = y_j + r_n$. Each term is weighted by $w_j e^{+y_j}$. The conversion of variables goes as

$$I_n(\xi_n) \equiv \int_{r_n}^\infty x \ln[D(x, \xi_n)]\mathrm{d}x = \sum_{j=1}^{J} w_j e^{+y_j} x_j \ln[D(x_j, \xi_n)]$$
$$= \sum_{j=1}^{J} w_j e^{+y_j}(y_j + r_n) \ln[D((y_j + r_n), \xi_n)]. \qquad \text{(L2.333)}$$

Summation of each of these integrals $I_n(\xi_n) = \int_{r_n}^\infty x \ln[D(x, \xi_n)]\mathrm{d}x$ over all sampling frequencies ξ_n for

$$G(l, T) = \frac{kT}{8\pi l^2} \sum_{n=0}^{\infty}{}' \int_{r_n}^\infty x \ln[D(x, \xi_n)]\mathrm{d}x = \frac{kT}{8\pi l^2} \sum_{n=0}^{\infty}{}' I_n(\xi_n) \qquad \text{(L2.334)}$$

is straightforward as long as the greatest ξ_n actually computed is much greater than all absorption frequencies in any of the dielectric response functions. This condition requires going to frequencies corresponding to $\hbar\xi_n$ of hundreds of electron volts. Recall that at room temperature $\hbar\xi_n = 0.159\, n$ eV, so that it may be necessary to sum several hundred terms.

Adding up the $I_n(\xi_n)$ only up to a limit $n = n_s$ and then converting the remainder of the summation into an integral over frequency may considerably shorten the summation. Because of the logarithmic nature of frequency, for large n the integrals $I_n(\xi_n)$ may vary slowly from term to term. In that case the sum over discrete n may be converted into an integral in continuously varying n and the variable of integration converted into a continuously varying frequency with $\mathrm{d}\xi = [(2\pi kT)/\hbar]\,\mathrm{d}n$:

$$\sum_{n=0}^{\infty}{}' I_n(\xi_n) \rightarrow \sum_{n=0}^{n_s}{}' I_n(\xi_n) + \int_{n_s+\frac{1}{2}}^\infty I_n(\xi_n)\mathrm{d}n = \sum_{n=0}^{n_s}{}' I_n(\xi_n) + \frac{\hbar}{2\pi kT}\int_{\xi_{n_s+\frac{1}{2}}}^\infty I(\xi)\mathrm{d}\xi. \qquad \text{(L2.335)}$$

Why $n_s + 1/2$, not $n_s + 1$, for the beginning of the integral?

Recall that the prime in summation indicates that the $n = 0$ term is to be given half weight, multiplied by 1/2 compared with the other terms in the sum. This is because the n summation itself originates in an integration [Level 3, Eqs. (L3.32)–(L3.47)] at which points are evaluated at integer n corresponding to ranges of integration from $n-(1/2)$ to $n+(1/2)$, except for $n=0$ for which the range is 0 to 1/2. The summation $\sum_{n=0}^{n_s}$, here in Eq. (L2.335), covers the frequency range corresponding to 0 to $n_s+(1/2)$. The integral $\int_{n_s+\frac{1}{2}}^{\infty} I_n(\xi_n)\mathrm{d}n$ picks up from there.

The integration over ξ is considerably shortened if we convert ξ to $\xi = 10^{\nu}$ and make ν a variable that goes from a finite value to "infinity." The integration in ν is itself effected as a summation of steps $\Delta\nu = 0.1$ (or whatever increment makes sense for the problem):

$$\int_{\xi_{n_s+\frac{1}{2}}}^{\infty} I(\xi)\mathrm{d}\xi \rightarrow \int_{\nu_{n_s+\frac{1}{2}}}^{\infty} I(10^{\nu})\mathrm{d}10^{\nu}$$

$$= 2.303\int_{\nu_{n_s+\frac{1}{2}}}^{\infty} I(10^{\nu})10^{\nu}\mathrm{d}\nu \rightarrow 2.303\Delta\nu\sum_{t=0}^{t_{max}} I(10^{\nu})10^{\nu},\ \nu = \nu_{n_s+\frac{1}{2}} + t\Delta\nu. \tag{L2.336}$$

Replacement of a sum by an integral is not a general procedure. It is necessary to determine a value of index n_s, or frequency $\xi_{n_s+\frac{1}{2}}$, at which to switch from summation to integration. It pays to shop for n_s big enough for accuracy yet small enough for convenience appropriate to the materials being examined and to the spectral information available about them. The reward is programs that will run orders of magnitude faster than procedures based on dogged summation. That degree of speed-up can give time to compute tedious integrands such as those encountered with inhomogeneous systems (e.g., Level 2, Tables P.7 and Level 3, Section L3.C.2).

PROBLEM L2.19: When can the discrete-sampling frequency summation be replaced with an integral over an imaginary frequency? Show that the condition

$$\frac{I(\xi_{n+1}) - 2I(\xi_n) + I(\xi_{n-1})}{24I(\xi_n)} \ll 1$$

does the trick.

L.2.4.C. Numerical conversion of full spectra into forces

Modern computation sticks to the numbers. The virtues of pure numerical conversion of spectra are best illustrated graphically. Figure L2.30 shows $\omega_R^2\varepsilon''(\omega_R)$ as $\mathrm{Re}[j_{cv}(\omega_R)]$ [Eq. (L2.299)] for crystals of AlN, Al_2O_3, MgO, SiO_2, water, and silicon.[14] To display the enormous amount of information in these spectra they are plotted two ways: vertically offset to be seen individually (left) and on the same vertical axis (right).

Fed into the transform $\varepsilon(i\xi) = 1 + \frac{2}{\pi}\int_0^{\infty}\{[\omega_R\varepsilon''(\omega_R)]/(\omega_R^2+\xi^2)\}\mathrm{d}\omega_R$ [Eq. (L2.275)] to create the functions $\varepsilon(i\xi)$ needed for computation, these spectra give strikingly featureless curves (see Fig. L2.31).[14]

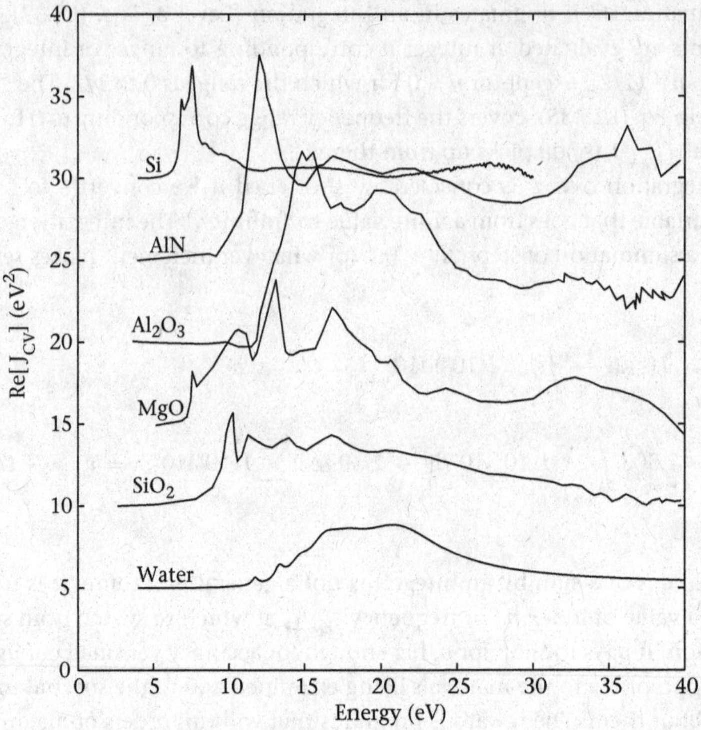

Figure L2.30

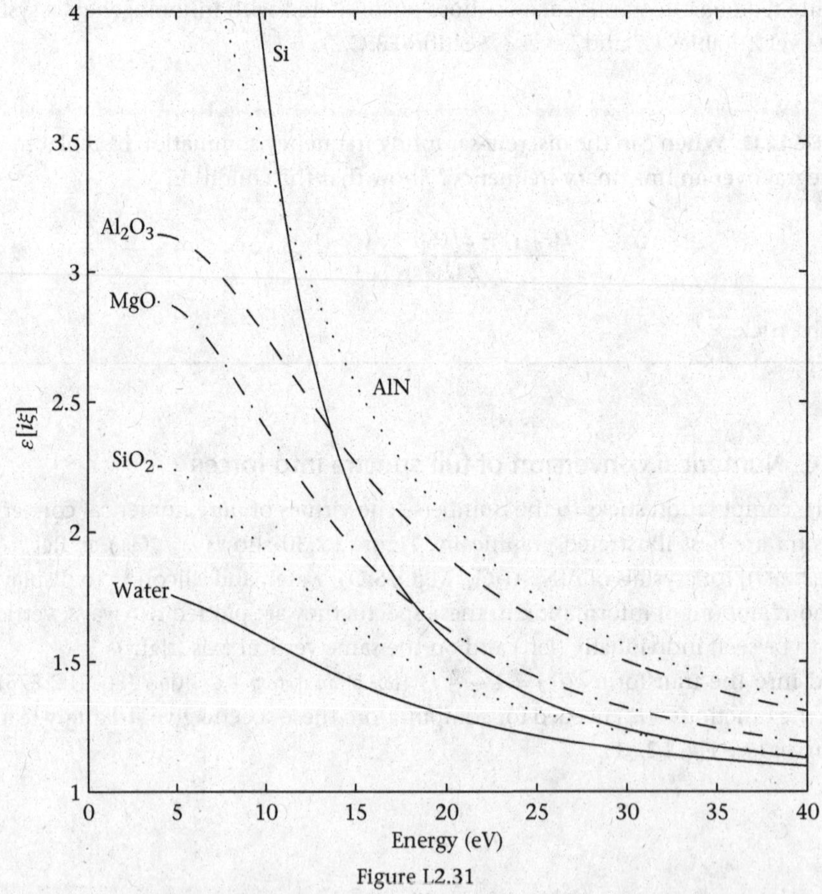

Figure L2.31

To show features at higher frequencies, the function $\mathrm{Re}[J_{cv}(\omega_R)]$ multiplies $\varepsilon''(\omega_R)$ by ω_R^2 whereas $\varepsilon(i\xi)$ depends on $\varepsilon''(\omega_R)$ weighted by the first power of ω_R. These $\varepsilon(i\xi)$ are then used to compute Hamaker coefficients (in 1 zJ = 10^{-21} J) in the limit of no retardation for attraction across a vacuum[13] or across water[14] as in this table:

$A_{AlN/water/AlN} = 102.2$ zJ	$A_{AlN/vac/AlN} = 228.5$ zJ
$A_{Al_2O_3/water/Al_2O_3} = 58.9$ zJ (27.5 zJ[15])	$A_{Al_2O_3/vac/Al_2O_3} = 168.7$ zJ (145zJ[15])
$A_{MgO/water/MgO} = 26.9$ zJ	$A_{MgO/vac/MgO} = 114.5$ zJ
$A_{SiO_2/water/SiO_2} = 6.0$ zJ (1.6zJ[15])	$A_{SiO_2/vac/SiO_2} = 66.6$ zJ (66 zJ[15])
$A_{Si/water/Si} = 112.5$ zJ	$A_{Si/vac/Si} = 212.6$ zJ

For two different reasons, $A_{AlN/vac/AlN}$ and $A_{Si/vac/Si}$ are the big winners here. Aluminum nitride has strong resonances at very high frequencies and therefore has an $\varepsilon(i\xi)$ that extends over a wider frequency range. Silicon shows relatively weak resonance at high frequency but is the strongest of all the substances shown in its response at lower frequencies. Because of the weighting in the construction of $\mathrm{Re}[J_{cv}(\omega_R)]$, this strong lower-frequency response is not so obvious until $\mathrm{Re}[J_{cv}(\omega_R)]$ is converted into $\varepsilon(i\xi)$.

Even exhaustive full-spectral computations have their frustrating uncertainties. Compare these tabulated Hamaker coefficients with those in parentheses,[15] quoted in the Prelude, which used earlier, slightly different, data and slightly different procedures[16] to create $\varepsilon(i\xi)$. The comparison reminds us to continue to search for the best data and to be aware of the unavoidable ambiguities due to limited data and to computational procedure.

Temperature comes into computation two ways. First, there is the way temperature affects electromagnetic fluctuations, how variable T is handled in formulae. Second, changes in temperature actually affect spectral response. By measuring response at different temperatures, we can determine both these consequences of varied temperature. Figure L2.32 shows the response of Al_2O_3 at different temperatures.[17] The nonretarded Hamaker coefficient for Al_2O_3 across vacuum goes from 145 zJ at 300 K to 152 zJ at 800 K and then down to 125 zJ at $T = 1925$ K.[14]

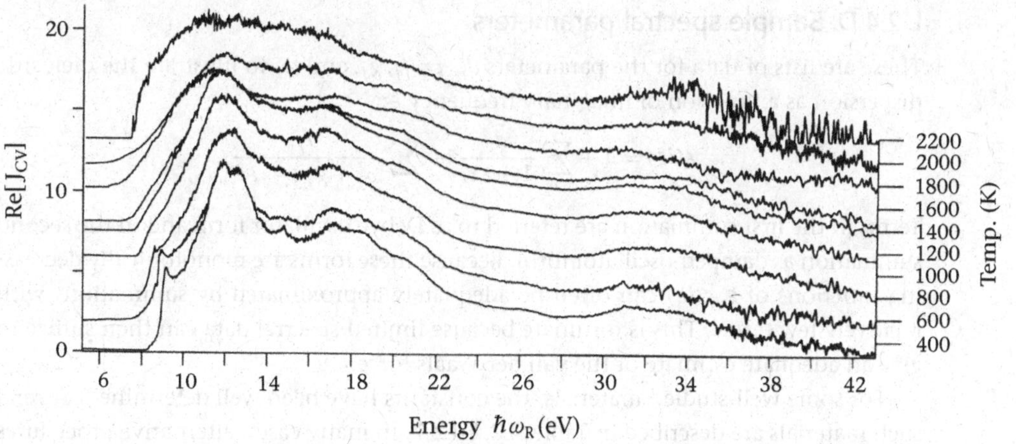

Figure L2.32

Table L2.1. Pure water[18,19]

1. Microwave frequencies: Debye dipolar-relaxation form[20]

$$d = 74.8,\ 1/\tau = 1.05 \times 10^{11}\ \text{rad/s} = 6.55 \times 10^{-5}\ \text{eV}.$$

2. Infrared frequencies: Damped-oscillator form[21,22]

ω_j, eV	f_j, (eV)^2	g_j, eV
2.07×10^{-2}	6.25×10^{-4}	1.5×10^{-2}
6.9×10^{-2}	3.5×10^{-3}	3.8×10^{-2}
9.2×10^{-2}	1.28×10^{-3}	2.8×10^{-2}
2.0×10^{-1}	5.44×10^{-4}	2.5×10^{-2}
4.2×10^{-1}	1.35×10^{-2}	5.6×10^{-2}

3. Ultraviolet frequencies: Damped-oscillator form[23,24]

ω_j, eV	f_j, (eV)^2	g_j, eV
8.25	2.68	0.51
10.0	5.67	0.88
11.4	12.0	1.54
13.0	26.3	2.05
14.9	33.8	2.96
18.5	92.8	6.26

4. *Alternative fitting to spectral data* without a constraint on parameters that fixes the value of the index of refraction; for details see note 24.

ω_j, eV	f_j, (eV)^2	g_j, eV
8.2	3.2	0.61
10.0	3.9	0.81
11.2	10.0	1.73
12.9	24.0	2.49
14.4	27.1	3.41
18.0	159.	9.90

L.2.4.D. Sample spectral parameters

These are lists of data for the parameters d_j, τ_j, f_j, g_j, and ω_j to illustrate the dielectric dispersion as a function of imaginary frequency ξ:

$$\varepsilon(i\xi) = 1 + \sum_j \frac{d_j}{1 + \xi\tau_j} + \sum_j \frac{f_j}{\omega_j^2 + g_j\xi + \xi^2}.$$

Terms in the first summation are referred to as Debye oscillator form and in the second summation as damped-oscillator form. Because these forms are monotonically decreasing functions of ξ, $\varepsilon(i\xi)$ can often be adequately approximated by summations with relatively few terms. This is fortunate because limited spectral data can then suffice to give an adequate estimate of the van der Waals force.

For some well-studied materials, the constants have been well determined. Several such materials are described in Tables L2.1–L2.7. In many cases, alternative procedures have been used to fit experimental spectra, and alternative tables are given. There is

Table L2.2. Tetradecane

1. Four-term fit[24]**:** Only UV frequencies needed

ω_j, eV	f_j, (eV)^2	g_j, eV
8.76	14.76	0.72
10.16	32.91	1.45
12.45	43.13	2.55
16.92	72.26	5.14

2. Four-term fit: *Without* index of refraction constraint, UV frequencies

ω_j, eV	f_j, (eV)^2	g_j, eV
8.71	16.83	0.82
10.15	42.26	1.82
12.78	64.18	3.72
18.70	146.06	9.65

3. Ten-term fit: UV frequencies

ω_j, eV	f_j, (eV)^2	g_j, eV
8.44	6.61	0.36
8.97	9.94	0.55
9.70	13.79	0.74
10.54	16.54	0.91
11.58	17.48	1.14
12.92	19.66	1.44
14.58	21.21	1.77
16.56	22.30	2.11
18.97	22.23	2.50
22.03	19.22	2.68

4. Ten-term fit: Without index of refraction constraint, UV frequencies

ω_j, eV	f_j, (eV)^2	g_j, eV
8.45	10.00	0.51
9.07	12.65	0.71
9.87	19.22	1.01
10.81	21.24	1.24
11.97	21.39	1.53
13.38	24.22	1.91
15.20	29.42	2.49
17.58	36.91	3.39
20.97	43.72	4.60
26.46	56.23	3.73

usually not a big difference in the forces computed from these different parameter sets. Still, it usually pays to look up the source of the data and to try different approximations to test the reliability of a computation.

In computation, try as much as possible to use the same kind of approximation in determining the $\varepsilon(i\xi)$*'s of all the materials involved.* Even in the simplest A|m|B computation,

Table L2.3. Polystyrene

Four-term fit[25]: Only UV frequencies needed

ω_j, eV	f_j, (eV)^2	g_j, eV
6.35	14.6	0.65
14.0	96.9	5.0
11.0	44.4	3.5
20.1	136.9	11.5

Table L2.4. Gold

1. Four-term fit[24] to absorption data[26]

ω_j, eV	f_j, (eV)^2	g_j, eV
—	9.7	3.21
2.9	4.95	0.67
4.0	41.55	2.22
8.9	207.76	8.50

2. Four-term fit to absorption data[27]

ω_j, eV	f_j, (eV)^2	g_j, eV
—	40.11	—
3.87	59.61	2.62
8.37	122.55	6.41
23.46	1031.19	27.57

3. Four-term fit to absorption data[28]

ω_j, eV	f_j, (eV)^2	g_j, eV
—	53.0	1.8
3.0	5.0	0.8
4.8	104.0	4.4

Table L2.5. Silver

1. Four-term fit[24] to absorption data[26]

ω_j, eV	f_j, (eV)^2	g_j, eV
—	56.3	—
5.6	54.5	2.7

2. Four-term fit to absorption and reflection data[27]

ω_j, eV	f_j, (eV)^2	g_j, eV
—	91.9	—
5.2	41.1	1.9
15.5	131.0	5.4
22.6	88.5	3.6
34.6	2688.4	94.2

Table L2.6. Copper

Four-term fit[24] to absorption and reflection data[27]

ω_j, eV	f_j, (eV)^2	g_j, eV
—	77.9	—
2.6	10.1	0.9
4.8	71.3	3.5
16.1	498.6	24.9
78.3	900.5	78.0

Table L2.7. Mica[29]

Data set a[30]

1. Microwave term: Debye form

$d = 1.36$, $1/\tau = 6.58 \times 10^{-5}$eV

2. Infrared term: Damped-oscillator form

ω_j, eV	f_j, (eV)^2	g_j, eV
8.4×10^{-2}	1.058×10^{-2}	0

3. Ultraviolet term: Damped-oscillator form

ω_j, eV	f_j, (eV)^2	g_j, eV
12.8	252.3	0

Data set b[31]

1. Microwave term: Debye form

$d = 0.4$, $1/\tau = 1.24 \times 10^{-6}$ eV

2. Infrared term: Damped-oscillator form

ω_j, eV	f_j, (eV)^2	g_j, eV
3.95×10^{-2}	0.312×10^{-2}	0

3. Ultraviolet term: Damped-oscillator form

ω_j, eV	f_j, (eV)^2	g_j, eV
10.33	157.93	0

Data set c[32]

1. Microwave term: Debye form

$d = 0.4$, $1/\tau = 1.24 \times 10^{-6}$ eV

2. Infrared term: Damped-oscillator form

ω_j, eV	f_j, (eV)^2	g_j, eV
3.95×10^{-2}	0.312×10^{-2}	0

3. Ultraviolet term: Damped-oscillator form

ω_j, eV	f_j, (eV)^2	g_j, eV
15.66	355.6	7.62

it is risky to use a detailed set of data for material A while making gross approximations for m and B. Better to treat A, m, and B in the same approximation.

The function $\varepsilon(i\xi)$ itself is dimensionless. Frequency ξ is in radians per second, but for compactness is often written in units of electron volts (the energy of a photon with that same radial frequency). If a quantity is tabulated in units of electron volts, it can be converted to radians per second by multiplying by 1.519×10^{15} (e.g., Level 1, the table on the frequency spectrum). This holds for the quantities ω_j and g_j, which are given in electron volts. The numerator f_j is given in electron volts squared [to keep $\varepsilon(i\xi)$ dimensionless] and can be converted to $(\text{radians/second})^2$ if one chooses to work in those units through multiplying by $(1.519 \times 10^{15})^2$. In the Debye form $d_j/(1 + \xi\tau_j)$, the numerator d_j is dimensionless and the inverse relaxation time $1/\tau_j$ is in electron volts.

The methods of fitting to spectral data to extract parameters are given in the cited references. One should be explicitly aware of rather different spectra observed on nominally the same material and of different parameter sets obtained from different kinds of data (reflectivity versus absorption, for example). It is always a good idea to use the different sets of parameters given in the tables to see the range of ambiguity in computed van der Waals forces. (And remember that before the Lifshitz theory, ambiguity in forces could be a factor of a thousand! The factor-of-two-or-three ambiguities that occur now are relatively small.)

L.2.4.E. Department of tricks, shortcuts, and desperate necessities

For many years people failed to take advantage of the modern theory of van der Waals forces on the grounds that there were no data sets for reliable computation. This same fear seems to hobble people even now because of limited spectroscopic information or experience. As it happens, even limited spectral information gives computations that are far more reliable than the use of formulae that are appropriate for gases rather than for solids or liquids.

It is always worth trying to see what numbers come out from the simplest or most rudimentary dielectric-dispersion data. As happens so often in physics, it is usually a good idea to try to make the *same crudeness of approximation* to the spectra of all materials in the computation. Similar assumptions have the happy tendency to compensate for the consequences of approximation. Dissimilar assumptions cause artificially big errors.

Approximate spectroscopic information can often be surprisingly useful. Perhaps the very simplest approximation of all is to use the index of refraction to estimate the dielectric permittivity of a material that is nearly transparent in the visible region. Then use the ionization potential, often tabulated in handbooks, to create a single absorption frequency. For example, consider several plastics that are nonpolar enough that we even ignore any significant terms from the microwave and IR regions.

The *Handbook of Chemistry and Physics* (CRC Press, Boca Raton, FL) gives the index of refraction n and the first ionization potential, a voltage I.P. corresponding to an energy $e \times$ I.P. required to rip the first electron off the material. The dielectric permittivity must equal n^2 at low frequencies. (This is the constraint referred to in the "with-constraint" parameters that fit more extensive data.) The single UV absorption frequency ω_{uv} corresponds to a photon energy $\hbar\,\omega_{\text{uv}} = e \times \text{I.P.}$

The general form for $\varepsilon(i\xi)$ will have only one term:

$$\varepsilon(i\xi) = 1 + \frac{f_{\mathrm{uv}}}{\omega_{\mathrm{uv}}^2 + \xi^2},$$

which can also be written in the equivalent form

$$\varepsilon(i\xi) = 1 + \frac{c_{\mathrm{uv}}}{1 + (\xi/\omega_{\mathrm{uv}}^2)},$$

where $c_{\mathrm{uv}} = f_{\mathrm{uv}}/\omega_{\mathrm{uv}}^2$ and is equal to $n_2 - 1$ in order that $\varepsilon(i\xi) = n^2$ at visible frequencies $\xi \ll \omega_{\mathrm{uv}}$.

If the ionization potential is in volts, as is usual, then the energy of ionization is the magnitude of electronic charge, $|e| = 1.6 \times 10^{-19}$ C times the I.P., that is, in *electron volts*; ω_{uv} is this energy divided by $\hbar = 1.0545 \times 10^{-34}$ J s. In practice it is far easier simply to express ξ and ω_{uv} directly in electron volts, as suggested in the worked examples. Then ω_{uv} simply has the numerical value of I.P. in volts.

Material	$\boldsymbol{\varepsilon(0) = n^2}$	$\boldsymbol{n}$[a]	$\boldsymbol{C_{\mathrm{uv}} = n^2 - 1}$	**I.P.**[b] **(eV)**	$\boldsymbol{\omega_{\mathrm{uv}}}$**(rad/s)**
Polyethylene	2.34	1.53	1.34	10.15[c]	1.54×10^{16}
Polypropylene	2.22	1.49	1.22	10.15[c]	1.54×10^{16}
Polytetra-fluoro-ethylene (Teflon)	1.96	1.40	0.96	10.15[c]	1.54×10^{16}
Polystyrene	2.53	1.59	1.53	8.47	1.29×10^{16}

[a] Refractive indices are from the *Handbook of Chemistry & Physics*, 50th ed.
[b] Ionization potentials I.P. are from R. W. Kiser, *Introduction to Mass Spectrometry* (Prentice-Hall, Englewood Cliffs, NJ, 1965).
[c] Here using I.P. of polyethylene.
NB: Rough numbers! Values vary from sample to sample and from handbook to handbook.
Source: Table modified from D. Gingell and V. A. Parsegian, "Prediction of van der Waals interactions between plastics in water using the Lifshitz theory," J. Colloid Interface Sci., **44**, 456–463 (1973).

L.2.4.F. Sample programs, approximate procedures

As approximate fits to spectra, oscillator models often miss essential details in the physics of the material response. Spectra of real samples reveal the consequences of composition, structure, doping, oxidation or reduction, multiplicity of phases, contaminant or introduced charges, etc., on electronic structure. These consequences from sample preparation can qualitatively affect intermolecular forces. To the extent possible, the best procedure is to use the best spectral data collected on the actual materials used in force measurement or materials designed for particular force properties. Given the present progress in spectroscopy, such coupling of spectra and forces may soon become routine.

Why then ever use simple oscillator models? For many materials, fits of such models to incomplete data are all that are available for computation. More important, to learn about the connection between spectra and forces, it helps to connect forces with analytic forms of the dielectric function. Although the forms themselves are approximate, they present a familiar language in which ever-more-detailed spectral information can be intuitively expressed. All of this goes with the caveat that these models allow only relatively crude estimates of the magnitudes and directions of the forces. In this spirit, this section tabulates parameters for $\varepsilon(i\xi)$ and presents some elementary programs.

Any linear dielectric response can be described as the sum (or integral) of damped harmonic oscillators in the form of Eq. (L2.318),

$$\varepsilon(\omega) = 1 + \sum_j \frac{d_j}{1 - i\omega\tau_j} + \sum_j \frac{f_j}{\omega_j^2 + g_j(-i\omega) + (-i\omega)^2}, \quad \text{(L2.337)}$$

which for $\omega = i\xi$ immediately becomes Eq. (L2.327):

$$\varepsilon(i\xi) = 1 + \sum_j \frac{d_j}{1 + \xi\tau_j} + \sum_j \frac{f_j}{\omega_j^2 + g_j\xi + \xi^2}. \quad \text{(L2.338)}$$

The slowly decreasing form of this last relation shows why incompleteness of spectral information need not always impede force computation. Even limited data can often give an adequate idea of the magnitude of forces.

Consider the simplest case, the interaction of two like materials across a planar layer, here water (Table L2.1 parameters) across hydrocarbon (Table L2.2): first a program for computation in its simplest form, then an annotated version of the same program explaining the physics and math behind each step.

Example: Computation of van der Waals force of water across a hydrocarbon film of thickness *l* (written as a MathCad program)

$$\mathrm{Ew}(z) := 1 + \frac{4.9 \times 10^{-3}}{z + 6.55 \times 10^{-5}}$$

$$\mathrm{Ew}(z) := \mathrm{Ew}(z) + \frac{6.3 \times 10^{-4}}{.021^2 + .015z + z^2} + \frac{3.5 \times 10^{-3}}{.069^2 + .038z + z^2}$$
$$+ \frac{1.3 \times 10^{-3}}{.092^2 + .028z + z^2} + \frac{5.44 \times 10^{-4}}{.2^2 + .025z + z^2} + \frac{1.4 \times 10^{-2}}{.42^2 + .056z + z^2}$$

$$\mathrm{Ew}(z) := \mathrm{Ew}(z) + \frac{2.68}{8.25^2 + .51z + z^2} + \frac{5.67}{10.^2 + .88z + z^2} + \frac{12.}{11.4^2 + 1.54z + z^2}$$
$$+ \frac{26.3}{13.^2 + 2.05z + z^2} + \frac{33.8}{14.9^2 + 2.96z + z^2} + \frac{92.8}{18.5^2 + 6.26z + z^2}$$

$$\mathrm{Eh}(w) := 1 + \frac{14.76}{8.76^2 + .72z + z^2} + \frac{32.91}{10.16^2 + 1.45z + z^2}$$
$$+ \frac{43.13}{12.45^2 + 2.55z + z^2} + \frac{72.26}{16.92^2 + 5.14z + z^2}$$

$$\mathrm{Fwh}(z) := \left[\frac{(\mathrm{Ew}(z) - \mathrm{Eh}(z))}{(\mathrm{Ew}(z) + \mathrm{Eh}(z))}\right]$$

$$\mathrm{r}(z) := \mathrm{p}(z)l; \qquad \mathrm{p}(z) := \frac{2\mathrm{Ew}(z)^{1/2}z}{3 \times 10^{10}} 1.5072 \times 10^{15}; \qquad \mathrm{R}(z) = (1 + r(z))^{*}\mathrm{e}^{-\mathrm{r}(z)}$$

$$\mathrm{N} := 1000$$

$$\mathrm{n} := 1..\mathrm{N}$$

$$\mathrm{Swh} := \sum_{\mathrm{n}} \mathrm{Fhw}(z)^2 \mathrm{R}(z).$$

$$\mathrm{Q} := 5$$

$$\mathrm{q} := 1..\mathrm{Q}$$

$$\text{Swh} := \text{Swh} + .5^* \sum_q \frac{\text{Fhw}(0)^{2q}}{q^3}$$

$$\text{Awh(l)} := -\frac{3}{2}\text{kT Swh;}$$

$$\text{Gwh(l)} := -\frac{\text{kT}}{8\pi l^2}\text{Swh}$$

Annotated version of this same program for computation of van der Waals force of water across a hydrocarbon film of thickness *l* (written as a MathCad program)

In this case materials A and B are the same; water is "w"; the medium m sandwiched between A and B is hydrocarbon "h". The approximate formula for the interaction energy is

$$G(l) = -\frac{A_{\text{wh/wh}}}{12\pi l^2},$$

where the coefficient $A_{\text{wh/wh}}$ is

$$A_{\text{wh/wh}} = \frac{3}{2}kT{\sum_n}' \overline{\Delta}_{\text{wh}}^2 R_n,$$

$$\overline{\Delta}_{\text{wh}} = \frac{\varepsilon_{\text{w}} - \varepsilon_{\text{h}}}{\varepsilon_{\text{w}} + \varepsilon_{\text{h}}}.$$

The screening factor now has the velocity of light in water, $c/\varepsilon_{\text{w}}^{1/2}$ at each frequency ξ_n with

$$R_n = (1 + r_n)e^{-r_n},\ r_n = \left(\frac{2l}{c/\varepsilon_{\text{w}}^{1/2}}\right) \Big/ \left(\frac{1}{\xi_n}\right).$$

Summation is from $n = 0$ to ∞ (remembering the factor 1/2 times the $n = 0$ term). The imaginary eigenfrequencies $\xi_n = [(2\pi kT)/\hbar]n$ can be in units of radians per second (most logically) or in units of electron volts for their corresponding photon energy $\hbar\xi_n$ (more convenient numbers for tabulating and computing ε's).

At $T = 20\,^\circ\text{C}$, the coefficient connecting the summation index n with the imaginary eigenfrequency ξ_n is

$$\frac{2\pi kT}{\hbar} = \frac{2 \times 3.14159 \times 1.38054 \times 10^{-16}\ (\text{ergs/K}) \times 293.15\ \text{K}}{1.0545 \times 10^{-27}}$$

so that $\xi_n = 2.411 \times 10^{14} n$ rad/s $= 0.159n$ eV.

To compute: **First**, we define the dielectric-permittivity functions

$$\varepsilon(i\xi) = 1 + \sum_j \frac{d_j}{1 + \xi\tau_j} + \sum_j \frac{f_j}{\omega_j^2 + g_j\xi + \xi^2}$$

by using the constants for d_j, τ_j, f_j, g_j, and ω_j given in Tables L2.1–L2.7.

Because computing programs don't usually allow the Greek letters in which formulae are written, write "z" for "ξ" and "E" for "ε" so that $\varepsilon_{\text{w}}(i\xi)$ becomes Ew(z), etc. Also, don't bother with subscripting. For water (with data copied from Table L2.1), we get

$$\text{Ew(z)} := 1 + \frac{4.9 \times 10^{-3}}{z + 6.55 \times 10^{-5}} \quad \text{(Debye, dipolar relaxation)}$$

$$\text{Ew(z)} := \text{Ew(z)} + \frac{6.3\times10^{-4}}{.021^2+.015z+z^2} + \frac{3.5\times10^{-3}}{.069^2+.038z+z^2} + \frac{1.3\times10^{-3}}{.092^2+.028z+z^2}$$

$$+\frac{5.44\times10^{-4}}{.2^2+.025z+z^2} + \frac{1.4\times10^{-2}}{.42^2+.056z+z^2} \quad \text{(infrared absorption frequencies)}$$

$$\text{Ew(z)} := \text{Ew(z)} + \frac{2.68}{8.25^2+.51z+z^2} + \frac{5.67}{10.^2+.88z+z^2} + \frac{12.}{11.4^2+1.54z+z^2}$$

$$+\frac{26.3}{13.^2+2.05z+z^2} + \frac{33.8}{14.9^2+2.96z+z^2} + \frac{92.8}{18.5^2+6.26z+z^2}$$

(ultraviolet absorption frequencies)

For hydrocarbon (with tetradecane data copied from Table L2.2.1), we get

$$\text{Eh(w)} := 1 + \frac{14.76}{8.76^2+.72z+z^2} + \frac{32.91}{10.16^2+1.45z+z^2}$$

$$+\frac{43.13}{12.45^2+2.55z+z^2} + \frac{72.26}{16.92^2+5.14z+z^2}$$

Second, we create the difference-over-sum of epsilons $\overline{\Delta}_{\text{Am}} = [(\varepsilon_{\text{A}} - \varepsilon_{\text{m}})/(\varepsilon_{\text{A}} + \varepsilon_{\text{m}})]$ where $\varepsilon_{\text{A}} = \varepsilon_{\text{water}}\ \varepsilon_{\text{m}} = \varepsilon_{\text{hydrocarbon}}$:

$$\text{Fwh(z)} := \left[\frac{(\text{Ew(z)} - \text{Eh(z)})}{(\text{Ew(z)} + \text{Eh(z)})}\right]$$

Third, (optional, for relativistic screening effects), we create the ratio of travel time to fluctuation lifetime,

$$r_n = \left(\frac{2l}{c/\varepsilon_w^{1/2}}\right) \Big/ \left(\frac{1}{\xi_n}\right),$$

$$\text{r(z)} := \text{p(z)}l$$

$$\text{p(z)} := \frac{2\text{Ew(z)}^{1/2}z}{3\times10^{10}} 1.5072\times10^{15}$$

Note here that I have taken the velocity of light, 3×10^{10} cm/s, and that l must be in centimeters in this case. The factor 1.5072×10^{15}, to convert electron volts to radians per second, is used because the imaginary frequency z in the summation below is easier to see in units of electron volts but frequency here has to be in radians per second.

The relativistic screening factor at each frequency z (that is, ξ_n) is then (by the approximate equal-light-velocities formula) [Eq. (L1.16), Fig. L1.12, Eq. (L2.26)]:

$$R(z) = [1 + r(z)]\, e^{-r(z)}.$$

Fourth, we do the computation itself, the summation of the product $\text{Fhw(z)}^2\text{R(z)}$, $\sum_{n=0}^{\prime\infty} \text{Fhw(z)}^2\text{R(z)}$, over all frequencies z (that is, ξ_n) remembering to treat the $z=0$ (a.k.a. $n=0$) term differently because of the factor of 1/2 (and because of higher order terms in Fhw(z)^2 that can also be important).

In principle, this summation $\sum_{n=0}^{\prime\infty}$ goes all the way to infinite frequency. In practice, the summation is sharply limited by two facts.

One, as frequency z approaches very high values, the relativistic retardation screening factor $R(z)$ goes to zero.

Two, the form of the dielectric functions is such that for large values of z the terms in the denominator $\omega_j^2 + g_j z + z^2$ (or $w_j^2 + g_j \xi + \xi^2$) are dominated by the z^2 term. Then each of the contributions in the dielectric susceptibilities Ew(z) and Eh(z) ($\varepsilon_{\mathrm{w}}(i\xi_{\mathrm{n}})$ and $\varepsilon_{\mathrm{h}}(i\xi_n)$) decreases as the square of z. The difference between susceptibilities that come into $\mathrm{Fhw(z)}^2$ decreases as the *fourth* power of z. {Formally, it would seem that the Debye term $[d/(1+\xi\tau)] = [d/(1+z\tau)]$ would die the most slowly, but in practice the coefficients d_j and τ_j are such that this term is for most purposes dead after the first few eigenfrequencies.}

To be sure about the reliability of computation, it is always a good idea to ask the computer to give results for different values of N, the upper limit in the number of terms:

$$\mathrm{N} := 1000$$

$$\mathrm{n} := 1..\mathrm{N}$$

$$\mathrm{Swh} := \sum_{\mathrm{n}} \mathrm{Fhw(z)}^2 \mathrm{R(z)}$$

To first approximation, the zero-frequency term requires an additional contribution $.5\mathrm{Fhw(0)}^2$. In fact (see the subsequent full derivation), this first approximation is not enough and this $n = 0$ term is in fact a power series in $\mathrm{Fhw(0)}^2 : \frac{1}{2}\sum_{\mathrm{q}=1}^{\infty} \mathrm{Fhw(0)}^{2\mathrm{q}}/\mathrm{q}^3$. Because $\mathrm{Fhw(0)}^2$ is less than 1 and because of the q^3 in the q denominator, this series converges very fast, usually within four or five terms at most. [When the medium is a salt solution with a Debye screening length $1/\kappa$, there is a screening of this low-frequency term that has the form $(1 + 2\kappa l)\, e^{-2\kappa l}$. This extra screening does not apply here in a hydrocarbon medium, but it can be very important in aqueous solutions.]

The summation in Swh is then completed with

$$\mathrm{Q} := 5$$

$$\mathrm{q} := 1..\mathrm{Q}$$

$$\mathrm{Swh} := \mathrm{Swh} + .5^* \sum_{\mathrm{q}} \frac{\mathrm{Fhw(0)}^{2\mathrm{q}}}{\mathrm{q}^3}$$

Once a value of Q and N are found that give a reliable estimate for this sum Swh, then it is only a matter of multiplying by $-\frac{kT}{8\pi l^2}$ or $-\frac{3}{2}kT$ to arrive at the energy of interaction Gwh(l) and Hamaker coefficient Awh(l):

$$\mathrm{Gwh(l)} := -\frac{\mathrm{kT}}{8\pi\, l^2}\,\mathrm{Swh} \quad \text{and} \quad \mathrm{Awh(l)} := -\frac{3}{2}\mathrm{kT\ Swh}$$

For good practice, it is probably a good idea to compute the Hamaker coefficient and the interaction free energy for several different film thicknesses to build an intuition about the magnitudes of forces and to see where retardation screening begins to be felt.

LEVEL THREE

Foundations

L3.1. Story, stance, strategy

As described in earlier sections, any two material bodies will interact across an intermediate substance or space. This interaction is rooted in the electromagnetic fluctuations—spontaneous, transient electric and magnetic fields—that occur in material bodies as well as in vacuum cavities. The frequency spectrum of these fluctuations is uniquely related to the electromagnetic absorption spectrum, the natural resonance frequencies of the particular material. In principle, electrodynamic forces can be calculated from absorption spectra.

Lifshitz's original formulation in 1954 (see Prelude, note 17) used a method, due to Rytov, to consider the correlation in electromagnetic fluctuations between two bodies separated by a vacuum gap. The force between the bodies is derived from the Maxwell stress tensor corresponding to the spontaneous electromagnetic fields that arise in the gap between boundary surfaces—the walls of the Planck–Casimir box. His result, for the case of two semi-infinite media separated by a planar slab gap, reduces in special limits to all previous valid results, specifically those of Casimir[1] and of Casimir and Polder[2] for, respectively, the interaction between two metal plates or between two point particles. In 1959, Dzyaloshinskii, Lifshitz, and Pitaevskii[3] (DLP) published a derivation that used diagram techniques of quantum field theory to allow the gap between the two bodies to be filled with a nonvacuous material.

The DLP result can be derived as well through an intuitive and heuristic method wherein the energy of the electromagnetic interaction is viewed as the energy of electromagnetic waves that fit between the dielectric boundaries of the planar gap. When the restrictions that the Planck–Casimir box be empty and that the walls be conductors are removed, it is possible to derive[4] the electromagnetic interaction between any two materials across a gap filled with a third substance by use of mode summation following a method introduced by van Kampen et al.[5] This method was put on a more stable foundation by Langbein[6] and elaborated by Mahanty and Ninham.[7] It has been granted the status of a rigorous theory by Barash and Ginsburg.[8] I believe that it is still only heuristic because of at least one shaky step, the assumption of pure oscillations even in regions of absorbing frequencies. No matter. Discussion of its rigor is secondary to its convenience in formulation and to its utility. The procedure of van Kampen et al. provides the same result as the more abstruse steps of the DLP method. I have decided to present it here because it clarifies the foundations of van der Waals forces in condensed media and lets us think creatively about many similar physical problems. Even in its more tedious steps, the complete working through of this one case explains puzzling features of van der Waals forces—the role of dielectric and magnetic susceptibilities, the need to think in terms of quanta, the use of imaginary frequencies, the emergence of eigenfrequencies, etc.—that can discourage the neophyte from taking full advantage of modern thinking. An expansion, an integration by parts, a contour integral, etc., that seem a gingerly traverse through a labyrinth become a set of steps to a higher-order view of electromagnetic fluctuations that create forces.

Once successfully demonstrated for the original Lifshitz result, the heuristic method can be immediately worked in more complicated geometries.

L3.2. Notation used in level 3 derivations

NB: Instead of using A and B to represent semi-infinite bodies, Level 3 derivations use L and R to orient "left" and "right" during equation solving.

L3.2.A. Lifshitz result

A_i, B_i	Coefficients of surface modes in i = L, m, or R
c	Velocity of light in vacuum; $c^2\varepsilon_0\mu_0 = 1$ in mks ("SI" or "Système International") units; $\varepsilon_0\mu_0$ is usually written as $1/c^2$ in the text
$D_E(\omega)$, $D_M(\omega)$ or $D_E(i\xi)$, $D_M(i\xi)$	Electric- and magnetic-mode dispersion relations for frequency ω or ξ
E, H	Electric and magnetic fields
$E_\eta = \left(\eta + \frac{1}{2}\right)\hbar\omega_j, \eta = 0, 1, 2, \ldots$	Oscillator-energy levels; η is also used locally in other contexts as index of summation
E_ω, H_ω	Fourier frequency components of $E(t)$ and $H(t)$; the ω subscript is dropped during derivation
$g(\omega_j) = -kT\ \ln[Z(\omega_j)]$	Free energy of mode ω_j
$G_l(\rho)$	Free energy of a surface wave of radial wave vector of magnitude ρ for a separation l between regions L and R
$G_{LmR}(l)$	Interaction free energy, compared with infinite separation, between L and R at a distance l
$(\mathbf{u}, \mathbf{v})$	Radial wave vectors in (x, y) direction: $\rho^2 = u^2 + v^2$; $\rho_i^2 = u^2 + v^2 - \frac{\varepsilon_i\mu_i\omega^2}{c^2} = \rho^2 - \frac{\varepsilon_i\mu_i\omega^2}{c^2}$
$Z(\omega_j)$	Partition function
$\overline{\Delta}_{ji}$, Δ_{ji}	Difference-over-sum functions for electric and magnetic modes

ε_i, μ_i	Relative electric and magnetic susceptibilities in regions i = L (left), i = m (middle), and i = R (right)
$\xi_n = \frac{2\pi kT}{\hbar} n, n = 0, \pm 1, \pm 2, \pm 3, \ldots$	Eigenfrequencies of summation
σ	Conductivity
ω	Radial frequency (real or complex)
$\{\omega_j\}$	Set of surface modes

$$p \equiv 2\rho_m l/r_n,\ s_i = \sqrt{p^2 - 1 + (\varepsilon_i\mu_i/\varepsilon_m\mu_m)}$$

$$r_n \equiv (2l\varepsilon_m^{1/2}\mu_m^{1/2}/c)\xi_n$$

$$x_i = 2l\rho_i,\ x_m = x = 2l\rho_m,$$

$$x_i^2 = x_m^2 + \left(\frac{2l\xi_n}{c}\right)^2 (\varepsilon_i\mu_i - \varepsilon_m\mu_m)$$

$$\rho_i^2 = \rho_m^2 + \frac{\xi_n^2}{c^2}(\varepsilon_i\mu_i - \varepsilon_m\mu_m)$$

L3.2.B. Layered systems

$l_{i/i+1}$	Position of interface between materials i and i + 1, i to the left of i + 1
$\mathbf{M}_{i+1/i}$	Matrix to convert surface-wave coefficients A_i, B_i in material i to A_{i+1}, B_{i+1} in material i + 1

$$\overline{\Delta}_{i+1/i} \equiv \left(\frac{\varepsilon_{i+1}\rho_i - \varepsilon_i\rho_{i+1}}{\varepsilon_{i+1}\rho_i + \varepsilon_i\rho_{i+1}}\right),$$

$$\Delta_{i+1/i} \equiv \left(\frac{\mu_{i+1}\rho_i - \mu_i\rho_{i+1}}{\mu_{i+1}\rho_i + \mu_i\rho_{i+1}}\right)$$

When there is no ambiguity the slash is omitted in the subscripts, as in $\overline{\Delta}_{Lm}$, $\mathbf{M}_{mL}$, or $\mathbf{M}_{Rm}^{eff}$ (between substrate R and medium m through intervening layers). For a finite number of layers, refer to material layers $A_1, A_2, \ldots, A_{j'}$ of thickness $a_1, a_2, \ldots, a_{j'}$ on half-space L; $B_1, B_2, \ldots, B_j$ of thickness $b_1, b_2, \ldots, b_{j'}$ on half-space R. Indices j or j′ count *away* from the central medium m. For repeating layers and multilayers, a single layer of material B′, thickness b′ on half-space R, is successively followed by N pairs of material B, thickness b, and B′, thickness b′.

$U_\nu(x)$	Chebyshev polynomial of the second kind.

L3.2.C. Ionic-fluctuation forces

ν	Ionic valence
n_ν	Mean number density of ions of valence ν
$\kappa_i^2 \equiv \frac{k_i^2}{\varepsilon_z^i}$	Debye constant in medium i = L, m or R
ρ_{ext}	External charge density
σ	Conductivity
ϕ	Electric potential

$$k_i^2 \equiv \frac{e^2}{\varepsilon_0 kT} \sum_{\nu=-\infty}^{\nu=\infty} \nu^2 n_\nu^i \text{ in mks units,}$$

$$k_i^2 \equiv \frac{4\pi e^2}{kT} \sum_{\nu=-\infty}^{\nu=\infty} \nu^2 n_\nu^i \text{ in cgs units}$$

$$p = \beta_m/\kappa_m, \quad s_i = \sqrt{p^2 - 1 + \kappa_i^2/\kappa_m^2}$$

$$x = 2\beta_m l, \quad x_i = \sqrt{x^2 - (\kappa_i^2 - \kappa_m^2)(2l)^2}$$

$$\beta_i^2 = \rho^2 + \kappa_i^2, \text{i} = \text{L, m or R}$$

$$\rho^2 = u^2 + v^2$$

L3.2.D. Anisotropic media

$\beta_i(\theta_i)$	Radial wave vector in medium i = L, m, or R; also written as $\beta_i(\theta_i) = \rho g_i(\theta_i - \psi)$ with a variable of integration ψ
$\varepsilon^m(\theta_m)$, $\varepsilon^R(\theta_R)$	Matrices of dielectric response of m and R in x, y, z directions, after rotation
ε_x^i, ε_y^i, ε_z^i	Relative dielectric response in directions x, y, z of materials i = L, m, or R; x, y parallel to planar surface
θ_m and θ_R	Rotation of principal axes of m or R with respect to principal axis of L ($\theta_R \equiv 0$)

L3.2.E. Anisotropic ionic media

$$\beta_i^2(\theta_i) = \rho^2 g_i^2(\theta_i - \psi) + \kappa_i^2$$

L3.3. A heuristic derivation of Lifshitz's general result for the interaction between two semi-infinite media across a planar gap

The scheme

Formally, the sum of random electromagnetic-field fluctuations in any set of bodies can be Fourier (frequency) decomposed into a sum of oscillatory modes extending through space. The "shaky step" in this derivation, already mentioned, is that we treat the modes extending over dissipative media as though they were pure sinusoidal oscillations. Implicitly this treatment filters all the fluctuations and dissipations to imagine pure oscillations; only then does the derivation transform these oscillations into the smoothed, exponentially decaying disturbances of random fluctuation.

Consider the original Lifshitz geometry of two half-spaces separated by a medium of thickness l (see Fig. L3.1).

We are specifically interested in the set $\{\omega_j\}$ of those modes, of radial frequency ω_j, which occur because of the location of boundary surfaces—"surface modes"—between different media: $\omega_j = \omega_j(l)$. We may determine these modes by direct solution of Maxwell's equations. This set $\{\omega_j(l)\}$ depends on the dielectric properties of each material L, R, and m, as well as on the spacing l. Each oscillation has a free energy $g(\omega_j)$; these energies are summed for a total free energy,

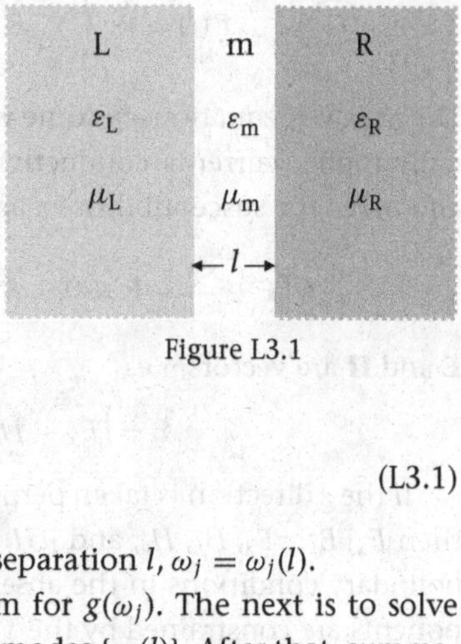

Figure L3.1

$$G(l) = \sum_{\{\omega_j\}} g(\omega_j). \tag{L3.1}$$

G depends on separation because ω_j is a function of separation l, $\omega_j = \omega_j(l)$.

The first step of the derivation is to find the form for $g(\omega_j)$. The next is to solve Maxwell's equations in order to find the set of surface modes $\{\omega_j(l)\}$. After that, summation over $g(\omega_j)$ leads to the general form of the van der Waals interaction. "L" and "R" (rather than "A" and "B" as used in the rest of the book) are used to designate materials on left and right for clarity in solving the wave equations.

Form of oscillator free energy $g(\omega_j)$

We often think of there being two separate quantum features of nonclassical oscillators, the change in energy levels in quantal units $h\nu$ (or $\hbar\omega$) and the finite zero-point energy

$\frac{1}{2}h\nu$ (or $\frac{1}{2}\hbar\omega$) of the oscillator in its lowest state. In fact, the two are connected. The zero-point fluctuation is an immediate consequence of the uncertainty principle. Observed for a time inverse to its frequency, an electromagnetic mode or degree of freedom has an uncertainty in its corresponding energy, an uncertainty proportional to the time of observation. The zero-point energy is also an unavoidable consequence of the fact that energy goes in and out in multiples of $\hbar\omega$. If the lowest energy state did not have energy $\frac{1}{2}\hbar\omega$, the oscillator energy would not go to the classical limit of kT at high temperature.[9]

The free energy $g(\omega_j)$ of an oscillator with energy levels

$$E_\eta = \left(\eta + \frac{1}{2}\right)\hbar\omega_j, \quad \eta = 0, 1, 2, \ldots, \tag{L3.2}$$

goes as the log of the partition function

$$Z(\omega_j) = \sum_{\eta=0}^{\infty} e^{-\hbar\omega_j\left(\eta+\frac{1}{2}\right)/kT}, \tag{L3.3}$$

$$g(\omega_j) = -kT\ \ln[Z(\omega_j)] = -kT\ \ln\left(e^{-\hbar\omega_j/2kT}\sum_{\eta=0}^{\infty} e^{-\hbar\omega_j\eta/kT}\right)$$
$$= -kT\ \ln[e^{-\hbar\omega_j/2kT}/(1 - e^{-\hbar\omega_j/kT})] = kT\ln[2\sinh(\hbar\omega_j/2kT)]. \tag{L3.4}$$

Finding the set of electromagnetic surface modes $\{\omega_j\}$

We look at the electric- and magnetic-field fluctuations in terms of Fourier components E_ω and H_ω such that, as functions of time, these fields are

$$E(t) = \mathrm{Re}\left(\sum_\omega E_\omega e^{-i\omega t}\right), \qquad H(t) = \mathrm{Re}\left(\sum_\omega H_\omega e^{-i\omega t}\right). \tag{L3.5}$$

The Maxwell equations become wave equations for E_ω and H_ω. In the absence of externally applied currents, conductivity, and externally inserted charges, with scalar electric and magnetic susceptibilities ε and μ that are constant in each region, we have[10]

$$\nabla^2\mathbf{E} + \frac{\varepsilon\mu\omega^2}{c^2}\mathbf{E} = 0, \quad \nabla\cdot\mathbf{E} = 0; \quad \nabla^2\mathbf{H} + \frac{\varepsilon\mu\omega^2}{c^2}\mathbf{H} = 0, \quad \nabla\cdot\mathbf{H} = 0. \tag{L3.6}$$

E and **H** are vectors, i.e.,

$$\mathbf{E} = \hat{\imath}E_x + \hat{\jmath}E_y + \hat{k}E_z, \quad \mathbf{H} = \hat{\imath}H_x + \hat{\jmath}H_y + \hat{k}H_z. \tag{L3.7}$$

If the z direction is taken perpendicular to the interface between different materials, then E_x, E_y, εE_z, H_x, H_y, and μH_z are continuous at each material boundary (Gaussian boundary conditions in the absence of extra charge or current). Also the x, y, z components are constrained by the $\nabla\cdot\mathbf{H} = 0$, $\nabla\cdot\mathbf{E} = 0$ conditions in Eqs. (L3.6).

Each component of the **E** and **H** fields is periodic in the x,y plane and has the general form $f(z)e^{i(ux+vy)}$, i.e.,

$$E_x = e_x(z)e^{i(ux+vy)}; E_y = e_y(z)e^{i(ux+vy)}; E_z = e_z(z)e^{i(ux+vy)}; \tag{L3.8a}$$

$$H_x = h_x(z)e^{i(ux+vy)}; H_y = h_y(z)e^{i(ux+vy)}; H_z = h_z(z)e^{i(ux+vy)}. \tag{L3.8b}$$

Put into the wave equation, this form gives

$$f''(z) = \rho_i^2 f(z), \tag{L3.9}$$

where, in each material region i,

$$\rho_i^2 = (u^2 + v^2) - \frac{\varepsilon_i \mu_i \omega^2}{c^2}. \tag{L3.10}$$

This yields six solutions of the form

$$f_i(z) = A_i e^{\rho_i z} + B_i e^{-\rho_i z}. \tag{L3.11}$$

If we keep the convention that $\mathrm{Re}(\rho_i) > 0$, the A and B coefficients must be restricted such that

$$A_R = 0 \text{ for } z > l \text{ (region R)},$$
$$B_L = 0 \text{ for } z < 0 \text{ (region L)}. \tag{L3.12}$$

There is also a constraint between the A and B coefficients for each kind of mode within each region:

$$\nabla \cdot \mathbf{E} = 0 = iue_x(z) + ive_y(z) + e_z'(z)$$
$$= (iuA_x + ivA_y + \rho A_z)e^{\rho z} + (iuB_x + ivB_y - \rho B_z)e^{-\rho z}, \tag{L3.13}$$

so that

$$A_z = -\frac{i}{\rho}(uA_x + vA_y), \quad B_z = \frac{i}{\rho}(uB_x + vB_y), \tag{L3.14}$$

and similarly for $\nabla \cdot \mathbf{H} = 0$.

The boundary conditions at $z = 0$ for the electric modes give

$$E_{Lx} = E_{mx} \rightarrow A_{Lx} = A_{mx} + B_{mx},$$
$$E_{Ly} = E_{my} \rightarrow A_{Ly} = A_{my} + B_{my},$$
$$\varepsilon_L E_{Lz} = \varepsilon_m E_{mz} \rightarrow \varepsilon_L A_{Lz} = \varepsilon_m A_{mz} + \varepsilon_m B_{mz}. \tag{L3.15}$$

Multiply the first of these equations by iu, the second by iv, and use the preceding $\nabla \cdot \mathbf{E} = 0$ condition to eliminate all the A_x, A_y, B_x, and B_y coefficients. They are irrelevant for what we want to do. Obtain

$$-A_{Lz}\rho_L = (-A_{mz} + B_{mz})\rho_m. \tag{L3.16}$$

At $z = l$ these same boundary conditions give

$$E_{Rx} = E_{mx} \rightarrow B_{Rx}e^{-\rho_R l} = A_{mx}e^{\rho_m l} + B_{mx}e^{-\rho_m l}, \tag{L3.17a}$$
$$E_{Ry} = E_{my} \rightarrow B_{Ry}e^{-\rho_R l} = A_{my}e^{\rho_m l} + B_{my}e^{-\rho_m l}, \tag{L3.17b}$$
$$\varepsilon_R E_{Rz} = \varepsilon_m E_{mz} \rightarrow \varepsilon_R B_{Rz}e^{-\rho_R l} = \varepsilon_m A_{mz}e^{\rho_m l} + \varepsilon_m B_{mz}e^{-\rho_m l}. \tag{L3.18}$$

Again eliminate all A_x, A_y, B_x, B_y to find

$$B_{Rz}e^{-\rho_R l}\rho_R = (-A_{mz}e^{\rho_m l} + B_{mz}e^{-\rho_m l})\rho_m. \tag{L3.19}$$

We now have four equations in the four A_z and B_z coefficients:

$$\varepsilon_L A_{Lz} = \varepsilon_m A_{mz} + \varepsilon_m B_{mz},$$
$$-A_{Lz}\rho_L = (-A_{mz} + B_{mz})\rho_m,$$
$$\varepsilon_R B_{Rz}e^{-\rho_R l} = \varepsilon_m A_{mz}e^{\rho_m l} + \varepsilon_m B_{mz}e^{-\rho_m l},$$
$$B_{Rz}e^{-\rho_R l}\rho_R = (-A_{mz}e^{\rho_m l} + B_{mz}e^{-\rho_m l})\rho_m. \tag{L3.20}$$

Eliminating the four coefficients gives

$$1-\left(\frac{\rho_L\varepsilon_m-\rho_m\varepsilon_L}{\rho_L\varepsilon_m+\rho_m\varepsilon_L}\right)\left(\frac{\rho_R\varepsilon_m-\rho_m\varepsilon_R}{\rho_R\varepsilon_m+\rho_m\varepsilon_R}\right)e^{-2\rho_m l}=0. \tag{L3.21}$$

This is the desired condition for finding the allowed electrical surface modes. The ε's and ρ's are functions of frequency. Whenever the frequency is such that Eq. (L3.21) is satisfied, it is a frequency that satisfies the conditions that its wave "fit" in the box, that it exist between the walls and then die away outside the walls (as an exponential in distance from either wall). It might be easier to think about this condition if we define a function

$$D_E(\omega)\equiv 1-\left(\frac{\rho_L\varepsilon_m-\rho_m\varepsilon_L}{\rho_L\varepsilon_m+\rho_m\varepsilon_L}\right)\left(\frac{\rho_R\varepsilon_m-\rho_m\varepsilon_R}{\rho_R\varepsilon_m+\rho_m\varepsilon_R}\right)e^{-2\rho_m l} \tag{L3.22}$$

and say that the frequencies $\{\omega_j(l)\}$ in which we are interested are those that occur when $D_E(\omega)=0$.

In addition to the electrical fluctuations, there are all the magnetic-field fluctuations that satisfy the same kind of condition. By inspection, we can see that these magnetic modes are to be determined in exactly the same way as the electrical modes, but we use the magnetic susceptibilities μ_m, μ_L, and μ_R, rather than ε_m, ε_L, and ε_R:

$$D_M(\omega)\equiv 1-\left(\frac{\rho_L\mu_m-\rho_m\mu_L}{\rho_L\mu_m+\rho_m\mu_L}\right)\left(\frac{\rho_R\mu_m-\rho_m\mu_R}{\rho_R\mu_m+\rho_m\mu_R}\right)e^{-2\rho_m l}, \tag{L3.23}$$

with which we can define a function

$$D(\omega)\equiv D_E(\omega)D_M(\omega) \tag{L3.24}$$

that has the property

$$D(\omega_j)=0 \tag{L3.25}$$

for each allowed surface mode.

Each set of frequencies $\{\omega_j\}$ is for a given pair of u, v radial wave components that occur in the composite radial wave vectors, ρ_L, ρ_m, ρ_R. We must sum over all possible radial wave vectors u, v as well as over all allowed frequencies at each u, v. For compactness define

$$\rho^2\equiv u^2+v^2, \tag{L3.26}$$

so that for each material

$$\rho_i^2=u^2+v^2-\frac{\varepsilon_i\mu_i\omega^2}{c^2}=\rho^2-\frac{\varepsilon_i\mu_i\omega^2}{c^2}; \tag{L3.27}$$

explicitly,

$$\rho_L^2=\rho^2-\frac{\varepsilon_L\mu_L\omega^2}{c^2},\quad \rho_m^2=\rho^2-\frac{\varepsilon_m\mu_m\omega^2}{c^2},\quad \rho_R^2=\rho^2-\frac{\varepsilon_R\mu_R\omega^2}{c^2}. \tag{L3.28}$$

For the existence of surface modes, i.e., those excitations that go to zero infinitely far away from boundary surfaces, we require that the real part of these ρ_i wave vectors be positive, $\mathrm{Re}(\rho_i)>0$, or

$$\rho^2>\mathrm{Re}\left(\frac{\varepsilon_i\mu_i\omega^2}{c^2}\right). \tag{L3.29}$$

In commonsense terms this inequality means that we are not allowed to use all values of u and v. The wave vectors must be large enough to ensure that the modes die away from the surface. This condition would go with the wavelength λ of the mode of radial frequency $\omega = 2\pi\nu$ if it were traveling in infinite space, $\lambda = (c/\sqrt{\varepsilon\mu})/\nu$. If the velocity of light c were infinite, so would be the wavelength; then all values of u, v would be allowed. But the velocity of light is finite. Too-small u, v leads to a ρ_i^2 that is negative, a composite wave vector ρ_i that is imaginary. Such a wave $e^{\pm\rho_i z}$ would not die away exponentially as required for a surface mode. This wave cannot be included in the set of modes that depend on the position of the boundary surfaces. This restriction in u, v, because of the finite velocity of light, is the source of the relativistic screening of van der Waals forces. Watch how it translates into a limit on the integrations for the total interaction free energy.

Summation of the free energies of the allowed surface modes

It should be clear that this summation–integration over all possibilities involves looking at all allowed $u^2 + v^2 = \rho^2$ as well as at all frequencies at each value of ρ. Think of a free energy $G_l(\rho)$ for the sum of free energies from the set of frequencies $\{\omega_j\}$ at one particular ρ, *then* add up these $G_l(\rho)$ over *all* allowed ρ.

Formally,

$$G_l(\rho) = \sum_{\{\omega_j\}} g(\omega_j) \tag{L3.30}$$

at each ρ. Then the total free energy $G_{\mathrm{LmR}}(l)$ can be defined as the real part of an integral over all ρ:

$$G_{\mathrm{LmR}}(l) = \frac{1}{(2\pi)^2}\mathrm{Re}\left\{\int_0^\infty 2\pi\rho\,[G_l(\rho) - G_\infty(\rho)]\mathrm{d}\rho\right\}. \tag{L3.31}$$

This integration in ρ uses a standard device in summation over wave vectors. Because u, v go with radial frequencies, $\omega = 2\pi\nu$, the units of u and v are 2π. Because $u^2 + v^2 = \rho^2$, we can combine all u, v that contribute to ρ in the range ρ to $\rho + \mathrm{d}\rho$. The number of these u, v combinations is the area of the circle $2\pi\rho\,\mathrm{d}\rho$ divided by the area per u, v combination $(2\pi)^2$. The lower bound in the integration is for those values of u, v allowed by susceptibilities of the medium m.

The remaining work is to effect the actual summation over $\{\omega_j\}$. This task is facilitated by two standard tricks of mode analysis.

Trick #1 Use the Cauchy integral theorem,

$$\sum_{\{\omega_j\}} g(\omega_j) = \frac{1}{2\pi i}\oint_C g(\omega)\frac{\mathrm{d}\ln[D(\omega)]}{\mathrm{d}\omega}\mathrm{d}\omega, \tag{L3.32}$$

where the contour of integration in the complex plane includes the zeros of $D(\omega)$.[11]

This integration over the entire complex plane of frequencies is where the combination of real and imaginary frequencies comes into the formulation of the van der Waals interaction. Recognize the frequency ω as a complex variable $\omega = \omega_{\mathrm{R}} + i\xi$ with real ω_R and imaginary ξ components that describe, respectively, oscillation $e^{i\omega_{\mathrm{R}}t}$ and exponential decay $e^{-\xi t}$ (see Fig. L3.2) (see also Level 2, Computation, Subsection L2.4.A).

To capture all possible positive frequencies ω_{R} that satisfy the condition $D(\omega) = 0$, take a contour of a semicircle of infinite radius centered at the origin and then a straight line from $\xi = +\infty$ to $\xi = -\infty$.

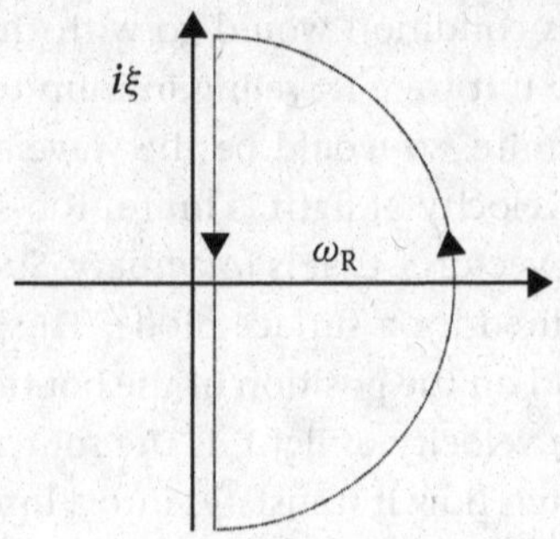

Figure L3.2

On the imaginary-frequency axis the free-energy function

$$g(\omega) = kT\ \ln[2\sinh(\hbar\omega/2kT)] = kT\ \ln\left(e^{\hbar\omega/2kT} - e^{-\hbar\omega/2kT}\right) \tag{L3.33}$$

has branch-point infinities at

$$\omega = i\xi = i\frac{2\pi kT}{\hbar}n, \quad n = 0, \pm 1, \pm 2, \pm 3, \ldots, \tag{L3.34}$$

or

$$\xi_n = \frac{2\pi kT}{\hbar}n. \tag{L3.35}$$

These points lie only on the path of integration. This property of the function $g(\omega)$ can be avoided by use of Trick #2.

Trick #2 Expand the logarithm as an infinite series[12]

$$g(\omega) = \frac{\hbar\omega}{2} - kT\sum_{\eta=1}^{\infty}\frac{e^{-(\hbar\omega/kT)\eta}}{\eta}. \tag{L3.36}$$

The contour integration can be taken as a line integral,

$$\begin{aligned} G_l(\rho) &= \sum_{\{\omega_j\}} g(\omega_j) = \frac{1}{2\pi i}\oint_C g(\omega)\frac{\mathrm{d}\ln[D(\omega)]}{\mathrm{d}\omega}\mathrm{d}\omega \\ &= \frac{1}{2\pi i}\int_{+i\infty}^{-i\infty} g(\omega)\frac{\mathrm{d}\ln[D(\omega)]}{\mathrm{d}\omega}\mathrm{d}\omega, \end{aligned} \tag{L3.37}$$

on the imaginary-frequency axis alone. This is because $\varepsilon(\omega) \to 1$, $\mu(\omega) \to 1$ as $|\omega| \to \infty$. This convergence to unity is an automatic consequence of the fact that no material can respond to an infinitely rapidly varying electric or magnetic field. In that limit all the ε's and μ's of all materials are equal to that of a vacuum; $D(\omega)$ is identically equal to 1. Its derivatives must equal zero.

For each radial wave vector ρ, the integration over frequency for free energy $G_l(\rho)$ can be done by parts. This turns out to be a clean separation into physically real and physically extraneous components:

$$\begin{aligned} G_l(\rho) &= \sum_{\{\omega_j\}} g(\omega_j) = \frac{-1}{2\pi i}\int_{-\infty}^{+\infty} g(i\xi)\frac{\mathrm{d}\ln[D(i\xi)]}{\mathrm{d}\xi}\mathrm{d}\xi \\ &= \sum_{\{\omega_j\}}\frac{1}{2}\hbar\omega_j + kT\frac{1}{2\pi i}\left[\sum_{\eta=1}^{\infty}\frac{e^{-(\hbar i\xi/kT)\eta}}{\eta}\ln D(i\xi)\Big|_{-\infty}^{+\infty}\right. \\ &\qquad \left. + \sum_{\eta=1}^{\infty}\frac{\hbar i}{kT}\int_{-\infty}^{+\infty} e^{-(\hbar i\xi/kT)\eta}\ln D(i\xi)\mathrm{d}\xi\right]. \end{aligned} \tag{L3.38}$$

The first term in [ ] is identically zero because, as just noted, $D(|\omega| \to \pm\infty) = 1$. The exponential in the remaining integral can be expanded into sine and cosine functions,

$$\sum_{\eta=1}^{\infty} e^{-(\hbar\xi/kT)\eta} = \sum_{\eta=1}^{\infty} \cos[(\hbar\xi/kT)\eta] - i\sum_{\eta=1}^{\infty} \sin[(\hbar\xi/kT)\eta], \tag{L3.39}$$

where we can use the transformation[13]

$$\sum_{\eta=1}^{\infty} \cos(\eta x) = \pi \sum_{\eta=-\infty}^{+\infty} \delta(x - 2\pi\eta) - \frac{1}{2} \tag{L3.40}$$

to create three integrals:

$$+\frac{\hbar}{2\pi}\left\{\int_{-\infty}^{\infty} \pi \sum_{\eta=-\infty}^{+\infty} \delta[(\hbar\xi/kT) - 2\pi\eta] \ln D(i\xi)\, d\xi \right.$$
$$\left. -\frac{1}{2}\int_{-\infty}^{+\infty} \ln D(i\xi)\, d\xi - i\sum_{\eta=1}^{\infty}\int_{-\infty}^{\infty} \sin[(\hbar\xi/kT)\eta] \ln D(i\xi)\, d\xi \right\}. \tag{L3.41}$$

Because we are interested in only the real part of the interaction energy $G_l(\rho)$, $i\sum_{\eta=1}^{\infty}\int_{-\infty}^{\infty} \sin[(\hbar\xi/kT)\eta] \ln D(i\xi)\, d\xi$, the third integral in { }, provides zero contribution to the energy. All that we require to know this is that susceptibilities $\varepsilon(i\xi)$ and $\mu(i\xi)$ are real on the imaginary-frequency axis. Then $D(i\xi)$ is a purely real quantity and $i \ln[D(i\xi)]$ is purely imaginary.

Taking the outside factors into the brackets, we take the first integral in { },

$$\frac{\hbar}{2}\int_{-\infty}^{\infty} \sum_{\eta=-\infty}^{+\infty} \delta[(\hbar\xi/kT) - 2\pi\eta] \ln D(i\xi)\, d\xi, \tag{L3.42}$$

by setting $x \equiv (\hbar\xi/kT)$ to write it as

$$\frac{kT}{2}\sum_{n=-\infty}^{\infty}\int_{-\infty}^{\infty} \delta(x - 2\pi n) \ln D(ikTx/\hbar)dx = \frac{kT}{2}\sum_{n=-\infty}^{\infty} \ln D(i\xi_n). \tag{L3.43}$$

This transformation reveals the source of the definition

$$\xi_n = \frac{2\pi kT}{\hbar} n \tag{L3.44}$$

for the imaginary sampling frequencies used in force and energy computation.

The second integral in { } is[14]

$$-\frac{1}{2}\frac{\hbar}{2\pi}\int_{-\infty}^{+\infty} \ln D(i\xi)\, d\xi = -\frac{\hbar}{2}\frac{1}{2\pi i}\int_{-i\infty}^{+i\infty} \ln D(\omega)\, d\omega$$
$$= -\frac{\hbar}{2}\frac{1}{2\pi i}\oint_C \omega \frac{d \ln D(\omega)}{d\omega} d\omega = -\sum_{\{\omega_j\}} \frac{1}{2}\hbar\omega_j, \tag{L3.45}$$

where we have gone back to a contour integral to collect the $\{\omega_j\}$ frequencies. This term cancels the first term in $G_l(\rho)$ (L 3.38).

The quantity $G_l(\rho)$ is then the tidy result

$$G_l(\rho) = \frac{kT}{2}\sum_{n=-\infty}^{\infty} \ln D(i\xi_n), \tag{L3.46}$$

where [Eq. (L3.44)] $\xi_n = [(2\pi kT)/\hbar]n$, $n = 0, \pm1, \pm2, \pm3, \ldots$.

It is worth repeating here that this function is evaluated on the imaginary-frequency axis at points where the oscillator free-energy function takes on infinite value, the δ

function in expression (L3.41). The values of the function $D(i\xi)$ are still a function of the susceptibilities $\varepsilon(i\xi_n)$ and $\mu(i\xi_n)$ at each of these points. It is common practice to assume that these susceptibilities are even functions of ξ. This is not correct, in principle, because they take on infinite values corresponding to resonance frequencies on only the lower half of the complex-frequency plane (see Level 2, Computation, Subsection L 2.4.A). Nevertheless, the positions of $i\xi_n$ are usually far enough from the positions of resonance frequencies to allow us to assume that $\varepsilon(i\xi_n) = \varepsilon(i|\xi_n|)$ and $\mu(i\xi_n) = \mu(i|\xi_n|)$ or at least that $D(i\xi_n) = D(i|\xi_n|)$. It is common practice then to write the summation in energy over only positive n:

$$G_l(\rho) = kT \sum_{n=0}^{\infty}{}' \ln D(i\xi_n), \tag{L3.47}$$

where the prime in summation reminds us to multiply the $n = 0$ term by 1/2.

Integration over all wave vectors for the total interaction free energy

Now that $G_l(\rho)$ is well defined, recall [Eq. (L3.31)]

$$G_{\mathrm{LmR}}(l) = \frac{1}{(2\pi)^2}\int_0^{\infty} 2\pi\rho\,[G_l(\rho) - G_{\infty}(\rho)]\mathrm{d}\rho, \tag{L3.48}$$

and [Eq. (L3.24)]

$$D(\omega) \equiv D_{\mathrm{E}}(\omega)D_{\mathrm{M}}(\omega)$$

with [Eqs. (L3.22) and (L3.23)]

$$D_{\mathrm{E}}(\omega) \equiv 1 - \left(\frac{\rho_{\mathrm{L}}\varepsilon_{\mathrm{m}} - \rho_{\mathrm{m}}\varepsilon_{\mathrm{L}}}{\rho_{\mathrm{L}}\varepsilon_{\mathrm{m}} + \rho_{\mathrm{m}}\varepsilon_{\mathrm{L}}}\right)\left(\frac{\rho_{\mathrm{R}}\varepsilon_{\mathrm{m}} - \rho_{\mathrm{m}}\varepsilon_{\mathrm{R}}}{\rho_{\mathrm{R}}\varepsilon_{\mathrm{m}} + \rho_{\mathrm{m}}\varepsilon_{\mathrm{R}}}\right)e^{-2\rho_{\mathrm{m}}l},$$

$$D_{\mathrm{M}}(\omega) \equiv 1 - \left(\frac{\rho_{\mathrm{L}}\mu_{\mathrm{m}} - \rho_{\mathrm{m}}\mu_{\mathrm{L}}}{\rho_{\mathrm{L}}\mu_{\mathrm{m}} + \rho_{\mathrm{m}}\mu_{\mathrm{L}}}\right)\left(\frac{\rho_{\mathrm{R}}\mu_{\mathrm{m}} - \rho_{\mathrm{m}}\mu_{\mathrm{R}}}{\rho_{\mathrm{R}}\mu_{\mathrm{m}} + \rho_{\mathrm{m}}\mu_{\mathrm{R}}}\right)e^{-2\rho_{\mathrm{m}}l},$$

and [Eqs. (L3.28)]

$$\rho_{\mathrm{L}}^2 = \rho^2 - \frac{\varepsilon_{\mathrm{L}}\mu_{\mathrm{L}}\omega^2}{c^2},\quad \rho_{\mathrm{m}}^2 = \rho^2 - \frac{\varepsilon_{\mathrm{m}}\mu_{\mathrm{m}}\omega^2}{c^2},\quad \rho_{\mathrm{R}}^2 = \rho^2 - \frac{\varepsilon_{\mathrm{R}}\mu_{\mathrm{R}}\omega^2}{c^2}.$$

We now know that these are to be evaluated at the imaginary frequencies $\omega = i\xi_n$, where $\omega^2 = -\xi_n^2$. The ρ_i's as used are purely real positive quantities

$$\rho_{\mathrm{L}}^2 = \rho^2 + \frac{\varepsilon_{\mathrm{L}}\mu_{\mathrm{L}}\xi_n^2}{c^2},\quad \rho_{\mathrm{m}}^2 = \rho^2 + \frac{\varepsilon_{\mathrm{m}}\mu_{\mathrm{m}}\xi_n^2}{c^2},\quad \rho_{\mathrm{R}}^2 = \rho^2 + \frac{\varepsilon_{\mathrm{R}}\mu_{\mathrm{R}}\xi_n^2}{c^2}. \tag{L3.49}$$

The variable of integration $\rho, 0 \le \rho < \infty$, can be changed to ρ_{m}, $[(\varepsilon_{\mathrm{m}}\mu_{\mathrm{m}}\xi_n^2)/c^2] \le \rho_{\mathrm{m}} < \infty$ with $\rho\mathrm{d}\rho = \rho_{\mathrm{m}}\mathrm{d}\rho_{\mathrm{m}}$. The total interaction free energy with summation over eigenfrequencies, switched in position with radial-vector integration, can then be written in a number of equivalent forms:

1. As an integral in ρ_{m}:

$$G_{\mathrm{LmR}}(l) = \frac{kT}{(2\pi)}\sum_{n=0}^{\infty}{}' \int_{\frac{\varepsilon_{\mathrm{m}}^{1/2}\mu_{\mathrm{m}}^{1/2}\xi_n}{c}}^{\infty} \rho_{\mathrm{m}} \ln\left[\left(1 - \overline{\Delta}_{\mathrm{Lm}}\overline{\Delta}_{\mathrm{Rm}}e^{-2\rho_{\mathrm{m}}l}\right)\left(1 - \Delta_{\mathrm{Lm}}\Delta_{\mathrm{Rm}}e^{-2\rho_{\mathrm{m}}l}\right)\right]\mathrm{d}\rho_{\mathrm{m}}, \tag{L3.50}$$

$$\overline{\Delta}_{\mathrm{ji}} = \frac{\rho_{\mathrm{i}}\varepsilon_{\mathrm{j}} - \rho_{\mathrm{j}}\varepsilon_{\mathrm{i}}}{\rho_{\mathrm{i}}\varepsilon_{\mathrm{j}} + \rho_{\mathrm{j}}\varepsilon_{\mathrm{i}}},\ \Delta_{\mathrm{ji}} = \frac{\rho_{\mathrm{i}}\mu_{\mathrm{j}} - \rho_{\mathrm{j}}\mu_{\mathrm{i}}}{\rho_{\mathrm{i}}\mu_{\mathrm{j}} + \rho_{\mathrm{j}}\mu_{\mathrm{i}}},\quad \rho_{\mathrm{i}}^2 = \rho_{\mathrm{m}}^2 + \frac{\xi_n^2}{c^2}(\varepsilon_{\mathrm{i}}\mu_{\mathrm{i}} - \varepsilon_{\mathrm{m}}\mu_{\mathrm{m}}). \tag{L3.51}$$

2. With a variable of integration $x = 2\rho_{\mathrm{m}} l$, and with

$$r_n \equiv \left(2l\varepsilon_{\mathrm{m}}^{1/2}\mu_{\mathrm{m}}^{1/2}/c\right)\xi_n \tag{L3.52}$$

for the minimum value of x so that $r_n \leq x < \infty$, $\rho_{\mathrm{m}}\,\mathrm{d}\rho_{\mathrm{m}} = (2l)^2 x\,\mathrm{d}x$, we have

$$G_{\mathrm{LmR}}(l) = \frac{kT}{8\pi l^2}\sum_{n=0}^{\infty}{}' \int_{r_n}^{\infty} x \ln\left[\left(1 - \overline{\Delta}_{\mathrm{Lm}}\overline{\Delta}_{\mathrm{Rm}}e^{-x}\right)\left(1 - \Delta_{\mathrm{Lm}}\Delta_{\mathrm{Rm}}e^{-x}\right)\right]\mathrm{d}x, \tag{L3.53}$$

$$\overline{\Delta}_{\mathrm{ji}} = \frac{x_{\mathrm{i}}\varepsilon_{\mathrm{j}} - x_{\mathrm{j}}\varepsilon_{\mathrm{i}}}{x_{\mathrm{i}}\varepsilon_{\mathrm{j}} + x_{\mathrm{j}}\varepsilon_{\mathrm{i}}},\ \Delta_{\mathrm{ji}} = \frac{x_{\mathrm{i}}\mu_{\mathrm{j}} - x_{\mathrm{j}}\mu_{\mathrm{i}}}{x_{\mathrm{i}}\mu_{\mathrm{j}} + x_{\mathrm{j}}\mu_{\mathrm{i}}},\ x_{\mathrm{i}}^2 = x_{\mathrm{m}}^2 + \left(\frac{2l\xi_n}{c}\right)^2 (\varepsilon_{\mathrm{i}}\mu_{\mathrm{i}} - \varepsilon_{\mathrm{m}}\mu_{\mathrm{m}}). \tag{L3.54}$$

3. As an integral in p, where

$$\rho_{\mathrm{m}} = \frac{\varepsilon_{\mathrm{m}}^{1/2}\mu_{\mathrm{m}}^{1/2}\xi_n}{c}p;\ x = 2\rho_{\mathrm{m}}\, l = r_n p, \tag{L3.55}$$

$$\rho_{\mathrm{m}}\mathrm{d}\rho_{\mathrm{m}} = \frac{\varepsilon_{\mathrm{m}}\mu_{\mathrm{m}}\xi_n^2}{c^2}\,p\,\mathrm{d}p \quad \text{or} \quad x\,\mathrm{d}x = r_n^2 p\,\mathrm{d}p = \frac{4\varepsilon_{\mathrm{m}}\mu_{\mathrm{m}}\xi_n^2 l^2}{c^2}p\,\mathrm{d}p, \quad 1 \leq p < \infty.$$

$$G_{\mathrm{LmR}}(l) = \frac{kT}{2\pi c^2}\sum_{n=0}^{\infty}{}' \varepsilon_{\mathrm{m}}\mu_{\mathrm{m}}\xi_n^2 \int_1^{\infty} p \ln\left[\left(1 - \overline{\Delta}_{\mathrm{Lm}}\overline{\Delta}_{\mathrm{Rm}}e^{-r_n p}\right)\left(1 - \Delta_{\mathrm{Lm}}\Delta_{\mathrm{Rm}}e^{-r_n p}\right)\right]\mathrm{d}p, \tag{L3.56}$$

$$\overline{\Delta}_{\mathrm{ji}} = \frac{s_{\mathrm{i}}\varepsilon_{\mathrm{j}} - s_{\mathrm{j}}\varepsilon_{\mathrm{i}}}{s_{\mathrm{i}}\varepsilon_{\mathrm{j}} + s_{\mathrm{j}}\varepsilon_{\mathrm{i}}},\ \Delta_{\mathrm{ji}} = \frac{s_{\mathrm{i}}\mu_{\mathrm{j}} - s_{\mathrm{j}}\mu_{\mathrm{i}}}{s_{\mathrm{i}}\mu_{\mathrm{j}} + s_{\mathrm{j}}\mu_{\mathrm{i}}},\ s_{\mathrm{i}} = \sqrt{p^2 - 1 + (\varepsilon_{\mathrm{i}}\mu_{\mathrm{i}}/\varepsilon_{\mathrm{m}}\mu_{\mathrm{m}})},\ s_m = p. \tag{L3.57}$$

This last form is convenient for differentiating $\ln[D(i\xi_n)]$ with respect to spacing l to obtain the force per unit area $-\mathrm{d}G_{\mathrm{LmR}}(l)/\mathrm{d}l$.

L3.4. Derivation of van der Waals interactions in layered planar systems

By modifying the function $D(i\xi_n)$, we can use the form

$$G_{\mathrm{LmR}}(l) = \frac{kT}{2\pi c^2} \sum_{n=0}^{\infty}{}' \varepsilon_{\mathrm{m}}\mu_{\mathrm{m}}\xi_n^2 \int_1^{\infty} p \ln\left[D(i\xi_n)\right] \mathrm{d}p \tag{L3.58}$$

to express the interaction between layered planar systems. The sums-over-differences $\overline{\Delta}_{\mathrm{Rm}}$, $\overline{\Delta}_{\mathrm{Lm}}$ and Δ_{Rm}, Δ_{Lm} are modified to become effective Δ's that include the boundaries between all layers.

The electromagnetic surface modes in each region still have the form $f(z) = \mathrm{A}e^{\rho z} + \mathrm{B}e^{-\rho z}$ [Eq. (L3.11)] with the A and B coefficients restricted in the semi-infinite spaces: to the left, $\mathrm{B_L} = 0$, and to the right $\mathrm{A_R} = 0$ [Eqs. (L3.12)]. As before, these restrictions ensure that we are looking at modes associated with the surfaces. For writing efficiency the procedure is described for the electric-field boundary conditions only.

As in the derivation of the Lifshitz expression, at each interface at a position $l_{\mathrm{i/i+1}}$ between adjacent materials, i and i + 1, the electric-field boundary conditions for continuous E_x, E_y, and εE_z create a connection between $\mathrm{A_i}$, $\mathrm{A_{i+1}}$ and $\mathrm{B_i}$, $\mathrm{B_{i+1}}$[15]:

$$\left(-\mathrm{A_{i+1}}e^{\rho_{\mathrm{i+1}}l_{\mathrm{i/i+1}}} + \mathrm{B_{i+1}}e^{-\rho_{\mathrm{i+1}}l_{\mathrm{i/i+1}}}\right)\rho_{\mathrm{i+1}} = \left(-\mathrm{A_i}e^{\rho_{\mathrm{i}}l_{\mathrm{i/i+1}}} + \mathrm{B_i}e^{-\rho_{\mathrm{i}}l_{\mathrm{i/i+1}}}\right)\rho_{\mathrm{i}},$$
$$\left(\mathrm{A_{i+1}}e^{\rho_{\mathrm{i+1}}l_{\mathrm{i/i+1}}} + \mathrm{B_{i+1}}e^{-\rho_{\mathrm{i+1}}l_{\mathrm{i/i+1}}}\right)\varepsilon_{\mathrm{i+1}} = \left(\mathrm{A_i}e^{\rho_{\mathrm{i}}l_{\mathrm{i/i+1}}} + \mathrm{B_i}e^{-\rho_{\mathrm{i}}l_{\mathrm{i/i+1}}}\right)\varepsilon_{\mathrm{i}}. \tag{L3.59}$$

This pair of equations can be written in matrix form,

$$\begin{pmatrix} \mathrm{A_{i+1}} \\ \mathrm{B_{i+1}} \end{pmatrix} = \mathbf{M}_{\mathrm{i+1/i}} \begin{pmatrix} \mathrm{A_i} \\ \mathrm{B_i} \end{pmatrix} \tag{L3.60}$$

to describe the transition between the coefficients describing the particular surface mode in layers i and i + 1. By convention, material i + 1 lies to the right of material i; i to the right of i − 1 (see Fig. L3.3).

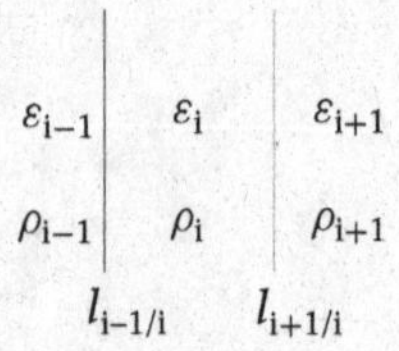

Figure L3.3

Aside from a multiplicative factor $[(\varepsilon_{\mathrm{i+1}}\rho_{\mathrm{i}} + \varepsilon_{\mathrm{i}}\rho_{\mathrm{i+1}})/(2\varepsilon_{\mathrm{i+1}}\rho_{\mathrm{i+1}})]$, matrix $\mathbf{M}_{\mathrm{i+1/i}}$ has the form[16]

$$\begin{bmatrix} e^{-\rho_{\mathrm{i+1}}l_{\mathrm{i/i+1}}}e^{+\rho_{\mathrm{i}}l_{\mathrm{i/i+1}}} & -\overline{\Delta}_{\mathrm{i+1/i}}e^{-\rho_{\mathrm{i+1}}l_{\mathrm{i/i+1}}}e^{-\rho_{\mathrm{i}}l_{\mathrm{i/i+1}}} \\ -\overline{\Delta}_{\mathrm{i+1/i}}e^{+\rho_{\mathrm{i+1}}l_{\mathrm{i/i+1}}}e^{+\rho_{\mathrm{i}}l_{\mathrm{i/i+1}}} & e^{+\rho_{\mathrm{i+1}}l_{\mathrm{i/i+1}}}e^{-\rho_{\mathrm{i}}l_{\mathrm{i/i+1}}} \end{bmatrix}, \tag{L3.61}$$

where

$$\overline{\Delta}_{i+1/i} \equiv \left(\frac{\varepsilon_{i+1}\rho_i - \varepsilon_i\rho_{i+1}}{\varepsilon_{i+1}\rho_i + \varepsilon_i\rho_{i+1}}\right) \tag{L3.62}$$

There is an equivalent transition matrix for magnetic modes with

$$\Delta_{i+1/i} \equiv \left(\frac{\mu_{i+1}\rho_i - \mu_i\rho_{i+1}}{\mu_{i+1}\rho_i + \mu_i\rho_{i+1}}\right). \tag{L3.63}$$

The derivations for added layers and multilayers use the following variables:

$$\rho_i^2 = \rho_m^2 + \frac{\xi_n^2}{c^2}(\varepsilon_i\mu_i - \varepsilon_m\mu_m), \tag{L3.64}$$

$$\overline{\Delta}_{ji} = \left(\frac{\varepsilon_j\rho_i - \varepsilon_i\rho_j}{\varepsilon_j\rho_i + \varepsilon_i\rho_j}\right), \quad \Delta_{ji} = \left(\frac{\mu_j\rho_i - \mu_i\rho_j}{\mu_j\rho_i + \mu_i\rho_j}\right); \tag{L3.65}$$

$$x_i = 2l\rho_i, \quad x_m = x = 2l\rho_m, \tag{L3.66}$$

$$x_i^2 = x_m^2 + \left(\frac{l\xi_n}{c}\right)^2(\varepsilon_i\mu_i - \varepsilon_m\mu_m), \tag{L3.67}$$

$$\overline{\Delta}_{ji} = \frac{x_i\varepsilon_j - x_j\varepsilon_i}{x_i\varepsilon_j + x_j\varepsilon_i}, \quad \Delta_{ji} = \frac{x_i\mu_j - x_j\mu_i}{x_i\mu_j + x_j\mu_i}; \tag{L3.68}$$

$$s_i = x_i/r_n, \quad s_m = p = x/r_n, \quad \rho_m = \frac{\varepsilon_m^{1/2}\mu_m^{1/2}\xi_n}{c}p, \tag{L3.69}$$

$$s_i = \sqrt{p^2 - 1 + (\varepsilon_i\mu_i/\varepsilon_m\mu_m)}, \tag{L3.70}$$

$$\overline{\Delta}_{ji} = \frac{s_i\varepsilon_j - s_j\varepsilon_i}{s_i\varepsilon_j + s_j\varepsilon_i}, \quad \Delta_{ji} = \frac{s_i\mu_j - s_j\mu_i}{s_i\mu_j + s_j\mu_i}. \tag{L3.71}$$

When material region i is a slab of finite thickness $(l_{i/i+1} - l_{i-1/i})$, we simplify the multiplication of matrices needed to incorporate additional layers by introducing factors $e^{+\rho_{i+1}l_{i/i+1}}$ and $e^{-\rho_{i+1}l_{i/i+1}}$, respectively multiplying A_{i+1}, B_{i+1}; then $e^{+\rho_i l_{i-1/i}}$ and $e^{-\rho_i l_{i-1/i}}$, respectively multiplying A_i, B_i. This transformation removes arbitrary additive reference points in the positions of the interfaces and allows us to focus on the physically important electric and magnetic events occurring at the interface.

Now for the transition between coefficients in material i + 1 and material i, we write[17]

$$\begin{pmatrix} A_{i+1} \\ B_{i+1} \end{pmatrix} = \mathbf{M}_{i+1/i}\begin{pmatrix} A_i \\ B_i \end{pmatrix}, \tag{L3.72}$$

$$\mathbf{M}_{i+1/i} = \begin{bmatrix} 1 & -\overline{\Delta}_{i+1/i}e^{-2\rho_i(l_{i/i+1}-l_{i-1/i})} \\ -\overline{\Delta}_{i+1/i} & e^{-2\rho_i(l_{i/i+1}-l_{i-1/i})} \end{bmatrix}. \tag{L3.73}$$

This simplification brings out the essential fact that it is the slab thickness that is the important measure of distance; it leaves unaffected the condition that the 1–1 element of the final matrix be equal to zero.

The Lifshitz result rederived

In the simplest L, m, R case, the connection among coefficients A_L, B_L, A_m, B_m, and A_R, B_R now reads (see Fig. L3.4)

$$\begin{pmatrix} A_R \\ B_R \end{pmatrix} = \mathbf{M}_{Rm}\mathbf{M}_{mL}\begin{pmatrix} A_L \\ B_L \end{pmatrix}. \tag{L3.74}$$

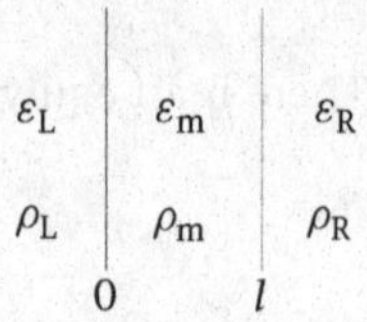

Figure L3.4

The surface-mode requirement $A_R = 0$, $B_L = 0$ is enforced by the condition that the 1–1 element of the matrix product $\mathbf{M}_{Rm}\mathbf{M}_{mL}$ be equal to zero. Formally,

$$\begin{pmatrix} 0 \\ B_R \end{pmatrix} = \mathbf{M}_{Rm}\mathbf{M}_{mL}\begin{pmatrix} A_L \\ 0 \end{pmatrix}. \tag{L3.75}$$

The matrix at the R/m interface brings in the slab thickness $(l_{i/i+1} - l_{i-1/i}) = (l_{m/R} - l_{L/m}) = l$:

$$\mathbf{M}_{Rm} = \begin{bmatrix} 1 & -\overline{\Delta}_{Rm}e^{-2\rho_m l} \\ -\overline{\Delta}_{Rm} & e^{-2\rho_m l} \end{bmatrix}. \tag{L3.76}$$

Because L is a semi-infinite medium, the matrix $\mathbf{M}_{mL}$ seems to have the ambiguity of an undefined $(l_{i/i+1} - l_{i-1/i})$. The surface-mode condition $B_L = 0$ immediately removes this nonphysical ambiguity through

$$\mathbf{M}_{mL}\begin{pmatrix} A_L \\ 0 \end{pmatrix} = \begin{bmatrix} 1 & -\overline{\Delta}_{mL}e^{-2\rho_i(l_{i/i+1}-l_{i-1/i})} \\ -\overline{\Delta}_{mL} & e^{-2\rho_i(l_{i/i+1}-l_{i-1/i})} \end{bmatrix}\begin{pmatrix} A_L \\ 0 \end{pmatrix} = \begin{pmatrix} 1 \\ -\overline{\Delta}_{mL} \end{pmatrix} A_L, \tag{L3.77}$$

so that

$$\begin{pmatrix} 0 \\ B_R \end{pmatrix} = \mathbf{M}_{Rm}\mathbf{M}_{mL}\begin{pmatrix} A_L \\ 0 \end{pmatrix} = \begin{pmatrix} 1 + \overline{\Delta}_{mL}\overline{\Delta}_{Rm}e^{-2\rho_m l} \\ -\overline{\Delta}_{Rm} - \overline{\Delta}_{mL}e^{-2\rho_m l} \end{pmatrix} A_L. \tag{L3.78}$$

Set equal to zero to satisfy $A_R = 0$, the element $1 + \overline{\Delta}_{mL}\overline{\Delta}_{Rm}e^{-2\rho_m l} = 1 - \overline{\Delta}_{Lm}\overline{\Delta}_{Rm}e^{-2\rho_m l}$ combines with the equivalent relation for magnetic terms to create the dispersion relation

$$D(i\xi_n) = (1 - \overline{\Delta}_{Lm}\overline{\Delta}_{Rm}e^{-2\rho_m l})(1 - \Delta_{Lm}\Delta_{Rm}e^{-2\rho_m l}) = 0, \tag{L3.79}$$

already derived for this simplest planar case [Eqs. (L3.22)–(L3.25)].

One singly coated surface

Generalization to layered structures is then a matter of successive matrix multiplications. Consider next the case in which there is a layer of thickness b_1 of material B_1 on material R (see Fig. L3.5). Here, for one layer on one side, and later for successive layerings on either side, we create matrix products of the form $\mathbf{M}^{eff}_{Rm}\mathbf{M}^{eff}_{mL}$, where the inner index m is the intervening medium and we seek interactions between the coated materials R and L. In the single-layer case of Fig. L3.5, matrix $\mathbf{M}_{Rm}$ is replaced with

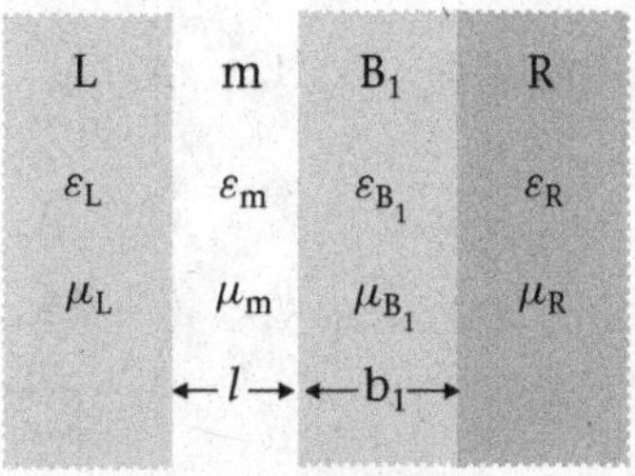

Figure L3.5

$\mathbf{M}_{\mathrm{Rm}}^{\mathrm{eff}} = \mathbf{M}_{\mathrm{RB}}\mathbf{M}_{\mathrm{B_1m}}$ with interface $l_{\mathrm{R/B_1}}$ at position $z = l + \mathrm{b}_1$. Doggedly substituting into the general form for the transition matrices, we obtain[18]

$$\mathbf{M}_{\mathrm{Rm}}^{\mathrm{eff}} = \mathbf{M}_{\mathrm{RB_1}}\mathbf{M}_{\mathrm{B_1m}}$$

$$= \left(1 + \overline{\Delta}_{\mathrm{RB_1}}\overline{\Delta}_{\mathrm{B_1m}}e^{-2\rho_{\mathrm{B_1}}\mathrm{b_1}}\right)\begin{bmatrix} 1 & -\dfrac{\overline{\Delta}_{\mathrm{RB_1}}e^{-2\rho_{\mathrm{B_1}}\mathrm{b_1}} + \overline{\Delta}_{\mathrm{B_1m}}}{1 + \overline{\Delta}_{\mathrm{RB_1}}\overline{\Delta}_{\mathrm{B_1m}}e^{-2\rho_{\mathrm{B_1}}\mathrm{b_1}}}e^{-2\rho_{\mathrm{m}}l} \\ -\dfrac{\overline{\Delta}_{\mathrm{RB_1}} + \overline{\Delta}_{\mathrm{B_1m}}e^{-2\rho_{\mathrm{B_1}}\mathrm{b_1}}}{1 + \overline{\Delta}_{\mathrm{RB_1}}\overline{\Delta}_{\mathrm{B_1m}}e^{-2\rho_{\mathrm{B_1}}\mathrm{b_1}}} & \dfrac{\overline{\Delta}_{\mathrm{RB_1}}\overline{\Delta}_{\mathrm{B_1m}} + e^{-2\rho_{\mathrm{B_1}}\mathrm{b_1}}}{1 + \overline{\Delta}_{\mathrm{RB_1}}\overline{\Delta}_{\mathrm{B_1m}}e^{-2\rho_{\mathrm{B_1}}\mathrm{b_1}}}e^{-2\rho_{\mathrm{m}}l} \end{bmatrix}. \tag{L3.80}$$

The surface-mode condition

$$\begin{pmatrix} 0 \\ \mathrm{B_R} \end{pmatrix} = \mathbf{M}_{\mathrm{Rm}}^{\mathrm{eff}}\mathbf{M}_{\mathrm{mL}}\begin{pmatrix} \mathrm{A_L} \\ 0 \end{pmatrix}$$

with

$$\mathbf{M}_{\mathrm{mL}}\begin{pmatrix} \mathrm{A_L} \\ 0 \end{pmatrix} = \begin{pmatrix} 1 \\ -\overline{\Delta}_{\mathrm{mL}} \end{pmatrix}\mathrm{A_L}$$

gives

$$1 - \overline{\Delta}_{\mathrm{Lm}}\overline{\Delta}_{\mathrm{Rm}}^{\mathrm{eff}}e^{-2\rho_{\mathrm{m}}l} = 0, \tag{L3.81}$$

where

$$\overline{\Delta}_{\mathrm{Rm}}^{\mathrm{eff}} = -\frac{\overline{\Delta}_{\mathrm{RB_1}}e^{-2\rho_{\mathrm{B_1}}\mathrm{b_1}} + \overline{\Delta}_{\mathrm{B_1m}}}{1 + \overline{\Delta}_{\mathrm{RB_1}}\overline{\Delta}_{\mathrm{B_1m}}e^{-2\rho_{\mathrm{B_1}}\mathrm{b_1}}}. \tag{L3.82}$$

The full dispersion relation including magnetic terms becomes

$$D_{\mathrm{LmB_1R}}(i\xi_n) = \left(1 - \overline{\Delta}_{\mathrm{Lm}}\overline{\Delta}_{\mathrm{Rm}}^{\mathrm{eff}}e^{-2\rho_{\mathrm{m}}l}\right)\left(1 - \Delta_{\mathrm{Lm}}\Delta_{\mathrm{Rm}}^{\mathrm{eff}}e^{-2\rho_{\mathrm{m}}l}\right) = 0. \tag{L3.83}$$

The functions $\overline{\Delta}_{\mathrm{Rm}}$, Δ_{Rm} in the simplest L/m/R case have been replaced with

$$\overline{\Delta}_{\mathrm{Rm}}^{\mathrm{eff}}(\mathrm{b_1}) = \frac{(\overline{\Delta}_{\mathrm{RB_1}}e^{-2\rho_{\mathrm{B_1}}\mathrm{b_1}} + \overline{\Delta}_{\mathrm{B_1m}})}{1 + \overline{\Delta}_{\mathrm{RB_1}}\overline{\Delta}_{\mathrm{B_1m}}e^{-2\rho_{\mathrm{B_1}}\mathrm{b_1}}}, \quad \Delta_{\mathrm{Rm}}^{\mathrm{eff}}(\mathrm{b_1}) = \frac{(\Delta_{\mathrm{RB_1}}e^{-2\rho_{\mathrm{B_1}}\mathrm{b_1}} + \Delta_{\mathrm{B_1m}})}{1 + \Delta_{\mathrm{RB_1}}\Delta_{\mathrm{B_1m}}e^{-2\rho_{\mathrm{B_1}}\mathrm{b_1}}}. \tag{L3.84}$$

As in that simplest LmR case, the full free energy has the form

$$G_{\mathrm{LmB_1R}}(l;\mathrm{b_1}) = \frac{kT}{2\pi c^2}\sum_{n=0}^{\infty}{}' \varepsilon_{\mathrm{m}}\mu_{\mathrm{m}}\xi_n^2\int_1^{\infty} p[\ln D_{\mathrm{LmB_1R}}(i\xi_n)]\mathrm{d}p \tag{L3.85}$$

except for the replacements $\overline{\Delta}_{\mathrm{Rm}}^{\mathrm{eff}}$, $\Delta_{\mathrm{Rm}}^{\mathrm{eff}}$.

Now the Rm interface is split into two interfaces, $\mathrm{RB_1}$ and $\mathrm{B_1m}$, where $\overline{\Delta}_{\mathrm{RB_1}}$ is reduced by a factor $e^{-2\rho_{\mathrm{B_1}}\mathrm{b_1}}$ for the thickness of the layer. When R and $\mathrm{B_1}$ have the same material properties, $\overline{\Delta}_{\mathrm{RB_1}} = 0$ so that $\overline{\Delta}_{\mathrm{Rm}}^{\mathrm{eff}}$ reverts to $\overline{\Delta}_{\mathrm{Rm}}$. When $\mathrm{b_1} \gg l$, $\overline{\Delta}_{\mathrm{Rm}}^{\mathrm{eff}}$ goes to $\overline{\Delta}_{\mathrm{B_1m}}$ as

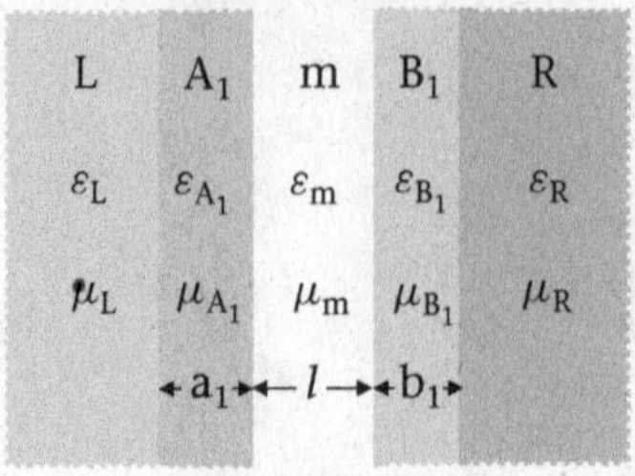

Figure L3.6

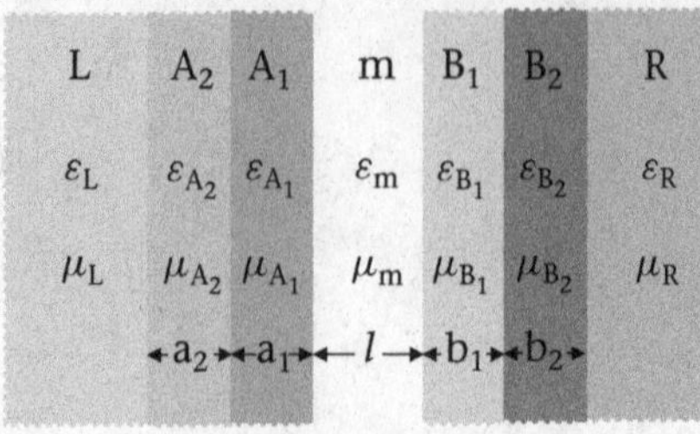

Figure L3.7

though material R were absent, the interaction is that between material B_1 and L across m at a distance l. When there are small differences in material properties, the $\overline{\Delta}$'s and Δ's are $\ll 1$; the products $\overline{\Delta}_{RB_1}\overline{\Delta}_{B_1m}$ and $\Delta_{RB_1}\Delta_{B_1m}$ in the denominators of $\overline{\Delta}_{Rm}^{eff}$ and Δ_{Rm}^{eff} can be ignored.

Two singly coated surfaces

It follows that the case in which each of the bodies is covered with a single layer (see Fig. L3.6) requires transformation of $\overline{\Delta}_{Lm}$ to

$$\overline{\Delta}_{Lm}^{eff} = \frac{(\overline{\Delta}_{LA_1}e^{-2\rho_{A_1}a_1} + \overline{\Delta}_{A_1m})}{1 + \overline{\Delta}_{LA_1}\overline{\Delta}_{A_1m}e^{-2\rho_{A_1}a_1}}, \tag{L3.86}$$

with

$$D_{LA_1mB_1R}(i\xi_n) = \left(1 - \overline{\Delta}_{Lm}^{eff}\overline{\Delta}_{Rm}^{eff}e^{-2\rho_m l}\right)\left(1 - \Delta_{Lm}^{eff}\Delta_{Rm}^{eff}e^{-2\rho_m l}\right) = 0. \tag{L3.87}$$

Adding layers

Further layering involves successive obvious substitutions for the $\overline{\Delta}_{LA_1}$, $\overline{\Delta}_{LB_1}$, Δ_{LA_1}, and Δ_{LB_1}. For a second layer on L, R, or both bodies, make the following substitutions.[19]

In $\overline{\Delta}_{Lm}^{eff}$ for a single layer,

$$\overline{\Delta}_{LA_1} \text{ is replaced by } \frac{(\overline{\Delta}_{LA_2}e^{-2\rho_{A_2}a_2} + \overline{\Delta}_{A_2A_1})}{1 + \overline{\Delta}_{LA_2}\overline{\Delta}_{A_2A_1}e^{-2\rho_{A_2}a_2}}; \tag{L3.88}$$

in $\overline{\Delta}_{Rm}^{eff}$,

$$\overline{\Delta}_{RB_1} \text{ is replaced by } \frac{\left(\overline{\Delta}_{RB_2}e^{-2\rho_{B_2}b_2} + \overline{\Delta}_{B_2B_1}\right)}{1 + \overline{\Delta}_{RB_2}\overline{\Delta}_{B_2B_1}e^{-2\rho_{B_2}b_2}} \tag{L3.89}$$

(see Fig. L3.7).

Proceed by induction. Say body L is coated with j′ layers and a j′ + 1st is added; or R by j layers and a j + 1st is added. Imagine slicing into L (or R) to create another layer of material $A_{j'+1}$ (or B_{j+1}) of thickness $a_{j'+1}$ (or b_{j+1}) (see Fig. L3.8).

The previous functions $\overline{\Delta}_{LA_{j'}}$ and $\overline{\Delta}_{RB_j}$ go to[20]

$$\frac{\left(\overline{\Delta}_{LA_{j'+1}}e^{-2\rho_{A_{j'+1}}a_{j'+1}} + \overline{\Delta}_{A_{j'+1}A_{j'}}\right)}{1 + \overline{\Delta}_{LA_{j'+1}}\overline{\Delta}_{A_{j'+1}A_{j'}}e^{-2\rho_{A_{j'+1}}a_{j'+1}}}, \frac{\left(\overline{\Delta}_{RB_{j+1}}e^{-2\rho_{B_{j+1}}b_{j+1}} + \overline{\Delta}_{B_{j+1}B_j}\right)}{1 + \overline{\Delta}_{RB_{j+1}}\overline{\Delta}_{B_{j+1}B_j}e^{-2\rho_{B_{j+1}}b_{j+1}}}. \tag{L3.90}$$

The compound $\overline{\Delta}_{Lm}^{eff}$, $\overline{\Delta}_{Rm}^{eff}$ and Δ_{Lm}^{eff}, Δ_{Rm}^{eff} that are built up in this way can be maddeningly messy but are mathematically manageable. They simplify mercifully when the

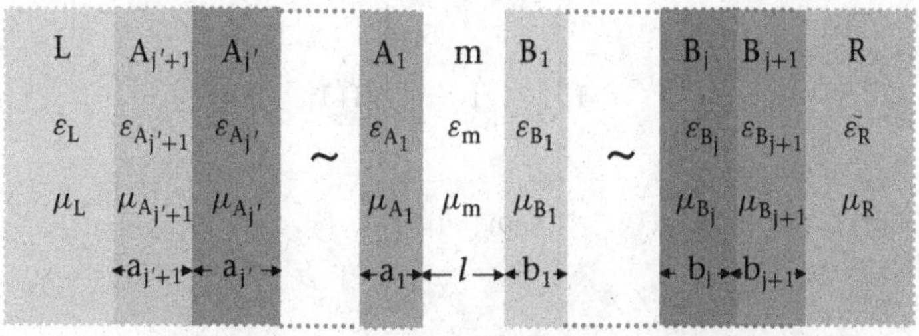

Figure L3.8

differences in ε's are small enough to let us ignore higher-power products of the $\overline{\Delta}$ sums-over-differences and to let us ignore differences in the ρ functions so that they may all be set equal to that of the medium ρ_m.

Multilayers

Multilayers are more fun. Imagine region R coated with N alternating layers of material B′ and B of thicknesses b′ and b with a final layer B′. Beyond this stack is a medium m of thickness l (see Fig. L3.9). How to construct the full transition matrix for this indefinitely extended system? The scheme for transition is still as in the simplest case:

$$\begin{pmatrix} A_R \\ B_R \end{pmatrix} = \mathbf{M}^{\mathrm{eff}}_{\mathrm{Rm}} \mathbf{M}^{\mathrm{eff}}_{\mathrm{mL}} \begin{pmatrix} A_L \\ B_L \end{pmatrix}. \tag{L3.91}$$

Matrix multiplication corresponds to the sequence of interfaces working from L to R. Because of the repeating structure, think of a succession of hops across each dielectric interface, written as matrices **D**, then traverses across layers written as matrices $\mathbf{T}^{21}$:

$$\begin{pmatrix} A_R \\ B_R \end{pmatrix} = \mathbf{D}_{\mathrm{RB'}} \left(\mathbf{T}_{\mathrm{B'}} \mathbf{D}_{\mathrm{B'B}} \mathbf{T}_{\mathrm{B}} \mathbf{D}_{\mathrm{BB'}}\right)^{N} \mathbf{T}_{\mathrm{B'}} \mathbf{D}_{\mathrm{B'm}} \mathbf{T}_{\mathrm{m}} \mathbf{D}_{\mathrm{mL}} \begin{pmatrix} A_L \\ B_L \end{pmatrix}. \tag{L3.92}$$

Decomposing the matrix

$$\mathbf{M}_{i+1/i} = \begin{bmatrix} 1 & -\overline{\Delta}_{i+1/i} e^{-2\rho_i (l_{i/i+1} - l_{i-1/i})} \\ -\overline{\Delta}_{i+1/i} & e^{-2\rho_i (l_{i/i+1} - l_{i-1/i})} \end{bmatrix} = \mathbf{D}_{i+1/i} \mathbf{T}_i \tag{L3.93}$$

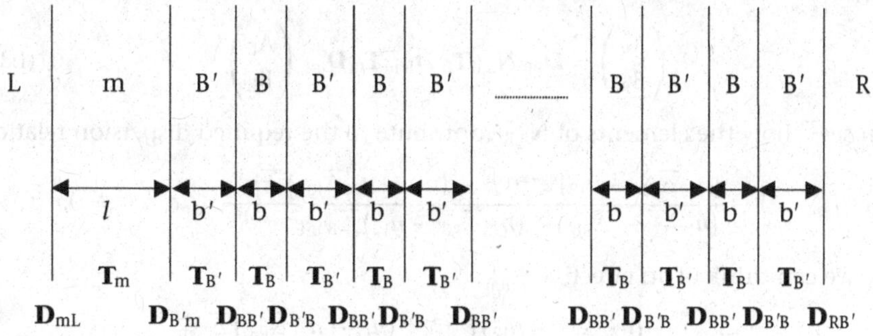

Figure L3.9

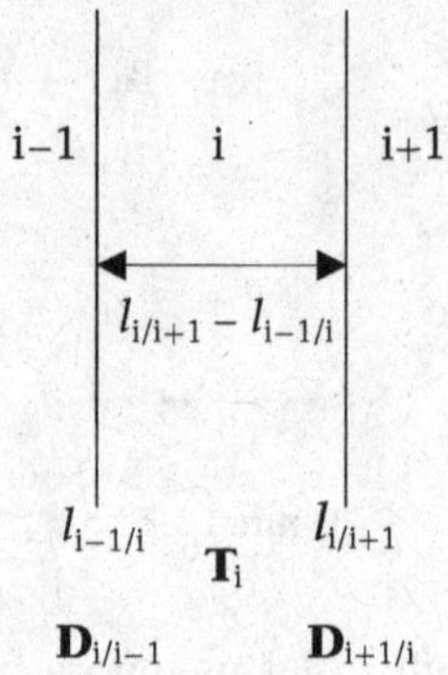

Figure L3.10

for the transition between regions i + 1 and i gives

$$\mathbf{D}_{i+1/i} = \begin{bmatrix} 1 & -\overline{\Delta}_{i+1/i} \\ -\overline{\Delta}_{i+1/i} & 1 \end{bmatrix}, \quad \mathbf{T}_i = \begin{bmatrix} 1 & 0 \\ 0 & e^{-2\rho_i(l_{i/i+1}-l_{i-1/i})} \end{bmatrix} \tag{L3.94}$$

(see Fig. L3.10).

For the N repeats of B/B′ pairs, $(\mathbf{T}_{B'}\mathbf{D}_{B'B}\mathbf{T}_B\mathbf{D}_{BB'})^N$, define

$$\mathbf{M}_{B'B} \equiv \mathbf{T}_{B'}\mathbf{D}_{B'B}\mathbf{T}_B\mathbf{D}_{BB'} \tag{L3.95}$$

but where

$$\mathbf{M}_{B'B} = \begin{bmatrix} m_{11} & m_{12} \\ m_{21} & m_{22} \end{bmatrix}$$

is normalized so that[22] $m_{11}m_{22} - m_{12}m_{21} = 1$:

$$m_{11} = \frac{1 - \overline{\Delta}_{B'B}^2 e^{-2\rho_B b}}{\left(1 - \overline{\Delta}_{B'B}^2\right) e^{-\rho_B b} e^{-\rho_{B'} b'}}, \qquad m_{12} = \frac{\overline{\Delta}_{B'B}\left(1 - e^{-2\rho_B b}\right)}{\left(1 - \overline{\Delta}_{B'B}^2\right) e^{-\rho_B b} e^{-\rho_{B'} b'}},$$

$$m_{21} = \frac{(e^{-2\rho_B b} - 1)\overline{\Delta}_{B'B} e^{-2\rho_{B'} b'}}{\left(1 - \overline{\Delta}_{B'B}^2\right) e^{-\rho_B b} e^{-\rho_{B'} b'}}, \qquad m_{22} = \frac{\left(e^{-2\rho_B b} - \overline{\Delta}_{B'B}^2\right) e^{-2\rho_{B'} b'}}{\left(1 - \overline{\Delta}_{B'B}^2\right) e^{-\rho_B b} e^{-\rho_{B'} b'}}. \tag{L3.96}$$

Once we have identified the elements of the transition matrix and normalized this matrix to satisfy the unitary condition, the dispersion relation for a full multilayer is immediately obtained. Then the Nth power of this matrix is

$$\mathbf{N}_{B'B} = \begin{bmatrix} m_{11} & m_{12} \\ m_{21} & m_{22} \end{bmatrix}^N \equiv \begin{bmatrix} n_{11} & n_{12} \\ n_{21} & n_{22} \end{bmatrix}. \tag{L3.97}$$

Through

$$\begin{pmatrix} A_R \\ B_R \end{pmatrix} = \mathbf{D}_{RB'}\mathbf{N}_{B'B}\mathbf{T}_{B'}\mathbf{D}_{B'm}\mathbf{T}_m\mathbf{D}_{mL}\begin{pmatrix} A_L \\ B_L \end{pmatrix} \tag{L3.98}$$

we can see[23] how the elements of $\mathbf{N}_{B'B}$ contribute to the required dispersion relation

$$D(i\xi_n) = 1 - \frac{\left(n_{22}\overline{\Delta}_{RB'} - n_{12}\right) e^{-2\rho_{B'} b'} + \left(n_{11} - n_{21}\overline{\Delta}_{RB'}\right)\overline{\Delta}_{B'm}}{\left(n_{11} - n_{21}\overline{\Delta}_{RB'}\right) + \left(n_{22}\overline{\Delta}_{RB'} - n_{12}\right)\overline{\Delta}_{B'm} e^{-2\rho_{B'} b'}}\overline{\Delta}_{Lm} e^{-2\rho_m l} = 0, \tag{L3.99}$$

where we can think of an effective $\overline{\Delta}_{Rm}^{\text{eff}}$:

$$\overline{\Delta}_{Rm}^{\text{eff}} = \frac{\left(n_{22}\overline{\Delta}_{RB'} - n_{12}\right) e^{-2\rho_{B'} b'} \left(n_{11} - n_{21}\overline{\Delta}_{RB'}\right)\overline{\Delta}_{B'm}}{\left(n_{11} - n_{21}\overline{\Delta}_{RB'}\right) + \left(n_{22}\overline{\Delta}_{RB'} - n_{12}\right)\overline{\Delta}_{B'm} e^{-2\rho_{B'} b'}}. \tag{L3.100}$$

For a few layers, N small, direct multiplication is practical. For an extended multilayer, tricks from optics and solid-state physics come in handy.

PROBLEM: Verify that, for $N = 0$, this formula [Eqs. (L3.99) and (L3.100)] reverts to the case [Eqs. (L3.82)–(L3.85)] of one layer on R and no layers on L.

SOLUTION: For no layers besides material B′ thickness b′,

$$\mathbf{N}_{\mathrm{B'B}} = \begin{bmatrix} 1 & 0 \\ 0 & 1 \end{bmatrix},$$

the identity matrix. The dispersion relation reduces to the familiar

$$D(i\xi_n) = 1 - \frac{\left(\overline{\Delta}_{\mathrm{RB'}} e^{-2\rho_{\mathrm{B'}} b'} + \overline{\Delta}_{\mathrm{B'm}}\right)}{\left(1 + \overline{\Delta}_{\mathrm{RB'}} \overline{\Delta}_{\mathrm{B'm}} e^{-2\rho_{\mathrm{B'}} b'}\right)} \overline{\Delta}_{\mathrm{Lm}} e^{-2\rho_{\mathrm{m}} l}$$

for a single layer of material B′ on R.

PROBLEM: Verify that, for B′ = B, Eqs. (L3.99) and (L3.100) become the case of one layer of thickness $[(N+1)(\mathrm{b}+\mathrm{b}') + \mathrm{b}']$ on half-space R with no layers on half-space L.

SOLUTION: When materials B′ and B are the same, there is effectively one large coating of $N+1$ layers of B′ and N layers of B. In this case $\overline{\Delta}_{\mathrm{B'B}} = 0$, $\rho_{\mathrm{B'}} = \rho_{\mathrm{B}}$:

$$m_{11} = e^{+\rho_{\mathrm{B'}}(\mathrm{b'}+\mathrm{b})}, \quad m_{12} = 0, \quad m_{21} = 0, \quad m_{22} = e^{-\rho_{\mathrm{B'}}(\mathrm{b'}+\mathrm{b})}.$$

The Nth power of the $\mathbf{M}$ matrix is

$$n_{11} = e^{+\rho_{\mathrm{B'}} N(\mathrm{b'}+\mathrm{b})}, \quad n_{12} = 0, \quad n_{21} = 0, \quad n_{22} = e^{-\rho_{\mathrm{B'}} N(\mathrm{b'}+\mathrm{b})},$$

so that

$$\overline{\Delta}_{\mathrm{Rm}}^{\mathrm{eff}} = \frac{\left(n_{22}\overline{\Delta}_{\mathrm{RB'}} - n_{12}\right) e^{-2\rho_{\mathrm{B'}} b'} + \left(n_{11} - n_{21}\overline{\Delta}_{\mathrm{RB'}}\right)\overline{\Delta}_{\mathrm{B'm}}}{\left(n_{11} - n_{21}\overline{\Delta}_{\mathrm{RB'}}\right) + \left(n_{22}\overline{\Delta}_{\mathrm{RB'}} - n_{12}\right)\overline{\Delta}_{\mathrm{B'm}} e^{-2\rho_{\mathrm{B'}} b'}}$$

$$= \frac{\overline{\Delta}_{\mathrm{RB'}} e^{-2\rho_{\mathrm{B'}} N(\mathrm{b'}+\mathrm{b})} e^{-2\rho_{\mathrm{B'}} b'} + \overline{\Delta}_{\mathrm{B'm}}}{1 + \overline{\Delta}_{\mathrm{RB'}} \overline{\Delta}_{\mathrm{B'm}} e^{-2\rho_{\mathrm{B'}} N(\mathrm{B'}+\mathrm{b})} e^{-2\rho_{\mathrm{B'}} \mathrm{B'}}},$$

where the thickness B′ of a single layer is replaced with $N(\mathrm{b} + \mathrm{b}') + \mathrm{b}'$ of that layer plus N additional layers with spacing b.

For multilayers we can take advantage of the fact that when a matrix $\mathbf{M}$ is unimodular, when $\det[\mathbf{M}] = 1$, its Nth power can be written as

$$\mathbf{M}^N = \begin{bmatrix} m_{11}U_{N-1} - U_{N-2} & m_{12}U_{N-1} \\ m_{21}U_{N-1} & m_{22}U_{N-1} - U_{N-2} \end{bmatrix}. \qquad \text{(L3.101)}$$

The m_{ij} are the elements of the original matrix $\mathbf{M}$, and U_N is a Chebyshev polynomial defined here by

$$U_\nu(x) = \frac{\sin\left[(\nu+1)\cos^{-1}(x)\right]}{(1-x^2)^{1/2}}, \quad \text{with } x = \frac{m_{11} + m_{22}}{2} \text{ for } \nu > 0, \qquad \text{(L3.102)}$$

$$U_{\nu=0}(x) = 1, \; U_{\nu<0}(x) = 0.$$

Because $\mathbf{M}_{\mathrm{B'B}}$ is unimodular by construction, there is a condition between elements:

$$m_{11} = (1 + m_{12}m_{21}) / m_{22}. \tag{L3.103}$$

It is sometimes convenient to define a variable ζ for the argument of the Chebyshev polynomial such that

$$x = \frac{m_{11} + m_{22}}{2} = \cosh(\zeta) \qquad \text{or} \qquad m_{11} + m_{22} = e^{+\zeta} + e^{-\zeta}. \tag{L3.104}$$

This definition of ζ together with the unimodular property gives a useful constraint:

$$(m_{11} - e^{-\zeta})(m_{22} - e^{-\zeta}) = m_{12}m_{21}. \tag{L3.105}$$

For $x > 1$, as must be the case here,

$$U_{N-1}(x) = \frac{\sinh(N\zeta)}{\sinh(\zeta)} = \frac{e^{+N\zeta} - e^{-N\zeta}}{e^{+\zeta} - e^{-\zeta}},$$

$$U_{N-2}(x) = \frac{\sinh[(N-1)\zeta]}{\sinh(\zeta)}. \tag{L3.106}$$

The matrix elements of

$$\mathbf{N}_{\mathrm{B'B}} = (\mathbf{T}_{\mathrm{B'}}\mathbf{D}_{\mathrm{B'B}}\mathbf{T}_{\mathrm{B}}\mathbf{D}_{\mathrm{BB'}})^N = \begin{bmatrix} n_{11} & n_{12} \\ n_{21} & n_{22} \end{bmatrix}$$

are then

$$n_{11} = m_{11}U_{N-1} - U_{N-2}, \qquad n_{12} = m_{12}U_{N-1},$$
$$n_{21} = m_{21}U_{N-1}, \qquad n_{22} = m_{22}U_{N-1} - U_{N-2}. \tag{L3.107}$$

Large-*N* limit

In the limit of large N, the dispersion relation reduces to[24]

$$D(i\xi_n) = 1 - \frac{m_{21}\overline{\Delta}_{\mathrm{B'm}} - (m_{22} - e^{-\zeta})e^{-2\rho_{\mathrm{B'}}b'}}{m_{21} - (m_{22} - e^{-\zeta})\,\overline{\Delta}_{\mathrm{B'm}}e^{-2\rho_{\mathrm{B'}}b'}}\overline{\Delta}_{\mathrm{Lm}}e^{-2\rho_m l} = 0. \tag{L3.108}$$

This is for the interaction between a half-space L and an infinitely layered half-space R. For the large limiting value of N used here, the right-hand half-space R disappears from the formulation.

By symmetry it is possible to generalize to multicoated L.

Interaction between two multilayer-covered surfaces

It is trivially easy to replace the half-space L with L coated with its own multilayer. Let this be successive layers of material A′, thickness a′, abutting L, then N_{L} layers of material A, thickness a, followed by material A′, thickness a′. The multilayer on R will have N_{R} b, b′ layers, where N is now replaced with N_{R} (see Fig. L3.11).

To see the generalization, recognize that, if R were uncovered and L covered with a multilayer, the interaction would be of the same form just derived. The formula would have N_{R} (formerly N) replaced by N_{L} for the number of additional layer pairs A′, A, of thicknesses a′, a, where there were layers B′, B of thickness b′, b.

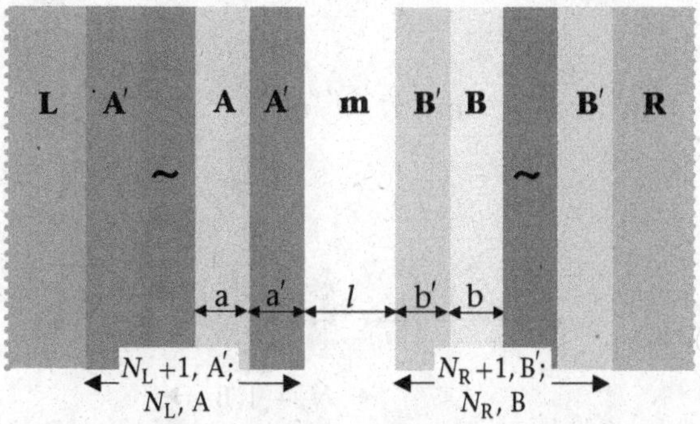

Figure L3.11

With this observation write the effective $\overline{\Delta}^{\mathrm{eff}}_{\mathrm{Lm}}$ and $\overline{\Delta}^{\mathrm{eff}}_{\mathrm{Rm}}$:

$$\overline{\Delta}^{\mathrm{eff}}_{\mathrm{Lm}} = \frac{\left[n_{22}^{(\mathrm{L})}\overline{\Delta}_{\mathrm{LA'}} - n_{12}^{(\mathrm{L})}\right] e^{-2\rho_{\mathrm{A'}}a'} + \left[n_{11}^{(\mathrm{L})} - n_{21}^{(\mathrm{L})}\overline{\Delta}_{\mathrm{LA'}}\right]\overline{\Delta}_{\mathrm{A'm}}}{\left[n_{11}^{(\mathrm{L})} - n_{21}^{(\mathrm{L})}\overline{\Delta}_{\mathrm{LA'}}\right] + \left[n_{22}^{(\mathrm{L})}\overline{\Delta}_{\mathrm{LA'}} - n_{12}^{(\mathrm{L})}\right]\overline{\Delta}_{\mathrm{A'm}} e^{-2\rho_{\mathrm{A'}}a'}},$$

$$\overline{\Delta}^{\mathrm{eff}}_{\mathrm{Rm}} = \frac{\left[n_{22}^{(\mathrm{R})}\overline{\Delta}_{\mathrm{RB'}} - n_{12}^{(\mathrm{R})}\right] e^{-2\rho_{\mathrm{B'}}b'} + \left[n_{11}^{(\mathrm{R})} - n_{21}^{(\mathrm{R})}\overline{\Delta}_{\mathrm{RB'}}\right]\overline{\Delta}_{\mathrm{B'm}}}{\left[n_{11}^{(\mathrm{R})} - n_{21}^{(\mathrm{R})}\overline{\Delta}_{\mathrm{RB'}}\right] + \left[n_{22}^{(\mathrm{R})}\overline{\Delta}_{\mathrm{RB'}} - n_{12}^{(\mathrm{R})}\right]\overline{\Delta}_{\mathrm{B'm}} e^{-2\rho_{\mathrm{B'}}b'}}. \tag{L3.109}$$

These go into the familiar dispersion relation

$$D(i\xi_n) = 1 - \overline{\Delta}^{\mathrm{eff}}_{\mathrm{Rm}}\overline{\Delta}^{\mathrm{eff}}_{\mathrm{Lm}} e^{-2\rho_{\mathrm{m}}l} = 0, \tag{L3.110}$$

where the elements $n_{\mathrm{ij}}^{(\mathrm{R})}$ and $n_{\mathrm{ij}}^{(\mathrm{L})}$ are constructed as before:

$$n_{11}^{(\mathrm{L})} = m_{11}^{(\mathrm{L})}U_{N_\mathrm{L}-1}(x_\mathrm{L}) - U_{N_\mathrm{L}-2}(x_\mathrm{L}), \qquad n_{11}^{(\mathrm{R})} = m_{11}^{(\mathrm{R})}U_{N_\mathrm{R}-1}(x_\mathrm{R}) - U_{N_\mathrm{R}-2}(x_\mathrm{R}),$$
$$n_{12}^{(\mathrm{L})} = m_{12}^{(\mathrm{L})}U_{N_\mathrm{L}-1}(x_\mathrm{L}), \qquad n_{12}^{(\mathrm{R})} = m_{12}^{(\mathrm{R})}U_{N_\mathrm{R}-1}(x_\mathrm{R}). \tag{L3.111}$$

$$n_{21}^{(\mathrm{L})} = m_{21}^{(\mathrm{L})}U_{N_\mathrm{L}-1}(x_\mathrm{L}), \qquad n_{21}^{(\mathrm{R})} = m_{21}^{(\mathrm{R})}U_{N_\mathrm{R}-1}(x_\mathrm{R}),$$
$$n_{22}^{(\mathrm{L})} = m_{22}^{(\mathrm{L})}U_{N_\mathrm{L}-1}(x_\mathrm{L}) - U_{N_\mathrm{L}-2}(x_\mathrm{L}), \qquad n_{22}^{(\mathrm{R})} = m_{22}^{(\mathrm{R})}U_{N_\mathrm{R}-1}(x_\mathrm{R}) - U_{N_\mathrm{R}-2}(x_\mathrm{R}). \tag{L3.112}$$

$$x_\mathrm{L} = \frac{m_{11}^{(\mathrm{L})} + m_{22}^{(\mathrm{L})}}{2}, \qquad x_\mathrm{R} = \frac{m_{11}^{(\mathrm{R})} + m_{22}^{(\mathrm{R})}}{2}, \tag{L3.113}$$

$$m_{11}^{(\mathrm{L})} = \frac{1 - \overline{\Delta}^2_{\mathrm{A'A}} e^{-2\rho_\mathrm{A}a}}{\left(1 - \overline{\Delta}^2_{\mathrm{A'A}}\right) e^{-\rho_\mathrm{A}a} e^{-\rho_{\mathrm{A'}}a'}}, \qquad m_{11}^{(\mathrm{R})} = \frac{1 - \overline{\Delta}^2_{\mathrm{B'B}} e^{-2\rho_\mathrm{B}b}}{\left(1 - \overline{\Delta}^2_{\mathrm{B'B}}\right) e^{-\rho_\mathrm{B}b} e^{-\rho_{\mathrm{B'}}b'}},$$
$$m_{12}^{(\mathrm{L})} = \frac{\overline{\Delta}_{\mathrm{A'A}}\left(1 - e^{-2\rho_\mathrm{A}a}\right)}{\left(1 - \overline{\Delta}^2_{\mathrm{A'A}}\right) e^{-\rho_\mathrm{A}a} e^{-\rho_{\mathrm{A'}}a'}}, \qquad m_{12}^{(\mathrm{R})} = \frac{\overline{\Delta}_{\mathrm{B'B}}\left(1 - e^{-2\rho_\mathrm{B}b}\right)}{\left(1 - \overline{\Delta}^2_{\mathrm{B'B}}\right) e^{-\rho_\mathrm{B}b} e^{-\rho_{\mathrm{B'}}b'}}. \tag{L3.114}$$

$$m_{21}^{(\mathrm{L})} = \frac{\left(e^{-2\rho_\mathrm{A}a} - 1\right)\overline{\Delta}_{\mathrm{A'A}} e^{-2\rho_{\mathrm{A'}}a'}}{\left(1 - \overline{\Delta}^2_{\mathrm{A'A}}\right) e^{-\rho_\mathrm{A}a} e^{-\rho_{\mathrm{A'}}a'}}, \qquad m_{21}^{(\mathrm{R})} = \frac{\left(e^{-2\rho_\mathrm{B}b} - 1\right)\overline{\Delta}_{\mathrm{B'B}} e^{-2\rho_{\mathrm{B'}}b'}}{\left(1 - \overline{\Delta}^2_{\mathrm{B'B}}\right) e^{-\rho_\mathrm{B}b} e^{-\rho_{\mathrm{B'}}b'}},$$
$$m_{22}^{(\mathrm{L})} = \frac{\left(e^{-2\rho_\mathrm{A}a} - \overline{\Delta}^2_{\mathrm{A'A}}\right) e^{-2\rho_{\mathrm{A'}}a'}}{\left(1 - \overline{\Delta}^2_{\mathrm{A'A}}\right) e^{-\rho_\mathrm{A}a} e^{-\rho_{\mathrm{A'}}a'}}, \qquad m_{22}^{(\mathrm{R})} = \frac{\left(e^{-2\rho_\mathrm{B}b} - \overline{\Delta}^2_{\mathrm{B'B}}\right) e^{-2\rho_{\mathrm{B'}}b'}}{\left(1 - \overline{\Delta}^2_{\mathrm{B'B}}\right) e^{-\rho_\mathrm{B}b} e^{-\rho_{\mathrm{B'}}b'}}. \tag{L3.115}$$

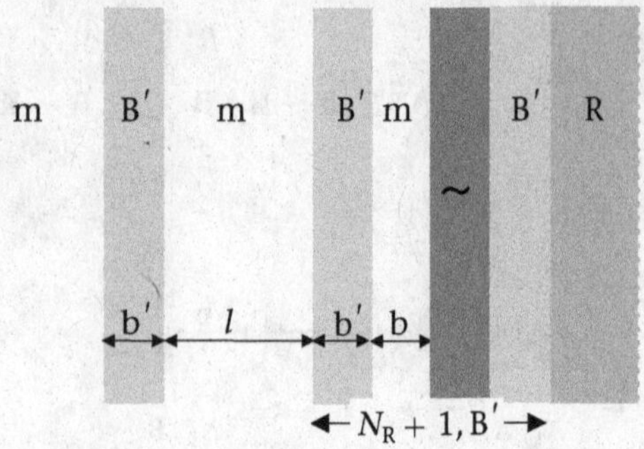

Figure L3.12

Layer of finite thickness adding on to a multilayer stack

The previous result can be immediately specialized to the case of a layer of finite thickness interacting with a previously existing stack of $N+1$ layers. Let half-space L as well as all materials B have the same dielectric properties as medium m. Let material A′ have the same properties as material B′ (see Fig. L3.12).

Then (Table P.6.c.1)

$$\overline{\Delta}^{\mathrm{eff}}_{\mathrm{Lm}} = \frac{\overline{\Delta}_{\mathrm{mB'}} e^{-2\rho_{\mathrm{B'}} \mathrm{b'}} + \overline{\Delta}_{\mathrm{B'm}}}{1 + \overline{\Delta}_{\mathrm{mB'}} \overline{\Delta}_{\mathrm{B'm}} e^{-2\rho_{\mathrm{B'}} \mathrm{b'}}} = \overline{\Delta}_{\mathrm{B'm}} \frac{1 - e^{-2\rho_{\mathrm{B'}} \mathrm{b'}}}{1 - \overline{\Delta}^2_{\mathrm{B'm}} e^{-2\rho_{\mathrm{B'}} \mathrm{b'}}}, \tag{L3.116}$$

$$\overline{\Delta}^{\mathrm{eff}}_{\mathrm{Rm}} = \frac{\left(n^{(\mathrm{R})}_{22} \overline{\Delta}_{\mathrm{RB'}} - n^{(\mathrm{R})}_{12}\right) e^{-2\rho_{\mathrm{B'}} \mathrm{b'}} + \left(n^{(\mathrm{R})}_{11} - n^{(\mathrm{R})}_{21} \overline{\Delta}_{\mathrm{RB'}}\right) \overline{\Delta}_{\mathrm{B'm}}}{\left(n^{(\mathrm{R})}_{11} - n^{(\mathrm{R})}_{21} \overline{\Delta}_{\mathrm{RB'}}\right) + \left(n^{(\mathrm{R})}_{22} \overline{\Delta}_{\mathrm{RB'}} - n^{(\mathrm{R})}_{12}\right) \overline{\Delta}_{\mathrm{B'm}} e^{-2\rho_{\mathrm{B'}} \mathrm{b'}}}, \tag{L3.117}$$

where now

$$m^{(\mathrm{R})}_{11} = \frac{1 - \overline{\Delta}^2_{\mathrm{B'm}} e^{-2\rho_{\mathrm{m}} \mathrm{b}}}{\left(1 - \overline{\Delta}^2_{\mathrm{B'm}}\right) e^{-\rho_{\mathrm{m}} \mathrm{b}} e^{-\rho_{\mathrm{B'}} \mathrm{b'}}}, \quad m^{(\mathrm{R})}_{12} = \frac{\overline{\Delta}_{\mathrm{B'm}} \left(1 - e^{-2\rho_{\mathrm{m}} \mathrm{b}}\right)}{\left(1 - \overline{\Delta}^2_{\mathrm{B'm}}\right) e^{-\rho_{\mathrm{m}} \mathrm{b}} e^{-\rho_{\mathrm{B'}} \mathrm{b'}}},$$

$$m^{(\mathrm{R})}_{21} = \frac{\left(e^{-2\rho_{\mathrm{m}} \mathrm{b}} - 1\right) \overline{\Delta}_{\mathrm{B'm}} e^{-2\rho_{\mathrm{B'}} \mathrm{b'}}}{\left(1 - \overline{\Delta}^2_{\mathrm{B'm}}\right) e^{-\rho_{\mathrm{m}} \mathrm{b}} e^{-\rho_{\mathrm{B'}} \mathrm{b'}}}, \quad m^{(\mathrm{R})}_{22} = \frac{\left(e^{-2\rho_{\mathrm{m}} \mathrm{b}} - \overline{\Delta}^2_{\mathrm{B'm}}\right) e^{-2\rho_{\mathrm{B'}} \mathrm{b'}}}{\left(1 - \overline{\Delta}^2_{\mathrm{B'm}}\right) e^{-\rho_{\mathrm{m}} \mathrm{b}} e^{-\rho_{\mathrm{B'}} \mathrm{b'}}}. \tag{L3.118}$$

L3.5. Inhomogeneous media

Dielectric properties need not change in sharp steps. Recognizing the possibility that they can change continuously reveals qualitatively new forms of force versus separation. Imagine, for example, regions where we can write $\varepsilon = \varepsilon(z)$. In the nonretarded limit, electric waves satisfy an equation

$$\nabla \cdot [\varepsilon(z) E(x, y, z)] = 0, \tag{L3.119}$$

or, with $E(x, y, z) = -\nabla\phi(x, y, z)$,

$$\nabla \cdot [\varepsilon(z)\nabla\phi(x, y, z)] = 0. \tag{L3.120}$$

Solutions of the form $\phi(x, y, z) = f(z)e^{i(ux+vy)}$ with $\rho^2 = u^2 + v^2$ create a differential equation,

$$f''(z) + \frac{\mathrm{d}\varepsilon/\mathrm{d}z}{\varepsilon(z)} f'(z) - \rho^2 f(z) = 0, \tag{L3.121}$$

where the primes here denote single or double differentiation by z.[25]

Several forms of $\varepsilon(z)$ admit exact solutions for $f(z)$ and the corresponding surface modes for van der Waals interaction. To examine these solutions, consider the interaction of two semi-infinite regions L and R of spatially unvarying dielectric response ε_L and ε_R, coated with slabs of thickness D_L and D_R within which there is spatially variable dielectric response $\varepsilon_a(z)$ and $\varepsilon_b(z)$, separated by a medium m of constant ε_m and variable thickness l (see Fig. L3.13).

Note that z is measured from the midpoint rather than from one of the material interfaces. Rather than create mathematically difficult forms of $\varepsilon(z)$ based on models of materials such as polymers, it is often more practical to use a lamplight strategy. Examine mathematically tractable forms of $(\mathrm{d}\varepsilon/\mathrm{d}z)/\varepsilon(z) = \mathrm{d}\ln[\varepsilon(z)]/\mathrm{d}z$.

Two such forms, for which formulae are given in the tables, are exponential (Table P.7.c):

$$\varepsilon(z) = \Gamma e^{-\gamma z}, z > 0, \quad \varepsilon(z) = \Gamma e^{+\gamma z}, z < 0,$$
$$\mathrm{d}\ln[\varepsilon(z)]/\mathrm{d}z = -\gamma,\ z > 0; \quad \mathrm{d}\ln[\varepsilon(z)]/\mathrm{d}z = +\gamma, z < 0, \tag{L3.122}$$

with $\Gamma > 0$, and power law (Table P.7.d)

$$\varepsilon(z) = (\alpha + \beta z)^n, z > 0, \quad \varepsilon(z) = (\alpha - \beta z)^n, z < 0,$$
$$\mathrm{d}\ln[\varepsilon(z)]/\mathrm{d}z = n\beta/(\alpha + \beta z), z > 0; \quad \mathrm{d}\ln[\varepsilon(z)]/\mathrm{d}z = -n\beta/(\alpha - \beta z), z < 0, \tag{L3.123}$$

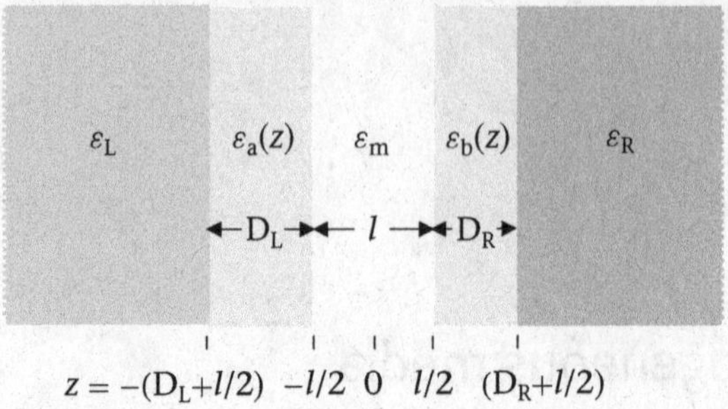

Figure L3.13

with α, β restricted only to maintain physically permissible $\varepsilon(z) \geq 1$ for positive imaginary frequencies where they are to be evaluated.

Many other forms of $\varepsilon(z)$ are probably tractable by use of experience from known solutions, e.g., for electromagnetic waves in the upper atmosphere. It turns out that instructive idiosyncrasies of inhomogeneity occur when ε or even $d\varepsilon/dz$ is continuous at the m|2 interfaces, $z = \pm l/2$. These idiosyncrasies become clear when results are compared with, for example, the two exponentials

$$\varepsilon(z) = \varepsilon_m e^{\gamma_e(z-l/2)}, z > l/2; \quad \varepsilon(z) = \varepsilon_m e^{-\gamma_e(z+l/2)}, z < -l/2, \tag{L3.124}$$

ε continuous at $z = \pm l/2$, and Gaussian

$$\varepsilon(z) = \varepsilon_m e^{\gamma_g^2(z-l/2)^2}, z > l/2; \quad \varepsilon(z) = \varepsilon_m e^{\gamma_g^2(z+l/2)^2}, z < -l/2,$$

both ε and $d\varepsilon/dz$ continuous at $z = \pm l/2$.

Here we work out several cases for which there is an arbitrary continuously varying $\varepsilon = \varepsilon(z)$ in a layer of fixed thickness between each outer half-space and the central medium of variable thickness l. Although there is no general closed-form solution for arbitrary $\varepsilon(z)$, it is possible to derive a mathematical procedure for evaluation. For clarity, consider successively more difficult situations: nonretarded interactions, symmetric and nonsymmetric geometry, retarded, nonsymmetric.

Arbitrary, continuous $\varepsilon(z)$ with discontinuities allowed at interfaces, nonretarded interactions

Two coated layers, symmetric configuration

Follow an outside–in strategy. Consider a semi-infinite body with permittivity ε_{out}, coated with an inhomogeneous layer of constant thickness D and permittivity $\varepsilon(z)$, facing a medium ε_m of variable thickness l. Subscripts for a, b, L, and R will be added later (Fig. L3.14).

To use what we know of forces between multiply layered structures, we first approximate the layer of variable $\varepsilon(z)$ by N slabs of equal thickness D/N bounded by $N+1$ interfaces, at

$$z_r = \frac{l}{2} + r\frac{D}{N} \text{ for } r = 0, 1, 2, \ldots, N. \tag{L3.125}$$

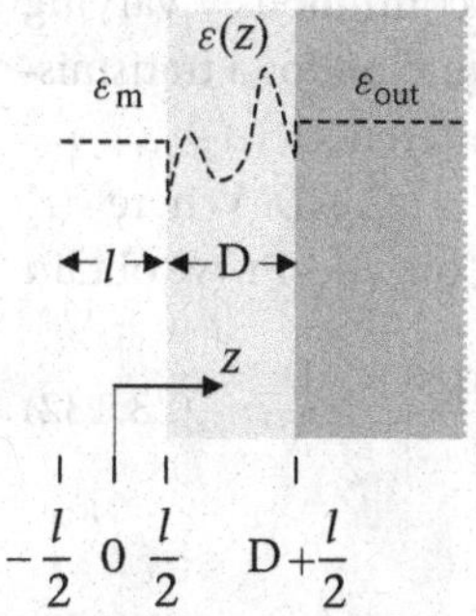

Figure L3.14

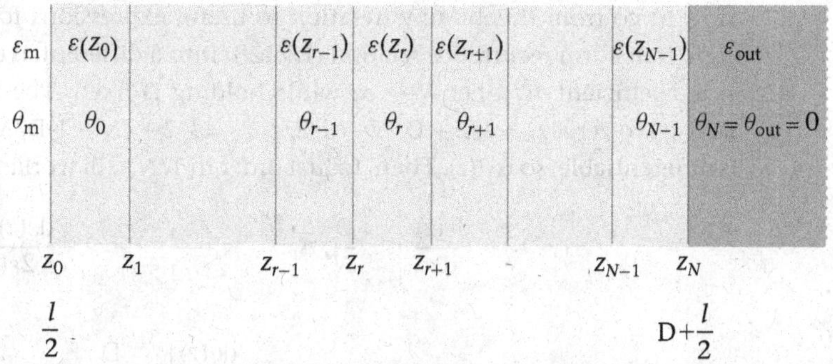

Figure L3.15

The interfaces of the variable region with the medium m and the outer half-space are, respectively, at $z_0 = (l/2)$ and $z_N = (l/2) + \mathrm{D}$ (see Fig. L3.15).

As usual, we seek solutions of the form $\mathrm{A}e^{+\rho z} + \mathrm{B}e^{-\rho z}$ to $f''(z) - \rho^2 f(z) = 0$ to the wave equation where [Eq. (L3.59)]

$$\begin{aligned} \mathrm{A}_{r-1}e^{\rho z_r} - \mathrm{B}_{r-1}e^{-\rho z_r} &= \mathrm{A}_r e^{\rho z_r} - \mathrm{B}_r e^{-\rho z_r}, \\ \varepsilon_{r-1}\left(\mathrm{A}_{r-1}e^{\rho z_r} + \mathrm{B}_{r-1}e^{-\rho z_r}\right) &= \varepsilon_r\left(\mathrm{A}_r e^{\rho z_r} + \mathrm{B}_r e^{-\rho z_r}\right), \\ \varepsilon_{r-1} \equiv \varepsilon\left(z_{r-1}\right), \varepsilon_r &\equiv \varepsilon\left(z_r\right). \end{aligned} \tag{L3.126}$$

We define

$$\theta_r \equiv \mathrm{A}_r/\mathrm{B}_r, \quad \overline{\Delta}_{r-1/r} \equiv \frac{\varepsilon_{r-1} - \varepsilon_r}{\varepsilon_{r-1} + \varepsilon_r} = -\overline{\Delta}_{r/r-1}, \tag{L3.127}$$

so that, for working outside-in,[26]

$$\theta_{r-1} = \frac{1}{e^{+2\rho z_r}}\left(\frac{\theta_r e^{+2\rho z_r} - \overline{\Delta}_{r-1/r}}{1 - \theta_r e^{+2\rho z_r}\overline{\Delta}_{r-1/r}}\right). \tag{L3.128}$$

At the right-most interface, index $r = N$, position $z_N = (l/2) + \mathrm{D}$, between material $\varepsilon(z_{N-1}) = \varepsilon_{N-1}$ and $\varepsilon_{\mathrm{out}} = \varepsilon_N$,

$$\theta_{N-1} = \frac{1}{e^{+\rho(l+2\mathrm{D})}}\left(\frac{\theta_N e^{+\rho(l+2\mathrm{D})} - \overline{\Delta}_{N-1/N}}{1 - \theta_N e^{+\rho(l+2\mathrm{D})}\overline{\Delta}_{N-1/N}}\right). \tag{L3.129}$$

Recall the condition for surface modes [Eqs. (L3.15)–(L3.18)] in the derivation of the original Lifshitz result (Subsection L3.2.A), the coefficient of $e^{+\rho z}$ must be zero. This means $\theta_N = \theta_{\mathrm{out}} = 0$, so that

$$\theta_{N-1} = -\frac{\overline{\Delta}_{N-1/\mathrm{out}}}{e^{+\rho(l+2\mathrm{D})}} = -\frac{\overline{\Delta}_{l/2+\mathrm{D/out}}}{e^{+\rho(l+2\mathrm{D})}}, \quad \overline{\Delta}_{(l/2+\mathrm{D})/\mathrm{out}} = \frac{\varepsilon(l/2+\mathrm{D}) - \varepsilon_{\mathrm{out}}}{\varepsilon(l/2+\mathrm{D}) + \varepsilon_{\mathrm{out}}}. \tag{L3.130}$$

At the interface between the variable region beginning at index $r = 0$, $z_0 = l/2$, and the medium m [again we apply Eq. (L3.128)],

$$\theta_{\mathrm{m}} = \frac{1}{e^{+\rho l}}\left(\frac{\theta_0 e^{+\rho l} - \overline{\Delta}_{\mathrm{m}/0}}{1 - \theta_0 e^{+\rho l}\overline{\Delta}_{\mathrm{m}/0}}\right) = \frac{1}{e^{+\rho l}}\left(\frac{\theta_0 e^{+\rho l} - \overline{\Delta}_{\mathrm{m}/(l/2)}}{1 - \theta_0 e^{+\rho l}\overline{\Delta}_{\mathrm{m}/(l/2)}}\right), \overline{\Delta}_{\mathrm{m}/(l/2)} = \frac{\varepsilon_{\mathrm{m}} - \varepsilon(l/2)}{\varepsilon_{\mathrm{m}} + \varepsilon(l/2)}. \tag{L3.131}$$

How to go from this bumpy iteration to useful expressions for continuously varying $\varepsilon(z)$? Transform recurrence relation (L3.128) into a differential equation for a transmission coefficient $\theta(z)$. Let $N \to \infty$ while holding D fixed. The differences $(z_r - z_{r-1}) = \mathrm{D}/N$ go to zero, $z_1 = l/2 + \mathrm{D}/N \to l/2$, $z_{N-1} = l/2 + (N-1)\mathrm{D}/N \to l/2 + \mathrm{D}$. Where $\varepsilon(z)$ is differentiable, so is $\theta(z)$. Then, to first order in $1/N$, still working outside–in, we obtain

$$\varepsilon_{r-1} \sim \varepsilon_r - \left.\frac{\mathrm{d}\varepsilon(z)}{\mathrm{d}z}\right|_{z=z_r} \frac{\mathrm{D}}{N}, \ \overline{\Delta}_{r-1/r} \equiv \frac{\varepsilon_{r-1} - \varepsilon_r}{\varepsilon_{r-1} + \varepsilon_r} \sim -\left.\frac{\mathrm{d}\varepsilon(z)/\mathrm{d}z}{2\varepsilon(z)}\right|_{z=z_r} \frac{\mathrm{D}}{N}, \qquad \text{(L3.132)}$$

$$\theta_{r-1} \sim \theta_r - \left.\frac{\mathrm{d}\theta(z)}{\mathrm{d}z}\right|_{z=z_r} \frac{\mathrm{D}}{N}. \qquad \text{(L3.133)}$$

Similarly, we expand Eq. (L3.128) to first order in $\overline{\Delta}_{r-1/r}$:

$$\theta_{r-1} \sim \left(\theta_r - \overline{\Delta}_{r-1/r}/e^{+2\rho z_r}\right)\left(1 + \theta_r e^{+2\rho z_r}\overline{\Delta}_{r-1/r}\right) \sim \theta_r - \overline{\Delta}_{r-1/r}\left(\frac{1}{e^{+2\rho z_r}} - e^{+2\rho z_r}\theta_r^2\right). \qquad \text{(L3.134)}$$

We equate approximations (L3.133) and (L3.134) for θ_{r-1}, introduce approximation (L3.132) for $\overline{\Delta}_{r-1/r}$,

$$\theta_{r-1} \sim \theta_r - \left.\frac{\mathrm{d}\theta(z)}{\mathrm{d}z}\right|_{z=z_r} \frac{\mathrm{D}}{N} \sim \theta_r + \left.\frac{\mathrm{d}\varepsilon(z)/\mathrm{d}z}{2\varepsilon(z)}\right|_{z=z_r} \frac{\mathrm{D}}{N}\left(\frac{1}{e^{+2\rho z_r}} - e^{+2\rho z_r}\theta_r^2\right), \qquad \text{(L3.135)}$$

and pass to the continuum limit to obtain a differential equation for $\theta(z)$:

$$\frac{\mathrm{d}\theta(z)}{\mathrm{d}z} = -\frac{e^{-2\rho z}}{2}\frac{\mathrm{d}\ln[\varepsilon(z)]}{\mathrm{d}z}\left[1 - e^{+4\rho z}\theta^2(z)\right]. \qquad \text{(L3.136)}$$

We define

$$u(z) \equiv e^{2\rho z}\theta(z) \qquad \text{(L3.137)}$$

to create an alternative, occasionally convenient, form,

$$\frac{\mathrm{d}u(z)}{\mathrm{d}z} = 2\rho u(z) - \frac{\mathrm{d}\ln[\varepsilon(z)]}{2\mathrm{d}z}\left[1 - u^2(z)\right]. \qquad \text{(L3.138)}$$

There is a boundary condition at the discontinuous step allowed to occur at $z = (l/2) + \mathrm{D}$, Eq. (L3.130). In the $N \to \infty$ limit,

$$\theta_{N-1} \to \theta\left(\frac{l}{2} + \mathrm{D}\right) = -\overline{\Delta}_{(l/2+\mathrm{D})/\mathrm{out}}e^{-\rho(l+2\mathrm{D})} \qquad \text{(L3.139a)}$$

or

$$u\left(\frac{l}{2} + \mathrm{D}\right) \equiv \theta\left(\frac{l}{2} + \mathrm{D}\right)e^{+\rho(l+2\mathrm{D})} = -\overline{\Delta}_{(l/2+\mathrm{D})/\mathrm{out}}. \qquad \text{(L3.139b)}$$

Beginning with this boundary condition for $\theta(\frac{l}{2} + \mathrm{D})$ or $u(\frac{l}{2} + \mathrm{D})$, we use differential equation (L3.136) or (L3.138) to propagate to the value of $\theta(l/2)$ or $u(l/2)$ needed at $z = (l/2)$. We apply this $\theta(l/2)$ or $u(l/2)$ to

$$\theta_{\mathrm{m}} = \left[\frac{\theta(l/2)e^{+\rho l} - \overline{\Delta}_{\mathrm{m}/(l/2)}}{1 - \theta(l/2)e^{+\rho l}\overline{\Delta}_{\mathrm{m}/(l/2)}}\right]e^{-\rho l} = \overline{\Delta}^{\mathrm{eff}}e^{-\rho l} \qquad \text{(L3.140a)}$$

with [Eq. (L3.127) with $r = 1$]:

$$\overline{\Delta}_{(l/2)/\mathrm{m}} = \frac{\varepsilon(l/2) - \varepsilon_{\mathrm{m}}}{\varepsilon(l/2) + \varepsilon_{\mathrm{m}}} = -\overline{\Delta}_{\mathrm{m}/(l/2)},$$

$$\overline{\Delta}^{\mathrm{eff}} \equiv \frac{\theta(l/2)e^{+\rho l} + \overline{\Delta}_{(l/2)/\mathrm{m}}}{1 + \theta(l/2)e^{+\rho l}\overline{\Delta}_{(l/2)/\mathrm{m}}} = \frac{u(l/2) + \overline{\Delta}_{(l/2)/\mathrm{m}}}{1 + u(l/2)\overline{\Delta}_{(l/2)/\mathrm{m}}}. \qquad \text{(L3.140b)}$$

What about the interaction of this coated semi-infinite body with another coated body, symmetric to it? To avoid unnecessary mathematics, think of the mirror image of the problem just solved. Speak of variables z_a and z_b in Fig. L3.16 as increasing, respectively, left and right from the midpoint in variable region of thickness l. These z_a, z_b of convenience connect with the "real" z as $z_a = -z$, $z_b = +z$. By symmetry $\varepsilon(z_a) = \varepsilon(z_b)$.

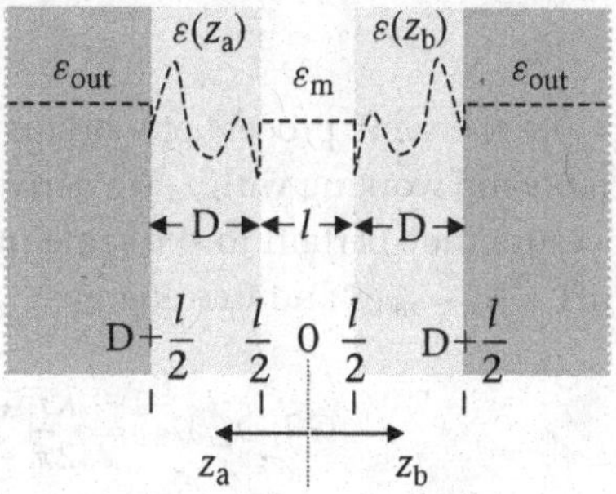

Figure L3.16

Everything is the same except that the role of θ is played by $1/\theta$. Why? Because in the right half-space the coefficient of $e^{+\rho z}$ in $\mathrm{A}e^{+\rho z} + \mathrm{B}e^{-\rho z}$ must go to zero but in the left half-space the coefficient of $e^{-\rho z}$ must be zero in order to have a surface mode.

The procedure for determining $1/\theta$ is to work from left to right. The result is $1/\theta_{\mathrm{m}}$ in the medium m. But this $1/\theta_{\mathrm{m}}$ defines the same $\mathrm{A_m}$ and $\mathrm{B_m}$ in the medium m as did the procedure for working from right to left. That is, in the middle we can write $\mathrm{A_{m(L)}}e^{+\rho z_a} + \mathrm{B_{m(L)}}e^{-\rho z_a}$ for coefficients determined from the left or $\mathrm{A_{m(R)}}e^{+\rho z_b} + \mathrm{B_{m(R)}}e^{-\rho z_b}$ determined from the right. Physically it is the same function. Because $z_a = -z_b$, $\mathrm{A_{m(L)}} = \mathrm{B_{m(R)}}$, $\mathrm{A_{m(R)}} = \mathrm{B_{m(L)}}$ to create the equality $1/\theta_{\mathrm{m}} = \theta_{\mathrm{m}}$. Written as $1 - \theta_{\mathrm{m}}^2 = 0$, this condition binding left and right generates the needed surface-mode dispersion relation, $D(l; \mathrm{D}) = \ln(1 - \theta_{\mathrm{m}}^2)$ and free energy:

$$G(l; \mathrm{D}) = \frac{kT}{2\pi}\sum_{n=0}^{\infty}{}' \int_0^{\infty} \rho \ln(1 - \theta_{\mathrm{m}}^2)\mathrm{d}\rho = \frac{kT}{2\pi}\sum_{n=0}^{\infty}{}' \int_0^{\infty} \rho \ln\left[1 - \overline{\Delta}^{\mathrm{eff}^2} e^{-2\rho l}\right]\mathrm{d}\rho, \qquad \text{(L3.141)}$$

with θ_{m} and $\overline{\Delta}^{\mathrm{eff}}$ as defined in Eqs. (L3.131) and (L3.140b).

PROBLEM: Is this obvious?

SOLUTION: Think about it until it is.

Two coated layers, nonsymmetric configuration, without retardation

Generalization to an asymmetric configuration is immediate. Here, the variation in ε in the region of fixed thickness $\mathrm{D_L}$ on the left is written as $\varepsilon_a(z_a)$ with the values $\varepsilon_a(l/2)$, $\varepsilon_a(\mathrm{D_L} + \frac{l}{2})$ at the inner and the outer interfaces respectively; in $\mathrm{D_R}$ on the right, a different function $\varepsilon_b(z_b)$, with $\varepsilon_b(l/2)$, $\varepsilon_b[\mathrm{D_R} + (l/2)]$; $\varepsilon_a[\mathrm{D_L} + (l/2)]$ need not equal ε_{L}; $\varepsilon_a(l/2)$ and $\varepsilon_b(l/2)$ need not equal ε_{m}; $\varepsilon_b[\mathrm{D_R} + (l/2)]$ need not equal ε_{R}. Discontinuities are allowed at the interfaces (see Fig. L3.17).

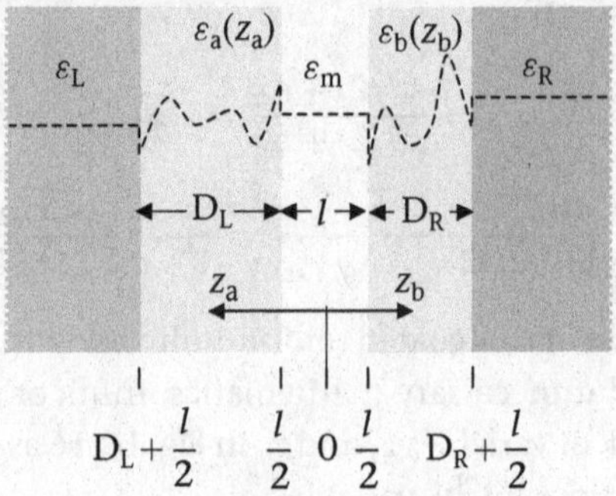

Figure L3.17

By the same process of working outside–inside with z_b, we derive a $\theta_{m(R)}$ from the right; and working with z_a we derive a $1/\theta_{m(L)}$ from the left. Again, these must be equal because they pertain to the same medium m. The dispersion relation is $D(l; D_L, D_R) = \ln[1 - \theta_{m(R)}\theta_{m(L)}]$ and free energy

$$G(l; D_L, D_R) = \frac{kT}{2\pi} \sum_{n=0}^{\infty}{}' \int_0^{\infty} \rho \ln\left[1 - \theta_{m(R)}\theta_{m(L)}\right] d\rho$$

$$= \frac{kT}{2\pi} \sum_{n=0}^{\infty}{}' \int_0^{\infty} \rho \ln\left[1 - \overline{\Delta}_{Lm}^{eff}\overline{\Delta}_{Rm}^{eff} e^{-2\rho l}\right] d\rho, \quad \text{(L3.142)}$$

where [Eqs. (L3.140a) and (L3.140b)]

$$\theta_{m(L)} = \left[\frac{\theta_a(l/2)e^{+\rho l} - \overline{\Delta}_{m/a}}{1 - \theta_a(l/2)e^{+\rho l}\overline{\Delta}_{m/a}}\right] e^{-\rho l} = \overline{\Delta}_{Lm}^{eff} e^{-\rho l}, \quad \text{(L3.143a)}$$

$$\theta_{m(R)} = \left[\frac{\theta_b(l/2)e^{+\rho l} - \overline{\Delta}_{m/b}}{1 - \theta_b(l/2)e^{+\rho l}\overline{\Delta}_{m/b}}\right] e^{-\rho l} = \overline{\Delta}_{Rm}^{eff} e^{-\rho l}, \quad \text{(L3.143b)}$$

$$\overline{\Delta}_{Lm}^{eff} \equiv \frac{\theta_a(l/2)e^{+\rho l} + \overline{\Delta}_{am}}{1 + \theta_a(l/2)e^{+\rho l}\overline{\Delta}_{am}} = \frac{u_a(l/2) + \overline{\Delta}_{am}}{1 + u_a(l/2)\overline{\Delta}_{am}}, \quad \overline{\Delta}_{am} = \frac{\varepsilon_a\left(\frac{l}{2}\right) - \varepsilon_m}{\varepsilon_a\left(\frac{l}{2}\right) + \varepsilon_m} = -\overline{\Delta}_{ma}, \quad \text{(L3.144a)}$$

$$\overline{\Delta}_{Rm}^{eff} \equiv \frac{\theta_b(l/2)e^{+\rho l} + \overline{\Delta}_{bm}}{1 + \theta_b(l/2)e^{+\rho l}\overline{\Delta}_{bm}} = \frac{u_b(l/2) + \overline{\Delta}_{bm}}{1 + u_b(l/2)\overline{\Delta}_{bm}}, \quad \overline{\Delta}_{bm} = \frac{\varepsilon_b\left(\frac{l}{2}\right) - \varepsilon_m}{\varepsilon_b\left(\frac{l}{2}\right) + \varepsilon_m} = -\overline{\Delta}_{mb}. \quad \text{(L3.144b)}$$

To derive $\theta_b(l/2)$ at the interface between the right-hand-side variable layer of thickness D_R and the medium m, begin as before with Eq. (L3.139a) at the outer interface of variable region with half-space R:

$$\theta_b\left(z_b = D_R + \frac{l}{2}\right) = -\overline{\Delta}_{bR} e^{-\rho(l+2D_R)}, \quad \overline{\Delta}_{bR} \equiv \frac{\varepsilon_b\left(\frac{l}{2} + D_R\right) - \varepsilon_R}{\varepsilon_b\left(\frac{l}{2} + D_R\right) + \varepsilon_R}. \quad \text{(L3.145)}$$

Then use this $\theta_b[D_R + (l/2)]$ as the boundary condition in Eq. (L3.136):

$$\frac{d\theta_b(z_b)}{dz_b} = -\frac{e^{-2\rho z_b}}{2} \frac{d \ln[\varepsilon_b(z_b)]}{dz_b} \left[1 - e^{+4\rho z_b}\theta_b^2(z_b)\right] \quad \text{(L3.146)}$$

or, using $u(z) \equiv e^{2\rho z}\theta(z)$, [Eqs. (L3.137) and (L3.138)], obtain

$$\frac{du_b(z_b)}{dz_b} = 2\rho u_b(z_b) - \frac{d\ln[\varepsilon_b(z_b)]}{2dz_b}\left[1 - u_b^2(z_b)\right] \qquad \text{(L3.147)}$$

to propagate—analytically or numerically—from $z_b = D_R + (l/2)$ to $z_b = (l/2)$. Feed the resulting $\theta_b(l/2)$ into $\theta_{m(R)}$ of Eq. (L3.143b) and then into the integration for the free energy $G(l; D_L, D_R)$ from Eq. (L3.142).

Proceed similarly from the left-hand side. For sanity, measure z_a leftward from the midpoint in medium m. Then the equations for L and a keep the same form as those for R and b. At the left-most boundary, echo Eq. (L3.145):

$$\theta_a\left(z_a = D_L + \frac{l}{2}\right) = -\overline{\Delta}_{aL}e^{-\rho(l+2D_L)}, \quad \overline{\Delta}_{aL} \equiv \frac{\varepsilon_a\left(\frac{l}{2}+D_L\right) - \varepsilon_L}{\varepsilon_a\left(\frac{l}{2}+D_L\right) + \varepsilon_L}. \qquad \text{(L3.148)}$$

Across the variable region a, echo Eqs. (L3.146) and (L3.147) previously derived for variable region b:

$$\frac{d\theta_a(z_a)}{dz_a} = -\frac{e^{-2\rho z_a}}{2}\frac{d\ln[\varepsilon_a(z_a)]}{dz_a}\left[1 - e^{+4\rho z_a}\theta_a^2(z_a)\right], \qquad \text{(L3.149)}$$

$$\frac{du_a(z_a)}{dz_a} = 2\rho u_a(z_a) - \frac{d\ln[\varepsilon_a(z_a)]}{2dz_a}\left[1 - u_a^2(z_a)\right]. \qquad \text{(L3.150)}$$

Again, feed the resulting $\theta_a(l/2)$ into $\theta_{m(L)}$ of Eq. (L3.143a) and into $G(l; D_L, D_R)$ of Eq. (L3.142).

Arbitrary, continuous $\varepsilon(z)$ with discontinuities allowed at interfaces, symmetric and asymmetric with finite velocity of light

To include the finite velocity of light, recall the boundary conditions on the electric and magnetic fields used to derive the Lifshitz L|m|R, added-layer, and multilayer interactions. The variable ρ now depends on the velocity of light in each layer. Specifically (Subsection L3.2.A),

$$\rho_r^2 = \rho^2 + \frac{\xi_n^2}{c^2}\varepsilon_r\mu_r = \rho_m^2 + \frac{\xi_n^2}{c^2}(\varepsilon_r\mu_r - \varepsilon_m\mu_m), \qquad \text{(L3.151)}$$

where we must now consider variable $\rho_r = \rho(z_r)$ as well as $\varepsilon_r = \varepsilon(z_r)$ (see Fig. L3.18).

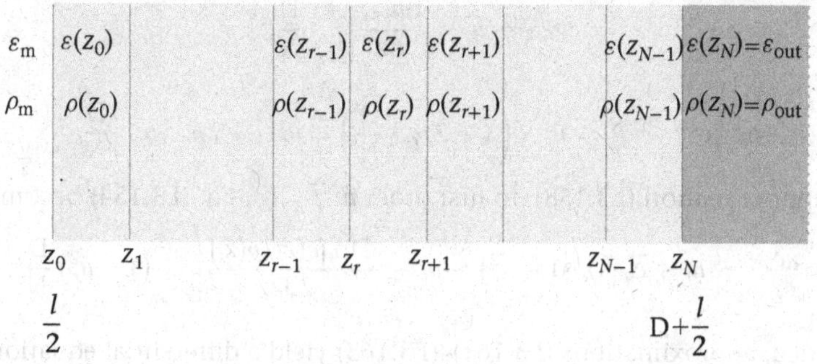

Figure L3.18

The relation between coefficients of the waves in successive layers has the form of Eqs. (L3.59):

$$\left(A_r e^{\rho_r z_r} + B_r e^{-\rho_r z_r}\right)\varepsilon_r = \left(A_{r-1} e^{\rho_{r-1} z_r} + B_{r-1} e^{-\rho_{r-1} z_r}\right)\varepsilon_{r-1},$$

$$\left(A_r e^{\rho_r z_r} - B_r e^{-\rho_r z_r}\right)\rho_r = \left(A_{r-1} e^{\rho_{r-1} z_r} - B_{r-1} e^{-\rho_{r-1} z_r}\right)\rho_{r-1} \tag{L3.152}$$

$\varepsilon(z_{r-1}) = \varepsilon_{r-1}$ | $\varepsilon(z_r) = \varepsilon_r$
$\rho(z_{r-1}) = \rho_{r-1}$ | $\rho(z_r) = \rho_r$
$\theta(z_{r-1}) = \theta_{r-1}$ | $\theta(z_r) = \theta_r$

z_r

Figure L3.19

at the boundary z_r between layers $r-1$ and r (see Fig. L3.19).

Now the connection between $\theta_r \equiv A_r/B_r$ and θ_{r-1} includes ρ_r,[27]

$$\theta_{r-1} e^{+2\rho_{r-1} z_r} = \left(\frac{\theta_r e^{+2\rho_r z_r} - \overline{\Delta}_{r-1/r}}{1 - \theta_r e^{+2\rho_r z_r}\overline{\Delta}_{r-1/r}}\right), \tag{L3.153}$$

$$u_{r-1} e^{2\rho_{r-1}(z_r - z_{r-1})} = \left(\frac{u_r - \overline{\Delta}_{r-1/r}}{1 - u_r \overline{\Delta}_{r-1/r}}\right), \tag{L3.154}$$

where

$$u_r = u(z_r) = \theta(z_r) e^{2\rho(z_r) z_r} = \theta_r e^{2\rho_r z_r}, \quad u_{r-1} = \theta_{r-1} e^{2\rho_{r-1} z_{r-1}} \tag{L3.155}$$

from the definition of the continuous function

$$u(z) \equiv e^{2\rho(z) z}\theta(z). \tag{L3.156}$$

For electric and, later, magnetic modes [see Eqs. (L3.62) and (L3.63)],

$$\overline{\Delta}_{r-1/r} \equiv \left(\frac{\varepsilon_{r-1}\rho_r - \varepsilon_r \rho_{r-1}}{\varepsilon_{r-1}\rho_r + \varepsilon_r \rho_{r-1}}\right), \quad \Delta_{r-1/r} \equiv \left(\frac{\mu_{r-1}\rho_r - \mu_r \rho_{r-1}}{\mu_{r-1}\rho_r + \mu_r \rho_{r-1}}\right). \tag{L3.157}$$

When the layer thickness $D/N = (z_r - z_{r-1}) \to 0$ as $N \to \infty$, we can expand as before for the nonretarded case [approximations (L3.132)–(L3.135) and Eqs. (L3.136)–(L3.138)]:

$$\overline{\Delta}_{r-1/r} = \frac{(\varepsilon_{r-1}/\rho_{r-1}) - (\varepsilon_r/\rho_r)}{(\varepsilon_{r-1}/\rho_{r-1}) + (\varepsilon_r/\rho_r)} \sim -\left.\frac{d\ln[\varepsilon(z)/\rho(z)\rho(z)]}{2dz}\right|_{z=z_r}\frac{D}{N}, \tag{L3.158}$$

$$\theta_{r-1} \sim \theta(z_r) - \left.\frac{d\theta(z)}{dz}\right|_{z=z_r}\frac{D}{N}, \tag{L3.159}$$

$$\rho_{r-1} \sim \rho_r - \left.\frac{d\rho(z)}{dz}\right|_{z=z_r}\frac{D}{N}, \tag{L3.160}$$

$$u_{r-1} \sim u_r - \left.\frac{du(z)}{dz}\right|_{z=z_r}\frac{D}{N}, \tag{L3.161}$$

$$u_{r-1} e^{2\rho_{r-1}(D/N)} \sim u_{r-1}\left(1 + 2\rho_{r-1}\frac{D}{N}\right) = u_{r-1} + 2\rho_{r-1} u_{r-1}\frac{D}{N}. \tag{L3.162}$$

By use of approximation (L3.158), to first order in $\overline{\Delta}_{r-1/r}$, Eq. (L3.154) becomes

$$u_{r-1} e^{2\rho_{r-1}(D/N)} \sim u_r - \overline{\Delta}_{r-1/r}\left(1 - u_r^2\right) \sim u_r + \left.\frac{d\ln[\varepsilon(z)/\rho(z)]}{2dz}\right|_{z=z_r}\left(1 - u_r^2\right)\frac{D}{N}. \tag{L3.163}$$

Combined, approximations (L3.161)–(L3.163) yield a differential equation[28]:

$$\frac{du(z)}{dz} = +2\rho(z)u(z) - \frac{d\ln[\varepsilon(z)/\rho(z)]}{2dz}[1 - u^2(z)]. \tag{L3.164a}$$

In terms of $\theta(z)$, by use of Eqs. (L3.155),

$$\frac{d\theta(z)}{dz} = -2z\frac{d\rho(z)}{dz}\theta(z) - \frac{d\ln[\varepsilon(z)/\rho(z)]}{2dz}e^{-2\rho(z)z}\left[1 - e^{+4\rho(z)z}\theta^2(z)\right]. \quad \text{(L3.164b)}$$

As in the nonretarded case, work from the outer boundaries toward the middle to solve these equations systematically. Although straightforward, this procedure merits explicit description. Begin with the form of Eq. (L3.142),

$$G(l; D_L, D_R) = \frac{kT}{2\pi}\sum_{n=0}^{\infty}{}' \int_0^{\infty} \rho \ln\left[\left(1 - \overline{\Delta}_{Lm}^{eff}\overline{\Delta}_{Rm}^{eff}e^{-2\rho l}\right)\left(1 - \Delta_{Lm}^{eff}\Delta_{Rm}^{eff}e^{-2\rho l}\right)\right]d\rho \quad \text{(L3.165)}$$

and build $\overline{\Delta}_{Lm}^{eff}, \overline{\Delta}_{Rm}^{eff}$ and $\Delta_{Lm}^{eff}, \Delta_{Rm}^{eff}$ outward–in from either side. These steps are spelled out for the dielectric terms and are then obvious for magnetic terms.

At $z_a = (l/2) + D_L$,

$$\theta_a\left(\frac{l}{2} + D_L\right)e^{+\rho_a(l/2+D_L)(l+2D_L)} = u_a\left(\frac{l}{2} + D_L\right) = +\overline{\Delta}_{La}, \quad \text{(L3.166a)}$$

at $z_b = (l/2) + D_R$,

$$\theta_b\left(\frac{l}{2} + D_R\right)e^{+\rho_b(l/2+D_R)(l+2D_R)} = +\overline{\Delta}_{Rb}, \quad \text{(L3.166b)}$$

so that

$$\overline{\Delta}_{La} = \frac{\varepsilon_L\rho_a\left(\frac{l}{2} + D_L\right) - \varepsilon_a\left(\frac{l}{2} + D_L\right)\rho_L}{\varepsilon_L\rho_a\left(\frac{l}{2} + D_L\right) + \varepsilon_a\left(\frac{l}{2} + D_L\right)\rho_L}, \quad \text{(L3.167a)}$$

$$\overline{\Delta}_{Rb} = \frac{\varepsilon_R\rho_b\left(\frac{l}{2} + D_R\right) - \varepsilon_b\left(\frac{l}{2} + D_R\right)\rho_R}{\varepsilon_R\rho_b\left(\frac{l}{2} + D_R\right) + \varepsilon_b\left(\frac{l}{2} + D_R\right)\rho_R}, \quad \text{(L3.167b)}$$

derived by use of Eqs. (L3.139a), (L3.139b), (L3.153), and (L3.154), with $r = N$ for half-space L or R, $\theta_N = \theta_L = \theta_R = 0$.

From $z_a = \frac{l}{2} + D_L$ to $z_a = \frac{l}{2}$

$$\frac{du_a(z_a)}{dz_a} = +2\rho_a(z_a)u_a(z_a) - \frac{d\ln[\varepsilon_a(z_a)/\rho(z_a)]}{2dz_a}\left[1 - u_a^2(z_a)\right]. \quad \text{(L3.168a)}$$

From $z_b = (l/2) + D_R$ to $z_b = (l/2)$,

$$\frac{du_b(z_b)}{dz_b} = +2\rho_b(z_b)u_b(z_b) - \frac{d\ln[\varepsilon_b(z_b)/\rho(z_b)]}{2dz_b}\left[1 - u_b^2(z_b)\right], \quad \text{(L3.168b)}$$

$$\frac{d\theta_a(z_a)}{dz_a} = -2z_a\frac{d\rho_a(z_a)}{dz_a}\theta_a(z_a) - \frac{d\ln[\varepsilon_a(z_a)/\rho_a(z_a)]}{2dz_a}e^{-2\rho_a(z_a)z_a}\left[1 - e^{+4\rho_a(z_a)z_a}\theta_a^2(z_a)\right], \quad \text{(L3.169a)}$$

$$\frac{d\theta_b(z_b)}{dz_b} = -2z_b\frac{d\rho_b(z_b)}{dz_b}\theta_b(z_b) - \frac{d\ln[\varepsilon_b(z_b)/\rho_b(z_b)]}{2dz_b}e^{-2\rho_b(z_b)z_b}\left[1 - e^{+4\rho_b(z_b)z_b}\theta_b^2(z_b)\right]. \quad \text{(L3.169b)}$$

by use of Eqs. (L3.164a) and (L3.164b) to find $\theta_a(\frac{l}{2})$, $\theta_b(\frac{l}{2})$ from $\theta_a(\frac{l}{2} + D_L)$, $\theta_b(\frac{l}{2} + D_R)$.

At $z_a = (l/2)$,

$$\theta_{m(L)} e^{+\rho_m l} = \left[\frac{\theta_a\left(\frac{l}{2}\right) e^{+\rho_a (l/2) l} - \overline{\Delta}_{ma}}{1 - \theta_a\left(\frac{l}{2}\right) e^{+\rho_a (l/2) l} \overline{\Delta}_{ma}} \right], \quad \text{(L3.170a)}$$

at $z_b = (l/2)$,

$$\theta_{m(R)} e^{+\rho_m l} = \left[\frac{\theta_b\left(\frac{l}{2}\right) e^{+\rho_b (l/2) l} - \overline{\Delta}_{mb}}{1 - \theta_b\left(\frac{l}{2}\right) e^{+\rho_b (l/2) l} \overline{\Delta}_{mb}} \right], \quad \text{(L3.170b)}$$

$$\overline{\Delta}_{Lm}^{eff} \equiv \left[\frac{\theta_a\left(\frac{l}{2}\right) e^{+\rho_a (l/2) l} + \overline{\Delta}_{am}}{1 + \theta_a\left(\frac{l}{2}\right) e^{+\rho_a (l/2) l} \overline{\Delta}_{am}} \right] = \left[\frac{u_a\left(\frac{l}{2}\right) + \overline{\Delta}_{am}}{1 + u_a\left(\frac{l}{2}\right) \overline{\Delta}_{am}} \right], \quad \text{(L3.171a)}$$

$$\overline{\Delta}_{Rm}^{eff} \equiv \left[\frac{\theta_b\left(\frac{l}{2}\right) e^{+\rho_b (l/2) l} + \overline{\Delta}_{bm}}{1 + \theta_b\left(\frac{l}{2}\right) e^{+\rho_b (l/2) l} \overline{\Delta}_{bm}} \right] = \left[\frac{u_b\left(\frac{l}{2}\right) + \overline{\Delta}_{bm}}{1 + u_b\left(\frac{l}{2}\right) \overline{\Delta}_{bm}} \right], \quad \text{(L3.171b)}$$

$$\overline{\Delta}_{am} = \frac{\varepsilon_a\left(\frac{l}{2}\right) \rho_m - \varepsilon_m \rho_a\left(\frac{l}{2}\right)}{\varepsilon_a\left(\frac{l}{2}\right) \rho_m + \varepsilon_m \rho_a\left(\frac{l}{2}\right)} = -\overline{\Delta}_{ma}, \quad \text{(L3.172a)}$$

$$\overline{\Delta}_{bm} = \frac{\varepsilon_b\left(\frac{l}{2}\right) \rho_m - \varepsilon_m \rho_b\left(\frac{l}{2}\right)}{\varepsilon_b\left(\frac{l}{2}\right) \rho_m + \varepsilon_m \rho_b\left(\frac{l}{2}\right)} = -\overline{\Delta}_{mb}, \quad \text{(L3.172b)}$$

again by use of Eqs. (L3.143a)–(L3.144b) and Eqs. (L3.153)–(L3.157), with $r = 0$ [$z_0 = (l/2)$] and m for $r - 1$.

PROBLEM L3.1: Show that this result converges to that for singly layered surfaces when $\varepsilon(z)$ is constant.

SOLUTION: $\varepsilon_a(z_a) = \varepsilon_a$, $\rho_a(z_a) = \rho_a$, and $\theta_a(z_a) = \theta_a$ are constant with position. At $z_a = (l/2) + D_L$,

$$\theta_a e^{+\rho_a (l + 2D_L)} = +\overline{\Delta}_{La},\ \theta_a e^{+\rho_a l} = +\overline{\Delta}_{La} e^{-2\rho_a D_L},\ \overline{\Delta}_{Lm}^{eff} = \left(\frac{\overline{\Delta}_{La} e^{-2\rho_a D_L} + \overline{\Delta}_{am}}{1 + \overline{\Delta}_{La} \overline{\Delta}_{am} e^{-2\rho_a D_L}} \right),$$

then similarly for other terms to recover the derived Eqs. (L3.145) and (L3.147) and tabulated (P.3.a.1) results.

L3.6. Ionic-charge fluctuations

Because they have so many unexpected features and because they are kind of a hybrid with electrostatic double-layer forces, ionic-charge-fluctuation forces deserve separate consideration.

In the language of dielectric response, how is one to regard the movement of mobile ions?

First, through conductivity. An applied electric field creates an electric current in a salt solution. Formally, a conductivity σ appears in the form that varies with frequency as $\sim \{i\sigma/[\omega(1-i\omega\tau)]\}$ in the dielectric permittivity $\varepsilon(\omega)$. In the limit of low frequency, $\omega\tau \to 0$, this diverges as $\sim (i\sigma/\omega)$. In that limit, a conducting material begins to appear as an infinitely polarizable medium, its mobile charges able to move indefinitely long distances.

In real life, we know this is not necessarily so. An electric field applied to a conducting medium can be maintained only as long as reactions or transfers of charges occur at the walls. The electrical outlet delivers and removes electrons; the electrodes react to remove or to produce ions. In real life we must recognize what goes on at the walls bounding a conducting medium. In the ideal "bad-electrode" limit there is no removing or producing a reaction at the walls. In that limit, under the action of a constant applied electric field, the charges pile up to create electrostatic double layers. When oscillating fields settle down at $\omega \to 0$ there is a spatially varying electric field across the space between the bounding walls.

Think in terms of a capacitor. With a pure, nonconducting dielectric material there is a constant electric field between plates (see Fig. L3.20). But across a salt solution between nonreactive, nonconducting, ideally bad electrodes (no chemical reactions at interfaces), there is a spatially varying electrostatic double-layer field set up by the electrode walls (see Fig. L3.21).

The nonlocal ionic response that depends on the location of the bounding interface is not the only complication encountered with ionic solutions. Electric-field fluctuations drive electric currents through the translation of mobile charges. These fluctuating currents in turn create fluctuating magnetic fields with their own interactions. In this text, by considering only fluctuations in electrostatic double layers, the electric-current complications are evaded. The ions are treated as external or "source" charges ρ_{ext} whereas the remaining material is treated as an ion-free dielectric. Because the characteristic times of the motion of ions are slow compared with the times of electric-signal

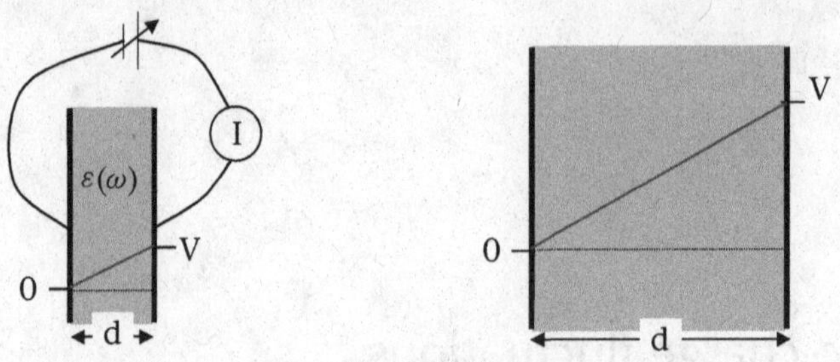

Figure L3.20

propagation and because effectively the velocity of light $c \to \infty$, relativistic considerations are also ignored.

The Maxwell equations reduce to[29]

$$\nabla \times \mathbf{H} = 0, \quad \nabla \times \mathbf{E} = 0, \quad \nabla \cdot \mathbf{H} = 0, \tag{L3.173}$$

and

$$\nabla \cdot (\varepsilon \mathbf{E}) = \rho_{ext}/\varepsilon_0 \text{ in mks units} \quad \text{or} \quad \nabla \cdot (\varepsilon \mathbf{E}) = 4\pi \rho_{ext} \text{ in cgs units.} \tag{L3.174}$$

The wave equation is built from $\nabla \cdot \mathbf{E} \propto \rho_{ext}/\varepsilon$. Because electrostatic double-layer equations are easier to think about in terms of potentials ϕ rather than electric fields $\mathbf{E} = -\nabla\phi$, we set up the problem of ionic-charge-fluctuation forces in terms of potentials. Charges ρ_{ext} come from the potential ϕ through the Boltzmann relation

$$\rho_{ext} = \sum_{\nu=-\infty}^{\nu=\infty} e\nu n_\nu e^{-e\nu\phi/kT} \approx \sum_{\nu=-\infty}^{\nu=\infty} e\nu n_\nu \left(1 - \frac{e\nu\phi}{kT}\right) = -\sum_{\nu=-\infty}^{\nu=\infty} \frac{e^2\nu^2 n_\nu}{kT}\phi, \tag{L3.175}$$

which is used for the Debye length $1/\kappa$ with

$$\kappa^2 \equiv \frac{e^2}{\varepsilon\varepsilon_0 kT}\sum_{\{\nu\}} n_\nu \nu^2 \text{ in mks units} \quad \text{or} \quad \kappa^2 \equiv \frac{4\pi e^2}{\varepsilon kT}\sum_{\{\nu\}} n_\nu \nu^2 \text{ in cgs units.} \tag{L3.176}$$

Here n_ν is the number density of ions of valence ν in the reference solution in which potential $\phi = 0$. The summation form $\sum_{\{\nu\}}$ simply means to take into account all mobile ions of all valences ν whereas the particular summation $\sum_{\{\nu\}} n_\nu \nu^2$ is proportional to the ionic strength of the bathing solution.

In terms of potential, the electrostatic double-layer "wave" equation is

$$\nabla^2 \phi = \kappa^2 \phi. \tag{L3.177}$$

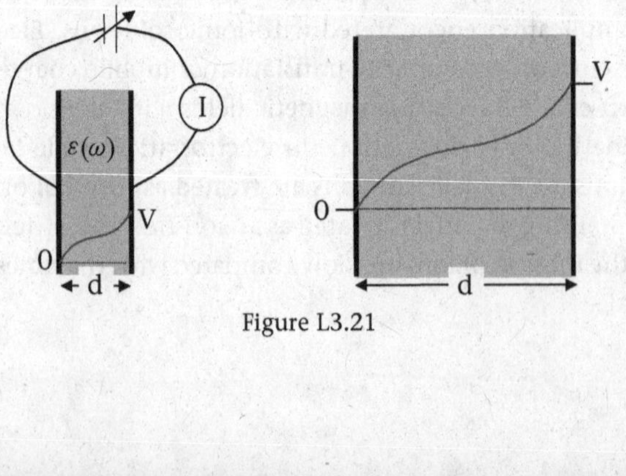

Figure L3.21

As in the L|m|R geometry of the Lifshitz interaction between planar half-spaces, fluctuations in potential ϕ have the form of waves in the x, y directions parallel to the surfaces and an exponential $f(z)$ that dies away from the surfaces.[30] The general form is like that used in the derivation of the Lifshitz result. For each radial wave vector $\mathbf{i}u + \mathbf{j}v$, the potential $\phi(x, y, z)$ has the form

$$\phi(x, y, z) = f(z)e^{i(ux+vy)}, \tag{L3.178}$$

with

$$f_i(z) = A_i e^{\beta_i z} + B_i e^{-\beta_i z}, \tag{L3.179}$$

where

$$\beta_i^2 = (u^2 + v^2) + \kappa_i^2 = \rho^2 + \kappa_i^2 \tag{L3.180}$$

in each region i = L, m, or R (see Fig. L3.22).

Compare the form $\beta_i^2 = \rho^2 + \kappa_i^2$ with $\rho_i^2 = \rho^2 - [(\varepsilon_i \mu_i \omega^2)/c^2]$ used for finite-frequency forces when we use imaginary frequency $\omega = i\xi : \rho_i^2 = \rho^2 + [(\varepsilon_i \mu_i \xi^2)/c^2]$. It already becomes clear that at least part of the action of ionic fluctuations is similar in form to the effect of retardation screening of finite-frequency fluctuations.

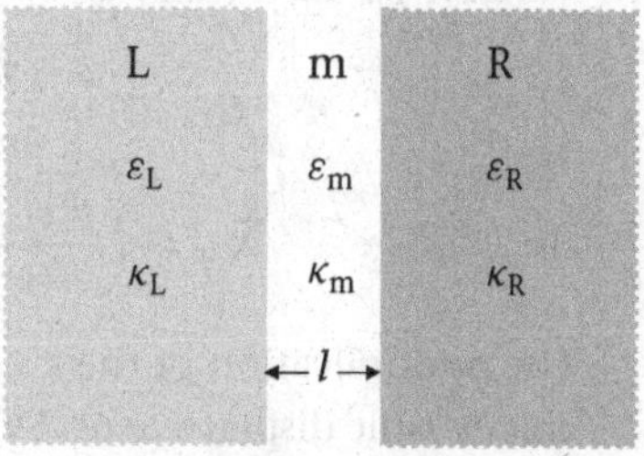

Figure L3.22

Restricting solutions of $f(z) = Ae^{\beta z} + Be^{-\beta z}$ only to those electric-field fluctuations affected by the location of the boundary surfaces, we must have A = 0 for region R and B = 0 for region L. In the absence of any additional charge on the interfaces, the potentials ϕ must be equal on both sides,

$$A_L = A_m + B_m, \quad B_R e^{-\beta_R l} = A_m e^{\beta_m l} + B_m e^{-\beta_m l}; \tag{L3.181}$$

as must the displacement vectors $\varepsilon \mathbf{E}_z = -\varepsilon \partial \phi / \partial z$,

$$\varepsilon_L A_L \beta_L = \varepsilon_m A_m \beta_m - \varepsilon_m B_m \beta_m, \quad -\varepsilon_R B_R \beta_R e^{-\beta_R l} = \varepsilon_m A_m \beta_m e^{\beta_m l} - \varepsilon_m B_m \beta_m e^{-\beta_m l}. \tag{L3.182}$$

This condition among ε's, β's, and l creates a dispersion relation of the same form of Eq. (L3.22) for the nonionic Lifshitz problem[31]:

$$D_{\text{ionic}}(\varepsilon_L, \varepsilon_m, \varepsilon_R, \kappa_L, \kappa_m, \kappa_R, l) \equiv 1 - \overline{\Delta}_{Lm} \overline{\Delta}_{Rm} e^{-2\beta_m l} = 0, \tag{L3.183}$$

where

$$\overline{\Delta}_{Lm} \equiv \left(\frac{\beta_L \varepsilon_L - \beta_m \varepsilon_m}{\beta_L \varepsilon_L + \beta_m \varepsilon_m} \right), \quad \overline{\Delta}_{Rm} \equiv \left(\frac{\beta_R \varepsilon_R - \beta_m \varepsilon_m}{\beta_R \varepsilon_R + \beta_m \varepsilon_m} \right). \tag{L3.184}$$

As in the original Lifshitz case [Eqs. (L3.27) and (L3.49)], we sum over modes with lateral wave vectors $\mathbf{i}u + \mathbf{j}v$ with u and v combining into wave-vector magnitude ρ as

$$\rho^2 \equiv (u^2 + v^2), \tag{L3.185}$$

so that

$$\begin{aligned} \beta_m^2 &= \rho^2 + \kappa_m^2; \\ \beta_L^2 &= \rho^2 + \kappa_L^2 = \beta_m^2 - \left(\kappa_m^2 - \kappa_L^2\right); \\ \beta_R^2 &= \rho^2 + \kappa_R^2 = \beta_m^2 - \left(\kappa_m^2 - \kappa_R^2\right). \end{aligned} \tag{L3.186}$$

Unlike the Lifshitz case, because only the zero-frequency ionic-charge fluctuations count, there is no summation over finite frequencies. There is only the integration over wave-vector magnitudes ρ to achieve the free energy $G_{LmR}(l)$ [as in Eq. (L3.31)]:

$$G_{LmR}(l) = \frac{1}{(2\pi)^2}\int_0^\infty 2\pi\rho\,[G_l(\rho) - G_\infty(\rho)]\mathrm{d}\rho, \tag{L3.187}$$

$$G_l(\rho) = (kT/2)\ln D_{\text{ionic}}, \tag{L3.188}$$

so that

$$G_{LmR}(l) = \frac{kT}{4\pi}\int_0^\infty \rho[\ln(D_{\text{ionic}})]\mathrm{d}\rho. \tag{L3.189}$$

The interaction free energy can be written in equivalent forms with different variables of integration:

1. Variable β_m such that $\beta_m^2 = \rho^2 + \kappa_m^2$, $\beta_m \mathrm{d}\beta_m = \rho\mathrm{d}\rho$, $\kappa_m \le \beta_m < \infty$:

$$G_{LmR}(l) = \frac{kT}{4\pi}\int_{\kappa_m}^\infty \beta_m \ln\left(1 - \overline{\Delta}_{Lm}\overline{\Delta}_{Rm}e^{-2\beta_m l}\right)\mathrm{d}\beta_m, \tag{L3.190}$$

$$\overline{\Delta}_{Lm} \equiv \left(\frac{\beta_L\varepsilon_L - \beta_m\varepsilon_m}{\beta_L\varepsilon_L + \beta_m\varepsilon_m}\right), \quad \overline{\Delta}_{Rm} \equiv \left(\frac{\beta_R\varepsilon_R - \beta_m\varepsilon_m}{\beta_R\varepsilon_R + \beta_m\varepsilon_m}\right). \tag{L3.191}$$

The multiplication of the ε's by β's creates an effective dielectric response that includes ionic displacement. Double-layer screening of zero-frequency fluctuations is through the exponential $e^{-2\beta_m l}$. The formal resemblance to retardation screening comes clear here and in subsequent similar factors.

2. Variable p such that $\beta_m = p\kappa_m$, $1 \le p < \infty$:

$$G_{LmR}(l) = \frac{kT\kappa_m^2}{4\pi}\int_1^\infty p\ln\left(1 - \overline{\Delta}_{Lm}\overline{\Delta}_{Rm}e^{-2\kappa_m l p}\right)\mathrm{d}p, \tag{L3.192}$$

$$\overline{\Delta}_{Lm} \equiv \left(\frac{s_L\varepsilon_L - p\varepsilon_m}{s_L\varepsilon_L + p\varepsilon_m}\right), \quad s_L = \sqrt{p^2 - 1 + \kappa_L^2/\kappa_m^2}, \tag{L3.193}$$

$$\overline{\Delta}_{Rm} \equiv \left(\frac{s_R\varepsilon_R - p\varepsilon_m}{s_R\varepsilon_R + p\varepsilon_m}\right), \quad s_R = \sqrt{p^2 - 1 + \kappa_R^2/\kappa_m^2} \tag{L3.194}$$

(s_L, p, s_R multiply ε_L, ε_m, ε_R, not as for the dipolar fluctuation formulae [Eqs. (L3.57)].

3. Variable x such that $x = 2\beta_m l$, $2\kappa_m l \le x < \infty$:

$$G_{LmR}(l) = \frac{kT}{16\pi l^2}\int_{2\kappa_m l}^\infty x\,[\ln(1 - \overline{\Delta}_{Lm}\overline{\Delta}_{Rm}e^{-x})]\mathrm{d}x, \tag{L3.195}$$

$$\overline{\Delta}_{Lm} \equiv \left(\frac{x_L\varepsilon_L - x\varepsilon_m}{x_L\varepsilon_L + x\varepsilon_m}\right), \quad x_L = 2\rho_L l = \sqrt{x^2 + (\kappa_L^2 - \kappa_m^2)(2l)^2}, \tag{L3.196}$$

$$\overline{\Delta}_{Rm} \equiv \left(\frac{x_R\varepsilon_R - x\varepsilon_m}{x_R\varepsilon_R + x\varepsilon_m}\right), \quad x_R = 2\rho_R l = \sqrt{x^2 + (\kappa_R^2 - \kappa_m^2)(2l)^2}. \tag{L3.197}$$

(x_L, x, x_R multiply ε_L, ε_m, ε_R, not as for the dipolar fluctuation formulae [Eqs. (L3.54)].

The choice among these equivalent forms is a matter of convenience. When the ε's are introduced at zero frequency, exclusive of ionic-conductance terms, the free energy is readily computed by numerical integration. Instructive features of these formulae emerge when the functions $\overline{\Delta}_{\mathrm{Lm}}, \overline{\Delta}_{\mathrm{Rm}}$ are reduced to special cases, as in the following examples:

1. Immerse L, m, and R in saltwater of uniform ionic strength so that $\kappa_{\mathrm{L}} = \kappa_{\mathrm{m}} = \kappa_{\mathrm{R}} = \kappa$. Then $\beta_{\mathrm{L}} = \beta_{\mathrm{m}} = \beta_{\mathrm{R}} = \beta$, and

$$\overline{\Delta}_{\mathrm{Lm}} \equiv \left(\frac{\varepsilon_{\mathrm{L}} - \varepsilon_{\mathrm{m}}}{\varepsilon_{\mathrm{L}} + \varepsilon_{\mathrm{m}}}\right), \quad \overline{\Delta}_{\mathrm{Rm}} \equiv \left(\frac{\varepsilon_{\mathrm{R}} - \varepsilon_{\mathrm{m}}}{\varepsilon_{\mathrm{R}} + \varepsilon_{\mathrm{m}}}\right). \tag{L3.198}$$

For $2\kappa l \gg 1$,[32]

$$G_{\mathrm{LmR}}(l) = \frac{kT}{4\pi}\int_{\kappa}^{\infty} \beta \ln\left(1 - \overline{\Delta}_{\mathrm{Lm}}\overline{\Delta}_{\mathrm{Rm}} e^{-2\beta l}\right) d\beta$$
$$\approx -\frac{kT}{16\pi l^2}\overline{\Delta}_{\mathrm{Lm}}\overline{\Delta}_{\mathrm{Rm}}(1 + 2\kappa l)e^{-2\kappa l}. \tag{L3.199}$$

There is an ionic-screening factor $(1 + 2\kappa l)e^{-2\kappa l} \leq 1$.

2. Let medium m be a salt solution, $\kappa_{\mathrm{m}} = \kappa$, and let L and R be pure dielectrics,[33] $\kappa_{\mathrm{L}} = \kappa_{\mathrm{R}} = 0$ with $\varepsilon_{\mathrm{m}} \gg \varepsilon_{\mathrm{L}}, \varepsilon_{\mathrm{R}}$.

   Then $\beta_{\mathrm{m}}^2 = \rho^2 + \kappa^2$, $\beta_{\mathrm{L}}^2 = \beta_{\mathrm{R}}^2 = \rho^2 = \beta_{\mathrm{m}}^2 - \kappa^2$, and

$$\overline{\Delta}_{\mathrm{Lm}} = \left(\frac{\varepsilon_{\mathrm{L}}\rho - \varepsilon_{\mathrm{m}}\sqrt{\rho^2 + \kappa^2}}{\varepsilon_{\mathrm{L}}\rho + \varepsilon_{\mathrm{m}}\sqrt{\rho^2 + \kappa^2}}\right),$$

$$\overline{\Delta}_{\mathrm{Rm}} = \left(\frac{\varepsilon_{\mathrm{R}}\rho - \varepsilon_{\mathrm{m}}\sqrt{\rho^2 + \kappa^2}}{\varepsilon_{\mathrm{R}}\rho + \varepsilon_{\mathrm{m}}\sqrt{\rho^2 + \kappa^2}}\right), \quad \overline{\Delta}_{\mathrm{Lm}}\overline{\Delta}_{\mathrm{Rm}} \approx 1, \tag{L3.200}$$

$$G_{\mathrm{LmR}}(l) = \frac{kT}{4\pi}\int_{\kappa}^{\infty} \beta_{\mathrm{m}} \ln\left(1 - \overline{\Delta}_{\mathrm{Lm}}\overline{\Delta}_{\mathrm{Rm}} e^{-2\beta_{\mathrm{m}} l}\right) d\beta_{\mathrm{m}} \approx -\frac{kT}{16\pi l^2}(1 + 2\kappa l)e^{-2\kappa l}. \tag{L3.201}$$

In addition to the screening factor $(1 + 2\kappa l)e^{-2\kappa l}$, ionic fluctuations create a larger $\overline{\Delta}_{\mathrm{Lm}}\overline{\Delta}_{\mathrm{Rm}}$.

3. Conversely, let L and R be salt solutions, $\kappa_{\mathrm{L}} = \kappa_{\mathrm{R}} = \kappa$, and let medium m be a pure dielectric, $\kappa_{\mathrm{m}} = 0$. Again $\overline{\Delta}_{\mathrm{Lm}}\overline{\Delta}_{\mathrm{Rm}} \approx 1$ so that[34]

$$G_{\mathrm{LmR}}(l) = \frac{kT}{4\pi}\int_{0}^{\infty} \beta_{\mathrm{m}} \ln\left(1 - \overline{\Delta}_{\mathrm{Lm}}\overline{\Delta}_{\mathrm{Rm}} e^{-2\beta_{\mathrm{m}} l}\right) d\beta_{\mathrm{m}}$$
$$\approx -\frac{kT}{16\pi l^2}\sum_{j=1}^{\infty}\frac{1}{j^3} \approx -\frac{1.202kT}{16\pi l^2}. \tag{L3.202}$$

There is the maximum coefficient of charge fluctuations because L and R are ionic solutions, but there is no additional double-layer screening of fluctuations correlated across the separation l.

L3.7. Anisotropic media

Correlated charge fluctuations between anisotropic bodies acting across an anisotropic medium create torques as well as attractions or repulsions. The formulae derived here for semi-infinite media can also be specialized to express the torque and force between anisotropic small particles or between long rodlike molecules. (For example, Table C.4 and Subsection L2.3.G.)

In this case the dielectric permittivity ε is a matrix rather than a scalar. When the principal axes of each material are perpendicular, this tensor can be written as

$$\boldsymbol{\varepsilon}^{\mathrm{i}} \equiv \begin{bmatrix} \varepsilon_x^{\mathrm{i}} & 0 & 0 \\ 0 & \varepsilon_y^{\mathrm{i}} & 0 \\ 0 & 0 & \varepsilon_z^{\mathrm{i}} \end{bmatrix}, \tag{L3.203}$$

where i = L, m, or R. For clarity we restrict ourselves to the case in which axis z for all materials is perpendicular to the interfaces. (Alleviation of this rectilinear restriction is algebraically tedious but no problem in principle.) As usual, each of the components $\varepsilon_x^{\mathrm{i}}$, $\varepsilon_y^{\mathrm{i}}$, and $\varepsilon_z^{\mathrm{i}}$ depends on frequency (see Fig. L3.23). Magnetic susceptibilities, not included, are easy to add.

These materials can be rotated, with respect to each other, about the z axis perpendicular to the interfaces. Let the angle of zero rotation be that mutual orientation at which the principal axes of the materials are in the x, y, and z directions. Then the effects of rotating materials m and R by amounts θ_{m} and θ_{R} with respect to θ_{L} (kept at $\theta_{\mathrm{L}} = 0$) can be written in terms of dielectric tensors $\boldsymbol{\varepsilon}^{\mathrm{m}}(\theta_{\mathrm{m}})$ and $\boldsymbol{\varepsilon}^{\mathrm{R}}(\theta_{\mathrm{R}})$, i = m or R:

$$\boldsymbol{\varepsilon}^{\mathrm{i}}(\theta_{\mathrm{i}}) = \begin{bmatrix} \varepsilon_x^{\mathrm{i}} + (\varepsilon_y^{\mathrm{i}} - \varepsilon_x^{\mathrm{i}})\sin^2(\theta_{\mathrm{i}}) & (\varepsilon_x^{\mathrm{i}} - \varepsilon_y^{\mathrm{i}})\sin(\theta_{\mathrm{i}})\cos(\theta_{\mathrm{i}}) & 0 \\ (\varepsilon_x^{\mathrm{i}} - \varepsilon_y^{\mathrm{i}})\sin(\theta_{\mathrm{i}})\cos(\theta_{\mathrm{i}}) & \varepsilon_y^{\mathrm{i}} + (\varepsilon_x^{\mathrm{i}} - \varepsilon_y^{\mathrm{i}})\sin^2(\theta_{\mathrm{i}}) & 0 \\ 0 & 0 & \varepsilon_z^{\mathrm{i}} \end{bmatrix} \tag{L3.204}$$

As in the derivation of the Lifshitz result it is necessary to delineate those electromagnetic modes that depend on the location of interfaces. For the nonretarded case, in which the finite velocity of light is ignored, the pertinent Maxwell equations can be written as

$$\nabla \cdot (\boldsymbol{\varepsilon}\mathbf{E}) = 0, \quad \nabla \times \mathbf{E} = 0. \tag{L3.205}$$

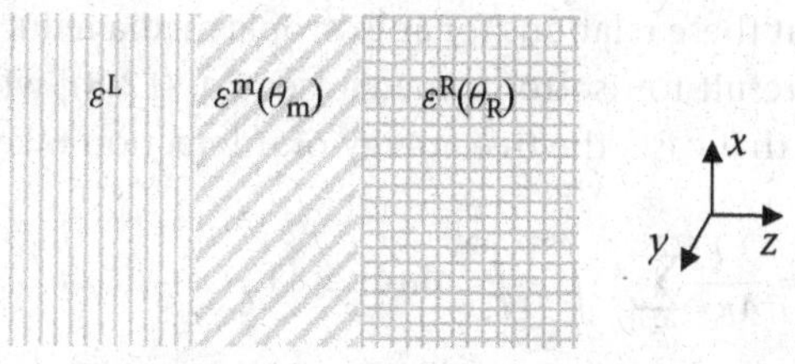

Figure L3.23

By the second of these equations, we know we can introduce a scalar potential ϕ such that $\mathbf{E} = -\nabla\phi$. This allows us to convert the first equation into

$$\nabla \cdot (\varepsilon \nabla \phi) = 0 \tag{L3.206}$$

which is to be solved subject to the usual conditions that E_x, E_y, and $(\varepsilon E)_z$ be continuous at the boundaries.

In each material the potential function $\phi_{\mathrm{i}}(x, y, z)$ will have the form of Eq. (L3.178)

$$\phi_{\mathrm{i}} = f_{\mathrm{i}}(z) e^{i(ux+vy)}, \ \mathrm{i} = \mathrm{L, m, or\ R}, \tag{L3.207}$$

to give an equation for $f_{\mathrm{i}}(z)$,

$$\varepsilon_z^{\mathrm{i}} f_{\mathrm{i}}''(z) - \left(\varepsilon_{11}^{\mathrm{i}} u^2 + 2\varepsilon_{12}^{\mathrm{i}} uv + \varepsilon_{22}^{\mathrm{i}} v^2\right) f_{\mathrm{i}}(z) = 0, \tag{L3.208}$$

where $\varepsilon_{pq}^{\mathrm{i}}$ is the pq (row/column) element of the *rotated* matrix given in Eq. (L3.204).

This differential equation can be more succinctly written as

$$f_{\mathrm{i}}''(z) - \beta_{\mathrm{i}}^2(\theta_{\mathrm{i}}) f_{\mathrm{i}}(z) = 0, \tag{L3.209}$$

with

$$\beta_{\mathrm{i}}^2(\theta_{\mathrm{i}}) = \frac{\varepsilon_x^{\mathrm{i}}}{\varepsilon_z^{\mathrm{i}}} (u\cos\theta_{\mathrm{i}} + v\sin\theta_{\mathrm{i}})^2 + \frac{\varepsilon_y^{\mathrm{i}}}{\varepsilon_z^{\mathrm{i}}} (v\cos\theta_{\mathrm{i}} - u\sin\theta_{\mathrm{i}})^2,$$

so that

$$f_{\mathrm{i}}(z) = \mathrm{A_i} e^{\beta_{\mathrm{i}} z} + \mathrm{B_i} e^{-\beta_{\mathrm{i}} z}. \tag{L3.210}$$

Because only surface modes have significance, we set $\mathrm{B_L} = \mathrm{A_R} = 0$.

The boundary conditions at $z = 0$ and l become

$$\begin{aligned} f_{\mathrm{L}}(0) &= f_{\mathrm{m}}(0), \\ \varepsilon_z^{\mathrm{L}} f_{\mathrm{L}}'(0) &= \varepsilon_z^{\mathrm{m}} f_{\mathrm{m}}'(0), \\ f_{\mathrm{m}}(l) &= f_{\mathrm{R}}(l), \\ \varepsilon_z^{\mathrm{m}} f_{\mathrm{m}}'(l) &= \varepsilon_z^{\mathrm{R}} f_{\mathrm{R}}'(l). \end{aligned} \tag{L3.211}$$

The solution for the remaining $\mathrm{A_i}$, $\mathrm{B_i}$ requires the dispersion relation

$$D(\xi_n, l, \theta_{\mathrm{m}}, \theta_{\mathrm{R}}) = 1 - \left[\frac{\varepsilon_z^{\mathrm{L}}\beta_{\mathrm{L}} - \varepsilon_z^{\mathrm{m}}\beta_{\mathrm{m}}(\theta_{\mathrm{m}})}{\varepsilon_z^{\mathrm{L}}\beta_{\mathrm{L}} + \varepsilon_z^{\mathrm{m}}\beta_{\mathrm{m}}(\theta_{\mathrm{m}})}\right]\left[\frac{\varepsilon_z^{\mathrm{R}}\beta_{\mathrm{R}}(\theta_{\mathrm{R}}) - \varepsilon_z^{\mathrm{m}}\beta_{\mathrm{m}}(\theta_{\mathrm{m}})}{\varepsilon_z^{\mathrm{R}}\beta_{\mathrm{R}}(\theta_{\mathrm{R}}) + \varepsilon_z^{\mathrm{m}}\beta_{\mathrm{m}}(\theta_{\mathrm{m}})}\right] e^{-2\beta_{\mathrm{m}}(\theta_{\mathrm{m}})l} = 0. \tag{L3.212}$$

It is easy to see that these relations for anisotropic media reduce immediately to the nonretarded Lifshitz result for isotropic media [see Eq. (L2.8)] where $\varepsilon_x^{\mathrm{i}} = \varepsilon_y^{\mathrm{i}} = \varepsilon_z^{\mathrm{i}} = \varepsilon_{\mathrm{i}}$, $\beta_{\mathrm{i}}^2(\theta) = u^2 + v^2$. As in that case, the free energy of interaction takes the form

$$G(l, \theta_{\mathrm{m}}, \theta_{\mathrm{R}}) = \frac{kT}{4\pi^2} \sum_{n=0}^{\infty}{}' \int_{-\infty}^{\infty} \int_{-\infty}^{\infty} \mathrm{d}u\mathrm{d}v \ \ln[1 - \overline{\Delta}_{\mathrm{Lm}}(\xi_n, u, v, \theta_{\mathrm{m}}) \overline{\Delta}_{\mathrm{Rm}} \times (\xi_n, u, v, \theta_{\mathrm{m}}, \theta_{\mathrm{R}}) \, e^{-2\beta_{\mathrm{m}}(\theta_{\mathrm{m}})l}], \quad \text{(L3.213)}$$

$$\overline{\Delta}_{\mathrm{Lm}}(\xi_n, u, v, \theta_{\mathrm{m}}) = \left[\frac{\varepsilon_z^{\mathrm{L}}\beta_{\mathrm{L}} - \varepsilon_z^{\mathrm{m}}\beta_{\mathrm{m}}(\theta_{\mathrm{m}})}{\varepsilon_z^{\mathrm{L}}\beta_{\mathrm{L}} + \varepsilon_z^{\mathrm{m}}\beta_{\mathrm{m}}(\theta_{\mathrm{m}})}\right], \quad \text{(L3.214)}$$

$$\overline{\Delta}_{\mathrm{Rm}}(\xi_n, u, v, \theta_{\mathrm{m}}, \theta_{\mathrm{R}}) = \left[\frac{\varepsilon_z^{\mathrm{R}}\beta_{\mathrm{R}}(\theta_{\mathrm{R}}) - \varepsilon_z^{\mathrm{m}}\beta_{\mathrm{m}}(\theta_{\mathrm{m}})}{\varepsilon_z^{\mathrm{R}}\beta_{\mathrm{R}}(\theta_{\mathrm{R}}) + \varepsilon_z^{\mathrm{m}}\beta_{\mathrm{m}}(\theta_{\mathrm{m}})}\right]. \quad \text{(L3.215)}$$

Through the β''s these expressions include the individual dependence of the dispersion relation on radial wave vectors u and v in the x and y directions.

By defining $u = \rho\cos\psi$ and $v = \rho\sin\psi$, the double integral in u and v can be transformed to integrals in polar coordinates ρ and ψ. Then

$$\beta_{\mathrm{i}}^2(\theta_{\mathrm{i}}) = \frac{\varepsilon_x^{\mathrm{i}}}{\varepsilon_z^{\mathrm{i}}}(u\cos\theta_{\mathrm{i}} + v\sin\theta_{\mathrm{i}})^2 + \frac{\varepsilon_y^{\mathrm{i}}}{\varepsilon_z^{\mathrm{i}}}(v\cos\theta_{\mathrm{i}} - u\sin\theta_{\mathrm{i}})^2 = \rho^2 g_{\mathrm{i}}^2(\theta_{\mathrm{i}} - \psi), \quad \text{(L3.216)}$$

where

$$g_{\mathrm{i}}^2(\theta_{\mathrm{i}} - \psi) \equiv \frac{\varepsilon_x^{\mathrm{i}}}{\varepsilon_z^{\mathrm{i}}} + \frac{(\varepsilon_y^{\mathrm{i}} - \varepsilon_x^{\mathrm{i}})}{\varepsilon_z^{\mathrm{i}}}\sin^2(\theta_{\mathrm{i}} - \psi) = \frac{\varepsilon_y^{\mathrm{i}}}{\varepsilon_z^{\mathrm{i}}} + \frac{(\varepsilon_x^{\mathrm{i}} - \varepsilon_y^{\mathrm{i}})}{\varepsilon_z^{\mathrm{i}}}\cos^2(\theta_{\mathrm{i}} - \psi). \quad \text{(L3.217)}$$

The ρ factors in the β's cancel in the $\overline{\Delta}$'s to give

$$\overline{\Delta}_{\mathrm{Lm}}(\xi_n, \theta_{\mathrm{m}}, \psi) = \left[\frac{\varepsilon_z^{\mathrm{L}} g_{\mathrm{L}}(-\psi) - \varepsilon_z^{\mathrm{m}} g_{\mathrm{m}}(\theta_{\mathrm{m}} - \psi)}{\varepsilon_z^{\mathrm{L}} g_{\mathrm{L}}(-\psi) + \varepsilon_z^{\mathrm{m}} g_{\mathrm{m}}(\theta_{\mathrm{m}} - \psi)}\right], \quad \text{(L3.218)}$$

$$\overline{\Delta}_{\mathrm{Rm}}(\xi_n, \theta_{\mathrm{m}}, \theta_{\mathrm{R}}, \psi) = \left[\frac{\varepsilon_z^{\mathrm{R}} g_{\mathrm{R}}(\theta_{\mathrm{R}} - \psi) - \varepsilon_z^{\mathrm{m}} g_{\mathrm{m}}(\theta_{\mathrm{m}} - \psi)}{\varepsilon_z^{\mathrm{R}} g_{\mathrm{R}}(\theta_{\mathrm{R}} - \psi) + \varepsilon_z^{\mathrm{m}} g_{\mathrm{m}}(\theta_{\mathrm{m}} - \psi)}\right] \quad \text{(L3.219)}$$

to let us write

$$G(l, \theta_{\mathrm{m}}, \theta_{\mathrm{R}}) = \frac{kT}{4\pi^2} \sum_{n=0}^{\infty}{}' \int_0^{2\pi} \mathrm{d}\psi \int_0^{\infty} \rho\mathrm{d}\rho \ \ln[1 - \overline{\Delta}_{\mathrm{Lm}}(\xi_n, \theta_{\mathrm{m}}, \psi) \times \overline{\Delta}_{\mathrm{Rm}}(\xi_n, \theta_{\mathrm{m}}, \theta_{\mathrm{R}}, \psi) \, e^{-2\rho g_{\mathrm{m}}(\theta_{\mathrm{m}} - \psi)l}]. \quad \text{(L3.220)}$$

By changing the variable of integration to $x \equiv 2\rho\, g_{\mathrm{m}}(\theta_{\mathrm{m}} - \psi)l$, we obtain

$$G(l, \theta_{\mathrm{m}}, \theta_{\mathrm{R}}) = \frac{kT}{16\pi^2 l^2} \sum_{n=0}^{\infty}{}' \int_0^{2\pi} \frac{\mathrm{d}\psi}{g_{\mathrm{m}}^2(\theta_{\mathrm{m}} - \psi)} \int_0^{\infty} x\mathrm{d}x \ln[1 - \overline{\Delta}_{\mathrm{Lm}}(\xi_n, \theta_{\mathrm{m}}, \psi) \times \overline{\Delta}_{\mathrm{Rm}}(\xi_n, \theta_{\mathrm{m}}, \theta_{\mathrm{R}}, \psi) \, e^{-x}]. \quad \text{(L3.221)}$$

$G(l, \theta_m, \theta_R)$ can be expanded in powers of $\overline{\Delta}_{Lm}\overline{\Delta}_{Rm} \leq 1$ to allow explicit integration in x:

$$G(l, \theta_m, \theta_R) = -\frac{kT}{16\pi^2 l^2} \sum_{n=0}^{\infty}{}' \sum_{j=1}^{\infty} \frac{1}{j^3} \int_0^{2\pi} \frac{\left[\overline{\Delta}_{Lm}(\xi_n, \theta_m, \psi)\,\overline{\Delta}_{Rm}(\xi_n, \theta_m, \theta_R, \psi)\right]^j \mathrm{d}\psi}{g_m^2(\theta_m - \psi)}. \tag{L3.222}$$

This free energy reduces to a familiar form (Eqs. [L2.8] & [P.1.a.3]) for isotropic materials.

Ion-containing anisotropic media (neglecting magnetic terms)

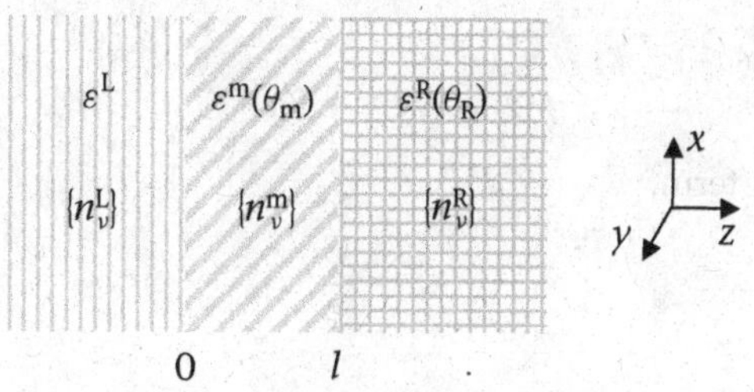

Figure L3.24

Begin with the Poisson equation but keep the ε matrix *inside* the divergence operation $\nabla \cdot (\varepsilon \nabla \phi) = -4\pi \rho_{ext}$ (see Fig. L3.24). The net electric-charge density ρ_{ext} at a given point depends on the magnitude of potential as in Debye–Hückel theory. As before in relation (L3.175),

$$\rho_{ext} = \sum_{\nu=-\infty}^{\nu=\infty} e\nu n_\nu^i e^{-e\nu\varphi/kT} \approx \sum_{\nu=-\infty}^{\nu=\infty} e\nu n_\nu^i \left(1 - \frac{e\nu\varphi}{kT}\right) = -\sum_{\nu=-\infty}^{\nu=\infty} \frac{e^2\nu^2 n_\nu^i}{kT}\varphi. \tag{L3.223}$$

Here n_ν^i is the mean number density of ions of valence ν of the solution's bathing regions i = L, m, or R. (By the net neutrality of salt solutions, with the summation over all mobile-ion valences, $\sum_{\nu=-\infty}^{\nu=\infty} \nu n_\nu^i = 0$.)

We define

$$k_i^2 \equiv \frac{e^2}{\varepsilon_0 kT} \sum_{\nu=-\infty}^{\nu=\infty} n_\nu \nu^2 \text{ in mks ("SI") units} \quad \text{or}$$

$$k_i^2 \equiv \frac{4\pi e^2}{kT} \sum_{\nu=-\infty}^{\nu=\infty} n_\nu \nu^2 \text{ in cgs ("Gaussian") units,} \tag{L3.224}$$

so that the equation to solve is of the form

$$\nabla \cdot (\varepsilon \nabla \phi) = \mathbf{k}^2 \phi. \tag{L3.225}$$

Notice that $\mathbf{k}^2$ differs from the Debye constant κ^2 by a factor ε. Because the dielectric permittivity is not a simple scalar quantity, it cannot be divided out of the $\nabla \cdot (\varepsilon \nabla \phi)$ on the left-hand side of the equation. Except for this difference, we can proceed as with ionic fluctuations in isotropic media.

In each region the potential can be Fourier-decomposed into periodic functions in the x and y directions:

$$\phi_i(x, y, z) = f_i(z) e^{i(ux+vy)}, \text{ i} = \text{L, m, R}. \tag{L3.226}$$

The differential equation for variation in the z direction perpendicular to the interfaces becomes

$$\varepsilon_z^{\mathrm{i}} f_{\mathrm{i}}''(z) - \left(\varepsilon_{11}^{\mathrm{i}} u^2 + 2\varepsilon_{12}^{\mathrm{i}} uv + \varepsilon_{22}^{\mathrm{i}} v^2\right) f_{\mathrm{i}}(z) = k_{\mathrm{i}}^2 f_{\mathrm{i}}(z) \tag{L3.227}$$

or

$$\varepsilon_z^{\mathrm{i}} f_{\mathrm{i}}''(z) - \beta_{\mathrm{i}}^2(\theta_{\mathrm{i}}) f_{\mathrm{i}}(z) = 0, \tag{L3.228}$$

with

$$\beta_{\mathrm{i}}^2(\theta_{\mathrm{i}}) = \frac{\varepsilon_x^{\mathrm{i}}}{\varepsilon_z^{\mathrm{i}}}(u\cos\theta_{\mathrm{i}} + v\sin\theta_{\mathrm{i}})^2 + \frac{\varepsilon_y^{\mathrm{i}}}{\varepsilon_z^{\mathrm{i}}}(v\cos\theta_{\mathrm{i}} - u\sin\theta_{\mathrm{i}})^2 + \frac{k_{\mathrm{i}}^2}{\varepsilon_z^{\mathrm{i}}}, \tag{L3.229}$$

to let us write

$$f_{\mathrm{i}}(z) = \mathrm{A_i} e^{\beta_{\mathrm{i}} z} + \mathrm{B_i} e^{-\beta_{\mathrm{i}} z}. \tag{L3.230}$$

Now there is an extra term,

$$\kappa_{\mathrm{i}}^2 \equiv \frac{k_{\mathrm{i}}^2}{\varepsilon_z^{\mathrm{i}}} \tag{L3.231}$$

in $\beta_{\mathrm{i}}^2(\theta_{\mathrm{i}})$,

$$\beta_{\mathrm{i}}^2(\theta_{\mathrm{i}}) = \frac{\varepsilon_x^{\mathrm{i}}}{\varepsilon_z^{\mathrm{i}}}(u\cos\theta_{\mathrm{i}} + v\sin\theta_{\mathrm{i}})^2 + \frac{\varepsilon_y^{\mathrm{i}}}{\varepsilon_z^{\mathrm{i}}}(v\cos\theta_{\mathrm{i}} - u\sin\theta_{\mathrm{i}})^2 + \kappa_{\mathrm{i}}^2, \tag{L3.232}$$

where κ_{i}^2 is built up from the mean number densities n_ν^{i} of ions of valence ν in regions $\mathrm{i} = \mathrm{L}$, m, or R:

$$\kappa_{\mathrm{i}}^2 \equiv \frac{e^2}{\varepsilon_0 \varepsilon_z^{\mathrm{i}} kT} \sum_{\nu=-\infty}^{\nu=\infty} \nu^2 n_\nu^{\mathrm{i}} \text{ in mks units} \qquad \kappa_{\mathrm{i}}^2 \equiv \frac{4\pi e^2}{\varepsilon_z^{\mathrm{i}} kT} \sum_{\nu=-\infty}^{\nu=\infty} \nu^2 n_\nu^{\mathrm{i}} \text{ in cgs units.} \tag{L3.233}$$

Set $u = \rho\cos\psi$ and $v = \rho\sin\psi$ to make

$$\beta_{\mathrm{i}}^2(\theta_{\mathrm{i}}) = \rho^2 g_{\mathrm{i}}^2(\theta_{\mathrm{i}} - \psi) + \kappa_{\mathrm{i}}^2, \tag{L3.234}$$

with

$$g_{\mathrm{i}}^2(\theta_{\mathrm{i}} - \psi) \equiv \frac{\varepsilon_x^{\mathrm{i}}}{\varepsilon_z^{\mathrm{i}}} + \frac{(\varepsilon_y^{\mathrm{i}} - \varepsilon_x^{\mathrm{i}})}{\varepsilon_z^{\mathrm{i}}}\sin^2(\theta_{\mathrm{i}} - \psi) = \frac{\varepsilon_y^{\mathrm{i}}}{\varepsilon_z^{\mathrm{i}}} + \frac{(\varepsilon_x^{\mathrm{i}} - \varepsilon_y^{\mathrm{i}})}{\varepsilon_z^{\mathrm{i}}}\cos^2(\theta_{\mathrm{i}} - \psi), \tag{L3.235}$$

as for the ion-free case.

Because the form of the functions $f_{\mathrm{i}}(z)$ is identical to that for the ion-free case, we can set $\mathrm{B_L} = \mathrm{A_R} = 0$ and apply the boundary conditions at $z = 0$ and l as before:

$$f_{\mathrm{L}}(0) = f_{\mathrm{m}}(0),\ \varepsilon_z^{\mathrm{L}} f'_{\mathrm{L}}(0) = \varepsilon_z^{\mathrm{m}} f'_{\mathrm{m}}(0),\ f_{\mathrm{m}}(l) = f_{\mathrm{R}}(l),\ \varepsilon_z^{\mathrm{m}} f'_{\mathrm{m}}(l) = \varepsilon_z^{\mathrm{R}} f'_{\mathrm{R}}(l). \tag{L3.236}$$

The solution for the remaining $\mathrm{A_i}$, $\mathrm{B_i}$ creates a dispersion relation similar to that in the ion-free case except for the κ_{i}^2 added to the $\beta_{\mathrm{r}}^2(\theta_{\mathrm{r}})$ functions:

$$\begin{aligned} D(l, \theta_{\mathrm{m}}, \theta_{\mathrm{R}}) &= 1 - \left[\frac{\varepsilon_z^{\mathrm{L}}\beta_{\mathrm{L}} - \varepsilon_z^{\mathrm{m}}\beta_{\mathrm{m}}(\theta_{\mathrm{m}})}{\varepsilon_z^{\mathrm{L}}\beta_{\mathrm{L}} + \varepsilon_z^{\mathrm{m}}\beta_{\mathrm{m}}(\theta_{\mathrm{m}})}\right]\left[\frac{\varepsilon_z^{\mathrm{R}}\beta_{\mathrm{R}}(\theta_{\mathrm{R}}) - \varepsilon_z^{\mathrm{m}}\beta_{\mathrm{m}}(\theta_{\mathrm{m}})}{\varepsilon_z^{\mathrm{R}}\beta_{\mathrm{R}}(\theta_{\mathrm{R}}) + \varepsilon_z^{\mathrm{m}}\beta_{\mathrm{m}}(\theta_{\mathrm{m}})}\right] e^{-2\beta_{\mathrm{m}}(\theta_{\mathrm{m}})l} \\ &= 1 - \left[\frac{\varepsilon_z^{\mathrm{L}}\beta_{\mathrm{L}} - \varepsilon_z^{\mathrm{m}}\beta_{\mathrm{m}}(\theta_{\mathrm{m}})}{\varepsilon_z^{\mathrm{L}}\beta_{\mathrm{L}} + \varepsilon_z^{\mathrm{m}}\beta_{\mathrm{m}}(\theta_{\mathrm{m}})}\right]\left[\frac{\varepsilon_z^{\mathrm{R}}\beta_{\mathrm{R}}(\theta_{\mathrm{R}}) - \varepsilon_z^{\mathrm{m}}\beta_{\mathrm{m}}(\theta_{\mathrm{m}})}{\varepsilon_z^{\mathrm{R}}\beta_{\mathrm{R}}(\theta_{\mathrm{R}}) + \varepsilon_z^{\mathrm{m}}\beta_{\mathrm{m}}(\theta_{\mathrm{m}})}\right] e^{-2\sqrt{\rho^2 g_{\mathrm{m}}^2(\theta_{\mathrm{m}} - \psi) + \kappa_{\mathrm{m}}^2}\, l} \\ &= 0. \end{aligned} \tag{L3.237}$$

For ionic-fluctuation forces, the ε's are now the dielectric constants in the limit of zero frequency ($\xi_n = 0$). The integration over wave vectors u, v can be converted to a ρ, ψ integration:

$$G_{n=0}(l, \theta_{\mathrm{m}}, \theta_{\mathrm{R}}) = \frac{kT}{8\pi^2} \int_0^{2\pi} \mathrm{d}\psi \int_0^{\infty} \rho \mathrm{d}\rho \ln\Big[1 - \overline{\Delta}_{\mathrm{Lm}}(\theta_{\mathrm{m}}, \psi)\, \overline{\Delta}_{\mathrm{Rm}}(\theta_{\mathrm{m}}, \theta_{\mathrm{R}}, \psi) \times e^{-2\sqrt{\rho^2 g_{\mathrm{m}}^2(\theta_{\mathrm{m}}-\psi)+\kappa_{\mathrm{m}}^2}\,l}\Big], \quad \text{(L3.238)}$$

$$\overline{\Delta}_{\mathrm{Lm}}(\theta_{\mathrm{m}}, \psi) = \left[\frac{\varepsilon_z^{\mathrm{L}}(0)\,\beta_{\mathrm{L}} - \varepsilon_z^{\mathrm{m}}(0)\,\beta_{\mathrm{m}}(\theta_{\mathrm{m}})}{\varepsilon_z^{\mathrm{L}}(0)\,\beta_{\mathrm{L}} + \varepsilon_z^{\mathrm{m}}(0)\,\beta_{\mathrm{m}}(\theta_{\mathrm{m}})}\right],$$

$$\overline{\Delta}_{\mathrm{Rm}}(\theta_{\mathrm{m}}, \theta_{\mathrm{R}}, \psi) = \left[\frac{\varepsilon_z^{\mathrm{R}}(0)\,\beta_{\mathrm{R}}(\theta_{\mathrm{R}}) - \varepsilon_z^{\mathrm{m}}(0)\,\beta_{\mathrm{m}}(\theta_{\mathrm{m}})}{\varepsilon_z^{\mathrm{R}}(0)\,\beta_{\mathrm{R}}(\theta_{\mathrm{R}}) + \varepsilon_z^{\mathrm{m}}(0)\,\beta_{\mathrm{m}}(\theta_{\mathrm{m}})}\right] \quad \text{(L3.239)}$$

The fecundity of this result emerges on examination of the specific cases considered in Level 2 and presented in the Tables of Formulae.

PROBLEM SETS

Problem sets for Prelude

Problem Pr.1: On average, how far apart are molecules in a dilute gas? Show that, for a gas at 1-atm pressure at room temperature, the average interparticle distance is ~30 Å.

Solution: Start with the ideal-gas law, $pV = NkT$. Take atmospheric pressure $p = 101.3$ kP $= 101.3 \times 10^3$ N/m^2 $= 1.013 \times 10^6$ ergs/cm^3, $N = N_{\text{Avogadro}} = 6.02 \times 10^{23}$, $kT = kT_{\text{room}} = 4.04 \times 10^{-21}$ J $= 4.04 \times 10^{-14}$ ergs. The volume for a mole under these "standard" conditions is then $V = 24 \times 10^3$ cm^3 $= 24 \times 10^{-3}$ m^3 $= 24$ liters.

Inversely, there are 24×10^{-3} m^3 $= 24 \times 10^{-3} \times 10^{+27}$ nm^3 $= 24 \times 10^{+24}$ nm^3 volume for these 0.602×10^{24} particles; hence 40 nm^3 $= (3.4\text{ nm})^3$ volume per particle or ~3 nm = 30 Å between particles, much bigger than the ~1–2-Å radius of an atom or small molecule.

Problem Pr.2: Calculate the effective Hamaker coefficient between the spherical atom and the gold surface.

Solution: The interaction energy between atom and surface goes as $-K_{\text{attr}}/z^3$ with $K_{\text{attr}} = 7.0 \times 10^{-49}$ J m^3. Introduce the point-particle-to-surface form $[-(2A_{\text{Ham}}/9)](\text{R}/z)^3$ in the table to translate K_{attr} into A_{Ham}, $K_{\text{attr}} = (2A_{\text{Ham}}/9)\text{R}^3$. Take ionic radius R ~ 2 Å $= 2 \times 10^{-10}$ m. $A_{\text{Ham}} = (9/2)(K_{\text{attr}}/\text{R}^3) \approx [(9 \times 7 \times 10^{-49}\text{ J m}^3)/(2 \times 8 \times 10^{-30}\text{ m}^3)] = 3.9 \times 10^{-19}$ J = 390 zJ.

Problem Pr.3: Show that the interaction between spheres separated by distances much greater than their radii will always be much less than thermal energy kT.

Solution: This weakness of inverse-sixth-power van der Waals forces between small particles is discussed at length in the main text. Its thermal triviality is easily seen. Begin with $[-(16/9)](\text{R}_6/z^6)A_{\text{Ham}}$ for the energy of interaction between two spheres of radius R and center-to-center separation z and ask what the A_{Ham} would have to be for the magnitude of this energy to be comparable with kT, $(16/9)(\text{R}_6/z^6)A_{\text{Ham}} = kT$ or $A_{\text{Ham}} = (9/16)(z^6/\text{R}^6)kT$. Even if the center-to-center separation z were equal to 4R, spheres separated by a distance equal to their diameter, R^6/z^6 would be $4^6 = 4096$. A_{Ham} would have to be a ridiculous $4096 \times (9/16)\, kT = 2304\, kT$ for there to be thermally significant attraction.

Problem Pr.4: Try something harder than spheres. Consider parallel cylinders of radius R, fixed length L, and surface separation l. Use the tabulated energy per unit length

$[-(A_{Ham}/24l^{3/2})]\ R^{1/2}$ to show that, for $A_{Ham} \approx 2\,kT_{room}$, a value typical of proteins (see table in preceding section), the energy is $\gg kT$ when

1. $R = L = 1\,\mu m \gg l = 10$ nm (dimensions of colloids), and
2. $R = 1$ nm, $L = 5$ nm $\gg l = 0.2$ nm (dimensions of proteins).

Solution:
$R = L = 1\,\mu m = 1000$ nm, $l = 10$ nm,

$$-\frac{A_{Ham}}{24l^{3/2}}R^{1/2}L = -\frac{2}{24}10^{+3}\,kT \approx -83\,kT;$$

$R = 1$ nm, $l = 0.2$ nm, $L = 5$ nm,

$$-\frac{A_{Ham}}{24l^{3/2}}R^{1/2}L = -\frac{2}{24}\frac{5}{0.089}kT \approx 5kT.$$

Problem Pr.5: Or easier than spheres. Consider a case of surface-shape complementarity imagined as two flat parallel surfaces. Show that the energy of interaction of two 1 nm × 1 nm patches 3 Å apart will yield an interaction energy $\sim kT$.

Solution:

$$\frac{A_{Ham}}{12\pi(3\times10^{-10}\,m)^2}(10^{-9}\ m)^2 = 2\,kT_{room}\frac{(10^{-9}\ m)^2}{12\pi(3\times10^{-10}\ m)^2} \approx \frac{kT_{room}}{2}.$$

Problem Pr.6: Show that van der Waals attraction at 100-Å separation is enough to hold up an ~2-cm cube even when $A_{Ham} = kT_{room}$.

Solution: Equate downward $F_{\downarrow} = F_{gravity} = \rho L^3\,g$ to upward $F_{\uparrow} = F_{vdW} = (A_{Ham}/6\pi l^3)(L^2)$ to find the $L = L_{bug}$ at which the forces balance. With $l = 100$ Å,

$$L_{bug} = \frac{A_{Ham}}{6\pi l^3\rho g} = \frac{4\times10^{-14}\ \text{ergs}}{6\pi(10^{-6}\ \text{cm})^3\times1\frac{\text{g}}{\text{cm}^3}\times980\frac{\text{dyn}}{\text{g}}} = \frac{4\times10^{-21}\ \text{J}}{6\pi(10^{-8}\ \text{m})^3\times1\frac{\text{kg}}{(0.1\ \text{m})^3}\times9.8\frac{\text{N}}{\text{kg}}}$$

to give $L_{bug} \sim 2$ cm = 0.02 m.

Problem Pr.7: Show how a change in shape makes a difference in the weight that can be maintained by a van der Waals force.

Solution: The force between a sphere and a nearby plane goes as the negative l derivative of interaction free energy, $[-(A_{Ham}/6)](R/l)$, or $F_{vdW} = F_{\uparrow} = (A_{Ham}/6)(R/l^2)$. This works against $F_{gravity} = F_{\downarrow} = \frac{4}{3}\pi R^3\rho g$. They balance (again with $l = 100$ Å) when

$$R^2_{bug} = \frac{A_{Ham}}{8\pi l^2\rho g} = \frac{4\times10^{-14}\text{ergs}}{8\pi(10^{-6}\ \text{cm})^2\times1\frac{\text{g}}{\text{cm}^3}\times980\frac{\text{dyn}}{\text{g}}} = \frac{4\times10^{-21}\ \text{J}}{8\pi(10^{-8}\ \text{m})^2\times1\frac{\text{kg}}{(0.1\ \text{m})^3}\times9.8\frac{\text{N}}{\text{kg}}}$$

to give $R_{bug} = 1.3\times10^{-3}$ cm $= 1.3\times10^{-5}$ m $= 13\,\mu$m.

Problem Pr.8: Show how van der Waals attraction can be a force for flattening a sphere against a flat surface.

Solution: The free energy of interaction between a perfect sphere and a flat surface goes as $[-(A_{Ham}/6)](R/l)$ whereas the interaction free energy between two flats goes as $-(A_{Ham}/12\pi l^2)$ per area. What if the sphere flattens slightly?
To first approximation there is negligible change in area or volume.

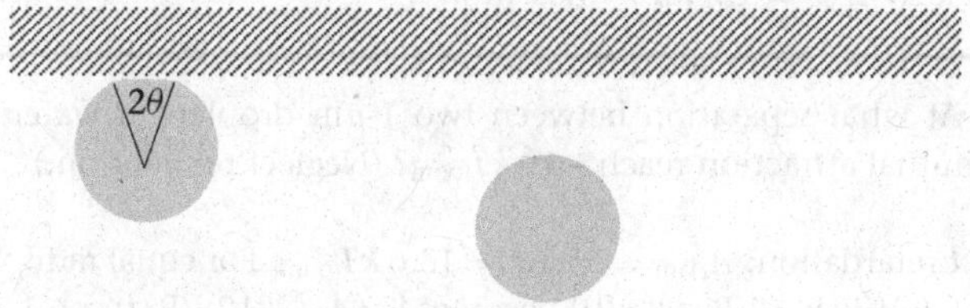

The loss of area from flattening of the sphere is

$$\int_0^{\theta} 2\pi R \sin\theta \, d(R\theta) = 2\pi R^2 \int_0^{\theta} \sin\theta \, d\theta = 2\pi R^2(1 - \cos\theta) \approx \pi R^2\theta^2,$$

whereas the flattened circular disk area is the same, $\pi(R\sin\theta)^2 \approx \pi R^2\theta^2$. Treating the energy of interaction at this flattened area as though it were the interaction between planes, and neglecting any additional interaction of the remaining curved parts of the sphere, we find that the interaction energy is

$$-\frac{A_{Ham}}{12\pi l^2}\pi R^2\theta^2 = -\frac{A_{Ham}}{12}\left(\frac{R}{l}\right)^2\theta^2.$$

Take $R = 1.3 \times 10^{-5}$m, $l = 100$ Å $= 10^{-8}$m, and $A_{Ham} = kT$ from the spherical bug statistics, so that $(A_{Ham}/12)(R/l)^2 \sim (kT/7)10^6$. If only 5% of the original spherical area were flattened $[(\pi R^2\theta^2)/(4\pi R^2)] = 0.05, \theta^2 = 0.2$, there would be an interaction energy $[-(A_{Ham}/12)](R/l)^2\theta^2 = -0.2(kT/7)10^6 = -3 \times 10^4\ kT$ compared with $[-(A_{Ham}/6)](R/l) = -1.6 \times 10^2\, kT$ for the undeformed sphere.

The first bending costs little bending energy but yields large attractive energy. Conversely, in practical situations, a small attractive force amplifies with bending. This amplification with induced bending deformation is also a reason that there can be confusion when one is interpreting forces between oppositely curved surfaces.

Warning: The preceding solution is not a complete solution of this problem; it is only just enough to show the magnitude of flattening force that emerges from weak van der Waals attraction.

Problem Pr.9: When does the van der Waals attraction between two spherical drops of water in air equal the gravity force between them? (Neglect retardation.)

Solution: Gravitational attraction goes as the product of masses and inverse-square distance:

$$F_{gravity}(z) = -G\frac{M_1M_2}{z^2}$$

(with a minus sign to remind us that the force is attractive). The gravitational constant $G = 6.673 \times 10^{-11}\ m^3/(s\,kg)$; the masses are $M_1 = M_2 = (4\pi/3)R^3\rho$ with density $\rho = 1\ g/cm^3 = 10^3\ kg/m^3$ so that $F_{gravity}(z) = -6.673 \times 10^{-11}\ m^3 s^{-2}\,kg^{-1}(4\pi/3)^2(R^6/z^2)10^6\ kg^2/m^2 = -1.17 \times 10^{-3}(R^6/z^2)$N, where we recall from force = mass × acceleration that 1 Newton force = kg mass × (meters/second)/second.

The van der Waals attraction force goes as the inverse-seventh power of separation, the negative derivative of $[-(16/9)](R^6/z^6)A_{Ham}$ (using the Table Pr. 1. long-distance form for the interaction): $F_{vdW}(z) = -6(16/9)(R^6/z^7)A_{Ham}$. Introducing $A_{Ham} \approx 55.1 \times 10^{-21}$ J (Table Pr.2.), we have $F_{vdW}(z) = -5.88 \times 10^{-19}(R^6/z^7)$N (J/m = N).

When we equate $F_{vdW}(z)$ with $F_{gravity}(z)$, $1.17 \times 10^{-3} = [(5.88 \times 10^{-19})/z^5]$, R^6 drops out. The forces are the same when $z = (5.88 \times 10^{-19}/1.17 \times 10^{-3})^{\frac{1}{5}} = 0.87$ mm. This strikingly macroscopic distance appears to be a fairly robust result. Even if $A_{Ham} = kT_{room} \approx 4 \times 10^{-21}$ J, ~7% of the strength previously assumed, the separation z at force equality would go down by a factor of $\sim(55.1/4)^{1/5} = 1.7$ for $z = .52$ mm.

Problem Pr.10: At what separation between two 1-μm droplets of water in air does the energy of their mutual attraction reach $-10\ kT_{room}$? (Neglect retardation.)

Solution: Neglect retardation: $A_{Ham} = 55.1$ zJ $= 13.6\ kT_{room}$. For equal radii, $R_1 = R_2 = R$, the tabulated $[-(A_{Ham}/6)\{(R_1R_2)/[(R_1+R_2)l]\}]$ becomes $[-(A_{Ham}/12)](R/l) \approx -1.1\ kT_{room}(R/l)$.

Equated to $-10\ kT_{room}$, this interaction energy is $-1.1\ kT_{room}\ (R/l) = -10\ kT_{room}$, or $l = 0.11\,R = 0.11\ \mu m = 110$ nm. This separation satisfies the requirement that separation l be much greater than the molecular detail of the sphere; it also satisfies the $l \ll R$ requirement for the formula used. The latter requirement would not be satisfied if we wanted to use this formula for an interaction energy -1 kT.

Problem Pr.11: Show that the forces "see" into the interacting bodies in proportion to separation.

Solution: Under the stated restrictions, as elaborated in the main text, we can think of the interaction between coated bodies as a sum of interactions between interfaces. Speak of the bodies B, coatings C, and medium m. Let the separation between coatings be l, and their thickness be c. Then there are four terms for the four pairs of interacting surfaces that have material m in between:

$$-\frac{A_{Cm/Cm}}{12\pi l^2} - \frac{A_{BC/Cm}}{12\pi(l+c)^2} - \frac{A_{BC/Cm}}{12\pi(l+c)^2} - \frac{A_{BC/BC}}{12\pi(l+2c)^2}.$$

The coefficient of each interaction is a sum over products of differences in dielectric responses. For example,

$$A_{BC/Cm} \sim \sum_{\substack{\text{sampling}\\ \text{frequencies}}} (\varepsilon_B - \varepsilon_C)(\varepsilon_C - \varepsilon_m)$$

when there are no large differences in the epsilons. (The convention throughout this text is that the outer material at an interface is written first in dielectric differences.) Near contact, when separation $l \ll$ thickness c, the first $-(A_{Cm/Cm}/12\pi l^2)$ term dominates.

When separation $l \gg$ thickness c, c is essentially zero, so all denominators are the same. We have a collection of numerators

$$\begin{aligned}
&(\varepsilon_C - \varepsilon_m)(\varepsilon_C - \varepsilon_m) + (\varepsilon_B - \varepsilon_C)(\varepsilon_C - \varepsilon_m) + (\varepsilon_B - \varepsilon_C)(\varepsilon_C - \varepsilon_m) + (\varepsilon_B - \varepsilon_C)(\varepsilon_B - \varepsilon_C)\\
&\quad = (\varepsilon_C^2 - 2\varepsilon_C\varepsilon_m + \varepsilon_m^2) + (\varepsilon_B\varepsilon_C - \varepsilon_B\varepsilon_m - \varepsilon_C^2 + \varepsilon_C\varepsilon_m)\\
&\qquad + (\varepsilon_B\varepsilon_C - \varepsilon_B\varepsilon_m - \varepsilon_C^2 + \varepsilon_C\varepsilon_m) + (\varepsilon_B^2 - 2\varepsilon_B\varepsilon_C + \varepsilon_C^2)\\
&\quad = \varepsilon_B^2 - 2\varepsilon_B\varepsilon_m + \varepsilon_m^2 = (\varepsilon_B - \varepsilon_m)(\varepsilon_B - \varepsilon_m),
\end{aligned}$$

so the interaction looks like $-(A_{Bm/Bm}/12\pi l^2)$. (Messy but yes!)

Problem Pr.12: Convince yourself that $A_{A-A} + A_{B-B} \geq 2A_{A-B}$.

Solution: Inequality follows from the mathematical identity $(\alpha - \beta)^2 \geq 0$ as $\alpha^2 + \beta^2 \geq +2\alpha\beta$. Associate A_{A-A} with a sum of terms of the form α^2, A_{B-B} with β^2, and A_{A-B} with $\alpha\beta$. There is no physics to this except that $\alpha \leftrightarrow [(\varepsilon_A - \varepsilon_m)/(\varepsilon_A + \varepsilon_m)]$ and $\beta \leftrightarrow [(\varepsilon_B - \varepsilon_m)/\ (\varepsilon_B + \varepsilon_m)]$ are mathematically real quantities.

Problem Pr.13: In vacuum, at what separation l does the travel time across and back equal the $\sim 10^{-14}$-s period of an IR frequency?

Solution: Take the velocity of light in vacuum, $c \approx 3 \times 10^8$ m/s $= 3 \times 10^{10}$ cm/s. The time is $(2l/c) = 10^{-14}$ s, $l = [(3 \times 10^8 \times 10^{-14})/2] = 1.5\,\mu$m.

For the $\sim 10^{-16}$-second period of a UV frequency, $l = 15$ nm.

Problem Pr.14: Show how a Hookean spring works against an inverse-power van der Waals interaction in a force balance of a sphere and a flat surface.

Solution: To think specifically, consider a sphere of radius R and a flat (or its Derjaguin-approximation equivalent, two perpendicular cylinders of radius R), and neglect all but nonretarded van der Waals forces. The force is the negative derivative of the free energy $-(A_{Ham}/6)(R/l)$,

$$F_{\text{interaction}}(l) = F_{\text{vdW}}(l) = -\frac{A_{\text{Ham}}}{6l^2}\text{R} \equiv -\frac{k_{\text{vdW}}}{l^2}$$

(which pulls to the left in the picture in the text), where l is the smallest distance between the two bodies. The spring force can be written as Hooke's law,

$$F_{\text{spring}}(x) = k_{\text{Hooke}}[(x - l) - x_0]$$

(which pulls to the right), where x is the separation set by the operator of the apparatus. At a balance point, $F_{vdW}(l) + F_{spring}(x) = 0$ creates a connection between x and (measured) l, $x = x_0 + l + [(k_{vdW}/k_{Hooke})/l^2]$. If we replace x with dx, we see l change by dl such that

$$\text{d}x = \frac{\text{d}x}{\text{d}l}\text{d}l = \left(1 - 2\frac{k_{\text{vdW}}/k_{\text{Hooke}}}{l^3}\right)\text{d}l.$$

For a stiff spring or for big l, dx = dl. Deviations from dx = dl reveal k_{vdW}. Most revealing is the instability that occurs when l is small enough that $\{1 - 2[(k_{vdW}/k_{Hooke})/l^3]\} = 0$. Then, when $l^3 = 2k_{vdW}/k_{Hooke}$, an imposed change dx provokes a jump in l.

Problem Pr.15: Convert an angle of deviation in a surface contour into an estimate of attractive energy across a film.

Solution: There are several ways to do this approximately. (A rigorous solution is beyond present scope.) Think of a thinned region as having a positive surface free energy per area $\gamma = \gamma' + \gamma_{vdW}$ that differs by a negative amount γ_{vdW} from the energy per area γ' where the film is no longer thinned. The angle θ reveals the balance between the two energies, each pulling along the line of the membrane: $\gamma' + \gamma_{vdW} = \gamma' \cos\theta \approx \gamma'[1 - (\theta^2/2)]$ so that $\gamma_{vdW} \approx -\gamma'(\theta^2/2)$. Because there are two interfaces to the film, the energy from the van der Waals interaction is twice this energy from one side.

Haydon and Taylor [Nature (London), **217**, 739–740 (1968)] reported an angle θ of $1°52' = 0.03258$ rad and spoke of an interfacial tension γ' of the bulk media 3.72 ergs/cm^2 and emerged with $\gamma_{vdW} \approx -\gamma'(\theta^2/2) = -0.00197$ ergs/cm^2 or a van der Waals energy of -0.00394 ergs/cm^2.

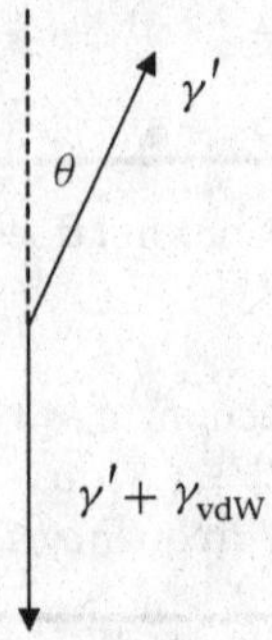

Problem Pr.16: What is the attractive energy that creates a flattening between two vesicles under tension $\overline{\mathbf{T}}$?

Solution: Speak of an angle θ between the rounded and flattened parts of the vesicles. Think of vectors $\overline{\mathbf{T}}$ for tension along the vesicles and an extra (negative!) free energy G per unit area gained by the flattening of two vesicles against each other. This energy acts as an extra pull on the junction point, a pull to increase the flattened area.

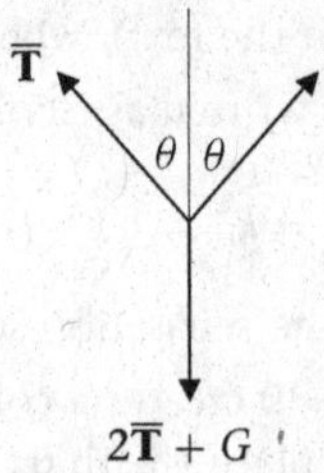

The vectors balance when $2\overline{\mathbf{T}} + G = 2\overline{\mathbf{T}}\cos\theta$ or $G = -2\overline{\mathbf{T}}(1 - \cos\theta)$.

Problem Pr.17: To gauge the difference between long-range and short-range charge-fluctuation forces, compute the van der Waals attraction free energy between two flat parallel regions of hydrocarbon across 3 nm of vacuum. Compare this long-range free energy with the ~20 mJ/m^2(= mN/m = erg/cm^2 = dyn/cm) surface tension of an oil.

Solution: Feed $A_{Ham} = 47 \times 10^{-21}$ J, the Hamaker coefficient for tetradecane across vacuum, into the interaction energy per area $-(A_{Ham}/12\pi l^2)$ of plane-parallel half-spaces of separation $l = 3 \times 10^{-9}$ m:

$$-\frac{A_{Ham}}{12\pi l^2} = -\frac{47 \times 10^{-21}\ \mathrm{J}}{12\pi(3 \times 10^{-9}\ \mathrm{m})^2} \approx 1.4 \times 10^{-4} = 0.14\ \mathrm{mJ/m^2}.$$

Problem Pr.18: To get an idea about the onset of graininess, consider the interaction between one point particle and a pair of point particles at a small separation **a**; show how the interaction becomes proportional to $\mathbf{a}^2/z^2$ when the distance z between point and pair becomes much greater than **a**.

Solution: Imagine a scene

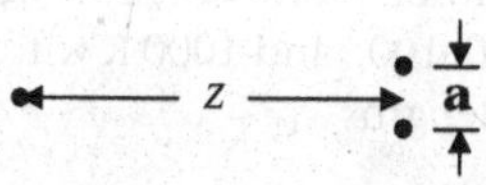

in which the particle on the left enjoys an inverse-sixth-power interaction with each of the particles on the right. Assume that the interactions are additive so that the energy goes as

$$\frac{2}{[z^2+(a/2)^2]^3} = \frac{2}{z^6[1+(a/2z)^2]^3}.$$

For $z \ll a$, expand $[1+(a/2z)^2]^3 \approx 1 + 3(a/2z)^2$ so that the sum of the interactions looks like

$$\frac{2}{z^6[1+(a/2z)^2]^3} \approx \frac{2}{z^6}[1-3(a/2z)^2].$$

Even for $a = 5z$, the a^2 term gives $3(a/2z)^2 = 0.03$.

From the other direction,

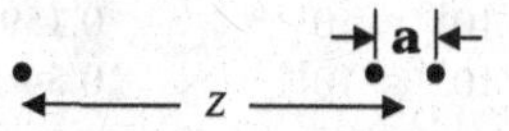

$$\frac{1}{\left(z-\frac{a}{2}\right)^6} + \frac{1}{\left(z+\frac{a}{2}\right)^6} \approx \frac{2}{z^6}\left[1+21\left(\frac{a}{2z}\right)^2\right].$$

Here, for $a = 5z$, the correction $21(a/2z)^2 = 0.21$; for $a = 10z$, $21(a/2z)^2 \approx 0.05$.

Problem Pr.19: Peel vs. Pull. Imagine a tape of width W with an adhesion energy G per area. Peeling off a length z removes an area of adhesion Wz and thereby incurs work GWz. Perpendicular lifting off a patch of area $A = 1\ \text{cm}^2$ costs GA.

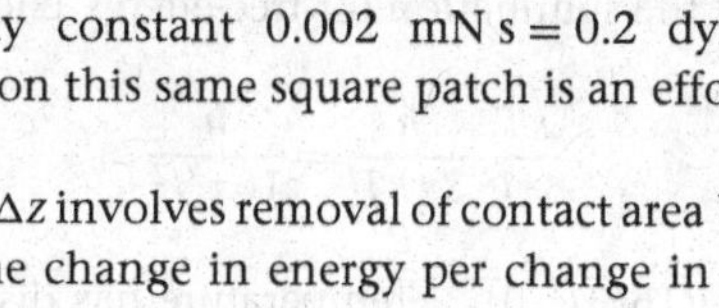

Assuming that adhesion comes from only a van der Waals attraction $G = -[A_{\text{Ham}}/(12\pi l^2)]$, neglecting any balancing forces or any elastic properties of the tape, show that when tape–surface separation $l = 0.5$ nm (5 Å), $W = z = 0.01$ m (1 cm), and $G = 0.2\ \text{mJ/m}^2$ ($0.2\ \text{erg/cm}^2$), the peeling force is a tiny constant 0.002 mN s = 0.2 dyn whereas the maximum perpendicular-pull-off force on this same square patch is an effortful $80\ \text{N} = 8 \times 10^6$ dyn.

Solution: Peeling a distance Δz involves removal of contact area $W\Delta z$ and a change in energy $GW\Delta z$. The force is GW, the change in energy per change in length, $GW = 0.2\ \text{mJ/m}^2 \times 0.01\ \text{m} = 0.002\ \text{mN} = 0.2\ \text{dyn}$.

Pulling perpendicular to the plane of contact incurs a pressure $P = -(\partial G/\partial l) = -(A_{\text{Ham}}/6\pi l^3) = (2/l)G$ and a force, pressure × area,

$$P \times W \times z = \frac{2Wz}{l}G = \frac{2(0.01\ \text{m})^2}{5\times10^{-10}\ \text{m}} 0.2\times10^{-3}\frac{\text{J}}{\text{m}^2} = 80\ \text{N} = 8\times10^{+6}\ \text{dyn}$$

here at its maximum when $l = 0.5$ nm.

Problem sets for Level 1

Problem L1.1: How important is temperature in determining which sampling frequencies act in the charge-fluctuation force? For $n = 1, 10,$ and 100, compute imaginary radial frequencies $\xi_1(T)$ at $T = 0.1, 1.0, 10, 100,$ and 1000 K with corresponding frequencies $\nu_1(T)$, photon energies $\hbar\xi_1(T)$, and wavelengths λ_1.

Solution: From $\xi_n(T) \equiv \{[(2\pi kT)/\hbar)]n\}$, with $k = 1.3807 \times 10^{-16}$ ergs/K $= 1.3807 \times 10^{-23}$ J/K $= 8.6173 \times 10^{-5}$ eV/K, $\hbar = 1.0546 \times 10^{-27}$ ergs s $= 1.0546 \times 10^{-34}$ J s $= 6.5821 \times 10^{-16}$ eV s, use $[(2\pi kT)/\hbar] = 8.22 \times 10^{+11}$ $T=$ for ξ_1 in radians per second and $2\pi kT = 5.4 \times 10^{-4}\ T$ for $\hbar\xi_1(T)$ in electron volts. The "wavelength" λ_1, really a decay distance when coupled with imaginary frequency, is $2\pi\lambda_1 = \xi_1/c$ with the velocity of light $c = 3 \times 10^{10}$ cm/s $= 3 \times 10^8$ m/s.

T(K)	ξ_1 (rad/s)	ν_1(Hz)	$\hbar\xi_1$(eV)	λ_1 (Å)
0.1	8.2×10^{10}	$1.3 \times 10^{10} \approx 10^{10.1}$	5.4×10^{-5}	2.3×10^8 Å $= 2.3$ cm
1.0	8.2×10^{11}	$1.3 \times 10^{11} \approx 10^{11.1}$	5.4×10^{-4}	2.3×10^7 Å $= 2.3$ mm
10	8.2×10^{12}	$1.3 \times 10^{12} \approx 10^{12.1}$	5.4×10^{-3}	2.3×10^6 Å $= .23$ mm
100	8.2×10^{13}	$1.3 \times 10^{13} \approx 10^{13.1}$	5.4×10^{-2}	2.3×10^5 Å $= 23.\ \mu$m
300	$25. \times 10^{13}$	$3.9 \times 10^{13} \approx 10^{13.6}$	0.159	7.7×10^4 Å $= 7.7\ \mu$m
1000	8.2×10^{14}	$1.3 \times 10^{14} \approx 10^{14.1}$	0.54	2.3×10^4 Å $= 2.3\ \mu$m

Is it clear how details of spectra at IR frequencies are progressively lost at higher temperatures? At $T = 1000$ K, the first finite sampling frequency, $\xi_{n=1} = 8.2 \times 10^{14}$ rad/s, occurs near the visible range whereas $\xi_{n=3} = 3 \times \xi_1 = 24.6 \times 10^{14} = 10^{15.4}$ rad/s is just at the boundary between the IR and the visible frequencies. The ξ_n summation leaps over everything from zero to the near-visible range.

Problem L1.2: If you take the kT factor too seriously, then it looks as though van der Waals interactions increase linearly with absolute temperature. Show that, for contributions from a sampling-frequency range $\Delta\xi$ over which $\overline{\Delta}$'s change little, there is little change in van der Waals forces with temperature, *except* for temperature-dependent changes in the component ε's themselves.

Solution: Although the contribution from each sampling frequency ξ_n enjoys a coefficient kT, the density of sampling frequencies goes down inversely with temperature. From $\xi_n = [(2\pi kT)/\hbar]n$, we know that a range $\Delta\xi$ includes $\Delta n = [\hbar/(2\pi kT)]\Delta\xi$ sampling frequencies. For that frequency range $\Delta\xi$ the summation for free energy encounters a product

$$\frac{kT}{8\pi l^2}\frac{\hbar}{2\pi kT} = \frac{\hbar}{16\pi^2 l^2}$$

for a contribution $\{-[(\hbar\Delta\xi)/(16\pi^2 l^2)]\}\overline{\Delta}^2$. Temperature has disappeared except for the dependence on temperature of the dielectric susceptibilities composing $\overline{\Delta}$. (*Warning:* Don't practice this maneuver without carefully verifying the condition that $\overline{\Delta}$ varies negligibly over the range. $\Delta\xi$.) For further discussion see Level 2, Section L2.3.A.

Problem L1.3: If the interaction is really a free energy versus separation, then it must have energetic and entropic parts. What is the entropy of a van der Waals interaction?

Solution: The temperature derivative of a work (free energy) as defined here requires that we take a derivative of the whole $G_{AmB}(l) \approx -\frac{kT}{8\pi l^2}\sum_{n=0}^{\prime\infty}\overline{\Delta}_{Am}\overline{\Delta}_{Bm}\,R_n(l)$. For sanity, ignore retardation [set $R_n(l) = 1$]; take material A = material B, medium m = vacuum, so that

$$G(l;T) \approx -\frac{kT}{8\pi l^2}\sum_{n=0}^{\infty}{}' \overline{\Delta}^2, \quad \overline{\Delta} = \frac{\varepsilon(i\xi_n)-1}{\varepsilon(i\xi_n)+1}, \quad \xi_n = \frac{2\pi kT}{\hbar}n = \xi_n(T), \quad \varepsilon(i\xi_n) = \varepsilon[T, i\xi_n(T)].$$

There are two kinds of temperature effects to think about:

1. If the photon energies are comparable with or less than thermal energy, $\hbar\xi_n(T) \leq kT$, the electromagnetic excitations are themselves stimulated by temperature.
2. If the epsilons change with temperature, so will the strength of contributions even at sampling frequencies for which $\hbar\xi_n(T) \gg kT$.

At high sampling frequencies, the contribution from the range $\Delta\xi$, $-\frac{\hbar\Delta\xi}{16\pi^2 l^2}\overline{\Delta}^2$, has an entropic component

$$-\frac{\hbar\Delta\xi}{16\pi^2 l^2}\frac{\partial\overline{\Delta}^2}{\partial T} = -\frac{\hbar\Delta\xi}{8\pi^2 l^2}\frac{\partial\overline{\Delta}}{\partial T} = -\frac{\hbar\Delta\xi}{8\pi^2 l^2}\frac{2}{(\varepsilon+1)^2}\frac{\partial\varepsilon}{\partial T}.$$

Both consequences of varied temperature are seen in the isolated $n = 0$ term, $-\frac{kT}{16\pi l^2}\overline{\Delta}^2(0)$, where

$$S_{n=0} = -\frac{\partial G_{n=0}}{\partial T} = \frac{\partial}{\partial T}\frac{kT}{16\pi l^2}\overline{\Delta}^2(0) = \frac{k}{16\pi l^2}\overline{\Delta}^2(0) + \frac{kT}{16\pi l^2}\frac{\partial\overline{\Delta}^2(0)}{\partial T},$$

with

$$\frac{\partial\overline{\Delta}^2}{\partial T} = 2\frac{2}{(\varepsilon+1)^2}\frac{\partial\varepsilon(0;T)}{\partial T}.$$

According to this derivative, if epsilon changes negligibly with temperature, then $G_{n=0} = -TS_{n=0}$. All entropy. Think about it.

Problem L1.4: For each sampling frequency ξ_n, or its corresponding photon energy $\hbar\xi_n$, what is the separation l_n at which $r_n = [(2l_n\varepsilon_m^{1/2}\xi_n)/c] = 1$?

Solution: Consider the medium to be a vacuum and find the answer that holds at room temperature. With $\xi_n = [(2\pi kT)/\hbar]n$, $[(2\pi kT_{room})/\hbar] = 2.411 \times 10^{14}$ rad/s = 0.159 eV,

$$l_n = \frac{c}{2\xi_n} \approx \frac{3\times 10^8\ \text{m/s}}{2\times 2.4\times 10^{14} n\ \text{rad/s}} \approx \frac{6.25\times 10^{-7}}{n}\text{m} = \frac{6.25\times 10^{+3}}{n}\ \text{Å}.$$

For $n = 100$, $\hbar\xi_{n=100} \approx 16$ eV, $l_{n=100} = 625$ Å; $n = 1000$, $\hbar\xi_{n=1000} \approx 160$ eV, $l_{n=1000} = 62.5$ Å. The result gives a nifty way to think about what sampling frequencies count at a given separation.

Problem L1.5: Show how this power law emerges from free energy $G_{AmB}(l)$.

Solution: To expose the effective power of l, put $G_{AmB}(l)$ in the form $G_{Am/Bm}(l) = [\mathrm{b}/l^{p(l)}]$, where all l dependence now resides in exponent $p(l)$. Take the logarithm, $\ln[G_{AmB}(l)] = \ln(\mathrm{b}) - p(l)\ln(l)$. It immediately becomes clear that the (negative) derivative of $\ln[G_{AmB}(l)]$ with respect to $\ln(l)$ gives us the power we seek.

Problem L1.6: How does convergence of the sum under the influence of only the retardation function $R_n(l;\xi_n)$ create the appearance of the $1/l^3$ variation of free energy?

Solution: Convergence of the sum is due to large values of $r_n(l;\xi_n)$ in $R_n(l;\xi_n) = [1 + r_n(l;\xi_n)]e^{-r_n(l;\xi_n)} \approx r_n(l;\xi_n)e^{-r_n(l;\xi_n)}$. Recall that $r_n = [(2l\varepsilon_{\mathrm{m}}^{1/2}\xi_n)/c]$ and that $\xi_n = [(2\pi kT)/\hbar]n$. For a brief decade of separations, the flat part of the curve in Fig. L1.18, all that matters is that r_n be proportional to $l \times n$, i.e, $r_n = \alpha ln$. The summation looks like $\sum_{n=0}^{\prime\infty} r_n(l;\xi_n)e^{-r_n(l;\xi_n)} = \sum_{n=0}^{\prime\infty}\alpha lne^{-\alpha ln}$ and smooths out into an integral $\int \alpha lne^{-\alpha ln}dn$. This is equivalent to $(1/\alpha l)\int xe^{-x}\,dx$, where the factor in front introduces a $1/l$ factor in the otherwise $1/l^2$ energy.

Problem L1.7: Take the l derivative of $G_{\mathrm{AmB}}(l)$, approximation (L1.5), in the equal-light-velocities approximation, to obtain $P_{\mathrm{AmB}}(l)$, approximation (L1.20).

Solution: Differentiate $G_{\mathrm{AmB}}(l) \approx -\frac{kT}{8\pi l^2}\sum_{n=0}^{\prime\infty}\overline{\Delta}_{\mathrm{Am}}\overline{\Delta}_{\mathrm{Bm}}\,\mathrm{R}_n(l) = -[kT/(8\pi l^2)]\sum_{n=0}^{\prime\infty}\overline{\Delta}_{\mathrm{Am}}\overline{\Delta}_{\mathrm{Bm}}$ $(1+r_n(l))e^{-r_n(l)}$ directly and through $r_n(l) = [(2\varepsilon_{\mathrm{m}}^{1/2}\xi_n)/c]l$, where

$$\frac{\partial[1+r_n(l)]e^{-r_n(l)}}{\partial l} = \frac{\partial[1+r_n(l)]e^{-r_n(l)}}{\partial r_n(l)}\frac{\partial r_n(l)}{\partial l} = [e^{-r_n} - (1+r_n)e^{-r_n}]\frac{r_n}{l} = -\frac{r_n^2}{l}e^{-r_n}.$$

$$\frac{\partial G_{\mathrm{AmB}}(l)}{\partial l} = -\frac{kT}{8\pi}\left[-\frac{2}{l^3}\sum_{n=0}^{\infty}{}'\,\overline{\Delta}_{\mathrm{Am}}\overline{\Delta}_{\mathrm{Bm}}\,(1+r_n)e^{-r_n} + \frac{1}{l^2}\sum_{n=0}^{\infty}{}'\,\overline{\Delta}_{\mathrm{Am}}\overline{\Delta}_{\mathrm{Bm}}\left(\frac{-r_n^2}{l}\right)e^{-r_n}\right]$$

$$= +\frac{kT}{4\pi l^3}\sum_{n=0}^{\infty}{}'\,\overline{\Delta}_{\mathrm{Am}}\overline{\Delta}_{\mathrm{Bm}}\left(1+r_n+\frac{r_n^2}{2}\right)e^{-r_n} = -P_{\mathrm{AmB}}(l).$$

Problem L1.8: Can there be van der Waals repulsion between bodies separated by a vacuum? (Far-fetched, zestfully discussed by Casimir cognoscenti.)

Solution: Yes, even across a vacuum, asymmetry can create van der Waals repulsion. In a vacuum, $\varepsilon_{\mathrm{m}} = \mu_{\mathrm{m}} = 1$. Now let, for example, $\varepsilon_{\mathrm{A}} > 1$, $\mu_{\mathrm{A}} = 1$, $\varepsilon_{\mathrm{B}} = 1$, $\mu_{\mathrm{B}} > 1$. To show that these bodies will repel, it suffices to show that members of the $\overline{\Delta}_{\mathrm{Am}}$, $\overline{\Delta}_{\mathrm{Bm}}$ and Δ_{Am}, Δ_{Bm} pairs have opposite signs. Here

$$s_{\mathrm{A}} = \sqrt{p^2 - 1 + \varepsilon_{\mathrm{A}}\mu_{\mathrm{A}}/\varepsilon_{\mathrm{m}}\mu_{\mathrm{m}}} = p\sqrt{1+\frac{\varepsilon_{\mathrm{A}}-1}{p^2}}, \qquad s_{\mathrm{B}} = p\sqrt{1+\frac{\mu_{\mathrm{B}}-1}{p^2}}$$

while, as usual, $1 \le p < \infty$.

The sign of $\overline{\Delta}_{\mathrm{Am}} = [(p\varepsilon_{\mathrm{A}} - s_{\mathrm{A}}\varepsilon_{\mathrm{m}})/(p\varepsilon_{\mathrm{A}} + s_{\mathrm{A}}\varepsilon_{\mathrm{m}})]$ is the sign of $\varepsilon_{\mathrm{A}} - \sqrt{1+\frac{\varepsilon_{\mathrm{A}}-1}{p^2}} \ge \varepsilon_{\mathrm{A}} - \sqrt{\varepsilon_{\mathrm{A}}} \ge 0$.

The sign of $\overline{\Delta}_{\mathrm{Bm}} = [(p\varepsilon_{\mathrm{B}} - s_{\mathrm{B}}\varepsilon_{\mathrm{m}})/(p\varepsilon_{\mathrm{B}} + s_{\mathrm{B}}\varepsilon_{\mathrm{m}})]$ is that of $1 - \sqrt{1+\frac{\mu_{\mathrm{B}}-1}{p^2}} \le 1 - \sqrt{\mu_{\mathrm{B}}} \le 0$. Similarly, $\Delta_{\mathrm{Am}} = [(p\mu_{\mathrm{A}} - s_{\mathrm{A}}\mu_{\mathrm{m}})/(p\mu_{\mathrm{A}} + s_{\mathrm{A}}\mu_{\mathrm{m}})] \le 0$; $\Delta_{\mathrm{Bm}} = [(p\mu_{\mathrm{B}} - s_{\mathrm{B}}\mu_{\mathrm{m}})/(p\mu_{\mathrm{B}} + s_{\mathrm{B}}\mu_{\mathrm{m}})] \ge 0$.

Problem L1.9: Using the result given in Table P.9.e in Level 2, derive free energy and torque [Eqs. (L1.24a) and (L1.24b)].

Solution: For weakly birefringent materials A and B, Table P.9.e gives the free energy of interaction with respect to infinite separation in the form

$$G(l,\theta) = -\frac{kT}{8\pi l^2}\sum_{n=0}^{\infty}{}'\left[\overline{\Delta}_{\overline{\mathrm{A}}\mathrm{m}}\overline{\Delta}_{\overline{\mathrm{B}}\mathrm{m}} + \overline{\Delta}_{\overline{\mathrm{A}}\mathrm{m}}\frac{\gamma_{\mathrm{B}}}{2} + \overline{\Delta}_{\overline{\mathrm{B}}\mathrm{m}}\frac{\gamma_{\mathrm{A}}}{2} + \frac{\gamma_{\mathrm{A}}\gamma_{\mathrm{B}}}{8}(1+2\cos^2\theta)\right].$$

For A and B of identical material properties, $\overline{\Delta}_{\mathrm{Am}} = \overline{\Delta}_{\mathrm{Bm}} = \overline{\Delta}$, $\gamma_{\mathrm{A}} = \gamma_{\mathrm{B}} = \gamma$,

$$G(l,\theta) = -\frac{kT}{8\pi l^2}\sum_{n=0}^{\infty}{}'\left[\overline{\Delta}^2 + \gamma\overline{\Delta} + \gamma^2(1+2\cos^2\theta)\right].$$

The derivative with respect to θ requires $\partial(2\cos^2\theta)/\partial\theta = -4\cos\theta\sin\theta = -2\sin(2\theta)$.

Problem L1.10: If $\sum_{n=0}^{\prime\infty}\gamma^2 = 10^{-2}$, how big an area L^2 of the two parallel-planar faces would suffer an energy change kT because of a 90° turn in mutual orientation?

Solution: Look only at the θ-dependent part of the free energy per area $G(l,\theta)$ multiplied by L^2,

$$-\frac{kT}{8\pi l^2}\mathrm{L}^2 2\cos^2\theta\sum_{n=0}^{\infty}{}'\gamma^2 = -\frac{kT}{4\pi l^2}\mathrm{L}^2\cos^2\theta\sum_{n=0}^{\infty}{}'\gamma^2$$

for $\theta = 0$ and $\theta = \pi/2$. For the latter, $\cos^2\theta = 0$; for the former, we have $-\frac{kT}{8\pi l^2}\mathrm{L}^2 2\cos^2\theta\sum_{n=0}^{\prime\infty}\gamma^2 = -\frac{kT}{4\pi l^2}\mathrm{L}^2 10^{-2}$ for $-\frac{kT}{4\pi l^2}\mathrm{L}^2 10^{-2} = -kT$ or $\frac{\mathrm{L}^2}{l^2} = 4\pi 10^{+2}$, $\mathrm{L} = 20\pi^{1/2}l \approx 35l$. For $l \sim 100$ Å, $\mathrm{L} \sim 0.35$ μm.

Problem L1.11: Neglecting retardation, show how the interaction between two coated bodies, Eq. (L1.29),

$$G(l;\mathrm{a}_1,\mathrm{b}_1) = -\frac{A_{\mathrm{A_1m/B_1m}}(l)}{12\pi l^2} - \frac{A_{\mathrm{A_1m/BB_1}}(l+\mathrm{b}_1)}{12\pi(l+\mathrm{b}_1)^2} - \frac{A_{\mathrm{AA_1/B_1m}}(l+\mathrm{a}_1)}{12\pi(l+\mathrm{a}_1)^2}$$
$$-\frac{A_{\mathrm{AA_1/BB_1}}(l+\mathrm{a}_1+\mathrm{b}_1)}{12\pi(l+\mathrm{a}_1+\mathrm{b}_1)^2}$$

can be converted into the interaction between two parallel slabs, Eq. (L1.30),

$$G(l;\mathrm{a}_1,\mathrm{b}_1) = -\frac{A_{\mathrm{A_1m/B_1m}}}{12\pi}\left[\frac{1}{l^2} - \frac{1}{(l+\mathrm{b}_1)^2} - \frac{1}{(l+\mathrm{a}_1)^2} + \frac{1}{(l+\mathrm{a}_1+\mathrm{b}_1)^2}\right].$$

Solution: Set A = m = B. Because

$$\overline{\Delta}_{\mathrm{A_1m}} = \frac{\varepsilon_{\mathrm{A_1}} - \varepsilon_{\mathrm{m}}}{\varepsilon_{\mathrm{A_1}} + \varepsilon_{\mathrm{m}}} = -\overline{\Delta}_{\mathrm{mA_1}}, \qquad \overline{\Delta}_{\mathrm{B_1m}} = \frac{\varepsilon_{\mathrm{B_1}} - \varepsilon_{\mathrm{m}}}{\varepsilon_{\mathrm{B_1}} + \varepsilon_{\mathrm{m}}} = -\overline{\Delta}_{\mathrm{mB_1}}$$

we can convert coefficients

$$A_{\mathrm{AA_1/B_1m}}(l+\mathrm{a}_1) \to A_{\mathrm{mA_1/B_1m}}(l+\mathrm{a}_1) = -A_{\mathrm{A_1m/B_1m}}(l+\mathrm{a}_1),$$
$$A_{\mathrm{A_1m/BB_1}}(l+\mathrm{b}_1) \to A_{\mathrm{A_1m/mB_1}}(l+\mathrm{b}_1) = -A_{\mathrm{A_1m/B_1m}}(l+\mathrm{b}_1),$$
$$A_{\mathrm{AA_1/BB_1}}(l+\mathrm{a}_1+\mathrm{b}_1) \to A_{\mathrm{mA_1/mB_1}}(l+\mathrm{a}_1+\mathrm{b}_1) = +A_{\mathrm{A_1m/B_1m}}(l+\mathrm{a}_1+\mathrm{b}_1).$$

Problem L1.12: Show how the nonretarded interaction between slabs goes from inverse-square to inverse-fourth-power variation.

Solution: Let $\eta = \mathrm{h}/w$. Expand

$$\left[1 - \frac{2w^2}{(w+\mathrm{h})^2} + \frac{w^2}{(w+2\mathrm{h})^2}\right] = 1 - \frac{2}{(1+\eta)^2} + \frac{1}{(1+2\eta)^2}$$
$$= 1 - \frac{2}{1+2\eta+\eta^2} + \frac{1}{1+4\eta+4\eta^2}$$

$$\approx 1 - 2\left[1 - (2\eta + \eta^2) + (2\eta + \eta^2)^2\right]$$
$$+ \left[1 - (4\eta + 4\eta^2) + (4\eta + 4\eta^2)^2\right] \approx 6\eta^2 = 6\left(\frac{h}{w}\right)^2,$$

so that

$$G(w; h) = -\frac{A_{HW/HW}}{12\pi w^2}\left[1 - \frac{2w^2}{(w+h)^2} + \frac{w^2}{(w+2h)^2}\right] = -\frac{A_{HW/HW} h^2}{2\pi w^4}.$$

Problem L1.13: Working in the nonretarded limit and in the limit of close approach $l \ll R_1, R_2$, compare the free energy of interaction *per unit area* of planar facing surfaces, $G_{1m/2m}(l)$, with the free energy *per interaction*, that is, the integral, $G_{ss}(l; R_1, R_2)$ of $F_{ss}(l; R_1, R_2)$. In particular, show that

$$G_{ss}(l; R_1, R_2) = G_{1m/2m}(l)\frac{2\pi R_1 R_2 l}{(R_1 + R_2)}.$$

It is as though the energy per interaction between spheres is the energy per area between planes of the same materials but multiplied by a continuously varying area $2\pi R_1 R_2 l/(R_1 + R_2)$ that goes to zero as the spheres are brought into contact.

Solution: Free energy is the integral of force, $G_{ss}(l; R_1, R_2) = -\int_{\infty}^{l} F_{ss}(l; R_1, R_2)dl$. From Eq. (L1.36), $F_{ss}(l; R_1, R_2) = [(2\pi R_1 R_2)/(R_1 + R_2)]G_{pp}(l)$, set $G_{pp}(l) = G_{1m/2m}(l) = \{-[A_{1m/2m}/(12\pi l^2)]\}$, and integrate with respect to l:

$$G_{ss}(l; R_1, R_2) = \left(-\frac{A_{1m/2m}}{12\pi}\right)\frac{2\pi R_1}{[1 + (R_1/R_2)]l} = G_{1m/2m}(l)\frac{2\pi R_1 l}{[1 + (R_1/R_2)]}.$$

Problem L1.14: Obtain Eq. (L1.42) for a sphere–plane interaction from Eq. (L1.40) for a sphere–sphere interaction.

Solution: Taking $z = R_1 + R_2 + l$, $z^2 - (R_1 + R_2)^2 = 2l(R_1 + R_2) + l^2$, $z^2 - (R_1 - R_2)^2 = 4R_1R_2 + 2l(R_1 + R_2) + l^2$. With $R_1 \to \infty$, $R_2 = R$, ignore terms that don't have a factor R_1 and ignore the difference between R_1 and $R_1 + R_2$:

$$\frac{R_1R_2}{z^2 - (R_1 + R_2)^2} = \frac{R_1R_2}{2l(R_1 + R_2) + l^2} \to \frac{R}{2l},$$

$$\frac{R_1R_2}{z^2 - (R_1 - R_2)^2} = \frac{R_1R_2}{4R_1R_2 + 2l(R_1 + R_2) + l^2} \to \frac{R}{4R + 2l},$$

$$\frac{z^2 - (R_1 + R_2)^2}{z^2 - (R_1 - R_2)^2} = \frac{2l(R_1 + R_2) + l^2}{4R_1R_2 + 2l(R_1 + R_2) + l^2} \to \frac{2l}{4R + 2l} = \frac{l}{2R + l}.$$

Problem L1.15: From Eq. (L1.40) obtain Eq. (L1.43) for sphere–sphere interactions in the limit of close approach, $l \ll R_1$ and R_2.

Solution: As in the previous problem, expand, but ignore all terms that do not diverge as $l \to 0$:

$$\frac{R_1R_2}{z^2 - (R_1 + R_2)^2} = \frac{R_1R_2}{2l(R_1 + R_2) + l^2} \to \frac{R_1R_2}{2l(R_1 + R_2)},$$

$$\frac{R_1R_2}{z^2 - (R_1 - R_2)^2} = \frac{R_1R_2}{4R_1R_2 + 2l(R_1 + R_2) + l^2} \to \frac{1}{4},$$

$$\ln \frac{z^2 - (R_1 + R_2)^2}{z^2 - (R_1 - R_2)^2} = \ln \frac{2l(R_1 + R_2) + l^2}{4R_1R_2 + 2l(R_1 + R_2) + l^2} \to \ln \frac{2l(R_1 + R_2)}{4R_1R_2} \to \ln l.$$

The close-approach limit is dominated by the $1/l$ term:

$$G_{ss}(z; R_1, R_2) \to -\frac{A_{1m/2m}}{3} \frac{R_1R_2}{2l(R_1 + R_2)} = -\frac{A_{1m/2m}}{6} \frac{R_1R_2}{(R_1 + R_2)l}.$$

Problem L1.16: Show that for $\tau = 1/1.05 \times 10^{11}$ radians/second (Table L2.1 in Level 2, Subsection L2.4.D), $\xi_{n=1}\tau \gg 1$ at room temperature.

Solution: From Level 1, the table in the section on the frequency spectrum, $\xi_{n=1} = 2.411 \times 10^{14}$ rad/s, $\xi_{n=1}\tau = 2.3 \times 10^3 \gg 1$.

Problem L1.17: Show how Eqs. (L1.59) emerge from the equation of Table S.6.a.

Solution: Set $\varepsilon_m = 1$, and $\alpha = \beta$ in Table S.6.a:

$$g_{ab}(r) = -\frac{6kT}{r^6} \sum_{n=0}^{\infty}{}' \left[\frac{\alpha(i\xi_n)\,\beta(i\xi_n)}{(4\pi\varepsilon_m(i\xi_n))^2} \right] \left(1 + r_n + \frac{5}{12}r_n^2 + \frac{1}{12}r_n^3 + \frac{1}{48}r_n^4\right) e^{-r_n};$$

then use Eqs. (L1.56):

$$\frac{\alpha(i\xi)}{4\pi} = \frac{\alpha_{ind}(i\xi)}{4\pi\varepsilon_0} \text{ (mks)}, \qquad \frac{\alpha(i\xi)}{4\pi} = \alpha_{ind}(i\xi) \text{ (cgs)}.$$

Problem L1.18: Derive Eqs. (L1.62) from Eqs. (L1.59).

Solution: Except for $r_{n=0} = 0$, at finite temperature all $r_n \to \infty$ when $z \to \infty$. Only the $n = 0$ term, with its factor of 1/2, endures in the summation. Ergo,

$$g_{London}(r \to \infty) = -\frac{3kT}{r^6} \left(\frac{\alpha_{ind}(0)}{4\pi\varepsilon_0} \right)^2 \text{ in mks}$$

$$g_{London}(r \to \infty) = -\frac{3kT}{r^6} \alpha_{ind}(0)^2 \text{ in cgs.}$$

Problem sets for Level 2

Problem L2.1: Show that this simple form of $\overline{\Delta}_{ji}$ and Δ_{ji} emerges immediately from assuming that the velocity of light is finite but everywhere equal.

Solution: Making believe that the velocity of light $c/\sqrt{\varepsilon_i\mu_i}$ is the same in all media is the same as saying that all $\varepsilon_i\mu_i$ are equal. Fed to $s_i = \sqrt{p^2 - 1 + (\varepsilon_i\mu_i/\varepsilon_m\mu_m)}$, this equality of $\varepsilon_i\mu_i$'s makes $s_i = p$, $\overline{\Delta}_{ji} = [(s_i\varepsilon_j - s_j\varepsilon_i)/(s_i\varepsilon_j + s_j\varepsilon_i)] \to [(\varepsilon_j - \varepsilon_i)/(\varepsilon_j + \varepsilon_i)]$, and $\Delta_{ji} = [(s_i\mu_j - s_j\mu_i)/(s_i\mu_j + s_j\mu_i)] \to [(\mu_j - \mu_i)/(\mu_j + \mu_i)]$. This is not the trivial result that trivial ease in derivation might suggest.

Problem L2.2: The limiting finite pressure in Eq. (L2.102) merits further consideration. Show that it comes (1) from the derivative of $G(l; a, b)$, Eq. (L2.99), in the $l \to 0$ limit and (2) from the integral for $P(l; a, b)$, Eq. (L2.101) in that same zero-l limit.

Solution:

1. The first term in {...} in Eq. (L2.99) goes to zero as $l^2 \ln l$ so that its derivative is zero as is the derivative of the constant, last term. We are left with the derivative of

$$\frac{1}{b^2}\ln\left(\frac{l+a+b}{l+a}\right)+\frac{1}{a^2}\ln\left(\frac{l+a+b}{l+b}\right)$$
$$=\left(\frac{1}{a^2}+\frac{1}{b^2}\right)\ln\left[(a+b)\left(1+\frac{l}{a+b}\right)\right]-\frac{1}{b^2}\ln\left[a\left(1+\frac{l}{a}\right)\right]-\frac{1}{a^2}\ln\left[b\left(1+\frac{l}{b}\right)\right]$$
$$\approx \text{constant}+\left[\left(\frac{1}{a^2}+\frac{1}{b^2}\right)\left(\frac{1}{a+b}\right)-\frac{1}{b^2a}-\frac{1}{ba^2}\right]l,$$

with the l derivative $-\frac{2}{ab(a+b)}$.

2. At small l the integral for pressure in Eq. (L2.101) becomes

$$\int_0^b\int_0^a \frac{z_b z_a}{(z_a+z_b)^3}dz_a dz_b = \int_0^b z_b dz_b \int_0^a \frac{z_a}{(z_a+z_b)^3}dz_a.$$

First, integrate over z_a:

$$\int_0^a \frac{z_a}{(z_a+z_b)^3}dz_a = \int_{z_b}^{a+z_b}\frac{q-z_b}{q^3}dq = -\frac{1}{a+z_b}+\frac{1}{2z_b}+\frac{z_b}{2(a+z_b)^2}.$$

Then integrate each of these three terms over z_b:

$$-\int_0^b \frac{z_b dz_b}{(a+z_b)} = -\int_a^{a+b}\frac{(u-a)du}{u} = -b+a\ln\left(\frac{a+b}{a}\right);+\int_0^b \frac{z_b dz_b}{2z_b} = \frac{b}{2};$$

$$\int_0^b z_b dz_b \frac{z_b}{2(a+z_b)^2} = \frac{1}{2}\int_a^{a+b}\frac{(u-a)^2}{u^2}du = \frac{1}{2}\int_a^{a+b}du - a\int_a^{a+b}\frac{du}{u}+\frac{a^2}{2}\int_a^{a+b}\frac{du}{u^2}$$
$$=\frac{b}{2}-a\ln\left(\frac{a+b}{a}\right)+\frac{a}{2}-\frac{a^2}{2(a+b)}.$$

Collect the three results:

$$-b+a\ln\left(\frac{a+b}{a}\right)+\frac{b}{2}+\frac{b}{2}-a\ln\left(\frac{a+b}{a}\right)+\frac{a}{2}-\frac{a^2}{2(a+b)}$$
$$=\frac{a}{2(a+b)}[(a+b)-a]=\frac{ab}{2(a+b)}.$$

Problem L2.3: Instead of the limiting form $\varepsilon \to 1+N\alpha$, use the Clausius–Mossoti expression $\varepsilon \approx [(1+2N\alpha/3)/(1-N\alpha/3)]$ [approximation (L2.135)] in expression (L2.138) for the interaction of two condensed gases across a vacuum $\varepsilon_m = 1$. Then, $\overline{\Delta}_{Am} = \overline{\Delta}_{Bm} = [(\varepsilon-1)/(\varepsilon+1)]$. Show that the result is a power series in density N in which the corrections to the $N^2\alpha^2$ leading term come in as successive factors $N\alpha/3$, and then $49N^2\alpha^2/288$.

Solution: It suffices to consider the form of $[(\varepsilon-1)/(\varepsilon+1)]^2$, where $[(\varepsilon-1)/(\varepsilon+1)] = [(N\alpha)/2][1/(1+N\alpha/6)]$ or

$$\left(\frac{\varepsilon-1}{\varepsilon+1}\right)^2=\left(\frac{N\alpha}{2+N\alpha/3}\right)^2=\left(\frac{N\alpha}{2}\right)^2\frac{1}{1+N\alpha/3+N^2\alpha^2/36}$$
$$=\left(\frac{N\alpha}{2}\right)^2\sum_{\nu=0}^{\infty}\left(\frac{N\alpha}{3}+\frac{N^2\alpha^2}{36}\right)^{\nu}$$

$$\approx \left(\frac{N\alpha}{2}\right)^2 \left(1 + \frac{N\alpha}{3} + \frac{N^2\alpha^2}{36} + \frac{N^2\alpha^2}{9} + \cdots\right)$$
$$= \left(\frac{N\alpha}{2}\right)^2 \left(1 + \frac{N\alpha}{3} + \frac{5N^2\alpha^2}{36} + \cdots\right)$$

for the $q = 1$ term in Table (P.1.a.3) and expression (L2.138) with successive correction factors $N\alpha/3$, then $5N^2\alpha^2/36$.

The $q = 2$ term contributes $\frac{1}{8}(\frac{\varepsilon-1}{\varepsilon+1})^4$ whose leading term in density is

$$\frac{1}{8}\left(\frac{N\alpha}{2}\right)^4 = \left(\frac{N\alpha}{2}\right)^2 \frac{N^2\alpha^2}{32}$$

for an additional correction factor $[(N^2\alpha^2)/32]$, so that the corrections come in as factors $N\alpha/3$, and then $49N^2\alpha^2/288$.

Problem L2.4: Show how thin-body formulae can often be derived either as expansions or as derivatives.

Solution: By expansion: Define $\eta \equiv \mathrm{b}/l \ll 1$, so that

$$E(l;\mathrm{b}) = -\frac{A_{\mathrm{Ham}}}{12\pi}\left[\frac{1}{l^2} - \frac{1}{(l+\mathrm{b})^2}\right] = -\frac{A_{\mathrm{Ham}}}{12\pi l^2}\left[1 - \frac{1}{(1+\eta)^2}\right];$$

to leading terms in η, $[\,] \approx 1 - \frac{1}{1+2\eta} \approx 1 - (1 - 2\eta) = 2\eta$ to yield $E(l;\mathrm{b}) \approx -[A_{\mathrm{Ham}}\mathrm{b}/(6\pi l^3)]$.

By differentiation: With $\mathrm{E}(l) = -[A_{\mathrm{Ham}}/(12\pi l^2)]$, take the difference in energy by means of a differential $-\mathrm{d}E(l) = E(l) - E(l + \mathrm{d}l) = -[\mathrm{d}E(l)/\mathrm{d}l]\mathrm{d}l = \{-[A_{\mathrm{Ham}}/(6\pi l^3)]\mathrm{d}l$ because of shifting the separation l by a relatively small amount $\mathrm{d}l = \mathrm{b}$.

Problem L2.5: Derive approximation (L2.145) by expansion of Eq. (L2.144) and by differentiation of $-[A_{\mathrm{Ham}}/(12\pi l^2)]$ for the interaction of half-spaces.

Solution: Expand

$$E(l;\mathrm{b}) = -\frac{A_{\mathrm{Ham}}}{12\pi}\left[\frac{1}{l^2} - \frac{2}{(l+\mathrm{b})^2} + \frac{1}{(l+2\mathrm{b})^2}\right] = -\frac{A_{\mathrm{Ham}}}{12\pi l^2}\left[1 - \frac{2}{(1+\eta)^2} + \frac{1}{(1+2\eta)^2}\right]$$

in small $\eta \equiv \mathrm{b}/l$:

$$[\,] \approx \frac{6\eta^2}{(1+\eta)^2(1+2\eta)^2} \approx 6\eta^2, \quad E(l;\mathrm{b}) \approx -\frac{A_{\mathrm{Ham}}\mathrm{b}^2}{2\pi l^4}.$$

Differentiate $-[A_{\mathrm{Ham}}/(12\pi l^2)]$ twice with respect to l and multiply by b^2.

Problem L2.6: Because of the number of ways they can be used elsewhere, it is worth exercising the manipulations used to extract Eqs. (L2.150) and (L2.151) from the general form for $G_{\mathrm{AmB}}(l, T)$.

1. Ignore differences in magnetic susceptibilities, feed $\varepsilon_{\mathrm{A}} = \varepsilon_{\mathrm{m}} + N_{\mathrm{A}}\alpha_{\mathrm{E}}$ and $\varepsilon_{\mathrm{B}} = \varepsilon_{\mathrm{m}} + N_{\mathrm{B}}\beta_{\mathrm{E}}$ to $s_{\mathrm{A}} = \sqrt{p^2 - 1 + (\varepsilon_{\mathrm{A}}/\varepsilon_{\mathrm{m}})}$, $s_{\mathrm{B}} = \sqrt{p^2 - 1 + (\varepsilon_{\mathrm{B}}/\varepsilon_{\mathrm{m}})}$; expand to lowest powers in number densities so as to verify approximations (L2.147).
2. Similarly, introduce approximations (L2.147) into $\overline{\Delta}_{\mathrm{Am}}, \overline{\Delta}_{\mathrm{Bm}}$ and $\Delta_{\mathrm{Am}}, \Delta_{\mathrm{Bm}}$ and expand in densities to verify Eq. (L2.148).

3. From here it is an easy trip, differentiating with respect to l and then integrating with respect to p so as to achieve Eqs. (L2.150) and (L2.151).

Solution:

1. $s_A = \sqrt{p^2 - 1 + (\varepsilon_A/\varepsilon_m)} = \sqrt{p^2 - 1 + [(\varepsilon_m + N_A\alpha)/\varepsilon_m]} \approx p(1 + \frac{1}{2}\frac{N_A\alpha}{p^2\varepsilon_m}) = p + \frac{N_A\alpha}{2p\varepsilon_m}$.

2. $\Delta_{Am} = (\frac{p-s_A}{p+s_A}) \approx (\frac{-N_A\alpha}{4\varepsilon_m})(\frac{1}{p^2})$ and $\Delta_{Bm} \approx (\frac{-N_B\beta}{4\varepsilon_m})(\frac{1}{p^2})$;
so that $\Delta_{Am}\Delta_{Bm} \approx N_A N_B[\frac{\alpha\beta}{(4\varepsilon_m)^2}](\frac{1}{p^4}) \ll 1$, and

$$\overline{\Delta}_{Am} = \left(\frac{p\varepsilon_A - s_A\varepsilon_m}{p\varepsilon_A + s_A\varepsilon_m}\right) \approx \left[\frac{N_A\alpha}{2\varepsilon_m} - \frac{N_A\alpha}{(2p)^2\varepsilon_m}\right] = \left(\frac{N_A\alpha}{4\varepsilon_m}\right)\frac{1}{p^2}(2p^2-1) \ll 1,$$

$$\overline{\Delta}_{Bm} \approx \left(\frac{N_B\beta}{4\varepsilon_m}\right)\frac{1}{p^2}(2p^2-1);\ \overline{\Delta}_{Am}\overline{\Delta}_{Bm} \approx N_A N_B\left(\frac{\alpha\beta}{(4\varepsilon_m)^2}\right)\frac{1}{p^4}(2p^2-1)^2 \ll 1.$$

Introduce these into the free-energy relation

$$G_{AmB}(l,T) \approx -\frac{kT}{2\pi c^2}\sum_{n=0}^{\infty}{}' \varepsilon_m\mu_m\xi_n^2\int_1^{\infty} p(\overline{\Delta}_{Am}\overline{\Delta}_{Bm} + \Delta_{Am}\Delta_{Bm})e^{-r_n p}\,dp,$$

where

$$(\overline{\Delta}_{Am}\overline{\Delta}_{Bm} + \Delta_{Am}\Delta_{Bm}) = N_A N_B\alpha\beta\left(\frac{1}{4\varepsilon_m}\right)^2\left(\frac{1}{p^4}\right)(4p^4 - 4p^2 + 2),$$

and $p = x/r_n$, which becomes Eq. (L2.148):

$$G_{AmB}(l,T) = -\frac{kT}{8\pi}N_A N_B\sum_{n=0}^{\infty}{}'\frac{\alpha\beta}{(4\varepsilon_m)^2}\left(2\varepsilon_m^{1/2}\mu_m^{1/2}\xi_n/c\right)^2$$
$$\times\int_1^{\infty}\frac{1}{p^3}(4p^4 - 4p^2 + 2)e^{-\left(2\varepsilon_m^{1/2}\mu_m^{1/2}\xi_n/c\right)pl}\,dp.$$

3. The third derivative of G with respect to l is the third derivative of

$$e^{-\left(2\varepsilon_m^{1/2}\mu_m^{1/2}\xi_n/c\right)pl}, \qquad -\left(2\varepsilon_m^{1/2}\mu_m^{1/2}\xi_n/c\right)^3 p^3 e^{-\left(2\varepsilon_m^{1/2}\mu_m^{1/2}\xi_n/c\right)pl} = -\frac{r_n^3}{l^3}p^3 e^{-r_n p}.$$

Taking the derivative cancels out the p^3; it is then much simpler to carry out the p integration

$$\int_1^{\infty}\{4p^4 - 4p^2 + 2\}e^{-r_n p}\,dp = 96\frac{e^{-r_n}}{r_n^5}\left(1 + r_n + \frac{5}{12}r_n^2 + \frac{1}{12}r_n^3 + \frac{1}{48}r_n^4\right).$$

[These underlying three integrals use the general form of $\alpha_n(z)$ functions $\alpha_n(z) \equiv \int_1^{\infty} t^n e^{-zt}\,dt = n!\,z^{-n-1}e^{-z}(1 + z + \frac{z^2}{2!} + \cdots + \frac{z^n}{n!})$ for all or any positive integer n. See, e.g., Eqs. 5.1.5 and 5.1.8 in Abramowitz and Stegun. M. Abramowitz and I. A. Stegun, *Handbook of Mathematical Functions, with Formulas, Graphs, and Mathematical Tables* (Dover, New York, 1965). Don't confuse the α and n in this definition of $\alpha_n(z)$, directly quoted from Abramowitz and Stegun, with the α and n in the force formulae!]

$$-G'''_{AmB}(l) = -\frac{kT}{8\pi}N_A N_B\sum_{n=0}^{\infty}{}'\frac{\alpha\beta}{(4\varepsilon_m)^2}\left(2\varepsilon_m^{1/2}\mu_m^{1/2}\xi_n/c\right)^5 96\frac{e^{-r_n}}{r_n^5}\left(1 + r_n + \frac{5}{12}r_n^2 + \frac{1}{12}r_n^3 + \frac{1}{48}r_n^4\right)$$
$$= -12\frac{kT}{\pi l^5}N_A N_B\sum_{n=0}^{\infty}{}'\frac{\alpha\beta}{(4\varepsilon_m)^2}e^{-r_n}\left[1 + r_n + \frac{5}{12}r_n^2 + \frac{1}{12}r_n^3 + \frac{1}{48}r_n^4\right],$$

where $(2\varepsilon_m^{1/2}\mu_m^{1/2}\xi_n/c)^5$ has been turned back into $(r_n^5)/(l^5)$.

Problem L2.7: Show that, in the highly idealized limit of zero temperature, the terms in the sum $\sum_{n=0}^{\prime\infty} e^{-r_n}(\cdots)$ [Eq. (L2.151) and expression (L2.153)] change so slowly with respect to index n that the sum can be approximated by an integral $\int_0^\infty (\;)e^{-r_n}\,dn$. In this limit, derive Eq. (L2.154) with its apparently-out-of-nowhere factor of 23.

Solution: With $r_n = (2l\varepsilon_m^{1/2}\mu_m^{1/2}\xi_n/c) = (2l\varepsilon_m^{1/2}\mu_m^{1/2}/c)(2\pi kT/\hbar)\,n$, the differential dn can be converted into $dr_n = (2l\varepsilon_m^{1/2}\mu_m^{1/2}/c)d\xi_n = (2l\varepsilon_m^{1/2}\mu_m^{1/2}/c)(2\pi kT/\hbar)\,dn$. Then the function in Eq. (L2.151) can be integrated as

$$\int_0^\infty \left(1 + r_n + \frac{5}{12}r_n^2 + \frac{1}{12}r_n^3 + \frac{1}{48}r_n^4\right)e^{-r_n}\,dr_n \Big/ \left[(2l\varepsilon_m^{1/2}\mu_m^{1/2}/c)\,(2\pi kT/\hbar)\right].$$

From $\int_0^\infty e^{-x}x^n\,dx = n!$, the integral over r_n becomes $1 + 1 + (10/12) + (6/12) + (24/48) = (23/6)$. The interaction $g_{\alpha\beta}(l)$ from expression (L2.153) then has the form

$$g_{\alpha\beta}(l) = -\left\{\frac{23}{6}\left(\frac{3kT}{8\pi^2 l^6}\right)\Big/\left[(2l\varepsilon_m^{1/2}\mu_m^{1/2}/c)\,(2\pi kT/\hbar)\right]\right\}\left[\frac{\alpha(0)\beta(0)}{\varepsilon_m^2(0)}\right]^2,$$

which becomes Eq. (L2.154).

Problem L2.8: Assuming the worst-case situation, a metallic sphere for which $\alpha = 4\pi a^3$, and using the center-to-center distance z between spheres as a measure of number density, N is one sphere per cubic volume z^3, show that the inequality condition $N\alpha \ll 3$ becomes $4\pi a^3 \ll 3z^3$.

For $z = 4a$, with a diameter's worth of separation between spheres, show that the inequality between $N\alpha$ and $(N\alpha)^2/3$ is a factor of $\sim 1/16$.

Solution: The ratio of terms is $[(N\alpha)^2/3]/N\alpha = N\alpha/3 = 4\pi a^3/3z^3$. For $z = 4a$, $4\pi a^3/3z^3 = 4\pi a^3/3(4a)^3 \sim (1/16)$.

Problem L2.9: Under the regime of weak fields such that $\mu_{\text{dipole}}E \ll kT$, show that the orientation polarization $\mu_{\text{dipole}}\cos\theta$, averaged over all angles and weighted by energies $\mu_{\text{dipole}}E\cos\theta$ in a Boltzmann distribution, is $(\mu_{\text{dipole}}^2/3kT)E$.

Solution: Multiply polarization $\mu_{\text{dipole}}\cos\theta$ at each angle by the Boltzmann factor $e^{+\mu_{\text{dipole}}E\cos\theta/kT}$ in energy $-\mu_{\text{dipole}}E\cos\theta$. For angles between θ and $\theta + d\theta$, weight by the solid angle $2\pi\sin\theta\,d\theta$.

Small E allows the expansion $e^{+\mu_{\text{dipole}}E\cos\theta/kT} \approx 1 + \mu_{\text{dipole}}E\cos\theta/kT$ so that the full average goes as

$$\frac{\int_0^\pi \mu_{\text{dipole}}\cos\theta\, e^{+\mu_{\text{dipole}}E\cos\theta/kT}\,2\pi\sin\theta\,d\theta}{\int_0^\pi e^{+\mu_{\text{dipole}}E\cos\theta/kT}\,2\pi\sin\theta\,d\theta}$$

$$\approx \frac{\mu_{\text{dipole}}^2}{kT}E\,\frac{\int_1^{-1}\cos^2\theta\,d(\cos\theta)}{\int_1^{-1}d(\cos\theta)} = \frac{\mu_{\text{dipole}}^2}{3kT}E.$$

Problem L2.10: Collecting Eqs. (L2.181), (L2.182), (L2.195), and (L2.196), expanding everything to lowest terms in particle number density, and by using Eqs. (L2.198) and (L2.199), derive Eqs. (L2.200), (L2.204), and (L2.206) for $g_{\text{D-D}}(l)$, $g_{\text{D-M}}(l)$, and $g_{\text{M-M}}(l)$, respectively.

Solution: Expansions yield

$$s \approx \left[p^2 + N\left(\frac{\Gamma_s}{n_m} - \frac{\alpha}{\varepsilon_m}\right)\right]^{\frac{1}{2}} \approx p + \frac{N}{2p}\left(\frac{\Gamma_s}{n_m} - \frac{\alpha}{\varepsilon_m}\right),\ \varepsilon_{susp} = \varepsilon_m\left(1 + N\frac{\alpha}{\varepsilon_m}\right),$$

$$\overline{\Delta}_{Sm} = \left(\frac{s\varepsilon_{susp} - p\varepsilon_m}{s\varepsilon_{susp} + p\varepsilon_m}\right) \approx \frac{N}{2}\left[\left(1 - \frac{1}{2p^2}\right)\frac{\alpha}{\varepsilon_m} + \frac{1}{2p^2}\frac{\Gamma_s}{n_m}\right].$$

For the integrand in Eq. (L2.198), take

$$p^2\overline{\Delta}_{Sm} \approx \frac{N}{2}\left[\frac{\alpha}{\varepsilon_m}\left(p^2 - \frac{1}{2}\right) + \frac{\Gamma_s}{n_m}\frac{1}{2}\right],$$

$$p^4\overline{\Delta}_{Sm}^2 \approx \frac{N^2}{4}\left[\left(\frac{\alpha}{\varepsilon_m}\right)^2\left(p^4 - p^2 + \frac{1}{4}\right) + \left(\frac{\alpha}{\varepsilon_m}\right)\left(\frac{\Gamma_s}{n_m}\right)\left(p^2 - \frac{1}{2}\right) + \frac{1}{4}\left(\frac{\Gamma_s}{n_m}\right)^2\right].$$

The factor N^2 allows us to write Eq. (L2.198) as

$$-G'''_{SmS}(l) = 2\pi l N^2 g_p(l) \approx -\frac{2kT\kappa_m^5}{\pi}\frac{N^2}{4}\int_1^\infty \left[\frac{\alpha}{\varepsilon_m}\left(p^2 - \frac{1}{2}\right) + \frac{\Gamma_s}{n_m}\frac{1}{2}\right]^2 e^{-2\kappa_m lp}\,dp.$$

For the point–particle interaction,

$$g_p(l) = -\frac{kT\kappa_m^5}{4\pi^2 l}\int_1^\infty [\,]e^{-2\kappa_m lp}\,dp = g_{D-D}(l) + g_{D-M}(l) + g_{M-M}(l),$$

the individual terms are separate integrals:

$$g_{D-D}(l) \approx -\frac{kT\kappa_m^5}{4\pi^2 l}\left(\frac{\alpha}{\varepsilon_m}\right)^2\int_1^\infty\left(p^4 - p^2 + \frac{1}{4}\right)e^{-2\kappa_m lp}\,dp,$$

$$g_{D-M}(l) \approx -\frac{kT\kappa_m^5}{4\pi^2 l}\left(\frac{\alpha}{\varepsilon_m}\right)\left(\frac{\Gamma_s}{n_m}\right)\int_1^\infty\left(p^2 - \frac{1}{2}\right)e^{-2\kappa_m lp}\,dp,$$

$$g_{M-M}(l) \approx -\frac{kT\kappa_m^5}{4\pi^2 l}\frac{1}{4}\left(\frac{\Gamma_s}{n_m}\right)^2\int_1^\infty e^{-2\kappa_m lp}\,dp.$$

Use

$$\alpha_n(z) = \int_1^\infty p^n e^{-zp}\,dp = \frac{n!}{z^{n+1}}e^{-z}\left(1 + z + \frac{z^2}{2!} + \frac{z^3}{3!} + \frac{z^4}{4!} + \cdots + \frac{z^n}{n!}\right),$$

$z\alpha_n(z) = e^{-z} + n\alpha_{n-1}(z)$ (e.g., M. Abramowitz and I. A. Stegun, *Handbook of Mathematical Functions, with Formulas, Graphs, and Mathematical Tables* (Dover, New York, 1965) Chapter 5, Eqs. 5.1.5, 5.1.8, and 5.1.15) to integrate

$$\int_1^\infty\left(p^4 - p^2 + \frac{1}{4}\right)e^{-zp}\,dp = 24\frac{e^{-z}}{z^5}\left(1 + z + \frac{5}{12}z^2 + \frac{1}{12}z^3 + \frac{1}{96}z^4\right)$$

for Eq. (L2.200);

$$\int_1^\infty\left(p^2 - \frac{1}{2}\right)e^{-zp}\,dp = 2\frac{e^{-z}}{z^3}\left(1 + z + \frac{1}{4}z^2\right)$$

for Eq. (L2.204), and

$$\int_1^\infty e^{-zp}\,dp = \frac{e^{-z}}{z}$$

for Eq. (L2.206).

Problem L2.11: Take the l derivative required for Eq. (L2.211).

Solution: Because $r_n = (2l\varepsilon_m^{1/2}/c)\xi_n = (2\varepsilon_m^{1/2}\xi_n/c)l$, all l dependence of $G_{AmB}(l, T)$ occurs inside the p integrand:

$$G'_{AmB}(l) = \frac{NkT}{8\pi}\sum_{n=0}^{\infty}{}' \left[\frac{\beta(i\xi_n)}{4\varepsilon_m(i\xi_n)}\right](2\varepsilon_m^{1/2}\xi_n/c)^3 \int_1^{\infty} p^2 \left[\overline{\Delta}_{Am}\left(\frac{2p^2-1}{p^2}\right) - \Delta_{Am}\frac{1}{p^2}\right] e^{-(2\varepsilon_m^{1/2}\xi_n/c)pl}\,dp$$

$$= \frac{NkT}{8\pi l^3}\sum_{n=0}^{\infty}{}' \left[\frac{\beta(i\xi_n)}{4\varepsilon_m(i\xi_n)}\right] r_n^3 \int_1^{\infty} (\overline{\Delta}_{Am}(2p^2-1) - \Delta_{Am})e^{-r_n p}\,dp = N\,g_p(l).$$

Problem L2.12: Beginning with Eq. (L2.211), convert summation to integration for the zero-temperature limit of Eqs. (L2.217) and (L2.218).

Solution: Replace r_n with $(2l\varepsilon_m^{1/2}/c)\xi_n$, replace summation over n with integration over ξ so as to introduce a factor $\hbar/(2\pi kT)$, and take the susceptibilities ε_m and β to be independent of frequency. The general form from Eq. (L2.211),

$$-\frac{kT}{32\pi l^3}\sum_{n=0}^{\infty}{}' \left[\frac{\beta(i\xi_n)}{\varepsilon_m(i\xi_n)}\right] r_n^3 \int_1^{\infty} [\overline{\Delta}_{Am}(2p^2-1) - \Delta_{Am}]e^{-r_n p}\,dp,$$

becomes

$$-\frac{\hbar}{8\pi^2 c^3}\varepsilon_m^{3/2}\left(\frac{\beta}{\varepsilon_m}\right)\int_0^{\infty} d\xi\,\xi^3 \int_1^{\infty} [\overline{\Delta}_{Am}(2p^2-1) - \Delta_{Am}]e^{-(2l\varepsilon_m^{1/2}\xi/c)p}\,dp.$$

Dispose of the ξ integration,

$$\int_0^{\infty} \xi^3 e^{-(2l\varepsilon_m^{1/2}p/c)\xi}\,d\xi = (c/2l\varepsilon_m^{1/2}p)^4 \int_0^{\infty} x^3 e^{-x}\,dx = 6\,(c/2l\varepsilon_m^{1/2}p)^4,$$

to leave the p integration of Eqs. (L2.217) and (L2.218):

$$-\frac{\hbar}{8\pi^2 c^3}\varepsilon_m^{3/2}\left(\frac{\beta}{\varepsilon_m}\right)6\left(\frac{c}{2l\varepsilon_m^{1/2}}\right)^4 \int_1^{\infty} \{[\overline{\Delta}_{Am}(2p^2-1) - \Delta_{Am}]/p^4\}\,dp$$

$$= -\frac{3\hbar c}{8\pi l^4}\left(\frac{\beta/4\pi}{\varepsilon_m^{3/2}}\right)\frac{1}{2}\int_1^{\infty} \{[\overline{\Delta}_{Am}(2p^2-1) - \Delta_{Am}]/p^4\}\,dp.$$

Problem L2.13: Expanding Eq. (L2.218) for small differences in $\varepsilon_A \approx \varepsilon_m$, show how the 23/30 comes into approximation (L2.200).

Solution: Let $\varepsilon_A/\varepsilon_m = 1 + \eta$, where $|\eta| \ll 1$ and retain terms linear in η:

$$\left(\frac{\varepsilon_A - \varepsilon_m}{\varepsilon_A + \varepsilon_m}\right) \approx \frac{\eta}{2},\ s_A = \sqrt{p^2 - 1 + (\varepsilon_A/\varepsilon_m)} \to p + (\eta/2p),\ \Delta_{Am} = \frac{p - s_A}{p + s_A} \to -\frac{\eta}{4p^2},$$

$$\overline{\Delta}_{Am} = \frac{p\varepsilon_A - s_A\varepsilon_m}{p\varepsilon_A + s_A\varepsilon_m} \to -\frac{\eta(1-2p^2)}{4p^2};$$

$$\Theta(\varepsilon_A/\varepsilon_m) \equiv \frac{1}{2}\int_1^{\infty} \{[\overline{\Delta}_{Am}(2p^2-1) - \Delta_{Am}]/p^4\}\,dp$$

$$= +\frac{\eta}{8}\int_1^{\infty} [(2 - 4p^2 + 4p^4)/p^6]\,dp$$

$$= \frac{\eta}{8}\left(\frac{2}{5} - \frac{4}{3} + 4\right) \approx \left(\frac{\varepsilon_A - \varepsilon_m}{\varepsilon_A + \varepsilon_m}\right)\left(\frac{23}{30}\right).$$

Problem L2.14: Beginning with the leading, $j = 1$, term in Eq. (L2.225) for $G_{\mathrm{AmB}}(l, \theta)$ and introducing lowest-order terms for $\overline{\Delta}_{\mathrm{Am}}(\xi_n, \psi)$ and $\overline{\Delta}_{\mathrm{Bm}}(\xi_n, \theta, \psi)$ from Eq. (L2.228), derive Eq. (L2.229).

Solution: Take

$$G_{\mathrm{AmB}}(l,\theta) \approx -\frac{kT}{16\pi^2 l^2}\sum_{n=0}^{\infty}{}' \int_0^{2\pi} \overline{\Delta}_{\mathrm{Am}}(\xi_n,\psi)\overline{\Delta}_{\mathrm{Bm}}(\xi_n,\theta,\psi)\mathrm{d}\psi$$

$$\to -\frac{kT}{16\pi^2 l^2}v^2\sum_{n=0}^{\infty}{}' \int_0^{2\pi}\left[\overline{\Delta}_{\perp} + \frac{(\overline{\Delta}_{\|} - 2\overline{\Delta}_{\perp})}{4}\cos^2(-\psi)\right]$$

$$\times\left[\overline{\Delta}_{\perp} + \frac{(\Delta_{\|} - 2\overline{\Delta}_{\perp})}{4}\cos^2(\theta-\psi)\right]\mathrm{d}\psi$$

$$= -\frac{kT}{16\pi^2 l^2}v^2\sum_{n=0}^{\infty}{}' \int_0^{2\pi}\{\}\,\mathrm{d}\psi,$$

where

$$\{\} = \left\{\overline{\Delta}_{\perp}^2 + \overline{\Delta}_{\perp}\frac{(\overline{\Delta}_{\|} - 2\overline{\Delta}_{\perp})}{4}[\cos^2(-\psi) + \cos^2(\theta-\psi)] + \left(\frac{\overline{\Delta}_{\|} - 2\overline{\Delta}_{\perp}}{4}\right)^2 \cos^2(-\psi)\cos^2(\theta-\psi)\right\}.$$

Use

$$\int_0^{2\pi}\cos^2(-\psi)\cos^2(\theta-\psi)\mathrm{d}\psi = 2\pi\left(\frac{1}{8} + \frac{\cos^2\theta}{4}\right),$$

so that

$$G_{\mathrm{AmB}}(l,\theta) \approx -\frac{kT}{8\pi l^2}v^2\sum_{n=0}^{\infty}{}'\left[\overline{\Delta}_{\perp}^2 + \frac{\overline{\Delta}_{\perp}}{4}(\overline{\Delta}_{\|} - 2\overline{\Delta}_{\perp}) + \left(\frac{\overline{\Delta}_{\|} - 2\overline{\Delta}_{\perp}}{4}\right)^2\left(\frac{1}{8} + \frac{\cos^2\theta}{4}\right)\right]$$

$$= -\frac{kT}{8\pi l^2}v^2\sum_{n=0}^{\infty}{}'\left[\overline{\Delta}_{\perp}^2 + \frac{\overline{\Delta}_{\perp}}{4}(\overline{\Delta}_{\|} - 2\overline{\Delta}_{\perp}) + \frac{2\cos^2\theta + 1}{2^7}(\overline{\Delta}_{\|} - 2\overline{\Delta}_{\perp})^2\right].$$

Problem L2.15: From Eq. (L2.229), putting Eq. (L2.230) into the form of an Abel transform $h(l) = \int_{-\infty}^{\infty} g(l^2 + y^2)\,\mathrm{d}y$, use the inverse Abel transform

$$g(l) = -\frac{1}{\pi}\int_l^{\infty}\frac{h'(y)}{\sqrt{y^2 + l^2}}\mathrm{d}y$$

to derive Eq. (L2.233) for the attraction of parallel thin rods.

Solution: The second derivative of G from Eq. (L2.229) gives us a form $\mathrm{d}^2G/\mathrm{d}l^2 = -6N^2C(\theta)/l^4$. From this it is clear that the role of $h(y)$ or $h(l)$ is played by $h(y) = -6C(\theta)/y^4$ so that $h'(y) = 24C(\theta)/y^5$. The problem here is too simple to merit complicated integrations. For parallel rods, $\theta = 0$, simply test the form $g(l^2 + y^2) = -c_{\|}/(y^2 + l^2)^{5/2}$. Use

$$\int_{-\infty}^{\infty}\frac{\mathrm{d}y}{(y^2 + l^2)^{5/2}} = 2\int_0^{\infty}\frac{\mathrm{d}y}{(y^2 + l^2)^{5/2}} = \frac{4}{3}\frac{1}{l^4}$$

(e.g., Gradshteyn and Ryzhik, Eq. 3.252.3). I. S. Gradshteyn and I. M. Ryzhik, *Table of Integrals, Series, and Products* (Academic, New York, 1965).

Putting everything together, we obtain

$$\frac{d^2G_{AmB}(l,\theta=0)}{dl^2} \approx -\frac{6N^2C(\theta=0)}{l^4} = N^2\int_{-\infty}^{\infty} g(\sqrt{l^2+y^2},\theta=0)dy$$

$$= -N^2c_\parallel \int_{-\infty}^{\infty}\frac{dy}{(l^2+y^2)^{\frac{5}{2}}} = -N^2c_\parallel\frac{4}{3l^4},$$

with

$$C(\theta=0) = \frac{kT}{8\pi}(\pi a^2)^2 \sum_{n=0}^{\infty}{}' \left[\overline{\Delta}_\perp^2 + \frac{\overline{\Delta}_\perp}{4}(\overline{\Delta}_\parallel - 2\overline{\Delta}_\perp) + \frac{3}{2^7}(\overline{\Delta}_\parallel - 2\overline{\Delta}_\perp)^2\right],$$

extract

$$c_\parallel = \frac{9kT}{16\pi}(\pi a^2)^2 \sum_{n=0}^{\infty}{}' \left[\overline{\Delta}_\perp^2 + \frac{\overline{\Delta}_\perp}{4}(\overline{\Delta}_\parallel - 2\overline{\Delta}_\perp) + \frac{3}{2^7}(\overline{\Delta}_\parallel - 2\overline{\Delta}_\perp)^2\right].$$

Problem L2.16: Show that, if $\alpha(\omega)$ for an isolated particle has the form of a resonant oscillator, $\alpha(\omega) = [f_\alpha/(\omega_\alpha^2 - \omega^2 - i\omega\gamma_\alpha)]$, so does $\varepsilon(\omega)$ when we use the Lorentz–Lorenz transform $\varepsilon(\omega) = \{[1+2N\alpha(\omega)/3]/[1-N\alpha(\omega)/3]\}$ for particles at number density N. The strength of response from the total number of particles is preserved through replacing f_α with Nf_α; the resonance frequency ω_α^2 is shifted to $\omega_\alpha^2 - Nf_\alpha/3$; the width parameter γ_α remains the same.

Solution: For one oscillator, test the form $\varepsilon(\omega) = 1 + [f_\varepsilon/(\omega_\varepsilon^2 - \omega^2 - i\omega\gamma_\varepsilon)]$:

$$\varepsilon(\omega) = \frac{1+2N\alpha(\omega)/3}{1-N\alpha(\omega)/3} = \frac{(\omega_\alpha^2-\omega^2-Nf_\alpha/3-i\omega\gamma_\alpha)+Nf_\alpha}{(\omega_\alpha^2-\omega^2-Nf_\alpha/3-i\omega\gamma_\alpha)}$$

$$= 1 + \frac{Nf_\alpha}{(\omega_\alpha^2-Nf_\alpha/3)-\omega^2-i\omega\gamma_\alpha} = 1 + \frac{f_\varepsilon}{\omega_\varepsilon^2-\omega^2-i\omega\gamma_\varepsilon},$$

where $f_\varepsilon = Nf_\alpha$, $\omega_\varepsilon^2 = \omega_\alpha^2 - Nf_\alpha/3$, and $\gamma_\varepsilon = \gamma_\alpha$.

Problem L2.17: How dilute is dilute? Use $\varepsilon(\omega) = \{[1+2N\alpha(\omega)/3]/[1-N\alpha(\omega)/3]\}$ to show how deviation from dilute-gas pairwise additivity of energies creeps in with increasing number density N. Ignoring retardation, imagine two like nondilute gases with $\varepsilon_A = \varepsilon_B = \varepsilon = [(1+2N\alpha/3)/(1-N\alpha/3)]$ interacting across a vacuum $\varepsilon_m = 1$. Expand this $\varepsilon(\omega)$ beyond the linear term in N, feed the result to the difference-over-sum $\overline{\Delta}^2 = [(\varepsilon-1)/(\varepsilon+1)]^2$ (Table P.1.a.4) used to compute forces.

Apply the result to metal spheres of radius a, $\alpha/4\pi = a^3$ [Table S.7 and Eqs. (L2.166)–(L2.169)] occupying an average volume $(1/N) = (4\pi/3)\rho^3$ per particle. Show that, for an average distance $z \sim 2\rho$ between particle centers, the condition of diluteness becomes $z \gg 2a$.

Solution: For compactness, write in terms of $\eta = N\alpha$ so that

$$\varepsilon - 1 = \frac{\eta}{1-\eta/3}, \qquad \varepsilon + 1 = \frac{2+\eta/3}{1-\eta/3}, \qquad \overline{\Delta}^2 = \left(\frac{\varepsilon-1}{\varepsilon+1}\right)^2 = \left(\frac{\eta}{2+\eta/3}\right)^2$$

$$= \left(\frac{\eta}{2}\right)^2\frac{1}{(1+\eta/6)^2} \approx \left(\frac{\eta}{2}\right)^2(1-\eta/3+\cdots).$$

The result is a factor $(1 - N\alpha/3 + \cdots)$ on the leading term. The measure of small $N\alpha$ is then $N\alpha/3 \ll 1$. For $\alpha/4\pi = a^3$ and $(1/N) = (4\pi/3)\rho^3$, $N\alpha = 3(a^3/\rho^3)$, the measure of low density becomes $(a^3/\rho^3) \ll 1$.

Problem L2.18: Show that, in the regime of pairwise summability, the continuum limit is violated by terms of the order of $(a/z)^2$ where, just here, a is atomic spacing.

Solution: (by example): Consider a point particle a minimum distance z from a "rod" of similar point particles spaced at intervals **a** along straight line. To see the effect of noncontinuum structure in the rod, calculate the difference in the particle–rod interaction (1) when the isolated particle is at opposite one of the particles on the rod or (2) when the particle sits opposite the midpoint between two particles in the rod. That is, case (2) is case (1), but the rod is shifted over by a distance a/2.

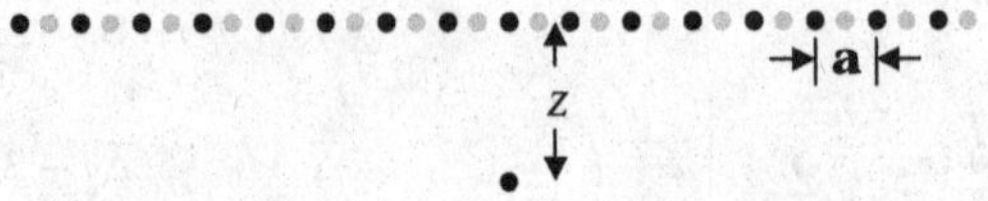

Index the positions of the particles along the rod such that their positions are at ja, where $-\infty \leq j \leq \infty$. Then the distance between the point particle and any particle on the rod is $r_j^2 = z^2 + j^2(a/2)^2$. It is easy to compare the energies in the two positions by comparing the interaction at distance r_j on one rod with the average of the interactions at displaced positions r_{j-1} and r_{j+1}, where $r_{j\pm1}^2 = z^2 + (j \pm 1)^2(a/2)^2 = r_j^2 + (1 \pm 2j)(a/2)^2$. Here we assume incremental interactions that go as $1/r^6$; use

$$\frac{1}{r_{j\pm1}^6} \approx \frac{1}{r_j^6}[1 - 3(1 \pm 2j)(a/2r_j)^2] = \frac{1}{r_j^6} - \frac{3}{r_j^6}\left(\frac{a}{2r_j}\right)^2 (1 \pm 2j)$$

to take a difference $\frac{1}{r_j^6} - \frac{1}{2}\left(\frac{1}{r_{j-1}^6} + \frac{1}{r_{j+1}^6}\right) = -\frac{6}{r_j^6}\left(\frac{a}{2r_j}\right)^2$.

From this difference we see that the deviation from effective inverse-sixth-power in summation is a factor that goes as $(a/r)^2$. To sum over all interactions, take the continuum limit as the basis of comparison. That is, integrate over $r^2 = z^2 + \rho^2$, as

$$\int_{-\infty}^{\infty} \frac{d\rho}{(z^2 + \rho^2)^{\nu/2}} = \frac{1}{z^{\nu-1}} \int_{-\infty}^{\infty} \frac{d(\rho/z)}{[1 + (\rho/z)^2]^{\nu/2}} \sim \frac{1}{z^{\nu-1}}$$

where $\nu = 6$ or 8. For $\nu = 6$ this gives the leading-term inverse-fifth-power interaction between a point and a rod (or the per-unit-length interaction between parallel thin rods, Table C.4.a). For $\nu = 8$ we have the correction term, a factor of the order of $(a/z)^2$.

Problem L2.19: When can the discrete-sampling frequency summation be replaced with an integral over an imaginary frequency? Show that the condition

$$\frac{I(\xi_{n+1}) - 2I(\xi_n) + I(\xi_{n-1})}{24I(\xi_n)} \ll 1$$

does the trick.

Solution: The summation is like a Simpson's rule, usually taken as an approximation for an integral. Think of evaluating discrete points at $\xi_{n=0} = 0, \xi_{n=1}, \xi_{n=2}, \xi_{n=3}, \ldots$, along a frequency axis for which the width of the abscissa about each point is $(2\pi kT)/\hbar$, extending by $(\pi kT)/\hbar$ each way except next to $\xi_0 = 0$, at which it extends only to the right. This summation over steps can be replaced with an integral only if the integrand $I(\xi)$ changes slowly over this range from $\xi_n - [(\pi kT)/\hbar]$ to $\xi_n + [(\pi kT)/\hbar]$. How slowly?

Expand $I(\xi)$ about ξ_n:

$$I(\xi) = I(\xi_n) + \left.\frac{\partial I(\xi)}{\partial \xi}\right|_{\xi_n}(\xi - \xi_n) + \left.\frac{\partial^2 I(\xi)}{\partial \xi^2}\right|_{\xi_n}\frac{(\xi - \xi_n)^2}{2} + \left.\frac{\partial^3 I(\xi)}{\partial \xi^3}\right|_{\xi_n}\frac{(\xi - \xi_n)^3}{6} + \cdots .$$

Integrate, seeing that odd powers yield zero,

$$\int_{\xi_n - \frac{\pi kT}{\hbar}}^{\xi_n + \frac{\pi kT}{\hbar}} I(\xi)\mathrm{d}\xi = I(\xi_n)\frac{2\pi kT}{\hbar} + 0 + \left.\frac{\partial^2 I(\xi)}{\partial \xi^2}\right|_{\xi_n}\left(\frac{\pi kT}{\hbar}\right)^3\frac{1}{3} + 0 + \cdots$$

$$= I(\xi_n)\frac{2\pi kT}{\hbar} + \left.\frac{\partial^2 I(\xi)}{\partial \xi^2}\right|_{\xi_n}\left(\frac{2\pi kT}{\hbar}\right)^3\frac{1}{24} + \cdots .$$

The relative deviation of this integrated version from the summation term $I(\xi_n)[(2\pi kT)/\hbar]$ over that same frequency range is

$$\left.\frac{[\partial^2 I(\xi)/\partial \xi^2]}{24 I(\xi_n)}\right|_{\xi_n}\left(\frac{2\pi kT}{\hbar}\right)^2 .$$

For a first estimate of $[\partial^2 I(\xi)/\partial \xi^2]|_{\xi_n}$, use the $I(\xi_n)$ functions themselves. Take $[\partial I(\xi)/\partial \xi]|_{\xi_n} \approx [I(\xi_{n+1}) - I(\xi_n)]/(2\pi kT/\hbar)$. Then

$$[\partial^2 I(\xi)/\partial \xi^2]|_{\xi_n} \approx (\{[I(\xi_{n+1}) - I(\xi_n)]/(2\pi kT/\hbar)\} - \{[I(\xi_n) - I(\xi_{n-1})]/(2\pi kT/\hbar)\})/(2\pi kT/\hbar)$$

$$= [I(\xi_{n+1}) - 2I(\xi_n) + I(\xi_{n-1})]/(2\pi kT/\hbar)^2,$$

so that the condition for a small difference becomes

$$\frac{I(\xi_{n+1}) - 2I(\xi_n) + I(\xi_{n-1})}{24 I(\xi_n)} \ll 1.$$

Typically the changes in ε's occur on a scale proportional to absorption frequency. Thus, when the difference in sampling frequencies is less than the absorption frequencies that are important in changing I's (or ε's), then the condition will be met. At room temperature, this means for n's $\gtrsim 10$.

NOTES

Prelude

1. *Entropic*: to do with disorder or uncertainty from multiple possibilities.
2. *Steric*: to do with three-dimensional structure, solidity; spatial arrangement of atoms in molecules.
3. From B. V. Derjaguin, "P. N. Lebedev's ideas on the nature of molecular forces," Sov. Phys. Usp., **10**, 108–11 (1967). The article traces the development of Lebedev's observations that led to the great work of H. B. G. Casimir and E. M. Lifshitz, who in turn made explicit the connection between polarizability and charge-fluctuation forces.
4. An excellent history of van der Waals theory and measurement is given in B. V. Derjaguin, N. V. Churaev, and V. M. Miller, *Surface Forces*, V. I. Kissin, trans., J. A. Kitchener, ed. (Consultants Bureau, Plenum, New York, London, 1987). A more succinct but instructive summary is given in J. Mahanty and B. W. Ninham, *Dispersion Forces* (Academic, London, New York, San Francisco, 1976). Summaries of more recent work can be found in J. N. Israelachvili, *Intermolecular and Surface Forces*, 2nd ed. (Academic, London (1992); L. Spruch, "Long-range (Casimir) interactions," Science, **272**, 1452–5 (1996), and M. Kardar and R. Golestanian, "The 'friction' of vacuum, and other fluctuation-induced forces," Rev. Mod. Phys., **71**, 1233–45 (1999). One measure of the breadth of the subject can be seen in the slight overlap between topics covered in different histories.

   The van der Waals interaction story is an excellent subject for scientific historians. Think of the elements:

   - Starts and stops in thinking: Lebedev's insight missed for decades, enduring fear of non-physicists to use easier, modern theory.
   - Personal relations: Derjaguin, Lebedev's stepson; Casimir, Verwey's son-in-law; van Kampen, the uncle of 't Hooft; Derjaguin, Lifshitz et al., working within close distance in a tightly controlled society.
   - Social scene: Nazis in Holland, Stalin's Russia.
   - Disjunction among disciplines: Physicists with their "Casimir effect," chemical engineers and physical chemists with their "DLVO theory" and terror in many of tackling abstruse physics, lack of interest by most parties in each others' motivating questions. (DLVO = Derjaguin–Landau–Verwey–Overbeek.)

5. H. C. Hamaker, "The London–van der Waals attraction between spherical particles," Physica, **4**, 1058–72 (1937).
6. For an affectionate appreciation of Hamaker and his work, see K. J. Mysels and P. C. Scholten, "H. C. Hamaker, more than a constant," Langmuir **7**(1): 209–11 (1991).
7. The Derjaguin–Landau contribution was published as an elegant sketch during the early years of World War II. B. Derjaguin and L. Landau, "Theory of the stability of strongly charged lyophobic sols and of the adhesion of strongly charged particles in solution of electrolytes,"

Acta Physicochim., URSS, **14**, 633–62 (1941). See also, Current Contents, 32: p. 20, August 10, 1987 for a brief reminiscence by B. Derjaguin.

8. The classic 1948 Verwey–Overbeek text is well worth studying even today. E. J. W. Verwey and J. Th. G. Overbeek, *Theory of the Stability of Lyophobic Colloids* (Dover, Mineola, NY, 1999; originally published by Elsevier, New York, 1948). In 1967 Verwey told me that their studies were done in secret while Nazi soldiers controlled the Philips Laboratories where he and Overbeek pretended to do assigned work. Because they could publish nothing during the war, the world was eventually blessed with a coherent monograph that has defined much of colloid research ever since. This text is especially valuable for its sensitive, systematic treatment of electrostatic double layers.
9. H. B. G. Casimir, "On the attraction between two perfectly conducting plates," Proc. Nederl. Akad. Wetensch., **B51**, 793–5 (1948).
10. H. B. G. Casimir, pp. 3–7 in "The Casimir Effect 50 Years Later," *Proceedings of the Fourth Workshop on Quantum Field Theory Under the Influence of External Conditions*, Michael Bordag, ed. (World Scientific, Singapore, 1999).
11. See D. Kleppner, "With apologies to Casimir" Phys. Today, **43**, 9–10 (October 1990), for a commentary on the Casimir derivations in the context of modern vacuum electrodynamics.
12. See, e.g., Section 44 in L. D. Landau and E. M. Lifshitz, *Quantum Mechanics (Non-Relativistic Theory)*, Vol. 3 of Course of Theoretical Physics Series, 3rd ed. (Pergamon, Oxford, 1991).
13. For a review see M. Kardar and R. Golestanian, "The 'friction' of vacuum, and other fluctuation-induced forces," Rev. Mod. Physics, **71**, 1233–45 (1999).
14. H. Wennerstrom, J. Daicic, and B. W. Ninham, "Temperature dependence of atom–atom interactions," Phys. Rev. A, **60**, 2581–4 (1999).
15. H. B. G. Casimir and D. Polder, "The influence of retardation on the London–van der Waals forces," Phys. Rev., **73**, 360–71 (1948).
16. Casimir's 1998 recollections open a collection of instructive essays in "The Casimir effect 50 years later," in *Proceedings of the Fourth Workshop on Quantum Field Theory Under the Influence of External Conditions*, Michael Bordag, ed. (World Scientific, Singapore, 1999). Another excellent recent text by K. A. Milton, *The Casimir Effect: Physical Manifestations of Zero-Point Energy* (World Scientific, Singapore, 2001), also gives a good idea of the huge amount of important physics that flowed from Casimir's early insight.
17. The original derivations are in E. M. Lifshitz, Dokl. Akad. Nauk. SSSR, **97**, 643 (1954); **100**, 879 (1955); "The theory of molecular attractive forces between solids," Sov. Phys., **2**, 73–83 (1956) [Zh. Eksp. Teor. Fiz., **29**, 94 (1955)]; the best source for study in the context of fluctuation theory is in Chapter VIII of E. M. Lifshitz and L. P. Pitaevskii, "Statistical physics," Part 2 in Vol. 9 of the Course of Theoretical Physics Series, L. D. Landau and E. M. Lifshitz, eds., Vol. 9, (Pergamon, New York, 1991).
18. The interaction of real metal plates is in fact far more complicated than what is derived assuming ideal infinite conductance. See B. W. Ninham and J. Daicic, "Lifshitz theory of Casimir forces at finite temperature," Phys. Rev. A, **57**, 1870–80 (1998), for an instructive essay that includes the effects of finite temperature, finite conductance, and electron-plasma properties. The nub of the matter is that the Casimir result is strictly correct only at zero temperature.
19. For a review see B. V. Derjaguin, "The force between molecules," Sci. Am., **203**, 47–53 (1960) and B. V. Derjaguin, I. I. Abrikosova, and E. M. Lifshitz, "Direct measurement of molecular attraction between solids separated by a narrow gap," Q. Rev. (London), **10**, 295–329 (1956). N. V. Churaev, "Boris Derjaguin, dedication," Adv. Colloid Interface Sci., **104**, ix–xiii (2003) summarizes a productive life.
20. The *Discussions of the Faraday Society*, Vol. 18 (1954), shows how bad it could get, scientifically and personally.
21. I. E. Dzyaloshinskii, E. M. Lifshitz, and L. P. Pitaevskii, "The general theory of van der Waals forces," Adv. Phys., **10**, 165 (1961).
22. B. V. Derjaguin, "Untersuchungen über die Reibung und Adhäsion, IV," Kolloid-Z., **69**, 155–64 (1934).

23. J. Blocki, J. Randrup, W. J. Swiatecki, and C. F. Tsang, "Proximity forces," Ann. Phys., **105**, 427–62 (1977).
24. B. M. Axilrod and E. Teller, "Interaction of the van der Waals type between three atoms," J. Chem. Phys., **11**, 299–300 (1943); see also the pedagogical article by C. Farina, F. C. Santos, and A. C. Tort, "A simple way of understanding the non-additivity of van der Waals dispersion forces," Am. J. Phys., **67**, 344–9 (1999) for the step from two-body to three-body interactions.
25. S. M. Gatica, M. M. Calbi, M. W. Cole, and D. Velegol, "Three-body interactions involving clusters and films," Phys. Rev., **68**, 205409 (1–8 November 2003).
26. V. A. Parsegian, "Long range van der Waals forces," in *Physical Chemistry: Enriching Topics From Colloid and Interface Science*, H. van Olphen and K. J. Mysels, eds. IUPAC I.6, Colloid and Surface Chemistry (Theorex, La Jolla, CA, 1975), pp. 27–73.
27. R. H. French, "Origins and applications of London dispersion forces and Hamaker constants in ceramics," J. Am. Ceram. Soc., **83**, 2117–46 (2000).
28. V. A. Parsegian and S. L. Brenner, "The role of long range forces in ordered arrays of tobacco mosaic virus," Nature (London), **259**, 632–5 (1976).
29. C. M. Roth, B. L. Neal, and A. M. Lenhoff, "Van der Waals interactions involving proteins," Biophys. J., **70**, 977–87 (1996).
30. L. Bergstrom, "Hamaker constants of inorganic materials," Adv. Colloid Interface Sci., **70**, 125–69 (1997), together with a brief tutorial on computation, contains a useful collection of interaction coefficients in the nonretarded limit.
31. H. D. Ackler, R. H. French, and Y.-M. Chiang, "Comparisons of Hamaker constants for ceramic systems with intervening vacuum or water: From force layers and physical properties," J. Colloid Interface Sci., **179**, 460–9 (1996).
32. R. R. Dagastine, D. C. Prieve, and L. R. White, "The dielectric function for water and its application to van der Waals forces," J. Colloid Interface Sci., **231**, 351–8 (2000).
33. V. A. Parsegian and G. H. Weiss, "Spectroscopic parameters for computation of van der Waals forces," J. Colloid Interface Sci., **81**, 285–9 (1981).
34. A. Shih and V. A. Parsegian, "Van der Waals forces between heavy alkali atoms and gold surfaces: Comparison of measured and predicted values," Phys. Rev. A, **12**, 835–41 (1975). See also several antecedent papers cited in this paper. With measured coefficient $K = 7 \times 10^{-36}$ ergs cm^3 and range of separations $5 \times 10^{-6} < r < 8 \times 10^{-6}$ cm, the magnitude of interaction energy is $5.6 \times 10^{-20} > K/r^3 > 1.4 \times 10^{-20}$ ergs $= 1.4 \times 10^{-27}$ J (vs. $kT_{\text{room}} \sim 4.1 \times 10^{-14}$ ergs $= 4.1 \times 10^{-21}$ J).
35. Interactions between crossed cylinders of mica in air, uncoated or coated with fatty acid monolayers, are described in J. N. Israelachvili and D. Tabor, "The measurement of van der Waals dispersion forces in the range 1.5 to 130 nm," Proc. R. Soc. London Ser. A, **331**, 19–38 (1972). An excellent review of this and related work is given in J. N. Israelachvili and D. Tabor, *Van der Waals Forces: Theory and Experiment*, Vol. 7 of Progress in Surface and Membrane Science Series (Academic Press, New York and London, 1973). Later reconciliation of theory and experiment required taking note of cylinder radius; L. R. White, J. N. Israelachvili, and B. W. Ninham, "Dispersion interaction of crossed mica cylinders: A reanalysis of the Israelachvili–Tabor experiments," J. Chem. Soc. Faraday Trans. 1, **72**, 2526–36 (1976).

    For measurements between crossed mica cylinders coated with phospholipid bilayers in water, see J. Marra and J. Israelachvili, "Direct measurements of forces between phosphatidylcholine and phosphatidylethanolamine bilayers in aqueous electrolyte solutions," Biochemistry, **24**, 4608–18 (1985). Interpretation in terms of expressions for layered structures and the connection to direct measurements between bilayers in water is given in V. A. Parsegian, "Reconciliation of van der Waals force measurements between phosphatidylcholine bilayers in water and between bilayer-coated mica surfaces," Langmuir, **9**, 3625–8 (1993). The bilayer–bilayer interactions are reported in E. A. Evans and M. Metcalfe, "Free energy potential for aggregation of giant, neutral lipid bilayer vesicles by van der Waals attraction," Biophys. J., **46**, 423–6 (1984).

36. See N. Mishchuk, J. Ralston, and D. Fornasiero, "Influence of dissolved gas on van der Waals forces between bubbles and particles," J. Phys. Chem. A, **106**, 689–96 (2002) for more fun with bubbles.
37. E. S. Sabisky and C. H. Anderson, "Verification of the Lifshitz theory of the van der Waals potential using liquid-helium films," Phys. Rev. A, **7**, 790–806 (1973).

    The story gets better. C. H. Anderson and E. S. Sabisky, "The absence of a solid layer of helium on alkaline earth fluoride substrates," J. Low Temp. Phys., **3**, 235–8 (1970), reported the thickness of helium liquid condensed from vapor onto ceramic substrates. Van der Waals attraction nicely explains film thickness vs. the chemical potential of helium in the vapor.
38. See, e.g., A. Muerk, P. F. Luckham, and L. Bergstrom, "Direct measurement of repulsive and attractive van der Waals forces between inorganic materials," Langmuir, **13**, 3896–9 (1997) and S.-W. Lee and W. M. Sigmund, "AFM study of repulsive van der Waals forces between Teflon AF^{TM} thin film and silica or alumina," Colloids Surf. A, **204**, 43–50 (2002), as well as references therein.
39. A good description of early measurements is in the book "*Surface Forces*" by Derjaguin et al. (see note 4). Measurements made with the crossed mica cylinders of a "surface force apparatus" are reviewed by J. N. Israelachvili, *Intermolecular and Surface Forces* (Academic, New York, 1992).
40. See, for example, A. M. Marvin and F. Toigo, "Van der Waals interaction between a point particle and a metallic surface. II. Applications," Phys. Rev. A, **25**, 803–15 (1982).
41. For example, S. Bhattacharjee, C.-H. Ko, and M. Elimelech, "DLVO interaction between rough surfaces," Langmuir, **14**, 3365–75 (1998).
42. V. B. Bezerra, G. L. Klimchitskaya, and C. Romero, "Surface impedance and the Casimir force," Phys. Rev. A, **65**, 012111–1 to 9 (2001), with references to several instructive texts such as V. M. Mostepanenko and N. N. Trunov, *The Casimir Effect and Its Applications* (Clarendon, Oxford, 1997) and P. W. Milonni, *The Quantum Vacuum* (Academic, San Diego, CA, 1994).
43. E. Elizalde and A. Romeo, "Essentials of the Casimir effect and its computation," Am. J. Phys., **59**, 711–19 (1991).
44. A. Ajdari, B. Duplantier, D. Hone, L. Peliti, and J. Prost, "Pseudo-Casimir effect in liquid-crystals" J. Phys. (Paris) II, **2**, 487–501 (1992).
45. See, e.g., T. G. Leighton, *The Acoustic Bubble* (Academic, San Diego, London, 1994), pp. 356–66.
46. C. I. Sukenik, M. G. Boshier, D. Cho, V. Sandoghdar, and E. A. Hinds, "Measurement of the Casimir–Polder force," Phys. Rev. Lett., **70**, 560–3 (1993).
47. S. K. Lamoreaux, "Demonstration of the Casimir force in the .6 to 6 μm range," Phys. Rev. Lett., **78**, 5–8 (1997); and "Erratum," Phys. Rev. Lett., **81**, 5475–6 (1998).
48. H. B. Chan, V. A. Aksyuk, R. N. Kleiman, D. J. Bishop, and F. Capasso, "Quantum mechanical actuation of microelectromechanical systems by the Casimir force," Science, **291**, 1941–4 (2001); "Nonlinear micromechanical Casimir oscillator," Phys. Rev. Lett., **87**, 211801 (2001).
49. F. Chen, U, Mohideen, G. L. Klimchitskaya, and V. M. Mostepanenko "Demonstration of the lateral Casimir force," Phys. Rev. Lett., **88**, 101801 (2002).
50. C. Argento and R. H. French, "Parametric tip model and force–distance relation for Hamaker constant determination from atomic force microscopy," J. Appl. Phys., **80**, 6081–90 (1996).
51. S. Eichenlaub, C. Chan, and S. P. Beaudoin, "Hamaker constants in integrated circuit metalization," J. Colloid Interface Sci., **248**, 389–97 (2002).
52. B. V. Derjaguin, I. I. Abrikosova, and E. M. Lifshitz, "Direct measurement of molecular attraction between solids separated by a narrow gap," Q. Rev. (London), **10**, 295–329 (1956); see also W. Arnold, S. Hunklinger, and K. Dransfeld, "Influence of optical absorption on the van der Waals interaction between solids," Phys. Rev. B, **19**, 6049–56 (1979), for more recent measurements with glasses.
53. E. A. Evans and W. Rawicz, "Entropy-driven tension and bending elasticity in condensed-fluid membranes," Phys. Rev. Lett., **64**, 2094–7 (1990); E. A. Evans, "Entropy-driven tension in vesicle membranes and unbinding of adherent vesicles," Langmuir, **7**, 1900–8 (1991).
54. L. J. Lis, M. McAlister, N. Fuller, R. P. Rand, and V. A. Parsegian, "Interactions between neutral phospholipids bilayer membranes," Biophys. J., **37**, 657–66 (1982).

55. J. Marra and J. N. Israelachvili, "Direct measurements of forces between phosphatidylcholine and phosphatidylethanolamine bilayers in aqueous electrolyte solutions," Biochemistry, **24**, 4608–18 (1985).
56. V. A. Parsegian, "Reconciliation of van der Waals force measurements between phosphatidylcholine bilayer in water and between bilayer-coated mica surfaces," Langmuir, **9**, 3625–8 (1993).
57. D. Gingell and J. A. Fornes, "Demonstration of intermolecular forces in cell adhesion using a new electrochemical technique," Nature (London), **256**, 210–11 (1975); D. Gingell and I. Todd, "Red blood cell adhesion. II. Interferometric examination of the interaction with hydrocarbon oil and glass," J. Cell Sci., **41**, 135–49 (1980).
58. See D. C. Prieve, "Measurement of colloidal forces with TIRM," Adv. Colloid Interface Sci., **82**, 93–125 (1999), for a clear description of technique as well as references; also S. G. Bike, "Measuring colloidal forces using evanescent wave scattering," Curr. Opin. in Colloid Interface Sci., **5**, 144–50 (2000).
59. Aerosols are treated in an expectably large literature. Only in a relatively few papers is the modern theory of van der Waals forces correctly used to analyze stability. For the first steps in using the modern theory, see W. H. Marlow, "Lifshitz—van der Waals forces in aerosol particle collisions. I. Introduction: Water droplets," J. Chem. Phys., **73**, 6288–95 (1980), and W. H. Marlow, "Size effects in aerosol particle interactions: The van der Waals potential and collision rates," Surf. Sci., **106**, 529–37 (1981); then the later work, V. Arunachalam, R. R. Lucchese, and W. H. Marlow, "Development of a picture of the van der Waals interaction energy between clusters of nanometer-range particles," Phys. Rev. E, **58**, 3451–7 (1998) and "Simulations of aerosol aggregation including long-range interactions," Phys. Rev. E, **60**, 2051–64 (1999).
60. There is a large and instructive literature on van der Waals forces in ceramics. The excellent review by R. H. French, "Origins and applications of London dispersion forces and Hamaker constants in ceramics," J. Am. Ceram. Soc., **83**, 2117–46 (2000), presents this work in the larger context of colloid and interface science. It is particularly useful for the explanations of how spectral data are gathered and converted to a form used in computation. The article by H. D. Ackler, R. H. French, and Y.-M. Chiang, "Comparisons of Hamaker constants for ceramic systems with intervening vacuum or water: From force laws and physical properties," J. Colloid Interface Sci., **179**, 460–9 (1996), gives many examples of interaction coefficients.
61. C. Eberlein, "Sonoluminescence as quantum vacuum radiation," Phys. Rev. Lett., **76**, 3842–5 (1996).
62. L. A. Crum, "Sonoluminescence," Phys. Today, **47**, 22–9 (September 1994). See also Section 5.2, T. G. Leighton, note 45.
63. See, e.g., Milton and Borlag texts listed in note 16.
64. D. Lohse, B. Schmitz, and M. Versluis, "Snapping shrimp make flashing bubbles," Nature (London), **413**, 477–8 (2001).
65. K. Autumn, W.-P. Chang, R. Fearing, T. Hsieh, T. Kenny, L. Liang, W. Zesch, and R. J. Full, "Adhesive force of a single gecko foot-hair," Nature (London), **405**, 681–5 (2000); K. Autumn, M. Sitti, Y. A. Liang, An. M. Peattie, W. R. Hansen, S. Sponberg, T. W. Kenny, R. Fearing, J. N. Israelachvili, and R. J. Full, "Evidence for van der Waals adhesion in gecko setai," Proc. Natl. Acad. Sci. USA, **99**, 12252–6 (2002).
66. A. K. Geim, S. V. Dubonos, I. V. Grigorieva, K. S, Novoselov, A. A. Zhukov, and S. Yu. Shapoval, "Microfabricated adhesive mimicking gecko foot-hair," *Nature Materials*, Vol. 2, pp. 461–3. 1 June 2003 doi:10:1038/nmat917.
67. M. Elbaum and M. Schick, "Application of the theory of dispersion forces to the surface melting of ice," Phys. Rev. Lett., **66**, 1713–16 (1991); relativistic retardation suppresses charge fluctuations at the higher frequencies at which $\varepsilon_{\text{ice}}(i\xi) > \varepsilon_{\text{water}}(i\xi)$. L. A. Wilen. J. S. Wettlaufer, M. Elbaum, and M. Schick, "Dispersion-force effects in interfacial premelting of ice," Phys. Rev. B, **52**, 12426–33 (1995); R. Bar-Ziv and S. A. Safran, "Surface melting of ice induced by hydrocarbon films," Langmuir, **9**, 2786–8 (1993).
68. P. Richmond, B. W. Ninham, and R. H. Ottewill, "A theoretical study of hydrocarbon adsorption on water surfaces using Lifshitz theory," J. Colloid Interface Sci., **45**, 69–80 (1973). See also I. M. Tidswell, T. A. Rabedeau, P. S. Pershan, and S. D. Kosowsky, "Complete wetting of

a rough surface: An x-ray study," Phys. Rev. Lett., **66**, 2108–11 (1991); E. Cheng and M. W. Cole, "Retardation and many-body effects in multilayer-film adsorption," Phys. Rev. B, **38**, 987–95 (1988); and M. O. Robbins, D. Andelman, and J.-F. Joanny, "Thin liquid films on rough or heterogeneous solids," Phys. Rev. A, **43**, 4344–54 (1991), for more recent descriptions and copious references to related work.

69. Actually, the situation may be worse than described. Not only must we worry about the nonadditivity of $1/r^6$ van der Waals energies acting in condensed systems, but there is good evidence that the $1/r^6$ form itself is inaccurate even for the interaction of two isolated particles when they are at the separations they would find themselves in in a condensed medium. See, e.g., T. C. Choy, "Van der Waals interaction of the hydrogen molecule: An exact implicit energy density functional," Phys. Rev. A, **62**, 012506 (2000); it appears that $1/r^6$, $1/r^8$, and $1/r^{10}$ terms comparably contribute to the energies at internuclear separations typical for condensed media. Such expansions were elaborated long ago; e.g., H. Margenau, "Van der Waals forces," Rev. Mod. Phys., **11**, 1–35 (1939). Interatomic potentials used in computer simulations are usually fit to carefully computed pairwise potentials that are transformed to practically convenient assumed forms for attractive and repulsive components, e.g., T. A. Halgren, "Representation of van der Waals (vdW) interactions in molecular mechanics force fields: Potential form, combination rules, and vdW parameters," J. Am. Chem. Soc., **114**, 7827–43 (1992). If we probe further into what happens at the short distances characteristic of condensed media, it turns out that "first-order perturbation" interactions cannot be neglected. See, e.g., footnote p. 341, L. D. Landau and E. M. Lifshitz, *Quantum Mechanics (Non-Relativistic Theory)*, Vol. 3 of Course of Theoretical Physics Series, 3rd ed. (Pergamon, Oxford, 1991) and K. Cahill and V. A. Parsegian "Rydberg-London potential for diatomic molecules and unbonded atom pairs," J. Chem. Phys. **121**, 10839–10842 (2004).
70. J. Mahanty and B. W. Ninham, *Dispersion Forces* (Academic, London, 1976), Chap. 4.
71. Two excellent texts, unfortunately out of print, explicitly on van der Waals forces from the point of view elaborated in the present book: J. Mahanty and B. W. Ninham, note 70, and D. Langbein, *Van der Waals Attraction* (Springer-Verlag, Berlin, 1974). There are several recent "Casimir Effect" texts, some of which have already been mentioned in specific connections: P. R. Berman, ed., *Cavity Quantum Electrodynamics* (Academic, Boston, 1994); "The Casimir Effect 50 Years Later," in *Proceedings of the Fourth Workshop on Quantum Field Theory Under the Influence of External Conditions*, Michael Bordag, ed. (World Scientific, Singapore, 1999); J. Feinberg, A. Mann, and M. Revzen, "Casimir effect: The classical limit," Ann. Phys., **288**, 103–36 (2001); M. Krech, *The Casimir Effect in Critical Systems* (World Scientific, Singapore, 1994); F. S. Levin and D. A. Micha, eds., *Long-Range Casimir Forces: Theory and Recent Experiments on Atomic Systems* (Plenum, New York, 1993); P. W. Milonni, *The Quantum Vacuum* (Academic, San Diego, CA, 1994); K. A. Milton, listed in note 16; V. M. Mostepanenko and N. N. Trunov, *The Casimir Effect and its Applications* (Clarendon, Oxford, 1997); and B. E. Sernelius, *Surface Modes in Physics* (Wiley, New York, 2001).

Level 1, Introduction

1. See, e.g., J. M. Seddon and J. D. Gale, *Thermodynamics and Statistical Mechanics* (The Royal Society of Chemistry, London 2001), for friendly background reading.
2. It is unfortunate that this macroscopic-continuum limitation is sometimes forgotten in overzealous application. The same limitation also holds in the theory of the electrostatic double layers for which we often make believe that the medium is a featureless continuum. Neglect of structure in double layers is equally risky, though, and even more common than in the computation of van der Waals forces.

   Creating spatially varying dielectric susceptibilities and solving the charge-fluctuation equations with these more detailed structures sometimes circumvents the macroscopic-continuum limitation.
3. O. Kenneth, I. Klich, A. Mann, and M. Revzen, "Repulsive Casimir forces," Phys. Rev. Lett., **89**, 033001 (2002); O. Kenneth and S. Nussinov, "Small object limit of the Casimir effect and the sign of the Casimir force," Phys. Rev. D, **65**, 095014 (2002).

4. V. A. Parsegian, see Prelude note 35.
5. See the seminal paper by B. W. Ninham and V. Yaminsky, "Ion binding and ion specificity: The Hofmeister effect and Onsager and Lifshitz theories," Langmuir, **13**, 2097–108 (1997), for the connection between solute interaction and van der Waals forces from the perspective of macroscopic continuum theory.
6. See J. E. Kiefer, V. A. Parsegian, and G. H. Weiss, "Model for van der Waals attraction between spherical particles with nonuniform adsorbed polymer," J. Colloid Interface Sci., **51**, 543–6 (1975), for numerical examples. See D. Prieve "Measurement of colloidal forces with TIRM," Adv. Colloid Interface Sci., **82**, 93–125 (1999), for a modern review of colloid measurements.
7. The point is prettily made in Fig. 2 of M. M. Calbi, S. M. Gatica, D. Velegol, and M. W. Cole, "Retarded and nonretarded van der Waals interactions between a cluster and a second cluster or a conducting surface," Phys. Rev. A, **67**, 033201 (2003). See also Level 2, Subsection L2.3.E, in this text.
8. See, e.g., G. D. Fasman, ed., *Handbook of Biochemistry and Molecular Biology* (CRC Press, Boca Raton, FL, 1975), pp. 372–82; W. J. Fredricks, M. C. Hammonds, S. B. Howard, and F. Rosenberger, "Density, thermal expansivity, viscosity and refractive index of lysozyme solutions at crystal growth concentrations," J. Cryst. Growth, **141**, 183–192, Tables 5 and 6 (1994).
9. For example, V. A. Parsegian, *Digest of Literature on Dielectrics* (National Academy of Sciences, USA, Washington, DC, 1970), Chap. 10.
10. C. M. Roth, B. L. Neal, and Am. M. Lenhof, "Van der Waals interactions involving proteins," Biophys. J., **70**, 977–87 (1996), in a more careful computation, give 3.1 kT_{room}.
11. Recall that a dipole moment is the product of the distance d between two charges $+q$ and $-q$ and the magnitude of these charges, $\mu_D = qd$. In cgs units, two electronic charges of magnitude 4.803×10^{-10} esu at a separation of 1 Å $= 10^{-8}$ cm would have a moment of 4.803×10^{-10} esu $\times\ 10^{-8}$ cm $= 4.803 \times 10^{18}$ esu cm = 4.803 Debye units. That is, 1 Debye unit $= 10^{-18}$ esu cm. [See Table S.8 and Eqs. (L2.170)–(L2.171) in Level 2, the subsection on "Atoms or molecules in a dilute gas."]
12. This is actually very rough thinking, perhaps excused only by the safety factor of 100 in the different characteristic times. The response of a collection of unbound ionic charges is like that in a metal. At low frequencies, it does depend on the viscous drag on individual ions, the same drag that shows up in the diffusion constant. However, as described in Level 2, the high-frequency limiting response of a collection of negligibly bound charges of mass m, charge e and number density n has a form $\varepsilon(\omega) = 1 - (\omega_p^2/\omega^2)$ or $\varepsilon(i\xi) = 1 + (\omega_p^2/\xi^2)$ or $\varepsilon(i\xi) = 1 + (\omega_p^2/\xi^2)$, where $\omega_p^2 = [(ne^2)/(\varepsilon_0 m)](\text{mks}) = [(4\pi ne^2)/m](\text{cgs})$ for yanking the massive particles back and forth. At a 1-M univalent ion density, $n = [(0.602 \times 10^{24})/(10^{-3}\text{m}^3)] = [(0.602 \times 10^{24})/(10^3\text{cm}^3)]$, of particles with an atomic mass 10, $m = [(10^{-2}\text{ kg})/(0.602 \times 10^{24})] = [10\text{ g}/(0.602 \times 10^{24})]$, and charge $e = 1.609 \times 10^{-19}$ C, $e = 4.803 \times 10^{-10}$ sc, the "plasma frequency" $\omega_P = 10.2 \times 10^{12} \approx 10^{13}$rad/s $\approx 1.6 \times 10^{12}$ Hz is much smaller than the first nonzero sampling frequency, $[(2\pi kT)/\hbar] = 2.41 \times 10^{14}$radians/s $= 3.84 \times 10^{13}$ Hz. Even at this molar concentration, mobile charges contribute negligibly to the dielectric response at first nonzero frequency: $\varepsilon(\omega) = 1 - (\omega_p^2/\omega^2) = 1 - [(10.2 \times 10^{12})^2/(2.41 \times 10^{14})^2] = 1 - 0.00179$ or $\varepsilon(i\xi_1) = 1 + (\omega_p^2/\xi_1^2) = 1.00179$.
13. J. G. Kirkwood and J. B. Shumaker, "Forces between protein molecules in solution arising from fluctuations in proton charge and configuration," Proc. Natl. Acad. Sci. USA, **38**, 863–71 (1952).

Level 2, Formulae

1. B. V. Derjaguin, Kolloid-Z., **69**, 155–64 (1934).
2. H. C. Hamaker, "The London–van der Waals attraction between spherical particles," Physica, **4**, 1058–72 (1937).
3. H. B. G. Casimir and D. Polder, "The influence of retardation on the London–van der Waals forces," Phys. Rev., **73**, 360–71 (1948).

4. For example, L. D. Landau and E. M. Lifshitz, *Electrodynamics of Continuous Media* (Pergamon, Oxford, 1984), Chap. 2, Section 9, p. 44, Eq. 9.7.
5. See, e.g., Chap. III, Section 5, in J. C. Slater and N. H. Frank, *Electromagnetism* (Dover, New York, 1947).
6. See e.g., Chap. IX, Section 3, of J. C. Slater and N. H. Frank, *Electromagnetism* (Dover, New York, 1947) and Section 2.3 of M. Born and E. Wolf, *Principles of Optics* (Pergamon, Oxford, 1970).
7. P. J. W. Debye, *Polar Molecules* (Dover, New York, reprint of 1929 edition) presents the fundamental theory with stunning clarity. See also, e.g., H. Fröhlich, "Theory of dielectrics: Dielectric constant and dielectric loss," in *Monographs on the Physics and Chemistry of Materials Series*, 2nd ed. (Clarendon, Oxford University Press, Oxford, June 1987). Here I have taken the zero-frequency response and multiplied it by the frequency dependence of the simplest dipolar relaxation. I have also put $\omega = i\xi$ and taken the sign to follow the convention for poles consistent with the form of derivation of the general Lifshitz formula. This last detail is of no practical importance because in the summation $\sum'$ over frequencies ξ_n only the first, $n = 0$, term counts. The relaxation time τ is such that permanent-dipole response is dead by ξ_1 anyway. The permanent-dipole response is derived in many standard texts.
8. As before convert the sum $\sum'$ into an integral over the index n, then convert variable n into variable frequency $\xi = (2\pi kT/\hbar)\, n$. The coefficient $(2\pi kT/\hbar)$ changes the coefficient of the interaction energy from kT units to $\hbar$ units.
9. $$\int_1^\infty (2p^2-1)e^{-r_n p}\,\mathrm{d}p = 2\int_1^\infty p^2 e^{-r_n p}\,\mathrm{d}p - \int_1^\infty e^{-r_n p}\,\mathrm{d}p = 2\frac{2!}{r_n^3}\left(1 + r_n + \frac{r_n^2}{2!}\right)e^{-r_n} - \frac{e^{-r_n}}{r_n}.$$
10. Lord Rayleigh (J. W. Strutt), "On the influence of obstacles arranged in rectangular order upon the properties of a medium," Philosoph. Mag., **42**, 481–502 (1892).
11. Because $\overline{\Delta}_{\mathrm{Am}}(\psi)\overline{\Delta}_{\mathrm{Bm}}(\theta-\psi) \ll 1$,

$$G_{\mathrm{AmB}}(l,\theta) = \frac{kT}{8\pi^2}\int_0^{2\pi}\mathrm{d}\psi\int_0^\infty \ln[D(\rho,\psi,l,\theta)]\rho\,\mathrm{d}\rho$$
$$\approx -\frac{kT}{8\pi^2}\int_0^{2\pi}\mathrm{d}\psi\int_0^\infty \overline{\Delta}_{\mathrm{Am}}(\psi,\rho)\overline{\Delta}_{\mathrm{Bm}}(\theta-\psi,\rho)e^{-2\sqrt{\rho^2+\kappa_{\mathrm{m}}^2}\,l}\rho\,\mathrm{d}\rho.$$

From the Level 3 derivation, Eqs. (L3.234) and (L3.239) specialized to the present case,

$$\overline{\Delta}_{\mathrm{Am}}(\psi,\rho) = \frac{\varepsilon_\perp\sqrt{\rho^2\left[1+\left(\frac{\varepsilon_\parallel-\varepsilon_\perp}{\varepsilon_\perp}\right)\cos^2\psi\right]+\kappa_{\mathrm{A}}^2} - \varepsilon_{\mathrm{m}}\sqrt{\rho^2+\kappa_{\mathrm{m}}^2}}{\varepsilon_\perp\sqrt{\rho^2\left[1+\left(\frac{\varepsilon_\parallel-\varepsilon_\perp}{\varepsilon_\perp}\right)\cos^2\psi\right]+\kappa_{\mathrm{A}}^2} + \varepsilon_{\mathrm{m}}\sqrt{\rho^2+\kappa_{\mathrm{m}}^2}},$$

$$\overline{\Delta}_{\mathrm{Bm}}(\theta-\psi,\rho) = \frac{\varepsilon_\perp\sqrt{\rho^2\left[1+\left(\frac{\varepsilon_\parallel-\varepsilon_\perp}{\varepsilon_\perp}\right)\cos^2(\theta-\psi)\right]+\kappa_{\mathrm{B}}^2} - \varepsilon_{\mathrm{m}}\sqrt{\rho^2+\kappa_{\mathrm{m}}^2}}{\varepsilon_\perp\sqrt{\rho^2\left[1+\left(\frac{\varepsilon_\parallel-\varepsilon_\perp}{\varepsilon_\perp}\right)\cos^2(\theta-\psi)\right]+\kappa_{\mathrm{B}}^2} + \varepsilon_{\mathrm{m}}\sqrt{\rho^2+\kappa_{\mathrm{m}}^2}}.$$

For $2v\overline{\Delta}_\parallel \ll 1$, $v\overline{\Delta}_\parallel \ll 1$, $\left(\frac{\varepsilon_\parallel-\varepsilon_\perp}{\varepsilon_\perp}\right) \approx v(\overline{\Delta}_\parallel - 2\overline{\Delta}_\perp) = N\pi a^2(\overline{\Delta}_\parallel - 2\overline{\Delta}_\perp) \ll 1$. For $N\Gamma_{\mathrm{c}} \ll n_{\mathrm{m}}$, $\kappa_{\mathrm{A}}^2 = \kappa_{\mathrm{B}}^2 \approx \kappa_{\mathrm{m}}^2[1 + N(\Gamma_{\mathrm{c}}/n_{\mathrm{m}} - 2\pi a^2\overline{\Delta}_\perp)]$. Then, by tedious expansion,

$$\varepsilon_\perp\sqrt{\rho^2\left[1+\left(\frac{\varepsilon_\parallel-\varepsilon_\perp}{\varepsilon_\perp}\right)\cos^2\psi\right]+\kappa_{\mathrm{A}}^2} \to \left(\varepsilon_{\mathrm{m}} + N\pi a^2\varepsilon_{\mathrm{m}}2\overline{\Delta}_\perp\right)$$
$$\times\sqrt{\rho^2\left[1+N\pi a^2\left(\overline{\Delta}_\parallel - 2\overline{\Delta}_\perp\right)\cos^2\psi\right] + \kappa_{\mathrm{m}}^2\left[1+N\left(\Gamma_{\mathrm{c}}/n_{\mathrm{m}} - 2\pi a^2\overline{\Delta}_\perp\right)\right]} \approx \varepsilon_{\mathrm{m}}\sqrt{\left(\rho^2+\kappa_{\mathrm{m}}^2\right)}$$
$$\times\left\{1 + N\left[2\pi a^2\overline{\Delta}_\perp + \frac{\rho^2\pi a^2\left(\overline{\Delta}_\parallel - 2\overline{\Delta}_\perp\right)\cos^2\psi + \kappa_{\mathrm{m}}^2\left(\Gamma_{\mathrm{c}}/n_{\mathrm{m}} - 2\pi a^2\overline{\Delta}_\perp\right)}{2\left(\rho^2+\kappa_{\mathrm{m}}^2\right)}\right]\right\},$$

so that

$$\overline{\Delta}_{\mathrm{Am}}(\psi,\rho) \approx N\left[\pi a^2\overline{\Delta}_{\perp} + \frac{\rho^2\pi a^2\left(\overline{\Delta}_{\|} - 2\overline{\Delta}_{\perp}\right)\cos^2\psi + \kappa_{\mathrm{m}}^2\left(\Gamma_{\mathrm{c}}/n_{\mathrm{m}} - 2\pi a^2\overline{\Delta}_{\perp}\right)}{4\left(\rho^2+\kappa_{\mathrm{m}}^2\right)}\right],$$

$$\overline{\Delta}_{\mathrm{Bm}}(\theta-\psi,\rho) \approx N\left[\pi a^2\overline{\Delta}_{\perp} + \frac{\rho^2\pi a^2\left(\overline{\Delta}_{\|} - 2\overline{\Delta}_{\perp}\right)\cos^2(\theta-\psi) + \kappa_{\mathrm{m}}^2\left(\Gamma_{\mathrm{c}}/n_{\mathrm{m}} - 2\pi a^2\overline{\Delta}_{\perp}\right)}{4\left(\rho^2+\kappa_{\mathrm{m}}^2\right)}\right].$$

In the limit at which κ_{m} goes to zero, these functions reduce to the correct form for dipolar fluctuations only. It is the product of these times $e^{-2\sqrt{\rho^2+\kappa_{\mathrm{m}}^2}\,l}$ that must be integrated over $d\psi$ and $\rho d\rho$ to obtain the interaction of the two arrays A and B. We know in advance that it is the second and third derivatives with respect to l that give us at-an-angle and parallel rod–rod interactions. The second derivative creates a factor $4\left(\rho^2+\kappa_{\mathrm{m}}^2\right)$.

For succinctness, temporarily define

$$C \equiv \pi a^2\overline{\Delta}_{\perp}, \quad D \equiv \pi a^2\left(\overline{\Delta}_{\|} - 2\overline{\Delta}_{\perp}\right),$$

$$K \equiv \kappa_{\mathrm{m}}^2\left(\Gamma_{\mathrm{c}}/n_{\mathrm{m}} - 2\pi a^2\overline{\Delta}_{\perp}\right) \quad \text{or} \quad K/\kappa_{\mathrm{m}}^2 \equiv \left(\Gamma_{\mathrm{c}}/n_{\mathrm{m}} - 2C\right),$$

so that

$$\overline{\Delta}_{\mathrm{Am}}(\psi) \approx N\left[C + \frac{\left(\rho^2 D\cos^2\psi + K\right)}{4\left(\rho^2+\kappa_{\mathrm{m}}^2\right)}\right],$$

$$\overline{\Delta}_{\mathrm{Bm}}(\theta-\psi) \approx N\left[C + \frac{\left(\rho^2 D\cos^2(\theta-\psi) + K\right)}{4\left(\rho^2+\kappa_{\mathrm{m}}^2\right)}\right],$$

or

$$\overline{\Delta}_{\mathrm{Am}}(\psi)\overline{\Delta}_{\mathrm{Bm}}(\theta-\psi) \approx N^2\left\{\left[C + \frac{K}{4\left(\rho^2+\kappa_{\mathrm{m}}^2\right)}\right] + \frac{\rho^2 D\cos^2(-\psi)}{4\left(\rho^2+\kappa_{\mathrm{m}}^2\right)}\right\}$$
$$\times\left\{\left[C + \frac{K}{4\left(\rho^2+\kappa_{\mathrm{m}}^2\right)}\right] + \frac{\rho^2 D\cos^2(\theta-\psi)}{4\left(\rho^2+\kappa_{\mathrm{m}}^2\right)}\right\}.$$

Its integral over ψ is

$$\int_0^{2\pi}\overline{\Delta}_{\mathrm{Am}}(\psi)\overline{\Delta}_{\mathrm{Bm}}(\theta-\psi)d\psi$$
$$= 2\pi N^2\left\{\left[C + \frac{K}{4\left(\rho^2+\kappa_{\mathrm{m}}^2\right)}\right]^2 + \left[C + \frac{K}{4\left(\rho^2+\kappa_{\mathrm{m}}^2\right)}\right]\left[\frac{\rho^2 D}{4\left(\rho^2+\kappa_{\mathrm{m}}^2\right)}\right]\right.$$
$$\left. + \left[\frac{\rho^2 D}{4\left(\rho^2+\kappa_{\mathrm{m}}^2\right)}\right]^2\left(\frac{1}{8} + \frac{\cos^2\theta}{4}\right)\right\}.$$

This product must be integrated over ρ as $\int_0^\infty \overline{\Delta}_{\mathrm{Am}}(\psi)\overline{\Delta}_{\mathrm{Bm}}(\theta-\psi)e^{-2\sqrt{\rho^2+\kappa_{\mathrm{m}}^2}\,l}\rho d\rho$. It can be simplified by changing the variable of integration to p such that

$$p^2 \equiv \left(\rho^2+\kappa_{\mathrm{m}}^2\right)/\kappa_{\mathrm{m}}^2$$

to create an integral of the form

$$G_{\mathrm{AmB}}(l,\theta) \approx -\frac{kT}{4\pi}N^2\kappa_{\mathrm{m}}^2\int_1^\infty f(p,\theta)e^{-2p\kappa_{\mathrm{m}}l}p\,dp$$

whose integrand $f(p,\theta)$ has terms

$$f(p,\theta) \equiv \left(C + \frac{K}{4p^2\kappa_{\mathrm{m}}^2}\right)^2 + \left(C + \frac{K}{4p^2\kappa_{\mathrm{m}}^2}\right)\left[\frac{(p^2-1)D}{4p^2}\right] + \left[\frac{(p^2-1)D}{4p^2}\right]^2\left(\frac{1}{8} + \frac{\cos^2\theta}{4}\right).$$

The required second derivative with respect to separation brings out another factor $(-2p\kappa_m)^2$:

$$\frac{d^2 G_{AmB}(l,0)}{dl^2} = -N^2\frac{kT}{\pi}\kappa_m^4 \int_1^\infty f(p,\theta)e^{-2p\kappa_m l}p^3 dp = N^2 \sin\theta\, g(l,\theta).$$

12. The required integration involves terms with different powers of p that emerge from $f(p,\theta)p^3$:

$$p^3 f(p,\theta) = \left[C^2 + \frac{CD}{4} + \frac{D^2}{16}\left(\frac{1}{8} + \frac{\cos^2\theta}{4}\right)\right]p^3$$
$$+\left[\frac{2KC}{4\kappa_m^2} - \frac{CD}{4} + \frac{KD}{16\kappa_m^2} - \frac{D^2}{8}\left(\frac{1}{8} + \frac{\cos^2\theta}{4}\right)\right]p$$
$$+\left[\frac{K^2}{16\kappa_m^4} - \frac{KD}{16\kappa_m^2} + \frac{D^2}{16}\left(\frac{1}{8} + \frac{\cos^2\theta}{4}\right)\right]p^{-1},$$

or, returning to our regular notation, by

$$K/\kappa_m^2 \equiv (\Gamma_c/n_m - 2C), \quad D \equiv \pi a^2(\overline{\Delta}_\parallel - 2\overline{\Delta}_\perp), \quad C \equiv \pi a^2\overline{\Delta}_\perp,$$

$$\frac{p^3 f(p,\theta)}{(\pi a^2)^2} = \left[\Delta_\perp^2 + \frac{\Delta_\perp(\Delta_\parallel - 2\Delta_\perp)}{4} + \frac{(\Delta_\parallel - 2\Delta_\perp)^2}{2^7}(1+2\cos^2\theta)\right]p^3$$

$$+\left[\begin{array}{l}\left(\frac{\Gamma_c}{\pi a^2 n_m}\right)\frac{\Delta_\perp}{2} + \left(\frac{\Gamma_c}{\pi a^2 n_m}\right)\frac{(\Delta_\parallel - 2\Delta_\perp)}{16} \\ -\Delta_\perp^2 - \frac{3\Delta_\perp(\Delta_\parallel - 2\Delta_\perp)}{8} - \frac{(\Delta_\parallel - 2\Delta_\perp)^2}{2^6}(1+2\cos^2\theta)\end{array}\right]p$$

$$+\left[\begin{array}{l}\left(\frac{\Gamma_c}{\pi a^2 n_m}\right)^2\frac{1}{16} - \left(\frac{\Gamma_c}{\pi a^2 n_m}\right)\frac{\Delta_\perp}{4} - \left(\frac{\Gamma_c}{\pi a^2 n_m}\right)\frac{(\Delta_\parallel - 2\Delta_\perp)}{16} \\ +\frac{\Delta_\perp^2}{4} + \frac{\Delta_\perp(\Delta_\parallel - 2\Delta_\perp)}{8} + \frac{(\Delta_\parallel - 2\Delta_\perp)^2}{2^7}(1+2\cos^2\theta)\end{array}\right]p^{-1}.$$

Here,

$$\int_1^\infty p^3 e^{-2p\kappa_m l}dp = \frac{6}{(2\kappa_m l)^4}e^{-2\kappa_m l}\left[1 + 2\kappa_m l + \frac{(2\kappa_m l)^2}{2} + \frac{(2\kappa_m l)^3}{6}\right].$$

The p term has a factor

$$\int_1^\infty p e^{-2p\kappa_m l}dp = \frac{1}{(2\kappa_m l)^2}e^{-2\kappa_m l}(1 + 2\kappa_m l).$$

And the last term has a factor

$$\int_1^\infty p^{-1}e^{-2p\kappa_m l}dp = E_1(2\kappa_m l),$$

the exponential integral. M. Abramowitz and I. A. Stegun, *Handbook of Mathematical Functions, with Formulas, Graphs, and Mathematical Tables* (Dover, New York, 1965), Chap. 5, Eqs. 5.1.1, 5.1.11, 5.1.12, and 5.1.51).

Together then,

$$\frac{d^2 G_{AmB}(l,\theta)}{dl^2} = -N^2\frac{kT}{\pi}\kappa_m^4\int_1^\infty f(p,\theta)e^{-2p\kappa_m l}p^3 dp$$
$$= N^2 \sin\theta\, g(l,\theta) = -N^2\frac{kT}{\pi}\kappa_m^4(\pi a^2)^2\{\,\}$$

or

$$g(l,\theta) = -\frac{kT\kappa_{\rm m}^4(\pi a^2)^2}{\pi \sin\theta}\{\},$$

where

$$\{\} = \left[\overline{\Delta}_\perp^2 + \frac{\overline{\Delta}_\perp(\overline{\Delta}_\parallel - 2\overline{\Delta}_\perp)}{4} + \frac{(\overline{\Delta}_\parallel - 2\overline{\Delta}_\perp)^2}{2^7}(1+2\cos^2\theta)\right]$$

$$\times \frac{6e^{-2\kappa_{\rm m}l}}{(2\kappa_{\rm m}l)^4}\left[1 + 2\kappa_{\rm m}l + \frac{(2\kappa_{\rm m}l)^2}{2} + \frac{(2\kappa_{\rm m}l)^3}{6}\right]$$

$$+\left[\begin{matrix}\left(\frac{\Gamma_{\rm c}}{\pi a^2 n_{\rm m}}\right)\frac{\overline{\Delta}_\perp}{2} + \left(\frac{\Gamma_{\rm c}}{\pi a^2 n_{\rm m}}\right)\frac{(\overline{\Delta}_\parallel - 2\overline{\Delta}_\perp)}{16}\\ -\overline{\Delta}_\perp^2 - \frac{3\overline{\Delta}_\perp(\overline{\Delta}_\parallel - 2\overline{\Delta}_\perp)}{8} - \frac{(\overline{\Delta}_\parallel - 2\overline{\Delta}_\perp)^2}{2^6}(1+2\cos^2\theta)\end{matrix}\right]$$

$$\times \frac{1}{(2\kappa_{\rm m}l)^2}e^{-2\kappa_{\rm m}l}(1+2\kappa_{\rm m}l)$$

$$+\left[\begin{matrix}\left(\frac{\Gamma_{\rm c}}{\pi a^2 n_{\rm m}}\right)^2\frac{1}{16} - \left(\frac{\Gamma_{\rm c}}{\pi a^2 n_{\rm m}}\right)\frac{\overline{\Delta}_\perp}{4} - \left(\frac{\Gamma_{\rm c}}{\pi a^2 n_{\rm m}}\right)\frac{(\overline{\Delta}_\parallel - 2\overline{\Delta}_\perp)}{16}\\ +\frac{\overline{\Delta}_\perp^2}{4} + \frac{\overline{\Delta}_\perp(\overline{\Delta}_\parallel - 2\overline{\Delta}_\perp)}{8} + \frac{(\overline{\Delta}_\parallel - 2\overline{\Delta}_\perp)^2}{2^7}(1+2\cos^2\theta)\end{matrix}\right]E_1(2\kappa_{\rm m}l).$$

13. As in the nonionic case, the interaction per unit area between thin slabs of A and B, the second derivative of $G_{\rm AmB}$, is the sum of parallel-rod interactions. Imagine, for example, one rod in the thin slab of A and its interactions with all the rods in the thin slab of B, an integration of $g(\sqrt{y^2+l^2}, \theta = 0) \equiv g_\parallel(\sqrt{y^2+l^2})$ at all distances $\sqrt{y^2+l^2}$. This integration has the form of the Abel transform, $h(l) = \int_{-\infty}^{\infty} g_\parallel(\sqrt{y^2+l^2})\,{\rm d}y$. [Section 8.11 in Alexander D. Poularikas, ed., *The Transforms and Applications Handbook* (CRC Press, Boca Raton, FL, 1996).] Here $h(l)$ is ${\rm d}^2G/{\rm d}l^2$.

We use the inverse transform

$$g_\parallel(l) = -\frac{1}{\pi}\int_l^\infty \frac{h'(x)}{\sqrt{x^2-l^2}}\,{\rm d}x.$$

To effect this inverse transform we take the next derivative of G,

$$\frac{{\rm d}^3 G_{\rm AmB}(l,\theta=0)}{{\rm d}l^3} = -N^2\frac{2kT}{\pi}\kappa_{\rm m}^5\int_1^\infty f(p,\theta=0)e^{-2p\kappa_{\rm m}l}\,p^4\,{\rm d}p,$$

write it as a function of variable x, and then integrate in x. Specifically, the spatially dependent part of ${\rm d}^3G_{\rm AmB}(l,\theta)/{\rm d}l^3$ is $e^{-2p\kappa_{\rm m}l}$, which becomes $e^{-2p\kappa_{\rm m}x}$. The inverse transform is then

$$-\frac{1}{\pi}\int_l^\infty \frac{e^{-2\kappa_{\rm m}px}}{\sqrt{x^2-l^2}}{\rm d}x = -\frac{1}{\pi}\int_1^\infty \frac{e^{-2\kappa_{\rm m}plt}}{\sqrt{t^2-1}}{\rm d}t = -\frac{1}{\pi}K_0(2\kappa_{\rm m}pl).$$

[I. S. Gradshteyn and I. M. Ryzhik, *Table of Integrals, Series, and Products* (Academic, New York, 1965) 8.432.2].

In this way, the energy of interaction per unit length is

$$N^2 g(l,\theta=0) = N^2 g_\parallel(l) = -N^2\frac{2kT}{\pi^2}\kappa_{\rm m}^5\int_1^\infty f(p,\theta=0)K_0(2\kappa_{\rm m}pl)p^4{\rm d}p$$

or

$$g_\parallel(l) = -\frac{2kT\kappa_{\rm m}^5}{\pi^2}\int_1^\infty p^4 f(p,\theta=0)K_0(2\kappa_{\rm m}pl){\rm d}p = -\frac{2kT\kappa_{\rm m}^5(\pi a^2)^2}{\pi^2}\{\},$$

where

$$\{\} = \left[\overline{\Delta}_{\perp}^2 + \frac{\overline{\Delta}_{\perp}(\overline{\Delta}_{\parallel} - 2\overline{\Delta}_{\perp})}{4} + \frac{3(\overline{\Delta}_{\parallel} - 2\overline{\Delta}_{\perp})^2}{2^7}\right]\int_1^{\infty} K_0(2\kappa_{\mathrm{m}} pl)p^4 \mathrm{d}p$$

$$+\left[\begin{array}{l}\left(\frac{\Gamma_{\mathrm{c}}}{\pi a^2 n_{\mathrm{m}}}\right)\frac{\overline{\Delta}_{\perp}}{2} + \left(\frac{\Gamma_{\mathrm{c}}}{\pi a^2 n_{\mathrm{m}}}\right)\frac{(\overline{\Delta}_{\parallel} - 2\overline{\Delta}_{\perp})}{16} \\ -\overline{\Delta}_{\perp}^2 - \frac{3\overline{\Delta}_{\perp}(\overline{\Delta}_{\parallel} - 2\overline{\Delta}_{\perp})}{8} - \frac{3(\overline{\Delta}_{\parallel} - 2\overline{\Delta}_{\perp})^2}{2^6}\end{array}\right]\int_1^{\infty} K_0(2\kappa_{\mathrm{m}} pl)p^2 \mathrm{d}p$$

$$+\left[\begin{array}{l}\left(\frac{\Gamma_{\mathrm{c}}}{\pi a^2 n_{\mathrm{m}}}\right)^2\frac{1}{16} - \left(\frac{\Gamma_{\mathrm{c}}}{\pi a^2 n_{\mathrm{m}}}\right)\frac{\overline{\Delta}_{\perp}}{4} - \left(\frac{\Gamma_{\mathrm{c}}}{\pi a^2 n_{\mathrm{m}}}\right)\frac{(\overline{\Delta}_{\parallel} - 2\overline{\Delta}_{\perp})}{16} \\ +\frac{\overline{\Delta}_{\perp}^2}{4} + \frac{\overline{\Delta}_{\perp}(\overline{\Delta}_{\parallel} - 2\overline{\Delta}_{\perp})}{8} + \frac{3(\overline{\Delta}_{\parallel} - 2\overline{\Delta}_{\perp})^2}{2^7}\end{array}\right]\int_1^{\infty} K_0(2\kappa_{\mathrm{m}} pl)\mathrm{d}p.$$

14. For large values of its argument, $K_0(2\kappa_{\mathrm{m}} pl) \sim \sqrt{\frac{\pi}{2}}\frac{e^{-2\kappa_{\mathrm{m}} pl}}{\sqrt{2\kappa_{\mathrm{m}} pl}}$ (Abramowitz and Stegun 9.7.2, p. 378). Integrals are dominated by their value near $p = 1$ with convergence because of the exponential:

$$\int_1^{\infty} K_0(2\kappa_{\mathrm{m}} pl)p^{2q}\,\mathrm{d}p \sim \sqrt{\frac{\pi}{2}}\int_1^{\infty}\frac{e^{-2\kappa_{\mathrm{m}} pl}}{\sqrt{2\kappa_{\mathrm{m}} pl}}p^{2q-\frac{1}{2}}\,\mathrm{d}p$$

$$\sim \sqrt{\frac{\pi}{2}}\frac{1}{(2\kappa_{\mathrm{m}} l)^{1/2}}\int_1^{\infty} e^{-2\kappa_{\mathrm{m}} lp}p^{2q-\frac{1}{2}}\,\mathrm{d}p.$$

The last integration defines a gamma function:

$$\alpha_n(x) = \int_1^{\infty} e^{-xt}t^n \mathrm{d}t = x^{-n-1}\Gamma(1+n, x)$$

M. Abramowitz and I. A. Stegun, *Handbook of Mathematical Functions, with Formulas, Graphs, and Mathematical Tables* (Dover, New York, 1965) p. 262, Eq. 6.5.10 and

$$\Gamma(a, x) \sim x^{a-1}e^{-x}\left[1 + \frac{a-1}{x} + \frac{(a-1)(a-2)}{x^2} + \cdots\right]$$

ibid., p. 263 (Eq. 6.5.32) with $x = 2\kappa_{\mathrm{m}} l$, $n = 2q - \frac{1}{2} = a - 1$,

$$\sqrt{\frac{\pi}{2}}\frac{1}{(2\kappa_{\mathrm{m}} l)^{1/2}}\int_1^{\infty} e^{-2\kappa_{\mathrm{m}} lp}p^n \mathrm{d}p$$

$$= \sqrt{\frac{\pi}{2}}\frac{(2\kappa_{\mathrm{m}} l)^{-n-1}}{(2\kappa_{\mathrm{m}} l)^{1/2}}(2\kappa_{\mathrm{m}} l)^n e^{-2\kappa_{\mathrm{m}} l}\left(1 + \frac{n}{2\kappa_{\mathrm{m}} l} + \cdots\right) \sim \sqrt{\frac{\pi}{2}}\frac{e^{-2\kappa_{\mathrm{m}} l}}{(2\kappa_{\mathrm{m}} l)^{3/2}}.$$

Level 2, Computation

1. The general derivation of these relations is given in standard texts. Think of the paraphrase here as "Landau and Lifshitz Lite." See, for example, Eq. 82.15, p. 281, in Section 82 of L. D. Landau, E. M. Lifshitz, and L. P. Pitaevskii, *Electrodynamics of Continuous Media*, 2nd ed., Vol. 8 of Course of Theoretical Physics Series (Pergamon, Oxford, 1993), as well as Eq. 123.19, p. 383, in Section 123 of L. D. Landau and E. M. Lifshitz, *Statistical Physics*, 3rd ed., Vol. 5 of Course of Theoretical Physics Series (Pergamon, Oxford, 1993). Much of what is given in the main text here is a paraphrase from these sources. F. Wooten's *Optical Properties of Solids* (Academic, New York, 1972) is a treasury of good teaching. Virtually any elementary electricity and magnetism text will suffice for background reading to the present text. M. Born and E. Wolf,

Principles of Optics: Electromagnetic Theory of Propagation, Interference and Diffraction of Light (Cambridge University Press, New York, 1999), is a personal favorite.

2. This correspondence was pointed out with stunning clarity by J. B. Johnson, "Thermal agitation of electricity in conductors," Nature (London), Vol. 119, 50–5 (1927). He begins a four-paragraph letter with: *Ordinary electric conductors are sources of spontaneous fluctuations of voltage which can be measured with sufficiently sensitive instruments. This property of conductors appears to be the result of thermal agitation of the electric charges in the material of the conductor*, and ends powerfully with: *The limit to the smallness of voltage which can be usefully amplified is often set, not by the vacuum tube, but by the very matter of which electrical circuits are built.*
3. This is a major result whose nontrivial derivation is given in several texts, e.g., L. D. Landau and E. M. Lifshitz, *Electrodynamics of Continuous Media* and *Statistical Physics*, already cited. See also Chap. 2 in Sh. Kogan, *Electronic Noise and Fluctuations in Solids* (Cambridge University Press, New York, 1996). For background, A. van der Ziel, *Noise* (Prentice-Hall, New York, 1954), systematically develops the classic theory. N. Wax, *Selected Papers on Noise and Stochastic Processes* (Dover, New York, 2003), is a good place to wander.
4. To see this we look ahead a few pages to where ε language is used to describe a material of conductivity σ. In the limit of low frequencies,

$$\varepsilon(\omega) = \frac{4\pi i\sigma}{\omega(1 - i\omega b)} \to \frac{4\pi i\sigma}{\omega} = i\varepsilon''(\omega).$$

With $\omega\varepsilon''(\omega) = 4\pi\sigma$ the current fluctuation $kT\{[\omega\varepsilon''(\omega)]/[(2\pi)^2 d]\}$ in this low frequency limit becomes $kT\{(4\pi\sigma)/[(2\pi)^2 d]\}$. The conductivity of the material between the plates can be converted into a resistance $R = d/(L^2\sigma) = (d/\sigma)$ for plates of unit area to let us write $kT\{(4\pi\sigma)/[(2\pi)^2 d]\} = (kT/\pi R)$. This is the density of current fluctuations over a range $d\omega$ in radial frequency. Because $\omega = 2\pi\nu$, for ordinary frequency ν measured in hertz (cycles/second), the density over a range in $d\nu$ is $(kT/\pi R)\,2\pi = (2kT/R)$. Because experiment cannot distinguish between "positive" and "negative" oscillatory frequencies, the measured density of fluctuations is that between ν and $\nu + d\nu$ as well as between $-\nu$ and $-(\nu + d\nu)$. The last expression must therefore be multiplied by 2 to give $\langle I^2\rangle_\nu = (4kT/R)$ or, with voltage $V = IR$, $\langle V^2\rangle_\nu = 4kTR$, the voltage noise given by H. Nyquist, Eq. 1, for the classical high-temperature low-frequency limit, in "Thermal agitation of electric charge in conductors," Phys. Rev., **32**, 110–13 (1928); Nyquist generalized it to include quantization of voltage fluctuation and showed a connection to fluctuations in a gas.
5. For example, L. Landau and E. M. Lifshitz, *Electrodynamics of Continuous Media*, 2nd ed. (Pergamon, Oxford, 1993), p. 396, Eq. 113.8.
6. See D. Y. Smith, "Dispersion theory, sum rules, and their application to the analysis of optical data," Chap. 3, and D. W. Lynch, "Interband absorption—mechanisms and interpretation," Chap. 10, both in *Handbook of Optical Constants of Solids* (Academic, New York, 1985), as well as references therein; also D. Y. Smith, M. Inokuti, and W. Karstens, "Photoresponse of condensed matter over the entire range of excitation energies: Analysis of silicon," Phys. Essays, **13**, 465–72 (2000).
7. See R. H. French, "Origins and applications of London dispersion forces and Hamaker constants in ceramics," J. Am. Ceram. Soc., **83**, 2117–46 (2000); K. van Benthem, R. H. French, W. Sigle, C. Elsässer, and M. Rühle, "Valence electron energy loss study of Fe-doped $SrTiO_3$ and a $\Sigma 13$ boundary: Electronic structure and dispersion forces," Ultramicroscopy, **86**, 303–18 (2001), and the extensive literature cited therein. Energy "E" in those papers is written as "$\hbar\omega$" here.
8. Designed primarily to describe electron response at optical and higher frequencies, J_{cv} has instructive powers rigorously described in Chap. 5 of F. Wooten, *Optical Properties of Solids* (Academic, New York, 1972).
9. E. Shiles, T. Sasaki, M. Inokuti, and D. Y. Smith, "Self-consistency and sum-rule tests in the Kramers–Kronig analysis of optical data: Applications to aluminum," Phys. Rev. B, **22**, 1612–28 (1980).

10. The text here does not give a proper derivation of this response, only a sketch of its form. See P. J. W. Debye, *Polar Molecules* (Dover, New York, reprint of 1929 edition), and H. Fröhlich, *Theory of Dielectrics: Dielectric Constant and Dielectric Loss*, in Monographs on the Physics and Chemistry of Materials Series (Clarendon, Oxford University Press, 1987).
11. See, for example, F. Buckley and A. A. Maryott, "Tables of Dielectric Dispersion Data for Pure Liquids and Dilute Solutions, NBS Circular 589 (National Bureau of Standards, Gaithersburg, MD, 1958).
12. See R. Podgornik, G. Cevc, and B. Zeks, "Solvent structure effects in the macroscopic theory of van der Waals forces," J. Chem. Phys., **87**, 5957–66 (1987), for a systematic exposition of problems in general formulation and solutions of specific cases involving $\varepsilon(\omega; \mathbf{k})$.
13. R. H. French, R. M. Cannon, L. K. DeNoyer, and Y.-M. Chiang, "Full spectral calculations of non-retarded Hamaker constants for ceramic systems from interband transition strengths," Solid State Ionics, **75**, 13–33 (1995). The figure here is a modified version of Fig. 2 in that paper; nonretarded Hamaker coefficients across a vacuum are from Table 1, column 3, of that paper.
14. Figures (L2.30)–(L2.32) and coefficients computed here are from Dr. Lin DeNoyer (personal communication, 2003) by the GRAMS program that sums n's to $\hbar\xi_n = 250$ eV and takes the radial vector integration by uniform spacing ("Simpson's rule"). Electronic Structure Tools, Spectrum Square Associates, 755 Snyder Hill Road, Ithaca, NY 14850, USA; GRAMS/32, Galactic Industries, 325 Main Street, Salem, NY 03079, USA. The Gecko Hamaker program is available for education and research at http://sourceforge.net/projects/geckoproj/.
15. H. D. Ackler, R. H. French, and Y.-M. Chiang, "Comparisons of Hamaker constants for ceramic systems with intervening vacuum or water: From force layers and physical properties," J. Colloid Interface Sci., **179**, 460–9 (1996).
16. R. H. French, H. Müllejans, and D. J. Jones, "Optical properties of aluminum oxide: Determined from vacuum ultraviolet and electron energy loss spectroscopies," J. Am. Ceram. Soc., **81**, 2549–57 (1998).
17. R. H. French, D. J. Jones, and S. Loughin, "Interband electronic structure of α-alumina up to 2167 K," J. Am. Chem. Soc., **77**, 412–22 (1994). Figure here quoted from Fig. 5a of that paper.
18. A numerical tabulation of recent data for water is given in R. R. Dagastine, D. C. Prieve, and L. R. White, "The dielectric function for water and its application to van der Waals forces," J. Colloid Interface Sci., **231**, 351–8 (2000). The tabulation, at *http://www.cheme.cmu.edu/jcis/*, gives $\varepsilon(i\xi_n)$ for ξ_n at room temperature together with a suggested procedure to compute $\varepsilon(i\xi_n)$ at other temperatures. This site also presents data tables for several other materials.
19. C. M. Roth and A. M. Lenhoff, "Improved parametric representation of water dielectric data for Lifshitz theory calculations," J. Colloid Interface Sci., **179**, 637–9 (1996), present another set of parameters for water.
20. F. Buckley and A. A. Maryott, "Tables of dielectric data for pure liquids and dilute solutions." NBS Circular 589 (National Bureau of Standards, Gaithersburg, MD, 1958).
21. L. D. Kislovskii, "Optical characteristics of water and ice in the infrared and radiowave regions of the spectrum," Opt. Spectr. (USSR), **7**, 201–6 (1959).
22. D. Gingell and V. A. Parsegian, "Computation of van der Waals interactions in aqueous systems using reflectivity data," J. Theor. Biol., **36**, 41–52 (1972).
23. J. M. Heller, R. N. Hamm, R. D. Birkhoff and L. R. Painter, "Collective oscillation in liquid water," J. Chem. Phys. **60**, 3483–86 (1974).
24. V. A. Parsegian and G. H. Weiss, "Spectroscopic parameters for computation of van der Waals forces," J. Colloid Interface Sci., **82**, 285–8 (1981).
25. V. A. Parsegian, Chap. 4 in *Physical Chemistry: Enriching Topics from Colloid and Interface Science*, H. van Olphen and K. J. Mysels, eds. (IUPAC, Theorex, La Jolla, CA, 1975).
26. G. B. Irani, T. Huen, and T. Wooten, J. Opt. Soc. Am., **61**, 128–9 (1971).
27. H.-J. Hagemann, W. Gudat, and C. Kunz, *Optical Constants from the Far Infrared to the X-Ray Regions: Mg, Al, Ca, Ag, Au, Bi, C and Al_2O_3* (Deutsches Electronen-Synchrotron, Hamburg, 1974). This book has a large amount of very useful data beyond what is quoted in the sample list here.

28. P. B. Johnson and R. W. Christy, Phys. Rev. B, **6**, 4370 (1972); parameters from fitting to imaginary part $\varepsilon''(\omega)$ of dielectric dispersion.
29. V. A. Parsegian, Langmuir, **9**, 3625–8 (1993); note that the form of the $\varepsilon(i\xi)$ function used in that paper differs from that used here. Resonance frequencies ω_1, ω_2, and ω_3 are the same. Numerator C_1 there is equal to the "Debye" d here; and $1/\tau = \omega_1$. Numerators C_2 and C_3 are converted to the f_j given here by multiplying by the respective $\omega_j{}^2$. The number g there is converted to the g_j given here by $g_j = (\omega_j^2/g)$.
30. Parameter set "a" is a "consensus set" from the Department of Applied Mathematics, IAS, Australian National University (Patrick Kekicheff, 1992 personal communication).
31. Parameter set "b" is from the text by J. Mahanty and B. W. Ninham, *Dispersion Forces* (Academic, London, 1976).
32. Parameter set "c." MW and IR as in set "b," UV term from D. Chan and P. Richmond, Proc. R. Soc. London Ser. A, **353**, 163–76 (1977).

Level 3, Foundations

1. H. B. G. Casimir, see Prelude note 9.
2. H. B. G. Casimir and D. Polder, "The influence of retardation on the London-van der Waals forces," Phys. Rev., **73**, 360–72 (1948).
3. I. E. Dzyaloshinskii, E. M. Lifshitz, and L. P. Pitaevskii, "The general theory of van der Waals forces," Adv. Phys., **10**, 165 (1961), for the method, though applied only to a vacuum gap; see also Chapter VIII, E. M. Lifshitz and L. P. Pitaevskii, *Statistical Physics*, Part 2 in Vol. 9 of Course of Theoretical Physics Series (Pergamon, Oxford, 1991); a systematic derivation of the full DLP result is given also in Chap. 6 of A. A. Abrikosov, L. P. Gorkov, & I. E. Dzyaloshinski, *Methods of Quantum Field Theory in Statistical Physics*, R. A. Silverman, trans. (Dover, New York, 1963).
4. B. W. Ninham, V. A. Parsegian, and G. H. Weiss, "On the macroscopic theory of temperature-dependent van der Waals forces," J. Stat. Phys., **2**, 323–8 (1970).
5. N. G. van Kampen, B. R. A. Nijboer, and K. Schram, "On the macroscopic theory of van der Waals forces," Phys. Lett., **26A**, 307 (1968).
6. D. Langbein, *Theory of van der Waals Attraction*, Vol. 72 of Springer Tracts in Modern Physics Series, G. Hohler, ed. (Springer-Verlag, Berlin Heidelberg, New York, 1974).
7. J. Mahanty and B. W. Ninham, *Dispersion Forces* (Academic, London, New York, San Francisco, 1976).
8. Yu. S. Barash, *Van der Waals Forces* (Nauka, Moscow, 1988) (in Russian); Yu. S. Barash and V. L. Ginzburg, "Electromagnetic fluctuations in matter and molecular (Van-der-Waals) forces between them," Usp. Fiz. Nauk, **116**, 5–40 (1975), English translation in Sov. Phys.-Usp. **18**, 305–22 (1975).
9. What if we ignored the additive constant? What if we said that the energy levels of the oscillator were only multiples of $\hbar\omega$, $E_\eta = \eta\hbar\omega$, $\eta = 0, 1, 2, \ldots$, with no additive $\hbar\omega/2$?

   Compute the average energy from a sum over states:

$$\overline{E} = \frac{\sum_{\eta=0}^{\infty} \hbar\omega\,\eta\, e^{-\hbar\omega\,\eta/kT}}{\sum_{\eta=0}^{\infty} e^{-\hbar\omega\,\eta/kT}} = \frac{\partial}{\partial\left(-\frac{1}{kT}\right)} \ln\left(\sum_{\eta=0}^{\infty} e^{-\hbar\omega\,\eta/kT}\right).$$

   If we define $x \equiv e^{-\hbar\omega/kT}$ so that $[\partial x/\partial(-\frac{1}{kT})] = \hbar\omega\, x$, then

$$\sum_{\eta=0}^{\infty} e^{-\hbar\omega\,\eta/kT} = \sum_{\eta=0}^{\infty} x^\eta = \frac{1}{1-x}.$$

   We have $\ln(\sum_{\eta=0}^{\infty} e^{-\hbar\omega\,\eta/kT}) = -\ln(1-x)$. The average energy is

$$\overline{E} = \frac{\partial x}{\partial\left(-\frac{1}{kT}\right)} \frac{d[-\ln(1-x)]}{dx} = \hbar\omega \frac{x}{1-x} = kT\frac{\hbar\omega}{kT}\frac{e^{-\hbar\omega/kT}}{1-e^{-\hbar\omega/kT}}.$$

At high temperature, such that $kT \gg \hbar\omega$, this expands to

$$\overline{E} = kT\frac{\hbar\omega}{kT}\frac{1-\frac{\hbar\omega}{kT}}{1-1+\frac{\hbar\omega}{kT}-\frac{1}{2}\left(\frac{\hbar\omega}{kT}\right)^2} = kT\frac{1-\frac{\hbar\omega}{kT}}{1-\frac{1}{2}\left(\frac{\hbar\omega}{kT}\right)}$$

$$\approx kT\left(1-\frac{1}{2}\frac{\hbar\omega}{kT}\right) = kT - \frac{1}{2}\hbar\omega.$$

From the equipartition theorem we know that there is energy $kT/2$ in each degree of freedom; for a bound oscillator with kinetic and potential energy, we know that at high temperature the average energy of must go exactly to kT. The average energy computed just above is too small by an additive constant $\frac{1}{2}\hbar\omega$. This means that the energy levels taken to be $E_\eta = \eta\hbar\omega$ were too small by $\frac{1}{2}\hbar\omega$. The right levels are the familiar

$$E_\eta = \left(\eta + \frac{1}{2}\right)\hbar\omega.$$

The assumption that energy changes in quantal photons is enough to say that energy cannot go to zero at $T = 0$. There is always a zero-point energy $\frac{1}{2}\hbar\omega$.

10. In the absence of conductivity σ, external charge ρ_{ext}, and externally applied currents j_{ext}, with spatially unvarying scalar relative ε and μ in each region, the Maxwell equations

$$\nabla\cdot\mathbf{D} = \nabla\cdot(\varepsilon\varepsilon_0\mathbf{E}) = \rho_{\text{ext}} \text{ in mks,} \qquad \nabla\cdot\mathbf{D} = \nabla\cdot(\varepsilon\mathbf{E}) = 4\pi\rho_{\text{ext}} \text{ in cgs;}$$

$$\nabla\cdot\mathbf{H} = 0, \qquad \nabla\cdot\mathbf{H} = 0;$$

$$\nabla\times\mathbf{E} = -\frac{\partial\mathbf{B}}{\partial t} = -\mu\mu_0\frac{\partial\mathbf{H}}{\partial t}, \qquad \nabla\times\mathbf{E} = -\frac{1}{c}\frac{\partial\mathbf{B}}{\partial t} = -\frac{\mu}{c}\frac{\partial\mathbf{H}}{\partial t};$$

$$\nabla\times\mathbf{H} = \varepsilon\varepsilon_0\frac{\partial\mathbf{E}}{\partial t} + \sigma\mathbf{E} + \mathbf{j}_{\text{ext}}, \qquad \nabla\times\mathbf{H} = \frac{\varepsilon}{c}\frac{\partial\mathbf{E}}{\partial t} + \frac{4\pi\sigma}{c}\mathbf{E} + \frac{4\pi}{c}\mathbf{j}_{\text{ext}};$$

become

$$\nabla\cdot\mathbf{E} = 0 \text{ in mks,} \qquad \nabla\cdot\mathbf{E} = 0 \text{ in cgs;}$$

$$\nabla\cdot\mathbf{H} = 0, \qquad \nabla\cdot\mathbf{H} = 0;$$

$$\nabla\times\mathbf{E} = -\mu\mu_0\frac{\partial\mathbf{H}}{\partial t}, \qquad \nabla\times\mathbf{E} = -\frac{\mu}{c}\frac{\partial\mathbf{H}}{\partial t};$$

$$\nabla\times\mathbf{H} = \varepsilon\varepsilon_0\frac{\partial\mathbf{E}}{\partial t}, \qquad \nabla\times\mathbf{H} = \frac{\varepsilon}{c}\frac{\partial\mathbf{E}}{\partial t}.$$

To reduce these to wave equations for $\mathbf{E}_\omega, \mathbf{H}_\omega$, use the identity $\nabla\times\nabla\times\mathbf{E} = -\nabla^2\mathbf{E} + \nabla(\nabla\cdot\mathbf{E})$ and drop the second term because $\nabla\cdot\mathbf{E} = 0$. Then use $\nabla\times\mathbf{E} = -\frac{\mu}{c}\frac{\partial\mathbf{H}}{\partial t}$ to create a wave equation for $\mathbf{E}$:

$$\nabla^2\mathbf{E} = -\nabla\times\nabla\times\mathbf{E} = -\nabla\times\left(-\frac{\mu}{c}\frac{\partial\mathbf{H}}{\partial t}\right) = \frac{\mu}{c}\frac{\partial(\nabla\times\mathbf{H})}{\partial t}$$

$$= \frac{\mu}{c}\frac{\partial\left(\frac{\varepsilon}{c}\frac{\partial\mathbf{E}}{\partial t}\right)}{\partial t} = \frac{\varepsilon\mu}{c^2}\frac{\partial^2\mathbf{E}}{\partial t^2}.$$

There is a similar reduction for $\mathbf{H}$ and $\nabla\times\mathbf{H} = \frac{\varepsilon}{c}\frac{\partial\mathbf{E}}{\partial t}$.

For $\mathbf{E}$, $\mathbf{H}$ of the form $\mathbf{E}(t) = \text{Re}\left[\sum_\omega \mathbf{E}_\omega e^{-i\omega t}\right]$, $\mathbf{H}(t) = \text{Re}\left[\sum_\omega \mathbf{H}_\omega e^{-i\omega t}\right]$, the second derivative introduces a factor $-\omega^2$:

$$\nabla^2\mathbf{E} + \frac{\varepsilon\mu\omega^2}{c^2}\mathbf{E} = 0; \qquad \nabla^2\mathbf{H} + \frac{\varepsilon\mu\omega^2}{c^2}\mathbf{H} = 0,$$

where the subscript ω has been dropped as understood here and from now on.

11. According to the Cauchy theorem, the value at $z = \text{a}$ of a complex function $g(z)$ of a complex variable z can be written as an integral $g(\text{a}) = \frac{1}{2\pi i}\oint_C \frac{g(z)}{(z-\text{a})}\,\text{d}z$.

The closed contour C goes around the point a in the complex-number plane. The essential point is that the quantity $(z - \text{a})$ in the denominator creates a mathematical pole of the "first order" [first power in $(z - \text{a})$] at position $(z = \text{a})$. In application of the theorem to the summation of free energies $g(\omega_j)$, the derivative $d\ \ln[D(\omega)]/d\omega$ automatically creates first-order poles to select $g(\omega_j)$ from the free energy form $g(\omega)$. It allows us to pluck exactly those frequencies that satisfy the surface-mode condition and to add up the free energies of all such modes. Think of $D(\omega)$ as a polynomial:

$$D(\omega) = \prod_{\{\omega_j\}} (\omega - \omega_j) \quad \text{so that} \quad \ln[D(\omega)] = \sum_{\{\omega_j\}} (\omega - \omega_j).$$

The derivative is

$$\frac{\mathrm{d}\ln[D(\omega)]}{\mathrm{d}\omega} = \sum_{\{\omega_j\}} \frac{1}{(\omega - \omega_j)},$$

so that

$$\frac{1}{2\pi i}\oint_C g(\omega)\frac{\mathrm{d}\ln[D(\omega)]}{\mathrm{d}\omega}\mathrm{d}\omega = \frac{1}{2\pi i}\oint_C g(\omega)\sum_{\{\omega_j\}}\frac{1}{(\omega-\omega_j)}\mathrm{d}\omega = \sum_{\{\omega_j\}} g(\omega_j).$$

12. $$kT\ln\left(e^{\hbar\omega/2kT} - e^{-\hbar\omega/2kT}\right) = kT\ln\left[e^{\hbar\omega/2kT}\left(1 - e^{-\hbar\omega/kT}\right)\right]$$
$$= kT\frac{\hbar\omega}{2kT} + kT\ln\left(1 - e^{-\hbar\omega/kT}\right) = \frac{\hbar\omega}{2} - kT\sum_{\eta=1}^{\infty}\frac{e^{-(\hbar\omega/kT)\eta}}{\eta}.$$

13. M. J. Lighthill, *An Introduction to Fourier Analysis and Generalised Functions* (Cambridge University Press, Cambridge, 1958).
14. The signs can be tricky, so spell it out:

$$-\frac{1}{2}\frac{\hbar}{2\pi}\int_{-\infty}^{+\infty}\ln D(i\xi)\,\mathrm{d}\xi = -\frac{\hbar}{2}\frac{1}{2\pi i}\int_{-i\infty}^{+i\infty}\ln D(\omega)\,\mathrm{d}\omega$$
$$= -\frac{\hbar}{2}\frac{1}{2\pi i}\left[\omega\ln D(\omega)\Big|_{-i\infty}^{+i\infty} - \int_{-i\infty}^{+i\infty}\omega\frac{\mathrm{d}\ln D(\omega)}{\mathrm{d}\omega}\mathrm{d}\omega\right] = +\frac{\hbar}{2}\frac{1}{2\pi i}\int_{-i\infty}^{+i\infty}\omega\frac{\mathrm{d}\ln D(\omega)}{\mathrm{d}\omega}\mathrm{d}\omega$$
$$= -\frac{\hbar}{2}\frac{1}{2\pi i}\int_{+i\infty}^{-i\infty}\omega\frac{\mathrm{d}\ln D(\omega)}{\mathrm{d}\omega}\mathrm{d}\omega = -\frac{\hbar}{2}\frac{1}{2\pi i}\oint_C\omega\frac{\mathrm{d}\ln D(\omega)}{\mathrm{d}\omega}\mathrm{d}\omega = -\sum_{\{\omega_j\}}\frac{1}{2}\hbar\omega_j.$$

15. Vector $\mathbf{E}$ has the form $\mathbf{E} = \hat{\imath}E_x + \hat{\jmath}E_y + \hat{k}E_z$, where E_x, E_y, and εE_z are continuous at each material boundary. E_x, E_y, and εE_z, are also constrained by $\nabla\cdot\mathbf{E} = 0$. Each component of the $\mathbf{E}$ and $\mathbf{H}$ fields is periodic in the x, y plane and has the general form $f(z)$ $f(z)e^{i(ux+vy)}$, or $E_x = e_x(z)e^{i(ux+vy)}$; $E_y = e_y(z)e^{i(ux+vy)}$; $E_z = e_z(z)e^{i(ux+vy)}$. The wave equation $f''(z) = \rho^2 f(z)$ has solutions $f(z) = \mathrm{A}e^{\rho z} + \mathrm{B}e^{-\rho z}$ for each field component. $\nabla\cdot\mathrm{E} = 0$ constrains the A_x, A_y, A_z, B_x, B_y, and B_z coefficients so that

$$iue_x(z) + ive_y(z) + e_z'(z) = (iu\mathrm{A}_x + iv\mathrm{A}_y + \rho\mathrm{A}_z)e^{\rho z} + (iu\mathrm{B}_x + iv\mathrm{B}_y - \rho\mathrm{B}_z)e^{-\rho z} = 0,$$

or

$$\mathrm{A}_z = -\frac{i}{\rho}(u\mathrm{A}_x + v\mathrm{A}_y), \quad \mathrm{B}_z = \frac{i}{\rho}(u\mathrm{B}_x + v\mathrm{B}_y).$$

E_x continuous requires that

$$\mathrm{A}_{(i+1)x}\,e^{\rho_{i+1}l_{i/i+1}} + \mathrm{B}_{(i+1)x}e^{-\rho_{i+1}l_{i/i+1}} = \mathrm{A}_{(i)x}\,e^{\rho_i l_{i/i+1}} + \mathrm{B}_{(i)x}\,e^{-\rho_i l_{i/i+1}}.$$

E_y continuous requires that

$$\mathrm{A}_{(i+1)y}\,e^{\rho_{i+1}l_{i/i+1}} + \mathrm{B}_{(i+1)y}e^{-\rho_{i+1}l_{i/i+1}} = \mathrm{A}_{(i)y}\,e^{\rho_i l_{i/i+1}} + \mathrm{B}_{(i)y}\,e^{-\rho_i l_{i/i+1}}.$$

Multiply the first of these equations by iu, the second by iv, then add and replace all A_x, A_y, B_x, B_y with A_z, B_z coefficients:

$$\left[-\mathrm{A}_{(i+1)z}e^{\rho_{i+1}l_{i/i+1}} + \mathrm{B}_{(i+1)z}e^{-\rho_{i+1}l_{i/i+1}}\right]\rho_{i+1} = \left[-\mathrm{A}_{(i)z}e^{\rho_i l_{i/i+1}} + \mathrm{B}_{(i)z}e^{-\rho_i l_{i/i+1}}\right]\rho_i.$$

$\varepsilon_R E_z$ continuous requires that

$$\left[A_{(i+1)z}e^{\rho_{i+1}l_{i/i+1}} + B_{(i+1)z}e^{-\rho_{i+1}l_{i/i+1}}\right]\varepsilon_{i+1} = \left[A_{(i)z}e^{\rho_i l_{i/i+1}} + B_{(i)z}e^{-\rho_i l_{i/i+1}}\right]\varepsilon_i.$$

The subscript z is dropped in the main text and in the subsequent endnotes.

16.
$$\left(-A_{i+1}e^{+\rho_{i+1}l_{i/i+1}} + B_{i+1}e^{-\rho_{i+1}l_{i/i+1}}\right)\rho_{i+1} = \left(-A_i e^{+\rho_i l_{i/i+1}} + B_i e^{-\rho_i l_{i/i+1}}\right)\rho_i,$$
$$\left(A_{i+1}e^{+\rho_{i+1}l_{i/i+1}} + B_{i+1}e^{-\rho_{i+1}l_{i/i+1}}\right)\varepsilon_{i+1} = \left(A_i e^{+\rho_i l_{i/i+1}} + B_i e^{-\rho_i l_{i/i+1}}\right)\varepsilon_i.$$

Adding and subtracting these two equations gives

$$A_{i+1} = e^{-\rho_{i+1}l_{i/i+1}}e^{+\rho_i l_{i/i+1}}\frac{1}{2}\left(\frac{\varepsilon_i}{\varepsilon_{i+1}} + \frac{\rho_i}{\rho_{i+1}}\right)A_i + e^{-\rho_{i+1}l_{i/i+1}}e^{-\rho_i l_{i/i+1}}\frac{1}{2}\left(\frac{\varepsilon_i}{\varepsilon_{i+1}} - \frac{\rho_i}{\rho_{i+1}}\right)B_i,$$

$$B_{i+1} = e^{+\rho_{i+1}l_{i/i+1}}e^{+\rho_i l_{i/i+1}}\frac{1}{2}\left(\frac{\varepsilon_i}{\varepsilon_{i+1}} - \frac{\rho_i}{\rho_{i+1}}\right)A_i + e^{+\rho_{i+1}l_{i/i+1}}e^{-\rho_i l_{i/i+1}}\frac{1}{2}\left(\frac{\varepsilon_i}{\varepsilon_{i+1}} + \frac{\rho_i}{\rho_{i+1}}\right)B_i.$$

Factoring out an uninteresting $[(\varepsilon_{i+1}\rho_i + \varepsilon_i\rho_{i+1})/(2\varepsilon_{i+1}\rho_{i+1})]$ yields

$$A_{i+1} = \left[\frac{(\varepsilon_{i+1}\rho_i + \varepsilon_i\rho_{i+1})}{2\varepsilon_{i+1}\rho_{i+1}}\right]^{-1}\left(e^{-\rho_{i+1}l_{i/i+1}}e^{+\rho_i l_{i/i+1}}A_i + \overline{\Delta}_{i/i+1}e^{-\rho_{i+1}l_{i/i+1}}e^{-\rho_i l_{i/i+1}}B_i\right),$$

$$B_{i+1} = \left[\frac{(\varepsilon_{i+1}\rho_i + \varepsilon_i\rho_{i+1})}{2\varepsilon_{i+1}\rho_{i+1}}\right]^{-1}\left(e^{+\rho_{i+1}l_{i/i+1}}e^{+\rho_i l_{i/i+1}}\overline{\Delta}_{i/i+1}A_i + e^{+\rho_{i+1}l_{i/i+1}}e^{-\rho_i l_{i/i+1}}B_i\right),$$

to create the matrix

$$\begin{bmatrix} e^{-\rho_{1+1}l_{i/i+1}}e^{+\rho_i l_{1/i+1}} & -\overline{\Delta}_{i/i+1}e^{-\rho_{1+1}l_{i/i+1}}/e^{-\rho_i l_{1/i+1}} \\ e^{+\rho_{1+1}l_{i/i+1}}e^{+\rho_i l_{1/i+1}}\overline{\Delta}_{i/i+1} & e^{+\rho_{1+1}l_{i/i+1}}e^{-\rho_i l_{1/i+1}} \end{bmatrix}$$

$$= \begin{bmatrix} e^{-\rho_{1+1}l_{i/i+1}}e^{+\rho_i l_{1/i+1}} & -\overline{\Delta}_{i+1/i}e^{-\rho_{1+1}l_{i/i+1}}e^{-\rho_i l_{1/i+1}} \\ -\overline{\Delta}_{i+1/i}e^{+\rho_{1+1}l_{i/i+1}}e^{+\rho_i l_{1/i+1}} & e^{+\rho_{1+1}l_{i/i+1}}e^{-\rho_i l_{1/i+1}} \end{bmatrix},$$

where $\overline{\Delta}_{i+1/i} \equiv [(\varepsilon_{i+1}\rho_i - \varepsilon_i\rho_{i+1})/(\varepsilon_{i+1}\rho_i + \varepsilon_i\rho_{i+1})]$.

17. The transition can be effected by going back to the previous derivation of the matrix and redefining the A's and B's. It is culturally instructive to see it also in terms of the matrix algebra that is used extensively in this section.

Begin with the old matrix and coefficients:

$$\mathbf{M}^{old}_{i+1/i} = \begin{bmatrix} e^{-\rho_{i+1}l_{i/i+1}}e^{+\rho_i l_{i/i+1}} & -\overline{\Delta}_{i+1/i}e^{-\rho_{i+1}l_{i/i+1}}e^{-\rho_i l_{i/i+1}} \\ -\overline{\Delta}_{i+1/i}e^{+\rho_{i+1}l_{i/i+1}}e^{+\rho_i l_{i/i+1}} & e^{+\rho_{i+1}l_{i/i+1}}e^{-\rho_i l_{i/i+1}} \end{bmatrix};$$

$$\begin{pmatrix} A^{old}_{i+1} \\ B^{old}_{i+1} \end{pmatrix} = \mathbf{M}^{old}_{i+1/i}\begin{pmatrix} A^{old}_i \\ B^{old}_i \end{pmatrix}$$

Then define the new coefficients:

$$\begin{pmatrix} A^{new}_{i+1} \\ B^{new}_{i+1} \end{pmatrix} = \begin{pmatrix} A^{old}_{i+1}e^{+\rho_{i+1}l_{i/i+1}} \\ B^{old}_{i+1}e^{-\rho_{i+1}l_{i/i+1}} \end{pmatrix} = \begin{bmatrix} e^{+\rho_{i+1}l_{i/i+1}} & 0 \\ 0 & e^{-\rho_{i+1}l_{i/i+1}} \end{bmatrix}\begin{pmatrix} A^{old}_{i+1} \\ B^{old}_{i+1} \end{pmatrix}$$

$$\begin{pmatrix} A^{new}_i \\ B^{new}_i \end{pmatrix} = \begin{pmatrix} A^{old}_i e^{+\rho_i l_{i-1/i}} \\ B^{old}_i e^{-\rho_i l_{i-1/i}} \end{pmatrix} = \begin{bmatrix} e^{+\rho_i l_{i-1/i}} & 0 \\ 0 & e^{-\rho_i l_{i-1/i}} \end{bmatrix}\begin{pmatrix} A^{old}_i \\ B^{old}_i \end{pmatrix}$$

or

$$\begin{pmatrix} A^{old}_i \\ B^{old}_i \end{pmatrix} = \begin{bmatrix} e^{-\rho_i l_{i-1/i}} & 0 \\ 0 & e^{+\rho_i l_{i-1/i}} \end{bmatrix}\begin{bmatrix} e^{+\rho_i l_{i-1/i}} & 0 \\ 0 & e^{-\rho_i l_{i-1/i}} \end{bmatrix}\begin{pmatrix} A^{old}_i \\ B^{old}_i \end{pmatrix}$$

$$= \begin{bmatrix} e^{-\rho_i l_{i-1/i}} & 0 \\ 0 & e^{+\rho_i l_{i-1/i}} \end{bmatrix}\begin{pmatrix} A^{new}_i \\ B^{new}_i \end{pmatrix}.$$

In these terms,

$$\begin{pmatrix} A_{i+1}^{new} \\ B_{i+1}^{new} \end{pmatrix} = \begin{bmatrix} e^{+\rho_{i+1} l_{i/i+1}} & 0 \\ 0 & e^{-\rho_{i+1} l_{i/i+1}} \end{bmatrix} \begin{pmatrix} A_{i+1}^{old} \\ B_{i+1}^{old} \end{pmatrix}$$

$$= \begin{bmatrix} e^{+\rho_{i+1} l_{i/i+1}} & 0 \\ 0 & e^{-\rho_{i+1} l_{i/i+1}} \end{bmatrix} \mathbf{M}_{i+1/i}^{old} \begin{pmatrix} A_{i}^{old} \\ B_{i}^{old} \end{pmatrix}$$

$$= \begin{bmatrix} e^{+\rho_{i+1} l_{i/i+1}} & 0 \\ 0 & e^{-\rho_{i+1} l_{i/i+1}} \end{bmatrix} \mathbf{M}_{i+1/i}^{old} \begin{bmatrix} e^{-\rho_i l_{i-1/i}} & 0 \\ 0 & e^{+\rho_i l_{i-1/i}} \end{bmatrix} \begin{pmatrix} A_{i}^{new} \\ B_{i}^{new} \end{pmatrix}$$

$$= \mathbf{M}_{i+1/i}^{new} \begin{pmatrix} A_{i}^{new} \\ B_{i}^{new} \end{pmatrix},$$

where

$$\mathbf{M}_{i+1/i}^{new} = \begin{bmatrix} e^{+\rho_{i+1} l_{i/i+1}} & 0 \\ 0 & e^{-\rho_{i+1} l_{i/i+1}} \end{bmatrix} \mathbf{M}_{i+1/i}^{old} \begin{bmatrix} e^{-\rho_i l_{i-1/i}} & 0 \\ 0 & e^{+\rho_i l_{i-1/i}} \end{bmatrix}$$

$$= \begin{bmatrix} e^{+\rho_{i+1} l_{i/i+1}} & 0 \\ 0 & e^{-\rho_{i+1} l_{i/i+1}} \end{bmatrix}$$

$$\times \begin{bmatrix} e^{-\rho_{i+1} l_{i/i+1}} e^{+\rho_i l_{i/i+1}} & -\overline{\Delta}_{i+1/i} e^{-\rho_{i+1} l_{i/i+1}} e^{-\rho_i l_{i/i+1}} \\ -\overline{\Delta}_{i+1/i} e^{+\rho_{i+1} l_{i/i+1}} e^{+\rho_i l_{i/i+1}} & e^{+\rho_{i+1} l_{i/i+1}} e^{-\rho_i l_{i/i+1}} \end{bmatrix} \begin{bmatrix} e^{-\rho_i l_{i-1/i}} & 0 \\ 0 & e^{+\rho_i l_{i-1/i}} \end{bmatrix}$$

$$= \begin{bmatrix} e^{+\rho_i (l_{i/i+1} - l_{i-1/i})} & -\overline{\Delta}_{i+1/i} e^{-\rho_i (l_{i/i+1} - l_{i-1/i})} \\ -\overline{\Delta}_{i+1/i} e^{+\rho_i (l_{i/i+1} - l_{i-1/i})} & e^{-\rho_i (l_{i/i+1} - l_{i-1/i})} \end{bmatrix}$$

$$= e^{+\rho_i (l_{i/i+1} - l_{i-1/i})} \begin{bmatrix} 1 & -\overline{\Delta}_{i+1/i} e^{-2\rho_i (l_{i/i+1} - l_{i-1/i})} \\ -\overline{\Delta}_{i+1/i} & e^{-2\rho_i (l_{i/i+1} - l_{i-1/i})} \end{bmatrix}$$

18.

$$\mathbf{M}_{i+1/i} = \begin{bmatrix} 1 & -\overline{\Delta}_{i+1/i} e^{-2\rho_i (l_{i/i+1} - l_{i-1/i})} \\ -\overline{\Delta}_{i+1/i} & e^{-2\rho_i (l_{i/i+1} - l_{i-1/i})} \end{bmatrix}$$

gives

$$\mathbf{M}_{RB_1} = \begin{bmatrix} 1 & -\overline{\Delta}_{RB_1} e^{-2\rho_{B_1} b_1} \\ -\overline{\Delta}_{RB_1} & e^{-2\rho_{B_1} b_1} \end{bmatrix},$$

$$\mathbf{M}_{B_1 m} = \begin{bmatrix} 1 & -\overline{\Delta}_{B_1 m} e^{-2\rho_m l} \\ -\overline{\Delta}_{B_1 m} & e^{-2\rho_m l} \end{bmatrix}$$

with their product

$$\mathbf{M}_{Rm}^{eff} = \mathbf{M}_{RB_1} \mathbf{M}_{B_1 m} = \begin{bmatrix} 1 & -\overline{\Delta}_{RB_1} e^{-2\rho_{B_1} b_1} \\ -\overline{\Delta}_{RB_1} & e^{-2\rho_{B_1} b_1} \end{bmatrix} \begin{bmatrix} 1 & -\overline{\Delta}_{B_1 m} e^{-2\rho_m l} \\ -\overline{\Delta}_{B_1 m} & e^{-2\rho_m l} \end{bmatrix}$$

$$= \begin{bmatrix} 1 + \overline{\Delta}_{RB_1} \overline{\Delta}_{B_1 m} e^{-2\rho_{B_1} b_1} & -\overline{\Delta}_{RB_1} e^{-2\rho_{B_1} b_1} e^{-2\rho_m l} - \overline{\Delta}_{B_1 m} e^{-2\rho_m l} \\ -\overline{\Delta}_{RB_1} - \overline{\Delta}_{B_1 m} e^{-2\rho_{B_1} b_1} & \overline{\Delta}_{RB_1} \overline{\Delta}_{B_1 m} e^{-2\rho_m l} + e^{-2\rho_{B_1} b_1} e^{-2\rho_m l} \end{bmatrix}.$$

19. Replace $\mathbf{M}_{Rm}^{eff} = \mathbf{M}_{RB_1 m}^{eff} = \mathbf{M}_{RB_1} \mathbf{M}_{B_1 m}$ with $\mathbf{M}_{Rm}^{eff} = \mathbf{M}_{RB_2 B_1 m}^{eff} = \mathbf{M}_{RB_2} \mathbf{M}_{B_2 B_1} \mathbf{M}_{B_1 m}$. In place of

$$\mathbf{M}_{RB_1} = \begin{bmatrix} 1 & -\overline{\Delta}_{RB_1} e^{-2\rho_{B_1} b_1} \\ -\overline{\Delta}_{RB_1} & e^{-2\rho_{B_1} b_1} \end{bmatrix}$$

we now have

$$\mathbf{M}_{\mathrm{RB_2}}\mathbf{M}_{\mathrm{B_2B_1}} = \begin{bmatrix} 1 & -\overline{\Delta}_{\mathrm{RB_2}}e^{-2\rho_{\mathrm{B_2}}\mathrm{b_2}} \\ -\overline{\Delta}_{\mathrm{RB_2}} & e^{-2\rho_{\mathrm{B_2}}\mathrm{b_2}} \end{bmatrix}\begin{bmatrix} 1 & -\overline{\Delta}_{\mathrm{B_2B_1}}e^{-2\rho_{\mathrm{B_1}}\mathrm{b_1}} \\ -\overline{\Delta}_{\mathrm{B_2B_1}} & e^{-2\rho_{\mathrm{B_1}}\mathrm{b_1}} \end{bmatrix}$$

$$= \begin{bmatrix} \left(1+\overline{\Delta}_{\mathrm{RB_2}}\overline{\Delta}_{\mathrm{B_2B_1}}e^{-2\rho_{\mathrm{B_2}}\mathrm{b_2}}\right) & -\left(\overline{\Delta}_{\mathrm{RB_2}}e^{-2\rho_{\mathrm{B_2}}\mathrm{b_2}}+\overline{\Delta}_{\mathrm{B_2B_1}}\right)e^{-2\rho_{\mathrm{B_1}}\mathrm{b_1}} \\ -\left(\overline{\Delta}_{\mathrm{RB_2}}+\overline{\Delta}_{\mathrm{B_2B_1}}e^{-2\rho_{\mathrm{B_2}}\mathrm{b_2}}\right) & \left(\overline{\Delta}_{\mathrm{RB_2}}\overline{\Delta}_{\mathrm{B_2B_1}}+e^{-2\rho_{\mathrm{B_2}}\mathrm{b_2}}\right)e^{-2\rho_{\mathrm{B_1}}\mathrm{b_1}} \end{bmatrix}.$$

The $A_R = 0$ condition uses only elements from the first row of this transition matrix. Hence the comparison between $-\overline{\Delta}_{\mathrm{RB_1}}e^{-2\rho_{\mathrm{B_1}}\mathrm{b_1}}$ and $\{[(\overline{\Delta}_{\mathrm{RB_2}}e^{-2\rho_{\mathrm{B_2}}\mathrm{b_2}}+\overline{\Delta}_{\mathrm{B_2B_1}})]/[(1+\overline{\Delta}_{\mathrm{RB_2}}\overline{\Delta}_{\mathrm{B_2B_1}}e^{-2\rho_{\mathrm{B_2}}\mathrm{b_2}})]\}e^{-2\rho_{\mathrm{B_1}}\mathrm{b_1}}$.

20. Replace $\mathbf{M}^{\mathrm{eff}}_{\mathrm{RB_j\ldots B_1m}}$ with $\mathbf{M}^{\mathrm{eff}}_{\mathrm{RB_{j+1}B_j\ldots B_1m}} = \mathbf{M}_{\mathrm{RB_{j+1}}}\mathbf{M}^{\mathrm{eff}}_{\mathrm{B_{j+1}B_j\ldots B_1m}}$ In place of

$$\mathbf{M}_{\mathrm{RB_j}} = \begin{bmatrix} 1 & -\overline{\Delta}_{\mathrm{RB_j}}e^{-2\rho_{\mathrm{B_j}}\mathrm{b_j}} \\ -\overline{\Delta}_{\mathrm{RB_j}} & e^{-2\rho_{\mathrm{B_j}}\mathrm{b_j}} \end{bmatrix}$$

we now have

$$\mathbf{M}_{\mathrm{RB_{j+1}}}\mathbf{M}_{\mathrm{B_{j+1}B_j}}$$

$$= \begin{bmatrix} 1 & -\overline{\Delta}_{\mathrm{RB_{j+1}}}e^{-2\rho_{\mathrm{B_{j+1}}}\mathrm{b_{j+1}}} \\ -\overline{\Delta}_{\mathrm{RB_{j+1}}} & e^{-2\rho_{\mathrm{B_{j+1}}}\mathrm{b_{j+1}}} \end{bmatrix}\begin{bmatrix} 1 & -\overline{\Delta}_{\mathrm{B_{j+1}B_j}}e^{-2\rho_{\mathrm{B_j}}\mathrm{b_j}} \\ -\overline{\Delta}_{\mathrm{B_{j+1}B_j}} & e^{-2\rho_{\mathrm{B_j}}\mathrm{b_j}} \end{bmatrix}$$

$$= \begin{bmatrix} 1+\overline{\Delta}_{\mathrm{RB_{j+1}}}\overline{\Delta}_{\mathrm{B_{j+1}B_j}}e^{-2\rho_{\mathrm{B_{j+1}}}\mathrm{b_{j+1}}} & -\overline{\Delta}_{\mathrm{B_{j+1}B_j}}e^{-2\rho_{\mathrm{B_j}}\mathrm{b_j}} - \overline{\Delta}_{\mathrm{RB_{j+1}}}e^{-2\rho_{\mathrm{B_{j+1}}}\mathrm{b_{j+1}}}e^{-2\rho_{\mathrm{B_j}}\mathrm{b_j}} \\ -\overline{\Delta}_{\mathrm{RB_{j+1}}} - \overline{\Delta}_{\mathrm{B_{j+1}B_j}}e^{-2\rho_{\mathrm{B_{j+1}}}\mathrm{b_{j+1}}} & \overline{\Delta}_{\mathrm{RB_{j+1}}}\overline{\Delta}_{\mathrm{B_{j+1}B_j}}e^{-2\rho_{\mathrm{B_j}}\mathrm{b_j}} + e^{-2\rho_{\mathrm{B_{j+1}}}\mathrm{b_{j+1}}}e^{-2\rho_{\mathrm{B_j}}\mathrm{b_j}} \end{bmatrix}$$

$$= \left(1+\overline{\Delta}_{\mathrm{RB_{j+1}}}\overline{\Delta}_{\mathrm{B_{j+1}B_j}}e^{-2\rho_{\mathrm{B_{j+1}}}\mathrm{b_{j+1}}}\right)$$

$$\times \begin{bmatrix} 1 & -\dfrac{\overline{\Delta}_{\mathrm{RB_{j+1}}}e^{-2\rho_{\mathrm{B_{j+1}}}\mathrm{b_{j+1}}}+\overline{\Delta}_{\mathrm{B_{j+1}B_j}}}{1+\overline{\Delta}_{\mathrm{RB_{j+1}}}\overline{\Delta}_{\mathrm{B_{j+1}B_j}}e^{-2\rho_{\mathrm{B_{j+1}}}\mathrm{b_{j+1}}}}e^{-2\rho_{\mathrm{B_j}}\mathrm{b_j}} \\ -\dfrac{\overline{\Delta}_{\mathrm{RB_{j+1}}}+\overline{\Delta}_{\mathrm{B_{j+1}B_j}}e^{-2\rho_{\mathrm{B_{j+1}}}\mathrm{b_{j+1}}}}{1+\overline{\Delta}_{\mathrm{RB_{j+1}}}\overline{\Delta}_{\mathrm{B_{j+1}B_j}}e^{-2\rho_{\mathrm{B_{j+1}}}\mathrm{b_{j+1}}}} & \dfrac{\overline{\Delta}_{\mathrm{RB_{j+1}}}\overline{\Delta}_{\mathrm{B_{j+1}B_j}}+e^{-2\rho_{\mathrm{B_{j+1}}}\mathrm{b_{j+1}}}}{1+\overline{\Delta}_{\mathrm{RB_{j+1}}}\overline{\Delta}_{\mathrm{B_{j+1}B_j}}e^{-2\rho_{\mathrm{B_{j+1}}}\mathrm{b_{j+1}}}}e^{-2\rho_{\mathrm{B_j}}\mathrm{b_j}} \end{bmatrix}.$$

The $A_R = 0$ condition uses elements from only the first row of this transition matrix. Hence the comparison between the 1–2 elements

$$-\overline{\Delta}_{\mathrm{RB_j}}e^{-2\rho_{\mathrm{B_j}}\mathrm{b_j}} \quad \text{and} \quad -\frac{\overline{\Delta}_{\mathrm{RB_{j+1}}}e^{-2\rho_{\mathrm{B_{j+1}}}\mathrm{b_{j+1}}}+\overline{\Delta}_{\mathrm{B_{j+1}B_j}}}{1+\overline{\Delta}_{\mathrm{RB_{j+1}}}\overline{\Delta}_{\mathrm{B_{j+1}B_j}}e^{-2\rho_{\mathrm{B_{j+1}}}\mathrm{b_{j+1}}}}e^{-2\rho_{\mathrm{B_j}}\mathrm{b_j}}.$$

21. The original derivation is given in B. W. Ninham and V. A. Parsegian, "Van der Waals interactions in multilayer systems" J. Chem. Phys. **53**, 3398–402 (1970). The details of a modified derivation are given in R. Podgornik, P. L. Hansen, and V. A. Parsegian, "On a reformulation of the theory of Lifshitz–van der Waals-interactions in multilayered systems, J. Chem. Phys, **119**, 1070–77 (2003).

22.
$$\mathbf{T}_{\mathrm{B'}} = \begin{bmatrix} 1 & 0 \\ 0 & e^{-2\rho_{\mathrm{B'}}\mathrm{b'}} \end{bmatrix}, \quad \mathbf{D}_{\mathrm{B'B}} = \begin{bmatrix} 1 & -\overline{\Delta}_{\mathrm{B'B}} \\ -\overline{\Delta}_{\mathrm{B'B}} & 1 \end{bmatrix},$$

$$\mathbf{T}_{\mathrm{B'}}\mathbf{D}_{\mathrm{B'B}} = \begin{bmatrix} 1 & -\overline{\Delta}_{\mathrm{B'B}} \\ -\overline{\Delta}_{\mathrm{B'B}}e^{-2\rho_{\mathrm{B'}}\mathrm{b'}} & e^{-2\rho_{\mathrm{B'}}\mathrm{b'}} \end{bmatrix},$$

$$\mathbf{T}_{\mathrm{B}} = \begin{bmatrix} 1 & 0 \\ 0 & e^{-2\rho_{\mathrm{B}}\mathrm{b}} \end{bmatrix}, \quad \mathbf{D}_{\mathrm{BB'}} = \begin{bmatrix} 1 & -\overline{\Delta}_{\mathrm{BB'}} \\ -\overline{\Delta}_{\mathrm{BB'}} & 1 \end{bmatrix},$$

$$\mathbf{T}_{\mathrm{B}}\mathbf{D}_{\mathrm{BB'}} = \begin{bmatrix} 1 & -\overline{\Delta}_{\mathrm{BB'}} \\ -\overline{\Delta}_{\mathrm{BB'}}e^{-2\rho_{\mathrm{B}}\mathrm{b}} & e^{-2\rho_{\mathrm{B}}\mathrm{b}} \end{bmatrix}.$$

$$\mathbf{M}_{\mathrm{B'B}} = \mathbf{T}_{\mathrm{B'}}\mathbf{D}_{\mathrm{B'B}}\mathbf{T}_{\mathrm{B}}\mathbf{D}_{\mathrm{BB'}} = \begin{bmatrix} 1 & -\overline{\Delta}_{\mathrm{B'B}} \\ -\overline{\Delta}_{\mathrm{B'B}}e^{-2\rho_{\mathrm{B'}}\mathrm{b'}} & e^{-2\rho_{\mathrm{B'}}\mathrm{b'}} \end{bmatrix}\begin{bmatrix} 1 & -\overline{\Delta}_{\mathrm{BB'}} \\ -\overline{\Delta}_{\mathrm{BB'}}e^{-2\rho_{\mathrm{B}}\mathrm{b}} & e^{-2\rho_{\mathrm{B}}\mathrm{b}} \end{bmatrix}$$

$$= \begin{bmatrix} 1-\overline{\Delta}^2_{\mathrm{B'B}}e^{-2\rho_{\mathrm{B}}\mathrm{b}} & \overline{\Delta}_{\mathrm{B'B}}(1-e^{-2\rho_{\mathrm{B}}\mathrm{b}}) \\ (e^{-2\rho_{\mathrm{B}}\mathrm{b}}-1)\overline{\Delta}_{\mathrm{B'B}}e^{-2\rho_{\mathrm{B'}}\mathrm{b'}} & (e^{-2\rho_{\mathrm{B}}\mathrm{b}}-\overline{\Delta}^2_{\mathrm{B'B}})e^{-2\rho_{\mathrm{B'}}\mathrm{b'}} \end{bmatrix}$$

The unimodular condition requires only a normalizing factor on the matrix. Each term must be divided by the square root of the determinant,

$$\left(1-\overline{\Delta}^2_{\mathrm{B'B}}e^{-2\rho_{\mathrm{B}}\mathrm{b}}\right)\left(e^{-2\rho_{\mathrm{B}}\mathrm{b}}-\overline{\Delta}^2_{\mathrm{B'B}}\right)e^{-2\rho_{\mathrm{B'}}\mathrm{b'}} - \overline{\Delta}_{\mathrm{B'B}}(1-e^{-2\rho_{\mathrm{B}}\mathrm{b}})$$
$$(e^{-2\rho_{\mathrm{B}}\mathrm{b}}-1)\overline{\Delta}_{\mathrm{B'B}}e^{-2\rho_{\mathrm{B'}}\mathrm{b'}} = \left(1-\overline{\Delta}^2_{\mathrm{B'B}}\right)^2 e^{-2\rho_{\mathrm{B}}\mathrm{b}}e^{-2\rho_{\mathrm{B'}}\mathrm{b'}},$$

i.e., divide by $(1-\overline{\Delta}^2_{\mathrm{B'B}})\,e^{-\rho_{\mathrm{B}}\mathrm{b}}e^{-\rho_{\mathrm{B'}}\mathrm{b'}}$ to create the required matrix elements m_{ij} in

$$\begin{bmatrix} \dfrac{1-\overline{\Delta}^2_{\mathrm{B'B}}e^{-2\rho_{\mathrm{B}}\mathrm{b}}}{\left(1-\overline{\Delta}^2_{\mathrm{B'B}}\right)e^{-\rho_{\mathrm{B}}\mathrm{b}}e^{-\rho_{\mathrm{B'}}\mathrm{b'}}} & \dfrac{\overline{\Delta}_{\mathrm{B'B}}\left(1-e^{-2\rho_{\mathrm{B}}\mathrm{b}}\right)}{\left(1-\overline{\Delta}^2_{\mathrm{B'B}}\right)e^{-\rho_{\mathrm{B}}\mathrm{b}}e^{-\rho_{\mathrm{B'}}\mathrm{b'}}} \\ \dfrac{(e^{-2\rho_{\mathrm{B}}\mathrm{b}}-1)\overline{\Delta}_{\mathrm{B'B}}e^{-2\rho_{\mathrm{B'}}\mathrm{b'}}}{\left(1-\overline{\Delta}^2_{\mathrm{B'B}}\right)e^{-\rho_{\mathrm{B}}\mathrm{b}}e^{-\rho_{\mathrm{B'}}\mathrm{b'}}} & \dfrac{\left(e^{-2\rho_{\mathrm{B}}\mathrm{b}}-\overline{\Delta}^2_{\mathrm{B'B}}\right)e^{-2\rho_{\mathrm{B'}}\mathrm{b'}}}{\left(1-\overline{\Delta}^2_{\mathrm{B'B}}\right)e^{-\rho_{\mathrm{B}}\mathrm{b}}e^{-\rho_{\mathrm{B'}}\mathrm{b'}}} \end{bmatrix}.$$

23. Use

$$\mathbf{D}_{\mathrm{mL}} = \begin{bmatrix} 1 & -\overline{\Delta}_{\mathrm{m/L}} \\ -\overline{\Delta}_{\mathrm{m/L}} & 1 \end{bmatrix} = \begin{bmatrix} 1 & -\overline{\Delta}_{\mathrm{mL}} \\ -\overline{\Delta}_{\mathrm{mL}} & 1 \end{bmatrix},\quad \mathbf{D}_{\mathrm{B'm}} = \begin{bmatrix} 1 & -\overline{\Delta}_{\mathrm{B'm}} \\ -\overline{\Delta}_{\mathrm{B'm}} & 1 \end{bmatrix},$$

$$\mathbf{T}_{\mathrm{m}} = \begin{bmatrix} 1 & 0 \\ 0 & e^{-2\rho_{\mathrm{m}}l} \end{bmatrix},\quad \mathbf{T}_{\mathrm{B'}} = \begin{bmatrix} 1 & 0 \\ 0 & e^{-2\rho_{\mathrm{B'}}\mathrm{b'}} \end{bmatrix},\quad \mathbf{T}_{\mathrm{B}} = \begin{bmatrix} 1 & 0 \\ 0 & e^{-2\rho_{\mathrm{B}}\mathrm{b}} \end{bmatrix}$$

$$\mathbf{D}_{\mathrm{BB'}} = \begin{bmatrix} 1 & -\overline{\Delta}_{\mathrm{BB'}} \\ -\overline{\Delta}_{\mathrm{BB'}} & 1 \end{bmatrix},\quad \mathbf{D}_{\mathrm{B'B}} = \begin{bmatrix} 1 & -\overline{\Delta}_{\mathrm{B'B}} \\ -\overline{\Delta}_{\mathrm{B'B}} & 1 \end{bmatrix},$$

$$\mathbf{D}_{\mathrm{RB'}} = \begin{bmatrix} 1 & -\overline{\Delta}_{\mathrm{RB'}} \\ -\overline{\Delta}_{\mathrm{RB'}} & 1 \end{bmatrix}$$

in

$$\begin{pmatrix} \mathrm{A_R} \\ \mathrm{B_R} \end{pmatrix} = \mathbf{D}_{\mathrm{RB'}}\mathbf{N}_{\mathrm{B'B}}\mathbf{T}_{\mathrm{B'}}\mathbf{D}_{\mathrm{B'm}}\mathbf{T}_{\mathrm{m}}\mathbf{D}_{\mathrm{mL}}\begin{pmatrix} \mathrm{A_L} \\ 0 \end{pmatrix}$$

$$= \begin{bmatrix} n_{11}-n_{21}\overline{\Delta}_{\mathrm{RB'}} & n_{12}-n_{22}\overline{\Delta}_{\mathrm{RB'}} \\ n_{21}-n_{11}\overline{\Delta}_{\mathrm{RB'}} & n_{22}-n_{22}\overline{\Delta}_{\mathrm{RB'}} \end{bmatrix}\begin{bmatrix} (1+\overline{\Delta}_{\mathrm{mL}}\overline{\Delta}_{\mathrm{B'm}}e^{-2\rho_{\mathrm{m}}l}) \\ -\left(\overline{\Delta}_{\mathrm{B'm}}+\overline{\Delta}_{\mathrm{mL}}e^{-2\rho_{\mathrm{m}}l}\right)e^{-2\rho_{\mathrm{B'}}\mathrm{b'}} \end{bmatrix}\mathrm{A_L}.$$

Element 11 of the product is set to zero to satisfy the condition that $\mathrm{A_R} = 0$:

$$(n_{11}-n_{21}\overline{\Delta}_{\mathrm{RB'}})(1+\overline{\Delta}_{\mathrm{mL}}\overline{\Delta}_{\mathrm{B'm}}e^{-2\rho_{\mathrm{m}}l}) - (n_{12}-n_{22}\overline{\Delta}_{\mathrm{RB'}})e^{-2\rho_{\mathrm{B'}}\mathrm{b'}}(\overline{\Delta}_{\mathrm{B'm}}+\overline{\Delta}_{\mathrm{mL}}e^{-2\rho_{\mathrm{m}}l})$$
$$= [(n_{11}-n_{21}\overline{\Delta}_{\mathrm{RB'}}) - (n_{12}-n_{22}\overline{\Delta}_{\mathrm{RB'}})\overline{\Delta}_{\mathrm{B'm}}e^{-2\rho_{\mathrm{B'}}\mathrm{b'}}]$$
$$+ [(n_{11}-n_{21}\overline{\Delta}_{\mathrm{RB'}})\overline{\Delta}_{\mathrm{B'm}} - (n_{12}-n_{22}\overline{\Delta}_{\mathrm{RB'}})e^{-2\rho_{\mathrm{B'}}\mathrm{b'}}]\overline{\Delta}_{\mathrm{mL}}e^{-2\rho_{\mathrm{m}}l}.$$

Because the physically important part of this dispersion relation is that which depends on variable separation l, factor out the physically irrelevant $(n_{11}-n_{21}\overline{\Delta}_{\mathrm{RB'}})$

$-\left(n_{12}-n_{22}\overline{\Delta}_{\mathrm{RB'}}\right)\overline{\Delta}_{\mathrm{B'm}}e^{-2\rho_{\mathrm{B'}}\mathrm{b'}}$ so that it will not play a role in the $\ln[D(i\xi_n)]$ function:

$$D(i\xi_n)=1-\frac{\left(n_{11}-n_{21}\overline{\Delta}_{\mathrm{RB'}}\right)\overline{\Delta}_{\mathrm{B'm}}-\left(n_{12}-n_{22}\overline{\Delta}_{\mathrm{RB'}}\right)e^{-2\rho_{\mathrm{B'}}\mathrm{b'}}}{\left(n_{11}-n_{21}\overline{\Delta}_{\mathrm{RB'}}\right)-\left(n_{12}-n_{22}\overline{\Delta}_{\mathrm{RB'}}\right)\overline{\Delta}_{\mathrm{B'm}}e^{-2\rho_{\mathrm{B'}}\mathrm{b'}}}\overline{\Delta}_{\mathrm{Lm}}e^{-2\rho_{\mathrm{m}}l}.$$

24.

$$n_{11}=m_{11}U_{N-1}-U_{N-2}\quad n_{12}=m_{12}U_{N-1},$$
$$n_{21}=m_{21}U_{N-1}\qquad n_{22}=m_{22}U_{N-1}-U_{N-2};$$
$$U_{N-1}(x)=\frac{e^{+N\zeta}-e^{-N\zeta}}{e^{+\zeta}-e^{-\zeta}}.\ U_{N-2}(x)=\frac{e^{+(N-1)\zeta}-e^{-(N-1)\zeta}}{e^{+\zeta}-e^{-\zeta}},$$
$$x=\frac{m_{11}+m_{22}}{2}=\cosh(\zeta);$$
$$n_{11}=m_{11}U_{N-1}-U_{N-2}\to m_{11}\frac{e^{+N\zeta}-e^{-N\zeta}}{e^{+\zeta}-e^{-\zeta}}-\frac{e^{+(N-1)\zeta}-e^{-(N-1)\zeta}}{e^{+\zeta}-e^{-\zeta}}$$
$$=\frac{e^{+N\zeta}}{e^{+\zeta}-e^{-\zeta}}[m_{11}(1-e^{-2N\zeta})-e^{-\zeta}+e^{-2N\zeta}e^{+\zeta}]$$
$$=\frac{e^{+N\zeta}}{e^{+\zeta}-e^{-\zeta}}[m_{11}-e^{-\zeta}-(m_{11}-e^{+\zeta})e^{-2N\zeta}],$$

$$n_{22}=m_{22}U_{N-1}-U_{N-2}$$
$$\to m_{22}\frac{e^{+N\zeta}-e^{-N\zeta}}{e^{+\zeta}-e^{-\zeta}}-\frac{e^{+(N-1)\zeta}-e^{-(N-1)\zeta}}{e^{+\zeta}-e^{-\zeta}}$$
$$=\frac{e^{+N\zeta}}{e^{+\zeta}-e^{-\zeta}}\left[m_{22}(1-e^{-2N\zeta})-e^{-\zeta}+e^{-2N\zeta}e^{+\zeta}\right],$$

$$n_{12}=m_{12}U_{N-1}=m_{12}\frac{e^{+N\zeta}}{e^{+\zeta}-e^{-\zeta}}(1-e^{-2N\zeta}),$$

$$n_{21}=m_{21}U_{N-1}=m_{21}\frac{e^{+N\zeta}-e^{-N\zeta}}{e^{+\zeta}-e^{-\zeta}}=m_{21}\frac{e^{+N\zeta}}{e^{+\zeta}-e^{-\zeta}}(1-e^{-2N\zeta}).$$

In the limit of large N, omitting a common factor $[e^{+N\zeta}/(e^{+\zeta}-e^{-\zeta})]$ in all elements yields

$$n_{11}\to\left(m_{11}-e^{-\zeta}\right),\quad n_{22}\to\left(m_{22}-e^{-\zeta}\right),\quad n_{12}\to m_{12},n_{21}\to m_{21}.$$

The ratio

$$\frac{\left(n_{11}-n_{21}\overline{\Delta}_{\mathrm{RB'}}\right)\overline{\Delta}_{\mathrm{B'm}}-\left(n_{12}-n_{22}\overline{\Delta}_{\mathrm{RB'}}\right)e^{-2\rho_{\mathrm{B'}}\mathrm{b'}}}{\left(n_{11}-n_{21}\overline{\Delta}_{\mathrm{RB'}}\right)-\left(n_{12}-n_{22}\overline{\Delta}_{\mathrm{RB'}}\right)\overline{\Delta}_{\mathrm{B'm}}e^{-2\rho_{\mathrm{B'}}\mathrm{b'}}}$$

in the general dispersion relation reduces to

$$\frac{\overline{\Delta}_{\mathrm{B'm}}-\frac{\left(n_{12}-n_{22}\overline{\Delta}_{\mathrm{RB'}}\right)}{\left(n_{11}-n_{21}\overline{\Delta}_{\mathrm{RB'}}\right)}e^{-2\rho_{\mathrm{B'}}\mathrm{b'}}}{1-\frac{\left(n_{12}-n_{22}\overline{\Delta}_{\mathrm{RB'}}\right)}{\left(n_{11}-n_{21}\overline{\Delta}_{\mathrm{RB'}}\right)}\overline{\Delta}_{\mathrm{B'm}}e^{-2\rho_{\mathrm{B'}}\mathrm{b'}}}$$

where

$$\frac{n_{12}-n_{22}\overline{\Delta}_{\mathrm{RB'}}}{n_{11}-n_{21}\overline{\Delta}_{\mathrm{RB'}}}=\frac{m_{12}-(m_{22}-e^{-\zeta})\overline{\Delta}_{\mathrm{RB'}}}{(m_{11}-e^{-\zeta})-m_{21}\overline{\Delta}_{\mathrm{RB'}}}$$
$$=\frac{(m_{22}-e^{-\zeta})}{m_{21}}\frac{m_{12}-(m_{22}-e^{-\zeta})\overline{\Delta}_{\mathrm{RB'}}}{m_{12}-(m_{22}-e^{-\zeta})\overline{\Delta}_{\mathrm{RB'}}}=\frac{(m_{22}-e^{-\zeta})}{m_{21}},$$

$$D(i\xi_n)=1-\frac{m_{21}\overline{\Delta}_{\mathrm{B'm}}-(m_{22}-e^{-\zeta})e^{-2\rho_{\mathrm{B'}}\mathrm{b'}}}{m_{21}-(m_{22}-e^{-\zeta})\overline{\Delta}_{\mathrm{B'm}}e^{-2\rho_{\mathrm{B'}}\mathrm{b'}}}\overline{\Delta}_{\mathrm{Lm}}e^{-2\rho_{\mathrm{m}}l}.$$

25. It is sometimes practical to rewrite $f''(z) + \frac{d\varepsilon/dz}{\varepsilon(z)} f'(z) - \rho^2 f(z) = 0$ by using $v(z) \equiv \varepsilon(z)^{1/2} f(z)$ to create an alternative differential equation:

$$v''(z) + \left[\frac{1}{4}\left(\frac{d\varepsilon/dz}{\varepsilon(z)}\right)^2 - \frac{d^2\varepsilon/dz^2}{2\varepsilon(z)} - \rho^2\right] v(z) = 0.$$

26. For the transition between layers use Eq. (L3.59) and its corresponding endnote:

$$\left(-A_{i+1}e^{\rho_{i+1}l_{i/i+1}} + B_{i+1}e^{-\rho_{i+1}l_{i/i+1}}\right)\rho_{i+1} = \left(-A_i e^{\rho_i l_{i/i+1}} + B_i e^{-\rho_i l_{i/i+1}}\right)\rho_i,$$
$$\left(A_{i+1}e^{\rho_{i+1}l_{i/i+1}} + B_{i+1}e^{-\rho_{i+1}l_{i/i+1}}\right)\varepsilon_{i+1} = \left(A_i e^{\rho_i l_{i/i+1}} + B_i e^{-\rho_i l_{i/i+1}}\right)\varepsilon_i.$$

Replace i with $r-1$ and set all ρ's equal (nonretarded limit!):

$$A_{r-1}e^{\rho z_r} - B_{r-1}e^{-\rho z_r} = A_r e^{\rho z_r} - B_r e^{-\rho z_r},$$
$$\varepsilon_{r-1}\left(A_{r-1}e^{\rho z_r} + B_{r-1}e^{-\rho z_r}\right) = \varepsilon_r\left(A_r e^{\rho z_r} + B_r e^{-\rho z_r}\right).$$

Divide the second equation by the first to obtain

$$\varepsilon_{r-1}\frac{\theta_{r-1}e^{+2\rho z_r}+1}{\theta_{r-1}e^{+2\rho z_r}-1} = \varepsilon_r\frac{\theta_r e^{+2\rho z_r}+1}{\theta_r e^{+2\rho z_r}-1}$$

or

$$\theta_{r-1}e^{+2\rho z_r}+1 = \theta_{r-1}e^{+2\rho z_r}\frac{\varepsilon_r}{\varepsilon_{r-1}}\frac{\theta_r e^{+2\rho z_r}+1}{\theta_r e^{+2\rho z_r}-1} - \frac{\varepsilon_r}{\varepsilon_{r-1}}\frac{\theta_r e^{+2\rho z_r}+1}{\theta_r e^{+2\rho z_r}-1}.$$

Factoring out $\theta_{r-1}e^{+2\rho z_r}$ gives

$$\theta_{r-1}e^{+2\rho z_r}\left(1 - \frac{\varepsilon_r}{\varepsilon_{r-1}}\frac{\theta_r e^{+2\rho z_r}+1}{\theta_r e^{+2\rho z_r}-1}\right) = -\left(1 + \frac{\varepsilon_r}{\varepsilon_{r-1}}\frac{\theta_r e^{+2\rho z_r}+1}{\theta_r e^{+2\rho z_r}-1}\right)$$

or

$$\theta_{r-1}e^{+2\rho z_r} = -\left[\frac{\varepsilon_{r-1}\left(\theta_r e^{+2\rho z_r}-1\right) + \varepsilon_r\left(\theta_r e^{+2\rho z_r}+1\right)}{\varepsilon_{r-1}\left(\theta_r e^{+2\rho z_r}-1\right) - \varepsilon_r\left(\theta_r e^{+2\rho z_r}+1\right)}\right]$$
$$= \frac{\theta_r e^{+2\rho z_r} - \overline{\Delta}_{r-1/r}}{1 - \theta_r e^{+2\rho z_r}\overline{\Delta}_{r-1/r}} \qquad \overline{\Delta}_{r-1/r} = \frac{\varepsilon_{r-1}-\varepsilon_r}{\varepsilon_{r-1}+\varepsilon_r}.$$

27. For $\theta_r = \theta(z_r)$ (Eq. L3.153), using Eq. (L3.152) and L3.59) (and corresponding footnote), divide

$$\varepsilon_{r-1}\left(A_{r-1}e^{\rho_{r-1}z_r} + B_{r-1}e^{-\rho_{r-1}z_r}\right) = \varepsilon_r\left(A_r e^{\rho_r z_r} + B_r e^{-\rho_r z_r}\right)$$

by

$$\rho_{r-1}\left(A_{r-1}e^{\rho_{r-1}z_r} - B_{r-1}e^{-\rho_{r-1}z_r}\right) = \rho_r\left(A_r e^{\rho_r z_r} - B_r e^{-\rho_r z_r}\right).$$

Let,

$$\alpha \equiv \frac{\varepsilon_r/\rho_r}{\varepsilon_{r-1}/\rho_{r-1}}$$

to solve

$$\alpha\frac{\theta_r e^{+2\rho_r z_r}+1}{\theta_r e^{+2\rho_r z_r}-1} = \frac{\theta_{r-1}e^{+2\rho_{r-1}z_r}+1}{\theta_{r-1}e^{+2\rho_{r-1}z_r}-1},$$

$$\alpha\left(\theta_r e^{+2\rho_r z_r}+1\right)\left(\theta_{r-1}e^{+2\rho_{r-1}z_r}-1\right) = \left(\theta_r e^{+2\rho_r z_r}-1\right)\left(\theta_{r-1}e^{+2\rho_{r-1}z_r}+1\right),$$

for

$$\theta_{r-1}e^{+2\rho_{r-1}z_r} = \frac{\theta_r e^{+2\rho_r z_r} - \overline{\Delta}_{r-1/r}}{1 - \theta_r e^{+2\rho_r z_r}\overline{\Delta}_{r-1/r}}.$$

formally as in the nonretarded case but, where now

$$\overline{\Delta}_{r-1/r} = \frac{\varepsilon_{r-1}\rho_r - \varepsilon_r\rho_{r-1}}{\varepsilon_{r-1}\rho_r + \varepsilon_r\rho_{r-1}}.$$

28. Introduced into approximation (L3.162), approximations (L3.160) and (L3.161) give

$$u_{r-1}e^{2\rho_{r-1}\frac{D}{N}} \sim u_{r-1} + 2\rho_{r-1}u_{r-1}\frac{D}{N}$$

$$\sim u_r - \left.\frac{du(z)}{dz}\right|_{z=z_r}\frac{D}{N} + 2\left[\rho_r - \left.\frac{d\rho(z)}{dz}\right|_{z=z_r}\frac{D}{N}\right]\left[u_r - \left.\frac{du(z)}{dz}\right|_{z=z_r}\frac{D}{N}\right]\frac{D}{N}$$

$$\sim u_r - \left.\frac{du(z)}{dz}\right|_{z=z_r}\frac{D}{N} + 2\rho_r u_r\frac{D}{N} \sim u_r + \left[2\rho_r u_r - \left.\frac{du(z)}{dz}\right|_{z=z_r}\right]\frac{D}{N}.$$

Equate to approximation (L3.163):

$$u_{r-1}e^{2\rho_{r-1}\frac{D}{N}} \sim u_r + \left.\frac{d\ln[\varepsilon(z)/\rho(z)]}{2dz}\right|_{z=z_r}\left(1-u_r^2\right)\frac{D}{N}$$

$$= u_r + \left[2\rho_r u_r - \left.\frac{du(z)}{dz}\right|_{z=z_r}\right]\frac{D}{N},$$

or

$$\left.\frac{d\ln[\varepsilon(z)/\rho(z)]}{2dz}\right|_{z=z_r}\left(1-u_r^2\right) = \left[2\rho_r u_r - \left.\frac{du(z)}{dz}\right|_{z=z_r}\right]$$

for

$$\frac{du(z)}{dz} = 2\rho(z)u(z) - \frac{d\ln[\varepsilon(z)/\rho(z)]}{2dz}[1-u^2(z)].$$

29. With conductivity σ, with external charge ρ_{ext}, and with spatially unvarying scalar relative ε and μ in each region, but without externally applied currents $\mathbf{j}_{\text{ext}}$, in the limit of infinite light velocity c, the Maxwell equations

$$\nabla\cdot\mathbf{D} = \nabla\cdot(\varepsilon\varepsilon_0\mathbf{E}) = \rho_{\text{ext}} \text{ in mks,} \qquad \nabla\cdot\mathbf{D} = \nabla\cdot(\varepsilon\mathbf{E}) = 4\pi\rho_{\text{ext}} \text{ in cgs;}$$

$$\nabla\cdot\mathbf{H} = 0, \qquad \nabla\cdot\mathbf{H} = 0,$$

$$\nabla\times\mathbf{E} = -\frac{\partial\mathbf{B}}{\partial t} = -\mu\mu_0\frac{\partial\mathbf{H}}{\partial t}, \qquad \nabla\times\mathbf{E} = -\frac{1}{c}\frac{\partial\mathbf{B}}{\partial t} = -\frac{\mu}{c}\frac{\partial\mathbf{H}}{\partial t},$$

$$\nabla\times\mathbf{H} = \varepsilon\varepsilon_0\frac{\partial\mathbf{E}}{\partial t} + \sigma + \mathbf{j}_{\text{ext}}, \qquad \nabla\times\mathbf{H} = \frac{\varepsilon}{c}\frac{\partial\mathbf{E}}{\partial t} + \frac{4\pi\sigma}{c} + \frac{4\pi}{c}\mathbf{j}_{\text{ext}},$$

become

$$\nabla\cdot\mathbf{E} = \rho_{\text{ext}}/\varepsilon_0\varepsilon \text{ in mks,} \qquad \nabla\cdot\mathbf{E} = 4\pi\rho_{\text{ext}}/\varepsilon \text{ in cgs,}$$

$$\nabla\cdot\mathbf{H} = 0, \qquad \nabla\cdot\mathbf{H} = 0,$$

$$\nabla\times\mathbf{E} = 0, \qquad \nabla\times\mathbf{E} = 0,$$

$$\nabla\times\mathbf{H} = 0, \qquad \nabla\times\mathbf{H} = 0.$$

30. $\nabla^2\phi = \kappa^2\phi = \left(\frac{\partial^2}{\partial x^2} + \frac{\partial^2}{\partial y^2} + \frac{\partial^2}{\partial z^2}\right)\phi$. Put $\phi(x,y,z) = f(z)e^{i(ux+vy)}$ to have

$$-(u^2+v^2)f(z) + f''(z) = \kappa^2 f(z) \text{ or } \beta^2 f(z) = f''(z)$$

where $\beta^2 \equiv (u^2+v^2) + \kappa^2$.

31. Eliminating A_L from

$$A_L = A_m + B_m \text{ and } \varepsilon_L A_L \beta_L = \varepsilon_m A_m \beta_m - \varepsilon_m B_m \beta_m$$

gives

$$A_m \overline{\Delta}_{Lm} = -B_m \text{ where } \overline{\Delta}_{Lm} \equiv \left(\frac{\beta_L\varepsilon_L - \beta_m\varepsilon_m}{\beta_L\varepsilon_L + \beta_m\varepsilon_m}\right).$$

Eliminating $B_R e^{-\beta_R l}$ from

$$B_R e^{-\beta_R l} = A_m e^{\beta_m l} + B_m e^{-\beta_m l},$$
$$-\varepsilon_R B_R \beta_R e^{-\beta_R l} = \varepsilon_m A_m \beta_m e^{\beta_m l} - \varepsilon_m B_m \beta_m e^{-\beta_m l}$$

gives

$$A_m = -B_m \overline{\Delta}_{Rm} e^{-2\beta_m l} \text{ where } \overline{\Delta}_{Rm} \equiv \left(\frac{\beta_R\varepsilon_R - \beta_m\varepsilon_m}{\beta_R\varepsilon_R + \beta_m\varepsilon_m}\right),$$

so that

$$1 - \overline{\Delta}_{Lm}\overline{\Delta}_{Rm} e^{-2\beta_m l} = 0.$$

32.
$$\begin{aligned} G_{LmR}(l) &= \frac{kT}{4\pi}\int_\kappa^\infty \beta \ln\left(1 - \overline{\Delta}_{Lm}\overline{\Delta}_{Rm} e^{-2\beta l}\right) d\beta \\ &= \frac{kT}{16\pi l^2}\int_{2\kappa l}^\infty x \ln\left(1 - \overline{\Delta}_{Lm}\overline{\Delta}_{Rm} e^{-x}\right) dx \\ &= -\frac{kT}{16\pi l^2}\sum_{j=1}^\infty \frac{\left(\overline{\Delta}_{Lm}\overline{\Delta}_{Rm}\right)^j}{j^3}(1 + 2\kappa l j)\, e^{-2\kappa l j} \\ &\approx -\frac{kT}{16\pi l^2}\overline{\Delta}_{Lm}\overline{\Delta}_{Rm}\,(1 + 2\kappa l)\, e^{-2\kappa l}. \end{aligned}$$

33.
$$\begin{aligned} G_{LmR}(l) &= \frac{kT}{4\pi}\int_\kappa^\infty \beta_m \ln\left(1 - \overline{\Delta}_{Lm}\overline{\Delta}_{Rm} e^{-2\beta_m l}\right) d\beta_m \\ &= \frac{kT}{16\pi l^2}\int_{2\kappa l}^\infty x \ln(1 - e^{-x})\, dx \\ &= -\frac{kT}{16\pi l^2}\sum_{j=1}^\infty \frac{(1 + 2\kappa l j)}{j^3} e^{-2\kappa l j} \approx -\frac{kT}{16\pi l^2}(1 + 2\kappa l)\, e^{-2\kappa l}. \end{aligned}$$

34.
$$\begin{aligned} G_{LmR}(l) &= \frac{kT}{4\pi}\int_0^\infty \beta_m \ln\left(1 - \overline{\Delta}_{Lm}\overline{\Delta}_{Rm} e^{-2\beta_m l}\right) d\beta_m \\ &= -\frac{kT}{16\pi l^2}\sum_{j=1}^\infty \frac{1}{j^3} = -\frac{kT}{16\pi l^2}\zeta(3) \approx -\frac{1.202kT}{16\pi l^2}. \end{aligned}$$

INDEX

Printed in the United States
By Bookmasters